The Real Number System

Real Numbers

Irrational Numbers: Any real number that is not rational, such as: $-\sqrt{2}$, $-\sqrt{3}$, $\sqrt{0.8}$, and π.

Rational Numbers: Real numbers that can be expressed in the form $\frac{a}{b}$, where a and b are integers and $b \neq 0$, such as: $-4\frac{3}{4}$, $-\frac{2}{3}$, 0.018, $0.\overline{3}$, and $\frac{5}{8}$.

Integers: ... , $-3, -2, -1, 0, 1, 2, 3, ...$

Whole Numbers: $0, 1, 2, 3, ...$

Natural Numbers: $1, 2, 3, ...$

Units of Measurement

Distance	Area	Volume	Capacity
Inches (in.)	Square inches (in.^{2})	Cubic inches (in.^{3})	Ounces (oz.)
Feet (ft.)	Square feet (ft.^{2})	Cubic feet (ft.^{3})	Cups (c.)
Yards (yd.)	Square yards (yd.^{2})	Cubic yards (yd.^{3})	Pints (pt.)
Miles (mi.)	Square miles (mi.^{2})	Cubic miles (mi.^{3})	Quarts (qt.)
			Gallons (gal.)
Centimeters (cm)	Square centimeters (cm.^{2})	Cubic centimeters (cm^3 or cc)	
Meters (m)	Square meters (m^2)	Cubic meters (km^3)	Milliliters (ml)
Kilometers (km)	Square kilometers (km^2)	Cubic kilometers (km^3)	Liters (l)
Light-years (lt-yr)	Square light-years (lt-yr^2)	Cubic light-years (lt-yr^3)	

Mass	Weight	Temperature	Time	Speed
Slugs	Ounces (oz.)	Degrees Fahrenheit (°F)	Years (yr.)	A distance unit over
	Pounds (lb.)	Degrees Celsius (°C)	Days (d.)	a time unit
Milligrams (mg)	Tons (T)	Kelvin (K)	Hours (hr.)	
Grams (g)			Seconds (sec.)	Miles per hour (mi./hr.)
Kilograms (kg)	Newtons (N)			Meters per second (m/sec.)
Metric tons (t)				

Prealgebra

SECOND EDITION

TOM CARSON
Midlands Technical College

Boston   San Francisco   New York
London   Toronto   Sydney   Tokyo   Singapore   Madrid
Mexico City   Munich   Paris   Cape Town   Hong Kong   Montreal

Publisher: Greg Tobin
Editor in Chief: Maureen O'Connor
Acquisitions Editor: Jennifer Crum
Managing Editor: Ron Hampton
Developmental Editor: Polina Sabinin
Text Design: Andrea Menza
Cover Design: Dennis Schaefer
Production Services: Pre-Press Company
Associate Editor: Katie Nopper
Editorial Assistant: Elizabeth Bernardi
Media Producer: Sharon Smith
Software Development: TestGen: Ted Hartman; MathXL: Jozef Kubit
Marketing Manager: Dona Kenly
Marketing Coordinator: Tracy Rabinowitz
Prepress Services Buyer: Caroline Fell
First Print Buyer: Hugh Crawford
Cover Photo: © James Lemass/Index Stock Imagery
Cover Image: Vaulted arch, Boston Harbor Hotel, Boston, Massachusetts

Photo Credits: p. 1: Digital Vision; p. 27: PhotoDisc; p. 71: NASA; p. 85: Digital Vision;
p. 102: AP Photo/Stephan Savoia; p. 111: Richard Cummins/Corbis; p. 138: NASA; p. 149:
Corbis; p. 233: PhotoDisc; p. 289: Corbis; p. 379: NASA; p. 399: PhotoDisc; p. 487:
PhotoDisc; p. 561: PhotoDisc; p. 627: Wolfgang Rattay/Reuters/Corbis

Library of Congress Cataloging-in-Publication Data
Carson, Tom, 1967–
      Prealgebra/Tom Carson.
         p. cm.
      Includes index.
      ISBN 0-321-23736-6
      1. Mathematics. I. Title.
   QA39.3.C39 2005
   510—dc22   2004061637

1 2 3 4 5 6 7   QWT   08 07 06 05

CONTENTS

Chapter 8   Percents   561

Chapter 9   Statistics and Graphs   627

Prealgebra, Second Edition, is designed for the student who needs a brush up in arithmetic and basic algebra concepts or for people who are encountering algebra for the first time. Written in a relaxed, nonthreatening style, this text takes great care to ensure that students who have struggled with math in the past will be comfortable with the subject matter. Explanations are carefully developed to provide a sense of *why* a mathematical process works the way it does, instead of just an explanation of *how* to follow the process. Problems from science, engineering, accounting, health fields, the arts, and everyday life link mathematics to the real world. The link to real-world problem solving is further developed through two project portfolios available in a separate workbook (see page xiii for a complete description of the *Project Portfolios and Extra Practice Workbook*). In addition, a complete study system with a Learning Styles Inventory and learning strategies provides further guidance for students. (See pages xv–xxii).

Upon completing the material in this text, a student should be able to proceed successfully in an introductory algebra course or a survey math course. This text is designed to be versatile enough for use in a standard lecture format, a self-paced lab, or even in an independent-study format. A strong ancillary package provides a wealth of supplemental resources for both instructors and students.

Key Features

Real, relevant, and interesting applications. Nearly every application problem is a real situation taken from science, engineering, health, finance, the arts, or just everyday life. The real-world applications not only illustrate the uses of basic arithmetic and algebra concepts, but they also expose students to the wonders of the world in which we live. Often the problems follow up with open-ended discussion questions where there is no "correct" answer. These questions help students to think beyond just getting a numeric answer by encouraging them to apply mathematical results to solve problems (see pages 18, 247, and 543).

Thorough explanations. This text explains not only how to do the math, but also why the math works the way it does, where it comes from, and how it is relevant to students' everyday lives. Knowing all of this helps the students remember the concept.

Study system. A study system is presented in the *To the Student* section on pages xv–xx, and is reinforced throughout the text. The system recommends using color codes for taking notes. The color codes are consistent in the text itself: red for definitions; blue for outlines, rules, and things to remember; and black for notes and examples. In addition, the study system presents a Learning Styles Inventory and strategies for succeeding in the course, and these strategies are revisited in the chapter openers throughout the text (see pages 85, 289 and 561) as well as in the Learning Strategy boxes (see pages 142, 180, and 320).

Learning Styles Inventory. A Learning Styles Inventory is presented on page xxi to help students assess their particular style of learning.

Learning Strategy boxes. Learning Strategy boxes appear where appropriate in the text to offer advice on how to effectively use the study system and how to study specific topics based on a student's learning style.

Problem-solving process.   On page 63 of Section 1.7, a problem-solving outline is developed with the following headings:

1. Understand
2. Plan
3. Execute
4. Answer
5. Check

Every application example throughout the rest of the text follows the steps given in this outline, presenting the headings to show the thinking process clearly (see pages 131, 253, and 629).

Of Interest boxes.   The Of Interests boxes are designed to enhance the learning process by making concepts fun and interesting (see pages 139, 245, and 585).

Connection boxes.   Connection boxes bridge concepts so that students see how the concepts build on each other and are interrelated (see pages 14, 235, and 673).

Calculator explanations and exercises.   The relevant functions on a scientific calculator are explained and illustrated throughout the text. An occasional calculator icon in the exercise sets indicates problems that are appropriate to be solved using a scientific calculator (see pages 29, 338, and 434). Note that the icon does not mean that a calculator is required for those problems.

Worktext format.   The worktext format leaves space for working problems in the margins and exercise sets, encouraging students to become actively involved in their learning.

Examples and Your Turn problems that correlate.   Examples have a corresponding Your Turn margin problem for students to complete as they read through the text so that they can practice concepts continually. Answers to all Your Turn problems are provided at the bottom of the page, giving students immediate feedback and confidence (see pages 203, 349, and 604).

Puzzle Problems.   These mathematical brainteasers appear at the end of selected exercise sets to encourage creative and critical thinking (see pages 258, 417, and 681).

Continuous Review Exercises.   Exercises that review previous concepts appear at the end of each exercise set (see pages 173, 417, and 573). These exercises also review concepts that are foundational to the discussion in the next section. Answers to these review exercises appear in the back of the text.

Chapter Summaries and Review Exercises.   A thorough summary at the end of each chapter provides a list of key terms referenced by section, a two-column summary of key concepts (procedures and rules on the left and examples on the right), and a list of key formulas. A set of review exercises is also provided (see pages 143, 390, and 552). Answers to all chapter review exercises appear in the back of the text.

Practice Tests.   A practice test follows each set of chapter review exercises (see pages 145, 394, and 481). The problem types in the practice tests correlate to the short-answer tests in the *Printed Test Bank*. This is especially comforting for people who are math-anxious or experience test anxiety. Answers to all practice test questions are provided in the back of the text.

Cumulative Reviews.   Beginning with Chapter 2, cumulative review exercises appear after each practice test (see pages 285, 483, and 698). These help students stay current with *all* the material they have learned. Answers to all cumulative review exercises appear in the back of the text.

What's New in This Edition?

Revising the first edition has proved to be a worthy endeavor. Our primary focus was on making the book even more accessible to students. We did this in several ways, including:

- Adding concept questions to the beginning of each exercise set to encourage students to think about the problems conceptually rather than by rote.
- Adding more low-level to mid-level exercises so students can gradually build up to more challenging problems.
- Decreasing the level of Chapter 3, focusing on single-variable polynomials.
- Reworking many explanations to be more concise and focused.
- Adding Learning Strategy boxes, which offer study strategies based on the student's learning style.
- Adding section numbers to all review exercises (at the end of exercise sets, in end-of-chapter review exercises, and in cumulative review exercises) and practice test exercises so that students can easily identify which section they need to review if they cannot remember how to solve a particular problem.
- Making Example numbers and Your Turn numbers correspond so that students and instructors can more easily determine which example goes with which Your Turn exercise.
- Updating and expanding the glossary; and
- Creating a new design that is more inviting to students while still retaining the pedagogical use of color as outlined in the Study Plan in the *To the Student* section of the Preface.

We also made the book more accessible to instructors by adding an Annotated Instructor's Edition and creating an *Adjunct Support Manual* that includes helpful teaching advice for first-time teachers, as well as useful suggestions for approaching each section.

In addition to making the book even more user friendly, we also increased and updated the applications. By doing so, the applications remain interesting to instructors and students alike.

Student Supplements

Project Portfolios and Extra Practice Exercises Workbook

- Includes two unique project portfolios—one on building a house and one on operating a business. Each portfolio contains problems keyed to the relevant sections of the text, so that as they work through each project over the course of the semester, students apply the skills and concepts they've learned in a realistic and integrated context.
- Additional practice exercises are provided for students who want additional practice with key or difficult concepts. Solutions to all extra practice exercises are provided in the back of the book.

ISBN: 0-321-33441-8

Instructor Supplements

Annotated Instructor's Edition

- Includes answers to all exercises, including puzzle problems, printed in bright blue near the corresponding problem.
- Useful teaching tips are printed in the margin.

ISBN: 0-321-23747-1

Instructor's Solutions Manual

- By Doreen Kelly, *Mesa Community College*
- Contains complete solutions to all even-numbered section exercises and puzzle problems.

ISBN: 0-321-23746-3

Student's Solutions Manual

- By Doreen Kelly, *Mesa Community College*
- Contains complete solutions to the odd-numbered section exercises and solutions to all of the section-level review exercises, chapter review exercises, practice tests, and cumulative review exercises.

ISBN: 0-321-23742-0

Videotapes

- A series of lectures correlates directly to the chapter content of the text. A video symbol at the beginning of each exercise set references the videotape or DVT (see Digital Video Tutor, below).
- Include a stop-the-tape feature that encourages students to stop the videotape, work through an example, and resume play to watch the video instructor work through the solution.

ISBN: 0-321-23741-2

Digital Video Tutor

- Complete set of digitized videos on CD-ROMs for student use at home or on campus.
- Ideal for distance learning or supplemental instruction.

ISBN: 0-321-23737-4

Addison-Wesley Math Tutor Center

The Addison-Wesley Math Tutor Center is staffed by qualified mathematics instructors who provide students with tutoring on examples and odd-numbered exercises from the textbook. Tutoring is available via toll-free telephone, toll-free fax, e-mail, or the Internet. White Board technology allows tutors and students to actually see problems worked while they "talk" in real time over the Internet during tutoring sessions.

www.aw-bc.com/tutorcenter

Printed Test Bank/Instructor's Resource Guide

- By Laura Hoye, *Trident Technical College*
- The *Printed Test Bank* portion contains one diagnostic test per chapter; four free-response test forms per chapter, one of which contains higher-level questions; one multiple-choice test per chapter; one free-response midterm exam; two free-response final exams; and one multiple-choice final exam. The *Instructor's Resource Guide* portion of the manual contains sample syllabi, chapter-by-chapter teaching strategies, extra practice worksheets for the algebra topics in the text, overhead-ready figures from the text, and answers and teaching suggestions for the project portfolios.

ISBN: 0-321-23748-X

TestGen with Quizmaster

- Enables instructors to build, edit, print, and administer tests.
- Features a computerized bank of questions developed to cover all text objectives.
- Available on a dual-platform Windows/Macintosh CD-ROM.

ISBN: 0-321-23738-2

Student and Instructor Supplements

MathXL® Tutorials on CD (ISBN 0-321-23739-0)

This interactive tutorial CD-ROM provides algorithmically generated practice exercises that are correlated to the exercises in the textbook. Every practice exercise is accompanied by an example and a guided solution designed to involve students in the solution process. Selected exercises may also include a video clip to help students visualize concepts. The software tracks student activity and scores and can generate printed summaries of students' progress.

MathXL: www.mathxl.com (ISBN 0-321-72611-4)

MathXL is a powerful, online homework, tutorial, and assessment system that accompanies your Addison-Wesley textbook in mathematics or statistics. With MathXL, instructors can create, edit, and assign online homework and tests using algorithmically generated exercises correlated to your textbook. All student work is tracked in MathXL's online gradebook. Students can take chapter tests in MathXL and receive personalized study plans based on their test results. The study plan diagnoses weaknesses and links students directly to tutorial exercises for the objectives they need to study and retest. Students can also access supplemental video clips directly from selected exercises.

MyMathLab

MyMathLab is a series of text-specific, easily customizable online courses for Addison-Wesley textbooks in mathematics and statistics. MyMathLab is powered by CourseCompass—Pearson Education's online teaching and learning environment—and by MathXL—our online homework, tutorial, and assessment system. MyMathLab gives instructors the tools they need to deliver all or a portion of their course online, whether students are in a lab setting or working from home. MyMathLab provides a rich and flexible set of course materials, featuring free-response exercises that are algorithmically generated for unlimited practice and mastery. Students can also use online tools, such as video lectures, animations, and a multimedia textbook, to independently improve their understanding and performance. Instructors can use MyMathLab's homework and test managers to select and assign online exercises correlated directly to the textbook, and they can import TestGen tests into MyMathLab for added flexibility. MyMathLab's online gradebook—designed specifically for mathematics and statistics—automatically tracks students' homework and test results and gives the instructor control over how to calculate final grades. Instructors can also add offline (paper-and-pencil) grades to the MathXL gradebook. MyMathLab is available to qualified adopters. For more information, visit our Web site at www.mymathlab.com or contact your Addison-Wesley sales representative.

PROJECT PORTFOLIOS AND EXTRA EXERCISES PRACTICE WORKBOOK

In order to bridge mathematical concepts and real-world problem solving, two projects have been developed to accompany the text. The *Project Portfolios and Extra Practice Exercises Workbook* can be bundled with student copies of the text. The two projects included are:

1. Building a House
2. Starting and Operating a Business

Both projects are correlated to the chapter content of the text and build and develop as students progress throughout the semester. Students are guided through a series of questions that reference relevant sections in the text, and in working through the problems, students are able to apply the skills and concepts they have learned in a realistic and integrated context. Following is a little more about each project.

Building a House

Students are given the following resources to get them started:
- A financial profile with relevant income, credit, and debt information
- A floor plan
- A pricing option sheet
- Closing costs

This project takes students through the process of building a house. As the project progresses, students will have to think through budget constraints, financing, and construction. The mathematical topics developed in this project include computations, geometry (area, perimeter, surface area, and the Pythagorean theorem), algebraic expressions, solving equations, ratios, percents, and graphs.

Starting and Operating a Business

Students are given the following resources to get them started:
- Profile with initial capital and start-up costs
- Monthly production information, including material costs, labor costs and plant lease

This project takes students through the process of starting and operating a business. As the project progresses, students will be asked to work through budgeting, accounting, and employee benefits and to follow the stock market. The mathematical topics developed in this project include computations, algebraic expressions, solving equations, ratios, percents, and graphs.

Note to instructors: The *Printed Test Bank/Instructor's Resource Guide* includes tips and strategies for incorporating the project portfolios into your course.

Acknowledgments

So many people have helped me in so many ways that I could write another book in saying thank you. Though the words of thanks to follow may be few, no amount of space can contain the genuine gratitude that I feel toward each and every person that gave of themselves to make this work the best that it can be.

I would like to thank the following people who gave of their time in reviewing the text. Their thoughtful input was vital to the development of the text.

Carla Ainsworth, *Salt Lake Community College*
Kevin Bodden, *Lewis and Clark Community College*
Gail Burkett, *Palm Beach Community College*
Susan Caldiero, *Cosumnes River College*
John Close, *Salt Lake Community College*
William Coe, *Montgomery College of Rockville*
Theresa Evans, *Odessa College*
Margaret Finster, *Erie Community College, South*
Randy Gallaher, *Lewis and Clark Community College*
Barbara Gardner, *Carroll Community College*
Nancy Johnson, *Manatee Community College*
Harold Mardones, *Community College of Denver*
Susan Olson, *Blinn College*
Jolene Rhodes, *Valencia Community College*
Hazel Ross, *Monterey Peninsula College*
Becki Saylor, *Sanford Brown College*
Julia Simms, *Southern Illinois University, Edwardsville*
Kristen Stoley, *Blinn College*
Jeff Tennant, *Santa Fe Community College*
Shae Thompson, *Montana State University*
David Whittlesey, *Valencia Community College*

I would like to extend special thanks to Aimee Tait, who encouraged me to pursue this endeavor and put me in touch with Jennifer Crum and Jason Jordan, who gave me the opportunity. I am forever indebted to Jennifer Crum for believing in me and encouraging me throughout the entire process. My heartfelt thanks to Kari Heen and Katie Nopper, who kept me on track, and to Polina Sabinin, whose keen eyes and superb advice were crucial during the development of the manuscript. I would be remiss without also saying thank you to John Hornsby, who answered all my questions about the business.

A very special thank-you to Pamela Watkins and Carol Nessmith for their inspiration and contribution to the development of the study system. Special thanks to Howard Troughton, Sharon O'Donnell, Susan Schroeder, Joanne Kendall, and Vince Koehler for accuracy checking the manuscript.

Thank you to Dennis Schaefer, who created the beautiful design, and to Ron Hampton, who supervised the production process. Thanks to Lisa Laing and all the folks with Pre-Press Company who put together the finished product.

To Ruth Berry, Sharon Smith, Elizabeth Bernardi, and all the people involved in the development of the supplements package, my deepest thanks. To Doreen Kelly, thank you for the exceptional quality of the solutions manuals, and thank you to Laura Hoye for the superb *Printed Test Bank*.

Finally, I'd like to dedicate this work in thanks to my parents, Tom and Janice, who gave me all that I am, and to my wife Laura for her unwavering support and love.

Tom Carson

Why do I have to take this course?

Often, this is one of the first questions students ask when they find out they must take a mathematics course, especially when they believe that they will never use the math again. You may think that you will not use algebra directly in daily life, and you may assume that you can get by knowing just enough arithmetic to balance a checkbook. So, what is the real point of education? Why don't colleges just train students for the jobs they want? The purpose of education is not just job training but also exercise—mental exercise. An analogy that illustrates this quite well is the physical training of athletes.

During the off-season, athletes usually develop an exercise routine that may involve weight lifting, running, swimming, aerobics, or maybe even dance lessons. Athletes often seek out a professional trainer to push them further than they might push themselves. The trainer's job is not to teach an athlete better technique in his or her sport, but to develop the athlete's raw material—to work the body for more strength, stamina, and balance. Educators are like physical trainers, and going to college is like going to the gym. An educator's job is to push students mentally and work the "muscle" of the mind. A college program is designed to develop the raw material of the intellect so the student can be competitive in the job market. After the athlete completes the off-season exercise program, he or she returns to the coach and receives specific technique training. Similarly, when students complete their college education and begin a job, they receive specific training to do that job. If the trainer or teacher has done a good job with hard-working clients, the coaching or job training should be absorbed easily.

Taking this analogy a step further, a good physical trainer finds the athlete's weaknesses and designs exercises that the athlete has never performed before, and then pushes him or her accordingly. Teachers do the same thing—their assignments are difficult in order to work the mind effectively. If you feel "brainstrained" as you go through your courses, that's a good sign that you are making progress, and you should keep up the effort.

The following study system is designed to help you in your academic workouts. As teachers, we find that most students who struggle with mathematics have never really studied math. A student may think, "Paying attention in class is all I need to do." When you watch a teacher do math, however, keep in mind that you are watching a pro. Going back to the sports analogy, you can't expect to shoot a score of 68 in golf by watching Tiger Woods. You have to practice golf yourself in order to learn and improve. The study system outlined in the following pages will help you get organized and make efficient use of your time so that you can maximize the benefits of your course work.

What do I need to do to succeed?

We believe you must have or acquire four prerequisites in order to succeed in college:

1. **Positive Attitude**
2. **Commitment**
3. **Discipline**
4. **Time**

A **Positive Attitude** is most important because commitment and discipline flow naturally from it. Consider Thomas Edison, inventor of the lightbulb. He tried more than 2000 different combinations of materials for the filament before he found the successful combination. When asked by a reporter about all his failed attempts, Edison replied, "I didn't fail once, I invented the lightbulb. It was just a 2000-step process." Recognize that learning can be uncomfortable and difficult, and mistakes are part of the process. Embrace the learning process with its discomforts and difficulties, and you'll see how easy it is to be committed and disciplined.

Commitment means giving everything you've got with no turning back. Consider Edison again. Imagine the doubts and frustrations he must have felt trying material after material for the filament of his lightbulb without success. Yet he forged ahead. Quitting was simply not an option. In Edison's own words, "Our greatest weakness lies in giving up. The most certain way to succeed is always to try just one more time."

Discipline means doing things you should be doing even when you don't want to. According to author W. K. Hope, "Self-discipline is when your conscience tells you to do something and you don't talk back." Staying disciplined can be difficult given all the distractions in our society. The best way to develop discipline is to create a schedule and stick to it.

Make sure you have enough **Time** to study properly, and make sure that you manage that time wisely. Too often, students try to fit school into an already full schedule. Take a moment to complete the exercise that follows under "How do I do it all?" to make sure you haven't overcommitted yourself. Once you have a sense of how much time school requires, read on about the study system that will help you maximize the benefits of your study time.

How do I do it all?

Now that we know a little about what it takes to be successful, let's make sure that you have enough time for school. In general, humans have a maximum of 60 hours of productivity per week. Therefore, as a guide, let's set the maximum number of work hours, which means time spent at your job(s) and at school combined, at 60 hours per week. Use the following exercise to determine the time you commit to your job and to school.

Exercise

Calculate the time that you spend at your job and at school.

1. Calculate the total hours you work in one week.
2. Calculate the number of hours you are in class each week.
3. Estimate the number of hours you should expect to spend outside of class studying. *A general rule is to double the number of hours spent in class.*
4. Add your work hours, in-class hours, and estimated out-of-class hours to get your total time commitment.
5. Evaluate the results. *See below.*

Evaluating the Results

a. If your total is greater than 60 hours, you may find yourself overwhelmed. It may not occur at first, but doing too much for an extended period of time will eventually catch up with you, and something may suffer. It is in your best interest to cut back on work or school until you bring your time commitment under 60 hours per week.
b. If your total is under 60 hours, good. Be sure you consider other elements in your life, such as your family's needs, health problems, commuting, or anything that could make demands on your time. If you do not have enough time for everything, consider what can be cut back. It is important to note that it is far better to pass fewer classes than to fail many.

How do I make the best of my time? How should I study?

We've seen many students who had been making D's and F's in mathematics transform their grades to A's and B's by using the study system that follows.

The Study System

Your Notebook

1. Get a loose-leaf binder so that you can put papers in and take them out without ripping any pages.
2. Organize the notebook into four parts:
   a. Class notes
   b. Homework
   c. Study sheets (a single piece of paper for each chapter onto which you will transfer procedures from your notes)
   d. Practice tests

In Class

Involve your mind completely.

1. **Take good notes.** Use three different colors while taking notes. Most students like using red, blue, and black (pencil).
   - Use the red pen to write *definitions*. Also, use this color to mark problems or items that the instructor indicates will be covered on a test.
   - Use the blue pen to write *procedures* and *rules*.
   - Use the pencil to write *problems* and *explanations*.

   When taking notes, don't just write the solutions to the problems that the instructor works out, but write the explanations as well. To the side of the problem, make notes about each step so that you remember the significance of the steps. Pay attention to examples or issues the instructor emphasizes: they often appear on a test, so make an effort to include them in your notes. Include common errors that the instructor points out, or any words of caution. If you find it is difficult to write and pay attention at the same time, ask your instructor if you can record the lectures with a tape recorder. If your instructor follows the text closely, when he or she points out definitions or procedures in the text, highlight them or write a page reference in your notes. You can then write these referenced items in their proper place in your notes after class.
2. **Answer the instructor's questions.** You should think through every question and answer in your mind, in your notes, or out loud.
3. **Ask questions.** You may find it uncomfortable to ask questions in front of other people, but keep in mind that if you have a question, it is very likely that someone else has the same question. If you still don't feel like asking in class, be sure to ask as soon as class is over. The main thing is to get that question answered as soon as possible.

After Class

Prepare for the next class meeting as if you were going to have a test on everything covered so far.

To make the most of your time, set aside a specific time that is reserved for math. Since there are often many distractions at home, study math while on campus in a quiet place such as the library or tutorial lab. Staying on campus also allows you to visit your instructor or tutorial services if you have a question that you cannot resolve. Here is a systematic approach to organizing your math study time outside of class:

Troubleshooting: For the problems that you do not get correct, first look for simple arithmetic errors. If you find no arithmetic errors, then make sure you followed the procedure rules correctly. If you followed the rules correctly, then you have likely interpreted something incorrectly, either with the problem or the rules. Read the instructions again carefully and try to find similar examples in your notes or in the book. If you still can't find the mistake, go on to something else for a while. Often after taking a fresh look you will see the mistake right away. If all these tips fail to resolve the problem, then mark it as a question for the next class meeting.

1. As soon as possible, go over your notes. Clarify any sentences that weren't quite complete. Fill in any page-referenced material.
2. Read through the relevant section(s) in the text again, and make sure you understand all the examples.
3. Transfer each new procedure or rule to your study sheet for that chapter. Write down important terms and their definitions. Make headings for each objective in the section(s) you covered that day and write the procedures and definitions in your own words.
4. Study the examples worked in class. Transfer each example (without the solution) to the practice test section of your notebook, leaving room to work it out later.
5. Use your study sheet to do the assigned practice problems. As soon as you finish each problem, check your answer in the back of the book or in the *Student's Solutions Manual*. If you did not get it correct, then immediately revisit the problem to determine your error (see the box on troubleshooting). If you are asked to do even-numbered problems, work odd-numbered problems that mirror the even problems. That way you can check your answers for the odd-numbered problems, and then work the even-numbered problems with confidence.
6. After completing the homework, prepare a quiz for yourself. Select one of each type of homework problem. Don't just pick the easy ones! Set the quiz aside for later.
7. After making the quiz, study your study sheet. To test your understanding, write the rules and procedures in your own words. Do not focus on memorizing the wording in the textbook.
8. Now it is time to begin preparing for the next class meeting. Read the next section(s) to be covered. Don't worry if you do not understand everything. The idea is to get some feeling for the topics to be discussed so that the class discussion will actually be the second time you encounter the material. While reading, you might mark points that you find difficult so that if the instructor does not clear them up, you can ask about them. Also, attempt to work through the examples. The idea is for you to do as much as possible on your own before class so that the in-class discussion merely ties together loose ends and solidifies the material.
9. After you have finished preparing for the next day, go back and take the quiz that you made. If you get all the answers correct, then you have mastered the material. If you have difficulty, return to your study sheet and repeat the exercise of writing explanations for each objective.

How do I ace the test?

Preparing for a Test

If you have followed all of the preceding suggestions, then preparing for a test should be quite easy.

1. **Read.** In one sitting, read through all of your notes on the material to be tested. In the same sitting, read through the book, observing what the instructor has highlighted in class. To guide your studies, look at any information or documents provided by your instructor that address what will be on the test. The examples given by the instructor will usually reflect the test content.

2. **Study.** Compare your study sheet to the summary in the book at the end of the chapter. Use both to guide you in your preparation, but keep in mind that the sheet you made reflects what the instructor has emphasized. Make sure you understand everything on your study sheet. Write explanations of the objectives until you eliminate all hesitation about how to approach an objective. The rules and procedures should become second nature.

3. **Practice.** Create a game plan for the test by writing the rule, definition, or procedure that corresponds to each problem on your practice tests. (Remember, one practice test is in your book and the other you made from your notes.) Next, work through the practice tests without referring to your study sheet or game plan.

4. **Evaluate.** Once you have completed the practice tests, check them. The answers to the practice test are in the back of the book. Check the practice test that you made using your notes.

5. **Repeat.** Keep repeating steps 2, 3, and 4 until you get the right answer for every problem on the practice tests.

Taking a Test

1. When the test hits your desk, don't look at it. Instead, do a memory dump. On paper, write down everything you think you might forget. Write out rules, procedures, notes to yourself, things to watch out for, special instructions from the instructor, and so on. This will help you to relax while taking the test.

2. If you get to a problem that you cannot figure out, skip it and move on to another problem. First do all the problems you are certain of and then return to the ones that are more difficult.

3. Use all the time given. If you finish early, check to make sure you have answered every problem. Even if you cannot figure out a problem, at least guess. Use any remaining time to check as many problems as possible by doing them over on separate paper.

If You Are Not Getting Good Results

Evaluate the situation.   What are you doing or not doing in the course? Are you doing all the homework and taking the time to prepare as suggested? Sometimes, people misjudge how well they have prepared. Just like an athlete, to excel, you will need to prepare beyond the minimum requirements. Here are some suggestions:

1. **Go to your instructor.** Ask your instructor for help to evaluate what is wrong. Make use of your instructor's office hours, because this is your opportunity for individual attention.

2. **Get a tutor.** If your school has a tutorial service, and most do, do your homework there so you can get immediate help when you have a question.

3. **Use Addison-Wesley's support materials.** Use the support materials that are available for your text, which include a *Student's Solutions Manual*, the Addison-Wesley Math Tutor Center, videotapes (available on CD-ROM as well), and tutorial software available on CD and online. Full descriptions of these supplements are provided on pages ix–xi of this book.

4. **Join a study group.** Meet regularly with a few people from class and go over material together. Quiz each other and answer questions. Meet with the group only after you have done your own preparation so you can then compare notes and discuss any issues that came up in your own work. If you have to miss class, ask the study group for the assignments and notes.

We hope you find this study plan helpful. Be sure to take the Learning Styles Inventory that follows to help determine your primary learning style. Good luck!

LEARNING STYLES INVENTORY

What is your personal learning style?

A learning style is the way in which a person processes new information. Knowing your learning style can help you make choices in the way you focus on and study new material. Below are fifteen statements that will help you assess your learning style. After reading each statement, rate your response to the statement using the scale below. There are no right or wrong answers.

3 = Often applies   **2** = Sometimes applies   **1** = Never or almost never applies

_____ 1. I remember information better if I write it down or draw a picture of it.

_____ 2. I remember things better when I hear them instead of just reading or seeing them.

_____ 3. When I receive something that has to be assembled, I just start doing it. I don't read the directions.

_____ 4. If I am taking a test, I can visualize the page of text or lecture notes where the answer is located.

_____ 5. I would rather have the professor explain a graph, chart, or diagram to me instead of just showing it to me.

_____ 6. When learning new things, I want to do it rather than hear about it.

_____ 7. I would rather have the instructor write the information on the board or overhead instead of just lecturing.

_____ 8. I would rather listen to a book on tape than read it.

_____ 9. I enjoy making things, putting things together, and working with my hands.

_____ 10. I am able to conceptualize quickly and visualize information.

_____ 11. I learn best by hearing words.

_____ 12. I have been called hyperactive by my parents, spouse, partner, or professor.

_____ 13. I have no trouble reading maps, charts, or diagrams.

_____ 14. I can usually pick up on small sounds like bells, crickets, or frogs, or distant sounds like train whistles.

_____ 15. I use my hands and I gesture a lot when I speak to others.

Write your score for each statement beside the appropriate statement number below. Then add the scores in each column to get a total score for that column.

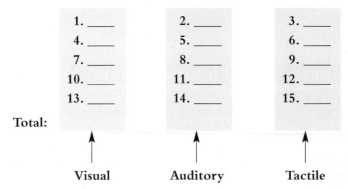

	1. _____	2. _____	3. _____
	4. _____	5. _____	6. _____
	7. _____	8. _____	9. _____
	10. _____	11. _____	12. _____
	13. _____	14. _____	15. _____
Total:	↑	↑	↑
	Visual	**Auditory**	**Tactile**

The largest total of the three columns indicates your dominant learning style.

Visual learners learn best by seeing. If this is your dominant learning style, then you should focus on learning strategies that involve seeing. The color coding in the study system (see page xvii) will be especially important. The same color coding is used in the text. Draw lots of diagrams, arrows, and pictures in your notes to help you see what is happening. Reading your notes, study sheets, and text repeatedly will be an important strategy.

Auditory learners learn best by hearing. If this is your dominant learning style, then you should use learning strategies that involve hearing. After getting permission from your instructor, bring a tape recorder to class to record the discussion. When you study your notes, play back the tape. Also, when you learn rules, say the rule over and over. As you work problems, say the rule before you do the problem. You may also find the videotapes to be beneficial because you can hear explanations of problems taken from the text.

Tactile (also known as kinesthetic) learners learn best by touching or doing. If this is your dominant learning style, you should use learning strategies that involve doing. Doing lots of practice problems will be important. Make use of the Your Turn exercises in the text. These are designed to give you an opportunity to do problems that are similar to the examples as soon as a topic is discussed. Writing out your study sheets and doing your practice tests repeatedly will be important strategies for you.

Note that the study system developed in this text is for all learners. Your learning style will help you decide what aspects and strategies in the study system to focus on, but being predominantly an auditory learner does not mean that you shouldn't read the textbook, do lots of practice problems, or use the color-coding system in your notes. Auditory learners can benefit from seeing and doing, and tactile learners can benefit from seeing and hearing. In other words, do not use your dominant learning style as a reason for not doing things that are beneficial to the learning process. Also, remember that the Learning Strategy boxes presented throughout the text provide tips to help you use your personal learning style to your advantage.

This Learning Styles Inventory is adapted from *Cornerstone: Building on Your Best* by Montgomery/Moody/Sherfield, © 2000. Reprinted by permission of Prentice-Hall, Inc., Upper Saddle River, NJ.

Whole Numbers

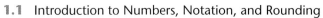

66 *The strokes of the pen need deliberation as much as the sword needs swiftness.* 99

—JULIA WARD HOWE,
POET

66 *As to the adjective, when in doubt, strike it out.* 99

—MARK TWAIN,
AMERICAN WRITER

66 *Good writing is clear thinking made visible.* 99

—BILL WHEELER,
ARTIST

Taking Notes

Good note taking is essential to success in college. If you have not read through the study system in the *To the Student* section at the beginning of this text, take a moment to do so now. You will find that this text follows the color-coding suggested in the note-taking discussion. Definitions will appear in red. Procedures and rules will appear in blue.

When taking notes, do your best to record not only what your instructor writes but also what your instructor says about an example. Outline the process involved in solving the problems. Note the instructor's cautions and warnings about common errors. If your instructor follows the text, be sure to include page references in your notes and highlight important points in the text.

Of course, it is difficult to try to understand everything and simultaneously write it all down. Try to write down the essence of the sentence and leave out nonessential words like "the." Also, develop codes for words that are used frequently. For example, $\therefore$ could mean "therefore" and + could mean "and." It is also helpful to tape lectures. (Ask your instructor before taping lectures, however.)

As soon as possible after class, tidy up your notes or even rewrite them. If you tape the lecture, you can listen to the tape as you rewrite the notes. This will allow you to fill in those nonessential words and replace any symbols with actual words, if needed.

Introduction to Numbers, Notation, and Rounding

OBJECTIVES

1 Name the digit in a specified place.

2 Write numbers in standard and expanded form.

3 Write the word name for a given number.

4 Use $<$, $>$, or $=$ to make a true statement.

5 Round numbers to a specified place.

What are **numbers**? They're not things we can see or touch. So what are they? What do we do with them?

DEFINITION   Numbers: Amounts or quantities.

Applications have been found for most mathematical ideas. There are some numbers and ideas in mathematics for which no practical use has yet been found. Many people merely play with mathematics for the sheer beauty and enjoyment of it. In this way, mathematics is like art.

In the applications in this text, we will use numbers to describe measurements, such as how much liquid is in a container or how fast an object is traveling or how much distance is between objects. If a number is used to describe a measurement, it will have a unit attached, such as feet (ft.) or meters (m) or seconds (sec.).

Let's be a little more specific about some numbers. Numbers are classified into groups called **sets.** In this first chapter, we'll focus on the set of **whole numbers.** The set of whole numbers contains 0 and all of the **natural numbers,** which are the counting numbers 1, 2, 3, and so on. Because every natural number is in the set of whole numbers, we say the set of natural numbers is a **subset** of the set of whole numbers.

DEFINITIONS   **Set:** A group of elements.

Subset: A set within a set. | Three periods, called an ellipsis, mean that the numbers continue forever. ▼

Natural numbers: 1, 2, 3, . . .

Whole numbers: 0, 1, 2, 3, . . .

Roman

Greek

Egyptian

Chinese

Hindu

Arabic

Babylonian

Although every natural number is a whole number, not every whole number is natural. The number 0 is a whole number but not a natural number. The diagram below shows the set of natural numbers as a subset of the set of whole numbers.

We use symbols called *numerals* or *digits* to write numbers. There are 10 numerals in our numeral system:

$$0, 1, 2, 3, 4, 5, 6, 7, 8, 9$$

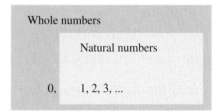

Whole numbers

Natural numbers

0,   1, 2, 3, ...

OF INTEREST

Numerals have been written many different ways throughout history. It is generally accepted that our modern numerals are derived from Hindu-Arabic forms. The Babylonians used wedge-shaped marks in clay tablets to represent numbers. The ancient Greeks used the letters of their alphabet to represent numbers. We still find Roman numerals used in clocks, in architecture, in books, and even in numbering the Super Bowls.

Since our numeral system has 10 numerals, we say it is a **base-10** system. We use combinations of numerals to represent numbers greater than 9. This is called a *place value system.*

OBJECTIVE 1 Name the digit in a specified place.

Our Place Value System

Figure 1–1 Place Values

Place Values															
Trillions period			Billions period			Millions period			Thousands period			Ones period			
Hundred trillions	Ten trillions	Trillions	Hundred billions	Ten billions	Billions	Hundred millions	Ten millions	Millions	Hundred thousands	Ten thousands	Thousands	Hundreds	Tens	Ones	

◄ **. . .**

Notice how the places are grouped in threes. These groups are called *periods*. When we write numbers using the place-value system, we say we are writing in *standard form*. When writing numbers in standard form, we separate the periods with commas, as in 3,243,972. The chart extends to the left indefinitely.

Note: The comma is optional in a four-digit number. For example, we can write 4,538 or 4538.

CONNECTION Think of the place-value chart as the money tray in a cash register. A digit in a particular place is like having that many bills in that tray of the register. For example, if we place a numeral like 7 in the thousands place, it's the same as having 7 one-thousand-dollar bills in the register, which is worth a total of $7,000.

Example 1 What digit is in the thousands place in 209,812?

Answer: 9

Explanation: The digit in the thousands place is the fourth digit from the right.

Do Your Turn 1 ▷

OBJECTIVE 2 Write numbers in standard and expanded form.

When we write a number in standard form such as 430, what are we saying?

Think of the money tray analogy:

The 4 in the hundreds place means 4 hundreds.
The 3 in the tens place means 3 tens.
The 0 in the ones place means 0 ones.

If we add those amounts together we have 430.

4 hundreds + 3 tens + 0 ones = 430

or

4 hundreds + 3 tens = 430

Note: We do not need to represent the 0 ones because it means there is a lack of bills in that slot of the money tray and mathematically the 0 will not affect the total.

Mathematically, 430 is the result of multiplying the 4 times 100, the 3 times 10, and the 0 times 1 and then adding those results together.

$$4 \times 100 + 3 \times 10 + 0 \times 1 \quad \text{or just} \quad 4 \times 100 + 3 \times 10$$

A number written this way is in *expanded notation* or *expanded form*.

PROCEDURE *To write a number in expanded form:*
1. Write each digit multiplied by its place value.
2. Express it all as a sum.

Your Turn 1

a. What digit is in the hundred thousands place in 62,407,981?

b. What digit is in the ten millions place in 417,290,006?

Answers to Your Turn 1
a. 4 b. 1

Write in expanded form.

a. 82,469

b. 7,082,049

c. 410,159,200

Example 2 Write the number in expanded form.

a. 57,483

Answer: $5 \times 10{,}000 + 7 \times 1000 + 4 \times 100 + 8 \times 10 + 3 \times 1$

Explanation: We wrote each digit multiplied by its respective place value and expressed it all as a sum. (Notice if we were to calculate, we'd get 57,483.)

b. 4,705,208

Answer: $4 \times 1{,}000{,}000 + 7 \times 100{,}000 + 5 \times 1000 + 2 \times 100 + 8 \times 1$

Explanation: We wrote each digit (except 0's) multiplied by its respective place value and expressed it all as a sum.

◁ **Do Your Turn 2**

Let's write expanded form in standard form.

PROCEDURE *To change a number from expanded form to standard form, write each digit in the place indicated by the corresponding place value.*

Write in standard form.

a. $2 \times 100{,}000 + 3 \times 10{,}000 + 1 \times 1000 + 5 \times 100 + 9 \times 10 + 8 \times 1$

b. 3 hundred thousands + 5 thousands + 9 tens + 7 ones

c. $9 \times 100{,}000{,}000 + 1 \times 10{,}000 + 4 \times 1000 + 7 \times 100$

Example 3 Write the number in standard form.

a. $6 \times 10{,}000 + 9 \times 1000 + 2 \times 100 + 5 \times 10 + 3 \times 1$

Answer: 69,253

Explanation: We wrote each digit in the place indicated by the corresponding place value.

b. $9 \times 1{,}000{,}000 + 2 \times 10{,}000 + 7 \times 100 + 9 \times 10 + 3 \times 1$

Answer: 9,020,793

Explanation: Notice that some places were skipped. There were no hundred thousands and no thousands in the expanded form, so we placed 0's in the hundred thousands and thousands places in the standard form.

◁ **Do Your Turn 3**

LEARNING STRATEGY

If you are using the note-taking system from the *To the Student* section of this text, remember to use a blue pen for procedures and rules, and a red pen for definitions.

OBJECTIVE 3 Write the word name for a given number.

The word name for a number is the way we say the number.

PROCEDURE *To write a word name, work from left to right through the periods.*
1. Write the name of the digits in the left-most period.
2. Write the period name followed by a comma.
3. Repeat steps 1 and 2 until you get to the ones period. We do not follow the ones period with its name.

Answers to Your Turn 2
a. $8 \times 10{,}000 + 2 \times 1000 + 4 \times 100 + 6 \times 10 + 9 \times 1$
b. $7 \times 1{,}000{,}000 + 8 \times 10{,}000 + 2 \times 1000 + 4 \times 10 + 9 \times 1$
c. $4 \times 100{,}000{,}000 + 1 \times 10{,}000{,}000 + 1 \times 100{,}000 + 5 \times 10{,}000 + 9 \times 1000 + 2 \times 100$

Answers to Your Turn 3
a. 231,598 b. 305,097
c. 900,014,700

WARNING Do not write the word *and* anywhere in a word name for a whole number. The word *and* takes the place of a decimal point. We'll see this in Chapter 6.

Example 4 The Earth is, on average, 92,958,349 miles from the Sun. Write the word name for 92,958,349.

Answer: ninety-two million, nine hundred fifty-eight thousand, three hundred forty-nine

Explanation: Starting at the left, we wrote 92 as "ninety-two" and then stated the period name, "million." Next we wrote 958 as "nine hundred fifty-eight" with its period name, "thousand." Finally, we wrote 349 as "three hundred forty-nine." Since this was the ones period we did not write the period name.

Do Your Turn 4▶

OBJECTIVE **4** Use <, >, or = to make a true statement.

When we use an equal sign to show that two amounts are equal, we call the statement an **equation.**

DEFINITION **Equation:** A mathematical statement that contains an equal sign.

For example, 12 = 12 is an equation. It is read as "twelve is equal to twelve." Because 12 is equal to 12, we say the equation is true. An equation can be false as well. 12 = 9 is an equation that is false because 12 and 9 are not equal.

When two amounts are not equal, we can use inequality symbols to write a true statement. A statement that contains an inequality symbol is an **inequality.**

DEFINITION **Inequality:** A mathematical statement that contains an inequality symbol.

We will consider two inequality symbols:

Greater than   >
Less than   <

In comparing 12 and 9, we can write a true statement by using an appropriate symbol.

We can say: 12 **is greater than** 9   or   9 **is less than** 12

Translation: 12 > 9   9 < 12

LEARNING STRATEGY

Notice that the < and > always open towards the greater value. If you are a visual learner, imagine the symbols as mouths that always open to eat the bigger meal.

Example 5 Use <, >, or = to make a true statement.

a. 159,208 ? 161,000

Answer: 159,208 < 161,000

Explanation: Because 159,208 is less than 161,000 we used the less than symbol.

b. 48,090 ? 12,489

Answer: 48,090 > 12,489

Explanation: Because 48,090 is greater than 12,489 we used the greater than symbol.

Do Your Turn 5▶

Your Turn 4

Write the word name.

a. 42,409 (2002 median family income; *Source:* U.S. Bureau of the Census)

b. 10,987,900,000,000 (2003 Gross Domestic Product; *Source:* U.S. Bureau of the Census)

c. 847,716 (Diameter of the Sun in miles)

Your Turn 5

Use < , > , or = to make a true statement.

a. 15,907 ? 15,906

b. 1,291,304 ? 1,291,309

c. 64,108 ? 64,108

d. 24,300 ? 25,300

Answers to Your Turn 4
a. forty-two thousand, four hundred nine
b. ten trillion, nine hundred eighty-seven billion, nine hundred million
c. eight hundred forty-seven thousand, seven hundred sixteen

Answers to Your Turn 5
a. 15,907 > 15,906
b. 1,291,304 < 1,291,309
c. 64,108 = 64,108
d. 24,300 < 25,300

OBJECTIVE 5 Round numbers to a specified place.

The speed of light in a vacuum is 299,792,458 meters per second (abbreviated m/sec.). However, because this value is rather tedious to say and work with, it is usually rounded to 300,000,000 meters per second. We round numbers to make them easier to communicate. We will see later that we can round numbers to quickly estimate calculations.

When we round, we must decide on a place value or have a place value specified. We then have to determine whether the given number is closer to the nearest value above or below it in the specified place. For example, suppose we decide to round 43,357 to the nearest thousand.

The nearest thousands to 46,357 are 46,000 and 47,000. We must determine which whole thousand it is closer to. Because 46,357 is below the halfway point (46,500), we rounded down to 46,000.

WARNING Be sure to include all 0's after the rounded place. Instead of 46, be sure to write 46,000. Think about money. $46 is very different from $46,000.

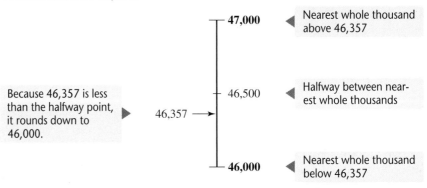

Notice that the digit to the right of the place to be rounded determines whether the number is above or below the halfway point. But suppose that digit is a 5, that is, at the halfway point. In those cases, we agree to always round up so that everyone rounds in a consistent way.

Our example suggests the following procedure:

PROCEDURE *To round a number to a given place value, consider the digit to the right of the desired place value.*
If this digit is 5 or greater, round up.
If this digit is 4 or less, round down.

Example 6 Round 43,572,991 to the specified place.

a. Hundred thousands

Solution: 43,5̸72,991

    Hundred   The digit to the right of 5 is 7, which is greater than 5, so we round up.
    thousands

Answer: 43,600,000

b. Ten thousands

Solution: 43,572,991

    Ten thousands   The digit to the right of 7 is 2, which is less than 4, so we round down.

Answer: 43,570,000

c. Millions

Solution: 43,572,991

↗

Millions   The digit to the right of 3 is 5, so we agree to round up.

Answer: 44,000,000

d. Hundreds

Solution: 43,572,991

↗

Hundreds   The digit to the right of 9 is 9, which is greater than 5, so we round up. Note that when we round the 9 in the hundreds place up, it becomes 10. Since 10 hundreds is 1000, we must increase the thousands place by 1 so that the 2 becomes a 3. This process of adding to the next place to the left is often called *carrying*.

Answer: 43,573,000

Do Your Turn 6 ▶

Your Turn 6

Round 602,549,961 to the specified place.

a. Ten thousands

b. Tens

c. Millions

d. Thousands

e. Hundreds

In the real world, we are not told what place to round to, so we round to a place that makes sense for the situation. Some questions to consider are

How precise must the numbers be in the situation?
How accurately can the amounts involved be measured?
Are others depending on what we do?
Are lives at stake?

For rough estimation purposes, we can round so that there is only one digit that is not 0. This means to round to the place farthest to the left.

Example 7

Round each so that there is only one nonzero digit.

a. 36,568

Solution: 36,568

↗

Farthest   The digit to the right of 3 is 6, which is greater
left   than 5, so we round up.

Answer: 40,000

b. 621,905

Solution: 621,905

↗

Farthest   The digit to the right of 6 is 2, which is less
left   than 5, so we round down.

Answer: 600,000

▲

Note: Remember, when rounding down, the digit in the rounded place remains the same.

Do Your Turn 7 ▶

Your Turn 7

Round each number so that there is only one nonzero digit.

a. 27,502,341

b. 6,128,200

c. 453,219

Answers to Your Turn 6
a. 602,550,000
b. 602,549,960
c. 603,000,000
d. 602,550,000
e. 602,550,000

Answers to Your Turn 7
a. 30,000,000
b. 6,000,000
c. 500,000

1.1 Exercises

For
Extra
Help

 Videotape
DVT

 Addison-Wesley
Tutor Center

 Math XL
Math XL

 MyMathLab

Student Solutions
Manual

1. List the numbers in the set of natural numbers.

2. List the numbers in the set of whole numbers.

3. Is every whole number a natural number? Explain.

4. How do you write a number in expanded notation?

5. How do you write the word name of a number?

6. Explain how to round a number.

For Exercises 7–10, name the digit in the requested place in the number 56,324,092.

7. Hundreds

8. Thousands

9. Ten millions

10. Hundred thousands

For Exercises 11–14, name the place held by the 7.

11. 2,457,502

12. 57,414

13. 9,706,541

14. 70,412,581

For Exercises 15–20, write the number in expanded form.

15. 24,319

16. 78,625

17. 5,213,304

18. 2,410,512

19. 93,014,008

20. 86,300,905

For Exercises 21–26, write the number in standard form.

21. 8 thousands + 7 hundreds + 9 tens + 2 ones

22. $9 \times 10,000 + 2 \times 1000 + 5 \times 100 + 8 \times 10 + 1 \times 1$

23. $6 \times 1,000,000 + 3 \times 10,000 + 9 \times 1000 + 2 \times 10$

24. 8 hundred millions + 7 thousands + 2 hundreds + 3 ones

25. $4 \times 10,000,000 + 9 \times 100,000 + 8 \times 10,000 + 1 \times 100 + 9 \times 1$

26. $7 \times 100,000,000 + 2 \times 1,000,000 + 5 \times 10,000 + 8 \times 10$

For Exercises 27–32, write the word name.

27. 7768 (diameter of Earth in miles)

28. 29,028 (height of the peak of Mt. Everest in feet)

29. 290,810,000 (estimated resident population of the United States as of July 1, 2003, according to the U.S. Bureau of the Census)

30. 6,914,000,000,000 (U.S. gross outstanding national debt in 2003)

31. 186,171 (speed of light in miles per second)

32. 299,792,458 (speed of light in meters per second)

For Exercises 33–38, use <, >, or = to make a true statement.

33. 599 ? 899

34. 88,332 ? 88,332

35. 4,299,308 ? 4,298,308

36. 89,900 ? 89,902

37. 609,001 ? 609,001

38. 9911 ? 9199

For Exercises 39–46, round 5,652,992,481 to the specified place.

39. Thousands

40. Hundred millions

41. Millions

42. Ten millions

43. Billions

44. Hundred thousands

45. Hundreds

46. Ten thousands

For Exercises 47–52, round each number so that there is only one nonzero digit.

47. 32,607

48. 281,506

49. 851,220

50. 4,513,541

51. 8723

52. 54,298

53. The distance from Earth to the Sun is 92,958,349 miles. Round the distance to a reasonable place. Why did you round to the place you chose?

54. The U.S. gross national debt was $6,914,000,000,000 in 2003. Round the debt to a reasonable place. Why did you round to the place you chose?

1.2 Adding, Subtracting, and Solving Equations with Whole Numbers

OBJECTIVES

1 Add whole numbers.

2 Estimate sums.

3 Solve applications involving addition.

4 Subtract whole numbers.

5 Solve equations containing a missing addend.

6 Solve applications involving subtraction.

7 Solve applications involving addition and subtraction.

OF INTEREST

In the 1400s, the Latin word *et*, which means "and," was used to indicate addition. The writers of the time would generally write the word quickly so that the e and t would run together, creating our modern + sign.

Source: D. E. Smith, *History of Mathematics.*

OBJECTIVE 1 Add whole numbers.

Let's examine the operation **addition.**

DEFINITION Addition: The arithmetic operation that combines amounts.

When we write an addition sentence, the *addends* are the parts that are added and the *sum* is the answer.

$$3 + 5 = 8$$

↑   ↑   ↑
Addends   Sum

Notice how the order of addends can be changed without affecting the sum.

$$5 + 3 \text{ and } 3 + 5 \text{ both make a sum of 8.}$$

This fact about addition is called the *commutative property*. The word *commutative* comes from the root word *commute*, which means to move from one place to another. That's exactly what happens when we move the addends—they commute.

COMMUTATIVE PROPERTY OF ADDITION

Changing the order of addends does not affect the sum.

In math language: $a + b = b + a$, where a and b are any numbers.

We add whole numbers according to place value. Most people prefer to stack the numbers aligning the like place values.

PROCEDURE *To add whole numbers:*
1. Stack with corresponding place values aligned.
2. Add the digits.

Sometimes when we add digits, the sum is more than 9, so we have to carry the extra to the next column.

Example 1 Add.

a. $423 + 64$

Solution:
$$\begin{array}{r} 423 \\ +64 \\ \hline 487 \end{array} \quad \text{or} \quad \begin{array}{r} 64 \\ +423 \\ \hline 487 \end{array}$$

▲   ▲

Note: The commutative property allows us to stack either way.

CONNECTION When we add the 2 and 6 digits we add 2 tens and 6 tens, which makes 8 tens. In expanded form:

4 hundreds + 2 tens + 3 ones
+ 6 tens + 4 ones
—————————————————
4 hundreds + 8 tens + 7 ones = 487

Explanation: We stacked the numbers with the ones and tens places aligned, then added the corresponding digits.

b. $5408 + 916$

Solution:
$$\begin{array}{r} \overset{11}{5408} \\ +916 \\ \hline 6324 \end{array}$$

Explanation: Notice that when we add the 8 and 6, we get 14. The 4 is placed in the ones place. The extra 10 in 14 is expressed as a 1 over the tens column of digits and is added with the other tens digits. The same thing happens with the 9 and 4 in the hundreds column. Finally, we added the 1 and 5 in the thousands column.

Do Your Turn 1 ▶

When we have three or more addends, we can group them any way we wish. Consider 2 + 3 + 4. We can add the 2 and 3 first, which is 5, and then add the 4 for a total of 9. Or, we can add the 3 and 4 first, which is 7, then add the 2 to again total 9.

This property of addition is called the *associative property* because to associate is to group. To write the associative property symbolically, we use parentheses to indicate the different ways we can group the addends.

$(2 + 3) + 4$ indicates add the 2 + 3 first, then add 4 to the result.
$2 + (3 + 4)$ indicates add the 3 + 4 first, then add that result to 2.

Either way, we get 9.

ASSOCIATIVE PROPERTY OF ADDITION

Grouping three or more addends differently does not affect the sum.

In math language: $(a + b) + c = a + (b + c)$, where a, b, and c are any numbers.

The commutative and associative properties allow for flexibility in the way we add. We can change the order of the addends or group the addends any way we wish without affecting the sum.

Example 2 Add. 35 + 60 + 18

Solution:
$$\begin{array}{r} \overset{1}{3}5 \\ 60 \\ +18 \\ \hline 113 \end{array}$$

CONNECTION

The addition properties tell us we can rearrange or group the addends in 35 + 60 + 18 differently and the sum will still be 113. Here are a couple of examples:

35 + 60 + 18	or	35 + 18 + 60
= 35 + 78		= 53 + 60
= 113		= 113

Explanation: We stacked the numbers by place value and added the digits.

Do Your Turn 2 ▶

OBJECTIVE 2 Estimate sums.

Often, we are not as much concerned about the actual answer to a calculation as we are about getting a quick approximation. This is called an *estimate*. We can estimate a calculation by first rounding the numbers and then adding the rounded numbers. How much we round depends on how close we want to be to the actual answer.

How do we decide what place to round the numbers to? First, it is important to be consistent. If you round one number to hundreds, then round the rest to the same place. Second, if we want an estimate that is close to the actual value, round only a

Your Turn 1

Add.

a. 527 + 41

b. 5802 + 549

c. 59,481 + 8574

Your Turn 2

Add.

a. 48 + 70 + 62

b. 114 + 85 + 30 + 79

c. 35,604 + 907 + 3215 + 42,008

Answers to Your Turn 1
a. 568 b. 6351 c. 68,055

Answers to Your Turn 2
a. 180 b. 308 c. 81,734

few places. If we just want a rough estimate, we can round the smallest number as much as possible, and then round the rest to the same place.

Your Turn 3

Estimate each sum by rounding, then find the actual sum.

a. 67,482 + 8190

b. 4586 + 62 + 871

Example 3 Estimate by rounding, then calculate the actual sum.

a. 68,214 + 4318

Estimate:	68,000	Actual:	68,214
	+4,000		+4,318
	72,000		72,532

Explanation: To estimate the sum, we first rounded the smallest number so that it has only one nonzero digit, so 4318 became 4000. Then, to be consistent, we rounded 68,214 to thousands as well and got 68,000. Last, we added the rounded numbers. To find the actual sum, we added the numbers as they were given. Notice that the estimate is quite close to the actual sum, which reassures us that the actual sum is reasonable.

b. 5896 + 539 + 72

Estimate:	5900	Actual:	5896
	540		539
	+70		+72
	6510		6507

Explanation: To estimate the sum, we began by rounding the smallest number, 72, to 70. Then to be consistent, we rounded the other two numbers to tens as well. Last, we added the rounded numbers. To find the actual sum, we added the numbers as they were given. Notice that the estimate affirms the actual sum.

Example 3 suggests the following procedure.

PROCEDURE *To estimate a calculation:*
1. Round the smallest number.
2. Round all other numbers to the same place as the smallest number.
3. Perform the calculation with rounded numbers.

◁ **Do Your Turn 3**

OBJECTIVE 3 Solve applications involving addition.

Often the words in a problem's statement can help us determine what arithmetic operation to use to solve the problem. We refer to those words as *key words*. Here are some key words and key questions that help us recognize addition:

KEY WORDS FOR ADDITION

Add, plus, sum, total, increased by, more than, in all, altogether, perimeter

Key Questions "How much in all?"
"How much altogether?"
"What is the total?"

Answers to Your Turn 3
a. Estimate: 75,000
   Actual: 75,672
b. Estimate: 5520
   Actual: 5519

Example 4 Bob and Kim find out that the basic cost of building the home they have chosen is $112,500. They decide they want to add some extra features. A wood-burning fireplace costs an additional $980. They also want to upgrade the fixtures and appliances at a cost of $2158. What will be the total price of the house?

Solution: Notice the word *total* in the question. Because the upgrades are costs in *addition* to the base price, we must add.

$$\begin{array}{r} {\scriptstyle 1\ 1} \\ 112,500 \\ 2,158 \\ +980 \\ \hline 115,638 \end{array}$$

Check by estimating:
$$\begin{array}{r} 113,000 \\ 2,000 \\ +1,000 \\ \hline 116,000 \end{array}$$

This is reasonably close to the actual sum.

Answer: The total price of the house is $115,638.

Do Your Turn 4 ▶

Suppose we want to construct a fence around some property. To get a sense of how much fencing material we need, we would measure or calculate the total distance around the space to be fenced. The total distance around a shape is called **perimeter**.

DEFINITION Perimeter: The total distance around a shape.

PROCEDURE *To find the perimeter of a shape, add the lengths of all the sides of the shape.*

Example 5 The Jensons want to put a border on the walls of their son's room at the ceiling. The room is a 10-foot-wide by 12-foot-long rectangle. How much border material must they buy?

Solution: To find the total amount of material, we need the total distance around the room. It is helpful to draw the picture.

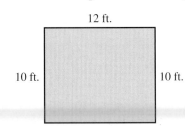

12 ft.

10 ft.   10 ft.

12 ft.

Perimeter = total distance around

$$P = 12 + 10 + 12 + 10$$
$$P = 44$$

Answer: The Jensons will need 44 ft. of border material.

Check by estimating: $10 + 10 + 10 + 10 = 40$ ft.

You probably noticed that we used the letter P in our example. Writing the word *perimeter* each time becomes tedious, so we abbreviate to P. A letter or symbol used to represent an unknown value is called a **variable.**

DEFINITION Variable: A symbol that can vary or change in value.

The word *variable* comes from the root word *vary*. The word was chosen because the same symbol may have different values in different problems or may even mean something totally different in a different context. For example, the letter P in a geometry context may mean perimeter while in a financial context it might mean principal.

Your Turn 4

Solve.

a. Below are the base price and prices for upgrades. Estimate the total cost of the house with the upgrades by rounding so that the numbers are easy to work with but are still fairly accurate. (Answers may vary.)

House base price = $125,480

Deck = $1280

Fireplace = $420

Kitchen upgrade = $675

b. Sarah is planning her budget for the week. She has $642 in her checking account and $283 in her savings account. She owes rent of $450. She knows groceries will be around $100, and she also needs to get her car serviced, which should be $125 according to the advertisement. Will she have enough money to cover these expenses?

Answers to Your Turn 4
a. $127,900 b. yes

Some symbols do not vary in value. We call these symbols **constants.**

> **DEFINITION** | **Constant:** Any symbol that does not vary in value.

All numerals are constants. For example, the symbol 3 always means three, so it is a constant.

◀ **Do Your Turn 5**

OBJECTIVE 4 Subtract whole numbers.

Subtraction is the *inverse* of addition. Addition and subtraction are inverse operations because they *undo* each other. If $5 + 4 = 9$, then we can undo the addition with subtraction like this: $9 - 4 = 5$. There are actually three ways of interpreting subtraction.

> **DEFINITION** | **Subtraction:** An operation of arithmetic that can be interpreted as
> 1. Take away
> 2. Difference
> 3. Missing addend

We interpret subtraction to mean *take away* when we remove an amount from another amount.

We interpret subtraction as *difference* when we must find the distance between two positions. For example, on a highway if you are at mile marker 42 and later you see you are at mile marker 60, you can use subtraction to find the difference between the two positions and thus determine how far you have traveled.

We interpret subtraction as a *missing addend* when an addend, which is part of an addition statement, is unknown. For example, in $(?) + 4 = 9$, the first addend is unknown. This corresponds to having $4 and finding how much more is needed to bring the total up to $9. We use subtraction to find that unknown addend.

When we write a subtraction sentence, the *subtrahend* is subtracted from the *minuend* and the *difference* is the answer.

$$9 - 4 = 5$$

Minuend Subtrahend Difference

CONNECTION

Since addition and subtraction are inverse operations we can check the accuracy of $9 - 4 = 5$ by adding $5 + 4 = 9$.

Let's look at how we subtract. We stack according to place value, just like in addition. However, subtraction is *not* commutative, which means we cannot change the order of the numbers in the minuend and subtrahend positions. In subtraction if we change the order of the minuend and subtrahend, we get different results.

$7 - 4$ is 3, but $4 - 7$ is not 3. When we discuss integers in Chapter 2, we'll learn that $4 - 7$ is -3.

Conclusion: To find the difference of two whole numbers, we stack the greater number on top, align the place values, then subtract the corresponding digits. For example, consider $795 - 32$.

Note: Because 795 is the greater number, it goes on top, and we subtract the digits.

$$\begin{array}{r} 795 \\ -32 \\ \hline 763 \end{array}$$

Check: $\begin{array}{r} 763 \\ +32 \\ \hline 795 \end{array}$

Note: We can use the inverse operation to check. Adding the difference, 763, to the subtrahend, 32, should equal the minuend, 795.

CONNECTION When we study negative numbers in Chapter 2, we will see that the greater number is not always the minuend.

When subtracting, if a digit in the top number is less than the digit directly below, we must rename the top number. Consider $8982 - 765$.

$$\begin{array}{r} ^{712} \\ 8\cancel{9}\cancel{8}2 \\ -765 \\ \hline 8217 \end{array}$$

Because the digit 2 is less than the 5 beneath it, we rename 2 by taking 1 ten from the 8 tens and adding that 10 to 2 to make 12. This process is often called *borrowing*. When we take 1 ten from the 8 tens it becomes 7 tens. Notice this does not change the value of 8982. We merely moved 1 ten from the tens place over to the ones place so that we could carry out the subtraction.

CONNECTION

Think about the cash register analogy. Suppose there are two $1 bills and eight $10 bills in the register, which amount to $82. To give more than two $1 bills out in change, the cashier must break one of the $10 bills. The manager gives the cashier ten $1 bills in exchange for one $10 bill. There are now seven $10 and twelve $1 in the register, which still amount to $82.

PROCEDURE *To subtract whole numbers:*

1. Stack the greater number on top of the smaller number, aligning the place values.
2. Subtract the digits in the bottom number from the digits directly above. If a digit in the top number is less than the digit beneath it, then rename the top digit.

Example 6 Subtract. $45,002 - 8,473$

Solution:
$$\begin{array}{r} ^{14\ \ 9\ 9} \\ ^{3\ \cancel{4}\ \cancel{10}\cancel{10}12} \\ 4\cancel{5},\cancel{0}\cancel{0}2 \\ -8,473 \\ \hline 36,529 \end{array}$$

Check by adding:
$$\begin{array}{r} ^{1\ 11} \\ 36,529 \\ +8,473 \\ \hline 45,002 \end{array}$$

Check by estimating:
$$\begin{array}{r} ^{315} \\ 4\cancel{5},000 \\ -8,000 \\ \hline 37,000 \end{array}$$

Explanation: The tens and hundreds places contain 0's so when we renamed the 2 we had to take 1 thousand from the 5 thousands to make 10 hundreds. Then we took 1 hundred from the 10 hundreds to make 10 tens. This, in turn, allowed us to take 1 ten from the 10 tens to make 12 in the ones place.

Do Your Turn 6 ▶

Your Turn 6

Estimate the difference by rounding, then calculate the actual difference.

a. $48,975 - 6241$

b. $5941 - 218$

c. $520,048 - 63,793$

OBJECTIVE 5 Solve equations containing a missing addend.

Suppose you decide to buy an appliance that costs $350. If you have $200, how much more do you need in order to have enough to make the purchase? This problem translates to an equation with a missing, or unknown, addend.

$$200 + (?) = 350$$

Because the additional amount needed is the difference between the $350 price and the $200 that you have, you can write a **related subtraction sentence** to calculate the needed amount.

Answers to Your Turn 6
a. Estimate: 43,000
  Actual: 42,734
b. Estimate: 5700
  Actual: 5723
c. Estimate: 460,000
  Actual: 456,255

We write the related subtraction for $200 + (?) = 350$ below.

$$(?) = 350 - 200$$
$$(?) = 150$$

Note: In the related subtraction sentence, we subtract the known addend, 200, from the sum, 350.

Answer: You will need $150.

PROCEDURE   *To find a missing addend, write a related subtraction sentence. Subtract the known addend from the sum.*

Earlier, we defined a variable as a symbol or letter that can vary in value, and we said that we use these symbols to represent unknown amounts. Because (?) indicates an unknown amount, it is a variable. However, it is more common to use a letter such as x, y, or t for a variable. From here on we will use letter variables.

Instead of: $200 + (?) = 350$ **we write:** $200 + x = 350$
$$(?) = 350 - 200 \qquad\qquad x = 350 - 200$$
$$(?) = 150 \qquad\qquad\qquad x = 150$$

The number 150 is the **solution** to the equation because it can replace the variable and make the equation true. The act of finding this solution is called *solving* the equation.

DEFINITION   Solution: A number that can replace the variable(s) in an equation and make the equation true.

The definition for *solution* suggests a method for checking. If 150 is correct, it should replace the x in $200 + x = 350$ and make the equation true.

Check: $200 + x = 350$
$$200 + 150 \stackrel{?}{=} 350$$
$$350 = 350$$

Note: The symbol $\stackrel{?}{=}$ is used to indicate that we are asking whether $200 + 150$ is equal to 350.

Because $200 + 150 = 350$, 150 is in fact the solution to $200 + x = 350$.

Your Turn 7

Solve and check.

a. $14 + x = 20$

b. $y + 32 = 60$

c. $140 + t = 216$

d. $m + 89 = 191$

Answers to Your Turn 7
a. $x = 6$ b. $y = 28$ c. $t = 76$
d. $m = 102$

Example 7   Solve and check. $45 + x = 73$

Solution: To solve for the missing addend, x, we write a related subtraction sentence.

$$x = 73 - 45 \qquad \text{Subtract the known addend, 45, from the sum, 73.}$$
$$x = 28$$

Check: $45 + x = 73$
$$45 + 28 \stackrel{?}{=} 73 \qquad \text{In the original equation, replace } x \text{ with 28.}$$
$$73 = 73 \qquad \text{It checks.}$$

Explanation: Think about the missing addend statement in money terms. If we have $45 and we want to end up with $73, we must subtract to find out how much more is needed.

◁ **Do Your Turn 7**

OBJECTIVE 6 Solve applications involving subtraction.

Like addition, there are key words and key questions that help us recognize subtraction situations.

KEY WORDS FOR SUBTRACTION

Subtract, minus, remove, decreased by, difference, take away, left, less than

Key Questions   "How much is left?" (take away)

"How much more/higher/warmer/colder/shorter?" (difference)

"How much more is needed?" (missing addend)

Example 8   Solve.

a. A small computer business has $5678 in the bank. During one month, $2985 is spent on parts for production. How much is left?

Solution: The word *spent,* indicating *to remove,* and the key question, *How much is left?* indicate we must subtract.

$$
\begin{array}{r}
\overset{15}{\underset{}{4}}\,\overset{.517}{5678} \\
-\ 2985 \\
\hline
2693
\end{array}
\qquad
\textbf{Check:}\quad
\begin{array}{r}
1\,1 \\
2693 \\
+2985 \\
\hline
5678
\end{array}
$$

Answer: $2693 is left.

b. A room-temperature (68°F) mixture of water and sugar is heated to boiling (212°F) so that more sugar may be added to supersaturate the mixture. How much did the temperature change?

Solution: Because we want to know the amount of change, we are looking for the difference so we must subtract.

$$
\begin{array}{r}
\overset{10}{\underset{}{1}}\overset{.012}{} \\
2\cancel{1}2 \\
-\ 68 \\
\hline
144
\end{array}
\qquad
\textbf{Check:}\quad
\begin{array}{r}
1\,1 \\
144 \\
+68 \\
\hline
212
\end{array}
$$

Answer: 144°F

c. A charity organization has collected $43,587 through a fundraiser. The goal for the year is to raise $125,000. How much more is needed?

Solution: Because we must increase the original amount of $43,587 by an unknown amount to end up with $125,000, this is a missing addend situation.

We can write $43,587 + x = $125,000.

To find that missing addend, we write a related sentence.

$$
\begin{array}{r}
\overset{9\ 9}{\underset{}{0\,12\,4\,\cancel{10}\,\cancel{10}\,\cancel{10}}} \\
\cancel{1}2\cancel{5},\cancel{0}\cancel{0}\cancel{0} \\
-43,587 \\
\hline
81,413
\end{array}
\qquad
\begin{aligned}
x &= 125,000 - 43,587 \\
x &= 81,413
\end{aligned}
\qquad \textbf{Subtract the known addend from the sum.}
$$

$$
\textbf{Check:}\quad
\begin{array}{r}
1\ 11 \\
81,413 \\
+43,587 \\
\hline
125,000
\end{array}
$$

Answer: $81,413 is needed.

Do Your Turn 8 ▶

> **Your Turn 8**
>
> *Solve.*
>
> a. A food bank has 3452 boxes of food. They distribute 928 boxes in one neighborhood. How many boxes do they have left?
>
> b. A plane is flying at an altitude of 24,500 feet. The plane experiences a downdraft, which causes it to abruptly drop to an altitude of 22,750 feet. How much altitude did the plane lose?
>
> c. Debbie sells cars at a very competitive dealership. As an incentive, the first salesperson to sell $150,000 worth of cars during the year will win a cruise to the Caribbean. Debbie has sold $78,520 so far. How much more does she need to sell to win? Can you see a flaw in the incentive idea?
>
> **Answers to Your Turn 8**
> a. 2524 boxes b. 1750 ft.
> c. $71,480

OBJECTIVE 7 Solve applications involving addition and subtraction.

Much of the time the problems we encounter in life require more than one step to find a solution. Let's consider some problems that involve addition and subtraction in the same problem.

Your Turn 9

Solve.

a. Find the missing current in the circuit shown.

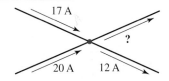

b. Lynda is a checkout clerk. At the end of her shift, the store manager makes sure the amount of money in the register equals her sales. The printout breaks the sales down into three categories: food, clothing, and nonperishable items. Lynda must count the money in the register and her credit card sales. Finally, $200 always stays in the register. The following chart shows the amounts. Based on the information in the chart, how did Lynda do? Is her register in balance? Does she have too little or too much and if so, how much?

Category	Amount
Food	$1587
Clothing	$2768
Nonperishables	$ 988

Totals	Amount
Register count	$4192
Credit cards	$1339

Example 9

a. In circuits, a wire connection is often referred to as a *node*. Current is a measure of electricity moving through a wire and is measured in amperes (or amps, A). One property of circuits is that all the current entering a node must equal all the current exiting the node. In the circuit diagram shown, how much must the unknown current be?

Solution: We can calculate the total current entering the node and then subtract the known current that is leaving the node.

Entering current: 9 + 15 = 24

If 24 amps are entering and we know 13 amps are leaving, then we can subtract 13 from 24 to find the unknown current.

Unknown current: 24 − 13 = 11

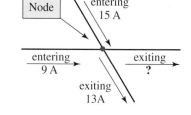

Answer: The unknown current is 11 A.

b. An accountant is given the following spreadsheet of expenses and income for a company. What is the final balance?

Description of expenses	Amount	Description of assets/income	Amount
Payroll	$10,548	Checking account	$ 5,849
Utilities	$ 329	Income	$19,538
Water	$ 87	Donations	$ 1,200
Waste disposal	$ 285		
New Inventory	$12,243		

Solution: We must calculate the total expenses and total assets/income, then find the difference of the amounts.

Expenses	Assets/Income
$10,548	$ 5,849
$ 329	$19,538
$ 87	+$ 1,200
$ 285	Total $26,587
− $12,243	
Total $23,492	

Now that we have the total expenses and income/assets figured, we need to find the difference to determine the final balance.

Balance
$26,587
− $23,492
$ 3,095

Answer: The final balance is $3095.

Answers to Your Turn 9
a. 25 A b. There should be $12 more in the register.

◁ **Do Your Turn 9**

1. What does the commutative property of addition say?

2. What does the associative property of addition say?

3. What is the perimeter of a shape?

4. Explain the difference between a constant and a variable.

5. How do you find an unknown addend?

6. What is a solution to an equation?

For Exercises 7–12, estimate each sum by rounding to the nearest thousand, then find the actual sum.

7. $6{,}051$
$+2{,}798$

8. $52{,}407$
$+31{,}596$

9. $91{,}512$
$+8{,}756$

10. $82{,}098$
$+7{,}971$

11. $10{,}516$
982
$+4{,}516$

12. $150{,}412$
258
$+6{,}239$

For Exercises 13–16, estimate each sum by rounding, then find the actual sum.

13. $9319 + 519 + 5408$

14. $6809 + 398 + 2087$

15. $43{,}210 + 135{,}569 + 2{,}088 + 516$

16. $128{,}402 + 4{,}480 + 93{,}095 + 98$

For Exercises 17–20, estimate by rounding to the nearest hundred, then find the actual difference.

17. 5873
-521

18. 9478
-253

19. $40{,}302$
$-6{,}141$

20. $510{,}304$
$-42{,}183$

For Exercises 21–26, estimate by rounding, then find the actual difference.

21. 50,016
−4,682

22. 42,003
−23,567

23. 51,980 − 25,461

24. 413,609 − 20,724

25. 210,007 − 43,519

26. 6,005,002 − 258,496

For Exercises 27–36, solve and check.

27. $8 + x = 12$

28. $12 + y = 19$

29. $t + 9 = 30$

30. $m + 16 = 28$

31. $17 + n = 35$

32. $r + 27 = 70$

33. $125 + u = 280$

34. $88 + a = 115$

35. $b + 76 = 301$

36. $470 + c = 611$

For Exercises 37–50, solve.

37. Tamika has to keep track of total ticket sales at a box office. There are four attendants selling tickets. The first attendant sold 548 tickets, the second sold 354, the third 481, and the fourth 427. How many tickets were sold in all?

38. Margo Incorporated produces a motor that turns the fan in air conditioners. The annual cost of materials last year was $75,348. The labor costs were $284,568. Utilities and operations costs for the plant were $54,214. The company received $548,780 in revenue from sales of the motors. What was the profit? (*Note:* Profit can be found by subtracting all costs from revenue.)

39. Mr. Grayson has just acquired some new land that he wants to fence in for a cow herd. To the right is a copy of the plot. How many feet of fencing will he need in all?

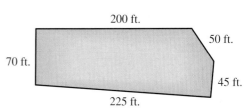

200 ft.
50 ft.
70 ft.
45 ft.
225 ft.

40. Jose is raising funds to run for a seat in the United States Senate. Currently, he has raised $280,540 and he intends to add $28,580 of his own money. His competitor has raised a total of $485,300. How much more does Jose need to raise to match his competitor's funds?

41. The photosphere, which is the outer layer of the Sun, has an average temperature of about 5500 K. A typical atomic blast at ground zero has a temperature of about 3400 K. How much warmer is the Sun's temperature?

42. The Sun is 92,958,349 miles from Earth. The Moon is 615,042 miles from Earth. During a solar eclipse, the Moon is directly between the Sun and Earth. How far is the Moon from the Sun during a solar eclipse?

43. In the morning, Serene had $47 in her wallet. She bought coffee for $2, lunch for $8, and filled her car's gas tank for $21. How much money did Serene have left when she got home?

44. Jesse's student loan and scholarship totaled $5976. He paid $4612 in tuition and fees, $676 for text books, and $77 for other school supplies. How much money does Jesse have left?

OF INTEREST

The symbol K represents kelvin temperature units. The scale was proposed by William Thomson Kelvin in 1848 and is used extensively in sciences. Increments on the Kelvin scale are the same as the Celsius scale. However, 0 K is equivalent to −273 °C.

45. An accountant is given the following spreadsheet of expenses and income for a company. What is the final balance?

Description of expenses	Amount	Description of assets/income	Amount
Payroll	$18,708	Checking account	$10,489
Utilities	$ 678	Income	$32,500
Water	$ 126		
Waste disposal	$ 1,245		
New inventory	$ 15,621		

46. An accountant is given the following spreadsheet of expenses and income for a company. What is the final balance?

Description of expenses	Amount	Description of assets/income	Amount
Payroll	$22,512	Checking account	$15,489
Utilities	$ 1,265	Income	$82,569
Water	$ 240		
Waste disposal	$ 1,350		
New inventory	$24,882		

47. Find the missing current in the circuit shown in the figure below.

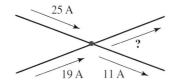

48. Find the missing current in the circuit shown in the figure below.

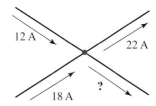

49. A family is purchasing a home in Florida that is said to be 2400 square feet. They later find out that this figure includes an unfinished bonus room over the garage and the garage itself. The garage is 440 square feet and the bonus room is 240 square feet. What is the area of the actual living space?

50. An animal rescue group had 37 cats in their shelter at the end of April. In May, 12 cats were returned to their owners, 21 went to new homes, and 26 new cats were brought in to the shelter. How many cats were in the shelter at the end of May?

PUZZLE PROBLEM Eight people are at a group-therapy session. Everyone hugs everyone once. How many hugs take place?

Review Exercises

[1.1] **1.** What digit is in the millions place in 456,028,549?

[1.1] **2.** Write the expanded form for 307,491,024.

[1.1] **3.** Write the word name for 1,472,359.

[1.1] **4.** Use < , > , or = to make a true statement.

    12,305 ? 12,350

[1.1] **5.** Round 23,405,172 to the nearest ten thousand.

1.3 Multiplying Whole Numbers and Exponents

OBJECTIVE 1 Multiply whole numbers.

OBJECTIVES

1 Multiply whole numbers.

2 Solve applications involving multiplication.

3 Evaluate numbers in exponential form.

4 Write in exponential form.

5 Solve applications.

Suppose we need to install five kitchen cabinet doors, each requiring four screws. Note that we could find the total number of screws by adding five 4's:

$$4 + 4 + 4 + 4 + 4 = 20$$

However, it is much faster to memorize that the sum of five 4's is always 20. This is the nature of **multiplication.**

DEFINITION **Multiplication:** Repeated addition of the same number.

When we write a multiplication sentence, the *factors* are the parts that are multiplied and the product is the answer.

$4 \times 5 = 20$

↑ ↑ ↑

Factors **Product**

We can also write: $4 \cdot 5 = 20$

$$(4)(5) = 20$$
$$4(5) = 20$$
$$(4)5 = 20$$

DISCUSSION When would we avoid using $\cdot$ or $\times$ to indicate multiplication?

A multiplication statement written with the product first, as in $20 = 4 \cdot 5$, is in *factored form.*

To make multiplication useful, you must memorize a certain number of multiplication facts. The more facts you memorize, the faster you'll be at multiplying. The table below has multiplication facts through 12.

Figure 1–2 Multiplication Table

×	0	1	2	3	4	5	6	7	8	9	10	11	12
0	0	0	0	0	0	0	0	0	0	0	0	0	0
1	0	1	2	3	4	5	6	7	8	9	10	11	12
2	0	2	4	6	8	10	12	14	16	18	20	22	24
3	0	3	6	9	12	15	18	21	24	27	30	33	36
4	0	4	8	12	16	20	24	28	32	36	40	44	48
5	0	5	10	15	20	25	30	35	40	45	50	55	60
6	0	6	12	18	24	30	36	42	48	54	60	66	72
7	0	7	14	21	28	35	42	49	56	63	70	77	84
8	0	8	16	24	32	40	48	56	64	72	80	88	96
9	0	9	18	27	36	45	54	63	72	81	90	99	108
10	0	10	20	30	40	50	60	70	80	90	100	110	120
11	0	11	22	33	44	55	66	77	88	99	110	121	132
12	0	12	24	36	48	60	72	84	96	108	120	132	144

Example 3 Estimate 42,109 × 7104 by rounding so that there is only one nonzero digit, then calculate the actual product.

Estimate: To estimate, we round 42,109 to 40,000 and 7104 to 7000. Because the numbers now contain lots of 0's, we can simply multiply the 7 times 4 mentally and write the total number of 0's from both numbers after the product.

$$
\begin{array}{r}
40{,}000 \leftarrow \textbf{Four 0's} \\
\times\ 7{,}000 \leftarrow \textbf{Three 0's} \\
\hline
280{,}000{,}000
\end{array}
$$

Think   **Total of seven 0's**
7 × 4

Actual:

$$
\begin{array}{r}
\overset{1}{}\ \ \overset{6}{\underset{3}{}} \\
42{,}109 \\
\times\ 7{,}104 \\
\hline
168\,436 \\
000\,00 \\
4\,210\,9 \\
+294\,763 \\
\hline
299{,}142{,}336
\end{array}
$$

Multiply 4 times 42,109.
Multiply 0 times 42,109.
Multiply 1 times 42,109.
Multiply 7 times 42,109.

Notice how the estimate and actual answer are reasonably close.

Do Your Turn 3 ▶

Estimate each product by rounding to only one nonzero digit, then calculate the actual product.

a. 56,045 × 6714

b. 24 × 365 × 8

OBJECTIVE 2 Solve applications involving multiplication.

There are certain key words you can look for that indicate multiplication. Since multiplication is repeated addition, you can expect to see words that indicate addition as well.

KEY WORDS FOR MULTIPLICATION

Multiply, times, product, each, of, by

Example 4 The human heart averages about 70 beats each minute.

a. How many times would the heart beat in an hour?

Solution: Since there are 60 minutes in an hour and each minute is 70 beats, we multiply 60 × 70 to find the number of heartbeats in an hour.

$$
\begin{array}{r}
60 \\
\times\ 70 \\
\hline
00 \\
+420 \\
\hline
4200
\end{array}
$$

Answer: The heart beats about 4200 times per hour.

Answers to Your Turn 3
a. Estimate: 420,000,000
   Actual: 376,286,130
b. Estimate: 80,000
   Actual: 70,080

1.3 Multiplying Whole Numbers and Exponents   25

Solve.

a. If the heart beats about 70 beats per minute, then how many times does the heart beat in a 30-day month?

b. A manager decides to monitor the copy machine use in her office. She finds that the people in her area are using two boxes of paper each week. Each box contains 10 reams of paper and each ream contains 500 pieces of paper. How many pages are being copied each week? Each month?

DISCUSSION What are some ways the manager could cut back on copying?

---

b. How many times would the heart beat in a week?

Solution: For the number of heartbeats in a week, we need to first figure out how many hours there are in a week. Since there are 24 hours in a day, and 7 days in a week, we multiply 24×7 to get the number of hours in a week.

$$\begin{array}{r} \overset{2}{2}4 \\ \times\ 7 \\ \hline 168 \end{array}$$

CONNECTION We can use the distributive property to do this calulation in our head.

$$7(24) = 7(20 + 4) = 7 \cdot 20 + 7 \cdot 4 = 140 + 28 = 168$$

There are 168 hours in a week. From part (a) we found there were 4200 beats in an hour. To calculate the number of beats in a week, we multiply 168×4200.

$$\begin{array}{r} 4200 \\ \times\ 168 \\ \hline 33\,600 \\ 252\,00 \\ +\,420\,0 \\ \hline 705{,}600 \end{array}$$

DISCUSSION What things affect the heart rate? Measure your own heart rate and do the same calculation to estimate the number of times your own heart beats in a week.

Answer: The heart beats about 705,600 times in a week (a truly amazing muscle).

◁ **Do Your Turn 4**

We can use multiplication to count objects arranged in a **rectangular array.**

DEFINITION Rectangular array: A rectangle formed by a pattern of neatly arranged rows and columns.

The buttons on a cell phone form a rectangular array.

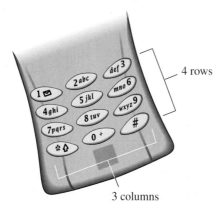

4 rows

3 columns

Four rows with three buttons in each row is a rectangular array. Since each row contains the same number of buttons, we can multiply to find the total number of buttons.

$$4 \cdot 3 = 12$$

Conclusion: To calculate the total number of items in a rectangular array, multiply the number of rows by the number of columns.

Answers to Your Turn 4
a. 3,024,000 times **b.** 10,000 each week; 40,000 each month

Example 5

Solve.

a. A section in the upper deck of a football stadium has 42 rows, each containing 40 seats. There are 8 sections in the upper deck. How many people can be seated in the upper deck altogether?

Solution: The 8 sections of 42 rows with 40 seats in each row form a large rectangular array, so we multiply $42 \times 40 \times 8$.

$$
\begin{array}{r}
42 \\
\times 40 \\
\hline
00 \\
+168 \quad\; \\
\hline
1680 \\
\end{array}
$$
↑
Number of seats in one section

$$
\begin{array}{r}
{\scriptstyle 5\;6} \\
1\,680 \\
\times 8 \\
\hline
13{,}440 \\
\end{array}
$$
↑
Number of seats in all 8 sections

CONNECTION We could have applied the commutative and associative properties and multiplied any two factors first.

$(42 \times 8) \times 40$ or $42 \times (40 \times 8)$
$= 336 \times 40$ $= 42 \times 320$
$= 13{,}440$ $= 13{,}440$

Answer: The upper deck can seat 13,440 people.

Explanation: Notice that when the example says "42 rows *each* containing 40 seats," the key word *each* tells us we have a repeated addition situation, so we multiply. Since there are 8 identical sections, we are once again repeatedly adding, so once again we multiply.

b. In order to get into a certain restricted building, a code must be entered. The code box has two digital windows (see illustration). The first window can contain any number from 1 to 5. The second window can contain any letter from A to Z (only uppercase letters appear). How many possible codes are there in all?

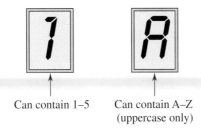

Can contain 1–5   Can contain A–Z
(uppercase only)

Solution: Each number in the first window could be paired with all 26 letters of the alphabet. In other words, the 1 could be paired with all 26 letters, then 2 with all 26 letters, and so on (see below). That means there could be a total of 5 sets of 26. Notice how we can make a rectangular array out of the combinations. Therefore, we multiply 5 times 26.

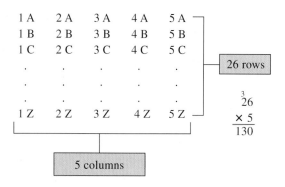

Your Turn 5

A certain combination lock has three dials. The first dial has the letters A–F inscribed. The second dial has the numbers 0–9. The third dial has the names of the planets. How many combinations are possible?

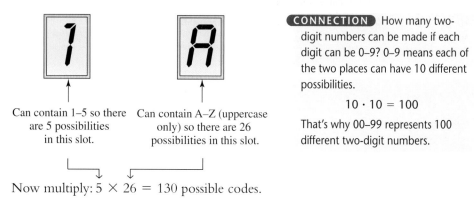

Answer: There are 130 possible codes.

Another way to analyze this problem is to think of each window as a place value or a *slot*, like on a slot machine. Determine the number of possible values each slot could contain and multiply.

Can contain 1–5 so there are 5 possibilities in this slot.

Can contain A–Z (uppercase only) so there are 26 possibilities in this slot.

CONNECTION How many two-digit numbers can be made if each digit can be 0–9? 0–9 means each of the two places can have 10 different possibilities.

$$10 \cdot 10 = 100$$

That's why 00–99 represents 100 different two-digit numbers.

Now multiply: $5 \times 26 = 130$ possible codes.

◄ **Do Your Turn 5**

The branch of mathematics that deals with counting total combinations and arrangements of items is called *combinatorics.*

PROCEDURE

To find the total number of possible combinations in a problem with multiple slots to fill, multiply the number of items that can fill the first slot by the number of items that can fill the next, and so on.

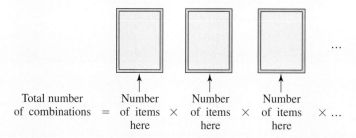

Total number of combinations = Number of items here × Number of items here × Number of items here × ...

OBJECTIVE 3 Evaluate numbers in exponential form.

A computer translates everything into *binary* code. This means that numbers are written so that each place value either contains a 1 or a 0 digit. If a certain computer chip has seven place values, then how many different numbers can be coded?

Answer to Your Turn 5
540 combinations

Suppose each box shown represents a bit or place value on the chip.

0 or 1 0 or 1 0 or 1 0 or 1 0 or 1 0 or 1 0 or 1

Since each box could have 2 different values, we must multiply seven 2's together.

$$2 \times 2 \times 2 \times 2 \times 2 \times 2 \times 2 = 128$$

Notice that we had to multiply the same number repeatedly. This suggests the need for a notation to indicate repeated multiplication. We use an **exponent** or **power** to indicate repeated multiplication of a **base** number.

DEFINITIONS **Exponent or Power:** A symbol written to the upper right of a base number that indicates how many times to use the base as a factor.

Base: The number that is repeatedly multiplied.

Notation: When we write a number with an exponent, we say it is in *exponential form*.

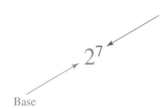

Exponent or power: Indicates the number of times the base is used as a factor.

2^7

Base

2^7 is read "two to the seventh power," or, simply, "two to the seventh."

We could answer the computer-coding question by saying the computer can code a total of 2^7 different numbers. To *evaluate* 2^7, we must multiply seven 2's together. It is helpful to first write the multiplication in factored form.

$$2^7 = 2 \cdot 2 \cdot 2 \cdot 2 \cdot 2 \cdot 2 \cdot 2 = 128 \qquad \textbf{WARNING} \quad 2^7 \text{ does not mean } 2 \cdot 7.$$

Exponential form   Factored form   Standard form

Example 6 Write 3^4 in factored form, then evaluate.

Solution: $3^4 = 3 \cdot 3 \cdot 3 \cdot 3$ The exponent 4 means the factored form has four 3's multiplied together.

$= 9 \cdot 3 \cdot 3$ Multiplying the first two 3's.

$= 27 \cdot 3$ Multiply 9 by the third 3.

$= 81$ Multiply 27 by the fourth 3.

Do Your Turn 6 ▶

OF INTEREST

In computer terminology, the place values are called *bits*. Bit is short for *binary dig*it.

 Calculator Tips

You will find several buttons on a scientific calculator for exponents.

$\boxed{x^2}$ is used when the exponent is 2.

Example: To calculate 34^2, enter $\boxed{3}$ $\boxed{4}$ $\boxed{x^2}$ $\boxed{\text{ENTER} \atop =}$.

Answer: 1156

The keys used for any exponent are $\boxed{y^x}$ or $\boxed{\wedge}$ depending on the calculator.

To use these keys, enter the base, press $\boxed{y^x}$ or $\boxed{\wedge}$, enter the exponent, and then press the equal sign.

Example: To calculate 9^8, type $\boxed{9}$ $\boxed{y^x}$ $\boxed{8}$ $\boxed{\text{ENTER} \atop =}$ or $\boxed{9}$ $\boxed{\wedge}$ $\boxed{8}$ $\boxed{\text{ENTER} \atop =}$.

Answer: 43,046,721

Your Turn 6

a. *Write each in factored form, then evaluate.*

1. 7^2

2. 10^4

3. 1^9

4. 0^4

 b. *Use a calculator to evaluate each of the following.*

1. 2^{12}

2. 5^7

Answers to Your Turn 6:
1. a. 49 b. 10,000 c. 1 d. 0
2. a. 4096 b. 78,125

OBJECTIVE 4 Write in exponential form.

Example 7 Write $5 \cdot 5 \cdot 5 \cdot 5 \cdot 5 \cdot 5$ in exponential form.

Answer: 5^6

Explanation: Since there are six 5's multiplied, the exponent is 6.

◀ **Do Your Turn 7**

Your Turn 7

Write each in exponential form.

a. $14 \cdot 14 \cdot 14 \cdot 14$

b. $2 \cdot 2 \cdot 2 \cdot 2 \cdot 2 \cdot 2 \cdot 2 \cdot 2 \cdot 2 \cdot 2$

c. $9 \cdot 9 \cdot 9 \cdot 9 \cdot 9 \cdot 9 \cdot 9$

d. How many different values can be represented by a 16-bit computer chip? Write your answer in exponential form, then use a calculator to evaluate.

When an exponential form has a base of 10 we say it is a *power* of 10. Consider the pattern:

$$10^2 = 10 \cdot 10 = 100$$
$$10^3 = 10 \cdot 10 \cdot 10 = 1{,}000$$
$$10^4 = 10 \cdot 10 \cdot 10 \cdot 10 = 10{,}000$$
$$10^5 = 10 \cdot 10 \cdot 10 \cdot 10 \cdot 10 = 100{,}000$$
$$10^6 = 10 \cdot 10 \cdot 10 \cdot 10 \cdot 10 \cdot 10 = 1{,}000{,}000$$

Conclusion: Because the exponent indicates the number of 10's to be multiplied, and each 10 contributes an additional 0 after the 1, the number of 0's in the product matches the exponent.

Notice that we can write expanded notation more simply with the powers of 10. Instead of writing all the zeros in the place values, we can simply express the place values with powers of 10.

Example 8 Write 1,239,405 in expanded notation using powers of 10.

Solution:

$$1 \times 1{,}000{,}000 + 2 \times 100{,}000 + 3 \times 10{,}000 + 9 \times 1{,}000 + 4 \times 100 + 0 \times 10 + 5 \times 1$$
$$= 1 \times 10^6 + 2 \times 10^5 + 3 \times 10^4 + 9 \times 10^3 + 4 \times 10^2 + 0 \times 10^1 + 5 \times 10^0$$
$$= 1 \times 10^6 + 2 \times 10^5 + 3 \times 10^4 + 9 \times 10^3 + 4 \times 10^2 + 5 \times 1$$

Note. We did not write the tens place in this expanded notation because it contained a 0 digit.

CONNECTION What does 10^0 mean?

Notice the pattern of exponents. Start with 10^3. If $10^3 = 1000$, $10^2 = 100$, and $10^1 = 10$, then 10^0 has to be 1 for the pattern to hold true.

$$10^0 = 1$$

Strange, but true.

Your Turn 8

Write each number in expanded form using powers of 10.

a. 9,210,984

b. 46,009

c. 1,204,918,205

Each power of 10 represents a different place value. Remember that the period names change every three places. So every third power of 10 will have a new name as the following table illustrates.

Figure 1–3 Powers of 10 and Period Names

Period Names		
10^3	$= 1{,}000$	= one thousand
10^6	$= 1{,}000{,}000$	= one million
10^9	$= 1{,}000{,}000{,}000$	= one billion
10^{12}	$= 1{,}000{,}000{,}000{,}000$	= one trillion
10^{15}	= (1 with 15 zeros)	= one quadrillion

The names continue using the preceding pattern. Some names are rather colorful.

10^{100}	= (1 with 100 zeros)	= googol
10^{googol}	= (1 with a googol of zeros)	= googolplex

◀ **Do Your Turn 8**

Answers to Your Turn 7

a. 14^4 b. 2^{10} c. 9^7
d. $2^{16} = 65{,}536$

Answers to Your Turn 8

a. $9 \times 10^6 + 2 \times 10^5 + 1 \times 10^4 + 9 \times 10^2 + 8 \times 10 + 4 \times 1$

b. $4 \times 10^4 + 6 \times 10^3 + 9 \times 1$

c. $1 \times 10^9 + 2 \times 10^8 + 4 \times 10^6 + 9 \times 10^5 + 1 \times 10^4 + 8 \times 10^3 + 2 \times 10^2 + 5 \times 1$

OBJECTIVE **5** Solve applications.

Suppose we need to know how much surface space is within the borders or edges of a shape. We measure the space in **square units** and call the measurement **area.**

DEFINITIONS **Square unit:** A 1 × 1 square.

Area: The total number of square units that completely fill a shape.

Examples of square units:

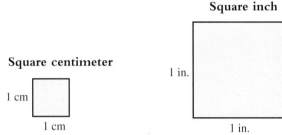

Square centimeter

1 cm

1 cm

A square centimeter is a 1 cm by 1 cm square and is represented mathematically by cm².

Square inch

1 in.

1 in.

A square inch is a 1 in. by 1 in. square and is represented mathematically by in.².

To *measure* the area of a rectangular room, we could place square tiles down until we completely cover the floor with tiles and then count the total number of tiles. In the following figure, we illustrate this approach for a 4-foot × 8-foot bathroom.

Note: If we tiled the floor with 1-foot × 1-foot square tiles, it would take 32 of those square feet to cover the floor. ▶

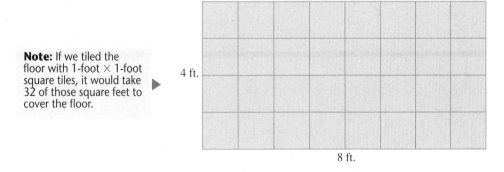

4 ft.

8 ft.

Notice the square units form a rectangular array so we can *calculate* the area by multiplying the length of the rectangle by the width, rather than measuring the area with square tiles. Calculating the area, we have $8 \cdot 4 = 32$ square feet, which agrees with our measurement.

Conclusion: Area of a rectangle = length × width
Using variables we can say $A = l \cdot w$.

The equation $A = lw$ is called a **formula.** ◀ **Note:** When variables are multiplied as in $A = l \cdot w$, the multiplication symbol is optional, so we can write $A = lw$.

DEFINITION **Formula:** An equation that describes a procedure.

A room measures 15 feet wide by 24 feet long. How many square feet of carpet will be needed to cover the floor of the room?

Example 9 A standard football field measures 100 yards by 50 yards, not counting the end zones. Find the area of a standard football field.

Solution: Since the field is a rectangle, to calculate the area, we multiply length times width.

$$A = lw$$
$$A = 100 \text{ yd.} \cdot 50 \text{ yd.}$$
$$A = 5000 \text{ yd.}^2$$

Note: This unit of area is read "yards squared" or "square yards." All area calculations will involve multiplying two distance measurements. Area units can always be expressed as the distance unit with an exponent of 2.

CONNECTION Notice that square units are written with an exponent of 2. It is as if yards times yards are multiplied to equal square yards.

Do Your Turn 9

Answer to Your Turn 9
360 ft.^{2}

1. What does the commutative property of multiplication say?

2. What does the associative property of multiplication say?

3. Express the distributive property in math language.

4. How do you calculate 3^4?

5. What is the area of a shape?

6. What is the formula $A = lw$ used for, and what do its variables represent?

For Exercises 7–10, find the product.

7. 352×1 **8.** 0×495 **9.** $8 \times 9 \times 0 \times 7$ **10.** $12 \times 3 \times 1$

For Exercises 11–16, estimate each product by rounding to a single nonzero digit, then find the actual product.

11. 24×18 **12.** 63×42

13. 54×91 **14.** 47×86

15. 246×381 **16.** 384×25

For Exercises 17–26, multiply.

17. 642×70 **18.** 408×92 **19.** 2065×482

20. 9402×608 **21.** $207 \times 41,308$ **22.** $361 \times 80,290$

23. $60,309 \times 4002$ **24.** $314 \times 12 \times 47$ **25.** $205 \times 23 \times 70$

26. $14 \times 1 \times 289 \times 307$

For Exercises 27–34, write each number in factored form and evaluate.

27. 2^4 **28.** 5^3 **29.** 1^6

30. 13^2 **31.** 3^5 **32.** 10^5

33. 10^7 **34.** 0^7

For Exercises 35–38, use a scientific calculator to evaluate.

35. 6536^2 **36.** 279^3 **37.** 42^5 **38.** 34^6

For Exercises 39–44, write in exponential form.

39. $9 \cdot 9 \cdot 9 \cdot 9$ **40.** $12 \cdot 12 \cdot 12$ **41.** $7 \cdot 7 \cdot 7 \cdot 7 \cdot 7$

42. $10 \cdot 10 \cdot 10 \cdot 10 \cdot 10 \cdot 10 \cdot 10$ **43.** $14 \cdot 14 \cdot 14 \cdot 14 \cdot 14 \cdot 14$ **44.** $2 \cdot 2 \cdot 2 \cdot 2 \cdot 2 \cdot 2 \cdot 2 \cdot 2$

For Exercises 45–50, write in expanded form using powers of 10.

45. 24,902   **46.** 604,057

47. 9,128,020   **48.** 10,945

49. 407,210,925   **50.** 3,029,408

For Exercises 51–70, solve.

51. In a certain city, avenues run north/south, and streets run east/west. If there are 97 streets and 82 avenues, how many intersections are there? If the city must put up 8 traffic lights at each intersection, then how many traffic lights are required in all?

52. A parking lot for a department store has 23 rows of spaces. Each row can hold 36 cars. What is the maximum number of cars that can be parked in the lot? Now, suppose the lot is full. Estimate the number of people in the store if we assume that there are two people in the store for each car in the lot.

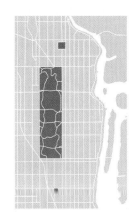

53. On average, the human heart beats 70 times in 1 minute. How many beats is this in a year? (Use 365 days in a year.) If the average male in the United States lives to be 74, estimate the number of heartbeats in an average male's lifetime. (*Source: National Vital Statistics Reports,* Vol. 52, No. 14, February 18, 2004, p. 33)

54. The average female life span in the United States is about 80 years. Using the information in Exercise 53, estimate the number of heartbeats in the average female's lifetime. (*Source: National Vital Statistics Reports,* Vol. 52, No. 14, February 18, 2004, p. 33)

55. A prescription indicates that a patient is to receive 10 milligrams of aminophylline per kilograms that the person weighs. If the person weighs 74 kilograms, then how many milligrams should the patient receive?

56. A nurse sets the drip rate for an IV at 26 drops each minute. How many drops does the patient receive in 1 hour?

57. Michael works 5 days every week and buys lunch each workday. On average, he spends $7 each day. If he has 10 vacation days each year, then how much does he spend on lunch in a week? In a year?

58. An advertisement indicates that a person can make up to $800 per week as a telemarketer. How much would a telemarketer earning $800 per week earn in one year?

DISCUSSION If you got the job and did not make $800 per week, can you argue that the flyer was false advertising? Why or why not?

59. It is recommended that a person drink eight 8-ounce glasses of water every day. If every one of the 290 million Americans drank eight 8-ounce glasses of water in the same day, how many ounces of water would be consumed in that day?

60. On the evening news you hear a conversation between the anchor and a commentator. The commentator says that a computer company sold nearly eight hundred thousand copies of a new software package at $40 each. The anchor then says, "So that's about $320 billion. Wow." Is the anchor's response accurate? Explain.

61. A combination lock is opened by a combination of three things. You must turn a key to one of two positions, left or right; set a dial to a day of the week; and set another dial to a letter from A to H. How many possible combinations are there?

 62. If a license plate consists of three capital letters and a three-digit number, how many license plates are possible? How many are possible in the entire United States if each state uses the same system of three letters and a three-digit number?

63. A certain computer chip processes 5-digit binary numbers, meaning each of the five places could contain a 1 or 0. How many different binary numbers can it process? If each letter, A–Z, and each numeral, 0–9, corresponds to a different binary number, would the 5-digit chip be able to recognize them all? Explain.

64. How many codes are possible using a 12-bit binary coder?

65. MIDI stands for musical instrument data interface and is the common coding for digital instruments, such as synthesizers. Most MIDI devices have a 7-bit binary memory chip. How many memory locations are possible?

66. A psychology quiz has 10 true-false questions. How many different answer keys are possible? Express the answer in exponential form, then calculate the actual number.

67. A man walks into a carpet store and says he has a 24-foot wide by 26-foot long room and therefore needs about 4000 square feet of carpet. Exactly how many square feet does he need? Is his estimate reasonable?

68. You wish to tile the kitchen floor in your home. The rectangular kitchen is 16 feet long by 12 feet wide. If you used 1-foot by 1-foot square tiles, how many would be needed?

69. Delia plans to stain a new wood deck. The deck is 15 feet wide by 20 feet long. What is the area of the deck?

70. Taylor plans to paint a wall in his living room. The wall is 18 feet long by 9 feet high. What is the area of the wall?

PUZZLE PROBLEM A certain bacterium reproduces by dividing itself every minute. This causes the total population to double every minute. If we begin with a single bacterium on a petri dish, how many will exist after 5 minutes? 10 minutes? Write a formula that describes the pattern.

Review Exercises

[1.1] **1.** Write the word name for 16,507,309.

[1.1] **2.** Write expanded notation for 23,506.

[1.2] **3.** Calculate. 54,391 + 2079 + 518

[1.2] **4.** Calculate. 901,042 − 69,318

[1.2] **5.** Solve and check. $n + 19 = 32$

1.4 Dividing, Square Roots, and Solving Equations with Whole Numbers

OBJECTIVES

1 Divide whole numbers.

2 Solve equations containing a missing factor.

3 Solve applications involving division.

4 Find the square root of a perfect square.

5 Solve applications involving square roots.

OBJECTIVE 1 Divide whole numbers.

Like addition and subtraction, multiplication and division are inverse operations. Because multiplication means to repeatedly add, it follows that **division** means to repeatedly subtract.

DEFINITION Division: Repeated subtraction of the same number.

When we write a division sentence, the *dividend* is divided by the *divisor* and the *quotient* is the answer.

Notation: $20 \div 5 = 4$

Dividend   Divisor   Quotient

CONNECTION Because multiplication and division are inverse operations, we can check $20 \div 5 = 4$ by multiplying: $4 \cdot 5 = 20$.

Like multiplication, there are several ways to indicate division. We can write $20 \div 5 = 4$ using a slash, fraction bar, or long division.

Slash: $20/5 = 4$

Fraction bar: $\dfrac{20}{5} = 4$

Long division: $5\overline{)20}$ with quotient 4

At one time, people divided by repeatedly subtracting the same number until they couldn't subtract anymore.

Let's calculate $20 \div 5$ by repeated subtraction. The quotient is the number of times we have to subtract 5 to get to 0 or a number between 0 and 5.

$$
\begin{array}{r}
20 \\
-5 \\
\hline
15 \\
-5 \\
\hline
10 \\
-5 \\
\hline
5 \\
-5 \\
\hline
0
\end{array}
$$

1st subtraction
2nd subtraction
3rd subtraction
4th subtraction ◄ Because we subtracted 5 from 20 four times to get 0, the quotient is 4.

The amount left after dividing is called the **remainder.**

DEFINITION Remainder: The amount left over after dividing two whole numbers.

In the case of $20 \div 5$, there are exactly four 5's in 20, so the remainder is 0. Sometimes the division does not end in 0. For example, if we had to split \$21 into \$5 bills we could only get four \$5 bills with \$1 left over. The \$1 left over is the remainder, and we would write $21 \div 5 = 4 \text{ r}1$, using the letter r to indicate the remainder.

Let's explore a few properties of division.

Like subtraction, division is neither commutative nor associative.

Division is not commutative: $20 \div 5 = 4$ while $5 \div 20 \neq 4$ ◀ **Note:** $\neq$ means "is not equal to."

Division is not associative:

$$(32 \div 8) \div 2 \quad \text{while} \quad 32 \div (8 \div 2)$$
$$= 4 \div 2 \qquad\qquad\qquad = 32 \div 4$$
$$= 2 \qquad\qquad\qquad\qquad = 8$$

Different results

What if the divisor is 1 as in $\dfrac{24}{1}$?

To check the quotient, we must be able to multiply the quotient by the divisor to get the dividend.

$$? \cdot 1 = 24$$

In Section 1.3, we learned that any number times 1 is the number. Therefore, $24 \cdot 1 = 24$, which means $\dfrac{24}{1} = 24$.

Conclusion: Any number divided by 1 is that number.

DIVISION PROPERTY

When 1 is the divisor, the quotient is equal to the dividend.

In math language: $n \div 1 = \dfrac{n}{1} = n$, when n is any number.

What if the divisor is 0 as in $\dfrac{14}{0}$?

To check the quotient, we must be able to multiply the quotient by the divisor to get the dividend.

$$? \cdot 0 = 14$$

This is impossible. Any number times 0 should make 0.

Conclusion: It is impossible to divide by 0.

When 0 is a divisor with a nonzero dividend, we say the quotient is *undefined*, so $\dfrac{14}{0}$ is undefined.

DIVISION PROPERTY

When 0 is the divisor with any dividend other than 0, the quotient is undefined.

In math language: $n \div 0$, or $\dfrac{n}{0}$, is undefined, when $n \neq 0$.

What if the dividend is 0 as in $\dfrac{0}{12}$?

Again, we must be able to multiply the quotient by the divisor to get the dividend.

$$? \cdot 12 = 0$$

Notice that the missing factor must be 0, therefore $\dfrac{0}{12} = 0$. In general, 0 divided by a nonzero number is 0. Before we formalize this rule, we must consider 0 divided by 0.

$$\frac{0}{0} = ?$$

To check we must be able to multiply the quotient by the divisor to get the dividend.

$$? \cdot 0 = 0$$

It is tempting to say the missing factor is still 0, but actually, any number multiplied by 0 equals 0, so we cannot determine a solution for $\dfrac{0}{0}$. We say $\dfrac{0}{0}$ is *indeterminate*.

Conclusion: When 0 is divided by any number other than 0, the quotient is 0. When 0 is divided by itself, the quotient is indeterminate.

DIVISION PROPERTIES

When 0 is the dividend, the quotient is 0 as long as the divisor is not also 0.

In math language: $0 \div n = \dfrac{0}{n} = 0$, when $n \neq 0$.

If both dividend and divisor are 0, the quotient is indeterminate.

In math language: $0 \div 0$ or $\dfrac{0}{0}$ is indeterminate.

What if we divide a number other than 0 by itself as in $\dfrac{65}{65}$?

Again, we must be able to check by multiplying the quotient times the divisor to get the dividend.

$$? \cdot 65 = 65$$

The only number that works here is 1 because $1 \cdot 65 = 65$, which means $\dfrac{65}{65} = 1$.

DIVISION PROPERTY

When a number (other than 0) is divided by itself, the quotient is 1.

In math language: $n \div n = \dfrac{n}{n} = 1$, when $n \neq 0$.

In division, when the remainder is 0, we say the divisor is an *exact divisor*. For example, because $20 \div 5 = 4$, we say that 5 is an exact divisor of 20 or that 20 is divisible by 5. We can use divisibility rules to determine if a number is an exact divisor for a given number without dividing the numbers. We call these *divisibility rules*. There are many divisibility rules, but the following divisibility rules are the most useful.

DIVISIBILITY RULES

1. 2 is an exact divisor for all even numbers. Even numbers have 0, 2, 4, 6, or 8 in the ones place.

   Example: 3596 is divisible by 2 because it is an even number. 3596 has the digit 6 in the ones place.

2. To determine whether 3 is an exact divisor for a given number:
   a. Add the digits in the dividend.
   b. If the resulting sum is a number that is divisible by 3, then so is the dividend.

   Example: 58,014 is divisible by 3 because the sum of the digits is
   $5 + 8 + 0 + 1 + 4 = 18$. Because 18 is divisible by 3, so is 58,014.

3. 5 is an exact divisor for numbers that have 0 or 5 in the ones place.

   Example: 91,285 is divisible by 5 because it has a 5 in the ones place.

Example 1 Use the divisibility rules to determine whether the given number is divisible by 2.

a. 45,091

Answer: 45,091 is not divisible by 2 because it is not even. It does not have a 0, 2, 4, 6, or 8 in the ones place.

b. 691,134

Answer: 691,134 is divisible by 2 because it is an even number.

Do Your Turn 1 ▶

Example 2 Use divisibility rules to determine whether the given number is divisible by 3.

a. 76,413

Solution: Add the digits. $7 + 6 + 4 + 1 + 3 = 21$

Because the sum, 21, is divisible by 3, so is the number.

Answer: 76,413 is divisible by 3.

b. 4256

Solution: Add the digits. $4 + 2 + 5 + 6 = 17$

Because the sum, 17, is not divisible by 3, neither is the number.

Answer: 4256 is not divisible by 3.

Do Your Turn 2 ▶

Example 3 Use divisibility rules to determine whether the given number is divisible by 5.

a. 2,380,956

Answer: 2,380,956 is not divisible by 5 because it does not end in 0 or 5.

b. 49,360

Answer: 49,360 is divisible by 5 because it ends in 0.

Do Your Turn 3 ▶

Your Turn 1

Use divisibility rules to determine whether the given number is divisible by 2.

a. 4,519,362

b. 5385

Your Turn 2

Use divisibility rules to determine whether the given number is divisible by 3.

a. 93,481

b. 412,935

Your Turn 3

Use divisibility rules to determine whether the given number is divisible by 5.

a. 6945

b. 42,371

Answers to Your Turn 1
a. yes b. no

Answers to Your Turn 2
a. no b. yes

Answers to Your Turn 3
a. yes b. no

Now that we have learned some properties and rules of division, we can discuss more complex divisions. Because repeated subtraction is tedious, the common method for dividing is *long division*.

In a completed long division, it can be difficult to comprehend everything that happens, so in our first example of long division, we will break the process into steps. Thereafter, we will only show the completed long division.

Your Turn 4

Divide.

a. $5028 \div 3$

b. $6409 \div 5$

c. $12{,}059 \div 23$

WARNING If a remainder is greater than the divisor, then the divisor can divide the dividend digits more times. For example, in our first step in Example 4, suppose we mistakenly thought 7 goes into 64 only 6 times.

$$
\begin{array}{r}
6 \\
7\overline{)6408} \\
-42 \\
\hline
22
\end{array}
$$

Since the remainder, 22, is greater than the divisor, 7, the quotient digit is not big enough. Further, since 7 goes into 22 three times, adding 3 to the incorrect quotient digit, 6, gives the correct digit:

$$3 + 6 = 9.$$

Example 4   Divide. $6408 \div 7$

Solution: $7\overline{)6408}$
↑
Because 6 is less than 7, we include the 4 with the 6. Because 64 is greater than 7, we are ready to divide.

$$
\begin{array}{r}
9 \\
7\overline{)6408} \\
-63 \\
\hline
1
\end{array}
$$

$$
\begin{array}{r}
9 \\
7\overline{)6408} \\
-63\downarrow \\
\hline
10
\end{array}
$$

$$
\begin{array}{r}
91 \\
7\overline{)6408} \\
-63 \\
\hline
10 \\
-7 \\
\hline
3
\end{array}
$$

$$
\begin{array}{r}
915 \\
7\overline{)6408} \\
-63 \\
\hline
10 \\
-7 \\
\hline
38 \\
-35 \\
\hline
3
\end{array}
$$

1. Write the problem in long division form.

2. Compare the left-most digit in the dividend with the divisor. If it is less than the divisor, then include the next digit. Continue in this way until the digits name a number that is greater than the divisor.

3. Divide 7 into 64 and place the quotient, which is 9, over the 4.

4. Multiply 9 times 7 to get 63, then subtract the 63 from 64 to get 1.

5. Write the next digit in the dividend, 0, beside the 1 remainder. It is common to say that we are *bringing down* the 0.

6. Divide 7 into 10 and place the quotient, 1, over the 0 digit in the dividend.

7. Multiply 1 times 7 to get 7. Subtract 7 from 10 to get 3.

8. Write the last digit, 8, beside the 3 to make the number 38.

9. Divide 7 into 38 to get 5 and write the 5 over the 8.

10. Multiply 5 times 7 to get 35, then subtract the 35 from 38 to get a remainder of 3.

Answer: 915 r3

Check: When checking results that have a remainder, we multiply the quotient by the divisor, then add the remainder to get the dividend.

$$
\begin{array}{r}
{}^{1\,3} \\
915 \quad \text{Quotient} \\
\times\ 7 \quad \text{Divisor} \\
\hline
6405 \\
+3 \quad \text{Remainder} \\
\hline
6408 \quad \text{Dividend}
\end{array}
$$

Answers to Your Turn 4
a. 1676 b. 1281 r4 c. 524 r7

◀ **Do Your Turn 4**

Example 5 Divide. $42{,}017 \div 41$

Solution:
$$
\begin{array}{r}
1{,}024 \\
41\overline{)42{,}017} \\
-41 \\
\hline
1\,0 \\
-0 \\
\hline
1\,01 \\
-82 \\
\hline
197 \\
-164 \\
\hline
33
\end{array}
$$

◀ Because 41 does not divide 10, we place a 0 in the quotient.

Answer: 1024 r33

We can check by reversing the process. We multiply the quotient by the divisor then add the remainder to the resulting product. We should get the dividend.

Check:
$$
\begin{array}{rl}
\overset{1}{1}\,024 & \text{Quotient} \\
\times 41 & \text{Divisor} \\
\hline
1\,024 & \\
40\,96 & \\
\hline
41{,}984 & \\
+33 & \text{Remainder} \\
\hline
42{,}017 & \text{Dividend}
\end{array}
$$

Do Your Turn 5 ▷

Your Turn 5

Divide.

a. $10{,}091 \div 97$

b. $2617 \div 65$

c. $493{,}231 \div 205$

OBJECTIVE 2 Solve equations containing a missing factor.

Suppose we have to design a room in a house, and we know we want the area to be 150 square feet and the length to be 15 feet. What must the width be?

We know the area of a rectangle is found by the formula $A = l \cdot w$, so we can write an equation with the width, w, as a missing factor:

$$15 \cdot w = 150$$

When we solved equations with a missing addend in Section 1.2, we used a related subtraction sentence. Similarly, to solve an equation with a missing factor, we use a related division sentence.

Related division: $w = 150 \div 15$ or $w = \dfrac{150}{15}$

$\phantom{\text{Related division:}}$ $w = 10$ $w = 10$

The width needs to be 10 feet. We can return to the formula for area to check.

$$A = 15 \cdot 10 = 150$$

Our example suggests the following procedure.

PROCEDURE *To solve for a missing factor, write a related division sentence, dividing the product by the known factor.*

Answers to Your Turn 5
a. 104 r3 **b.** 40 r17 **c.** 2406 r1

Find the missing factor.

a. $6 \cdot x = 54$

b. $n \cdot 18 = 378$

c. $28 \cdot a = 0$

d. $r \cdot 0 = 37$

Example 6 Solve and check.

a. $x \cdot 16 = 208$

Solution: $x = 208 \div 16$

$x = 13$

$$\begin{array}{r} 13 \\ 16\overline{)208} \\ -16 \\ \hline 48 \\ -48 \\ \hline 0 \end{array}$$

Check: In the original equation, replace x with 13 and verify that the equation is true.

$$x \cdot 16 = 208$$
$$13 \cdot 16 \stackrel{?}{=} 208 \quad \text{Replace } x \text{ with 13.}$$
$$208 = 208 \quad \begin{array}{l}\text{Multiply. The equation is true,}\\ \text{so 13 is the solution.}\end{array}$$

$$\begin{array}{r} \overset{1}{16} \\ \times 13 \\ \hline 48 \\ +16 \\ \hline 208 \end{array}$$

b. $14 \cdot n = 0$

Solution: $n = \dfrac{0}{14}$

$n = 0$

Explanation: The related division statement has 0 as a dividend. When 0 is a dividend with a divisor other than 0, the quotient is 0.

Check:

$$14 \cdot n = 0$$
$$14 \cdot 0 \stackrel{?}{=} 0 \quad \text{Replace } n \text{ with 0.}$$
$$0 = 0 \quad \text{Multiply. The equation is true, so 0 is the solution.}$$

c. $y \cdot 0 = 5$

Solution: $y = \dfrac{5}{0}$, which is undefined. There is no solution to this equation.

Check: We cannot check this problem because there is no number that can replace the missing factor to make the statement true.

◀ **Do Your Turn 6**

OBJECTIVE 3 Solve applications involving division.

Like the other operations, division problems have key words or phrases that indicate to divide.

KEY WORDS FOR DIVISION

Divide, distribute, each, split, quotient, into, per, over

Answers to Your Turn 6
a. $x = 9$ b. $n = 21$ c. $a = 0$
d. no solution

Example 7 An egg farmer has 4394 eggs to distribute into packages of a dozen each. How many packages can be made? How many eggs will be left?

Solution: We must split up the total number of eggs into groups of a dozen.

$$
\begin{array}{r}
366 \\
12\overline{)4394} \\
-36 \\
\hline
79 \\
-72 \\
\hline
74 \\
-72 \\
\hline
2
\end{array}
$$

Answer: 366 packages, each holding a dozen eggs, can be made with 2 eggs left over.

Check:
$$
\begin{array}{r}
366 \\
\times 12 \\
\hline
732 \\
+366 \\
\hline
4392 \\
+2 \\
\hline
4394
\end{array}
$$

Do Your Turn 7 ▶

OBJECTIVE 4 Find the square root of a perfect square.

Suppose we want to design a fenced area for our pet dog. The veterinarian tells us that the dog needs 400 square feet of area to stay healthy. If we make the fenced area a square, how long must each side be?

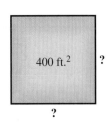

400 ft.²   ?

?

A **square** is a special rectangle in which the length and width are the same. Recall that to get the area of a rectangle we must multiply the length by the width. In the case of a square, since the length and width are the same, we would have to multiply a number by itself, or *square* the number to get the area. Note the two meanings of the word *square*.

DEFINITIONS Square: **a.** Geometric: A rectangle with all sides equal in length.

b. Algebraic: To multiply a number by itself.

So in designing the fence, to find the length of each side, we must find a number that can be squared to equal 400.

$$(\; ? \;)^2 = 400$$

We say this unknown number is a **square root** of 400.

Your Turn 7

Solve.

a. Alicia wants to place a wooden fence along the back of her property. She measures the distance to be 180 feet. She decides to buy prefabricated 8-foot sections of fencing. How many sections must she purchase?

b. Juan is to set an IV drip so that a patient receives 300 units of heparin, an anticoagulant medication, in D/W solution throughout 1 hour. How many units should the patient receive each minute?

OF INTEREST

D/W is an abbreviation for dextrose in water. Dextrose is a type of sugar in animal and plant tissues.

Answers to Your Turn 7
a. 23 whole sections **b.** 5 units

CONNECTION Exponents and roots are inverse operations, just as multiplication and division are inverse operations and addition and subtraction are inverse operations.

DEFINITION Square root: A base number that can be squared to equal a given number.

Since $20^2 = 400$, the square root of 400 is 20. Our fenced area should be 20 feet by 20 feet.

The symbol for square root is the *radical sign*. The number we wish to find the square root of is called the *radicand*.

$$\text{Radical sign} \rightarrow \sqrt{\text{Radicand}}$$

In this section, we will only consider radicands that are **perfect squares.** Perfect squares are numbers that have whole-number square roots.

CONNECTION In Chapter 2, we will learn that every natural number actually has *two* square roots.

DEFINITION Perfect square: A number that has a whole-number square root.

It is helpful to memorize a certain number of perfect squares. Table 1.1 lists the first 20 perfect squares and their roots.

Calculator Tips

To evaluate a square root, some calculators have a $\boxed{\sqrt{x}}$ key and some have a $\sqrt{}$ function, which is accessed by pressing $\boxed{2^{nd}}$ $\boxed{x^2}$. For example, to evaluate $\sqrt{169}$ using a $\boxed{\sqrt{x}}$ key, type $\boxed{1}$ $\boxed{6}$ $\boxed{9}$ $\boxed{\sqrt{x}}$. To evaluate $\sqrt{169}$ using a $\sqrt{}$ function, type $\boxed{2^{nd}}$ $\boxed{x^2}$ $\boxed{1}$ $\boxed{6}$ $\boxed{9}$ $\boxed{\text{ENTER} =}$

Table 1.1   Roots and Their Squares

Root	Perfect square	Root	Perfect square
0	$0^2 = 0$	10	$10^2 = 100$
1	$1^2 = 1$	11	$11^2 = 121$
2	$2^2 = 4$	12	$12^2 = 144$
3	$3^2 = 9$	13	$13^2 = 169$
4	$4^2 = 16$	14	$14^2 = 196$
5	$5^2 = 25$	15	$15^2 = 225$
6	$6^2 = 36$	16	$16^2 = 256$
7	$7^2 = 49$	17	$17^2 = 289$
8	$8^2 = 64$	18	$18^2 = 324$
9	$9^2 = 81$	19	$19^2 = 361$

DISCUSSION There is a pattern to the ones places in the perfect squares. Do you see it?

HINT:

$0^2 = \mathbf{0}$ and $10^2 = 10\mathbf{0}$

$1^2 = \mathbf{1}$ and $11^2 = 12\mathbf{1}$

$2^2 = \mathbf{4}$ and $12^2 = 14\mathbf{4}$

What is the pattern and why is it like that?

PROCEDURE *To find a square root of a given number, find a number that can be squared to equal the given number.*

Your Turn 8

Find each square root.

a. $\sqrt{144}$

b. $\sqrt{324}$

Example 8 $\sqrt{169}$

Answer: 13

Explanation: The square root of 169 is 13 because $13^2 = 169$.

Check: Verify that 13×13 is 169.

$$\begin{array}{r} 13 \\ \times 13 \\ \hline 39 \\ +13 \\ \hline 169 \end{array}$$

Answers to Your Turn 8
a. 12 b. 18

◁ Do Your Turn 8

Example 9 In building your new home, you decide to include an office. The plans can be developed with an office that has an area between 200 square feet and 250 square feet. What size perfect-square office would fit within the specified area?

Solution: The perfect square between 200 and 250 is 225, so the dimensions of the square office will be the square root of 225.

$$\sqrt{225} = 15$$

Answer: To make the office a perfect square with an area in the specified range, the dimensions should be 15 ft. × 15 ft.

Check: Verify that $15^2 = 225$

$$
\begin{array}{r}
\overset{2}{15} \\
\times\ 15 \\
\hline
75 \\
+\ 15 \\
\hline
225
\end{array}
$$

Do Your Turn 9 ▷

Your Turn 9

Solve.

a. A package of grass seed will cover 2500 square feet. What would be the dimensions of a square area that could be covered by the whole package?

b. A grid is drawn on a map with a scale where 1 square inch = 196 square miles. What are the dimensions in miles of each square in the grid?

Answers to Your Turn 9
a. 50 ft. × 50 ft.
b. 14 mi. × 14 mi.

For Extra Help

Videotape DVT

Tutor Center
Addison-Wesley Tutor Center

Math XL
Math XL

MyMathLab

Student Solutions Manual

1. When a number is divided by 1, the quotient is equal to _____.

2. When 0 is the divisor with any dividend other than 0, the quotient is _____.

3. When a number (other than 0) is divided by itself, the quotient is _____.

4. Explain how to find an unknown factor.

5. What is a square root of a given number?

6. List four perfect squares.

For Exercises 7–14, determine the quotient and explain your answer.

7. $26 \div 26$ **8.** $381 \div 1$ **9.** $0 \div 49$ **10.** $29 \div 0$

11. $0 \div 0$ **12.** $462 \div 462$ **13.** $22 \div 0$ **14.** $0 \div 95$

For Exercises 15–20, use divisibility rules to determine whether the given number is divisible by 2.

15. 19,761   **16.** 24,978   **17.** 143,706

18. 801,907   **19.** 431,970   **20.** 8445

For Exercises 21–26, use divisibility rules to determine whether the given number is divisible by 3.

21. 19,704   **22.** 1101   **23.** 26,093

24. 450,917   **25.** 98,757   **26.** 241,080

For Exercises 27–32, use divisibility rules to determine whether the given number is divisible by 5.

27. 27,005   **28.** 148,070   **29.** 64,320

30. 704,995   **31.** 4,195,786   **32.** 319,424

For Exercises 33–50, divide.

33. $3834 \div 9$ **34.** $2166 \div 6$ **35.** $1038 \div 5$

36. $24,083 \div 4$ **37.** $3472 \div 16$ **38.** $10,836 \div 28$

39. $6399 \div 26$ **40.** $9770 \div 19$ **41.** $12,592 \div 41$

42. $\dfrac{19{,}076}{38}$

43. $\dfrac{27{,}600}{120}$

44. $\dfrac{21{,}600}{270}$

45. $\dfrac{235{,}600}{124}$

46. $\dfrac{3{,}912{,}517}{93}$

47. $\dfrac{14{,}780}{0}$

48. $\dfrac{174{,}699}{87}$

49. $\dfrac{331{,}419}{207}$

50. $\dfrac{1{,}031{,}096}{514}$

For Exercises 51–62, solve and check.

51. $9 \cdot x = 54$

52. $y \cdot 4 = 40$

53. $m \cdot 11 = 88$

54. $15 \cdot a = 45$

55. $24 \cdot t = 0$

56. $b \cdot 21 = 105$

57. $17 \cdot n = 119$

58. $u \cdot 0 = 409$

59. $v \cdot 2 \cdot 13 = 1092$

60. $16 \cdot 5 \cdot k = 560$

61. $29 \cdot 6 \cdot h = 3480$

62. $18 \cdot c \cdot 41 = 6642$

For Exercises 63–76, solve.

63. Dedra's gross annual salary is $34,248. She gets paid once per month. What is her gross monthly salary?

64. A financial planner is asked to split $16,800 evenly among 7 investments. How much does she put into each investment?

65. Carl is to set an IV drip so that a patient receives 840 milliliters of 5% D/W solution in an hour. How many milliliters should the patient receive each minute?

66. The federal government grants a state $5,473,000 to be distributed equally among the 13 technical colleges in the state. How much does each college receive?

67. An employee of a copying company needs to make small fliers for a client. He can make 4 fliers out of each piece of paper. The client needs 500 fliers. How many pieces of paper must be used?

68. A printing company is asked to make 800 business cards for a client. It can print 12 cards on each sheet of card stock. How many sheets will be used?

69. How many 37¢ stamps can be bought with $15? Explain.

70. A long-distance company charges 19¢ per minute. How long can you talk for $5? Explain.

71. A cereal factory produces 153,600 ounces of cereal each day. Each box is to contain 16 ounces. The boxes are packaged in bundles of 8 boxes per bundle, then stacked on pallets, 24 bundles to a pallet. Finally, the pallets are loaded onto trucks, 28 pallets to each truck.
  a. How many boxes of cereal are produced each day?
  b. How many bundles are produced?
  c. How many pallets are needed?
  d. How many trucks are needed? Explain.

72. A bottling company produces 48,000 bottles each day. The bottles are packaged first in six-packs, then bundled into cases of 4 six-packs per case. The cases are then loaded onto pallets, 36 cases to a pallet.
  a. How many six-packs are produced?
  b. How many cases?
  c. How many pallets are needed? Explain.

73. In designing a room for a house, the desired length is 24 feet. The desired area is 432 square feet. What must the width be?

74. A landscaper needs to cover a 4000 square foot yard with sod. Sod is delivered in pallets. Each pallet holds enough sod to cover 504 square feet. How many pallets must be purchased if the company that sells the pallets will only sell whole pallets?

75. Barry decides to use prefabricated sections of wooden fencing to fence his back yard. Each section is 8 feet in length. How many sections must be purchased to complete 170 feet?

76. Building code says that electrical outlets must be spaced no more than 8 feet apart along the wall perimeter in each room. If a room has a perimeter of 60 feet, then how many electrical outlets will be needed to meet the code requirements?

For Exercises 77–88, find the square root. You should have most of these perfect squares and their roots memorized.

77. $\sqrt{100}$ **78.** $\sqrt{81}$ **79.** $\sqrt{169}$ **80.** $\sqrt{121}$

81. $\sqrt{0}$ **82.** $\sqrt{289}$ **83.** $\sqrt{1}$ **84.** $\sqrt{25}$

85. $\sqrt{49}$ **86.** $\sqrt{144}$ **87.** $\sqrt{196}$ **88.** $\sqrt{225}$

For Exercises 89–92, use a calculator to find the square root.

89. $\sqrt{1369}$ **90.** $\sqrt{7056}$ **91.** $\sqrt{45{,}796}$ **92.** $\sqrt{21{,}316}$

For Exercises 93–96, solve.

93. A machinist is hired to design a square plate cover for an engine. The hole to be covered is 36 square inches. What must be the dimensions of the plate?

94. A search-and-rescue team is flying in a helicopter over the ocean. They must search a square area of 784 square miles. What are the dimensions of this square?

95. The Jacksons wish to build a deck onto the back of their house. The builder charges $13 per square foot. They have $1872 in savings to spend.
a. How many square feet can they afford?
b. If they build a square deck, what would be the dimensions?

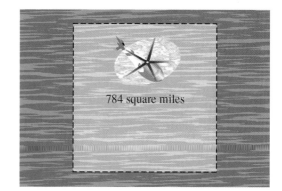

784 square miles

96. Michael is building a concrete patio. He has purchased enough concrete to cover an area of 324 square feet. If he built a square patio, what would be its dimensions?

Review Exercises

[1.2]   **1.** Solve and check. $184 + t = 361$

[1.2]   **2.** Connie is placing a decorative border strip near the ceiling in her bedroom. The room is 14 feet wide by 16 feet long. How much will she need?

[1.3]   **3.** Estimate 42×367 by rounding to the nearest ten, then find the actual product.

[1.3]   **4.** Evaluate. 5^4

[1.3]   **5.** Write 49,602 in expanded notation with powers of 10.

1.5 Order of Operations

OBJECTIVE 1 Simplify numerical expressions by following the order of operations agreement.

OBJECTIVE

1 Simplify numerical expressions by following the order of operations agreement.

Now let's simplify expressions like $16 - 3 \cdot 2$ that combine several different operations. Notice that we get different answers depending on the order in which we perform the operations.

If we subtract first, we get:

$$16 - 3 \cdot 2$$
$$= 13 \cdot 2$$
$$= 26$$

If we multiply first, we get:

$$16 - 3 \cdot 2$$
$$= 16 - 6$$
$$= 10$$

Which is correct? It turns out that 10 is the correct answer because we must multiply before subtracting.

The correct order is determined by an agreement. Because the specific order of operations that we follow is an agreement, it cannot be derived or proven.

ORDER OF OPERATIONS AGREEMENT

Perform operations in the following order:

1. Grouping symbols. These include parentheses ( ), brackets [ ], braces { }, absolute value | |, radicals $\sqrt{}$, and fraction bars ——.
2. Exponents.
3. Multiplication or division from left to right, in order as they occur.
4. Addition or subtraction from left to right, in order as they occur.

Note: We will not see absolute value symbols until Chapter 2.

Many people use the word $\overrightarrow{GEM}\overrightarrow{DAS}$ to help remember the order of operations. The arrows over MD and AS are there to remind us that we perform multiplication or division from left to right and then addition or subtraction from left to right.

LEARNING STRATEGY

If you are an auditory learner, try using a sentence like "Go eat my dad's anchovy sandwiches" to remember the order of operations.

Go	Eat	My	Dad's	Anchovy	Sandwiches.
r	x	u	i	d	u
o	p	l	v	d	b
u	o	t	i	i	t
p	n	i	s	t	r
i	e	p	i	i	a
n	n	l	o	o	c
g	t	i	n	n	t
	s	c			i
s		a			o
y		t			n
m		i			
b		o			
o		n			
l					
s					

Be careful, as the sentence seems to imply that multiplication comes before division and addition before subtraction, which is not necessarily true.

Some people use the word PEMDAS to remember the order of operations. Here, *P* stands for *parentheses*, which are only one type of grouping symbols and may be confusing.

WARNING Some people make the mistake of thinking that multiplication always comes before division and addition always comes before subtraction. However, proper order is to multiply or divide from *left to right* in the order they occur. Then add or subtract from *left to right* in the order they occur.

Your Turn 1

Simplify. Show your steps.

a. $15 + 4 \cdot 6$

b. $20 - 18 \div 3$

c. $12 + 14 \div 2 \cdot 5$

Example 1 Simplify. $8 + 6 \cdot 2$

Solution: $8 + 6 \cdot 2$

$= 8 + 12$ Multiply $6 \cdot 2$ first to get 12.

$= 20$ Add $8 + 12$ to get 20.

Note: We write each new step underneath the previous step instead of outwards to the right. This is an algebraic form of writing mathematics, which makes steps easy to follow.

DISCUSSION   Explain the mistake: $8 + 6 \cdot 2$
$$= 14 \cdot 2$$
$$= 28$$

◀ Do Your Turn 1

Your Turn 2

Simplify. Show your steps.

a. $7 - \dfrac{3^2}{9}$

b. $2^2 - \dfrac{12}{4}$

c. $5^2 \cdot 3 - 10$

Example 2 Simplify. $\dfrac{8}{2^2} + 5$

Solution: $\dfrac{8}{2^2} + 5$

$= \dfrac{8}{4} + 5$ Evaluate the exponential form. $2^2 = 4$

$= 2 + 5$ Divide. $\dfrac{8}{4} = 2$

$= 7$ Add. $2 + 5 = 7$

◀ Do Your Turn 2

Your Turn 3

Simplify. Show your steps.

a. $12 + 2^4 - 9 \cdot 2$

b. $2 \cdot 6 \div 4 + \sqrt{49}$

c. $3^3 - 5 \cdot 3 + \sqrt{64} - \dfrac{24}{3}$

Example 3 Simplify. $16 - 3^2 + 15 \div \sqrt{25}$

Solution: $16 - 3^2 + 15 \div \sqrt{25}$

$= 16 - 9 + 15 \div 5$ Evaluate the square and square root. $3^2 = 9$ and $\sqrt{25} = 5$

$= 16 - 9 + 3$ Divide. $15 \div 5 = 3$

$= 7 + 3$ Subtract. $16 - 9 = 7$

$= 10$ Add. $7 + 3 = 10$

◀ In $16 - 9 + 3$ our choices are subtraction or addition. We are to add or subtract from left to right, so we subtract first.

◀ Do Your Turn 3

Your Turn 4

Simplify. Show your steps.

a. $7(9 - 5) + 16 \div (3 + 5)$

b. $3^4 - 5(8 - 2) \div 6 + (7 - 3)^2$

Example 4 Simplify. $3(12 - 9) + 5^2 - (18 + 2) \div 5$

Solution:

A number next to parentheses means multiply.

$3(12 - 9) + 5^2 - (18 + 2) \div 5$

$= 3 \cdot 3 + 5^2 - 20 \div 5$ Perform operations within parentheses.

$= 3 \cdot 3 + 25 - 20 \div 5$ Evaluate the square. $5^2 = 25$

$= 9 + 25 - 4$ Multiply and divide from left to right. $3 \cdot 3 = 9$ and $20 \div 5 = 4$

$= 30$ Add and subtract from left to right. $9 + 25 = 34$ and $34 - 4 = 30$

Note: Once the calculations within parentheses, brackets, or braces have been completed, we can drop the parentheses, brackets, or braces.

Note: Once you get used to the order, you'll find that you can do some of the operations in the same step. For example, we could have performed the operations within the parentheses and evaluated 5^2 in one step. That would have saved writing an extra line.

◀ Do Your Turn 4

Answers to Your Turn 1
a. 39 b. 14 c. 47

Answers to Your Turn 2
a. 6 b. 1 c. 65

Answers to Your Turn 3
a. 10 b. 10 c. 12

Answers to Your Turn 4
a. 30 b. 92

When grouping symbols appear within other grouping symbols, we say they are *embedded*. When we have embedded grouping symbols, we must work from the innermost symbols outward, performing the inner computations first.

Example 5 Simplify. $[26 - 3(2 + 5)] + 30 \div \sqrt{9 + 16}$

Solution: $[26 - 3(2 + 5)] + 30 \div \sqrt{9 + 16}$

$= [26 - 3 \cdot 7] + 30 \div \sqrt{9 + 16}$	Add within the innermost parentheses. $2 + 5 = 7$
$= [26 - 21] + 30 \div \sqrt{9 + 16}$	Multiply within the brackets. $3 \cdot 7 = 21$
$= 5 + 30 \div \sqrt{9 + 16}$	Subtract within the brackets. $26 - 21 = 5$
$= 5 + 30 \div \sqrt{25}$	Add within the radical.
$= 5 + 30 \div 5$	Find the square root. $\sqrt{25} = 5$
$= 5 + 6$	Divide. $30 \div 5 = 6$
$= 11$	Add. $5 + 6 = 11$

Be careful, $\sqrt{9 + 16}$ is not the same as $\sqrt{9} + \sqrt{16}$.

$\sqrt{9 + 16}$ means to find the square root of the sum of 9 and 16, whereas $\sqrt{9} + \sqrt{16}$ means to find the sum of the square root of 9 and the square root of 16.

Look at the difference in answers.

$$\sqrt{9} + \sqrt{16} \qquad \sqrt{9 + 16}$$
$$= 3 + 4 \qquad\quad = \sqrt{25}$$
$$= 7 \qquad\qquad\quad = 5$$

Conclusion: When finding the square root of a sum or difference, we must add or subtract first, then find the root of the sum or difference.

Example 6 Simplify. $5[(12 - 8) \div 2] + 2\sqrt{16 \cdot 9}$

Solution: $5[(12 - 8) \div 2] + 2\sqrt{16 \cdot 9}$

$= 5[4 \div 2] + 2\sqrt{16 \cdot 9}$	Innermost parentheses first. $12 - 8 = 4$
$= 5 \cdot 2 + 2\sqrt{16 \cdot 9}$	Divide within the brackets. $4 \div 2 = 2$
$= 10 + 2\sqrt{16 \cdot 9}$	Multiply. $5 \cdot 2 = 10$
$= 10 + 2\sqrt{144}$	Multiply inside the radical. $16 \cdot 9 = 144$
$= 10 + 2 \cdot 12$	Find the square root. $\sqrt{144} = 12$
$= 10 + 24$	Multiply. $2 \cdot 12 = 24$
$= 34$	Add. $10 + 24 = 34$

A number next to a radical sign means multiply.

We saw in Example 5 that $\sqrt{9 + 16}$ is not the same as $\sqrt{9} + \sqrt{16}$. What about $\sqrt{16 \cdot 9}$ compared to $\sqrt{16} \cdot \sqrt{9}$? Let's have a look:

Multiplying first: $\sqrt{16 \cdot 9}$ **Roots first:** $\sqrt{16} \cdot \sqrt{9}$
$$= \sqrt{144} \qquad\qquad\qquad\qquad = 4 \cdot 3$$
$$= 12 \qquad\qquad\qquad\qquad\quad = 12$$

Notice the answer is the same either way. This is always true with square roots of products or quotients.

Conclusion: If multiplication or division occurs under a radical, we can multiply or divide first, then find the square root of the product or quotient. Or we can find the square roots first, then multiply or divide the roots.

Do Your Turn 5 ▷

Calculator Tips

Scientific calculators are programmed to follow the proper order of operations. This means you merely type in the expression as it is written.

Example: $(6 - 2)^3 + 4[9 + (8 - 3)]$

Type: | (| 6 | − |
2	)	^	3	+
4	(	9	+
(	8	−	3
)	)	ENTER =

Answer: 120

Your Turn 5

Simplify.

a. $\{5 + 2[18 \div (2 + 7)]\} + 5\sqrt{25 \cdot 4}$

b. $5^3 - 3[(6 + 14) \div (3 + 2)] + \sqrt{36 \div 9}$

c. $4\sqrt{169 - 25} + \{[30 - 3(5 + 2)] + [70 - (3 + 5)^2]\}$

Answers to Your Turn 5
a. 59 b. 115 c. 63

The fraction bar can be used as a grouping symbol. For example, in $\dfrac{7^2 - 4}{29 - 3(8 - 6)^3}$, the fraction line means to divide the result of the top expression by the result of the bottom expression. In fraction terminology the top is called the *numerator* and the bottom is called the *denominator*. Think of $\dfrac{7^2 - 4}{29 - 3(8 - 6)^3}$ as having implied parentheses in the numerator and denominator. In other words, the order of operations is the same as $(7^2 - 4) \div [29 - 3(8 - 6)^3]$.

Your Turn 6

Simplify using order of operations.

a. $\dfrac{12^2 - 24}{42 - 3(9 - 7)^2}$

b. $\dfrac{10 + 2(12 - 5)}{(3 + 5)2 - (7 - 5)^3}$

Example 7 Simplify. $\dfrac{7^2 - 4}{29 - 3(8 - 6)^3}$

Solution: Calculate the numerator and denominator separately, then divide.

$\dfrac{7^2 - 4}{29 - 3(8 - 6)^3}$

$= \dfrac{49 - 4}{29 - 3(2)^3}$ In the top: $7^2 = 49$.
At the bottom: $8 - 6 = 2$.

$= \dfrac{45}{29 - 3 \cdot 8}$ In the top: $49 - 4 = 45$.
At the bottom: $2^3 = 8$.

$= \dfrac{45}{29 - 24}$ Multiply in the bottom. $3 \cdot 8 = 24$

$= \dfrac{45}{5}$ Subtract at the bottom. $29 - 24 = 5$

$= 9$ Divide. $45/5 = 9$

◀ **Do Your Turn 6**

Answers to Your Turn 6
a. 4 b. 3

1.5 Exercises

For Extra Help

Videotape DVT

Addison-Wesley Tutor Center

Math XL

MyMathLab

Student Solutions Manual

1. List the four stages of the order of operations.

2. What is the first step in simplifying the expression $15 - 3 \times 4$? Explain.

3. What is the first step in simplifying the expression $8 + 2(10 - 4)$? Explain.

4. What is the first step in simplifying the expression $13 + 20 \div 5 \times 2$? Explain

For Exercises 5–50, simplify using the order of operations. Show your steps.

5. $7 + 5 \cdot 3$

6. $35 - 3 \cdot 7$

7. $18 + 36 \div 4 \cdot 3$

8. $20 \div 2 \cdot 5 + 6$

9. $12^2 - 6 \cdot 4 \div 8$

10. $7^2 + 5 \cdot 6 \div 10$

11. $36 - 3^2 \cdot 4 + 14$

12. $16 - 48 \div 4^2 + 5$

13. $\dfrac{39}{13} + \sqrt{36} - 2^3$

14. $29 - 4^2 + 21 \div \sqrt{49}$

15. $25 + 2(14 - 9)$

16. $43 - (2 + 4)^2$

17. $12 + 8\sqrt{25} \div 4$

18. $3\sqrt{36} - \dfrac{15}{3}$

19. $7\sqrt{64} - 40 \div 5 \cdot 2 + 9$

20. $4^2 - 6 \cdot 2 + \sqrt{81} - \dfrac{42}{7}$

21. $2^6 - 18 \div 3 \cdot 5 - \sqrt{100}$

22. $24 \div 8 \cdot 6 + \sqrt{121} - 5^2$

23. $3^2 \cdot 4 \div 6 - 5 + \sqrt{16}$

24. $5^2 \cdot 3 \div 15 + 4 - \sqrt{49}$

25. $2(14 + 3) - 8 \cdot 2 + (9 - 2)$

26. $5(13 - 7) + 12 \div (4 + 2)$

27. $(2 + 9)(19 - 16)^2$

28. $(5 - 3)(2 + 4)^2$

29. $2^5 \div (14 - 6) + 7(23 - 12)$

30. $2(7 - 4) + 8^2 - \dfrac{(16 + 5)}{7}$

31. $(13 - 8)3^2 - 48 \div (15 - 9)$

32. $\dfrac{15}{3} - 2^2 + \dfrac{(19 - 11)}{4}$

33. $58 \div (2\sqrt{49} - 6 \cdot 2)$

34. $2(4\sqrt{25} - 3 \cdot 5)$

35. $(3 + 4)\sqrt{121} + 3(6 - 4)$

36. $(19 - 16)\sqrt{169} + 54 \div 9(7 - 5)$

37. $[14 - (3 + 2)] \div 3 + 4(17 - 6)$

38. $31 - 3[(20 - 6) - 3 \cdot 2] + 2^4$

39. $\{18 - 4[21 \div (3 + 4)]\} + 3\sqrt{16 \cdot 4}$

40. $\sqrt{25 \cdot 9} + 2\{[15 - (3 + 9)] \cdot 6\}$

41. $4^3 - 5[(28 - 4) \div 2^3] + \sqrt{100 \div 25}$

42. $2[(2 + 9)\sqrt{36 \div 9}] + 28 \div 7$

43. $12 \div 4 \cdot 3 + 1^5 - \sqrt{17 - 1}$

44. $12 \div (4 \cdot 3) + 1^2 + \sqrt{29 - 4}$

45. $\dfrac{9^2 - 21}{56 - 2(4 + 1)^2}$

46. $\dfrac{22 + 3^3}{3(14 - 6) - (8 + 9)}$

47. $\dfrac{38 - 4(15 - 12)}{(3 + 5)^2 - 2(39 - 8)}$

48. $\dfrac{(12 - 5)^2 + 2^3}{10 \div 2 - (11 - 9)}$

49. $[485 - (68 + 39)] + 4^5 - 24 \cdot 16 \div 8$

50. $(25 - 9)^3 + 420 \div (28 - 7) + \sqrt{2209}$

For Exercises 51–54, explain the mistake, then work the problem correctly.

51. $48 - 6(9 - 4)$
$= 48 - 6(5)$
$= 42(5)$
$= 210$

52. $23 + \sqrt{100 - 64}$
$= 23 + 10 - 8$
$= 33 - 8$
$= 25$

53. $(3 + 5)^2 - 2\sqrt{49}$
$= 9 + 25 - 2\sqrt{49}$
$= 9 + 25 - 2(7)$
$= 9 + 25 - 14$
$= 34 - 14$
$= 20$

54. $[12 - 2 \cdot 3] + 4(3)^2$
$= [12 - 6] + 4(3)^2$
$= 6 + 4(3)^2$
$= 6 + 12^2$
$= 6 + 144$
$= 150$

Review Exercises

[1.2]   **1.** Estimate $42{,}320 + 25{,}015$ by rounding to the nearest ten thousand.

For Exercises 2 and 3, perform the indicated operation.

[1.3]   **2.** $498{,}503 \times 209$

[1.4]   **3.** $21{,}253 \div 17$

[1.3]   **4.** A 12-foot wide by 16-foot long room is to have ceiling molding installed. How much molding is needed? If the carpenter charges $2 per foot to install the molding, how much will it cost?

[1.3]   **5.** The floor of the same 12-foot wide by 16-foot long room is to be covered with square 1-foot by 1-foot tiles. How many tiles must be purchased?

1.6 More with Variables, Formulas, and Solving Equations

In Section 1.2, we defined *variable* and *constant*. In Section 1.3, we used variables to write a *formula* for the area of a rectangle. We have also learned how to solve equations with a missing addend (Section 1.2) or a missing factor (Section 1.4). In Section 1.5, we learned the order of operations.

In this section, we will bundle all of those topics together. We will derive more formulas, then use them to solve problems. When solving the problems, our formulas may contain numerical expressions that must be simplified using proper order of operations or they may lead to missing addend or missing factor equations that we must solve.

OBJECTIVE 1 Use the formula $P = 2l + 2w$ to find the perimeter of a rectangle.

First, let's derive a formula for the perimeter of a rectangle. Look at the rectangle below. We use the variable l to represent length and w for width. In Section 1.2, we learned that to find perimeter we add the lengths of all the sides, which suggests the following formula.

$$\text{perimeter of a rectangle} = \text{length} + \text{width} + \text{length} + \text{width}$$
$$P = \quad l \quad + \quad w \quad + \quad l \quad + \quad w$$

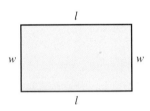

Notice that we can simplify the relationship even further. Since we are repeatedly adding two lengths and two widths, we can multiply the length by 2 and multiply the width by 2 then add. We now have a formula for the perimeter of a rectangle.

$$P = 2l + 2w$$

We can use the formula to find the perimeter of any rectangle if given values for its length and width.

PROCEDURE *To use a formula:*

1. Replace the variables with the corresponding given values.
2. Solve for the missing variable.

Example 1 A school practice field is to be enclosed with a chain-link fence. The field is 400 feet long by 280 feet wide. How much fencing is needed?

Solution: Because we must find the total distance around a rectangular field, we use the formula for the perimeter of a rectangle.

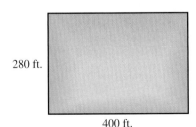

$P = 2l + 2w$ **Replace *l* with 400 and *w* with 280.**
$P = 2(400) + 2(280)$
$P = 800 + 560$ **Multiply.**
$P = 1360$ **Add.**

Answer: 1360 ft. of fencing is needed.

DISCUSSION What would happen if we replaced *l* with 280 feet and *w* with 400 feet? Would the answer be different?

Do Your Turn 1 ▶

Your Turn 1

Solve.

a. A rectangle has a length of 19 meters and width of 42 meters. Find the perimeter.

b. A rectangle has a length of 7 miles and a width of 6 miles. Find the perimeter.

Answers to Your Turn 1
a. 122 m b. 26 mi.

In Section 1.3, we discussed that area is measured in square units and developed a formula for calculating the area of a rectangle, $A = lw$. Rectangles are special forms of a more general class of figures called **parallelograms**, which have two pairs of **parallel** sides.

DEFINITIONS **Parallel lines:** Straight lines that never intersect.

Parallelogram: A four-sided figure with two pair of parallel sides.

Examples of parallelograms:

To find the area of a parallelogram, we need the length of the base and the height. The height is measured along a line that makes a 90° angle with the base. An angle that measures 90° is called a **right angle.**

DEFINITION **Right angle:** An angle that measures 90°.

We will use the letter b for the length of the base and h for the height.

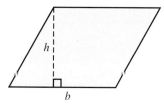

> **Note:** The right angle is indicated by the small square where the height line meets the base.

Notice that if we cut along the height line we would cut off a triangle. If we slide that triangle around to the right side and place the parallel sides together, the resulting figure is a rectangle.

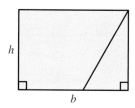

The rectangle we have made has the same area as the original parallelogram. The length of this rectangle is b and the width is h. So the formula to calculate the area of a parallelogram is $A = bh$.

> **Note:** Because squares and rectangles are special parallelograms, we can use $A = bh$ for them as well.

Example 2 Find the area of the parallelogram.

Solution: We use the formula for the area of a parallelogram.

$A = bh$

$A = (15)(9)$ **Replace b with 15 and h with 9.**

$A = 135$ **Multiply. Since the base and height are measured in feet, the area unit is square feet.**

Answer: The area is 135 ft.^{2}.

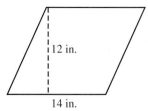
9 ft.
15 ft.

DISCUSSION What would happen if we replaced b with 9 feet and h with 15 feet? Would the answer be different?

Do Your Turn 2 ▷

Your Turn 2

Use the formula $A = bh$ to find the area.

a.

12 in.

14 in.

b.

8 ft.

4 ft.

OBJECTIVE 3 Use the formula $V = lwh$ to find the volume of a box.

Now let's consider **volume**, which is a measure of the amount of space inside a three-dimensional object. We measure volume using **cubic units.**

DEFINITIONS **Cubic unit:** A 1 × 1 × 1 cube.

Volume: The total number of cubic units that completely fill an object.

Examples of cubic units:

Cubic centimeter

1 cm
1 cm
1 cm

A cubic centimeter is a 1 cm by 1 cm by 1 cm cube and is represented mathematically by cm^3.

Cubic inch

1 in.
1 in.
1 in.

A cubic inch is a 1 in. by 1 in. by 1 in. cube and is represented

Let's develop a formula for calculating the volume of a box. Consider a box that is 3 meters long by 2 meters wide by 4 meters high. Let's *measure* its volume by filling it with cubic meters.

Note: If we fill the box with cubic meters, we can fit 6 cubic meters in a layer at the bottom. Because each layer is 1 meter high, we can stack a total of four layers inside the box. This means a total of 24 cubic meters fill the box.

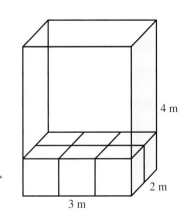

4 m

2 m

3 m

OF INTEREST

In mathematics, we write cubic centimeters as cm^3. Another common abbreviation is cc, which is used in medical dosage or in describing engine size.

Answers to Your Turn 2
a. 168 in.^{2} b. 32 ft.^{2}

Notice that we can *calculate* the volume by multiplying the length, width, and height.

volume of a box = length · width · height

volume = 3 m · 2 m · 4 m = 24 m³

> This is read *cubic meters* or *meters cubed*. All volume calculations will involve multiplying three distance measurements. Therefore, volume units will always be expressed as the distance unit with an exponent of 3.

So the formula for calculating the volume of a box is $V = lwh$.

Your Turn 3

Find the volume of each box.

a.

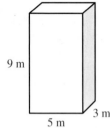

9 m

3 m

5 m

b.

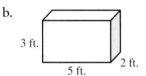

3 ft.

2 ft.

5 ft.

Example 3 An engine block is 25 centimeters long by 20 centimeters wide by 10 centimeters high. What is the volume of the engine block?

Solution: Because the engine block is a box, we use the formula for the volume of a box.

$$V = lwh$$
$$V = (25)(20)(10) \quad \text{Replace } l \text{ with 25, } w \text{ with 20, and } h \text{ with 10.}$$
$$V = 5000 \quad \text{Multiply. Since the length, width, and height are in centimeters, the volume is in cubic centimeters.}$$

Answer: The volume is 5000 cm³.

DISCUSSION What would happen if we interchanged the length value with the width value (or height)? Would the answer be different?

◄ **Do Your Turn 3**

OBJECTIVE 4 Solve for a missing number in a formula.

Sometimes when we substitute numbers into a formula, the resulting equation contains a missing factor.

Your Turn 4

Find the missing number.

a. The area of a parallelogram is 72 square inches. If the height is 9 inches, find the base.

b. The area of a rectangle is 60 square meters. If the width is 5 meters, find the length.

Example 4 The area of a parallelogram is 40 square feet and the base is 8 feet. Find the height.

Solution: Use the formula $A = bh$.

CONNECTION $40 = 8h$ is a missing factor equation. We learned to solve missing factor equations in Section 1.4 by dividing the product by the known factor.

$$A = bh$$
$$40 = 8h \quad \text{Replace } A \text{ with 40 and } b \text{ with 8.}$$
$$40 \div 8 = h \quad \text{Write a related division statement.}$$
$$5 = h \quad \text{Divide 40 by 8.}$$

Answer: The height is 5 ft.

Check: Verify that a parallelogram with a base of 8 feet and height of 5 feet has an area of 40 square feet.

$$A = bh$$
$$A = (8)(5) \quad \text{Replace } b \text{ with 8 and } h \text{ with 5.}$$
$$A = 40 \quad \text{Multiply. The height is correct.}$$

Answers to Your Turn 3
a. 135 m³ b. 30 ft.³

◄ **Do Your Turn 4**

Answers to Your Turn 4
a. 8 in. b. 12 m

Example 5 The volume of a box is 480 cubic centimeters. If the length is 10 centimeters and the width is 8 centimeters, then find the height.

Solution: Use the formula $V = lwh$.

$$V = lwh$$
$$480 = (10)(8)h \qquad \text{Replace } V \text{ with 480, } l \text{ with 10, and } w \text{ with 8.}$$
$$480 = 80h \qquad \text{Multiply 8 by 10.}$$
$$480 \div 80 = h \qquad \text{Write a related division statement.}$$
$$6 = h \qquad \text{Divide 480 by 80.}$$

Answer: The width is 6 cm.

Check: Verify that a box with a length of 10 centimeters, width of 8 centimeters, and height of 6 centimeters has a volume of 480 cubic centimeters.

$$V = (10)(8)(6) \qquad \text{Replace } l \text{ with 10, } w \text{ with 8, and } h \text{ with 6.}$$
$$V = 480 \qquad \text{Multiply. The height is correct.}$$

Do Your Turn 5 ▶

Your Turn 5

Solve for the missing number.

a. The volume of a box is 336 cubic centimeters. If the length is 12 centimeters and the width is 7 centimeters, find the height.

b. The volume of a box is 810 cubic inches. If the width is 9 inches and the height is 5 inches, find the length.

Answers to Your Turn 5
a. 4 cm b. 18 in.

For
Extra
Help

 Videotape
DVT

Tutor Center
Addison-Wesley
Tutor Center

 Math XL

 MyMathLab

 Student Solutions
Manual

1. In general, how do you use a formula?

2. What is the formula $A = bh$ used for, and what do its variables represent?

3. What is volume?

4. What is the formula $V = lwh$ used for, and what do its variables represent?

For Exercises 5–10, find the perimeter.

5.
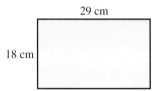
29 cm
18 cm

6.
10 ft.
15 ft.

7.
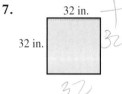
32 in.
32 in.
32
32

8.
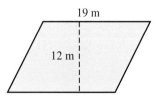
40 m
7 m
90

9.
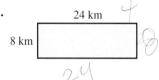
24 km
8 km
24

10.
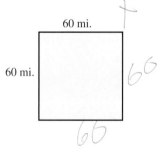
60 mi.
60 mi.
60
60

For Exercises 11–16, find the area.

11.
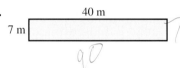
19 m
12 m

12.
9 ft.
13 ft.

13.
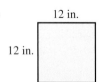
12 in.
12 in.

14.
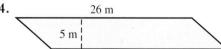
26 m
5 m

15.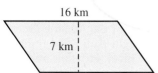
16 km
7 km

16.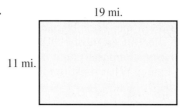
19 mi.
11 mi.

For Exercises 17–22, find the volume.

17.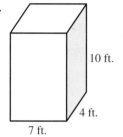
10 ft.
4 ft.
7 ft.

18.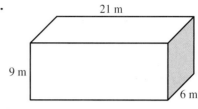
21 m
9 m
6 m

19.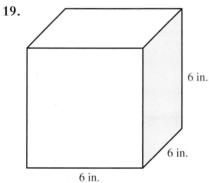
6 in.
6 in.
6 in.

20.
8 ft.
2 ft.   1 ft.

21.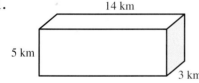
14 km
5 km
3 km

22.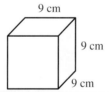
9 cm
9 cm
9 cm

For Exercises 23–34, find the missing number.

23. A parallelogram has an area of 144 square meters. If the base is 16 meters, find the height.

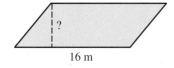
?
16 m

24. A parallelogram has an area of 399 square feet. If the height is 19 feet, find the base.

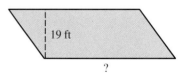
19 ft
?

25. A parallelogram has an area of 1922 square inches. If the height is 31 inches, find the base.

26. A parallelogram has an area of 2408 square centimeters. If the base is 56 centimeters, find the height.

27. A canning company feeds sheets of metal into a machine that molds them into cans. The area of a rectangular piece of sheet metal must be 180 square inches. To fit into the machine the width of a sheet of metal must be 9 inches. What must the length be?

28. A search-and-rescue helicopter is assigned to search an 3584-square-mile area of the Atlantic Ocean. It has flown north in a straight path of 64 miles and will now turn east (90° turn). How far must it travel along this new course to make a rectangle that will have the desired area?

29. A box has a volume of 12,320 cubic feet. If the length is 22 feet and the width is 35 feet, find the height.

30. A box has a volume of 1368 cubic meters. If the width is 19 meters and the height is 3 meters, find the length.

31. A box has a volume of 22,320 cubic inches. If the length is 31 inches and the height is 40 inches, find the width.

32. A box has a volume of 30,600 cubic centimeters. If the height is 60 centimeters and the length is 15 centimeters, find the width.

33. An engineer is designing a refrigerator. The desired volume is to be 24 cubic feet. If the length must be 3 feet and width must be 2 feet, what must the height be?

34. The research and development department for an electronics company has calculated that 18 cubic feet of storage space is needed to house a new high-resolution TV. The space for the TV in a common entertainment center measures 2 feet by 3 feet by 3 feet. Find the volume of the space for the TV in the entertainment center. Is it adequate for the new TV design?

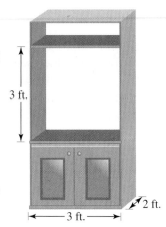

3 ft.

3 ft.

2 ft.

Review Exercises

[1.2] **1.** An accountant is given the following spreadsheet of expenses and income for a company. What is the final balance?

Description of expense	Amount	Description of assets/income	Amount
Payroll	$16,980	Checking account	$ 2,359
Utilities	$ 1,250	Income	$41,300
Water	$ 158		
Waste disposal	$ 97		
New inventory	$12,341		

[1.6] **2.** Find the area of a parallelogram with a base of 14 feet and a height of 16 feet.

[1.3] **3.** A contractor charges $16 per square foot to put in hardwood floors. How much would it cost to have the contractor cover a 32-square-foot entry space with hardwood flooring?

[1.3] **4.** To open a lock, you must use a combination of three things. You must turn a key to one of two positions, left or right; set a dial to a digit 0–9; and set another dial to a letter from A to F. How many possible combinations are there?

[1.4] **5.** An employee of a copying company needs to make leaflets for a client. She can make 6 leaflets out of each piece of paper. The client needs 1800 leaflets. How many pieces of paper must be used?

1.7 Problem Solving and Applications

OBJECTIVE 1 Use a problem-solving process to solve problems involving several operations.

OBJECTIVE

1 Use a problem-solving process to solve problems involving several operations.

As the problems we solve get more complex, it is helpful to follow a problem-solving outline. The purpose of the outline is to suggest strategies for solving problems and to offer insight into the process of problem solving. Throughout the rest of the text, all application problems will be solved using the outline.

PROCEDURE *Problem-Solving Outline*

1. **Understand** the problem.
   a. Read the question(s) (not the whole problem, just the question at the end) and note what it is you are to find.
   b. Now read the whole problem, underlining the key words.
   c. If possible and useful, make a list or table, simulate the situation, or search for a related example problem.
2. **Plan** your solution strategy by searching for a formula or translating the key words to an equation.
3. **Execute** the plan by solving the formula or equation.
4. **Answer** the question. Look at your note about what you were to find and make sure you answered that question. Include appropriate units.
5. **Check** the results.
   a. Try finding the solution in a different way, reversing the process, or estimating the answer and make sure the estimate and actual answer are reasonably close.
   b. Make sure the answer is reasonable.

DISCUSSION Why are units important in your answer? What are the pros and cons of each different method of checking?

LEARNING STRATEGY

If you are a visual learner, focus on visual strategies such as drawing pictures or making lists. If you are an auditory learner, try talking through the situation, expressing your ideas verbally. If you are a tactile learner, try to simulate or model the situation using pencil, paper, or any item nearby.

Let's use the problem-solving outline to solve problems involving *composite shapes*. Composite shapes are shapes formed by putting together two or more fundamental shapes, like rectangles or parallelograms. Or, a composite shape can be a shape within a shape.

Example 1 Shown here is a base plan for a new building. What is the total area occupied by the building?

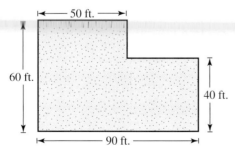

Understand: We are to find the total area occupied by the building.

The key words that we should underline are *total* and *area*. In Section 1.2, we learned that *total* indicates we must add. In Section 1.3, we learned that *area* indicates that we must multiply.

The building is a composite shape. There are several ways we can look at this composite shape.

We can view the area of the building as two rectangles put together. To do this, we can picture a horizontal line (left) or a vertical line (right) separating the shape into two rectangles:

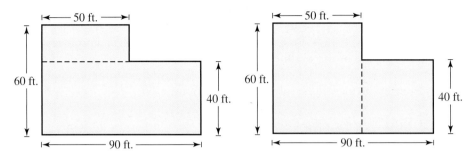

Or we could look at it as a rectangle in the upright corner that was removed from a larger rectangle.

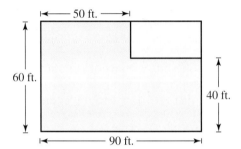

Visualize the rectangles in the way that is easiest for you. We will use the second figure from above.

Plan: Find the *total* area by adding the area of each rectangle. To find the area of each rectangle, we use the formula $A = lw$ (or $A = bh$).

Execute: First, we will find the area of the larger rectangle, which is 50 feet by 60 feet.

$$\text{area of the larger rectangle} = (50)(60) = 3000 \text{ ft.}^2$$

Now we need the area of the smaller rectangle. Notice that we were not given its length. We can use the other dimensions in the figure to determine that unknown length.

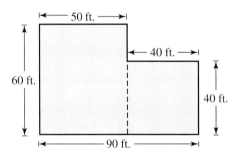

Note: Since the bottom is 90 feet across, the top must also be 90 feet across, so the unknown length of the smaller rectangle must be $90 - 50 = 40$ feet.

So the smaller rectangle is actually a 40-foot by 40-foot square. Now we can find its area:

$$\text{area of the square} = (40)(40) = 1600 \text{ ft.}^2$$

The total area of the building is the sum of the two areas.

$$\text{total area} = 3000 + 1600 = 4600 \text{ ft.}^2$$

Answer: The total area is 4600 ft.^{2}.   ◀   **Note:** Be sure to include the units.

Check: We will solve the problem again using the third figure, in which we visualized a smaller rectangle removed from the corner of a larger rectangle.

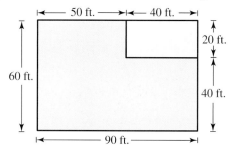

Since the smaller rectangle is *removed* from the larger one, we subtract its area from the area of the larger rectangle to find the area of the building.

area of building = area of larger rectangle − area of small rectangle

To calculate the area of a rectangle, we multiply length by width. The large rectangle is 60 feet by 90 feet.

$$\text{area of large rectangle} = (60)(90) = 5400 \text{ ft.}^2$$
$$\text{area of small rectangle} = (20)(40) = 800 \text{ ft.}^2$$
$$\text{area of the building} = 5400 - 800 = 4600 \text{ ft.}^2$$

Because we get the same answer, we can feel secure that our calculation for the total area of this shape is correct.

Do Your Turn 1 ▷

Example 2 A builder is required to sod the yard of a new home. The lot is a rectangle 75 feet wide by 125 feet long. The house occupies 2500 square feet of the lot. Sod must be bought in pallets of 500 square feet that cost $85 each. How much will it cost to sod the yard?

Understand: We are to find the total cost to sod a yard. Let's underline key words and important information.

A builder is required to sod the yard of a new home. The lot is a <u>rectangle 75 feet wide by 125 feet long</u>. The house <u>occupies 2500 square feet of the lot</u>. Sod must be bought in <u>pallets of 500 square feet</u> that cost <u>$85 each</u>. How much will it cost to sod the yard?

Because the problem involves rectangles, let's draw a picture.

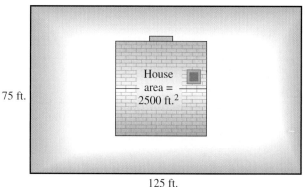

Your Turn 1

Solve.

a. The picture shows two bedrooms and a connecting hallway that are to be carpeted. How many square feet of carpet will be needed?

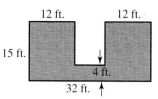

b. The picture shows the driveway and walk of a new house. The builder needs to know how much concrete to plan for. What is the total area of the driveway and walk?

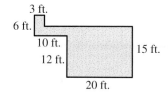

Answers to Your Turn 1

a. 392 ft.^{2} b. 339 ft.^{2}

To develop a plan, we will ask ourselves questions starting with the question in the problem.

What do we need to know in order to find the cost of the sod?

> Answer: If each pallet costs $85, then we need to know how many pallets will be used.

How can we figure the number of pallets required?

> Answer: If each pallet covers 500 ft.^{2}, then we need to know how many square feet of the lot will be covered with sod.

How can we figure the number of square feet to be covered?

> Answer: Because the whole lot, except for the house, is to be covered, we need to subtract the area occupied by the house from the total area of the lot.

We were given the area occupied by the house, but not the lot. How can we find the area of the lot?

> Answer: Because the lot is a rectangle and the area of a rectangle is found by the formula $A = lw$, we need to multiply the length and width of the lot.

Plan: Now we have our starting point. We'll find the area of the lot, then subtract the area occupied by the house, and this will give us the area to be covered with sod. From there, we can calculate how many pallets will be needed. Once we have the number of pallets, we can determine the cost.

Execute: Calculate the area of the lot:

$$A = lw$$
$$A = (125)(75)$$
$$A = 9375 \text{ ft.}^2$$

Now we can calculate how much area will be covered with sod by subtracting the area occupied by the house.

$$\text{area to sod} = 9375 - 2500 = 6875 \text{ ft.}^2$$

Next we need to determine how many pallets will be needed to cover 6875 square feet. Because each pallet covers 500 square feet, we need to divide 500 into 6875 to see how many sets of 500 are in 6875.

$$\text{number of pallets} = 6875 \div 500 = 13 \text{ r}375$$

This means 13 whole pallets are needed with 375 square feet uncovered by sod. Because partial pallets cannot be purchased, the builder must buy one more pallet to cover that 375 square feet, even though not all the sod will be used from that extra pallet. So, a total of 14 pallets must be purchased.

Because each of the 14 pallets costs $85, we must multiply to find the total cost.

$$\text{total cost} = (14)(85) = \$1190$$

Answer: The total cost to sod the yard is $1190.

Check: Let's do a quick estimate of the area. We'll round the length to 100 feet and the width to 80 feet and calculate the area.

$$\text{estimate} = 100 \cdot 80 = 8000 \text{ ft.}^2$$

The house occupies 2500 square feet, so let's round that to 3000 square feet. We can now estimate the amount of land to be covered in sod by subtracting the area of the house from our estimate for the whole lot.

approximate area of coverage = 8000 − 3000 = 5000 ft.^{2}

If each pallet covers 500 square feet, then about 10 pallets would be needed to cover 5000 square feet.

If we round the cost of each pallet to about $90, then the total cost is $90 · 10 pallets = $90. So we should expect the cost to be somewhere in the neighborhood of $900. Our actual calculation of $1190 is reasonable given the amount of rounding we did in our estimates.

Do Your Turn 2 ▶

Your Turn 2

Solve.

a. A 2000-square-foot house is built on a 9000-square-foot lot. The lot is to be land-scaped with 1800 square feet of flower beds and then the remaining area covered with grass seed. Each bag of grass seed covers 1000 square feet and costs $18. How much will it cost to seed the lot?

Example 3   Approximate the volume of the automobile shown.

Understand: *Volume* indicates we are to find the total number of cubic units that completely fill the car. Since we only need an approximation, we can view the automobile as two boxes stacked as shown. The formula for the volume of a box is $V = lwh$.

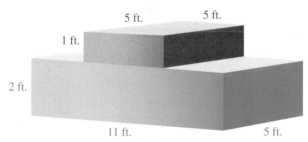

b. An 8-foot long by 4-foot wide entryway in a new house is to have a tile landing and the rest of the floor covered in hardwood. The tile landing is to be 4 feet long by 2 feet wide. How many square feet of hardwood will be needed? The cost of the tile is $7 per square foot and the cost of the hardwood is $13 per square foot. What will be the total cost of the floor in the entryway?

CONNECTION Two boxes combined is similar to two rectangles combined. To find the area made by the two combined rectangles, we add the areas of the rectangles. To find the volume of two boxes combined, we add the volumes of the two boxes.

Plan: To find the total volume, we add the volumes of the two boxes.

Execute: total volume = volume of bottom box + volume of top box

$$
\begin{aligned}
V &= \quad l \cdot w \cdot h \quad + \quad l \cdot w \cdot h \\
V &= \quad (11)(5)(2) \quad + \quad (5)(5)(1) \\
V &= \quad 110 \quad\quad + \quad 25 \\
V &= \quad 35
\end{aligned}
$$

Answer: The volume of the car is about 135 ft.^{3}

Check: Let's reverse the process. We start with 135 cubic feet and see if we can work our way back to the original dimensions. The last step in the preceding process was 110 + 25 = 135, so we can check by subtracting.

135 − 25 = 110   **This is true, so our addition checks.**

Answers to Your Turn 2
a. $108 (must purchase 6 bags of seed) **b.** 24 ft.^{2}; $368

Engineers are designing a clean room for a computer chip manufacturer. The room is to be **L**-shaped (see diagram). In designing the air recycling and filter system, the engineers need to know the volume of the room. Find the volume of the room if the whole room has a 10-foot ceiling.

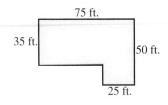

75 ft.

35 ft.

50 ft.

25 ft.

Now let's check 110 cubic feet. Since we multiplied (11)(5)(2), we can check by dividing $110 \div 2 \div 5$ to see if we get 11.

$$110 \div 2 \div 5$$
$$= 55 \div 5$$
$$= 11 \qquad \text{It checks.}$$

Now check 25 cubic feet. Since we multiplied (5)(5)(1) to get 25, we can check by dividing $25 \div 1 \div 5$ to see if we get 5.

$$25 \div 1 \div 5$$
$$= 25 \div 5$$
$$= 5 \qquad \text{It checks.}$$

◀ **Do Your Turn 3**

Examples 2 and 3 suggest the following procedure.

PROCEDURE *When solving for the area or volume of a composite form:*
1. Identify the shapes or objects that make up the composite form.
2. If those shapes or objects combine to make the composite form, add all their areas or volumes. If one shape or object is contained within another, subtract the area/volume of the inner shape/object from the area/volume of the outer shape/object.

Answer to Your Turn 3
30,000 ft.³

1.7 Exercises

For
Extra
Help

Videotape
DVT

Tutor Center
Addison-Wesley
Tutor Center

Math XL
Math XL

MyMathLab

Student Solutions
Manual

1. What is the first step in solving a problem?

2. What is the last step in the problem-solving process?

3. How would you calculate the area of a shape that is a combination of two or more shapes?

4. How would you calculate the area of a shape that is formed by cutting away a portion of a larger shape?

For Exercises 5–24, solve.

5. Crown molding is to be installed in a 16-foot-wide by 20-foot-long living room. The subcontractor charges $4 per foot to install the molding. What will be the total cost of installing molding for the room?

6. Angela plans to put a wallpaper border just below the ceiling in her bedroom. The room is 12 feet wide by 14 feet long. If the border costs $2 per foot, what will be the total cost?

7. Baseboard molding is to be placed where the walls connect to the floor in the room shown at the right. An 8-foot section of baseboard molding costs $4. If no partial sections are sold, how much will the total baseboard cost?

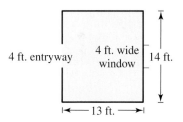

4 ft. entryway   4 ft. wide window   14 ft.

|← 13 ft. →|

8. Chair railing is to be placed on the walls around the room mentioned in Exercise 7. The railing will be at a height of 3 feet from the floor. If the railing costs $5 for an 8-foot section and no partial sections are sold, how much will the chair railing cost for the room?

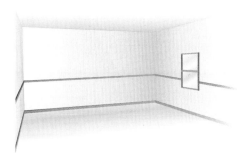

9. Mr. Williams wishes to fence a large pasture. The plot plan for the land is shown here. The fencing company charges $5 per foot for fencing and $75 for the 10-foot gate. What will be the total cost?

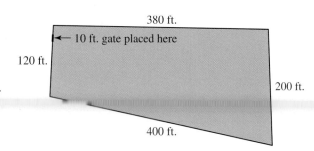

380 ft.

|← 10 ft. gate placed here

120 ft.

200 ft.

400 ft.

10. The Henry family wishes to put a fence around their backyard. The plan is shown here. The fencing company charges $8 per foot for the wood picket fence and $50 for each 4-foot-wide gate. What will be the total cost?

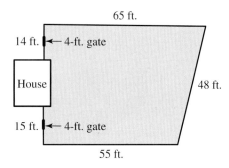

65 ft.

14 ft. |← 4-ft. gate

House

48 ft.

15 ft. |← 4-ft. gate

55 ft.

11. Mr. Williams decides to investigate the cost of building the fence in Exercise 9 himself.

   a. If he places wood posts every 8 feet (except where the gate goes), then how many posts must he purchase? The posts cost $2 each. What will be the cost for all the posts?

   b. He will nail four rows of barbed wire along the posts (except where the gate goes). How many feet of barb wire must he have in all? A roll of barbed wire contains 1320 feet of wire and costs $45. How many rolls will be needed? What will be the cost of barbed wire?

   c. Because he'll be nailing four rows of wire to each post, he'll need 4 U-nails for each post. How many nails will be needed? If the nails come in boxes of 100, how many boxes will he need? If the boxes cost $3 each, how much will the nails cost?

   d. The gate costs $75. What will be the total cost of materials?

   e. Compare the total cost of materials for Mr. Williams to build the fence himself with the total cost of having the fencing company build it. What should Mr. Williams do? What factors should he consider?

12. Mr. Henry from Exercise 10 decides to investigate the cost of building the fence himself. He decides that 8-foot prefabricated panels are the least expensive option. Each 8-foot panel costs $25. Posts are placed at each end of each section and at each end of each gate. The posts cost $3 each. Rental of an auger to dig the post holes costs $50. The gates cost $45 each. What will be the total cost? (*Hint:* Draw the plan for post placement.)

13. Find the area of the metal (shaded) in the plate shown. The hole in the center is an 8-centimeter by 8-centimeter square.

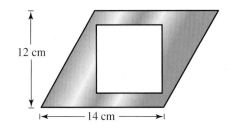

14. An 84-foot wide by 72-foot long building is on a lot that is 245 feet long by 170 feet wide. The entire area surrounding the building is to be paved for parking. Calculate the area that will be paved.

15. The floor of a house is to be made from 4-foot-wide by 8-foot-long sheets of plywood. The floor plan is shown here. How many sheets of plywood will be needed? If each sheet of wood costs $15, what will be the total cost?

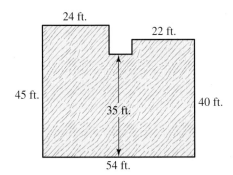

16. The area shown is to be carpeted. How many square feet will be needed in all? If the chosen carpet costs $3 for each square foot, then how much will it cost for carpet for the area?

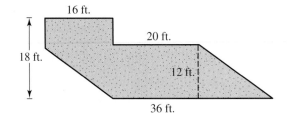

17. The walls of a 24-foot-long by 15-foot-wide room with a 9-foot ceiling are to be painted. The room has a 6-foot-wide by 7-foot-long entryway and two windows that each measure 3 feet-high by 2 feet-wide. What is the area to be painted? If each gallon of paint covers 400 square feet how many gallons will be needed?

18. The walls in nine offices in an office building are to be painted. Each office is 10 feet long by 9 feet wide and has 8-foot ceilings. Each office has a 3-foot-wide by 7-foot-high doorway and a 3-foot-high by 2-foot-wide window.
  a. What is the area to be painted?
  b. If a gallon of paint will cover 400 square feet, how many gallons will be needed?
  c. If each gallon costs $14, what will be the total cost of the paint?

19. Three floors in an office building are to have new carpet installed. To estimate the amount of carpet needed, the manager decides to use the dimensions of the building, which is 60 feet wide by 80 feet long. How many square feet of carpet will be needed? If the carpet company charges $2 per square foot, then what will be the cost of installing the carpet? What are the problems with his estimate?

20. The outside of a large building is to be covered with 4-foot by 8-foot glass panels. The building is a box that is 72 feet long by 60 feet wide by 160 feet high. How many glass panels will be needed? (*Note:* The roof will not be covered with the panels.)

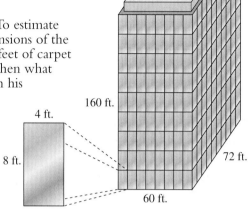

21. The assembling building where NASA constructs its rockets and shuttles is 120 feet wide by 140 feet high by 500 feet long. Find the volume of the building.

22. A bus is a box that is 35 feet long by 9 feet high by 8 feet wide. What is the volume of a bus?

23. The Antonov AN 124 is the world's largest commercial cargo aircraft. Its cargo compartment is approximately box shaped and measures 120 feet long, 21 feet wide, and 14 feet tall. What is the approximate volume of the cargo compartment?

24. Ziggurats are terraced pyramids. Find the volume of the ziggurat shown if the bottom level is 297 feet wide by 297 feet long by 12 feet high, the middle level is 260 feet wide by 260 feet wide by 10 feet high, and the top level is 145 feet wide by 145 feet wide by 9 feet high.

PUZZLE PROBLEM An *equilateral triangle* is a triangle with all three sides equal in length. Suppose we construct an equilateral triangle out of three equal length pencils like so:

We can construct larger equilateral triangles out of the smaller version like so:

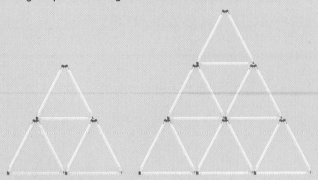

How many pencils would be needed to construct an equilateral triangle like those shown that has 15 pencils along each side?

Review Exercises

[1.3] **1.** Write 2,408,073 in expanded form using powers of 10.

[1.2] **2.** Solve and check. $x + 175 = 2104$

[1.3] **3.** Estimate $452 \cdot 71{,}203$ by rounding so that there is only one nonzero digit, and then calculate the actual product.

[1.4] **4.** Solve and check. $17 \cdot y = 3451$

[1.5] **5.** Simplify using the order of operations. $5^3 - \left[(9 + 3)^2 - 2^6\right] + \sqrt{100 - 36}$

Defined Terms

Review the following terms, and for those you do not know, study its definition on the page number next to it.

Section 1.1
Numbers *(p. 2)*
Set *(p. 2)*
Subset *(p. 2)*
Natural numbers *(p. 2)*
Whole numbers *(p. 2)*
Equation *(p. 5)*
Inequality *(p. 5)*

Section 1.2
Addition *(p. 10)*
Perimeter *(p. 13)*

Variable *(p. 13)*
Constants *(p. 14)*
Subtraction *(p. 14)*
Related sentence *(p. 16)*
Solution *(p. 16)*

Section 1.3
Multiplication *(p. 22)*
Rectangular array *(p. 26)*
Exponent or power
  (p. 29)
Base *(p. 29)*

Square unit *(p. 31)*
Area *(p. 31)*
Formula *(p. 31)*

Section 1.4
Division *(p. 36)*
Remainder *(p. 36)*
Square *(p. 43)*
Square root *(p. 44)*
Perfect square *(p. 44)*

Section 1.6
Parallel lines *(p. 56)*
Parallelogram *(p. 56)*
Right angle *(p. 56)*
Cubic unit *(p. 57)*
Volume *(p. 57)*

Our Place-Value System

Place Values														
Trillions period			Billions period			Millions period			Thousands period			Ones period		
Hundred trillions	Ten trillions	Trillions	Hundred billions	Ten billions	Billions	Hundred millions	Ten Millions	Millions	Hundred thousands	Ten Thousands	Thousands	Hundreds	Tens	Ones

◄ ···

Arithmetic Summary Diagram

Each operation has an inverse operation. In the following diagram, the operations build from the top down. Addition leads to multiplication, which leads to exponents. Subtraction leads to division, which leads to roots.

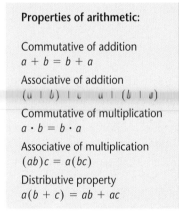
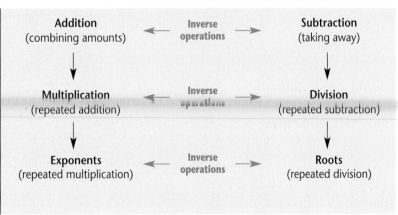

Properties of arithmetic:

Commutative of addition
$a + b = b + a$

Associative of addition
$(a + b) + c = a + (b + c)$

Commutative of multiplication
$a \cdot b = b \cdot a$

Associative of multiplication
$(ab)c = a(bc)$

Distributive property
$a(b + c) = ab + ac$

Addition
(combining amounts) → Inverse operations → Subtraction
(taking away)

Multiplication
(repeated addition) → Inverse operations → Division
(repeated subtraction)

Exponents
(repeated multiplication) → Inverse operations → Roots
(repeated division)

Procedures, Rules, and Key Examples

<table>
<tr><td>Procedures/Rules</td><td>Key Example(s)</td></tr>
</table>

Section 1.1 Introduction to Numbers, Notation, and Rounding

To write a number in expanded form:

1. Write each digit multiplied by its place value.
2. Express it all as a sum.

Example 1: Write 70,916 in expanded form.
Answer: $7 \times 10,000 + 9 \times 100 + 1 \times 10 + 6 \times 1$

To change a number from expanded form to standard form:
Write each digit in the place indicated by the corresponding place value.

Example 2: Write $2 \times 1,000,000 + 8 \times 10,000 + 5 \times 1000 + 4 \times 10 + 7$ in standard form.
Answer: 2,085,047

To write a word name, work from left to right through the periods.

1. Write the name of the digits in the left-most period.
2. Write the period name followed by a comma.
3. Repeat steps 1 and 2 until you get to the ones period. We do not follow the ones period with its name.

Warning: Do not write the word *and* anywhere in a word name for a whole number. The word *and* takes the place of a decimal point.

Example 3: Write the word name for 14,907,156.
Answer: fourteen million, nine hundred seven thousand, one hundred fifty-six

To round a number to a given place value, consider the digit to the right of the desired place value.

If this digit is 5 or greater, round up.

If this digit is 4 or less, round down.

Example 4: Round to the nearest thousand.
a. 54,259   **Answer:** 54,000
b. 127,502   **Answer:** 128,000
c. 4801   **Answer:** 5000

Section 1.2 Adding, Subtracting, and Solving Equations with Whole Numbers

To add whole numbers:

1. Stack by place value.
2. Add the digits.

Example 1: Add. $5408 + 913$

Solution:
$$\begin{array}{r} \overset{1\ \ 1}{5408} \\ +913 \\ \hline 6321 \end{array}$$

To estimate addition and subtraction:

1. Round the smallest number.
2. Round all other numbers to the same place as the smallest number.
3. Perform the calculation with rounded numbers.

Example 2: Estimate $12,591 + 4,298$ by rounding.

$$\begin{array}{rcl} 12,591 & \longrightarrow & 13,000 \\ +4,298 & \longrightarrow & +4,000 \\ & & \text{Estimate: } 17,000 \end{array}$$

To find the perimeter of a shape:
Add the lengths of all the sides of the shape.

Example 3: Find the perimeter.

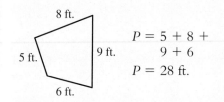

$P = 5 + 8 + 9 + 6$
$P = 28$ ft.

To subtract whole numbers:

1. Stack the larger number on top of the smaller number, aligning the place values.
2. Subtract the digits in the bottom number from the digits directly above. If digit in the top number is smaller than the digit beneath it, then rename the top digit.

To find a missing addend: write a related subtraction sentence, subtracting the known addend from the sum.

Example 4: Subtract. $8982 - 765$

Solution:
$$\begin{array}{r} {\scriptstyle 71} \\ 89\overset{}{8}2 \\ -765 \\ \hline 8217 \end{array}$$

Example 5: Solve. $23 + x = 57$

Related Subtraction: $\quad x = 57 - 23$
$$x = 34$$

Section 1.3 Multiplying Whole Numbers and Exponents

To multiply two whole numbers: stack them and apply the distributive property.

Example 1: Multiply. 503×62

Solution:

$$\begin{array}{r} {\scriptstyle 01} \\ 503 \\ \times 62 \\ \hline 1\,006 \\ +30\,18 \\ \hline 31,186 \end{array}$$

Multiply 2 times 503.

Multiply 6 times 503.

Add.

Multiplicative Property of 0: The product of 0 and a number is always 0.

In math language: $0 \cdot n = 0$ and $n \cdot 0 = 0$, when n is any number.

Multiplicative Property of 1: The product of 1 and a number is always the number.

In math language: $1 \cdot n = n$ and $n \cdot 1 = n$, when n is any number.

Example 2: Multiply.

a. $0 \cdot 15 = 0$

b. $271 \cdot 0 = 0$

c. $1 \cdot 94 = 94$

d. $68 \cdot 1 = 68$

Exponential form when the exponent is a natural number:

Repeated multiplication of the base.

Example 3: Write 2^5 in factored form, then calculate.

Solution: $2^5 = 2 \cdot 2 \cdot 2 \cdot 2 \cdot 2$
$$= 32$$

Example 4: Write $7 \cdot 7 \cdot 7 \cdot 7$ in exponential form.

Answer: 7^4

To find the total number of possible combinations in a problem with multiple slots to fill, multiply the number of items that can fill the first slot by the number of items that can fill the next, and so on.

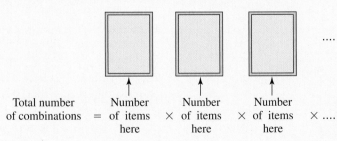

Example 5: In order to get into a certain restricted building, a code must be entered. The code box has two digital windows. The first window can contain any number from 1 to 6. The second window can contain any letter from A to J (only capitals appear). How many possible codes are there in all?

Solution:

number of items in first window		number of items in second window
6	$\times$	10	$= 60$

Section 1.4 Dividing, Square Roots, and Solving Equations with Whole Numbers

Division Properties:

When 1 is the divisor, the quotient is equal to the dividend.

> In math language: $n \div 1 = \dfrac{n}{1} = n$, when n is any number.

When 0 is the divisor with any dividend other than 0, the quotient is undefined.

> In math language: $n \div 0$ or $\dfrac{n}{0}$, is undefined, when $n \neq 0$.

When 0 is the dividend, the quotient is 0 as long as the divisor is not also 0.

> In math language: $0 \div n = \dfrac{0}{n} = 0$, when $n \neq 0$.

If both dividend and divisor are 0, the quotient is indeterminate.

> In math language: $0 \div 0$ or $\dfrac{0}{0}$, is indeterminate.

When a number (other than 0) is divided by itself, the quotient is 1.

> In math language: $n \div n = \dfrac{n}{n} = 1$, when $n \neq 0$.

Example 1: Divide.

a. $\dfrac{78}{1} = 78$

b. $\dfrac{25}{0}$ is undefined (because no number checks).

c. $\dfrac{0}{19} = 0$

d. $\dfrac{0}{0}$ is indeterminate (because every number checks).

e. $\dfrac{245}{245} = 1$

Divisibility Rules:

1. 2 is an exact divisor for all even numbers. Even numbers have 0, 2, 4, 6, or 8 in the ones place.

Example 2: Determine whether the number is divisible by 2.

a. 24,806 is divisible by 2 because it is an even number. It has a 6 in the ones place.

b. 617 is not divisible by 2 because it is not an even number. It has a digit other than 0, 2, 4, 6, or 8 in the ones place.

2. To determine whether 3 is an exact divisor for a given number:
   a. Add the digits in the dividend.
   b. If the resulting sum is a number that is divisible by 3, then so is the dividend.

Example 3: Determine whether the number is divisible by 3.

a. 5142 is divisible by 3 because the sum of its digits is divisible by 3.
$$5 + 1 + 4 + 2 = 12$$

b. 215 is not divisible by 3 because the sum of its digits is not divisible by 3.
$$2 + 1 + 5 = 8$$

3. 5 is an exact divisor for numbers that have 0 or 5 in the ones place.

Example 4: Determine whether the number is divisible by 5.

a. 945 and 260 are both divisible by 5 because they both have 0 or 5 in the ones place.

b. 612 is not divisible by 5 because it has a digit other than 0 or 5 in the ones place.

To do long division:

1. Write the problem in long division form.
2. Compare the left-most digit in the dividend with the divisor. Is it larger? If not, then include the next digit. Continue in this way until the digits name a number that is larger than the divisor.
3. Ask yourself how many times the divisor goes into those digits, and place that result over the last digit you included from step 2.
4. Multiply the result by the divisor and align the product under the digits from step 3.
5. Subtract. The remainder you get should be less than the divisor. If not, then add 1 to your quotient digit and repeat steps 4 and 5.
6. Write the next digit in the quotient directly beside the last digit in the remainder from step 5.
7. Divide the divisor into the number made by the remainder and the digit you brought down.
8. Repeat steps 3–7 until you have brought down and divided the digit in the ones place of the dividend. When you subtract this last time, if you get zero, then we say the number has no remainder. If you get a remainder other than zero, then list it as a remainder in your answer.

Example 5: Divide.

a. $350 \div 2$

$$
\begin{array}{r}
175 \\
2\overline{)350} \\
-2 \\
\hline
15 \\
-14 \\
\hline
10 \\
-10 \\
\hline
0
\end{array}
$$

Answer: 175

b. $6409 \div 7$

$$
\begin{array}{r}
915 \\
7\overline{)6409} \\
-63 \\
\hline
10 \\
-7 \\
\hline
39 \\
-35 \\
\hline
4
\end{array}
$$

Answer: 915 r 4

To solve for a missing factor:

Write a related division sentence, dividing the product by the known factor.

Example 6: Solve. $x \cdot 12 = 180$

Related Multiplication: $x = 180 \div 12$

$x = 15$

To find the square root of a given number:

Find a number that can be squared to equal the given number.

Example 7: Evaluate. $\sqrt{81}$

$\sqrt{81} = 9$ because $(9)^2 = 81$.

Section 1.5 Order of Operations

Order of operations agreement:

Perform operations in the following order:

1. Grouping symbols. These include parentheses ( ), brackets [ ], braces { }, absolute value ||, radicals $\sqrt{}$, and fraction bars —.

2. Exponents
3. Multiplication or division from left to right, in order as they occur.
4. Addition or subtraction from left to right, in order as they occur.

Note: Use $\overrightarrow{\text{GEM}}\overrightarrow{\text{DAS}}$ to remember the order.

Example 1: Simplify.

$$
\begin{aligned}
&26 - [19 - 2(15 - 9)] + 4^2 \\
&= 26 - [19 - 2(6)] + 4^2 \\
&= 26 - [19 - 12] + 4^2 \\
&= 26 - 7 + 4^2 \\
&= 26 - 7 + 16 \\
&= 19 + 16 \\
&= 35
\end{aligned}
$$

Section 1.6 More with Variables, Formulas, and Solving Equations

To use a formula:

1. Replace the variables with the corresponding given values.
2. Solve for the missing variable.

Example 1: Find the perimeter of a rectangle with a length of 24 centimeters and a width of 17 centimeters.

$$P = 2l + 2w$$
$$P = 2(24) + 2(17)$$
$$P = 48 + 34$$
$$P = 82 \text{ cm}$$

Section 1.7 Problem Solving and Applications

Problem-Solving Outline:

1. **Understand** the problem.
   a. Read the question(s) (not the whole problem, just the question at the end), and note what it is you are to find.
   b. Now read the whole problem, underlining the key words.
   c. If possible and useful, make a list or table, simulate the situation, or search for a related example problem.
2. **Plan** your solution strategy by searching for a formula or translating the key words to an equation.
3. **Execute** the plan by solving the formula or equation.
4. **Answer** the question. Look at your note about what you were to find and make sure you answered that question. Include appropriate units.
5. **Check** the results.
   a. Try finding the solution in a different way, reversing the process, or estimating the answer and make sure the estimate and actual answer are reasonably close.
   b. Make sure the answer is reasonable.

When solving for the area or volume of a composite form:

1. Identify the shapes or objects that make up the composite form.
2. If those shapes or objects combine to make the composite form, add all their areas or volumes. If one shape or object is contained within another, subtract the area/volume of the inner shape/object from the area/volume of the outer shape/object.

Example 1: Find the area of the shaded region.

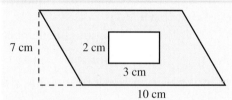

Understand: We are to find the area of the shaded region, which is a parallelogram with a rectangle removed.

Plan: Subtract the area of the rectangle from the area of the parallelogram.

Execute:

area = area of the parallelogram
    − area of the rectangle
$$A = (10)(7) - (3)(2)$$
$$A = 70 - 6$$
$$A = 64$$

Answer: The area of the shaded region is 64 cm².

Check: Verify the calculations. We will leave the check to the reader.

Key Words for Operations

Addition: Add, plus, sum, total, increased by, more than, in all, altogether, perimeter
Subtraction: Subtract, minus, difference, remove, decreased by, take away, less, less than, left
Multiplication: Multiply, times, product, each, of, by
Division: Divide, distribute evenly, each, split, quotient, into, per, over

Formulas

Perimeter of a rectangle:	$P = 2l + 2w$
Area of a parallelogram:	$A = bh$
Volume of a box:	$V = lwh$

For Exercises 1–6, answer true or false.

[1.1] **1.** 0 is a natural number.

[1.1] **2.** All natural numbers are also whole numbers.

[1.2] **3.** $16 - 2 = 2 - 16$.

[1.3] **4.** $2(3 + 5) = 2 \cdot 3 + 2 \cdot 5$

[1.4] **5.** 3 is an exact divisor for 6201.

[1.4] **6.** $\dfrac{0}{0} = 0$

For Exercises 7–10, solve as indicated.

[1.2] **7.** Explain in your own words how to check that 8 is the solution for $x + 9 = 17$.

[1.4] **8.** Explain or show why $\dfrac{14}{0}$ is undefined.

[1.5] **9.** Explain the mistake in the problem.

$$17 + 3(9 - 2)$$
$$= 17 + 3(7)$$
$$= 20(7)$$
$$= 140$$

[1.4] **10.** Explain in your own words the difference between the square of a number and the square root of a number.

[1.1] *For Exercises 11 and 12, write in expanded form.*

11. 5,680,901

12. 42,519

[1.1] *For Exercises 13–15, write in standard form.*

13. $9 \times 10,000 + 8 \times 1000 + 2 \times 100 + 7 \times 10 + 4 \times 1$

14. $8 \times 1,000,000 + 2 \times 10,000 + 9 \times 10 + 6 \times 1$

15. $7 \times 10^8 + 9 \times 10^5 + 2 \times 10^4 + 8 \times 10^3 + 6 \times 1$

[1.1] *For Exercises 16–18, write the word name.*

16. 47,609,204

17. 9421

18. 123,405,600

[1.1] *For Exercises 19 and 20, use $<$ or $>$ to make a true statement.*

19. 14 ? 19

20. 2930 ? 2899

[1.1] *For Exercises 21 and 22, round to the specified place.*

21. 5,689,412 to the nearest ten thousand

22. 2,512,309 to the nearest million

[1.2] *For Exercises 23 and 24, estimate by rounding, then calculate the actual sum.*

23. $45,902 + 6819$

24. $545 + 9091 + 28 + 30,009$

[1.2] *For Exercises 25 and 26, estimate by rounding, then calculate the actual difference.*

25. $541,908 - 56,192$

26. $8002 - 295$

[1.2] *For Exercises 27 and 28, solve and check.*

27. $29 + x = 54$

28. $y + 14 = 203$

[1.2] *For Exercises 29 and 30, solve.*

29. In the morning, Dionne's checking account balance was $349. During her lunch break she deposits her paycheck of $429 and pays a $72 phone bill. What is her new balance?

30. Find the missing current in the circuit shown.

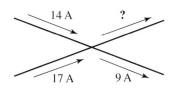

[1.3] *For Exercises 31 and 32, estimate by rounding so that there is only one nonzero digit, then calculate the actual product.*

31. 4591×307

32. $16,140 \times 25$

[1.3] *For Exercises 33 and 34, evaluate.*

33. 2^7

34. 5^3

[1.3] *For Exercises 35 and 36, write in exponential form.*

35. $10 \cdot 10 \cdot 10 \cdot 10 \cdot 10$

36. $7 \cdot 7 \cdot 7 \cdot 7 \cdot 7 \cdot 7 \cdot 7 \cdot 7$

[1.3] *For Exercises 37 and 38, solve.*

37. A computer translates everything into binary code. This means that numbers are written so that each place value either contains a 1 or a 0 digit. If a certain computer chip has 6 place values (bits), then how many different numbers can be coded?

38. A room measures 16 feet wide by 18 feet long. How many square feet of carpet will be needed to cover the floor of the room?

[1.4] *For Exercises 39 and 40, estimate by rounding so that there is only one nonzero digit, then calculate the actual quotient.*

39. $78,413 \div 19$

40. $\dfrac{83,451}{26}$

41. $19 \cdot b = 456$

42. $8 \cdot k = 2448$

[1.4] *For Exercises 43 and 44, solve.*

43. Jana is to administer 120 milliliters of Pitocin in an IV drip over the course of an hour. How many milliliters of Pitocin should the patient receive each minute?

44. Andre has a gross annual salary of $29,256. What is his gross monthly salary?

[1.4] *For Exercises 45 and 46, find the square root.*

45. $\sqrt{196}$

46. $\sqrt{225}$

[1.5] *For Exercises 47–50, simplify.*

47. $24 \div 8 \cdot 6 + \sqrt{121} - 5^2$

48. $4^3 - \left[(28 - 4) \div 2^2\right] + \sqrt{100 \div 25}$

49. $3\sqrt{25 - 16} + \left[25 + 2(14 - 9)^2\right]$

50. $\dfrac{9^2 - 21}{56 - 2(4 + 1)^2}$

[1.6] 51. Find the area of the figure.

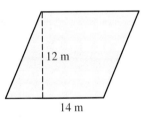

[1.6] 52. Find the volume of a box that is 7 feet wide, 6 feet high, and 9 feet long.

[1.7] *For Exercises 53–56, solve.*

53. In designing a room for a house, the desired length is 18 feet and the desired area is 288 square feet. What must the width be?

54. The Dobsons are told that a 225-square-foot patio is included in the price of the house they are having built. What are the dimensions of the patio?

55. A desk manufacturer has an order to make 500 desks (shown here). Each desk will have plastic trim placed along the edge. How many feet of trim must be purchased for the entire order?

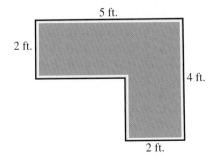

56. The picture shows the plans for a department store and parking lot. What is the area of the parking lot?

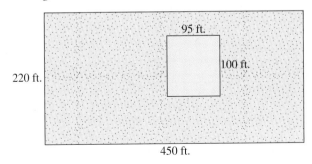

1. _____

[1.1] **1.** What digit is in the ten thousands place in 1,258,987?

2. _____

[1.3] **2.** Write 48,210,907 in expanded form using powers of 10.

3. _____

[1.1] **3.** Use $<$ or $>$ to make a true statement.

$$19,304 \quad ? \quad 19,204$$

4. _____

[1.1] **4.** Write the word name for 67,194,210.

5. _____

[1.2] **5.** Estimate the sum by rounding to the nearest hundred thousand.

$$2,346,502 + 481,901$$

6. _____

[1.2] **6.** Add. $63,209 + 4981$

7. _____

[1.3] **7.** Subtract. $480,091 - 54,382$

8. _____

[1.3] **8.** Solve and check. $27 + x = 35$

9. _____

[1.3] **9.** Multiply. 357×6910

10. _____

[1.3] **10.** Evaluate. 4^3

11. _____

[1.4] **11.** Divide. $45,019 \div 28$

12. _____

[1.4] **12.** Solve and check. $7 \cdot y = 126$

For Exercises 13–18, evaluate.

13. _____

[1.4] **13.** $\sqrt{196}$

14. _____

[1.5] **14.** $24 + \dfrac{40}{4}$

15. _____

[1.5] **15.** $14 + 2^5 - 8 \cdot 3$

16. _____

[1.5] **16.** $20 - 6(3 + 2) \div 15$

[1.5] **17.** $\dfrac{(14 + 26)}{8} - \sqrt{25 - 9}$

17. _____

[1.5] **18.** $\dfrac{5^2 + 27}{(4 + 2)^2}$

18. _____

[1.6] **19.** Calculate the area of the shape.

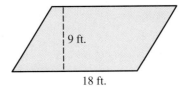

19. _____

[1.2] **20.** Nelly's account balance was $172. If she deposited a $27 check, bought a $156 money order, and withdrew $25 cash, what is her new balance?

20. _____

[1.3] **21.** Participants in a walk-run race are given identification cards that have a letter, A–Z, paired with a digit, 0–9. How many identification cards can be made in all?

21. _____

[1.4] **22.** A college admits 2487 new students. All new students must enroll in a college orientation course. If the maximum number of students allowed in a single class is 30, how many classes should be offered to accommodate all the new students?

22. _____

[1.6] **23.** A rectangular playground measures 65 feet wide by 80 feet long. What is the perimeter of the playground? If fence material costs $6 per foot, how much would fencing cost for the playground?

23. _____

[1.6] **24.** A walk-in refrigerator at a restaurant is a box measuring 8 feet long, 6 feet wide, and 7 feet high. What is the volume?

24. _____

[1.7] **25.** A wall 16 feet wide and 9 feet high is to be painted. This wall has an open doorway that is 3 feet wide by 7 feet high. How many square feet must be painted?

25. _____

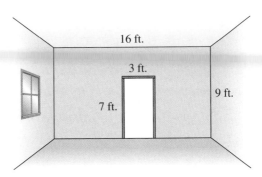

Integers

" He is educated who knows how to find out what he doesn't know."

—GEORGE SIMMEL,
SOCIOLOGIST

" Seeing, hearing, and feeling are miracles, and each part and tag of me is a miracle."

—WALT WHITMAN,
AMERICAN POET

" Let your performance do the thinking."

—H. JACKSON BROWN JR.,
AUTHOR

Learning Styles

Every person has a learning style. Some people learn best through seeing (visual learners), some through touching or doing (tactile learners), and some through hearing (auditory learners). If you do not know your learning style, then take a moment to complete the *Learning Styles Inventory* in the *To the Student* section at the beginning of the book.

In Chapter 2, we will be exploring positive and negative numbers. The operations of arithmetic used with these numbers require that you know and understand rules. Knowing your learning style can guide you in learning the rules and procedures. Also, look for Learning Strategy boxes throughout the text, which offer suggestions on how to study based on your learning style.

2.1 Introduction to Integers

OBJECTIVES

1 Identify integers.

2 Graph integers on a number line.

3 Use < or > to make a true statement.

4 Find the absolute value.

5 Find the additive inverse.

OF INTEREST

Negative numbers took some time to gain acceptance. The first known mention of negative numbers is found in the work by the Indian mathematician Brahmagupta (about A.D. 628). He called them "negative and affirmative quantities." There have been many different ways of indicating negative numbers. The Chinese would use colors, black for negative numbers and red for positive. In sixteenth-century Europe, it was popular to use the letter m for minus, to indicate negative numbers, as in m:7 to indicate −7.

OBJECTIVE 1 Identify integers.

In this chapter, we will learn about positive and negative numbers.

A positive number is written with a plus sign or no sign, as in +5 or 5. We read +5 (or 5) as "positive five," "plus five," or "five." All of the natural numbers (1, 2, 3, . . .) are positive numbers. A negative number is written with a minus sign, as in −6, which is read as "negative six" or "minus six." The number 0 has no sign. Zero is neither positive nor negative.

The whole numbers {0, 1, 2, 3, . . . } along with all the negative natural numbers {. . . , −3, −2, −1} form a set of numbers called the **integers.**

DEFINITION Integers: . . . , − 3, − 2, − 1, 0, 1, 2, 3, . . .

The following diagram shows how the set of integers contains the set of whole numbers and the set of natural numbers.

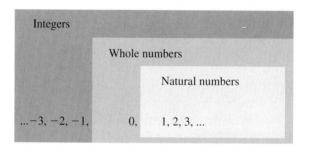

Positive and negative numbers can be used to describe such things as money, temperature, elevation, or direction. In the table to the right, we list

Situation	Negative Number	Positive Number
Money	Debt/Liability	Credit/Asset
Temperature	Below zero	Above zero
Elevation	Below sea level	Above sea level
Direction	Reverse/go down	Forward/go up

words that indicate a negative number or a positive number in each of the listed situations. Relating integers to familiar situations will help us understand arithmetic with signed numbers.

Example 1 Express each amount as a positive or negative quantity.

a. Angela owes $480 on her credit card.

Answer: − 480

Explanation: Because she owes the money, it is a debt and, therefore, is a negative amount.

b. It is 29°C outside.

Answer: 29

Explanation: Because 29°C is above zero, it is a positive amount. Notice when writing a positive number, we do not have to write +29. We can just write 29.

c. A research submarine is at a depth of 245 feet.

Answer: -245

Explanation: Because the submarine is below sea level, its position is a negative amount.

Do Your Turn 1 ▷

OBJECTIVE 2 Graph integers on a number line.

Many people find a number line helpful in doing calculations with integers. A number line is like a ruler. It is a line with marks evenly spaced and with each mark representing a number.

PROCEDURE *To graph a number on a number line, we draw a solid dot or point on the mark for that number.*

Example 2 Graph each given number on a number line.

a. 3

Solution:

b. -2

Solution:

Do Your Turn 2 ▷

OBJECTIVE 3 Use < or > to make a true statement.

We can use number lines to compare numbers. Since numbers increase from left to right on a number line, when comparing two integers, the integer that is farther to the right on a number line is the greater integer.

Example 3 Use < or > to make a true statement.

a. 9 ? -14

Answer: $9 > -14$

Explanation: The greater number is the number farther to the right on a number line. Graphing 9 and -14 on a number line, we see that 9 is farther to the right.

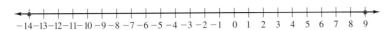

Or, think about elevation. If zero is sea level, then positive 9 is 9 feet *above* sea level and -14 is 14 feet *below* sea level. Obviously, 9 feet above sea level is higher.

b. -12 ? -7

Answer: $-12 < -7$

Your Turn 1

Express each amount as a positive or negative quantity.

a. Ian has $3250 in a savings account.

b. According to the U.S. National Debt Clock, on 4/1/04 the national debt was $7,145,177,977,986.

c. The melting point of gold is 1063°C.

d. A submarine is at a depth of 250 feet.

Your Turn 2

Graph each number on a number line.

a. -2

b. 0

c. 5

Answers to Your Turn 1
a. 3250
b. $7,145,177,977,986$
c. 1063
d. -250

Answers to Your Turn 2
a.

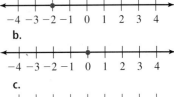
b.

c.

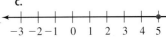

Your Turn 3

Use < or > to make a true statement.

a. 15 ? −26

b. −19 ? −17

c. −14 ? −2

d. 0 ? −9

Explanation: The greater number, in this case −7, is farther right on a number line.

$$\text{−12 −11 −10 −9 −8 −7 −6 −5 −4 −3 −2 −1 \quad 0}$$

In terms of elevation, −7 is 7 feet below sea level and −12 is 12 feet below sea level. Since −7 is closer to the surface, it is the greater number.

Conclusion: When both numbers are negative, the integer closer to zero is the greater integer.

◀ **Do Your Turn 3**

OBJECTIVE 4 Find the absolute value.

We also can use number lines to understand the *absolute value* of a number.

DEFINITION Absolute value: A given number's distance from zero on a number line.

Your Turn 4

Find the absolute value of each given number.

a. 14

b. −27

c. −1347

d. 0

Example 4 Find the absolute value of each given number.

a. 5

Answer: 5

Explanation: Because 5 is five steps from zero on a number line its absolute value is 5.

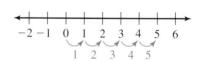

b. −6

Answer: 6

Explanation: Because −6 is six steps from zero on a number line, its absolute value is 6. Notice that it doesn't matter which direction we move, we just count the steps. As an analogy, imagine a size 6 shoe. It doesn't matter which way you point the shoe or where it is located, it's still a size 6 shoe.

◀ **Do Your Turn 4**

LEARNING STRATEGY

If you are a tactile learner, think of absolute value as physically walking from 0 to the given number and counting the steps you take to get to that number.

The symbols for absolute value are two vertical lines. For example, the absolute value of 5 is written $|5|$.

Answers to Your Turn 3
a. 15 > −26 b. −19 < −17
c. −14 < −2 d. 0 > −9

Answers for Your Turn 4
a. 14 b. 27 c. 1347 d. 0

RULES

The absolute value of a number is always positive or zero.
The absolute value of a negative number is always positive.
The absolute value of a positive number is always positive.
The absolute value of zero is zero.

Example 5 Simplify.

a. $|-8|$

◀ **Note:** In words $|-8|$ translates to "the absolute value of negative eight."

Answer: 8

Explanation: Because -8 is eight steps away from zero, its absolute value is 8.

b. $|27|$

Answer: 27

Explanation: Because 27 is 27 steps away from zero, its absolute value is 27.

Do Your Turn 5 ▷

Your Turn 5

Simplify.

a. $|-19|$

b. $|90|$

c. $|0|$

OBJECTIVE 5 Find the additive inverse.

Notice that 5 and -5 have the same absolute value because they are both the same distance from zero. However, because 5 is positive and -5 is negative, they are on opposite sides of zero on a number line. We say that 5 and -5 are **additive inverses.** They are called *additive* inverses because their sum is 0.

$$5 + (-5) = 0$$

DEFINITION Additive inverses: Two numbers whose sum is zero.

Example 6

a. Find the additive inverse of 3.

Answer: -3

Explanation: Since $3 + (-3) = 0$, 3 and -3 are additive inverses.

b. Find the additive inverse of -4.

Answer: 4

Explanation: Since $-4 + 4 = 0$, -4 and 4 are additive inverses.

c. Find the additive inverse of 0.

Answer: 0

Explanation: Since $0 + 0 = 0$, 0 is its own additive inverse.

d. Find the additive inverse of the additive inverse of -2.

Answer: -2.

Explanation: There are two additive inverses here. The additive inverse of -2 is 2, and the additive inverse of 2 is -2.

Note: When we find the additive inverse of the additive inverse of a number, the outcome is the original number.

Do Your Turn 6 ▷

Your Turn 6

a. Find the additive inverse of -6.

b. Find the additive inverse of 9.

c. Find the additive inverse of the additive inverse of 7.

Answers to Your Turn 5
a. 19 b. 90 c. 0

Answers to Your Turn 6
a. 6 b. -9 c. 7

The symbol for additive inverse is the minus sign. It may seem strange that the minus sign can mean subtraction, a negative number, and additive inverse. But consider that in the English language, there are many cases of the same word having different meanings. For example, consider the word *can*.

> If we say, "Open a can of paint," the word *can* is a noun.
> If we say, "I can do this," the word *can* acts as a verb.

We determine a word's meaning by context; that is, we look at how it is used in the sentence. The same is true for a minus sign.

> If we write $5 - 2$, we mean "five subtract two."
> If we write -2, we mean "negative two."
> If we write $-(2)$, we mean "the additive inverse of two."

Subtracting 2 is the same as adding the additive inverse of 2, which is negative 2. That's why the same symbol can be used for all three.

Your Turn 7

Simplify.

a. $-(8)$

b. $-(-(7))$

c. $-(-(-(11)))$

d. $-(-(-(-17)))$

Note:

In an additive inverse problem with a nonzero number, you can determine the sign of the outcome by counting the total number of minus signs.

If there are an even number of minus signs, then the outcome is positive because each pair of minus signs gives a positive.

$$-(-(5)) = 5$$
$$-(-(-(-8))) = 8$$

If there are an odd number of minus signs, then the outcome is negative because if we pair all of the minus signs, there will be an extra minus sign left.

$$-(6) = -6$$
$$-(-(-4)) = -4$$

Example 7 Simplify.

a. $-(3)$

CONNECTION Example 7a is the symbolic version of Example 6a.

Answer: -3

Explanation: $-(3)$ means "the additive inverse of 3," which is -3.

b. $-(-4)$

CONNECTION Example 7b is the symbolic version of Example 6b.

Answer: 4

Explanation: $-(-4)$ means, "the additive inverse of -4," which is 4. Notice that a pair of minus signs translates to positive.

c. $-(0)$

CONNECTION Example 7c is the symbolic version of Example 6c.

Answer: 0

Explanation: $-(0)$ means, "the additive inverse of 0," which is 0.

d. $-(-(-2))$

CONNECTION Example 7d is the symbolic version of Example 6d.

Answer: -2

Explanation: $-(-(-2))$ means, "the additive inverse of the additive inverse of -2," which is -2.

Do Your Turn 7

Answers to Your Turn 7
a. -8 b. 7 c. -11 d. 17

Example 8   Translate $-|-18|$ to words, then simplify.

Translation: The additive inverse of the absolute value of negative eighteen.

Answer: -18

Explanation: We must first find the absolute value of -18, which is 18. Then we must find the additive inverse of that result. The additive inverse of 18 is -18.

> **CONNECTION** Absolute value symbols are treated like parentheses. Operations within absolute value symbols are to be computed before operations outside of the absolute value symbols.

Do Your Turn 8 ▷

Your Turn 8

Translate to words, then simplify.

a. $-|20|$

b. $|-(-4)|$

c. $-(-(-(14)))$

Answers to Your Turn 8
a. The additive inverse of the absolute value of 20 is -20.
b. The absolute value of the additive inverse of -4 is 4.
c. The additive inverse of the additive inverse of the additive inverse of 14 is -14.

2.1 Exercises

For Extra Help

 Videotape DVT

 Addison-Wesley Tutor Center

 Math XL

 MyMathLab

 Student Solutions Manual

1. Write a list indicating the numbers in the set of integers.

2. Draw a number line showing the integers from -5 to 5.

3. The absolute value of every positive number is _____.

4. The absolute value of every negative number is _____.

5. The absolute value of zero is _____.

6. The additive inverse of every positive number is _____.

7. The additive inverse of every negative number is _____.

8. The additive inverse of zero is _____.

For Exercises 9–22, express each amount as a positive or negative integer.

9. Sonja receives a paycheck of $450.

10. Holly owes her friend $35.

11. Adam's business made a profit of $215,000 last year.

12. The *Titanic* is lying on the bottom of the North Atlantic Ocean at a depth of approximately 13,200 feet.

13. The tunnel connecting Britain to France is approximately 220 meters below the surface of the English Channel.

14. In the last century, the population of Bengal tigers in India decreased by 37,000.

15. A college football team gains 40 yards in one play.

16. The countdown to launch time of a space shuttle is stopped with 20 minutes to go.

17. Water boils at 100°C.

18. The Earth accelerates objects toward its surface at a rate of 32 feet per second every second.

19. Mt. Everest is the tallest mountain in the world. Its peak is 29,028 feet above sea level.

20. Carbon dioxide (CO_2) freezes at 78°C below zero.

21. A financial planner analyzes the Smiths' financial situation and finds they have a total debt of $75,243.

22. A plane's cruising altitude is 35,000 feet above sea level.

For Exercises 23–38, graph each integer on a number line.

23. 8   24. 2

25. 7   26. 5

OF INTEREST

The English Channel Tunnel is approximately 31 miles long, 23 miles of which are under water. The rubble from excavating the tunnel increased the size of Britain by 90 acres.

OF INTEREST

Mt. Everest is in the central Himalayan range on the border of Tibet and Nepal. Its summit is the highest elevation in the world and was first reached in 1953 by Edmund Hillary from New Zealand and Tenzing Norgay, a Nepalese guide.

Source: Microsoft Bookshelf '98

27. -9

28. -2

29. -6

30. -8

31. 0

32. 9

33. -4

34. -7

35. -5

36. 6

37. 10

38. -1

For Exercises 39–52, use $<$ or $>$ to make a true statement.

39. -20 ? 17

40. 26 ? -18

41. 20 ? 16

42. 12 ? 19

43. -30 ? -16

44. -12 ? -19

45. -9 ? -12

46. -14 ? -11

47. 0 $>$ -16

48. -4 ? -5

49. -19 ? -17

50. 0 ? 5

51. -8 ? -10

52. -2 ? -7

For Exercises 53–62, complete the sentence.

53. The absolute value of 26 is 26 .

54. The absolute value of 19 is 19 .

55. The absolute value of 21 is 21 .

56. The absolute value of 9 is 9 .

57. The absolute value of -18 is 18 .

58. The absolute value of -4 is 4 .

59. The absolute value of -10 is 10 .

60. The absolute value of -16 is 16 .

61. The absolute value of 0 is 0 .

62. The absolute value of -7 is 7 .

For Exercises 63–76, simplify.

63. $|14| = 14$

64. $|242|$ 242

65. $|-18|$ 18

66. $|-9|$ 9

67. $|-2004| = 2004$

68. $|12|$ 12

69. $|8|$ 8

70. $|-35|$ 35

71. $|0|$ 0

72. $|-200|$ 200

73. $|47|$ 47

74. $|97|$ 97

75. $|-377|$ 377

76. $|-84|$ 84

For Exercises 77–86, complete the sentence.

77. The additive inverse of -18 is 18 .

78. The additive inverse of 42 is -42 .

79. The additive inverse of 61 is -61 .

80. The additive inverse of -540 is 540 .

81. The additive inverse of 0 is 0 .

82. The additive inverse of -17 is 17 .

83. The additive inverse of the additive inverse of -8 is -8 .

84. The additive inverse of the additive inverse of -16 is -16 .

85. The additive inverse of the additive inverse of 43 is 43 .

86. The additive inverse of the additive inverse of 75 is 75 .

For Exercises 87–100, simplify.

87. $-(6)$ **88.** $-(9)$ **89.** $-(-14)$

90. $-(-19)$ **91.** $-(0)$ **92.** $-(-75)$

93. $-(63)$ **94.** $-(29)$ **95.** $-(-4)$

96. $-(-(7))$ **97.** $-(-(-12))$ **98.** $-(-(-(37)))$

99. $-(-(-(-87)))$ **100.** $-(-(-(-(42))))$

For Exercises 101–110, translate the expression to words, then simplify.

101. $|-4|$ **102.** $|42|$

103. $-(-8)$ **104.** $-(140)$

105. $-(-(14))$ **106.** $-(-(-28))$

107. $-|-5|$ **108.** $-|16|$

109. $-|-(-4)|$ **110.** $|-(-(-17))|$

Review Exercises

For Exercises 1–4, perform the indicated operation.

[1.2]   **1.** $84,759 + 9506$

[1.2]   **2.** $900,406 - 35,918$

[1.3]   **3.** 457×609

[1.4]   **4.** $85,314 \div 42$

[1.2]   **5.** Amelia begins a month with $452 in savings. Following is a list of her deposits and withdrawals for the month. What is her balance at the end of the month?

Deposits	Withdrawals
$45	$220
$98	$25
$88	$10
$54	

2.2 Adding Integers

OBJECTIVE 1 Add integers with like signs.

Let's examine some realistic situations in order to discover how we add signed numbers.

Example 1

a. Suppose you have $200 in your checking account and you deposit $40. What would be the balance?

Solution: This is a standard addition situation.

$$200 + 40 = 240$$

Explanation: Because the $40 deposit increases the $200 balance, we add the values to end up with a balance of $240.

We can also use a number line. Think of the first addend, 200, as the starting position. The absolute value of the second addend indicates the distance to move from the starting position, and its sign indicates which direction. Because 40 is positive, we move to the right 40 steps. The final position, 240, is the answer.

+ 40 means move 40 steps to the right

200   240
Starting position   Final position

b. Suppose you have a debt of $200 on a credit card and you purchase $40 worth of merchandise using the same credit card. What is the balance after the purchase?

Solution: Because the $200 is a debt, we write it as −200. A purchase using a credit card is a debt, so the $40 is written as −40. The addition statement looks like this:

$$-200 + (-40) = -240$$

Note: In math sentences, parentheses are placed around negative numbers that *follow* operation symbols. They actually are not needed, but are there to clarify the meaning of the plus and minus signs.

Explanation: By using that credit card to purchase merchandise, you are simply increasing your debt, so we add the money you already owe to the new debt to get a total debt.

On a number line, −200 + (−40) means start at −200 and move 40 steps in the negative direction, which is to the left.

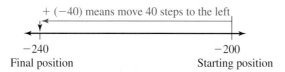

+ (−40) means move 40 steps to the left

−240   −200
Final position   Starting position

Notice that in Example 1 we added the absolute values because we were either increasing assets or increasing debt. On a number line, we were starting positive and moving more positive or starting negative and moving more negative.

Conclusion: When we add two numbers that have the same sign, we add their absolute values and keep the same sign.

Do Your Turn 1 ▷

OBJECTIVES

1 Add integers with like signs.

2 Add integers with different signs.

3 Add integers.

4 Solve applications involving addition of integers.

Your Turn 1

Add.

a. 19 + 74

b. −42 + (−39)

c. −15 + (−18)

d. 20 + 13

Answers to Your Turn 1
a. 93 b. −81 c. −33 d. 33

Your Turn 2

Add.

a. $-18 + 6$

b. $-20 + 28$

c. $30 + (-26)$

d. $14 + (-24)$

e. $15 + (-15)$

OBJECTIVE 2 Add integers with different signs.

Addition with signed numbers does not always mean we add.

Example 2

a. Suppose we have a debt of $200 on a credit card and we make a payment of $80. What is the balance?

Solution: The $200 debt is written as -200, and the $80 payment is a positive number added to the account. The addition statement looks like this:

$$-200 + 80 = -120$$

Explanation: Because the payment decreases the debt, we subtract the values, and since the debt exceeds the payment, we still owe money.

On a number line, we start at -200 and move 80 steps to the right, which is not enough to move past 0; so the result is still negative.

+ 80 means move 80 steps to the right

-200
Starting position

-120
Final position

Also notice that because of the commutative law of addition, we can write the statement as

$$80 + (-200) = -120$$

It doesn't matter which way it is arranged, a $200 debt with $80 in assets results in a debt of $120.

b. Suppose we have a debt of $200 on the credit card and we make a payment of $240. What is the balance?

Solution: Again, the $200 debt is written as -200 and the payment of $240 is a positive number added to the account towards that debt, so it is a positive amount. The addition statement looks like this:

$$-200 + 240 = 40$$

Explanation: In this case, we paid more than we owe, so we have a credit of $40.

On a number line, we start at -200 and move 240 steps to the right. Moving 200 steps to the right puts us at 0, so 240 steps moves 40 steps past 0, making the result 40.

+240 means move 240 steps to the right

-200
Starting position

0   40
Final position

From Example 2 we can draw the following conclusion.

Conclusion: When we add two numbers that have different signs, we subtract their absolute values and keep the sign of the number that has the greater absolute value.

◁ **Do Your Turn 2**

Answers to Your Turn 2

a. -12 b. 8 c. 4 d. -10 e. 0

OBJECTIVE 3 Add integers.

We can summarize what we learned in Examples 1 and 2 with the following two rules for adding integers. It is helpful to relate each addition situation to reconciling debts and credits.

PROCEDURE *To add two integers:*

If they have the same sign, we add their absolute values and keep the same sign. (Think: two credits or two debts.)

If they have different signs, we subtract their absolute values and keep the sign of the number with the greater absolute value. (*Think:* Payment toward a debt.)

Example 3 Explain each situation in terms of debts and credits, then add.

a. $-40 + (-21)$

Solution: We are to add two debts, which increases the debt, so we add and the result is still a debt. Using the rules, since the numbers have the same sign, we add the absolute values and keep the negative sign.

$$-40 + (-21) = -61$$

b. $28 + (-16)$

Solution: We have a $28 payment towards a $16 debt. We subtract the amount and since the payment exceeds the debt, the result is a credit. Using the rules, since the numbers have different signs, we subtract their absolute values, and because the positive integer, 28, has the greater absolute value, the result is positive.

$$28 + (-16) = 12$$

c. $10 + (-24)$

Solution: We have a $10 payment towards a $24 debt. We subtract the amounts and since the debt exceeds the payments, the result is a debt. Using the rules, since the numbers have different signs, we subtract their absolute values, and because the negative integer, -24, has the greater absolute value, the result is negative.

$$10 + (-24) = -14$$

Do Your Turn 3 ▶

Now let's consider how to add more than two integers. There are two popular approaches: 1) We can add from left to right, or 2) we can add all the positives and negatives separately, then reconcile the result.

Example 4 Add. $-14 + 26 + (-18) + 15 + (-32)$

Solution: First, let's add from left to right.

$$\begin{aligned}
&-14 + 26 + (-18) + 15 + (-32) \\
=\ & \quad\ 12 + (-18) + 15 + (-32) \qquad \text{Add } -14 \text{ and } 26 \text{ to get } 12. \\
=\ & \qquad\qquad -6 + 15 + (-32) \qquad \text{Add } 12 \text{ and } -18 \text{ to get } -6. \\
=\ & \qquad\qquad\qquad\quad 9 + (-32) \qquad \text{Add } -6 \text{ to } 15 \text{ to get } 9. \\
=\ & \qquad\qquad\qquad\qquad\quad -23 \qquad \text{Add } 9 \text{ and } -32 \text{ to get } -23.
\end{aligned}$$

Your Turn 4

Add.

a. $-18 + (-14) + 30 + (-2) + 15$

b. $-38 + 64 + 40 + (-100)$

c. $28 + (-14) + 5 + (-12) + (-10) + 3$

Calculator Tips

To add integers using a calculator, type the addition sentence as you read it from left to right.

Example:
$-42 + 17 + (-28)$

Type (−) 4 2

+ 1 7 +

(−) 2 8   ENTER =

Answer: -53

On some calculators, to enter a negative number, you first enter the value, then press a +/− key.

Your Turn 5

The following table lists the assets and debts for the Cromwell family. Calculate their net worth.

Assets

Savings = $3528
Checking = $1242
Furniture = $21,358
Stocks = $8749

Debts

Credit card balance = $3718
Mortgage = $55,926
Automobile 1 = $4857
Automobile 2 = $3310

Answers to Your Turn 4
a. 11 b. −34 c. 0

Answer to Your Turn 5
− $ 32,934

Now let's add all the positives and negatives separately, then reconcile the result. The commutative and associative properties of addition make this method possible because they allow us to rearrange and regroup addition any way we wish.

$$-14 + 26 + (-18) + 15 + (-32)$$

$$= 26 + 15 + (-14) + (-18) + (-32)$$ Rearrange the addends so the positives and negatives are grouped separately.

$$= \quad 41 \quad + \quad (-64)$$ Add the positives and negatives separately.

$$= -23$$ Reconcile the positive and negative sums.

▲ Think: The total credit is 41 and the total debt is 64. Because the debt outweighs the credit, the result is a debt of 23.

◀ **Do Your Turn 4**

OF INTEREST

Adding positives and negatives separately is the method banks use to calculate balances on their statements. They add the credits and debts separately, then reconcile those two sums.

OBJECTIVE 4 Solve applications involving addition of integers.

The key to solving addition problems that involve integers is to simulate the situation.

Example 5 Jane's credit card statement showed a closing balance of −$240. Since that time she made the following transactions. What is her balance now?

CHARGES	
THE BREAKFAST PLACE	$15
PACY'S	$53
CORNER VIDEO	$12
FINANCE CHARGE	$13
PAYMENT	**$300**

Understand: The balance is the sum of the closing balance, all charges, and all payments.

Plan: Write an addition statement adding the closing balance, all charges, and the payment.

Execute: net worth = $-240 + (-15) + (-53) + (-12) + (-13) + 300$

$$= -333 + 300$$
◀ **Note:** Adding all the negatives computes the total debt, $333.
$$= -33$$

Answer: Because the total debt of $333 has a greater absolute value than the total payment of $300, Jane's current balance is a debt of $33.

Check: Instead of adding the total debt and payments, we could compute the sum from left to right. We will leave this to the reader.

◀ **Do Your Turn 5**

Example 6 A research submarine is at a depth of 458 feet. The submarine's sonar detects that the ocean floor is another 175 feet down. What will be the depth if the submarine descends to the ocean floor? The submarine is designed to withstand the pressure from the surrounding water up to a depth of 600 feet. Is it safe to allow the submarine to descend to the ocean floor in this case?

Understand: We must determine if it is safe for the submarine to proceed from its current depth to the ocean floor.

Plan: We must first calculate the depth of the ocean floor. We can then assess whether it is beyond the submarine's depth limit of 600 feet. Because -458 feet is its starting position and it is to proceed down another 175 feet, we can add to get the depth of the ocean floor.

Execute: ocean floor depth $= -458 + (-175) = -633$

Answer: The depth of the ocean floor is -633 feet, which is beyond the submarine's 600-foot depth limit. Therefore, it is not safe to allow the submarine to descend to the ocean floor.

Check: We can use a vertical number line.

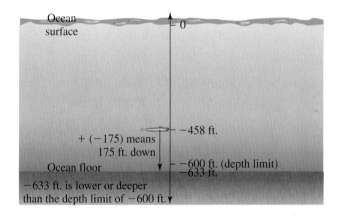

Do Your Turn 6 ▷

Your Turn 6

At sunset one cold winter evening in North Dakota, the temperature was $-5°F$. By midnight, the temperature had dropped another $15°F$. What was the temperature at midnight?

Example 7 We can use integers to analyze the forces acting on an object. The resultant force is the sum of the forces acting on the object. A steel beam weighs 250 pounds and has two wires attached to hold it up. Each wire has an upward force measuring 140 pounds. (See the following diagram.) Find the resultant force on the beam. What does this force tell you?

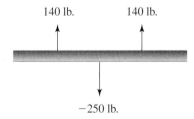

Understand: Because the weight is a force due to gravity pulling downward on objects, it has a negative value. The wires are pulling upward against gravity, so they have positive values.

Plan: To find the resultant force, we compute the sum of the forces.

Answer to Your Turn 6
$-20°F$

A 500-pound concrete slab is suspended by four ropes tied to the corners. Each rope has an upward force of 110 pounds. What is the resultant force? What does the resultant force tell you?

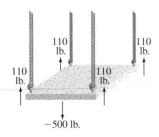

Execute: resultant force = −250 + 140 + 140

= −250 + 280   **Add the positive numbers.**

= 30

Answer: The resultant force is 30 lb. Because the resultant force is positive, it means the beam is traveling upward.

Check: Compute −250 + 140 + 140 from left to right.

−250 + 140 + 140

= −110 + 140   **Add −250 and 140.**

= 30

◁ **Do Your Turn 7**

Answer to Your Turn 7

−60 lb. Because the resultant force is negative, the slab is moving downward.

2.2 Exercises

For Extra Help

Videotape DVT

Addison-Wesley Tutor Center

Math XL

MyMathLab

Student Solutions Manual

1. Explain $-24 + (-6)$ in terms of debts or credits.

2. Explain $-20 + 12$ in terms of debts or credits.

3. When adding two numbers that have the same sign, we _____ their absolute values and _____.

4. When adding two numbers that have different signs, we _____ their absolute values and _____.

For Exercises 5–14, add.
Key thought: Adding integers with the same sign is similar to increasing assets or increasing debts.

5. $14 + 9$

6. $74 + 13$

7. $-15 + (-8)$

8. $-14 + (-16)$

9. $-9 + (-6)$

10. $-31 + (-9)$

11. $12 + 18 + 16$

12. $4 + 15 + 41$

13. $-38 + (-17) + (-21)$

14. $-60 + (-19) + (-32)$

For Exercises 15–24, add.
Key thought: Adding integers with different signs is similar to reconciling assets and debts to get net worth.

15. $28 + (-18)$

16. $17 + (-5)$

17. $-34 + 20$

18. $-25 + 16$

19. $-21 + 35$

20. $-35 + 47$

21. $35 + (-53)$

22. $16 + (-29)$

23. $-24 + 80$

24. $75 + (-43)$

For Exercises 25–46, solve by explaining each in terms of assets and debts.

25. $48 + 90$

26. $35 + 16$

27. $-45 + (-27)$

28. $-68 + (-42)$

29. $81 + (-23)$

30. $68 + (-15)$

31. $-15 + 42$

32. $-54 + 65$

33. $-81 + 60$

34. $-45 + 21$

35. $37 + (-58)$

36. $62 + (-92)$

37. $-45 + 45$

38. $64 + (-64)$

39. $48 + (-18) + 16 + (-12)$

40. $69 + (-20) + (-15) + 11$

41. $-36 + (-17) + 94 + (-9)$

42. $-42 + 25 + (-61) + (-3)$

43. $25 + (-17) + (-33) + 19 + 6$

44. $32 + 16 + (-51) + 3 + (-8)$

45. $-75 + (-14) + 38 + 9 + (-17)$

46. $-60 + 18 + (-40) + (-12) + 93$

For Exercises 47–58, solve.

47. Jason's checking account shows a balance of $24. Unfortunately, he forgot about a check for $40 and it clears. The bank then charges $17 for insufficient funds. What is his new balance?

48. Charlene has a current balance of −$1243 on her credit card. During the month, she makes the transactions shown to the right. What will be her balance at the end of the month?

Charges:	
Truman's	$58
Dave's Diner	$13
Fuel'n Go	$15
Finance charge	$18
Payment:	$150

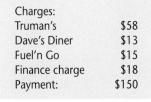

49. The following table lists the assets and debts for the Smith family. Calculate their net worth.

Assets	Debts
Savings = $1498	Credit card balance = $1841
Checking = $2148	Mortgage = $74,614
Furniture = $18,901	Automobile = $5488
Jewelry = $3845	

50. The following table lists the assets and debts for the Jones family, who just financed a new car. Calculate their net worth.

Assets	Debts
Savings = $214	Student loans = $15,988
Checking = $1242	Automobile 1 = $4857
Furniture = $21,358	Automobile 2 = $23,410

51. Lea's current credit card balance is −$125. During the month, she makes the following purchases with her card. If she was charged $9 in interest and made a $325 payment, what will be her new balance at the end of the month?

PURCHASES	
JACK'S CAFÉ	$12
MINUTE DRY-CLEAN	$8
TIRES ARE US	$120
PET COMFORT	$22

52. Greg's January closing balance on his credit card was −$1320. In February, he made a payment of $450 and another of $700. His finance charges were $19. Greg made purchases at Tommy's for $32 and Petromart for $27. What is his February closing balance?

53. A scientist working during the summer at the South Pole station notes that the temperature is about −21°F. During the winter she can expect the temperature to be as much as 57°F colder. What temperature might she expect during the winter?

OF INTEREST

Because of the tilt of the Earth, the South Pole experiences daylight for 6 months (summer) and then night for 6 months (winter). The sun rises in mid-September and sets in mid-March.

54. The Inuit people in the arctic regions of North America sometimes build igloos of snow blocks. During winter months, the temperatures can be as low as −58°F. A small heater can raise the internal temperature of a snow-block igloo by 108°F. What would be the temperature inside the igloo?

55. A research submarine is attached to a crane on a ship. The submarine is currently at a depth of 147 feet. The crane raises the submarine 69 feet and then must stop to be repaired. What is the elevation of the submarine?

56. A mining team is working 147 feet below ground. During the day, they dig down another 28 feet. What is their new position?

57. An elevator weighs 745 pounds. When the elevator is on the first floor of a 12-story building, the steel cable connected to the top of it adds another 300 pounds. Three people enter the elevator. One person weighs 145 pounds, the second weighs 185 pounds, and the third 168 pounds. The motor exerts an upward force of 1800 pounds. What is the resultant force?

58. A concrete block is suspended by two cables. The block weighs 500 pounds. Each cable is exerting 250 pounds of upward force. What is the resultant force? What does this mean?

PUZZLE PROBLEM   A snail is at the bottom of a well that is 30 feet deep. The snail climbs 3 feet per hour then stops to rest for an hour. During each rest period, the snail drops back down 2 feet. If the snail began climbing at 8 A.M., what time will it be when the snail gets out of the well?

Review Exercises

[1.1]   **1.** Write 42,561,009 in expanded form.

[1.1]   **2.** Write the word name for 2,407,006.

[1.7]   **3.** Becky plans to place a fence around her property. To the right is a plot plan of the property. She plans to have two 10-foot gates. The fence company charges $7 per foot for the fencing and $80 for each gate. What will be the total cost?

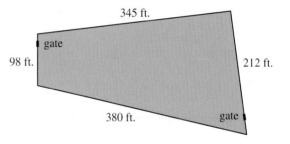

[1.2]   **4.** Subtract. 60,041 − 4596

[1.2]   **5.** Solve and check. 15 + *n* = 28

2.3 Subtracting Integers and Solving Equations

OBJECTIVES

1 Write subtraction statements as equivalent addition statements.

2 Solve equations containing a missing addend.

3 Solve applications involving subtraction of integers.

OBJECTIVE 1 Write subtraction statements as equivalent addition statements.

In Section 2.2, we learned that when we add two numbers with different signs, we actually subtract their values. This suggests that addition and subtraction are closely related. In fact, any subtraction statement can be written as an equivalent addition statement. This can be quite useful because some subtraction problems involving signed numbers are more challenging to analyze than their equivalent addition statements. Let's consider some examples to determine how to write subtraction as addition.

Example 1

a. Marty has $30 in his checking account. He writes a check for $42. What is his new balance?

Solution: Because the check amount of $42 is a deduction, we subtract it from the initial balance of $30. Since the deduction is more than the amount in the account, the result is negative.

$$30 - 42 = -12$$

In Section 2.2, we would have solved this problem using the following addition:

$$30 + (-42) = -12$$

This suggests that $30 - 42$ and $30 + (-42)$ are equivalent. Notice two things changed going from the subtraction to the addition: 1) the operation symbol changed from $-$ to $+$, and 2) the subtrahend, 42, changed to its additive inverse, -42.

The operation changes from $-$ to $+$. $\quad\quad \dfrac{30 - \quad 42}{}$ The subtrahend changes to its additive inverse.

$$= 30 + (-42)$$
$$= -12$$

b. Marty has a balance of $-$$12 and writes a check for $15. What is his new balance?

Solution: As in part a, we subtract the check amount from the initial balance. In this case, however, the initial balance is negative, so the deduction of $15 makes the balance more negative.

$$-12 - 15 = -27$$

Note that this subtraction is equivalent to the following addition:

$$-12 + (-15) = -27$$

Also, like in part a, the same two things changed going from the subtraction to the addition: 1) the operation symbol changed from $-$ to $+$, and 2) the subtrahend changed to its additive inverse (15 changed to -15).

The operation changes from $-$ to $+$. $\quad\quad \dfrac{-12 - \quad 15}{}$ The subtrahend changes to its additive inverse.

$$= -12 + (-15)$$
$$= -27$$

CONNECTION Subtracting a positive number is equivalent to adding a debt.

Example 1 suggests the following procedure.

PROCEDURE *To write a subtraction statement as an equivalent addition statement,*
1. Change the operation symbol from a minus sign to a plus sign.
2. Change the subtrahend (second number) to its additive inverse.

Do Your Turn 1 ▶

What if we subtract a negative number? We will see that our procedure still applies.

Example 2

a. Sarah has a balance of $28 in her checking account. The bank discovers she was charged a $5 service charge by mistake. The bank tells Angela the charge will be corrected. What will be his new balance?

Solution: The service charge is a debt, so it is written as −5. To correct the mistake, the bank must subtract that debt. Subtracting the debt means $5 is deposited back into the account, so the resulting balance is $33. As a subtraction statement, we would write:

$$28 - (-5) = 33$$

Notice that it is easier to think of this problem as an addition statement:

$$28 + 5 = 33$$

So, $28 - (-5)$ and $28 + 5$ are equivalent. Notice that the same two things changed in writing the subtraction as an addition: 1) the minus sign changed to a plus sign, and 2) the subtrahend changed to its additive inverse (−5 changed to 5).

The operation changes from − to +. $\underbrace{28 -}\ \underbrace{(-5)}$ The subtrahend changes to its additive inverse.
$$= 28 + \quad 5$$
$$= 33$$

b. Now suppose Sarah owes a friend $50 and, out of kindness, her friend decides to cancel (or remove) $30 of that debt. What is Sarah's new balance?

Solution: The $50 debt is written as −50. The $30 is also debt, so it is written as −30. Because her friend is canceling, or removing that debt, we subtract. If $30 of the $50 debt is removed, then she now only owes her friend $20, so her balance is −$20.

$$-50 - (-30) = -20$$

This problem is easier to think through in terms of an addition statement.

$$-50 + 30 = -20$$

Again, we changed two things: 1) the operation changed from − to +, and 2) the subtrahend changed to its additive inverse (−30 changed to 30).

The operation changes from − to +. $\underbrace{-50 -}\ \underbrace{(-30)}$ The subtrahend changes to its additive inverse.
$$= -50 + \quad 30$$
$$= -20$$

CONNECTION Subtracting a negative number is equivalent to making a deposit.

Do Your Turn 2 ▶

Your Turn 1

Write each subtraction as an equivalent addition statement, then evaluate.

a. $30 - 48$

b. $24 - 82$

c. $-16 - 25$

d. $-33 - 44$

Your Turn 2

Write each subtraction as an equivalent addition, then evaluate.

a. $42 - (-21)$

b. $-48 - (-10)$

c. $-36 - (-40)$

Answers to Your Turn 1
a. $30 + (-48) = -18$
b. $24 + (-82) = -58$
c. $-16 + (-25) = -41$
d. $-33 + (-44) = -77$

Answers to Your Turn 2
a. $42 + 21 = 63$
b. $-48 + 10 = -38$
c. $-36 + 40 = 4$

OBJECTIVE 2 Solve equations containing a missing addend.

In Section 1.2, we learned how to solve equations involving a missing addend. Let's recall the procedure.

PROCEDURE *To find a missing addend, write a related subtraction sentence. Subtract the known addend from the sum.*

Your Turn 3

Solve and check.

a. $18 + y = 12$

b. $c + 24 = 9$

c. $-15 + d = -23$

d. $k + (-21) = -6$

e. $9 + t = -17$

Example 3 Solve and check.

a. $25 + x = 10$

Solution: Write a related sentence.

$$25 + x = 10$$
$$x = 10 - 25 \qquad \text{Subtract the known addend from the sum.}$$
$$x = 10 + (-25) \qquad \text{Write the subtraction as an equivalent addition.}$$
$$x = -15 \qquad \text{Since we are adding two numbers with different signs, we subtract their absolute values and keep the sign of the number with the greatest absolute value.}$$

Check: Verify that -15 can replace x in $25 + x = 10$, and make the equation true.

$$25 + (-15) \overset{?}{=} 10$$
$$10 = 10 \qquad \text{True, so it checks.}$$

Explanation: Think about the original addition statement. If we start with 25 and end with 10, then a deduction has taken place. The only way this happens with addition is if a negative number is added.

b. $-14 + x = -20$

Solution: Write a related sentence.

$$-14 + x = -20$$
$$x = -20 - (-14) \qquad \text{Subtract the known addend from the sum.}$$
$$x = -20 + 14 \qquad \text{Write the subtraction as an equivalent addition.}$$
$$x = -6 \qquad \text{Since we are adding two numbers with different signs, we subtract their absolute values and keep the sign of the number with the greatest absolute value.}$$

Check: Verify that -6 can replace x in $-14 + x = -20$, and make the equation true.

$$-14 + (-6) \overset{?}{=} -20$$
$$-20 = -20 \qquad \text{True, so it checks.}$$

Explanation: Think through the original addition statement. Starting at -14, to end up at -20, we must continue 6 more steps in the negative direction. Therefore the second addend must be -6.

◄ **Do Your Turn 3**

Answers to Your Turn 3
a. -6 b. -15 c. -8 d. 15
e. -26

OBJECTIVE 3 Solve applications involving subtraction of integers.

In Chapter 1, we learned that subtraction problems can involve taking one amount away from another, finding the difference between amounts, or finding an unknown addend. First, let's consider a situation involving take away. Calculating a business's **net** involves subtracting its total **costs** from its total **revenue.** If the net is positive, we say it is a **profit**; if the net is negative, we say it is a **loss.**

DEFINITIONS **Net:** Money remaining after subtracting costs from revenue (money made minus money spent).

Cost: Money spent on production, operation, labor, and debts.

Revenue: Income (money made).

Profit: A positive net (when revenue is greater than cost).

Loss: A negative net (when revenue is less than cost).

The definition of net suggests the following formula.

$$\text{net} = \text{revenue} - \text{cost}$$
$$N = R \qquad - C$$

Example 4 The financial report for a business indicates that the total revenue for 2004 was $2,453,530 and the total costs were $2,560,000. What was the net? Did the business experience a profit or loss?

Understand: We must find the net given the revenue and cost. The formula for net is

$$N = R - C.$$

Plan: Replace R with 2,453,530 and C with 2,560,000 in $N = R - C$, then subtract.

Execute: $N = R - C$
$\qquad N = 2,453,530 - 2,560,000$
$\qquad N = -106,470$

Answer: The net for 2004 was $-\$106,470$, which means the business had a loss of $106,470.

Check: Reverse the process.

$$-106,470 + 2,560,000 \overset{?}{=} 2,453,530$$
$$2,453,530 = 2,453,530$$

Do Your Turn 4 ▶

Now, let's consider difference problems. In these problems, we must find the amount between two given numbers. To find the difference, we subtract the smaller number from the larger number.

Your Turn 4

The revenue for one month for a small business was $45,382, and the total costs were $42,295. What was the net? Was it a profit or loss?

Answer to Your Turn 4
$3087; profit

Solve.

a. The surface temperature on the Moon at lunar noon can reach 120°C. During the lunar night, the temperature can drop to −190°C. How much of a change in temperature occurs from day to night on the Moon?

b. A submarine goes from a depth of −78 feet to a depth of −34 feet. What is the difference in depth?

Example 5   On the evening news the meteorologist says the current temperature is 25°F. She then says that the evening low could get down to −6°F. How much of a change is this from the current temperature?

Understand: Because we are asked to find the amount of change, we must calculate a difference.

Plan: Find the difference between 25 and −6.

Execute: temperature difference = 25 − (−6)

$$= 25 + 6$$

Write the subtraction as an equivalent addition statement.

$$= 31$$

Answer: From 25°F to −6°F is a change of 31°F.

Check: Use a number line or picture.

CONNECTION   A thermometer is essentially a vertical number line.

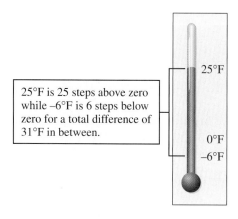

25°F is 25 steps above zero while –6°F is 6 steps below zero for a total difference of 31°F in between.

25°F

0°F
–6°F

◁ **Do Your Turn 5**

In missing addend problems, we are given a starting amount and must find how much more is needed to reach a new amount.

Example 6   A submarine is at −126 feet. How much must the submarine climb to reach −40 feet?

Understand: The word *climb* indicates that we must add an amount to −126 feet to reach a value of −40 feet so this is a missing addend situation. A picture is helpful.

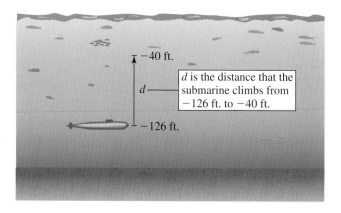

−40 ft.

d

d is the distance that the submarine climbs from −126 ft. to −40 ft.

−126 ft.

Answers to Your Turn 5
a. 310°C b. 44 ft.

Plan: Write a missing addend equation, then solve.

Execute: $-126 + d = -40$

$$d = -40 - (-126)$$ To solve, subtract the known addend, −126, from the sum −40.

$$d = -40 + 126$$ Write the subtraction as an equivalent addition.

$$d = 86$$

CONNECTION In the diagram, we can see that the missing addend amount is the same as the difference between -40 and -126. This explains why the missing addend statement, $-126 + d = -40$, becomes the difference statement, $d = -40 - (-126)$.

Answer: The submarine must climb 86 feet to reach a depth of -40 feet.

Check: Verify that if the submarine climbs 86 feet from -126 feet, it will be at -40 feet.

$$-126 + 86 \stackrel{?}{=} -40$$
$$-40 = -40 \quad \text{It checks.}$$

Do Your Turn 6 ▷

⌐Your Turn 6

Solve.

a. Barbara currently has a balance of $-\$58$ in her checking account. To avoid any further charges, she must have a minimum balance of $25. How much should she deposit to have the minimum balance?

b. A mining team is on an elevator platform at a depth of -148 feet. They are to ascend to a depth of -60 feet in order to dig a side tunnel from the main shaft. How much does the elevator need to rise to get them to the appropriate depth?

Answers to Your Turn 6
a. $83 b. 88 feet

For Extra Help

Videotape DVT

Addison-Wesley Tutor Center

Math XL

MyMathLab

Student Solutions Manual

1. Explain how to write a subtraction statement as an equivalent addition statement.

2. Do you always have to rewrite subtraction statements in order to find the correct difference? Explain.

3. Give an example of a subtraction problem that is equivalent to an addition of two positive numbers.

4. Give an example of a subtraction problem that is equivalent to an addition of two negative numbers.

For Exercises 5–16, write as an equivalent addition statement, then evaluate.

5. $18 - 25$

6. $30 - 54$

7. $-15 - 18$

8. $-21 - 6$

9. $20 - (-8)$

10. $33 - (-12)$

11. $-14 - (-18)$

12. $-20 - (-30)$

13. $-15 - (-8)$

14. $-26 - (-14)$

15. $0 - (-5)$

16. $0 - 16$

For Exercises 17–30, write as an equivalent addition statement as needed, then evaluate.

17. $-21 - 19$

18. $-15 - (-12)$

19. $-4 - (-19)$

20. $26 - (-6)$

21. $31 - 44$

22. $-13 - 28$

23. $35 - (-10)$

24. $-19 - (-24)$

25. $-28 - (-16)$

26. $9 - 15$

27. $0 - 18$

28. $-32 - 14$

29. $-27 - (-16)$

30. $0 - (-8)$

For Exercises 31–42, solve and check.

31. $18 + t = 12$

32. $21 + c = 16$

33. $d + (-6) = 9$

34. $a + (-14) = 15$

35. $-28 + u = -15$

36. $-22 + p = -19$

37. $-14 + m = 0$

38. $16 + n = 0$

39. $h + (-35) = -40$

40. $r + (-29) = -38$

41. $-13 + k = -25$

42. $-36 + x = -53$

For Exercises 43–56, solve.

43. Belinda has a balance of $126 in her bank account. A check written against her account for $245 arrives at the bank.
  a. What is her balance?
  b. Because Belinda has insufficient funds to cover the check amount, the bank assesses a service charge of $20. What is her balance after the service charge?

44. Brian has a credit of $86 on his account with an audio-video store. He uses his credit account to purchase some equipment at a total cost of $585. What is his new balance after the purchase?

45. The revenue for an insurance company in 2004 was $3,568,250 and their total cost were $1,345,680. What was the net? Was it a profit or loss?

46. As of 2004, *Titanic* was the top-grossing movie of all time, grossing $1,835,400,000 in total box office sales in the world. If the budget for making the film was $200,000,000, what was the profit? (*Source: Box Office Report*, June 2004)

47. Florence spent a total of $26,458 for her car, including the cost of financing and the cost of maintenance. Five years later, she sells the car for $4500. What is her net? Is it a profit or loss?

48. Gary bought a Fender Stratocaster guitar in 1960 for $150. In 2004, he takes the guitar to a guitar show and sells it for $12,000. What was the net? Is this a profit or loss?

DISCUSSION How would inflation affect the answer to Exercises 43 and 44?

49. The temperature at sunset was reported to be 19°F. By midnight it is reported to be −27°F. What is the amount of the decrease?

50. Liquid nitrogen has a temperature of $-208°C$. An orange is placed into the liquid nitrogen that causes the liquid to boil at $-196°C$. What is the amount of the increase in temperature?

51. Molecular motion is minimized at $-273°C$. Oxygen goes from liquid to solid at $-218°C$. How much colder does liquid oxygen need to get for its molecular motion to be minimized?

52. A container of liquid hydrogen fluoride is at a temperature of $-90°C$. The boiling point for hydrogen fluoride is $19°C$. How much must the temperature rise for the hydrogen fluoride to boil?

53. Derrick has a balance of $-\$37$ in his bank account. To avoid further charges, he must have a balance of $30. What is the minimum he can deposit to avoid further charges?

54. Danielle owns a business that is in financial trouble. Her accountant informs her that her net worth is $-\$5267$. Her net worth should be at least $2500 by the end of the month, otherwise she will be forced to close her business. What is the minimum profit she must clear in order to remain in business?

55. The lowest level of the Sears Tower in Chicago is at an elevation of -43 feet. The skydeck is at an elevation of 1353 feet. What is the distance between the lowest level and the skydeck?

56. As a result of initial expenses, a new business has a net worth of $-\$45,000$. The goal for the first year is to make enough profit to have a net worth of $15,800. How much must the business make in that first year to achieve its goal?

Review Exercises

[1.3] **1.** Multiply. $(145)(209)$

[1.3] **2.** Evaluate. $(3)^5$

[1.7] **3.** Jacqueline owns 20 square miles of land that borders a national park. The government wishes to purchase an 8-mile-long by 2-mile-wide rectangular strip of her land. How much land will she have left if she sells?

[1.4] **4.** Find the missing factor: $3 \cdot y = 24$

[1.4] **5.** Evaluate. $\sqrt{225}$

Multiplying, Dividing, and Solving Equations with Integers

OBJECTIVE 1 Multiply integers.

To discover the rules for multiplying and dividing signed numbers, we will look for a pattern in the following list of products. We'll keep the first factor the same and gradually decrease the other factor until it becomes negative. What happens to the product as we decrease the factor?

OBJECTIVES

1 Multiply integers.

2 Evaluate numbers in exponential form.

3 Divide integers.

4 Solve equations containing a missing factor.

5 Evaluate square roots.

6 Solve applications involving multiplication and division.

$2 \cdot 4 = 8$ ◀ **Note:** From our work with whole numbers in Chapter 1, we know that when we multiply two positive numbers, the product is positive.

$2 \cdot 3 = 6$ ◀ Each time we decrease the second factor by 1, we lose one of the repeatedly added 2's so that the product decreases by 2.

$2 \cdot 2 = 4$

$2 \cdot 1 = 2$ When we decrease the 1 factor to 0, the product must continue to decrease by 2. This is why $2 \cdot 0$ must equal 0, which affirms the rule that any number multiplied by 0 is 0.

$2 \cdot 0 = 0$ ◀

$2 \cdot (-1) = -2$ ◀ As we continue, the pattern must continue. Notice that when we decrease the 0 factor to -1, we must continue to decrease the product by 2. This explains why multiplying $2 \cdot (-1)$ must equal -2, and so on.

$2 \cdot (-2) = -4$

From this pattern, we can conclude that multiplying a positive number by a negative number gives a negative product. The commutative property of multiplication tells us that rearranging the factors does not affect the product.

Both $2 \cdot (-2)$ and $(-2) \cdot 2$ equal -4.

Therefore, multiplying a negative number times a positive number also gives a negative product.

RULE

When multiplying two numbers that have different signs, the product is negative.

Example 1 Multiply.

a. $8 \cdot (-6)$

Answer: -48

b. $-12 \cdot 5$

Answer: -60

Do Your Turn 1 ▷

Your Turn 1

Multiply.

a. $9 \cdot (-7)$

b. $-8 \cdot 10$

c. $-4(5)$

d. $6(-3)$

Answers to Your Turn 1
a. -63 b. -80 c. -20
d. -18

Now let's consider multiplying two numbers with the same sign. We have already seen that the product of two positive numbers is positive. To discover the rule for multiplying two negative numbers, we will look for a pattern in another list of products. We begin with a negative number times a positive number and gradually decrease the positive factor until it becomes negative.

$$(-2) \cdot 4 = -8$$ ◀ We already have established that the product of a negative number and a positive number is negative.

$$(-2) \cdot 3 = -6$$ ◀ As we decrease the positive factor from 4 to 3, the product actually increases from -8 to -6 which is an increase of 2.

$$(-2) \cdot 2 = -4$$

$$(-2) \cdot 1 = -2$$

$$(-2) \cdot 0 = 0$$

$$(-2) \cdot (-1) = 2$$ ◀ To continue the same pattern, when we decrease the 0 factor to -1, we must continue to increase the product by 2. This means the product must become positive.

$$(-2) \cdot (-2) = 4$$

We already knew that when we multiply two positive numbers, the product is positive. Now we've discovered that when we multiply two negative numbers, the product is positive as well.

RULE

When multiplying two numbers that have the same sign, the product is positive.

Your Turn 2

Multiply.

a. $-5 \cdot (-8)$

b. $(-4)(-9)$

c. $-10(-13)$

Example 2 Multiply.

a. $-6 \cdot (-8)$

Answer: 48

b. $(-9)(-7)$

Answer: 63

◀ **Do Your Turn 2**

We can now develop rules for multiplying more than two signed numbers. To discover those rules, let's look at some examples where we use the order of operations agreement, which tells us to multiply from left to right.

Example 3 Multiply.

a. $-2(-3)(-4)$

Solution: We can work our way from left to right.

$$-2(-3)(-4)$$
$$= 6(-4) \qquad \text{Multiply } -2 \text{ by } -3 \text{ to get positive 6.}$$
$$= -24 \qquad \text{Multiply 6 by } -4 \text{ to get } -24.$$

Answer: -24

b. $-1(-1)(-1)(-5)$

Solution: We can work our way from left to right.

$$-1(-1)(-1)(-5)$$
$$= 1(-1)(-5) \qquad \text{Multiply } -1 \text{ by } -1 \text{ to get positive 1.}$$
$$= -1(-5) \qquad \text{Multiply 1 by } -1 \text{ to get } -1.$$
$$= 5 \qquad \text{Multiply } -1 \text{ by } -5 \text{ to get positive 5.}$$

Answer: 5

CONNECTION Multiplying by -1 is like finding additive inverses. For example, finding the additive inverse of $-(-(-(-5)))$ is the same as the multiplication problem $-1(-1)(-1)(-5)$. Just as we counted the total number of negative signs to work through the longer additive inverse problems, we do the same thing when multiplying.

Answers to Your Turn 2
a. 40 b. 36 c.130

Example 3 suggests the following rules.

RULES *When multiplying signed numbers, count the total number of negative factors:*

If there are an *even* number of negative factors, then the product is *positive*.

If there are an *odd* number of negative factors, then the product is *negative*.

Do Your Turn 3 ▶

OBJECTIVE 2 Evaluate numbers in exponential form.

Because exponents mean repeated multiplication, we can now extend what we've learned to simplifying exponential forms with negative bases.

Example 4 Evaluate.

a. $(-5)^2$

Solution: $(-5)^2 = (-5)(-5) = 25$

Explanation: The exponent 2 means that we multiply the base -5 by itself. Because we are multiplying two negative numbers, the product is positive.

Conclusion: Any time the exponent is an even number with a negative base, we will be multiplying an even number of negative factors so the product will be positive.

b. $(-2)^3$

Solution: $(-2)^3 = (-2)(-2)(-2) = -8$

Explanation: The exponent 3 means that we multiply the base -2 as a factor three times. Multiplying the first two -2's, gives positive 4, which multiplies the last -2 to get -8. Because we have an odd number of negative factors, the result is negative.

Conclusion: Any time the exponent is an odd number with a negative base, we will be multiplying an odd number of negative factors so the product will be negative.

RULES *When evaluating an exponential form that has a negative base:*

If the exponent is even, the product is positive.

If the exponent is odd, the product is negative.

Do Your Turn 4 ▶

Your Turn 3

Multiply.

a. $(-4)(-6)(-2)$

b. $(-1)(-1)(-2)(-5)$

c. $-5(-7)(-2)(-1)$

d. $-3(-5)(2)$

e. $4(-4)(3)(-1)(-1)$

Your Turn 4

Evaluate.

a. $(-2)^5$

b. $(-2)^6$

Answers to Your Turn 3
a. -48 b. 10 c. 70 d. 30
e. -48

Answers to Your Turn 4
a. -32 b. 64

WARNING The expressions $(-2)^4$ and -2^4 are different.

The parentheses in $(-2)^4$ indicate that the base number is a negative number.

$(-2)^4$ is read as "negative 2 raised to the fourth power."

Solution: $(-2)^4 = (-2)(-2)(-2)(-2) = 16$

In -2^4, we view the sign as an additive inverse. We must calculate 2^4 first with the positive base, then find the opposite of the result!

-2^4 is read as "the additive inverse of the fourth power of 2."

Solution: $-2^4 = -[2 \cdot 2 \cdot 2 \cdot 2] = -16$

Alternatively, we can think of additive inverses as multiplying by -1, so we can think of -2^4 as $-1 \cdot 2^4$. The order of operations agreement says we should compute exponents before multiplying. So we calculate 2^4 first, which is 16, then multiply by -1 to get its additive inverse, -16.

<table>
<tr><td>

Your Turn 5

Evaluate.

a. -7^2

b. -4^3

</td><td>

Example 5 Evaluate. -3^4

Solution: $-3^4 = -[3 \cdot 3 \cdot 3 \cdot 3] = -81$

Explanation: -3^4 means find "the additive inverse of the fourth power of 3." 3 raised to the fourth power is 81, and the additive inverse of 81 is -81.

◁ **Do Your Turn 5**

</td></tr>
</table>

OBJECTIVE 3 Divide integers.

The sign rules for division are the same as for multiplication. We can see this by looking at the relationship between division and multiplication. Recall that we can use division to find missing factors. Consider the following statement.

$$-2n = 6$$

◀ **Note:** Remember $-2n$ means $-2 \cdot n$.

We have learned that a negative number multiplied by another negative number gives a positive product. This means that the unknown factor must be negative, that is, -3.

We can write the missing factor statement as a related division statement.

$$-2n = 6 \quad \text{can be written as} \quad n = 6 \div (-2) = -3 \quad \text{or} \quad n = \frac{6}{-2} = -3$$

Note: We will usually use the form $n = \dfrac{6}{-2}$

Conclusion: When dividing a positive number by a negative number, the quotient is negative.

What about a negative number divided by a positive number? Let's consider another missing factor statement.

$$3x = -12$$

We have learned that multiplying two numbers with different signs gives a negative product. Since our known factor is positive 3, the unknown factor must be negative, that is, -4.

Answers to Your Turn 5
a. -49 b. -64

Once again, let's write the missing factor statement as a division statement.

$$3x = -12$$
$$x = \frac{-12}{3}$$
$$x = -4$$

Conclusion: When dividing a negative number by a positive number the quotient is negative.

What if the dividend and divisor are both negative? Consider the following missing factor statement.

$$-4y = -20$$

We know by the rules of multiplication that a negative factor must be multiplied by a positive factor in order to get a negative product. This means that the missing factor here must be positive 5. Let's look at the related division sentence.

$$-4y = -20$$
$$y = \frac{-20}{-4}$$
$$y = 5$$

Conclusion: When dividing two negative numbers, the quotient is positive.

In summary, we can conclude that the rules for dividing signed numbers are the same as for multiplying.

| **RULES** *When multiplying or dividing two integers:* |
| If they have the same sign, the result is positive. |
| If they have different signs, the result is negative. |

Your Turn 6

Divide.

a. $36 \div (-4)$

b. $-40 \div 8$

c. $\dfrac{-28}{-7}$

d. $\dfrac{-54}{9}$

e. $\dfrac{-72}{-12}$

f. $\dfrac{-18}{0}$

> **Example 6**   Divide.
>
> a. $-45 \div 9$
>
> **Answer:** -5
>
> **Explanation:** We are dividing two numbers with different signs, so the quotient is negative.
>
> b. $\dfrac{-18}{-6}$
>
> **Answer:** 3
>
> **Explanation:** We are dividing two numbers with the same sign, so the quotient is positive.
>
> c. $-15 \div 0$
>
> **Answer:** undefined.
>
> **Explanation:** When the divisor is 0, the quotient is undefined.

Do Your Turn 6 ▷

Answers to Your Turn 6
a. -9 b. -5 c. 4 d. -6
e. 6 f. undefined

OBJECTIVE 4 Solve equations containing a missing factor.

In Section 1.4, we learned how to solve equations involving a missing factor. Let's recall the procedure.

PROCEDURE *To solve for a missing factor, write a related division sentence, dividing the product by the known factor.*

Example 7 Solve and check.

a. $9x = -63$

Solution: Write a related division sentence.

$$9x = -63$$
$$x = \frac{-63}{9}$$
$$x = -7$$

Note: Because we are dividing two numbers with different signs, the quotient is negative.

Check: Verify that -7 can replace x in $9x = -63$ and make the equation true.

$$9(-7) \stackrel{?}{=} -63$$
$$-63 = -63 \quad \textbf{True, so } -7 \textbf{ checks.}$$

b. $-30y = -60$

Solution: Write a related division sentence.

$$-30y = -60$$
$$y = \frac{-60}{-30}$$
$$y = 2$$

Note: Because we are dividing two numbers with the same sign, the quotient is positive.

Check: Verify that 2 can replace y in $-30y = -60$ and make the equation true.

$$-30(2) \stackrel{?}{=} -60$$
$$-60 = -60 \quad \textbf{True, so 2 checks.}$$

c. $0n = 19$

Solution: Write a related division sentence.

$$0n = 19$$
$$n = \frac{19}{0}, \text{ which is undefined.}$$

Because $\dfrac{19}{0}$ is undefined, there is no solution.

◀ Do Your Turn 7

OBJECTIVE 5 Evaluate square roots.

In Chapter 1, we learned that the square root of a given number is a number that, when squared, equals the given number. For example, a square root of 25 is 5 because $5^2 = 25$. But there is a second square root. Notice $(-5)^2 = 25$ also, so 25 actually has two square roots, 5 and -5. This suggests the following rule for square roots.

RULE

Every positive number has two square roots, a positive root and a negative root.

Your Turn 7

Solve and check.

a. $7n = -42$

b. $-12k = -36$

c. $-8h = 96$

d. $0m = -14$

Answers to Your Turn 7
a. -6 b. 3 c. -12
d. no solution

Example 8 Find all square roots of the given number.

a. 81

Answers: 9 and -9

> **Note:** We can express both the positive and negative root more concisely as ± 9.

Explanation: There are two square roots because $9^2 = 81$ and $(-9)^2 = 81$.

b. -16

Answer: no integer solution

Explanation: There is no integer that can be squared to equal a negative number because squaring any number, whether positive or negative, always produces a positive product.

> **DISCUSSION** If a negative number has no integer square root, why not say the square root of a negative number is undefined?

Do Your Turn 8 ▷

Your Turn 8

Find all square roots of the numbers.

a. 100

b. 25

c. -36

Now let's revisit the radical sign. In Section 1.4, we learned that the radical symbol means to find the *principal* square root, which is the positive square root. For example, $\sqrt{49} = 7$ because 7 is the positive square root of 49.

To indicate the negative square root of a number using a radical sign, we place a minus sign in front of the radical. For example, $-\sqrt{49}$ means to find the additive inverse of the principle square root of 49, so $-\sqrt{49} = -(7) = -7$.

Example 9 Simplify.

a. $\sqrt{144}$

Answer: 12

Explanation: The radical sign means the positive square root.

b. $-\sqrt{9}$

Answer: -3

Explanation: The negative sign in front of the radical sign indicates to find the additive inverse of the positive square root of 9, so $-\sqrt{9} = -(3) = -3$

c. $\sqrt{-36}$

Answer: $\sqrt{-36}$ is not an integer.

Explanation: There is no positive number that can be squared to equal -36.

The following rules summarize what we have learned about using the radical sign.

> **RULES**
>
> $\sqrt{n}$ means to find the principal square root of n, which is its positive square root.
> $-\sqrt{n}$ means to find the additive inverse of the principal square root of n, which is the negative square root of n.
> The square root of a negative number is not an integer.

Your Turn 9

Simplify.

a. $\sqrt{121}$

b. $-\sqrt{225}$

c. $\sqrt{-81}$

Answers to Your Turn 8
a. ± 10 b. ± 5
c. no integer solution

Answers to Your Turn 9
a. 11 b. -15
d. no integer solution

Do Your Turn 9 ▷

An oceanographer programs a small submarine to take readings every −18 feet. After seven readings, what will be the submarine's position?

Example 10 Jean went traveling for three months. Each month while she was away, −$19 in finance charges was posted to her account. What were her total finance charges upon her return?

Understand: Since Jean was charged the same amount each month for three months, we are repeatedly adding the same amount. Repeated addition means we can multiply.

Plan: Multiply.

Execute: $-19 \cdot 3 = -57$

Answer: Jean's total finance charges are $57.

Check: Reverse the process.

$$-57 \div 3 \stackrel{?}{=} -19$$
$$-19 = -19$$

◀ Do Your Turn 10

A ceiling grid is suspended from the roof of a building by 50 cables so that its weight (downward force) of 6000 pounds is evenly distributed among the 50 cables. What is the force on each cable?

Example 11 Rick's total debt is $12,600. If he makes one payment of equal amount each month for five years, what is his monthly debt payment?

Understand: Since the $12,600 is debt, we write it as $-12,500$. Making one payment each month for five years means he makes $5(12) = 60$ payments over the five years.

Plan: Since he is splitting the amount into equal payments, we divide.

Execute: $-12,600 \div 60 = -210$

Answer: Rick will pay $210 each month.

Check: Reverse the process.

$$-210 \cdot 60 \stackrel{?}{=} -12,600$$
$$-12,600 = -12,600$$

◀ Do Your Turn 11

Answer to Your Turn 10
−126 ft.

Answer to Your Turn 11
−120 lb.

1. When multiplying or dividing two numbers that have the same sign, the result is_____.

2. When multiplying or dividing two numbers that have different signs, the result is_____.

3. If the base of an exponential form is a negative number and the exponent is even, then the product is_____.

4. If the base is a negative number and the exponent is odd, the product is_____.

5. Every positive integer has _____ square roots.

6. What does $\sqrt{n}$ mean?

7. What does $-\sqrt{n}$ mean?

8. The square root of a negative number is _____.

For Exercises 9–34, multiply.

9. $-2 \cdot 16$

10. $-4 \cdot 11$

11. $-7 \cdot 9$

12. $-12 \cdot 6$

13. $4 \cdot (-2)$

14. $6 \cdot (-4)$

15. $13 \cdot (-5)$

16. $15 \cdot (-3)$

17. $0 \cdot (-9)$

18. $0 \cdot (-5)$

19. $(-1) \cdot (-32)$

20. $(-4) \cdot (-17)$

21. $-13 \cdot 0$

22. $-4 \cdot 0$

23. $15 \cdot (-1)$

24. $-12 \cdot (-2)$

25. $-21 \cdot (-8)$

26. $-9 \cdot (-18)$

27. $19(-20)$

28. $(-14)(31)$

29. $(-1)(-5)(-7)$

30. $(-3)(-6)(-1)$

31. $(-1)(-1)(-6)(-9)$

32. $(-8)(7)(-2)(-1)(-1)$

33. $(-20)(-2)(-1)(3)(-1)$

34. $(5)(-4)(-2)(-1)(3)$

For Exercises 35–54, evaluate.

35. $(-1)^2$ **36.** $(-1)^3$ **37.** $(-3)^2$ **38.** $(-2)^2$

39. $(-7)^2$ **40.** $(-12)^2$ **41.** $(-4)^3$ **42.** $(-5)^3$

43. $(-3)^4$ **44.** $(-2)^4$ **45.** $(-2)^6$ **46.** $(-3)^5$

47. -8^2 **48.** -11^2 **49.** -10^6 **50.** -10^4

51. -1^2 **52.** -1^3 **53.** -3^2 **54.** -2^2

For Exercises 55–72, divide.

55. $36 \div (-3)$ **56.** $-48 \div 8$ **57.** $-81 \div 27$ **58.** $40 \div (-10)$

59. $-32 \div (-4)$ **60.** $-100 \div (-20)$ **61.** $0 \div (-2)$ **62.** $0 \div 12$

63. $31 \div (-1)$ **64.** $-14 \div (-1)$ **65.** $\dfrac{65}{-13}$ **66.** $\dfrac{-96}{8}$

67. $\dfrac{-41}{0}$ **68.** $\dfrac{27}{0}$ **69.** $\dfrac{-124}{-4}$ **70.** $\dfrac{-91}{-7}$

71. $-\dfrac{28}{4}$ **72.** $-\dfrac{42}{7}$

For Exercises 73–92, solve and check.

73. $4x = 12$ **74.** $7p = 28$ **75.** $9x = -18$

76. $5x = -125$ **77.** $-12t = 48$ **78.** $-14g = 42$

79. $-6m = -54$ **80.** $-8n = -32$ **81.** $-2a = -24$

82. $-5k = -35$ **83.** $0v = -14$ **84.** $0b = 28$

85. $-1c = 17$ **86.** $-1c = -12$ **87.** $-18m = 0$

88. $-25b = 0$ **89.** $-2(-5)d = -50$ **90.** $3(-8)f = 72$

91. $-1(-1)(-7)g = -63$ **92.** $(-9)(-1)(2)w = -90$

For Exercises 93–103, evaluate each radical.

93. $\sqrt{81}$ **94.** $\sqrt{36}$ **95.** $\sqrt{64}$

96. $\sqrt{100}$ **97.** $\sqrt{49}$ **98.** $\sqrt{25}$

99. $\sqrt{-169}$ **100.** $\sqrt{-144}$ **101.** $-\sqrt{121}$

102. $-\sqrt{256}$ **103.** $\sqrt{0}$ **104.** $\sqrt{1}$

For Exercises 105–110, solve.

105. Rohini had a balance of $-\$214$ on her credit card in June. By November, her balance had tripled. What was her balance in November?

106. Michael has five credit accounts. Three of the accounts have a balance of $-\$100$ each. The other two accounts each have a balance of $-\$258$. What is Michael's total debt?

107. During one difficult year, the Morrisons had insufficient funds in their checking account on seven occasions, and each time, they were assessed a charge appearing as $-\$17$ on their statement. What were the total insufficient funds charges that year?

108. An oil drilling crew estimated the depth to an oil pocket to be -450 feet. After drilling to three times that depth, they finally found oil. At what depth did they find oil?

109. Alicia borrowed $1656 from her friend and promised to pay her back with equal monthly payments for 3 years. How much should each payment be?

110. Liam borrowed $25,200 from his parents to pay for college. After graduation he will make monthly payments for 6 years. How much should each of his payment be?

Review Exercises

[1.2] **1.** A wallpaper border is to be placed on the walls just below the ceiling around a room that is 14 feet wide by 16 feet long. How much border paper is needed?

[1.7] **2.** Sherry wishes to paint a wall in her house. The wall is 15 feet long and 10 feet high and has a window that is 4 feet long by 3 feet wide. How many square feet must she be prepared to paint?

[1.7] **3.** What is the volume of the object shown?

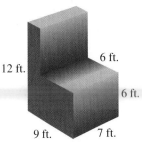

12 ft.   6 ft.   6 ft.   9 ft.   7 ft.

[1.5] **4.** Simplify. $19 - 6 \cdot 3 + 2^3$

[1.5] **5.** Simplify. $7\sqrt{64} - 2(5 + 3)$

2.5 Order of Operations

OBJECTIVE

1 Simplify numerical expressions by following the order of operations agreement.

OBJECTIVE 1 Simplify numerical expressions by following the order of operations agreement.

The same order of operations agreement applies to all arithmetic. Of course, with integers, it is important to remember the sign rules while following the proper order of operations. Let's recall the order of operations.

ORDER OF OPERATIONS AGREEMENT

Perform operations in the following order:

1. Grouping Symbols: parentheses ( ), brackets [ ], braces { }, absolute value | |, radicals $\sqrt{}$, and fraction bars.
2. Exponents.
3. Multiplication or division from left to right, in order as they occur.
4. Addition or subtraction from left to right, in order as they occur.

Your Turn 1

Simplify.

a. $-24 + (-9) \cdot 3$

b. $32 - 16 \div (-4)$

c. $-28 \div 4 + (-15)$

Your Turn 2

Simplify.

a. $18 - 48 \div (-6) \cdot (-3)$

b. $20 + (-3) \cdot 4 - (-6)$

c. $-7 \cdot 5 - (-30) \div 5$

d. $12 \cdot (-4) \div (-6) + 9 - 27$

Your Turn 3

Simplify.

a. $4 - (3 - 5) \cdot 7$

b. $12 \div [4 - (-2)] + (-4)$

c. $-4 + 5(7 - 4)$

Answers to Your Turn 1
a. -51 b. 36 c. -22

Answers to Your Turn 2
a. -6 b. 14 c. -29 d. -10

Answers to Your Turn 3
a. 18 b. -2 c. 11

Example 1 Simplify. $-8 + 6 \cdot (-2)$

Solution: $-8 + 6 \cdot (-2)$

$\qquad = -8 + (-12)$ Multiply first. $6(-2) = -12$

$\qquad = -20$ Add. $-8 + (-12) = -20$

> **Note.** Write steps below the equal sign instead of outward to the right.

CONNECTION Compare Example 1 with the first example in Section 1.5, and you will see that it is essentially the same problem. The only difference here is that some of the numbers are negative.

◁ **Do Your Turn 1**

Example 2 Simplify. $-16 - 20 \div (-4) \cdot 6$

Solution: $-16 - 20 \div (-4) \cdot 6$

$\qquad = -16 - (-5) \cdot 6$ Divide. $20 \div (-4) = -5$

$\qquad = -16 - (-30)$ Multiply. $-5 \cdot 6 = -30$

$\qquad = -16 + 30$ Write the subtraction as an equivalent addition.

$\qquad = 14$ Add. $-16 + 30 = 14$

> **Note:** We divide before multiplying here because we multiply or divide from left to right in the order they occur.

◁ **Do Your Turn 2**

Example 3 Simplify. $(12 - 4) \div (-2) + 1$

Solution: $(12 - 4) \div (-2) + 1$

$\qquad = 8 \div (-2) + 1$ Subtract inside the parentheses. $12 - 4 = 8$

$\qquad = -4 + 1$ Divide. $8 \div (-2) = -4$

$\qquad = -3$ Add. $-4 + 1 = -3$

◁ **Do Your Turn 3**

Example 4 Simplify. $-|-24 - 5(-2)|$

◀ **Note:** Absolute value symbols are grouping symbols.

Solution: $-|-24 - 5(-2)|$

$= -|-24 - (-10)|$ Multiply inside the absolute value symbols. $5(-2) = -10$

$= -|-24 + 10|$ Write the subtraction as an equivalent addition.

$= -|-14|$ Add. $-24 + 10 = -14$

$= -14$ Find the absolute value of -14, which is 14. Then find the additive inverse of 14, which is -14.

Do Your Turn 4 ▷

Your Turn 4

Simplify.

a. $|-4 - (6)(-5)|$

b. $-|25 + (-34)|$

c. $-2|48 \div (-6)|$

Example 5 Simplify. $(-1 + 6) - 2^3 + (-2)$

Solution: $(-1 + 6) - 2^3 + (-2)$

$= 5 - 2^3 + (-2)$ Add inside the parentheses. $-1 + 6 = 5$

$= 5 - 8 + (-2)$ Evaluate the expression with exponents. $2^3 = 8$

$= -3 + (-2)$ Subtract. $5 - 8 = -3$

$= -5$ Add. $-3 + (-2) = -5$

Do Your Turn 5 ▷

Your Turn 5

Simplify.

a. $(-4)^2 + (2 - 7) - (-9)$

b. $5^2 \div (-1 + 6) + 1$

c. $2 - [9 - (-2)] + (-1)^3$

Example 6 Simplify. $8 - (-3)^2 + (-15) \div \sqrt{25}$

Solution: $8 - (-3)^2 + (-15) \div \sqrt{25}$

$= 8 - 9 + (-15) \div 5$ Evaluate the exponential and root expressions. $(-3)^2 = 9$ and $\sqrt{25} = 5$.

$= 8 - 9 + (-3)$ Divide. $-15 \div 5 = -3$

$= -1 + (-3)$ Subtract. $8 - 9 = -1$

$= -4$ Add. $-1 + (-3) = -4$

Do Your Turn 6 ▷

Your Turn 6

Simplify.

a. $14 + (-2)^3 - 8 \cdot (-3)$

b. $(-5) \cdot 6 \div (-3) - \sqrt{49}$

c. $3^3 - 4 \cdot (-8) + \sqrt{81} - |24 \div (-8)|$

Note: A number next to parentheses means multiply.

Note: Recall that to write a subtraction as an equivalent addition, change the operation from $-$ to $+$ and change the subtrahend to its additive inverse.

Example 7 Simplify. $\sqrt{36} \div 3 + 2(-1 - 4)$

Solution: $\sqrt{36} \div 3 + 2(-1 - 4)$

$= \sqrt{36} \div 3 + 2(-1 + (-4))$ In the parentheses, write subtraction as equivalent addition.

$= \sqrt{36} \div 3 + 2(-5)$ Add in the parentheses. $-1 + (-4) = -5$

$= 6 \div 3 + 2(-5)$ Evaluate the root. $\sqrt{36} = 6$

$= 2 + (-10)$ Multiply and divide working from left to right. $6 \div 3 = 2$ and $2(-5) = -10$

$= -8$ Add. $2 + (-10) = -8$

Do Your Turn 7 ▷

Your Turn 7

Simplify.

a. $[-2 - (-5)] + [1 + (-3)] \cdot \sqrt{4}$

b. $\sqrt{16} \div 4 - 2(4 - 7)$

c. $12 \div \sqrt{9} + 3(-2 + 4)$

Answers to Your Turn 4
a. 26 b. -9 c. -16

Answers to Your Turn 5
a. 20 b. 6 c. -10

Answers to Your Turn 6
a. 30 b. 3 c. 65

Answers to Your Turn 7
a. -1 b. 7 c. 10

Simplify.

a. $4(9 - 15) + 16 \div [2 + (-10)]$

b. $-20 + 6(3 - 8) - [8 + (-5)]^2$

c. $|19 - 29| \div 2 + [14 - 2(-6)] \div (-13)$

Example 8 Simplify. $2(8 - 12) + (-2)^3 - [20 + (-2)] \div (-6)$

Solution: $2(8 - 12) + (-2)^3 - [20 + (-2)] \div (-6)$

$= 2(-4) + (-2)^3 - 18 \div (-6)$	**Calculate within parentheses first.** $8 - 12 = -4$ **and** $20 + (-2) = 18$
$= 2(-4) + (-8) - 18 \div (-6)$	**Evaluate the expression with the exponent.** $(-2)^3 = -8$
$= -8 + (-8) - (-3)$	**Multiply and divide, working from left to right.** $2(-4) = -8$ **and** $18 \div (-6) = -3$
$= -16 - (-3)$	**Add.** $-8 + (-8) = -16$
$= -16 + 3$	**Write the subtraction as an equivalent addition.**
$= -13$	**Add.** $-16 + 3 = -13$

◀ **Do Your Turn 8**

Example 9 Simplify.

a. $\{15 - 4[2 + (-5)]\} - 40 \div \sqrt{9 + 16}$

Solution: $\{15 - 4[2 + (-5)]\} - 40 \div \sqrt{9 + 16}$

$= \{15 - 4[-3]\} - 40 \div \sqrt{9 + 16}$	**Work inside the innermost grouping symbols first.** $2 + (-5) = -3$
$= \{15 - [-12]\} - 40 \div \sqrt{9 + 16}$	**Multiply within the bracket.** $4 \cdot -3 = -12$
$= \{15 + 12\} - 40 \div \sqrt{9 + 16}$	**Within the bracket, write subtraction as addition.**
$= 27 - 40 \div \sqrt{9 + 16}$	**Finish the bracket.** $15 + 12 = 27$
$= 27 - 40 \div \sqrt{25}$	**Add within the square root.** $9 + 16 = 25$
$= 27 - 40 \div 5$	**Evaluate the root.** $\sqrt{25} = 5$
$= 27 - 8$	**Divide.** $40 \div 5 = 8$
$= 19$	**Subtract.** $27 - 8 = 19$

Note: To find the square root of a sum or difference, we must add or subtract first, then find the square root of the sum or difference.

Simplify.

a. $\{-8 + 3[24 \div (4 + (-10))]\} + (-3)\sqrt{25 \cdot 4}$

b. $(-5)^3 - 4[(-6 + 14) \div (-7 + 6)] + \sqrt{81 \div 9}$

c. $\{[16 - 3(5 + 3)] + [26 - (3 - 9)^2]\} - 4\sqrt{25 - 9}$

d. $\{|-16(-3) - 40| + 7\} \div 3[-6 + \sqrt{121}]^2$

b. $5[(9 - 15) \div (-2)] - \sqrt{16 \cdot 9}$

Solution: $5[(9 - 15) \div (-2)] - \sqrt{16 \cdot 9}$

$= 5[(-6) \div (-2)] - \sqrt{16 \cdot 9}$	**Work with innermost grouping symbols.** $9 - 15 = -6$
$= 5[3] - \sqrt{16 \cdot 9}$	**Divide.** $(-6) \div (-2) = 3$
$= 5[3] - \sqrt{144}$	**Multiply in the radical.** $16 \cdot 9 = 144$
$= 5[3] - 12$	**Evaluate the radical.** $\sqrt{144} = 12$
$= 15 - 12$	**Multiply.** $5[3] = 15$
$= 3$	**Subtract.** $15 - 12 = 3$

Answers to Your Turn 8
a. -26 b. -59 c. 3

Note: We could have performed some of the calculations in the same step. For example, we could have computed $5[3]$ and $16 \cdot 9$ in the same step. However, combining steps increases the likelihood of making careless mistakes.

Answers to Your Turn 9
a. -50 b. -90 c. -34 d. 125

◀ **Do Your Turn 9**

Example 10 Simplify. $\dfrac{8^2 - (-16)}{8 - 4(4 - 6)^3}$

Note: Recall that the fraction bar means to divide. We must simplify the top (numerator) and bottom (denominator) separately before dividing.

Solution: $\dfrac{8^2 - (-16)}{8 - 4(4 - 6)^3}$

$= \dfrac{64 - (-16)}{8 - 4(-2)^3}$ Work the top and bottom separately. In the top: $8^2 = 64$. At the bottom: $4 - 6 = -2$.

$= \dfrac{64 + 16}{8 - 4(-8)}$ In the top: $64 - (-16) = 64 + 16$. At the bottom: $(-2)^3 = -8$.

$= \dfrac{80}{8 - (-32)}$ In the top: $64 + 16 = 80$. At the bottom: $4(-8) = -32$.

$= \dfrac{80}{8 + 32}$ Write subtraction as equivalent addition. $8 - (-32) = 8 + 32$

$= \dfrac{80}{40}$ Add. $8 + 32 = 40$

$= 2$ Divide. $80 \div 40 = 2$

Do Your Turn 10 ▷

Your Turn 10

Simplify.

a. $\dfrac{(2 - 14) + 4^3}{(-3)(-4) - 5^2}$

b. $\dfrac{2 \cdot (-4) - 7^2}{9 \cdot 5 - 3(3 - 7)^2}$

Answers to Your Turn 10
a. -4 b. 19

For Extra Help

Videotape DVT

Tutor Center
Addison-Wesley Tutor Center

Math XL

MyMathLab

Student Solutions Manual

1. List the four stages of the order of operations.

2. What is the first step in simplifying the expression $12 + 5 \cdot 2$? Explain.

3. What is the first step in simplifying the expression $2(3 - 8) + 9$? Explain.

4. What is the first step in simplifying the expression $12 \div (-3) \cdot 4 - 20$? Explain.

For Exercises 5–84, simplify using order of operations.

5. $3 - 10 \div 2$

6. $-8 \div 2 + 4$

7. $9 + 4 \cdot (-6)$

8. $15 - 5 \cdot 4$

9. $3 - 2 \cdot 4 + 11$

10. $2 + 16 \div 4 - 9$

11. $9 + 20 \div (-4) \cdot 3$

12. $-30 \div 2 \cdot (-3) + 21$

13. $3 - (-2)^2$

14. $(-3)^2 - 8$

15. $-5 + (-4)^2 - 1$

16. $5 - (-5)^2 + 13$

17. $9^2 - 6 \cdot (-4) \div 8$

18. $-14 - (-3)^2 \cdot 4 + 5$

19. $3 + (4 - 6^2)$

20. $-2 - (5 - 2^2)$

21. $3 + (4 - 6)^2$

22. $-2 - (5 - 2)^2$

23. $\sqrt{25 - 9}$

24. $\sqrt{16 + 9}$

25. $\sqrt{25} - \sqrt{9}$

26. $\sqrt{16} + \sqrt{9}$

27. $(-2)^4$

28. -2^4

29. -2^5

30. $(-2)^5$

31. $-(-3)^2$

32. $-(-2)^3$

33. $[-(-3)]^2$

34. $[-(-2)]^3$

35. $9 \div 3 - (-2)^2$

36. $4^3 - 3(-2)$

37. $3 - (-2)(-3)^2$

38. $2 - (-5)(-4)^2$

39. $|18 - 5(-4)|$

40. $|21 \div (-3) + 12|$

41. $-|-43 + 6 \cdot 4|$

42. $-|-16 - (-3)(-7)|$

43. $-5|26 \div 13 - 7 \cdot 2|$

44. $-4|3^2 + (-5)(6)|$

45. $(-3)^2 - 2[3 - 5(1 + 4)]$

46. $5[4 + 3(-2 + 1)] - (-4)^2$

47. $28 \div (-7) + \sqrt{49} + (-3)^3$

48. $25 - 6^2 + (-52) \div \sqrt{169}$

49. $12 + 8\sqrt{81} \div (-6)$

50. $-5\sqrt{36} + 40 \div (-5) \cdot 2 - 14$

51. $4^2 - 7 \cdot 5 + \sqrt{121} - 21 \div (-7)$

52. $(-2)^3 + 3\sqrt{25} - 18 \div 3$

53. $15 + (-3)^3 + (-5)|14 + (-2)9|$

54. $-28 \div 4 + |-3(-2)(-5)| - 2^5$

55. $(-15 + 12) + 5(-6) - (9 - (-12)) \div 3$

56. $39 \div 3 + (24 - 30) - 5^2 + (-21 - (-13))$

57. $(-2) \cdot (5 - 8)^2 \div 6 + (4 - 2)$

58. $(5 - 7)^2 \div (-2) - (-3) \div 3$

59. $-|38 - 14 \cdot 2| + 44 \div (5 - (-6)) - 2^4$

60. $-3\sqrt{49} + |16 \div (-8)| - (28 - 40) \div (-3)$

61. $[19 - 2(4 + (-9))] - 18 \div \sqrt{25 - 16}$

62. $[12(-2 - (-5)) - 40] \div [17 + (-21)]$

63. $[(-14 + 2) + 5] \div 0 + 9^4 \cdot 7$

64. $[19(-2) - (-18)] \div [15 - 5(2 - (-1))]$

65. $\{(-3)^2 + 4[(8 - 20) \div (-2)]\} + (-5)\sqrt{25 \cdot 4}$

66. $4\sqrt{16 \cdot 9} - \{(-4)^3 + 2[18 \div (-2) + (4 - (-2))]\}$

67. $[30 - 3(-4)] \div [(15 - 6) + (-2)] + (5 - 3)^4$

68. $\{[13 - 3(1 + 6)] + [18 - (5 - 2)^2]\} - 6\sqrt{25 - 9}$

69. $\{|-12(-5) - 38| + 2\} \div 3[-9 + \sqrt{49}]^3$

70. $-3\{14 - 2|20 - 7(4)|\} + [(3)(-9) - (-21)]^2$

71. $\{2[14 - 11] + \sqrt{(4)(-9)}\} - [12 - 6 \cdot 9]$

72. $5^3 - [19 + 4\sqrt{16 - 25}] + |41 - 50|^2$

73. $\dfrac{6(-4) + 16}{(4 + 7) - 9}$

74. $\dfrac{34 - 5(2)}{7 + (9 - 4)}$

75. $\dfrac{3(-12) + 1}{3^2 - 2}$

76. $\dfrac{20 + 12(-3)}{3^2 - 1}$

77. $\dfrac{(5 - 19) + 3^3}{(-2)(-6) - 5^2}$

78. $\dfrac{2^4 + 3(7 + 19)}{8^2 - (2 \cdot 8 + 1)}$

79. $\dfrac{3 \cdot (-6) - 6^2}{2(2 \quad 5)^2}$

80. $\dfrac{(-2)^3 + 5(7 - 15)}{4 \cdot 11 - 2(5 - 9)^2}$

81. $\dfrac{2[5(3 - 7) + (-3)^3] - 6}{15 \cdot 5 - (12 - 7)^2}$

82. $\dfrac{[6(12 - 15) + 5(-6)] - 4^2}{10 - 2(6 - 9)^2}$

83. $\dfrac{3[14 - 2(9)] + (5 - 11)^2}{(9 - 3)^2 - 4(10 - 1)}$

84. $\dfrac{-|14 - 3^2|^2}{3(-12) + 11}$

85. Explain the difference between $(-2)^4$ and -2^4.

86. Explain the difference between $-(-3)^2$ and $[-(3)]^2$.

87. Why do $(-2)^5$ and -2^5 both simplify to -32?

88. Why is the answer to Exercise 64 undefined?

For Exercises 89–96, explain the mistake in each problem, then work it correctly.

89. $28 - 5(24 - 30)$

$= 28 - 5(-6)$

$= 23(-6)$

$= -138$

90. $24 \div 2 \cdot 3 + 5$

$= 24 \div 6 + 5$

$= 4 + 5$

$= 9$

91. $4 - (9 - 4)^2$

$= 4 - (9 - 16)$

$= 4 - (-7)$

$= 4 + 7$

$= 11$

92. $19 - (-2)^5$

$= 19 - 32$

$= -13$

93. $34 - [3 \cdot 5 - (14 + 8)]$

$= 34 - 3 \cdot 5 - 22$

$= 34 - 15 - 22$

$= 19 - 22$

$= -3$

94. $\dfrac{3^2 - (25 - 4^2)}{2(-7)}$

$= \dfrac{3^2 - (25 - 16)}{-14}$

$= \dfrac{3^2 - 9}{-14}$

$= \dfrac{9 - 9}{-14}$

$= \dfrac{0}{-14}$, which is undefined.

95. $\sqrt{169 - 25}$

$= 13 - 5$

$= 8$

96. $-2[16(3 - 5) + 7]$

$= -2[16(-2) + 7]$

$= -2[32 + 7]$

$= -2[39]$

$= -78$

Review Exercises

[2.2] **1.** Felicia has a balance of $185 in her checking account. If she writes checks for $45, $68, and $95 what is her new balance?

[2.3] **2.** Tina has a television that she bought for $245. She sells it at a yard sale for $32. What was the net? Was it a profit or loss?

[2.3] **3.** A submarine at -98 feet ascends to -25 feet. How much did the submarine ascend?

[2.4] **4.** Jeff's credit card account has a balance of $0. If he uses the card to purchase three dining chairs that cost $235 each, what is the new balance?

2.6 Applications and Problem Solving

OBJECTIVE 1 Solve problems involving net.

Recall that the net profit or loss is calculated using the formula $N = R - C$, where N represents the net amount, R represents revenue, and C represents cost.

OBJECTIVES

1 Solve problems involving net.

2 Solve problems involving force.

3 Solve problems involving voltage.

4 Solve problems involving average rate.

Example 1 Cassif puts $2000 down on a car, then makes monthly payments of $275 for 5 years (60 months). After 6 years, he sells the car for $5200. If he spent a total of $1500 in routine maintenance and minor repairs, what is his net? Is it a profit or loss?

Understand: To calculate the net we need total revenue and total cost. The formula for net is $N = R - C$. Cassif's total revenue is the amount that he sold the car for, which was given to be $5200. His total cost is the sum of all the money he spent on the car.

Note: Recall that we developed the problem-solving process in Section 1.7.

Plan: Calculate total cost, then subtract the cost from the total revenue to get the net.

Execute: cost = amount down + total of all payments + maintenance costs

(60 payments of $275)

$$
\begin{aligned}
\text{cost} &= 2000 &+& 60(275) &+& 1500 \\
\text{cost} &= 2000 &+& 16,500 &+& 1500 \\
\text{cost} &= \$20,000
\end{aligned}
$$

Now use $N = R - C$ to get the net.

$N = R - C$

$N = 5200 - 20,000$ **Replace *R* with 5200 and *C* with 20,000.**

$N = -14,800$

Answer: Cassif's net is $-$14,800, which is a loss of $14,800.

Check: Reverse the process. The net added to cost should produce the revenue.

$$-14,800 + 20,000 \overset{?}{=} 5200$$
$$5200 = 5200 \quad \textbf{The net amount checks.}$$

Subtracting the maintenance costs and total of all payments from the total cost should produce the down payment.

$$20,000 - 1500 - 16,500 \overset{?}{=} 2000$$
$$2000 = 2000 \quad \textbf{The total cost checks.}$$

Dividing the total of all the payments by 60 should produce the amount of each payment.

$$16,500 \div 60 \overset{?}{=} 275$$
$$275 = 275 \quad \textbf{The total payments check.}$$

Do Your Turn 1 ▷

Your Turn 1

Shawna put $1200 down when she bought her car. She made 48 payments of $350 and spent $2500 in maintenance and repairs. Four years after paying off the car she sells it for $4400. What is her net? Is it a profit or loss?

OBJECTIVE 2 Solve problems involving force.

Many situations can be described mathematically using formulas. Force situations and voltage situations can be described with formulas that involve multiplication.

Answer to Your Turn 1

$-$$16,100; loss

Isaac Newton discovered a relationship for forces. Objects experience a force when they are accelerated. The formula for force is

$$F = ma.$$

F represents force, m represents mass, and a represents acceleration. In the American system of measurement, force is measured in pounds (lb), mass in slugs, and acceleration in feet per second per second ($ft./sec.^2$). In the metric system, force is measured in newtons (N), mass is measured in kilograms (kg), and acceleration is measured in meters per second per seconds ($m/sec.^2$).

Mass is a measure of how much material or matter (atoms and molecules) make up an object. Acceleration is a measure of how quickly an object increases or decreases its speed. For example, when you measure how quickly a car can go from 0 miles per hour to 60 miles per hour, you are measuring its acceleration.

A force that we all experience is the force due to gravity. Galileo Galilei discovered that all objects fall at the same rate. This seems counter to what we'd expect; however, it is easy to demonstrate.

Try this: Hold your pencil or pen in one hand and your book in the other at the same height above the floor. It should be pretty obvious to you that the book weighs more. Now drop them at the same time. Notice that they both hit the floor at the same time!

The Earth accelerates all objects downward at the same rate, about 32 feet per second per second. Because this acceleration is downward we write the acceleration as -32 $ft./sec.^2$, which means, in each second that an object falls toward the ground, gravity increases its speed by 32 ft./sec. So when you drop objects like a book or pencil, after 1 second they have a speed of 32 ft./sec., after 2 seconds they have a speed of 64 ft./sec., and so on.

The weight you feel when holding an object is the force due to gravity trying to accelerate the object toward the ground. Just think, you're holding back all the gravitational force that Earth can muster! Newton's law tells us that to calculate that force, we multiply the mass of the object by the acceleration constant -32 $ft./sec.^2$. The metric value of the acceleration due to gravity is about -10 $m/sec.^2$, which means -10 meters per second per second.

Example 2   A person has a mass of 5 slugs. What is his weight?

Understand: Weight is a force so we can use Newton's formula $F = ma$. The acceleration due to gravity is about -32 $ft./sec.^2$.

Plan: Replace m with 5 and a with -32 in $F = ma$, then multiply.

Execute: $F = ma$
$F = (5)(-32)$
$F = -160$

Note: The units for force could be represented as the product of mass units and the acceleration units. In the American system, slug · $ft./sec.^2$ is called pound (lb.). In the metric system, kg · $m/sec.^2$ is called newton (N). We will learn more about these units in Chapter 7.

Answer: The person weighs 160 lb.; the negative sign means that the force is directed downward.

Check: Reverse the process. Divide the force, -160 pounds, by the acceleration, -32 $ft./sec.^2$, to see if we get the mass, 5 slugs.

$$-160 \div -32 \overset{?}{=} 5$$
$$5 = 5$$

DISCUSSION   What if the man were on another planet? Would his mass change? Would his weight change?

Your Turn 2

The blue whale is the largest animal on Earth. The mass of an average blue whale is about 2625 slugs. Find the weight.

Answer to Your Turn 2
84,000 lb.

◁ **Do Your Turn 2**

Example 3 A dumbbell is measured to have a weight of 20 newtons. What is its mass?

Understand: Because weight is a force directed downward, we write 20 newtons as -20 newtons. We use Newton's law, where $F = ma$. Since newton is the metric unit of force, we must remember to use the metric value for the acceleration due to gravity, which is about -10 m/sec.^{2}.

Plan: Replace F with -20 and a with -10 in $F = ma$, then solve for m.

Execute:

$$F = ma$$
$$-20 = m(-10)$$
$$-20 \div (-10) = m$$
$$2 = m$$

Note: We have a missing factor equation. To find the unknown factor, m, we must divide the product, -20, by the known factor, -10.

Answer: The dumbbell has a mass of 2 kg.

Check: Verify that a dumbbell with a mass of 2 kilograms would have a weight of -20 newtons.

$$F = ma$$
$$-20 \stackrel{?}{=} (2)(-10)$$
$$-20 = -20 \qquad \text{It checks.}$$

Do Your Turn 3 ▶

Your Turn 3

Find the mass of a 17,800-newton elephant.

OF INTEREST

Everything that has mass also has gravity. The more massive an object, the more gravitational pull it has. The Moon is less massive than Earth, so it does not have as much gravitational pull as Earth. This is why you would weigh less on the Moon than on Earth.

Example 4 An elevator has a mass of 38 slugs and holds 4 people with the following masses: 4 slugs, 5 slugs, 6 slugs, and 3 slugs. What is the total force exerted on the cable?

Understand: We must calculate the total force on the cable exerted by the elevator and its occupants. The formula for force is $F = ma$. To calculate the downward force or weight, we need the total mass of the elevator and all its occupants. The acceleration due to gravity is -32.

Plan: Find the total mass of the elevator and all its occupants, then use $F = ma$ to calculate the force.

Execute:
$$\text{total mass} = \text{elevator} + \text{all occupants}$$
$$\text{total mass} = 38 + 4 + 5 + 6 + 3$$
$$\text{total mass} = 56 \text{ slugs}$$

To find the force or weight, replace m with 56 and a with -32 in $F = ma$, then multiply.

$$F = ma$$
$$F = (56)(-32)$$
$$F = -1792$$

Answer: The force is -1792 lb.

Check: Reverse the process. The force, -1792 pounds, divided by acceleration due to gravity, -32 ft./sec.^{2}, should equal the total mass, which is 56 slugs.

$$-1792 \div -32 \stackrel{?}{=} 56$$
$$56 = 56 \qquad \text{The force checks.}$$

Subtracting each person's mass from the total mass should equal the mass of the elevator.

$$56 - 3 - 6 - 5 - 4 \stackrel{?}{=} 38$$
$$38 = 38 \qquad \text{The total mass checks.}$$

Do Your Turn 4 ▶

Your Turn 4

A car has a mass of 1080 kilograms. Two passengers are in the car with masses of 68 kilograms and 80 kilograms. What is the total weight of the car with the passengers?

Answer to Your Turn 3
1780 kg

Answer to Your Turn 4
$-12,280$ N

OBJECTIVE 3 Solve problems involving voltage.

Georg Simon Ohm was a German scientist who studied electricity. He is credited with the discovery of the relationship among voltage, current, and resistance. His famous formula is $V = ir$.

The V stands for voltage, which is the electrical pressure in a circuit measured in volts (V). The i stands for current, a measure of electricity moving through a wire, which is measured in amperes or amps (A). The r stands for resistance to the flow of electricity through a wire, which is measured in ohms (named after Mr. Ohm in honor of his discovery). The symbol for ohms is the Greek letter Ω (omega).

Voltage and current can be negative numbers. A voltage source like a battery has $+$ and $-$ connections called terminals. A voltage meter measures voltage and has two probes, one for $+$ and one for $-$ terminals/connections. If we connect each probe to the appropriate terminal on the battery, we measure a positive voltage. If we reverse the probes, we measure a negative voltage. Voltage is positive or negative depending on the orientation of the measuring device.

Current is a measure of the flow of electricity. A current meter is essentially the same as a voltage meter. If we connect the probes, and the meter indicates negative current, the current is flowing in the opposite direction of the orientation of the probes from the meter.

Your Turn 5

The current in a circuit is measured to be 12 amps through a 20-ohm resistor. Find the voltage.

Note: You do not necessarily need to understand what a formula relates to (although it makes the formula more interesting) in order to work with the mathematics. In Example 5, you do not have to know that V means voltage, i means current, and r means resistance. As long as you are given some indication as to where each number goes, you can work with the formula.

Example 5 Calculate the voltage in a circuit that has a resistance of 20 ohms and a current of -9 amps.

Understand: We must calculate voltage given resistance and current. The formula that relates voltage, current, and resistance is $V = ir$.

Plan: Replace i with -9 and r with 20 in $V = ir$, then multiply.

Execute:
$$V = ir$$
$$V = (-9)(20)$$
$$V = -180$$

CONNECTION The formulas $F = ma$, $V = ir$, and $A = lw$ are all mathematically the same. They all involve the product of two factors. Many different phenomena have the same mathematical relationship.

Answer: The voltage is -180 V.

Check: Reverse the process. Divide the voltage, -180 volts, by the resistance, 20 ohms, to see if we get the current, 9 amps.

$$-180 \div 20 \overset{?}{=} -9$$
$$-9 = -9 \quad \text{It checks.}$$

◄ **Do Your Turn 5**

Example 6 The voltage in a circuit is measured to be 120 volts across a 40-ohm resistor. What is the current?

Understand: We must calculate current given voltage and resistance. We use the formula $V = ir$.

Plan: Replace V with 120 and r with 40 in $V = ir$, then solve for i.

Answer to Your Turn 5
240 V

Execute: $V = ir$
          $120 = i(40)$ ◀ Notice that we have a missing factor
     $120 \div 40 = i$ equation. To find the unknown factor, i,
              $3 = i$ we divide the product, 120, by the known
                              factor, 40.

Answer: The current is 3 A.

Check: Use $V = ir$ to verify that a circuit with a current of 3 amps and a resistance of 40 ohms has a voltage of 120 volts.

$$V = (3)(40)$$ In $V = ir$, replace i with 3 and r with 40.
$$V = 120$$ Multiply. It checks.

Do Your Turn 6 ▶

Your Turn 6

The voltage is measured to be -9 volts across a 3-ohm resistor. Find the current.

OBJECTIVE 4 Solve problems involving average rate.

Another situation that has the same mathematical relationship as $F = ma$, $V = ir$, and $A = lw$ is the formula that describes the relationship among distance, rate, and time.

If we drive a car 60 miles per hour for 2 hours, how far do we travel?

60 miles per hour means that we travel 60 miles for each hour we drive, so in 2 hours, we'll go 120 miles. Notice that to find distance, we multiply the rate by the time. In formula form we can say

$$\text{distance} = \text{rate} \cdot \text{time}$$
$$d = rt$$

Consider the fact that in reality we cannot drive a car at an exact speed for any lengthy period of time. Our rate will vary slightly as a result of hills, wind resistance, or even our inability to keep our foot pressure exactly the same on the accelerator. Because rates can vary we usually consider **average rate**. It is a measure of how quickly an object is able to travel a total distance in a total amount of time.

DEFINITION   **Average rate:** A measure of the rate at which an object travels a total distance in a total amount of time.

Example 7   A commuter train travels at an average rate of 30 miles per hour for 2 hours. How far does the train travel?

Understand: The rate was given to be 30 miles per hour. The time was given to be 2 hours. The formula that relates distance, rate, and time is $d = rt$.

Plan: Replace r with 30 and t with 2, then solve for d.

Execute: $d = rt$
          $d = (30)(2)$
          $d = 60$

Answer: The train travels 60 mi.

Check: Reverse the process. Divide the distance, $60 \div 2 \overset{?}{=} 30$
60 miles, by the time, 2 hours, to see if you get the $30 = 30$ It checks.
rate, 30 miles per hour.

Do Your Turn 7 ▶

Your Turn 7

A research submarine is lowered at an average speed of 8 feet per second. What will be the submarine's depth after 45 seconds?

Answer to Your Turn 6
-3 A

Answer to Your Turn 7
360 ft.

On a vacation trip, Candice leaves at 10 A.M. and travels 85 miles then takes a 30-minute break. She then travels another 120 miles and stops for another 30-minute break. Finally, she travels 70 miles and arrives at her destination at 4 P.M. What was her average rate?

Example 8 A bus leaves at 9 A.M. and travels 40 miles then stops for an hour. It then travels 110 miles and arrives at its final destination at 1 P.M. Find the average rate.

Understand: To find average rate we must consider the total distance of the trip and the total time. The trip was broken into two parts. We were given the distances for both parts. We were also given a departure time and arrival time, but there was a 1-hour stop that we must take into account. The formula that relates distance, rate, and time is $d = rt$.

Plan: Find the total distance of the trip and total time of the trip, then use the formula $d = rt$ to solve for r.

Execute: The total distance is $40 + 110 = 150$ miles.

The total time from 9 A.M. to 1 P.M. is 4 hours. However, the bus stopped for an hour so the actual travel time was 3 hours.

We now have the total distance of 150 miles and total time of 3 hours, so we can use $d = rt$.

$$d = rt$$
$$150 = r(3) \qquad \text{Replace } d \text{ with 150 and } h \text{ with 3.}$$
$$150 \div 3 = r \qquad \text{Solve for the missing factor by dividing.}$$
$$50 = r$$

Answer: The average rate was 50 mph.

Check: Use $d = rt$ to verify that if the bus traveled an average of 50 miles per hour for 3 hours, it would travel 150 miles.

$$d = (50)(3) \qquad \text{In } d = rt, \text{ replace } r \text{ with 50 and } t \text{ with 3.}$$
$$d = 150 \qquad \text{Multiply. It checks.}$$

◀ **Do Your Turn 8**

Answer to Your Turn 8
55 mph

For
Extra
Help

Videotape
DVT

Addison-Wesley
Tutor Center

Math XL

MyMathLab

Student Solutions
Manual

1. In the formula $N = R - C$, what does each variable represent?

2. What do positive and negative net amounts indicate?

3. In the formula $F = ma$, what does each variable represent?

4. What are the constants of the acceleration due to gravity for both the American system and metric system?

5. In the formula $d = rt$, what does each variable represent?

6. What is an average rate?

For Exercises 7–38, solve.

 7. Darwin put $800 down when he bought a Buick Century. He made 60 payments of $288 and spent $950 in maintenance and repairs. Three years after paying off the car he sells it for $4000. What is his net? Is it a profit or loss?

 8. In 1955, Malvin put $800 down on a new Chevrolet Bel-Air. He made 24 payments of $18. In 2000, he spends $2500 restoring the car for a classic car show. At the show, someone offers him $20,000 for the car. If he paid approximately $3000 for maintenance over the years, what would be his net if he accepted the offer? Is it a profit or loss?

9. Lynn takes out a loan to buy a fixer-upper house. She then spends $4500 in repairs and improvements. She sells the house for $80,560. If the payoff amount for the loan that she took out to buy the house is $71,484, what is her net? Is it a profit or loss?

10. Scott takes out a loan to buy a fixer-upper house. He spends $5475 in repairs and improvements. He sells the house for $94,200 but pays the new owners $3750 in closing costs. If the payoff for the loan that Scott took out to buy the house is $80,248, what is his net? Is it a profit or loss?

11. A car with a mass of 88 slugs is placed on a hydraulic lift. In the trunk of the car are golf clubs with a mass of 1 slug and a tool box with a mass of 2 slugs. What force must be applied to lift the truck?

12. Four concrete slabs each with a mass of 20 slugs are to be lifted by a crane. What force must be applied to lift the slabs?

13. A ski lift has a seat suspended from a wire above. Three skiers with masses of 70 kilograms, 82 kilograms, and 20 kilograms are riding the lift. The seat itself has a mass of 17 kilograms. What is the total force exerted on the wire above?

14. A car engine with a mass of 157 kilograms is hanging from a lift by a chain. What is the force on the chain?

15. A cable that can hold a maximum of 2500 pounds is to be used to lift 5 steel beams that each have a mass of 17 slugs. Will the rope hold?

16. A concrete slab has a mass of 25 slugs. The slab is being lifted by three cables, each able to support a maximum of 275 pounds. Will these cables be able to lift the slab?

For Exercises 17–21, use the chart of accelerations due to gravity for bodies in our solar system.

Accelerations Due to Gravity for Bodies in the Solar System

Body	Meters per second every second	Feet per second every second
Moon	−2	−5
Sun	−275	−900
Mercury	−4	−13
Venus	−9	−29
Mars	−4	−13
Jupiter	−26	−87
Saturn	−12	−39
Uranus	−10	−32
Neptune	−10	−32
Pluto	−4?	−13? (uncertain)

17. How much would an astronaut with a mass of 6 slugs weigh on Mars?

18. An astronaut has a mass of 54 kilograms. How much would she weigh on Jupiter?

19. How much would a 4-slug space probe weigh on Saturn?

20. How much would a 40-kilogram space probe weigh on Venus?

21. Neil Armstrong weighed 172 pounds on Earth at the time of the Apollo mission to the Moon in 1969. His space suit and backpack added another 180 pounds. How much did Armstrong weigh when he took that "one small step" onto the surface of the Moon?

OF INTEREST

On July 16, 1969, *Apollo 11* carrying Neil Armstrong, Edwin "Buzz" Aldrin, and Michael Collins was launched. On July 20, 1969, Armstrong and Aldrin handed the lunar module on the Moon while Collins piloted the command module in orbit around Moon. Three-and-a-half hours after landing, Neil Armstrong climbed out of the lunar module to become the first human to set foot on the Moon. His famous words upon setting foot on the Moon were: "That's one small step for man, one giant leap for mankind."

22. The *Viking 1* was a probe that landed on the surface of Mars in 1976. The lander weighed about 1280 pounds on Earth. What was its weight on Mars?

23. The blue whale is the largest animal on Earth. The largest blue whale ever caught weighed approximately 1,658,180 newtons. What was the mass of this blue whale?

24. The Eiffel Tower contains approximately 437,500 slugs of iron. What is the weight of the iron in the Eiffel Tower?

OF INTEREST

There were two Viking spacecraft, *Viking 1* and *Viking 2*. *Viking 1* was launched on August 20, 1975, and landed on the surface of Mars on July 20, 1976, becoming the first man-made object to touch the surface. *Viking 2* landed on September 3, 1976. The latest landings were of the Twin Mars Exploration Rovers *Spirit* (January 4, 2004) and *Opportunity* (January 25, 2004), which found evidence of water having existed on Mars at some prior time.

OF INTEREST

The Eiffel Tower, located in Paris, France, is 984 feet tall. It was designed and built by French engineer Alexandre Gustave Eiffel for the 1889 World's Fair.

25. The largest bear on Earth is the polar bear. These animals can weigh up to 8000 newtons, which is a downward force of -8000 newtons. What is the mass of one of these bears?

26. The *Titanic* weighed approximately 92,656,000 pounds, which is a downward force of $-92,656,000$ pounds. What was the mass of the *Titanic*?

> **OF INTEREST**
>
> Upon completion in 1912, the R.M.S *Titanic* became the largest passenger ship of its day at a length of 882 feet. On its maiden voyage from England to the United States, it struck an iceberg in the North Atlantic, which opened a 300-foot gash in its side. The ship sank in less than 3 hours. Only 711 of the 2224 on board survived. The wreck was discovered in 1985.

27. An F-16 pilot weighs 185 pounds. The pilot's plane goes into a sharp turn that causes him to experience six times the force of Earth's gravity (6 g's). How much does the pilot weigh during the turn?

> **OF INTEREST**
>
> The acceleration due to Earth's gravity exerts a force of 1 g. However, if we ride a roller coaster or a jet plane or anything that accelerates more than 32 feet per second every second, we experience more than 1 g. If we accelerate at 64 feet per second every second, we say we experience 2 g's. This means we experience a force that is twice Earth's gravity. If that force is directed downward, then a human body weighs twice as much as normal. If the human body experiences more than about 5 g's, the blood is accelerated out of the brain, causing a blackout. This is why jet pilots wear g-suits. The suit reacts to the g forces and squeezes the pilot's legs and lower body, squeezing the blood back into the brain so that the pilot does not black out.

28. Madeline weighs 130 pounds. During a tight turn on a roller coaster ride she experiences 3 g's. What does she weigh during the turn?

29. An electrical circuit has a resistance of 7 ohms and a current of -9 amps. Find the voltage.

30. An electrical circuit has a resistance of 100 ohms and a current of -12 amps. Find the voltage.

31. The voltage in an electrical circuit measures -220 volts. If the resistance is 5 ohms, find the current.

32. It is suspected that an incorrect resistor was put into a circuit. The correct resistor should be 12 ohms. A voltage measurement is taken and found to be 60 volts. The current is measured to be 10 amps. Is the resistor correct?

33. The elevator in the Empire State Building travels from the lobby to the 80th floor, a distance of about 968 feet, in about 44 seconds. What is the average rate of the elevator?

> **OF INTEREST**
>
> Completed in November 1930, the Empire State Building became the tallest man-made structure at a height of 381 meters with 102 floors. The current tallest structure is the Taipei Financial Center, in Taiwan, built in 2004. It is 509 meters tall and has 101 floors above ground and 5 below.

34. A research submarine is lowered at an average rate of 7 feet per second. What will be the submarine's depth after 29 seconds?

35. The space shuttle travels at a rate of about 17,060 miles per hour while in orbit. How far does the shuttle travel in 3 hours?

36. A ship travels at an average rate of 20 miles per hour. How far will the ship travel in 4 hours?

37. On a vacation trip, Devin leaves at 11 A.M. and travels 105 miles then takes a 30-minute break. He then travels another 140 miles and stops for another 30-minute break. Finally, he travel 80 miles and arrives at his destination at 5 P.M. What was his average speed?

38. On a vacation trip, Corrine leaves at 7 A.M. and travels 152 miles, then takes a 15-minute break. She then travels another 145 miles and stops for 45 minutes to eat lunch. Finally, she travels 135 miles and arrives at her destination at 2 P.M. What was her average speed?

PUZZLE PROBLEM A boat is in a harbor at low tide. Over the side hangs a ladder with its bottom step 6 inches below the surface of the water. The ladder steps are 12 inches apart. If the tide rises at a rate of 8 inches per hour, how many steps will be under water after 5 hours?

Review Exercises

[2.1] **1.** Graph $-(-3)$ on a number line.

For Exercises 2 and 3, simplify.

[2.1] **2.** $-(-(-5))$

[2.1] **3.** $|15|$

For Exercises 4 and 5, solve and check.

[2.3] **4.** $k + 76 = -34$

[2.4] **5.** $-14m = 56$

Defined Terms

Review the following terms, and for any you do not know, study its definition on the page number next to it.

Section 2.1
Integers *p. 00*
Absolute value *p. 00*
Additive inverses *p. 00*

Section 2.3
Net *p. 000*
Cost *p. 000*
Revenue *p. 000*

Profit *p. 000*
Loss *p. 000*

Section 2.6
Average rate *p. 000*

Procedures, Rules, and Key Examples

Procedures/Rules	Key Example(s)

Section 2.1 Introduction to Integers

To graph a number on a number line:
Draw a solid dot or point on the mark for that number.

Example 1: Graph -3 on a number line.

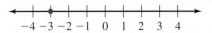
$$-4 \;-3 \;-2 \;-1 \;\;0 \;\;1 \;\;2 \;\;3 \;\;4$$

The absolute value of a number is always positive or zero.
The absolute value of a negative number is always positive.
The absolute value of a positive number is always positive.
The absolute value of zero is zero.

Example 2: Find the absolute value.
a. $|-9| = 9$
b. $|15| = 15$
c. $|0| = 0$

Section 2.2 Adding Integers

To add two integers:
If they have the same sign, we add their absolute values and keep the same sign.

If they have different signs, we subtract their absolute values and keep the sign of the number that has the larger absolute value.

Example 1: Add.
a. $5 + 9 = 14$
b. $-5 + (-9) = -14$
c. $-5 + 9 = 4$
d. $5 + (-9) = -4$

Section 2.3 Subtracting Integers and Solving Equations

To write a subtraction statement as an equivalent addition statement:
1. Change the operation symbol from a minus sign to a plus sign.
2. Change the subtrahend (second number) to its additive inverse.

Example 1: Subtract.
a. $5 - 9 = 5 + (-9) = -4$
b. $-5 - 9 = -5 + (-9) = -14$
c. $5 - (-9) = 5 + 9 = 14$
d. $-5 - (-9) = -5 + 9 = 4$

To find a missing addend:
Write a related subtraction sentence. Subtract the known addend from the sum.

Example 2: Solve.
$$x + 30 = 18$$
$$x = 18 - 30$$
$$x = -12$$

Section 2.4 Multiplying, Dividing, and Solving Equations with Integers

When multiplying or dividing two integers:
If they have the same sign, the result is positive.
If they have different signs, the result is negative.

Example 1: Multiply or divide.
a. $4 \cdot 6 = 24$ e. $18 \div 9 = 2$
b. $(-4)(-6) = 24$ f. $-18 \div (-9) = 2$
c. $4(-6) = -24$ g. $18 \div (-9) = -2$
d. $(-4)(6) = -24$ h. $-18 \div 9 = -2$

When evaluating an exponential form that has a negative base:

If the exponent is even, the product is positive.

If the exponent is odd, the product is negative.

To solve for a missing factor:

Write a related division sentence, dividing the product by the known factor.

Rules for square roots:

Every positive number has two square roots, a positive root and a negative root.

$\sqrt{n}$ means to find the principal square root of n, which is its positive square root.

$-\sqrt{n}$ means to find the additive inverse of the principal square root of n, which is the negative square root of n.

The square root of a negative number is not an integer.

Section 2.5 Order of Operations

Order of operations agreement:

Perform operations in the following order:

1. Grouping symbols: parentheses ( ), brackets [ ], braces { }, absolute value | |, radicals $\sqrt{}$, and fraction bar.

2. Exponents.

3. Multiplication or division from left to right, in order as they occur.

4. Addition or subtraction from left to right, in order as they occur.

Note: Use $\overrightarrow{\text{GEMDAS}}$ to remember the order.

Example 2: Evaluate.

a. $(-2)^4 = (-2)(-2)(-2)(-2) = 16$

b. $(-2)^5 = (-2)(-2)(-2)(-2)(-2)$
$= -32$

c. Tricky case: -2^4 means find the additive inverse of 2 raised to the 4th power.

$-2^4 = -[2 \cdot 2 \cdot 2 \cdot 2] = -16$

Example 3: Solve $6x = -24$

$x = -24 \div 6$

$x = -4$

Example 4: Find all square roots of 64.

Answer: 8 and -8 (or ± 8)

Example 5: Evaluate.

a. $\sqrt{81} = 9$

b. $\sqrt{-25} =$ not an integer

Example 1: Simplify.

$\{(-4)^2 + 5[(9 - 15) \div (-2)]\} +$
$\quad (-3)\sqrt{16 \cdot 9}$
$= \{(-4)^2 + 5[(-6) \div (-2)]\} +$
$\quad (-3)\sqrt{144}$
$= \{16 + 5[3]\} + (-3) \cdot 12$
$= \{16 + 15\} + (-36)$
$= 31 + (-36)$
$= -5$

Formulas

Net:	$N = R - C$
Voltage:	$V = ir$
Force:	$F = ma$
Distance:	$d = rt$

LEARNING STRATEGY

When studying rules, if you are a visual learner, read the rules over and over. If you are an auditory learner, record yourself *saying* the rules (or the rhymes or songs you've made out of them). Then listen to your recording over and over. If you are a tactile learner, write the rules repeatedly.

For Exercises 1–6, answer true or false.

[2.1] **1.** All whole numbers are integers.

[2.1] **2.** All integers are whole numbers.

[2.1] **3.** The additive inverse of any number is always a negative number.

[2.1] **4.** The absolute value of any number is always positive or zero.

[2.2] **5.** The sum of two negative numbers is a positive number.

[2.4] **6.** The product of two negative numbers is a positive number.

For Exercises 7–10, complete the rule.

[2.2] **7.** When two numbers that have different signs are added, we _____ _____ .

[2.3] **8.** To write a subtraction statement as an equivalent addition statement, we _____ _____ _____ .

[2.4] **9.** When two numbers that have different signs are multiplied, the product is _____ .

[2.4] **10.** When two numbers that have different signs are divided, the quotient is _____ .

For Exercises 11 and 12, express each amount as an integer.

[2.1] **11.** The *Titanic* rests at a depth of approximately 13,000 feet.

[2.1] **12.** Water boils at 212°F.

[2.1] **13.** Graph each integer on a number line.

 a. -8

 b. 5

[2.1] **14.** Use $<$ or $>$ to make a true statement.

 a. -15 ? 0

 b. -26 ? -35

 c. 12 ? -41

[2.1] **15.** The absolute value of 41 is ____ .

[2.1] **16.** $|16|$

[2.1] **17.** The additive inverse of 27 is ____ .

[2.1] **18.** $-(-17)$

[2.1] **19.** $(-(-26))$

[2.1] **20.** $-|-12|$

For Exercises 21–44, calculate.

[2.2] **21.** $-16 + 25$

[2.2] **22.** $-12 + (-14)$

[2.2] **23.** $18 + (-30)$

[2.3] **24.** $24 - 31$

[2.3] **25.** $-20 - 17$

[2.3] **26.** $-13 - (-19)$

[2.3] **27.** $-27 - (-22)$

[2.4] **28.** $6 \cdot (-9)$

[2.4] **29.** $-5(-12)$

[2.4] **30.** $(-1)(-3)(-6)(-2)$

[2.4] **31.** $(-5)(-2)(-3)$

[2.4] **32.** $(-3)^4$

[2.4] **33.** $(-10)^3$

[2.4] **34.** $\dfrac{-48}{8}$

[2.4] **35.** $\dfrac{-63}{-9}$

[2.4] 36. $\dfrac{-12}{0}$

[2.4] 37. $\sqrt{-36}$

[2.4] 38. $21 - 4 \cdot 7$

[2.5] 39. $-16 - 28 \div (-7)$

[2.5] 40. $-|26 - 3(-2)|$

[2.5] 41. $13 + 4(6 - 15) + 2^3$

[2.5] 42. $-3\sqrt{49} + 4(2 - 6)^2$

[2.5] 43. $[-12 + 4(15 - 10)] + \sqrt{100 - 36}$

[2.5] 44. $\dfrac{4 + (5 + 3)^2}{3(-13) + 5}$

For Exercises 45–48, solve and check.

[2.3] 45. $14 + x = -27$

[2.3] 46. $n + (-12) = -7$

[2.4] 47. $7k = -63$

[2.4] 48. $(-10)h = -80$

For Exercises 49–59, solve.

[2.2] 49. William's credit card balance on the statement was $-\$1100$. He has made the following transactions. What is his current balance?

CHARGES	
MUSIC SHOP	$23
AUTOMART	$14
BURGER HEAVEN	$9
FINANCE CHARGE	$15
PAYMENT	$900

[2.2] 50. A concrete block is suspended by two cables. The block weighs 600 pounds. Each cable is exerting 300 pounds of upward force. What is the resultant force? What does this mean?

[2.3] 51. The financial report for a business indicates that the total revenue for 2004 was $1,648,200 and the total costs were $928,600. What was the net? Did the business have a profit or loss?

[2.3] 52. The temperature at sunset was reported to be 12°F. By midnight it was reported to be -19°F. What is the amount of the decrease?

[2.3] 53. Arturo has a balance of $-\$45$ in his bank account. To avoid further charges he must have a balance of $25. What is the minimum he must deposit to avoid further charges?

[2.6] 54. A large air-conditioning unit with a mass of 18 slugs is placed on the roof of a building. What is the force on the roof?

[2.6] 55. A circuit has a resistance of 12 ohms and a current of 3 amps. Calculate the voltage.

[2.4] 56. Branford has a debt of $4272. He agrees to make monthly payments for 2 years to repay the debt. How much is each payment?

[2.6] 57. A boulder is estimated to have a mass of 75 slugs. A bulldozer can move up to 2500 pounds of material. Will the bulldozer be able to move the boulder?

[2.6] 58. Jacquelyn drives at an average rate of 65 miles per hour for 3 hours. How far does she travel?

[2.6] 59. Steve begins a trip at 7 A.M. and drives 150 miles. After a 30-minute break, he drives another 110 miles. After another 30-minute break, he drives 40 miles. If he arrives at 1:00 P.M., what was his average rate of speed?

[2.1]   **1.** Graph -5 and 8 on a number line.

1. _____

For Exercises 2–8, calculate.

[2.1]   **2.** $|26|$

2. _____

[2.1]   **3.** $-(-(-18))$

3. _____

[2.2]   **4.** $17 + (-29)$

4. _____

[2.2]   **5.** $-31 + (-14)$

5. _____

[2.3]   **6.** $20 - 34$

6. _____

[2.3]   **7.** $-16 - 19$

7. _____

[2.3]   **8.** $-30 - (-14)$

8. _____

[2.3]   **9.** Solve and check. $-19 + k = 25$

9. _____

[2.3]   **10.** Below is Allison's credit card statement. What is her new balance?

10. _____

Beginning balance = −$487	
Transactions	Amount
Clothing boutique	−$76
Payment	$125
Finance charges	−$14

For Exercises 11–13, calculate.

[2.4]   **11.** $9(-12)$

11. _____

[2.4]   **12.** $(-2)(-6)(-7)$

12. _____

[2.4]   **13.** $\dfrac{-48}{-12}$

13. _____

[2.4]   **14.** Solve and check. $6n = -54$

14. _____

[2.4]   **15.** Evaluate. $(-4)^3$

15. _____

[2.4]   **16.** Evaluate. -2^2

16. _____

17. _____

18. _____

19. _____

20. _____

21. _____

22. _____

23. _____

24. _____

25. _____

26. _____

27. _____

[2.4] **17.** Find all square roots of 81.

[2.4] **18.** During a very difficult month, Natasha overdrafted her account five times. Each overdraft charge appears as $-\$14$ on her statement. What were the total overdraft charges that month?

[2.6] **19.** Howard put \$12,000 down when he bought a Cooper Mini. He made 55 payments of \$89 and spent \$347 on maintenance and repair. Three years later, he sold the car for \$15,500. What is his net? Is it a profit or a loss?

[2.6] **20.** A block of concrete weighing 1380 newtons is lifted by a crane. What is the mass of the block?

[2.6] **21.** Lashanda drives at an average rate of 62 miles per hour. How far will she travel in 2 hours?

For Exercises 22–27, evaluate.

[2.5] **22.** $28 - 6 \cdot 5$

[2.5] **23.** $19 - 4(7 + 2) + 2^4$

[2.5] **24.** $|28 \div (2 - 6)| + 2(4 - 5)$

[2.5] **25.** $[18 \div 2 + (4 - 6)] - \sqrt{49}$

[2.5] **26.** $\dfrac{5^2 + 3(-12)}{36 \div 2 - 18}$

[2.5] **27.** $\dfrac{(-7)^2 + 11}{(-2)(5) + \sqrt{64}}$

For Exercises 1–6, answer true or false.

[2.1] **1.** 0 is an integer.

[2.1] **2.** -6 is a whole number.

[2.4] **3.** $-12 \div 0 = 0$

[1.3] **4.** $3 \cdot 4 + 3 \cdot 5 = 3(4 + 5)$

[2.2] **5.** The sum of a negative number and a positive number is always negative.

[2.3] **6.** $-9 - 12 = -9 + (-12)$

[1.5] **7.** List the order in which we perform operations according to the order of operations agreement.
1.
2.
3.
4.

[2.5] **8.** Explain the mistake in the problem.
$$-14(2) + \sqrt{16 + 9}$$
$$= -14(2) + 4 + 3$$
$$= -28 + 4 + 3$$
$$= -24 + 3$$
$$= -21$$

[1.3] **9.** $9 \cdot (4 \cdot 3) = (9 \cdot 4) \cdot 3$ is an illustration of the _____ property of _____.

[1.2] **10.** Explain the commutative property of addition in your own words and give an example.

[1.1] **11.** Write 5,680,901 in expanded form

[1.3] **12.** Write $5 \times 10^7 + 8 \times 10^5 + 3 \times 10^4 + 6 \times 10^3 + 9 \times 1$ in standard form.

[1.1] **13.** Write the word name for 409,254,006.

[2.1] **14.** Graph -4 on a number line.

For Exercises 15 and 16, use $<$ or $>$ to make a true statement.

[2.1] **15.** 135   ? -450

[2.1] **16.** -930 ? -932

[1.1] [2.1] **17.** Round $-23,410,512$ to the nearest million.

[1.2] **18.** Estimate $49,902 + 6519$ by rounding to the nearest thousand.

For Exercises 19–36, simplify.

[2.1] **19.** $|16|$

[2.1] **20.** $-(-5)$

[2.1] **21.** $-|-8|$

[2.1] **22.** $-(-(-(4)))$

[2.2] **23.** $287 + 48 + (-160) + (-82)$

[2.3] **24.** $-19 - 24$

[2.3] **25.** $-64 - (-14)$

[2.4] **26.** $-14(6)$

[2.4] **27.** $-12(-8)$

[2.4] **28.** $-1(12)(-6)(-2)$

[2.4] **29.** $(3)^5$

[2.4] **30.** $(-2)^6$

[2.4] **31.** -2^6

[1.4] **32.** $\dfrac{1208}{6}$

[2.4] 33. $-105 \div 7$

[2.4] 34. $\sqrt{121}$

[2.5] 35. $-15 - 2[36 \div (3 + 15)]$

[2.5] 36. $-4(2)^3 + [18 - 12(2)] - \sqrt{25 \cdot 4}$

For Exercises 37–40, solve and check.

[1.2] 37. $29 + x = 54$

[2.3] 38. $y + 60 = 23$

[1.4] 39. $26a = 2652$

[2.4] 40. $-18c = 126$

[1.2]
[1.6] 41. Find the perimeter and area.

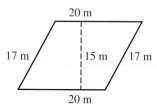

[1.6] 42. Find the volume.

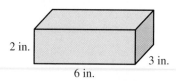

For Exercises 43–50, solve.

[1.2] 43. Shanisse's checking account balance is $45. On the way home from work, she fills up her car for $23, stops by the bank to deposit a check for $62, and buys $57 of groceries. What is her new balance?

[2.6] 44. Sonya put $1500 down on a new car. She made 60 payments of $276 and spent $1450 on maintenance and repairs. Three years after paying off the car, she sold it for $4300. What was her net? Was it a profit or loss?

[1.7] 45. A landscaper needs to know the area of the yard. A plot plan showing the position of the house on the lot is shown. Find the area of the yard.

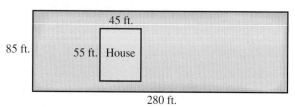

[1.3] 46. A computer translates everything into binary code. This means that numbers are written so that each place value either contains a 1 or a 0 digit. If a certain computer chip has 7 place values (bits), how many different numbers can be coded?

[1.4] 47. Jerry is to administer 300 milliliters of Heparin in an IV drip over the course of an hour. How many milliliters of Heparin should the patient receive each minute?

[2.6] 48. The eider duck generally covers 360 miles in one day of migration from the Arctic to New England. Its average speed is 45 miles per hour. How many hours does the eider duck fly on a migration day?

[2.6] 49. A cargo box that weighs 60,000 pounds is to be loaded onto a ship. What is the mass of the box?

[1.3] 50. Shares in a publishing company sell for $15 each. How much will it cost you to buy 350 shares?

Expressions and Polynomials

66 *The formula for success is simple: Practice and concentration, then more practice and more concentration.* 99

—BABE DIDRIKSON ZAHARIAS,
ATHLETE

66 *It's better to look ahead and prepare, than to look back and regret.* 99

—JACKIE JOYNER-KERSEE,
ATHLETE

66 *No one has ever drowned in sweat.* 99

—LOU HOLTZ,
FOOTBALL COACH

Practice

One of the most important elements to success is practice. To get the most out of your practice or study time, develop a plan and a routine (see the *To the Student* section for a detailed study system). Build into your daily routine blocks of time devoted to practice or study in an environment where you are not likely to be distracted. Some good study places are the library, a tutorial center, or an empty classroom.

To begin your practice or study sessions, the first thing you should do is tidy up your notes (see the Chapter 1 opener). Next, with your notes handy for reference, work through the assigned exercises. Once you've completed the assigned exercises, shift your focus to studying. Study your notes until you know the rules and procedures. Prepare as if every day is a test day.

3.1 Evaluating Expressions

OBJECTIVE 1 Differentiate between an expression and an equation.

OBJECTIVES

1 Differentiate between an expression and an equation.

2 Evaluate expressions.

In Chapters 1 and 2, we began developing concepts of algebra such as variables, constants, equations, and formulas. Recall the definition of an **equation.**

DEFINITION Equation: A mathematical relationship that contains an equal sign.

The formulas we developed, such as $F = ma$ and $V = lwh$, are equations because they have equal signs. Notice that the equal sign separates an equation into two sides, the left side and the right side. If we focus attention on the right side of $F = ma$, we see ma. Because ma by itself does not contain an equal sign, we cannot call it an equation. Instead, we call ma an **expression.**

DEFINITION Expression: A collection of constants, variables, and operations.

$F = ma$ is an equation, whereas ma is an expression.
$4 + 7 = 11$ is an equation, whereas $4 + 7$ is an expression.
$V = lwh$ is an equation, whereas lwh is an expression.

Note: Expressions or equations that contain only constants are called *numeric*, whereas those that contain variables are called *algebraic*.

Think of equations as complete sentences and expressions as phrases or incomplete sentences. In grammar, we learn that a complete sentence must have a subject and a verb. The verb in an equation is the equal sign. Because expressions do not have an equal sign, they do not have a verb, and therefore are incomplete.

Your Turn 1

Determine whether each of the following is an expression or equation and explain why.

a. $E = mc^2$

b. $19 - 2(3 + 8)$

c. $\dfrac{\sqrt{b^2 - 4ac}}{2a}$

d. $y = mx + b$

Example 1 Determine whether each of the following is an expression or equation and explain why.

a. $16 \cdot 5 = 80$

Solution: $16 \cdot 5 = 80$ is an equation because it has an equal sign. More specifically, it is a *numeric* equation because it does not contain variables.

b. $d = rt$

Solution: $d = rt$ is an equation because it contains an equal sign. More specifically, it is an *algebraic* equation because it contains variables.

c. rt

Solution: rt is an expression because it does not contain an equal sign. More specifically, it is an *algebraic* expression because it contains variables.

◀ Do Your Turn 1

Answers to Your Turn 1
a. equation; it contains an equal sign
b. expression; it does not contain an equal sign
c. expression; it does not contain an equal sign
d. equation; it contains an equal sign

OBJECTIVE 2 Evaluate expressions.

There are two actions we can perform with an expression.

1. Evaluate

2. Rewrite

In this section, we'll focus on evaluating. The rest of this chapter will deal with rewriting.

PROCEDURE *To evaluate an expression:*

1. Replace the variables with the corresponding given values.
2. Calculate using the order of operations agreement.

Example 2 Evaluate $x^2 + 5xy - 7$ when $x = -3$ and $y = 2$.

Solution: $x^2 + 5 \ x \ y - 7$

$(-3)^2 + 5(-3)(2) - 7$ — Replace *x* with −3 and *y* with **2**.

$= 9 + 5(-3)(2) - 7$ — Calculate. Simplify the exponential form first.

$= 9 + (-30) - 7$ — Multiply.

$= -21 - 7$ — Add. 9 + (−30) = −21

$= -28$ — Subtract. −21 − 7 = −28

WARNING When we replace a variable with a value, it is important to keep the operations the same. The most consistent way to ensure this is to always use parentheses.

Let's look at x^2. If we replace *x* with −3, we have $(-3)^2$, which, when simplified, becomes $(-3)(-3) = 9$. If we did not use the parentheses, we would have -3^2, which means $-[3 \cdot 3] = -9$.

Now, let's look at $5xy$. Using parentheses when we replace *x* with −3 and *y* with 2 gives us $5(-3)(2)$. Notice that if we did not use parentheses, we would have $5 - 32$.

Do Your Turn 2 ▷

Example 3 Evaluate $-x + 5(x + y)$ when $x = -3$ and $y = 1$.

Solution: $-(-3) + 5[(-3) + (1)]$ — Replace *x* with −3 and *y* with **1** using parentheses.

$= -(-3) + 5[-2]$ — Add inside the brackets. −3 + 1 = −2

$= 3 + (-10)$ — Simplify. −(−3) = 3 and 5U−2V = −10

$= -7$ — Add.

Do Your Turn 3 ▷

Example 4 Evaluate $|r^2 - p| + 4r$ when $r = -2$ and $p = 6$.

Solution: $|(-2)^2 - (6)| + 4(-2)$ — Replace *r* with −2 and *p* with **6**, using parentheses.

$= |4 - (6)| + 4(-2)$ — Evaluate exponents inside the absolute value. $(-2)^2 = 4$

$= |-2| + 4(-2)$ — Calculate inside the absolute value. 4 − (6) = −2

$= 2 + 4(-2)$ — Evaluate the absolute value. |−2| = 2

$= 2 + (-8)$ — Multiply. 4(−2) = −8

$= -6$ — Add.

Do Your Turn 4 ▷

Your Turn 2

Evaluate.

a. $5x - 7y + 2; x = 4,$ $y = -3$

b. $x^2 - 4x + y^3; x = -2,$ $y = -1$

c. $2mn - m^2 + 3np;$ $m = 3, n = -4,$ $p = -2$

Your Turn 3

Evaluate.

a. $2(x + y) - y; x = 4,$ $y = -3$

b. $-x + y(x - y);$ $x = -6, y = 2$

Your Turn 4

Evaluate.

a. $-3x + |2x - xy|;$ $x = 5, y = 4$

b. $-2n^2 - |6np|; n = -3,$ $p = 2$

Answer to Your Turn 2
a. 43 b. 11 c. −9

Answers to Your Turn 3
a. 5 b. −10

Answers to Your Turn 4
a. −5 b. −54

Evaluate.

a. $-4a + \sqrt{2a - b}$; $a = 9$, $b = 2$

b. $2xy + \sqrt{x - y}$; $x = 3$, $y = 7$

Example 5 Evaluate $7\sqrt{m + n} + 5mn$ when $m = -6$ and $n = 2$.

Solution: $7\sqrt{-6 + 2} + 5(-6)(2)$ Replace *m* with −6 and *n* with 2.

$= 7\sqrt{-4} + (-30)(2)$ Add within the radical.

Because the expression contains the square root of a negative number, the answer is not an integer.

Note: When a radical contains a sum or difference we treat the sum or difference as if there were parentheses around them.

◀ **Do Your Turn 5**

Your Turn 6

Evaluate.

a. $\dfrac{12 - x}{5 - x}$; $x = -2$

b. $\dfrac{x + y}{2x - y}$; $x = 3$, $y = 6$

c. $\dfrac{2x - y}{3x - 4}$; $x = -2$, $y = -4$

Example 6 Evaluate $\dfrac{3h^2}{k + 5}$ when $h = 2$ and $k = -5$.

Solution: $\dfrac{3(2)^2}{(-5) + 5}$ Replace *h* with 2 and *k* with −5, using parentheses.

$= \dfrac{3(4)}{0}$ Calculate the top and bottom expressions separately.

$= \dfrac{12}{0}$, which is undefined

Explanation: When we replace k with -5, the divisor expression equals zero. Any time zero is a divisor the expression is undefined or indeterminate (0 dividend).

◀ **Do Your Turn 6**

Recall the division properties of zero that we developed in Chapter 1.

DIVISION PROPERTIES

When zero is the divisor with any dividend other than zero, the quotient is undefined.

In math language: $n \div 0$, or $\dfrac{n}{0}$, is undefined, when $n \neq 0$.

When zero is the dividend, the quotient is zero as long as the divisor is not also zero.

In math language: $0 \div n = \dfrac{0}{n} = 0$, when $n \neq 0$.

If both dividend and divisor are zero, the quotient is indeterminate.

In math language: $0 \div 0$, or $\dfrac{0}{0}$, is indeterminate.

Your Turn 7

Find the value(s) that can replace the variable and cause the expression to be undefined.

a. $\dfrac{2x + 1}{7 - x}$

b. $\dfrac{-9}{(x + 2)(x - 3)}$

Example 7 Find the value(s) that can replace the variable and cause $\dfrac{2x}{x + 6}$ to be undefined.

Solution: In a division expression, if the divisor ever becomes 0, then the whole expression is undefined. In this case the divisor is $x + 6$, so we must find the value for x that causes $x + 6$ to become 0. If we replace x with -6, then $x + 6$ becomes $-6 + 6$, which is 0. Therefore, if $x = -6$, the expression is undefined.

◀ **Do Your Turn 7**

Answers to Your Turn 5
a. -32 b. not an integer

Answers to Your Turn 6
a. 2 b. undefined c. 0

Answers to Your Turn 7
a. $x = 7$ b. $x = -2$ or $x = 3$

For
Extra
Help

Videotape
DVT

Addison-Wesley
Tutor Center

Math XL

MyMathLab

Student Solutions
Manual

1. What is the differnce between an equation and an expression?

2. How do you evaluate an expression?

3. When is a division expression undefined?

4. When is a division expression indeterminate?

For Exercises 5–14, indicate whether each of the following is an expression or an equation.

5. $19 + 3 \cdot 2 = 25$

6. $|14 - 29| = 15$

7. $13 + 5\sqrt{169}$

8. $\dfrac{19 + 9}{4 - 2^3}$

9. $25 - x^2$

10. $x^2 + 5xy - 2 = 7$

11. $x + y = 9$

12. $\sqrt{1 - v^2/c^2}$

13. $9 + 2(x + 3) = 14 - x$

14. $mx + b$

For Exercises 15–52, evaluate the expression using the given values.

15. $4a - 9;\ a = 4$

16. $3b + 12;\ b = 2$

17. $3x + 8;\ x = -5$

18. $5y - 7;\ y = -2$

19. $-2m + n;\ m = 4,\ n = 7$

20. $3z - 2w;\ z = 5,\ w = 2$

21. $2x - 4y;\ x = 3,\ y = -2$

22. $3p - 2q;\ p = 4,\ q = -3$

23. $3y - 5(y + 2);\ y = 4$

24. $-2x + 3(x - 4);\ x = -2$

25. $t^2 - 2t + 4;\ t = 3$

26. $p^2 + 4p - 2;\ p = 2$

27. $a^2 + 9a - 2;\ a = -2$

28. $c^2 - 3c + 7;\ c = -3$

29. $m^2 + 5n - 3;\ m = -3,\ n = 1$

30. $a^2 + 4b + 9;\ a = -1,\ b = 2$

31. $3t^2 - 4u + 1;\ t = -2,\ u = 4$

32. $2r^2 + 4s - 4;\ r = -3,\ s = 5$

33. $r^3 - 4rt;\ r = -3,\ t = -1$

34. $6xy - x^3;\ x = -2,\ y = 3$

35. $b^2 - 4ac;\ b = -4,\ a = -3,\ c = 2$

36. $b^2 - 4ac;\ b = 3,\ a = -5,\ c = -1$

37. $|x^2 + y|$; $x = 2$, $y = -10$

38. $|r - p^3|$; $r = -7$, $p = -2$

39. $-|4a + 7b|$; $a = -5$, $b = 2$

40. $-|5x| + |y^3|$; $x = 6$, $y = -3$

41. $-6m - |n^3|$; $m = -3$, $n = -2$

42. $-3|c - 9cd|$; $c = 3$, $d = -1$

43. $\sqrt{m} + \sqrt{n}$; $m = 144$, $n = 25$

44. $\sqrt{x} - \sqrt{y}$; $x = 25$, $y = 16$

45. $\sqrt{m + n}$; $m = 144$, $n = 25$

46. $\sqrt{x - y}$; $x = 25$, $y = 16$

47. $\sqrt{x^2 + y^2}$; $x = -3$, $y = 4$

48. $\sqrt{mn} - 5n^2$; $m = 27$, $n = 3$

49. $\dfrac{3x + 5}{7 - 2y}$; $x = 10$, $y = 1$

50. $\dfrac{2p - 6}{3 + 4r}$; $p = -8$, $r = 2$

51. $\dfrac{-7m + n^2}{4m + 2n}$; $m = -5$, $n = 1$

52. $\dfrac{-6uv + 14}{3u - v^2}$; $u = -1$, $v = 4$

For Exercises 53–58, find the value(s) that can replace the variables and cause the expression to be undefined.

LEARNING STRATEGY

If you find that you are unsure of how to work a particular problem, look for examples in the text and your notes relating to that problem. If you still cannot figure it out, mark the problem and move on. When you've completed all that you can on your own, seek help to complete the exercises.

53. $\dfrac{x + 3}{10 - x}$

54. $\dfrac{12}{t + 9}$

55. $\dfrac{16}{y}$

56. $\dfrac{-11}{(x + 4)(x - 5)}$

57. $\dfrac{a}{(4 + a)(7 - a)}$

58. $\dfrac{n + 2}{(n - 6)(n - 1)}$

Review Exercises

[2.3] **1.** Rewrite $3 - 7 - 9$ as addition.

[1.1] **2.** What digit is in the thousands place in 472,603?

[1.3] **3.** Write $3 \cdot 10^5$ in standard form.

[1.1] **4.** Write 279 in expanded form.

[1.3] **5.** What is the exponent for 7? Explain.

[1.3] **6.** Evaluate 2^6.

3.2 Introduction to Polynomials

OBJECTIVE 1 Identify monomials.

OBJECTIVES

1. Identify monomials
2. Identify the coefficient and degree of a monomial.
3. Identify like terms.
4. Identify polynomials and their terms.
5. Identify the degree of a polynomial.
6. Write polynomials in descending order.

As we saw in Section 3.1, there are many types of expressions. We will now focus on expressions known as *polynomials*. Polynomials are algebraic expressions that are similar to whole numbers written in expanded notation.

For example, in expanded form, we see that 279 is like the expression $2x^2 + 7x + 9$.

Expanded form, written with base 10: $\quad 279 = 2 \cdot 10^2 + 7 \cdot 10 + 9$

$\qquad\qquad\qquad\qquad\qquad\qquad\qquad\quad \downarrow \qquad\quad \downarrow$

Polynomial form, written with base x: $\qquad 2x^2 \quad + \quad 7x \quad + 9$

The expression $2x^2 + 7x + 9$ is a *polynomial*. In order to define polynomial, we must first define **monomial** or **term**. Think about the prefixes involved. *Poly* means many and *mono* means one.

DEFINITION **Monomial** or **term:** An algebraic expression that is a constant, or a product of a constant and variables that are raised to whole-number powers.

Example 1 Is the expression a monomial? Explain.

a. $5x$

Explanation: $5x$ is a monomial because it is a product of a constant, 5, and a variable, x, that has an exponent of 1, which is a whole number.

From Example 1a, we can make the following conclusion:

Conclusion: Any number or variable with no apparent exponent has an understood exponent of 1.

$$5x = 5^1 x^1$$

We write $5x$ instead of $5^1 x^1$ because it is simpler; that is, it has fewer symbols.

b. $-4xy^3$

Explanation: $-4xy^3$ is a monomial because the constant -4 and variables xy^3 are all multiplied and the variables all have whole-number exponents. As we saw in Example 1a, we could write $-4xy^3$ as $(-4)^1 x^1 y^3$. Both of the variables' exponents, 1 and 3, are whole numbers.

c. x^6

Explanation: x^6 is a monomial. It might seem that x^6 is not a monomial because no numerical factor is visible. However, x^6 is the same as $1x^6$. There is no need to write the 1 because a product of 1 and an amount is the amount.

From Example 1c, we can make the following conclusion:

Conclusion: If a numerical factor is not apparent in a monomial, it is understood to be 1.

$$x^6 = 1x^6$$

We write x^6 instead of $1x^6$ because it is simpler.

d. 8

Explanation: 8 is a monomial because it is a constant.

Is the given expression a monomial? Explain.

a. $4x^7$

b. $-9mn$

c. -6

d. $\dfrac{3x^2}{y}$

e. $9a + 5$

e. $\dfrac{5}{xy}$

Explanation: $\dfrac{5}{xy}$ is not a monomial because it is not a product of a constant and variables. Rather, it is a quotient of a constant with variables. The variables make up the divisor.

f. $3x^2 + 5x - 4$

Explanation: It is not a monomial because it is not a product of a constant with variables. Rather, addition and subtraction are involved.

Note: Although $3x^2 + 5x - 4$ is not a monomial, it is a polynomial. Remember, monomials must be a constant or a product of constants and variables. Monomials do not have variables in a divisor, nor do they contain addition or subtraction.

◀ **Do Your Turn 1**

OBJECTIVE 2 Identify the coefficient and degree of a monomial.

Two important parts of a monomial are its **coefficient** and its **degree**.

Identify the coefficient and degree of each monomial.

a. m^6

b. $-7x^2y^3$

c. $14ab$

d. -10

DEFINITIONS **Coefficient:** The numerical factor in a monomial.

Degree of a monomial: The sum of the exponents on all variables in a monomial.

Example 2 Identify the coefficient and degree of each monomial.

a. $-7x^2$

Answer: Coefficient: -7

Degree: 2

Explanation: The coefficient is -7, because it is the numerical factor. The degree is 2, because it is the exponent for the variable x.

b. $-xy^3$

Answer: Coefficient: -1

Degree: $1 + 3 = 4$

Explanation: From the conclusions we made in Examples 1b and 1c, we can express $-xy^3$ as $-1x^1y^3$. In this alternative form, we can see that the numerical factor is -1, which is the coefficient. We can also see that the variables' exponents are 1 and 3. Because degree is the sum of the variables' exponents, we add 1 and 3, which equals 4.

c. 17

Answer: Coefficient: 17

Degree: 0

Explanation: The monomial 17 is equivalent to $17x^0$. They are equivalent because $x^0 = 1$, which means $17x^0 = 17(1) = 17$. In the alternative form, it is apparent that 17 is the coefficient because it is the numerical factor, and 0 is the degree because it is the variable's exponent.

◀ **Do Your Turn 2**

Answers to Your Turn 1
a. $4x^7$ is a monomial because it is a product of a constant, 4, and a variable, x, with a whole-number exponent, 7. b. $-9mn$ is a monomial because it is a product of a constant, -9, and variables, m and n, that both have the whole number 1 as their exponent. c. -6 is a monomial because -6 can be expressed as $-6x^0$. d. $\dfrac{3x^2}{y}$ is not a monomial because it has a variable, y, as a divisor. e. $9a + 5$ is not a monomial because addition is involved.

Answers to Your Turn 2
a. coefficient: 1, degree: 6
b. coefficient: -7, degree: 5
c. coefficient: 14, degree: 2
d. coefficient: -10, degree: 0

OBJECTIVE 3 Identify like terms.

Remember that monomials are also referred to as terms. In Section 3.3, we will learn how to simplify polynomials that contain **like terms.**

DEFINITION **Like terms:** Monomials that have the same variables raised to the same exponents.

Notice that the definition does not say anything about the coefficient, so the coefficients can be different. However, the variables and their exponents must match *exactly.*

Example 3 Determine whether the given monomials are like terms.

a. $5x^2$ and $9x^2$

Answer: $5x^2$ and $9x^2$ are like terms because they both have x^2. They have the same variable raised to the same exponent.

b. $9xyz$ and $5xy$

Answer: $9xyz$ and $5xy$ are not like terms because the variables are different. $9xyz$ has a z while $5xy$ does not.

c. $7xy^2$ and $7x^2y$

Answer: $7xy^2$ and $7x^2y$ are not like terms because the variables are not raised to the same exponents.

d. $5a$ and $9A$

Answer: $5a$ and $9A$ are not like terms. In mathematics, upper- and lowercase letters are considered different symbols, even if they are of the same letter. An uppercase A is considered a different variable than a lowercase a.

Do Your Turn 3 ▷

Your Turn 3

Determine whether the given monomials are like terms.

a. $-8x^3$ and $5x^3$

b. $12mn$ and $12m$

c. $-5y^3z$ and y^3z

d. $9t$ and $12T$

e. $4ab^3c$ and $3a^3bc$

OBJECTIVE 4 Identify polynomials and their terms.

Now we can define a **polynomial.**

DEFINITION **Polynomial:** A monomial or an expression that can be written as a sum of monomials.

All of the following expressions are polynomials.

$$2x^3 + 5x + 8$$
$$5x + 7$$
$$9x^3 - 4x^2 + 8x - 6$$
$$2x^2 + 5xy + 8y$$

In $9x^3 - 4x^2 + 8x - 6$, it would seem that the subtraction signs go against the word *sum* in the definition. However, in Chapter 2, we learned that subtraction can be written as addition, so $9x^3 - 4x^2 + 8x - 6$ can be expressed as a sum.

$$9x^3 - 4x^2 + 8x - 6 = 9x^3 + (-4x^2) + 8x + (-6)$$

Mathematicians prefer to write expressions with the fewest symbols possible. Because $9x^3 - 4x^2 + 8x - 6$ has fewer symbols than $9x^3 + (-4x^2) + 8x + (-6)$, it is the preferred form. An expression written with the fewest symbols possible is said to be in **simplest form.**

Answers to Your Turn 3
a. yes **b.** no **c.** yes **d.** no **e.** no

Notice that x is the only variable in the polynomial $9x^3 - 4x^2 + 8x - 6$. We can say that $9x^3 - 4x^2 + 8x - 6$ is a **polynomial in one variable,** x. The polynomial $2x^2 + 5xy + 8y$ has a mixture of two variables, x and y. We call $2x^2 + 5xy + 8y$ a **multivariable polynomial.**

DEFINITIONS **Polynomial in one variable:** A polynomial with only one variable.

Multivariable polynomial: A polynomial with more than one variable.

$p^2 - 5p + 2$ is a polynomial in one variable.
$q^2 + 8qt$ is a multivariable polynomial.

In this text, we will focus on polynomials in one variable.

Remember that monomials are often referred to as *terms*. This is especially true when we talk about the individual monomials that make up a polynomial.

Your Turn 4

Identify the terms in the given polynomial and their coefficients.

a. $12x^2 - 8x + 9$

b. $4m^3 + 5m^2 - 10m - 11$

Example 4 Identify the terms in the polynomial and their coefficients.

a. $9x^3 - 4x^2 + 8x - 6$

Answers: First term: $9x^3$ Coefficient: 9

Second term: $-4x^2$ Coefficient: -4

Third term: $8x$ Coefficient: 8

Fourth term: -6 Coefficient: -6

Explanation: We saw that we can express $9x^3 - 4x^2 + 8x - 6$ as $9x^3 + (-4x^2) + 8x + (-6)$. Notice the dual role of a minus sign. It indicates to subtract the term that follows it and that the coefficient of that term is negative. Similarly, a plus sign indicates to add the term that follows it and that the coefficient of that term is positive.

From Example 4a, we can make the following conclusion:

Conclusion: The sign of the term is the sign to the left of the term.

b. $4y^4 - 8y^3 - y + 2$

This minus sign not only means subtract y but also that the term's coefficient is -1.

Answers:

$4y^4 - 8y^3 - y + 2$

This minus sign not only means subtract $8y^3$ but also that the term's coefficient is -8.

This plus sign not only means add 2 but also that the term's coefficient is positive 2.

First term: $4y^4$ Coefficient: 4

Second term: $-8y^3$ Coefficient: -8

Third term: $-1y$ Coefficient: -1

Fourth term: 2   Coefficient: 2

◄ **Do Your Turn 4**

Answers to Your Turn 4
a. $12x^2$, coefficient 12;
$-8x$, coefficient -8;
9, coefficient 9
b. $4m^3$, coefficient 4;
$5m^2$, coefficient 5;
$-10m$, coefficient -10;
-11, coefficient -11

Some polynomials have special names that are based on the number of terms in the polynomial.

> *Mono*mial: A **single**-term polynomial.
> *Bi*nomial: A polynomial that has exactly **two** terms.
> *Tri*nomial: A polynomial that has exactly **three** terms.

Note: The prefix of each special name corresponds to the number of terms in the polynomial.

Polynomials that have more than three terms have no special name.

Example 5 Classify each expression as a monomial, binomial, trinomial, or none of these.

a. $5x^3$

Answer: $5x^3$ is a monomial because it is a single term.

b. $-3x^2 + 5x$

Answer: $-3x^2 + 5x$ is a binomial because it is a sum of two terms.

c. $4m^2 + 6m - 9$

Answer: $4m^2 + 6m - 9$ is a trinomial because it is a sum of three terms.

d. $x^3 + 4x^2 - 9x + 2$

Answer: The answer is none of these. Since $x^3 + 4x^2 - 9x + 2$ has more than three terms, it has no special name. Mathematicians merely call it a *polynomial*.

> **DISCUSSION** If we were to give $x^3 + 4x^2 - 9x + 2$ a special name, what do you think that name would be?

Do Your Turn 5 ▶

Your Turn 5

Classify each expression as a monomial, binomial, trinomial, or none of these.

a. $5x + 9$

b. $4x^3 - 2x^2 + 5x - 6$

c. $8x^3$

d. $x^2 + 2x - 9$

OBJECTIVE 5 Identify the degree of a polynomial.

Earlier in this section, we learned that the degree of a monomial is the sum of the variables' exponents. For a polynomial with more than one term, the degree is the greatest degree of all its terms.

> **DEFINITION** **Degree of a multiple-term polynomial:** The greatest degree of all the terms that make up the polynomial.

Example 6 Identify the degree of each polynomial.

a. $5x^3 + 9x^6 - 10x^2 + 8x - 2$

Answer: Degree: 6

Explanation: Look at the degree of each term.

$$\underset{\substack{\uparrow \\ \text{degree} \\ \text{is } 3}}{5x^3} + \underset{\substack{\uparrow \\ \text{degree} \\ \text{is } 6}}{9x^6} - \underset{\substack{\uparrow \\ \text{degree} \\ \text{is } 2}}{10x^2} + \underset{\substack{\uparrow \\ \text{degree} \\ \text{is } 1}}{8x} - \underset{\substack{\uparrow \\ \text{degree} \\ \text{is } 0}}{2}$$

The greatest degree of all the terms is 6.

b. $12a^7 - a^5 - 3a^9 + 4a - 16$

Answer: Degree: 9

Explanation: The greatest degree of all the terms is 9.

Do Your Turn 6 ▶

Your Turn 6

Identify the degree of each polynomial.

a. $-2m^5 + 13m^7 - 8m^3 + 4m + 17$

b. $3x^{12} + 5x^6 + 8x^2 - 10x - 7$

Answers to Your Turn 5
a. binomial **b.** none of these
c. monomial **d.** trinomial

Answers to Your Turn 6
a. 7 **b.** 12

OBJECTIVE 6 Write polynomials in descending order.

The commutative property of addition tells us that rearranging the order of the addends in a sum does not affect the sum. Since polynomials are sums of terms, rearranging the order of the terms in a polynomial does not change the polynomial.

For example, if we move the -9 term in $-9 + 3x^3 - 6x^2 + 5x$ to the end of the polynomial, we have $3x^3 - 6x^2 + 5x - 9$ which is equivalent to the original expression. Notice that the minus sign moved with the term. Also, when $+3x^3$ became the first term, the plus sign was no longer necessary because $3x^3$ is understood to have a positive coefficient.

It is customary to arrange the terms in a polynomial in descending order of degree.

PROCEDURE *To write a polynomial in descending order of degree, place the term with the greatest degree first, then the term with the next greatest degree, and so on.*

Your Turn 7

Write each polynomial in descending order of degree.

a. $9m^4 + 7m^9 - 3m^2 + 15 - 6m^7 + 2m$

b. $-14y^2 - 3y + 17 + 9y^3 - 13y^5$

c. $24 + 9x^2 - 15x + 8x^5 - 7x^3 - 4x^4$

Example 7 Write the polynomial in descending order of degree.

a. $5x^3 + 4x^2 - 3x^5 - 7 + 9x$

Answer: $-3x^5 + 5x^3 + 4x^2 + 9x - 7$

Explanation: The term with the greatest degree is $-3x^5$ so it is written first. Notice that the minus sign stays with that term. The term with the next greatest degree is $5x^3$. Note that when this positive term is no longer the first term, it is written with a plus sign to its left. Continuing the descending order, we have the degree-two term next, $4x^2$, then degree-one, $9x$, and finally degree-zero, which is -7.

b. $-16b^3 + 5b^6 + 10b^2 - 13b^4 + 19 - 12b$

Answer: $5b^6 - 13b^4 - 16b^3 + 10b^2 - 12b + 19$

Explanation: The term with the greatest degree is $5x^6$ so it is written first. Since its coefficient is positive, a plus sign is not needed when it becomes the first term. The term with the next greatest degree is $-13x^4$. Continuing the descending order, we have the degree-three term next, $-16x^3$, then degree-two, $10x^2$, then degree-one, $-12b$, and finally degree-zero, which is 19.

◀ **Do Your Turn 7**

Answers to Your Turn 7
a. $7m^9 - 6m^7 + 9m^4 - 3m^2 + 2m + 15$
b. $-13y^5 + 9y^3 - 14y^2 - 3y + 17$
c. $8x^5 - 4x^4 - 7x^3 + 9x^2 - 15x + 24$

1. What is a monomial? Give three examples.

2. What is the coefficient of a monomial?

3. What is the degree of a monomial?

4. What are like terms? Give an example of a pair of like terms.

5. What is a polynomial? Give an example.

6. What does it mean for an expression to be in simplest form?

7. What is the degree of a polynomial?

8. How do you write a polynomial in descending order of degree?

For Exercises 9–20, determine whether the expression is a monomial. Explain.

9. $5x^3$

10. $7y^2$

11. $8m + 5$

12. $3t^2 + 5t - 6$

13. $\dfrac{4x^2}{5y}$

14. $\dfrac{-mn}{k}$

15. $-9x^2y$

16. 19

17. $-x$

18. $5m^3n^7p$

19. -7

20. $x^3 - 9xy$

For Exercises 21–32, identify the coefficient and degree of each monomial.

21. $3x^8$

22. $5t^2$

23. $-9x$

24. $-m$

25. 8

26. -7

27. xy^2

28. $-2t^3u^4$

29. $-a^2b^3c$

30. $15mnp$

31. -1

32. k

For Exercises 33–44, determine whether the given monomials are like terms.

33. $8x$ and $-5x$

34. $3y^2$ and y^2

35. $-5m$ and $-7n$

36. $2a$ and $2b$

37. $4mn^2$ and $7mn^2$

38. $8h^3$ and h^3

39. 14 and -6

40. T and $-T$

41. $5n$ and $9N$

42. $6xY$ and $5xy$

43. x^3y^2z and $-4y^2x^3z$

44. 1 and -5

For Exercises 45–50, identify the terms in the polynomial and their coefficients.

45. $5x^2 + 8x - 7$

46. $4y^3 - 2y - 9$

47. $6t - 1$

48. $9n^4 - 8n^2 + 7n - 1$

49. $-6x^3 + x^2 - 9x + 4$

50. $-9x^2 - x + 5$

For Exercises 51–62, determine whether the expression is a monomial, binomial, trinomial, or none of these.

51. $5x + 2$

52. $9y^2 + 5y - 6$

53. $18x^2$

54. 2

55. $6x^3 - 14x^2 + 9x - 8$

56. $2a - 6b$

57. $16t^2 - 8t - 5$

58. $a^5 - 6a^3 + 7a^2 - a + 2$

59. -5

60. $b^3 + 7b^2 - 12b$

61. $y^3 - 8$

62. $m^4 - 5m^3 + m^2 + 6m - 9$

For Exercises 63–68, identify the degree of each polynomial.

63. $6x^3 - 9x^2 + 8x + 2$

64. $8y^4 + 2y - 5y^6 + y^3 + 4$

65. $-m^3 - 8m^5 + 3m^2 - 2m^6$

66. $5 + 6a + 3a^3 + 4a^2 - a^5$

67. $19t^9 + 4t^3 + 25t^{12} - 11t^5$

68. $-b^2 - 16b^8 + 5b^3 + 2b^4 - 22$

For Exercises 69–74, write each polynomial in descending order of degree.

69. $14t^6 + 9t^3 + 5t^2 - 8t^4 - 1$

70. $6x + 19 - 4x^3 + 3x^2$

71. $9 - 18y^3 + 12y - 10y^2 + y^5$

72. $21m - 6m^7 + 9m^2 + 13 - 8m^5$

73. $7a^3 + 9a^5 - 6 + 2a^2 - a$

74. $2n^4 - 5n^3 - n + 2 + n^6 - 7n^8$

Review Exercises

For Exercises 1–8, evaluate.

[2.2]   **1.** $-7 + 5$

[2.3]   **2.** $-9 - 7$

[2.3]   **3.** $8 - 15$

[2.3]   **4.** $-15 + 8 - 21 + 2$

[2.3]   **5.** Is $3 - 5$ the same as $5 - 3$? Explain.

[2.3]   **6.** Is $-5 + 3$ the same as $3 + (-5)$? Explain.

[2.4]   **7.** Is $(-3)(5)$ the same as $(5)(-3)$? Explain.

[1.6]   **8.** Find the perimeter of the shape.

3.3 Simplifying Polynomials

OBJECTIVE 1 Simplify polynomials in one variable by combining like terms.

OBJECTIVE

1 Simplify polynomials in one variable by combining like terms.

We have learned that there are two actions that we can perform on an expression: 1. evaluate it, or 2. rewrite it. In Section 3.2, we learned one way to rewrite a polynomial expression by arranging its terms in descending order of degree. In this section, we will learn how to rewrite polynomials that contain like terms.

Recall from Section 3.2 that like terms have the same variables raised to the same powers. For example, $3x$ and $2x$ are like terms because they have the same variable, x, raised to the same power, 1.

Now, suppose we have a polynomial that contains those like terms, such as $3x + 2x$. To understand how we combine those like terms, let's think about what each term means. Since multiplication is repeated addition, $3x$ means we have three x's added together and $2x$ means we have two x's added together. If we rewrite our polynomial as a sum of all those x's, we have the following:

$$\overbrace{3x}^{} \quad + \quad \overbrace{2x}^{}$$
$$= \underbrace{x + x + x \ + \ x + x}_{}$$
$$= 5x$$

Note: We have a total of five x's repeatedly added, so we can represent that sum as $5x$.

Notice that the coefficient of the result, 5, is the sum of the coefficients of the like terms, $3 + 2 = 5$. Because $5x$ has fewer symbols than $3x + 2x$, we say that we have *simplified* $3x + 2x$. In fact, $5x$ is in *simplest form* because it cannot be written with fewer symbols.

Our example suggests the following procedure.

> **PROCEDURE** *To combine like terms:*
> 1. Add or subtract the coefficients.
> 2. Keep the variables and their exponents the same.

CONNECTION We can affirm that $3x + 2x = 5x$ by evaluating $3x + 2x$ and $5x$ using the same value of x. For example, let's evaluate $3x + 2x$ when $x = 4$.

$$3(4) + 2(4) = 12 + 8 = 20$$

Evaluating $5x$ when $x = 4$ also gives a result of 20.

$$5(4) = 20$$

Equivalent expressions always yield the same result when we evaluate them using the same variable values.

Example 1   Combine like terms.

a. $4x + 9x$

Answer: $13x$

Explanation: We added the coefficients and kept the variable the same.

> **Note:** Remember that when adding two numbers that have the same sign, we add their absolute values and keep the same sign.
>
> When adding two numbers that have different signs, we subtract their absolute values and keep the sign of the number with the larger absolute value.

b. $-12y^2 + 7y^2$

Answer: $-5y^2$

Explanation: We subtracted the coefficients and kept the variable and its exponent the same.

c. $-4m + 4m$

Answer: 0

Explanation: Because $-4m$ and $4m$ are additive inverses, their sum is 0.

d. $-2y^3 - 5y^3$

Answer: $-7y^3$

Explanation: We added the coefficients and kept the variables and their exponents the same.

CONNECTION Adding measurements that have the same unit is like combining like terms. Suppose we have to calculate the perimeter of the rectangle shown.

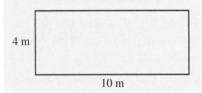

$P = 4\text{ m} + 10\text{ m} + 4\text{ m} + 10\text{ m}$
$P = 28\text{ m}$

This perimeter problem is similar to a problem in which we must combine like terms in a polynomial where all the variables are m.

$$4m + 10m + 4m + 10m = 28m$$

Notice that if the variables were different, like $4cm + 10cm + 4m + 10m$, we would not be able to combine the terms with different variables. In a similar way, we cannot combine measurements with different units. If the width were 4 cm and the length 10 m, then the perimeter is no longer 28 m because cm and m represent different distance measurements.

$$4\text{ cm} + 10\text{ m} + 4\text{ cm} + 10\text{ m} \neq 28$$

You may be wondering why we simplify expressions. One reason is to make evaluating expressions easier. Suppose we were asked to evaluate $-12y^2 + 7y^2$ when $y = 2$. Before we learned how to simplify, we would have evaluated the expression as shown below:

$$-12(2)^2 + 7(2)^2 \quad \text{Replace } y \text{ with 2 in } -12y^2 + 7y^2.$$
$$= -12(4) + 7(4) \quad \text{Square 2.}$$
$$= -48 + 28 \quad \text{Multiply.}$$
$$= -20 \quad \text{Add.}$$

However, because the simplified expression is equivalent to the original, we can evaluate the simplified expression using the same variable value and get the same result. We saw in Example 1b that $-12y^2 + 7y^2 = -5y^2$. Let's evaluate $-5y^2$ when $y = 2$.

$$-5(2)^2 \quad \text{Replace } y \text{ with 2 in } 5y^2.$$
$$= -5(4) \quad \text{Square 2.}$$
$$= -20 \quad \text{Multiply.}$$

Since the simplified expression has fewer symbols, it is much easier to evaluate.

◀ **Do Your Turn 1**

Sometimes there are several like terms in the same polynomial.

Example 2 Combine like terms and write the resulting polynomial in descending order of degree.

$$9x^2 + 5 + 7x + 2x^2 - 8 - 7x$$

Solution:

$$= 9x^2 + 2x^2 + 7x - 7x + 5 - 8 \quad \text{Collect like terms as needed.}$$
$$= 11x^2 + 0 - 3 \quad \text{Notice that } 7x \text{ and } -7x \text{ have a sum of 0.}$$
$$= 11x^2 - 3 \quad \text{We can drop 0 and bring the terms together.}$$

Your Turn 1

Combine like terms.

a. $6x + 9x$

b. $4y^2 - 20y^2$

c. $12x^3 - 12x^3$

d. $-m^2 - 12m^2$

e. $-x^2 + 4x^2 - 2x^2$

Answers to Your Turn 1
a. $15x$ **b.** $-16y^2$ **c.** 0
d. $-13m^2$ **e.** x^2

Explanation: First, we collected the like terms using the commutative property of addition to rearrange the polynomial so that the like terms were next to each other. This was optional. Some people prefer to combine like terms without first collecting them, as shown next. Most people mark out the terms as they go.

Alternative Solution:

$$9x^2 + 5 + 7x + 2x^2 - 8 - 7x$$

$$= 11x^2 + \quad 0 \quad - 3$$

$$= 11x^2 - 3$$

THINK	$9x^2 + 2x^2 = 11x^2$
	$7x + (-7x) = 0$
	$5 + (-8) = -3$
Mark out the terms as you combine.	

LEARNING STRATEGY

When combining like terms, mark through the terms that you combine so that you know what you've combined and what you have left. Also, combine the terms in order of degree, combining the highest-degree terms first, and so on. This will put the resulting polynomial in descending order of degree.

Example 3 Combine like terms and write the resulting polynomial in descending order of degree.

$$6x^2 + 9x^3 - 7x^4 + 15 + 2x^2 - 9x^3 + x - 18$$

Solution: We will combine the terms in descending order, marking through the terms as we go.

$$6x^2 + 9x^3 - 7x^4 + 15 + 2x^2 - 9x^3 + x - 18$$

$$= -7x^4 \quad - \quad 0 \quad + \quad 8x^2 \quad + x - 3$$

$$= -7x^4 + 8x^2 + x - 3$$

Explanation: We brought down $-7x^4$ because it had no like term. Also, it was the first term because it had the greatest degree.

The degree-three terms are additive inverses.

$$9x^3 + (-9x^3) = 0$$

Next, we combined the degree-two terms.

$$6x^2 + 2x^2 = 8x^2$$

We brought down the x term because it had no like term.

Last, we combined the degree-zero terms.

$$15 + (-18) = -3$$

Do Your Turn 2 ▷

For
Extra
Help

Videotape
DVT

Addison-Wesley
Tutor Center

Math XL

MyMathLab

Student Solutions
Manual

1. How do you combine like terms?

2. What conditions cause the sum of two like terms to be 0?

For Exercises 3–30, simplify by combining like terms.

3. $3x + 9x$

4. $5a + 2a$

5. $3y + y$

6. $b + 7b$

7. $12m - 7m$

8. $10x - 2x$

9. $n - 6n$

10. $5w - 14w$

11. $-6x + 8x$

12. $-5y + 2y$

13. $-15a - a$

14. $-9m - 3m$

15. $9t - 12t$

16. $7k - 18k$

17. $-q - 6q$

18. $-r - 8r$

19. $8a^2 + 5a^2$

20. $7x^3 - 2x^3$

21. $2y^2 - 9y^2$

22. $m^4 + 13m^4$

23. $j^3 + j^3$

24. $6n^8 - 19n^8$

25. $2b^2 - 2b^2$

26. $17c^4 - c^4$

27. $-3y^3 + 9y^3$

28. $-9x^2 + 9x^2$

29. $-4m^5 - 4m^5$

30. $-10t^2 + t^2$

For Exercises 31–46, simplify and write the resulting polynomial in descending order of degree.

31. $9x^2 + 5x + 2x^2 - 3x + 4$

32. $7y^2 + 4y - 3y^2 - y$

33. $7a - 8a^3 - a + 4 - 5a^3$

34. $3b - 9b^4 + 4b^4 - 7b$

35. $-9c^2 + c + 4c^2 - 5c$

36. $4m^7 - 9m^5 - 12m^5 + 7m^7$

37. $-3n^2 - 2n + 7 - 4n^2$

38. $p^3 + 9p^2 - 17 + 3p^2$

39. $12m^2 + 5m^3 - 11m^2 + 3 + 2m^3 - 7$

40. $4y + 15 - y^2 - 16 + 5y + 6y^2$

41. $-7x + 5x^4 - 2 + 3x^4 + 7x + 10 - 3x^2$

42. $-10b^3 + 7b^4 + 1 + b^3 - 2b^2 - 12 + b^4$

43. $t^5 + 7t^2 - 20 + 6t^4 - 7t^2 + 9t^3 + 18 - 4t - 8t^5$

44. $2x^3 - x^7 + 3x + 10x^3 - 3x^7 - 3x + x^3 - 5 + 12x^2 + 4$

45. $-3y^2 + 15y - 13 - 6y^2 + 9y^4 - 8y^3 - 12y^4 + 13 + 3y - 10y^4$

46. $-8m + 17m^6 + 12m^3 - 19 - 14m - 10 - m^2 + 5m^4 - 10m^6 + 3m^2$

Review Exercises

[1.1] **1.** Write 472 in expanded form.

[1.7] **2.** A border strip is to be glued to the edge of an office desk. How much border material is required?

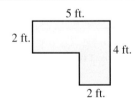

For Exercises 3–6, evaluate.

[2.1] **3.** $-(-6)$

[1.2] **4.** $392 + 149$

[2.2] **5.** $-285 + 173$

[2.3] **6.** $-41 - (-382)$

[1.3] **7.** What property is illustrated by $2(3 + 4) = 2 \cdot 3 + 2 \cdot 4$?

3.4 Adding and Subtracting Polynomials

OBJECTIVE 1 Add polynomials in one variable.

We can add and subtract expressions in the same way that we add and subtract numbers. However, instead of adding digits in like place values, we add like terms. Consider the following comparison:

OBJECTIVES

1 Add polynomials in one variable.

2 Write an expression for the perimeter of a given shape.

3 Subtract polynomials in one variable.

Numeric addition:

$352 + 231 = 583$

Polynomial addition:

$(3x^2 + 5x + 2) + (2x^2 + 3x + 1) = 5x^2 + 8x + 3$

We can stack like place values to find the sum:

$$
\begin{array}{r}
352 \\
+\ 231 \\
\hline
583
\end{array}
$$

We can stack like terms to find the sum:

$$
\begin{array}{r}
3x^2 + 5x + 2 \\
+\ 2x^2 + 3x + 1 \\
\hline
5x^2 + 8x + 3
\end{array}
$$

Actually, we do not have to stack polynomials in order to add them as long as we combine all the like terms.

PROCEDURE *To add polynomials, combine like terms.*

CONNECTION As we saw in Section 3.2, a polynomial like $3x^2 + 5x + 2$ can be tranformed to the number 352 by replacing x with 10.

$3x^2 + 5x + 2$

$3(10)^2 + 5(10) + 2$

$= 300 + 50 + 2$

$= 352$

Example 1 Add. $(5x^3 + 9x^2 + 2x + 3) + (3x^3 + 5x + 1)$

Solution: $(5x^3 + 9x^2 + 2x + 3) + (3x^3 + 5x + 1)$

$= 8x^3 + 9x^2 + 7x + 4$

Explanation: We combined like terms in order of degree.

First, we combined the degree-three terms: $5x^3 + 3x^3 = 8x^3$.

There was no degree-two term in the second polynomial, so we brought down the $9x^2$.

Next, we combined the degree-one terms: $2x + 5x = 7x$.

Last, we combined the degree-zero terms: $3 + 1 = 4$.

CONNECTION If we stack the polynomials in Example 1, we must leave a blank space under the $9x^2$ because there is no degree-two term in the second polynomial. It's like adding $5923 + 3051$.

$$
\begin{array}{r}
5x^3 + 9x^2 + 2x + 3 \\
+\ 3x^3 \qquad\ \ + 5x + 1 \\
\hline
8x^3 + 9x^2 + 7x + 4
\end{array}
\qquad
\begin{array}{r}
5923 \\
+\ 3051 \\
\hline
8974
\end{array}
$$

As the polynomials get more complex, these blanks become more common, which is why stacking is not the best method for all cases of adding polynomials.

Do Your Turn 1 ▶

Your Turn 1

Add.

a. $(2x + 8) + (6x + 1)$

b. $(3x^2 - 2x + 4) + (2x^2 + 5x - 9)$

c. $(y^4 + 3y^2 + 4) + (4y^4 + 7y^3 + 3)$

d. $(6t^3 + 4t + 2) + (2t^4 + t + 1)$

Answers to Your Turn 1
a. $8x + 9$ b. $5x^2 + 3x - 5$
c. $5y^4 + 7y^3 + 3y^2 + 7$
d. $2t^4 + 6t^3 + 5t + 3$

Your Turn 2

Add.

a. $(9x^4 - 12x^2 + 3x - 14) + (7x^4 + 2x^3 - 8x - 3)$

b. $(12x^5 + 7x^3 - 13x - 15) + (8x^4 - 7x^3 - 5x + 19)$

c. $(9m^3 + 4m - 8 + 14m^2) + (2m^3 - 7m - 8 + 3m^2)$

Example 2   Add. $(4x^5 - 9x - 6) + (2x^5 - 3x - 2)$

Solution: $(4x^5 - 9x - 6) + (2x^5 - 3x - 2)$

$$= 6x^5 - 12x - 8$$

Explanation: First, we combined the degree-five terms: $4x^5 + 2x^5 = 6x^5$.

Next, we combined degree-one terms: $-9x - 3x = -12x$.

Last, we combine degree-zero terms: $-6 + (-2) = -8$.

> **Note:** We determine the sign in between terms based on the outcome of combining.
>
> If the outcome is positive, as with $6x^5$, we place a plus sign to the left of the term. But since $6x^5$ is the first term in the answer, no plus sign is needed.
>
> If the outcome is negative, as with $-12x$, we place a minus sign to the left of the term.

◀ **Do Your Turn 2**

OBJECTIVE 2 Write an expression for the perimeter of a given shape.

Your Turn 3

Write an expression in simplest form for the perimeter of the shape shown.

$$
\begin{array}{c}
x+1 \qquad\qquad x+1 \\
\triangle \\
3x-2
\end{array}
$$

Example 3   Write an expression in simplest form for the perimeter of the rectangle shown.

Solution: To find the perimeter, we need to find the total distance around the shape. Instead of numerical measurements for the length and width, we have expressions. In this case, the length is the expression $3x - 1$ and the width is the expression $2x + 5$. We can write only an expression for the perimeter; we cannot get a numerical answer until we have a value for the variable.

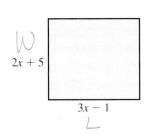

$$
\begin{aligned}
\text{perimeter} &= \text{length} + \text{width} + \text{length} + \text{width} \\
&= (3x - 1) + (2x + 5) + (3x - 1) + (2x + 5) \\
&= 3x + 2x + 3x + 2x - 1 + 5 - 1 + 5 \qquad \text{Collect like terms.} \\
&= \qquad 10x \qquad\qquad + 8 \qquad\qquad \text{Combine like terms.}
\end{aligned}
$$

Answer: The expression for the perimeter is $10x + 8$.

> **WARNING** $10x + 8$ is in simplest form. It might be tempting to try to write it as $18x$, but $10x$ and 8 are not like terms and cannot be combined.

◀ **Do Your Turn 3**

OBJECTIVE 3 Subtract polynomials in one variable.

We subtract polynomials just as we subtract numbers. However, polynomials are tricky because they have signs involved, like the integers. Recall that when we subtracted integers we found it simpler to write the subtraction statements as equivalent addition statements. We can apply this same principle to subtracting polynomials.

Let's compare polynomial subtraction to numeric subtraction.

Numeric subtraction:	**Polynomial subtraction:**
$896 - 254$	$(8x^2 + 9x + 6) - (2x^2 + 5x + 4)$

Answers to Your Turn 2
a. $16x^4 + 2x^3 - 12x^2 - 5x - 17$
b. $12x^5 + 8x^4 - 18x + 4$
c. $11m^3 + 17m^2 - 3m - 16$

Answer to Your Turn 3
$5x$

Like addition, we can align the like place values or like terms in a stacked form.

$$
\begin{array}{r}
896 \\
-\ 254 \\
\hline
642
\end{array}
\qquad\qquad
\begin{array}{r}
8x^2 + 9x + 6 \\
-\ 2x^2 + 5x + 4 \\
\hline
6x^2 + 4x + 2
\end{array}
$$

In the numeric subtraction, we subtracted digits in the like place value. In the polynomial subtraction, we subtracted like terms. Note that subtracting 4 from 6 is the same as adding 6 and -4, subtracting $5x$ from $9x$ is the same as adding $9x$ and $-5x$, and subtracting $2x^2$ from $8x^2$ is the same as adding $8x^2$ and $-2x^2$. Therefore, we would write the following equivalent addition.

$$
\begin{array}{r}
8x^2 + 9x + 6 \\
+\ -2x^2 - 5x - 4 \\
\hline
6x^2 + 4x + 2
\end{array}
$$

or in sentence form:

$$(8x^2 + 9x + 6) + (-2x^2 - 5x - 4) = 6x^2 + 4x + 2$$

Conclusion: We can write polynomial subtraction as equivalent polynomial addition by changing the signs of each term in the subtrahend (second) polynomial.

$$(8x^2 + 9x + 6) - (\ \ 2x^2 + 5x + 4)$$

Change the minus sign to a plus sign.   ↓   ↓   ↓   ↓   **Change all signs in the subtrahend.**

$$= (8x^2 + 9x + 6) + (-2x^2 - 5x - 4)$$
$$= 6x^2 + 4x - 2$$

CONNECTION   We use the distributive property to change all the signs in a polynomial. If we disregard the first polynomial, we have:

$$-(2x^2 + 5x + 4)$$
$$= -1(2x^2 + 5x + 4)$$

The distributive property tells us we can distribute the -1 (or minus sign) to each term inside the parentheses.

$$= -1 \cdot 2x^2 - 1 \cdot 5x - 1 \cdot 4$$
$$= -2x^2 - 5x - 4$$

Our exploration suggests the following procedure.

PROCEDURE   *To subtract polynomials:*

1. Write the subtraction statement as an equivalent addition statement.
   a. Change the operation symbol from a minus sign to a plus sign.
   b. Change the subtrahend (second polynomial) to its additive inverse. To get the additive inverse, we change *all* the signs in the polynomial.
2. Combine like terms.

Example 4   Subtract.

a. $(2x + 3) - (7x - 2)$

Solution: Write an equivalent addition statement, then combine like terms.

$$(2x + 3) - (7x - 2)$$

Change minus sign to a plus sign.   ↓   ↓   ↓   **Change all signs in the subtrahend.**

$$= (2x + 3) + (-7x + 2)$$

$$= -5x + 5 \qquad\qquad \text{Combine like terms.}$$

Explanation: To write the equivalent addition, we changed the minus sign to a plus sign and changed all signs in the subtrahend. After writing the equivalent addition, we combined $2x$ and $-7x$ to get $-5x$.

Finally, we combined 3 and 2 to get 5.

b. $(7x^2 + 8x + 5) - (3x^2 + 7x + 1)$

Solution: Write an equivalent addition statement, then combine like terms.

$$(7x^2 + 8x + 5) - \quad (3x^2 + 7x + 1)$$

Change the minus sign to a plus sign.   ↓ ↓ ↓ ↓   **Change all signs in the subtrahend.**

$$= (7x^2 + 8x + 5) + (-3x^2 - 7x - 1)$$

$$= 4x^2 + x + 4 \qquad \text{Combine like terms.}$$

Explanation: After writing the equivalent addition, we combined $7x^2$ and $-3x^2$ to get $4x^2$.

Next, we combined $8x$ and $-7x$ to get $1x$ or just x.

Finally, we combined 5 and -1 to get 4.

CONNECTION The polynomial subtraction in part b is equivalent to the following numeric subtraction:

$$\begin{array}{r} 785 \\ - \; 371 \\ \hline 414 \end{array}$$

Notice that the numeric result 414 corresponds to the polynomial result $4x^2 + x + 4$.

Although we could use the stacking method for subtracting polynomials, it is not the best method because we sometimes have missing terms or terms that are not like terms.

◀ **Do Your Turn 4**

Now let's subtract polynomials that contain minus signs.

Example 5 Subtract. $(12x^3 + 10x - 9) - (4x^3 - 5x + 3)$

Solution: Write an equivalent addition statement, then combine like terms.

$$(12x^3 + 10x - 9) - (4x^3 \quad - 5x + 3)$$

Change the minus sign to a plus sign.   ↓ ↓ ↓ ↓   **Change all signs in the subtrahend.**

$$= (12x^3 + 10x - 9) + (-4x^3 + 5x - 3)$$

$$= 8x^3 + 15x - 12 \qquad \text{Combine like terms.}$$

Explanation: After writing the equivalent addition, we combined $12x^3$ with $-4x^3$ to get $8x^3$.

Next, we combined $10x$ with $5x$ to get $15x$.

Finally, we combined -9 with -3 to get -12.

◀ **Do Your Turn 5**

Your Turn 4

Subtract.

a. $(5x^2 + 7x + 3) - (2x^2 + 3x + 1)$

b. $(8t^2 + 9t + 3) - (7t^2 + t + 2)$

c. $(9y^4 + 4y^2 + 7) - (y^4 + 3y^2 + 2)$

Your Turn 5

Subtract.

a. $(9x^2 - 7x + 5) - (2x^2 - 3x + 8)$

b. $(17x^4 + 9x^3 - 10x^2 - 12) - (13x^4 + 9x^3 - 3x^2 + 2)$

Answers to Your Turn 4
a. $3x^2 + 4x + 2$
b. $t^2 + 8t + 1$
c. $8y^4 + y^2 + 5$

Answers to Your Turn 5
a. $7x^2 - 4x - 3$
b. $4x^4 - 7x^2 - 14$

For Extra Help

 Videotape DVT

 Addison-Wesley Tutor Center

 Math XL

 MyMathLab

 Student Solutions Manual

1. Explain how to add two polynomials.

2. Explain how to write a subtraction of two polynomials as an equivalent addition.

For Exercises 3–28, add and write the resulting polynomial in descending order.

3. $(2x + 5) + (3x + 1)$

4. $(5x + 2) + (3x + 7)$

5. $(7y - 4) + (2y - 1)$

6. $(9m - 3) + (4m - 6)$

7. $(10x + 7) + (3x - 9)$

8. $(7p + 4) + (9p - 7)$

9. $(4t - 11) + (t + 13)$

10. $(3g - 4) + (g + 2)$

11. $(2x^2 + x + 3) + (5x + 7)$

12. $(7x^2 + 2x + 1) + (6x + 5)$

13. $(9n^2 - 14n + 7) + (6n^2 + 2n - 4)$

14. $(8t^2 + 12t - 10) + (2t^2 - 8t - 2)$

15. $(4p^2 - 7p - 9) + (3p^2 + 4p + 5)$

16. $(9a^2 + 3a - 5) + (5a^2 - 7a - 6)$

17. $(-3b^3 + 2b^2 - 9b) + (-5b^3 - 3b^2 + 11)$

18. $(9c^3 - 4c^2 + 5) + (-7c^2 - 10c + 7)$

19. $(3a^3 - a^2 + 10a - 4) + (6a^2 - 9a + 2)$

20. $(9x^3 + 4x^2 + x - 11) + (3x^2 - 8x - 5)$

21. $(5x^3 + 7x^2 - 8x + 3) + (-3x^3 - 4x^2 + 8x - 5)$

22. $(7y^3 - 9y^2 + 5y - 1) + (-4y^3 + 9y^2 + y + 8)$

23. $(t^3 - 4t^2 - 9t + 5) + (8t^3 - t^2 + 10t - 1)$

24. $(m^3 + 10m^2 - 8m - 2) + (9m^3 - 11m^2 + 3m - 6)$

25. $(5m^4 + 6m^2 - 8m + 12) + (3m^3 + 9m - 15)$

26. $(7x^5 - 3x^3 + 5x^2 - 18) + (x^4 - 5x^3 - 3x^2 + 8)$

27. $(x^5 + 5x^4 - 9x^2 + 11) + (-2x^5 - 14x^4 - 6x^3 + 5)$

28. $(10a^6 + 7a^4 - 9a - 5) + (-6a^6 - 2a^4 + 11a^3 - 9)$

For Exercises 29–32, write an expression for the perimeter in simplest form.

29.

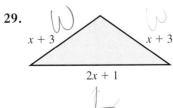

30.

$2x - 7$

$5x + 1$

31.

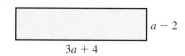

$a - 2$

$3a + 4$

32.

$m - 3$ $7m + 8$

$6m + 5$

For Exercises 33–36, find and explain the mistake(s) then work the problem correctly.

33. $(9y^3 + 4y^2 - 6y) + (y^3 - 2y^2)$

$= 10y^3 - 2y^2 - 6y$

34. $(10a^4 + 5a^3 - 11) + (5a^4 + 3a^2 - 4)$

$= 15a^4 + 3a^2 - 15$

35. $(8x^6 - 9x^4 + 15x^3 - 1) + (x^5 - 8x^3 - 3)$

$= 9x^{11} - 9x^4 + 7x^3 - 4$

36. $(n^3 + 2n^2 - 19) + (n^3 - 2n + 4)$

$= (2n^3 - 15)$

For Exercises 37–52, subtract and write the resulting polynomial in descending order of degree.

37. $(5x + 9) - (3x + 2)$

38. $(7m + 4) - (2m + 1)$

39. $(8t + 3) - (5t + 3)$

40. $(6a + 7) - (6a + 5)$

41. $(10n + 1) - (2n + 7)$

42. $(9x + 8) - (3x + 10)$

43. $(5x - 6) - (2x + 1)$

44. $(2m + 5) - (7m - 3)$

45. $(6x^2 + 8x - 9) - (9x + 2)$

46. $(8n^2 - 7n - 5) - (2n - 3)$

47. $(8a^2 - 10a + 2) - (-4a^2 + a + 8)$

48. $(y^2 + 3y - 9) - (-6y^2 + 8y - 1)$

49. $(7x^2 + 5x - 8) - (7x^2 + 9x + 2)$

50. $(2t^2 - t + 7) - (3t^2 - t - 6)$

51. $(-8m^3 + 9m^2 - 17m - 2) - (-8m^3 + 5m^2 + m - 6)$

52. $(5u^3 - 4u^2 + u - 10) - (-2u^3 + 4u^2 + u + 7)$

For Exercises 53–56, find and explain the mistake, then work the problem correctly.

53. $(5x^2 - 6x) - (3x^2 - 4x)$

$= (5x^2 - 6x) + (-3x^2 - 4x)$

$= 2x^2 - 10x$

54. $(10y^2 - 5y) - (y^2 + 3y)$

$= (10y^2 - 5y) + (-y^2 - 3y)$

$= 9y^2 + 8y$

55. $(-9t^2 + 4t + 11) - (5t^3 + 9t^2 + 6)$

$= (-9t^2 + 4t + 11) + (-5t^3 - 9t^2 - 6)$

$= -5t^3 + 4t + 5$

56. $(x^4 - 6x^2 + 3x) - (x^4 + 2x^3 - 4x)$

$= (x^4 - 6x^2 + 3x) + (-x^4 - 2x^3 + 4x)$

$= -2x^4 - 2x^3 - 6x^2 + 7x$

PUZZLE PROBLEM A trog is a mythical creature that reproduces in a peculiar way. An adult trog will produce one newborn every hour. A newborn trog spends its first hour growing to an adult. If we place one newborn trog in a cage, how many trogs will exist after six hours? Seven hours? Explain the pattern.

Review Exercises

For Exercises 1–4, evaluate.

[2.4]   **1.** $2 \cdot (-341)$

[2.4]   **2.** $(-3)^4$

[2.4]   **3.** $(-2)^5$

[1.3]   **4.** $2^3 \cdot 2^4$

[1.6]   **5.** A rectangular swimming pool is 25 meters long by 15 meters wide. What is the surface area of the water?

[1.3]   **6.** A lunch menu has a choice of five main dishes and four side dishes. How many different meals can a customer order?

3.5 Multiplying Polynomials

OBJECTIVE 1 Multiply monomials.

In order to multiply monomials, we need to understand how to multiply exponential forms that have the same base as in $2^3 \cdot 2^4$. One way to simplify $2^3 \cdot 2^4$, is to use the order of operations agreement.

$$2^3 \cdot 2^4 = 8 \cdot 16 \quad \text{Evaluate exponential forms.}$$
$$= 128 \quad \text{Multiply.}$$

However, there is an alternative. We can express the result in exponential form by expanding each of 2^3 and 2^4 into their factored forms.

$$\overbrace{2^3}^{} \cdot \overbrace{2^4}^{}$$
$$= \underbrace{2 \cdot 2 \cdot 2}_{} \cdot \underbrace{2 \cdot 2 \cdot 2 \cdot 2}_{}$$
$$= 2^7 \qquad \blacktriangleleft \quad \text{Because there are a total of seven 2's multiplied, we can express the product as } 2^7.$$

Note that $2^7 = 128$. Notice in the alternative method that the resulting exponent is the *sum* of the original exponents.

$$2^3 \cdot 2^4 = 2^{3+4} = 2^7$$

Our example suggests the following rule.

DISCUSSION Why do the exponential forms have to have the same base for us to add the exponents?

RULE *When multiplying exponential forms that have the same base, we can add the exponents and keep the same base.*

In math language: $n^a \cdot n^b = n^{a+b}$

Your Turn 1

Multiply.

a. $x^3 \cdot x^4$

b. $(n^6)(n^2)$

c. $m^5(m)$

Example 1 Multiply. $(x^2)(x^3)$

Solution: $(x^2)(x^3) = x^{2+3} = x^5$

Explanation: Because the bases were the same, we added the exponents and kept the same base.

Check: We can expand the exponential forms into their factored forms.

$$(x^2)(x^3)$$

x^2 means two x's. $\qquad \swarrow \qquad \searrow \qquad$ x^3 means three x's.

$$= \underbrace{(x \cdot x)}_{} \; \underbrace{(x \cdot x \cdot x)}_{} \qquad \text{There are a total of five } x\text{'s multiplied together, so we can express the product as } x^5.$$

$$= x^5$$

◀ **Do Your Turn 1**

Answers to Your Turn 1
a. x^7 b. n^8 c. m^6

Let's apply this rule to monomials that have coefficients such as $(2x^4)(3x^6)$. We can use the commutative property of multiplication to separate the coefficients and variables. We can then multiply the coefficients and use our rule for exponents.

$$(2x^4)(3x^6) = 2 \cdot 3 \cdot x^4 \cdot x^6$$
$$= 6x^{4+6} \qquad \text{Multiply coefficients and add exponents of the like bases, } x.$$
$$= 6x^{10}$$

We can confirm our result by expanding the original expression so that the exponential forms are in factored form.

$$(2x^4)(3x^6) = 2 \cdot x \cdot x \cdot x \cdot x \cdot 3 \cdot x \cdot x \cdot x \cdot x \cdot x \cdot x$$
$$= 2 \cdot 3 \cdot x \cdot x \cdot x \cdot x \cdot x \cdot x \cdot x \cdot x \cdot x \cdot x$$
$$= 6x^{10}$$

◄ **Note:** There are indeed a total of 10 factors of x.

CONNECTION Multiplying monomials like $(2x^4)(3x^6)$ is similar to multiplying numbers in expanded form. If we replace the x's with the number 10, then we have expanded form.

$$(2\mathbf{x}^4)(3\mathbf{x}^6)$$
$$(2 \cdot \mathbf{10}^4)(3 \cdot \mathbf{10}^6)$$

Multiplying numbers in expanded form must work the same way as multiplying monomials. We can multiply the $2 \cdot 3$ to get 6 and add the exponents of the like bases.

$$2 \cdot 3 \cdot 10^{4+6} = 6 \cdot 10^{10}$$

To confirm our result, we can write the expanded forms in standard form then multiply.

$$(2 \cdot 10^4)(3 \cdot 10^6) = (20{,}000)(3{,}000{,}000)$$
$$= 60{,}000{,}000{,}000$$
$$= 6 \cdot 10^{10} \qquad \text{The product has a 6 followed by ten 0's.}$$

Example 2 Multiply. $-3x^5(4x^2)(5x)$

Solution: $-3x^5(4x^2)(5x) = -3 \cdot 4 \cdot 5x^{5+2+1}$ Multiply coefficients and add exponents of the like bases, x.

$$= -60x^8$$

Check: We can expand the original expression so that the exponential forms are in factored form.

$$-3x^5(4x^2)(5x) = -3 \cdot x \cdot x \cdot x \cdot x \cdot x \cdot 4 \cdot x \cdot x \cdot 5 \cdot x$$
$$= -3 \cdot 4 \cdot 5 \cdot x \cdot x \cdot x \cdot x \cdot x \cdot x \cdot x \cdot x \qquad \text{Use the commutative property of multiplication.}$$
$$= -60x^8 \qquad \text{Multiply the numbers and count the factors of } x, \text{ which we write as an exponent, 8.}$$

Our examples so far suggest the following procedure.

PROCEDURE *To multiply monomials:*
1. Multiply coefficients.
2. Add the exponents of the like bases.

Do Your Turn 2 ▶

Your Turn 2

Multiply.

a. $4x^5 \cdot 2x^8$

b. $(4 \cdot 10^5)(2 \cdot 10^8)$

c. $12m^2(m^8)$

d. $-2y^2(4y)(9y^4)$

e. $2b^3(-4b^2)(3b^5)$

Answers to Your Turn 2
a. $8x^{13}$
b. $8 \cdot 10^{13}$
c. $12m^{10}$
d. $-72y^7$
e. $-24b^{10}$

CONNECTION Multiplying distance measurements in area and volume is like multiplying monomials.

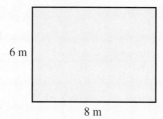

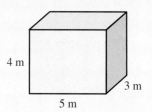

The area of the rectangle shown is

$$A = bh$$
$$A = (8 \text{ m})(6 \text{ m})$$
$$A = 48 \text{ m}^2$$

The area calculation is similar to multiplying two monomials with m as the variable.

Multiply: $(8m)(6m) = 48m^2$

The volume of the box shown is

$$V = lwh$$
$$V = (5 \text{ m})(3 \text{ m})(4 \text{ m})$$
$$V = 60 \text{ m}^3$$

The volume calculation is similar to multiplying three monomials with m as the variable.

Multiply: $(5m)(3m)(4m) = 60m^3$

OBJECTIVE 2 Simplify monomials raised to a power.

In order to simplify a monomial raised to a power, we need to understand how to simplify an exponential form raised to a power as in $(2^3)^2$. To simplify $(2^3)^2$, we could use the order of operations agreement and evaluate the exponential form within the parentheses first.

$$(2^3)^2 = (2 \cdot 2 \cdot 2)^2$$
$$= 8^2$$
$$= 64$$

However, there is an alternative. The outside exponent, 2, indicates to multiply 2^3 by itself.

$$(2^3)^2 = 2^3 \cdot 2^3$$
$$= 2^{3+3}$$
$$= 2^6$$

◀ **Note:** Because this is a multiplication of exponential forms that have the same base, we can add the exponents.

Note that $2^6 = 64$ and that the exponent, 6, in the result is the *product* of the original exponents, 3 and 2.

$$(2^3)^2 = 2^{3 \cdot 2} = 2^6$$

Our example suggests the following rule.

RULE *To simplify an exponential form raised to a power, we can multiply the exponents and keep the same base.*

In math language: $(n^a)^b = n^{a \cdot b}$

How might we simplify a monomial raised to a power, as in $(2x^3)^4$? The exponent, 4, means that $2x^3$ is a factor 4 times.

$$(2x^3)^4 = 2x^3 \cdot 2x^3 \cdot 2x^3 \cdot 2x^3$$
$$= 2 \cdot 2 \cdot 2 \cdot 2 \cdot x^{3+3+3+3}$$
$$= 16x^{12}$$

Note: $(2x^3)^4$ is read, "Two x to the third power, raised to the fourth power." Both 3 and 4 are exponents and are referred to as 3rd *power* and 4th *power*. However, in an effort to make the explanations easier to follow, we will use the word *power* to refer to the exponent outside the parentheses and the word *exponent* to refer to the variable's exponent.

Notice that the final coefficient is the result of evaluating 2^4, and the final exponent is the product of the power and the variable's initial exponent, so we could write

$$(2x^3)^4 = 2^4 x^{3 \cdot 4} = 16x^{12}.$$

Our example suggests the following procedure.

PROCEDURE *To simplify a monomial raised to a power:*

1. Evaluate the coefficient raised to that power.
2. Multiply each variable's exponent by the power.

Example 3 Simplify. $(-4x^2)^3$

Solution: $(-4x^2)^3 = (-4)^3 x^{2 \cdot 3}$

$$= -64x^6$$

Do Your Turn 3 ▷

OBJECTIVE 3 Multiply a polynomial by a monomial.

To multiply numbers like $2 \cdot 34$, we stack and use the distributive property to multiply each digit in the 34 by 2.

$$\begin{array}{r} 34 \\ \times\, 2 \\ \hline 68 \end{array}$$

Now let's multiply $2 \cdot 34$ without stacking. We can write 34 as $30 + 4$ so that the expression looks more like the distributive property form.

$$2(34) = 2(30 + 4)$$

Using the distributive property to distribute the 2 to the 30 and the 4, we have

$$2(30 + 4) = 2 \cdot 30 + 2 \cdot 4$$
$$= 60 + 8$$
$$= 68$$

Now let's look at a product involving a polynomial such as $2(3x + 4)$. We use the distributive property just as we did for $2(30 + 4)$. Compare the numeric and algebraic versions:

Numeric: $2(30 + 4) = 2 \cdot 30 + 2 \cdot 4$ Algebraic: $2(3x + 4) = 2 \cdot 3x + 2 \cdot 4$
$\qquad\qquad\quad = 60 + 8$ $\qquad\qquad\qquad\qquad\qquad = 6x + 8$
$\qquad\qquad\quad = 68$

Notice, if we let $x = 10$ in the algebraic version, we get the numeric version.

PROCEDURE *To multiply a polynomial by a monomial, use the distributive property to multiply each term in the polynomial by the monomial.*

Your Turn 3

Simplify.

a. $(3x^2)^4$

b. $(5a^6)^3$

c. $(-2m^4)^5$

d. $(-2t^3)^6$

CONNECTION In Chapter 1, we learned the distributive property is
$a(b + c) = ab + ac.$

Answers to Your Turn 3
a. $81x^8$
b. $125a^{18}$
c. $-32m^{20}$
d. $64t^{18}$

Multiply.

a. $2x(4x^2 + 3)$

b. $5a^2(6a^2 - 7a + 2)$

c. $-4n^3(8n - 9)$

d. $-7m^2(4m^2 - m + 3)$

Example 4 Multiply.

a. $2x(3x^2 + 4x + 1)$

Solution: We multiply each term in the polynomial by $2x$.

$$2x \quad (3x^2 \quad + \quad 4x \quad + \quad 1)$$
$$2x \cdot 3x^2$$
$$2x \cdot 4x$$
$$2x \cdot 1$$

Note:

$2x = 2x^1$ and $4x = 4x^1$, so
$2x \cdot 3x^2 = 2x^1 \cdot 3x^2 = 6x^{1+2} = 6x^3$
and $2x \cdot 4x = 2x^1 \cdot 4x^1 = 8x^{1+1} = 8x^2$.

$$= 2x \cdot 3x^2 + 2x \cdot 4x + 2x \cdot 1$$
$$= 6x^3 + 8x^2 + 2x$$

CONNECTION Multiplying $2x(3x^2 + 4x + 1)$ is essentially the same as this numeric problem:

$$20(300 + 40 + 1) = 20 \cdot 300 + 20 \cdot 40 + 20 \cdot 1$$
$$= 6000 + 800 + 20$$

If you think of x^3 as thousands place, x^2 as hundreds, and x as tens, then the algebraic result is the same.

$$2x(3x^2 + 4x + 1) = 6x^3 + 8x^2 + 2x$$

b. $-3x^2(4x^2 + 5x - 6)$

Solution: Multiply each term in the polynomial by $-3x^2$. Watch the signs! We are multiplying by a negative so the signs are affected.

$$-3x^2 \quad (4x^2 \quad + \quad 5x \quad - \quad 6)$$
$$-3x^2 \cdot 4x^2$$
$$-3x^2 \cdot 5x$$
$$-3x^2 \cdot (-6)$$

Note: Remember, the sign to the left of a term not only indicates addition or subtraction, but also indicates the sign of the term. For example, $+5x$ not only means to add $5x$ but also that $5x$ is positive. Similarly, -6 means not only subtract 6 but also negative 6.

$$= -3x^2 \cdot 4x^2 - 3x^2 \cdot 5x - 3x^2 \cdot (-6)$$
$$= -12x^4 - 15x^3 + 18x^2$$

Note: When we multiply a polynomial by a negative monomial, the signs of the resulting polynomial will be the opposite of the signs in the original polynomial.

$$-3x^2 \ (4x^2 \ + \ 5x \ - \ 6)$$
$$= \ -12x^4 \ - \ 15x^3 \ + \ 18x^2$$

◀ **Do Your Turn 4**

OBJECTIVE 4 Multiply polynomials.

Multiplying two binomials is like multiplying a pair of two-digit numbers. For example, $(x + 2)(x + 3)$ is like $(12)(13)$. To see how to multiply polynomials, let's consider the procedure for multiplying $(12)(13)$.

$$\begin{array}{r} 12 \\ \times \ 13 \\ \hline 36 \\ + \ 12 \\ \hline 156 \end{array}$$

Note: We think to ourselves "3 times 2 is 6, then 3 times 1 is 3," which creates 36. We then move to the 1 in the tens place of the 13 and do the same thing. Because this 1 digit is in the tens place it really means 10, so when we multiply this 10 times 12, it makes 120. We usually omit writing the 0 in the ones place and write 12 in the next two places.

Answers to Your Turn 4
a. $8x^3 + 6x$
b. $30a^4 - 35a^3 + 10a^2$
c. $-32n^4 + 36n^3$
d. $-28m^4 + 7m^3 - 21m^2$

Remember, we are using the distributive property. We multiply each digit in one number by each digit in the other number and shift underneath as we move to each new place.

The same process applies to the binomials. We can stack them like we did the numbers above. However, we'll only stack to see how the method works. You'll find that as the polynomials get more complex, the stacking method becomes too tedious.

$$
\begin{array}{r}
x + 2 \\
x + 3 \\
\hline
3x + 6 \\
+\ x^2 + 2x \\
\hline
x^2 + 5x + 6
\end{array}
$$

Note: We think "3 times 2 is 6, then 3 times x is $3x$." Now move to the x and think "x times 2 is $2x$, then x times x is x^2." Notice how we shifted so that the $2x$ and $3x$ line up. This is because they are like terms. It is the same as lining up the tens column when we work with whole numbers. Note that the numeric result, 156, is basically the same as the algebraic result, $x^2 + 5x + 6$.

Our examples suggest the following procedure.

PROCEDURE *To multiply two polynomials:*

1. Multiply every term in the first polynomial by every term in the second polynomial.
2. Combine like terms.

Example 5 Multiply.

a. $(x + 5)(x + 1)$

Solution: Multiply each term in $x + 1$ by each term in $x + 5$.

$$(x + 5)\ (x + 1)$$

$$
\begin{aligned}
&= x \cdot x + x \cdot 1 + 5 \cdot x + 5 \cdot 1 \\
&= x^2 \quad + x \quad\ \ + 5x + 5 \\
&= x^2 \quad + \quad 6x \qquad + 5 \qquad \text{Combine like terms. } x + 5x = 6x
\end{aligned}
$$

Explanation: First, we multiplied $x \cdot x$ to get x^2.

Then, we multiplied $x \cdot 1$ to get x.

Next, we multiplied $5 \cdot x$ to get $5x$.

Finally, we multiplied $5 \cdot 1$ to get 5.

To finish, we combined like terms x and $5x$ to get $6x$.

b. $(x + 4)(x - 3)$

Solution: Multiply each term in $x - 3$ by each term in $x + 4$.

$$(x + 4)\ (x - 3)$$

$$
\begin{aligned}
&= x \cdot x + x \cdot (-3) + 4 \cdot x + 4 \cdot (-3) \\
&= x^2 \quad -3x \quad\ \ + 4x \qquad - 12 \qquad \text{Combine like terms.} \\
&= x^2 \quad + \qquad x \qquad\qquad\ - 12
\end{aligned}
$$

Multiply.

a. $(x + 4)(x + 2)$

b. $(t + 3)(4t - 1)$

c. $(2y - 5)(3y + 1)$

d. $(n - 6)(7n - 3)$

Explanation: First, we multiplied $x \cdot x$ to get x^2.

Then, we multiplied $x \cdot (-3)$ to get $-3x$.

Next, we multiplied $4 \cdot x$ to get $4x$.

Finally, we multiplied $4 \cdot (-3)$ to get -12.

To finish, we combined the like terms $-3x$ and $4x$ to get x.

◀ **Do Your Turn 5**

LEARNING STRATEGY

Depending on your learning style, you may use different techniques to remember the procedure for multiplying two polynomials.

Tactile learners might imagine a party where two groups of people (two polynomials) meet each other. Every member of one group must shake hands with (multiply) each member of the second group.

Visual learners might want to draw arrows between the terms that represent multiplication, as we did in the two examples.

Auditory learners might want to use the word FOIL, a popular way to remember the process of multiplying two binomials. FOIL stands for **F**irst **O**uter **I**nner **L**ast. We can use Example 5a to demonstrate.

First terms: $x \cdot x = x^2$

$(x + 5)(x + 1)$

Outer terms: $x \cdot 1 = x$

$(x + 5)(x + 1)$

Inner terms: $5 \cdot x = 5x$

$(x + 5)(x + 1)$

Last terms: $5 \cdot 1 = 5$

$(x + 5)(x + 1)$

WARNING FOIL makes sense only when multiplying two binomials. When we multiply larger polynomials, there are too many inner terms. In such cases, we must return to the original procedure of multiplying *every term* of the second polynomial by *every term* of the first polynomial.

State the conjugate of the given binomial.

a. $x + 7$

b. $2a - 5$

c. $-9t + 1$

d. $-6h - 5$

Now let's multiply special binomials called **conjugates.**

DEFINITION Conjugates: Binomials that differ only in the sign separating the terms.

Examples of conjugates $x + 9$ and $x - 9$

$-5y - 7$ and $-5y + 7$

◀ **Note:** In conjugates, the first terms are the same and the second terms are additive inverses.

Answers to Your Turn 5
a. $x^2 + 6x + 8$
b. $4t^2 + 11t - 3$
c. $6y^2 - 13y - 5$
d. $7n^2 - 45n + 18$

Answers to Your Turn 6
a. $x - 7$ b. $2a + 5$
c. $-9t - 1$ d. $-6h + 5$

Example 6 State the conjugate of the given binomial.

a. $x - 2$

Answer: $x + 2$

b. $3x + 5$

Answer: $3x - 5$

c. $-4n + 3$

Answer: $-4n - 3$

◀ **Do Your Turn 6**

When we multiply conjugates, a special pattern occurs in the product.

Your Turn 7

Multiply.

a. $(x + 9)(x - 9)$

b. $(4n - 1)(4n + 1)$

c. $(-2y + 5)(-2y - 5)$

> **Example 7** Multiply. $(3x + 5)(3x - 5)$
>
> **Solution:** Multiply every term in $3x - 5$ by every term in $3x + 5$.
>
> $$(3x + 5)(3x - 5) = 3x \cdot 3x + 3x \cdot (-5) + 5 \cdot (3x) + 5 \cdot (-5) \quad \text{Distribute. (Use FOIL.)}$$
> $$= 9x^2 - 15x + 15x - 25 \quad \text{Multiply.}$$
> $$= 9x^2 + 0 - 25 \quad \begin{array}{l}\text{Combine like terms:}\\ -15x + 15x = 0.\end{array}$$
> $$= 9x^2 - 25$$
>
> **Note:** Since the like terms are additive inverses, their sum is 0, making the result a difference of squares.
>
> $$\underset{\substack{\uparrow \\ \text{Square} \\ \text{of } 3x}}{\underbrace{9x^2}} \overset{\text{Difference}}{-} \underset{\substack{\uparrow \\ \text{Square} \\ \text{of } 5}}{\underbrace{25}}$$

When we multiply two conjugates we will always have like terms that are additive inverses, so the product will always be a difference of squares. We can write the following rule.

> **RULE** *The product of two conjugates is a difference of two squares.*
>
> **In math language:** $(a + b)(a - b) = a^2 - b^2$

Do Your Turn 7 ▷

Your Turn 8

Multiply.

a. $(x + 3)(4x^2 - 5x + 1)$

b. $(2n - 1)(3n^2 - 7n - 5)$

> **Example 8** Multiply. $(x + 4)(2x^2 + 5x - 3)$
>
> **Solution:** We multiply every term in $2x^2 + 5x - 3$ by every term in $x + 4$.
>
> **WARNING** FOIL does *not* make sense here because there are too many terms. Remember, FOIL only describes how to multiply two binomials.
>
> $$(x + 4) \quad (2x^2 + 5x - 3)$$
>
> $$= x \cdot 2x^2 + x \cdot 5x + x \cdot (-3) + 4 \cdot 2x^2 + 4 \cdot 5x + 4 \cdot (-3)$$
> $$= 2x^3 + 5x^2 - 3x + 8x^2 + 20x - 12$$
> $$= 2x^3 + 13x^2 + 17x - 12$$

Explanation: We first multiplied each term in the trinomial by x. We then multiplied each term in the trinomial by 4. Finally, we combined like terms. Notice that we had two pairs of like terms. The $5x^2$ and $8x^2$ combined to equal $13x^2$ and $-3x$ and $20x$ combined to equal $17x$.

Answers to Your Turn 7
a. $x^2 - 81$ b. $16n^2 - 1$
c. $4y^2 - 25$

Answers to Your Turn 8
a. $4x^3 + 7x^2 - 14x + 3$
b. $6n^3 - 17n^2 - 3n + 5$

Do Your Turn 8 ▷

CONNECTION One of the ways to quickly check if you have the correct number of terms in your resulting unsimplified polynomial is to use the knowledge we developed in Chapter 1. Recall that to get the total number of possible combinations in a problem with multiple slots to fill, we can multiply the number of items that can fill the first slot by the number of items that can fill the next, and so on.

Let's look at the resulting unsimplified polynomial from Example 7.

Slot 1 contains one of the two terms from $x + 4$.

$$x \cdot 2x^2 + x \cdot 5x + x \cdot (-3) + 4 \cdot 2x^2 + 4 \cdot 5x + 4 \cdot (-3)$$

Slot 2 contains one of the three terms from $2x^2 + 5x - 3$.

Each term has two slots, one for a term from the binomial $(x + 4)$ and the other for a term from the trinomial $(2x^2 + 5x - 3)$. The binomial has two possible terms that can fill the first slot; the trinomial has three. To find the total number of possible combinations, we multiply $2 \cdot 3$ and get 6, the number of terms in the resulting unsimplified polynomial.

If we were to multiply a trinomial by a trinomial, we should have $3 \cdot 3 = 9$ terms in the resulting polynomial before simplifying.

Your Turn 9

Multiply.

a. $2x(3x - 2)(x + 1)$

b. $-a^2(a + 3)(2a - 1)$

c. $t^3(t - 1)(t + 1)$

d. $-2p^4(2p - 3)(3p + 1)$

Example 9 Multiply. $3x^2(2x - 3)(x + 4)$

Solution: $3x^2(2x - 3)(x + 4)$
$= [3x^2 \cdot 2x - 3x^2 \cdot 3](x + 4)$ Multiply $2x - 3$ by $3x^2$.
$= [6x^3 - 9x^2](x + 4)$ Simplify.
$= 6x^3 \cdot x + 6x^3 \cdot 4 - 9x^2 \cdot x - 9x^2 \cdot 4$ Multiply $6x^3 - 9x^2$ and $x + 4$.
$= 6x^4 + 24x^3 - 9x^3 - 36x^2$ Simplify.
$= 6x^4 + 15x^3 - 36x^2$ Combine like terms.

Explanation: We had to multiply three expressions. When we multiply three numbers, we multiply two of them first and then multiply the third number by the result. For example, for $2 \cdot 3 \cdot 4$ we multiply $2 \cdot 3$ to get 6, and then $6 \cdot 4$ for the final answer of 24. Same thing here, we multiply the first binomial by the monomial and then multiply the remaining binomial by the result.

Check: To check we can multiply the two binomials first, then multiply the result by the monomial.

$3x^2(2x - 3)(x + 4)$
$= 3x^2[2x \cdot x + 2x \cdot 4 - 3 \cdot x - 3 \cdot 4]$ Multiply $2x - 3$ and $x + 4$ first.
$= 3x^2[2x^2 + 8x - 3x - 12]$ Simplify.
$= 3x^2[2x^2 + 5x - 12]$ Combine like terms.
$= 3x^2 \cdot 2x^2 + 3x^2 \cdot 5x - 3x^2 \cdot 12$ Multiply $2x^2 + 5x - 12$ by $3x^2$.
$= 6x^4 + 15x^3 - 36x^2$ Simplify.

We get the same result.

◀ **Do Your Turn 9**

Answers to Your Turn 9
a. $6x^3 + 2x^2 - 4x$
b. $-2a^4 - 5a^3 + 3a^2$
c. $t^5 - t^3$
d. $-12p^6 + 14p^5 + 6p^4$

OBJECTIVE 5 Write an expression for the area of a rectangle or the volume of a box.

Example 10 Write an expression for the area of the rectangle shown.

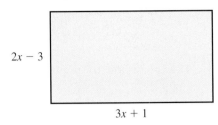

$2x - 3$

$3x + 1$

Your Turn 10

Write an expression for the area of the rectangle shown.

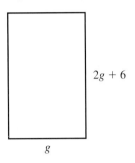

$2g + 6$

g

Solution: To find the area of a rectangle, we multiply length times width. In this case, the length and width are expressions, so we can write only an expression for the area. We cannot get a numerical answer until we are given a value for x.

$$\begin{aligned} \text{area} &= \text{length} \cdot \text{width} \\ &= (3x + 1)(2x - 3) \\ &= 3x \cdot 2x + 3x \cdot (-3) + 1 \cdot 2x + 1 \cdot (-3) \\ &= 2x \cdot 3x + 2x \cdot 1 - 3 \cdot 3x - 3 \cdot 1 \qquad \text{Multiply all terms.} \\ &= 6x^2 - 9x + 2x - 3 \\ &= 6x^2 - 7x - 3 \qquad \text{Combine like terms.} \end{aligned}$$

Answer: The expression for the area is $6x^2 - 7x - 3$.

Do Your Turn 10 ▷

Example 11 Write an expression for the volume of the box shown.

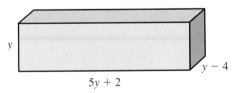

y

$y - 4$

$5y + 2$

Your Turn 11

Write an expression for the volume of the box shown.

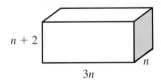

$n + 2$

$3n$

n

Solution: To find the volume of a box, we multiply the length, width, and height. As in Example 9, these dimensions are expressions, so we can write only an expression for the volume.

$$\begin{aligned} \text{volume} &= \text{length} \cdot \text{width} \cdot \text{height} \\ &= (5y + 2)(y - 4)(y) \\ &= (5y + 2)(y \cdot y - 4 \cdot y) \qquad \text{Multiply } y - 4 \text{ by } y. \\ &= (5y + 2)(y^2 - 4y) \qquad \text{Simplify.} \\ &= 5y \cdot y^2 + 5y \cdot (-4y) + 2 \cdot y^2 + 2 \cdot (-4y) \qquad \text{Multiply } 5y + 2 \text{ and } y^2 - 4y. \\ &= 5y^3 - 20y^2 + 2y^2 - 8y \qquad \text{Simplify.} \\ &= 5y^3 - 18y^2 - 8y \qquad \text{Combine like terms.} \end{aligned}$$

◀ **Note:** When multiplying three expressions we use the associative property, multiplying two of them first and then multiplying the result by the third.

Answer: The expression for the volume is $5y^3 - 18y^2 - 8y$.

Do Your Turn 11 ▷

Answer to Your Turn 10
$2g^2 + 6g$

Answer to Your Turn 11
$3n^3 + 6n^2$

For Extra Help

Videotape DVT

Addison-Wesley Tutor Center

Math XL

MyMathLab

Student Solutions Manual

1. When multiplying exponential forms that have the same base, we _____ the exponents and keep the same base.

2. Explain how to multiply monomials.

3. When an exponential form is raised to a power, we _____ the exponents and keep the same base.

4. Explain how to simplify a monomial raised to a power.

5. Explain how to multiply a polynomial by a monomial.

6. Explain how to multiply two binomials.

7. What are conjugates? Give an example of a pair of conjugates.

8. The product of two binomials that are conjugates is a _____ of squares.

For Exercises 9–26, multiply.

9. $x^3 \cdot x^4$

10. $y^3 \cdot y^5$

11. $t \cdot t^9$

12. $m^4 \cdot m$

13. $7a^2 \cdot 3a$

14. $9b^3 \cdot 4b$

15. $2n^3 \cdot 5n^4$

16. $4x^5 \cdot 3x^3$

17. $-8u^4 \cdot 3u^2$

18. $9x^2(-4x)$

19. $-y^5(-8y^2)$

20. $-3n^4(-7n^6)$

21. $5x(9x^2)(2x^4)$

22. $2a^2(5a)(3a)$

23. $3m(-3m^2)(4m)$

24. $u^3(7u^4)(-8u^3)$

25. $3y^4(-5y^3)(-4y^2)$

26. $-6k^5(2k)(-4k^2)$

For Exercises 27 and 28, find and explain the mistakes, then work the problems correctly.

27. $9x^3 \cdot 5x^4$
$= 45x^{12}$

28. $-3m^2 \cdot 10m^5$
$= -30m^{10}$

For Exercises 29–42, simplify.

29. $(2x^3)^2$

30. $(3a^4)^3$

31. $(-7m^5)^2$

32. $(-4n^2)^2$

33. $(-2y^6)^5$

34. $(-3t^7)^3$

35. $(-4x^5)^4$

36. $(-5y^3)^4$

37. $(5y^6)^2$

38. $(6a^4)^3$

39. $(-5v^6)^3$

40. $(-2j^6)^5$

41. $(3x)(2x^3)^4$

42. $7a^3 \cdot (6a)^2$

For Exercises 43 and 44, find and explain the mistakes, then work the problems correctly.

43. $(3x^4)^4$
$= 3x^8$

44. $(-2b^4)^3$
$= -6b^7$

For Exercises 45–64, multiply.

45. $4(2x + 3)$ **46.** $9(3x + 1)$ **47.** $5(3x - 7)$ **48.** $8(2t - 3)$

49. $-3(5t + 2)$ **50.** $-6(4y + 5)$ **51.** $-4(7a - 9)$ **52.** $-2(5m - 6)$

53. $6u(3u + 4)$ **54.** $3x(7x - 1)$ **55.** $-2a(3a + 1)$ **56.** $-5y(2y - 4)$

57. $-2x^3(5x - 8)$ **58.** $7y^2(4y - 3)$ **59.** $8x(2x^2 + 3x - 4)$ **60.** $2n(7n^2 - 5n + 3)$

61. $-x^2(5x^2 - 6x + 9)$ **62.** $-t^3(3t^2 + t - 8)$ **63.** $-2p^2(3p^2 + 4p - 5)$ **64.** $-3a^2(4a^2 - a + 3)$

For Exercises 65–78, multiply.

65. $(x + 4)(x + 2)$ **66.** $(a + 3)(a + 1)$

67. $(m - 3)(m + 5)$ **68.** $(t + 7)(t - 2)$

69. $(y - 8)(y + 1)$ **70.** $(u + 4)(u - 9)$

71. $(3x + 2)(4x - 5)$ **72.** $(5t - 4)(3t + 1)$

73. $(4x - 1)(3x - 5)$ **74.** $(2a - 7)(a - 6)$

75. $(3t - 5)(4t + 1)$ **76.** $(h + 2)(4h - 3)$

77. $(a - 7)(2a - 5)$ **78.** $(2x - 5)(3x - 1)$

For Exercises 79–86, state the conjugate of the given binomial.

79. $x - 7$ **80.** $y + 2$ **81.** $2x + 5$ **82.** $3t - 8$

83. $-2 + 8b$ **84.** $-4 - u$ **85.** $9 - n$ **86.** $2 + 5k^2$

For Exercises 87–92, multiply.

87. $(x + 3)(x - 3)$ **88.** $(a - 4)(a + 4)$ **89.** $(5t + 6)(5t - 6)$

90. $(-6x - 1)(-6x + 1)$ **91.** $(2m + n)(2m - n)$ **92.** $(-h + 2k)(-h - 2k)$

For Exercises 93 and 94, find and explain the mistakes, then work the problems correctly.

93. $(2x - 5)(3x + 1)$
$= 6x + 2x - 15x + 5$
$= -7x + 5$

94. $(a - 6)(a - 6)$
$= a^2 + 36$

For Exercises 95–104, multiply.

95. $(x + 5)(2x^2 - 3x + 1)$ **96.** $(a + 2)(5a^2 + 3a - 4)$

97. $(2y - 3)(4y^2 + y - 6)$ **98.** $(3m - 7)(m^2 - 5m - 2)$

99. $2a(a - 1)(a + 2)$

100. $3c(c + 4)(c - 2)$

101. $-x^2(2x + 3)(x + 1)$

102. $-p^3(p + 2)(4p + 1)$

103. $-3q^2(3q - 1)(3q + 1)$

104. $-2t^3(5t + 2)(5t - 2)$

For Exercises 105 and 106, write an expression for the area.

105.

8y

3y + 1

106.

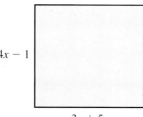

4x − 1

3x + 5

For Exercises 107 and 108, write an expression for the volume.

107.

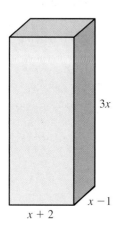

3x

x − 1

x + 2

108.

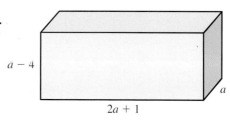

a − 4

2a + 1

a

PUZZLE PROBLEM If there are 10^{11} stars in a galaxy and about 10^{11} galaxies in the visible universe, about how many stars are in the visible universe?

Review Exercises

[1.4] **1.** Divide. $846 \div 2$

[2.4] **2.** Find the missing factor. $-6 \cdot (?) = 24$

[1.6] **3.** A rectangular room has an area of 132 square feet. If one wall is 12 feet long, what must the length of the other wall be?

[1.6] **4.** The packing crate shown has a volume of 210 cubic feet. Find the missing length.

[3.3] **5.** Combine like terms. $3x - 7x$

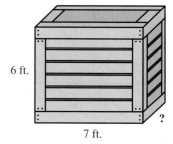

6 ft.

7 ft.

?

3.6 Prime Numbers and GCF

OBJECTIVE 1 Determine if a number is prime, composite, or neither.

In this section, we learn about special numbers called **prime numbers.**

DEFINITION **Prime number:** A natural number other than 1 that has exactly different factors, 1 and the number itself.

Let's develop a list of primes.

Is 0 a prime? No, because 0 is not a natural number. (0 is a whole number.)

Is 1 a prime? No, because a prime must have exactly two *different* factors and 1's factors are the same 1 and 1.

Is 2 a prime? Yes, because 2 is a natural number whose only factors are 1 and 2 (itself).

Is 3 a prime? Yes, because 3 is a natural number whose only factors are 1 and 3 (itself).

Is 4 a prime? No, because 4 has factors other than 1 and 4 (2 is also a factor of 4).

If we continue this line of thinking, we get the following list of prime numbers:

$$2, 3, 5, 7, 11, 13, 17, 19, 23, 29, 31, 37, 41, 43, \ldots$$

This list of prime numbers continues without end. However, as we progress to larger numbers, it becomes more difficult to determine whether a given number is prime.

Note that once we establish a prime number, such as 2, no number that is divisible by *that* prime number can be a prime number. Because every even number is divisible by 2, we can say that no even number larger than 2 is prime. Or, in the case of the prime number 3, no number that is divisible by 3 can be prime. Numbers like 4, 6, and 9 that are not prime are called **composite numbers.**

DEFINITION **Composite number:** A natural number that has factors other than 1 and itself.

Is 0 composite? No, because it is not a natural number. Remember, 0 is a whole number.

Is 1 composite? No, because its only factor is itself.

Conclusion: 0 and 1 are neither prime nor composite.

The first composite number is 4. Here is a list of composites:

$$4, 6, 8, 9, 10, 12, 14, 15, 16, 18, 20, 21, 22, 24, 25, \ldots$$

DISCUSSION Is every prime number an odd number?
Is every odd number a prime number?
Is every even number a composite?

Conclusion: Every composite number is divisible by at least one prime number.

OBJECTIVES

1 Determine if a number is prime, composite, or neither.

2 Find the prime factorization of a given number.

3 Find all possible factors of a given number.

4 Find the greatest common factor of a given set of numbers by listing.

5 Find the greatest common factor of a given set of numbers using prime factorization.

6 Find the greatest common factor of a set of monomials.

This conclusion gives us a way to determine if a given number is prime or composite. If a given number is composite, it is divisible by a prime. All we need to do is divide a given number by each prime on the list of prime numbers, searching for a prime that divides the given number without a remainder. But what if we don't find a prime that divides the given number without a remainder? How far down the list do we need to go before we conclude a given number is itself a prime? Let's consider a couple of cases.

First, let's determine whether 91 is prime or composite. We divide 91 by each of the prime numbers. It is helpful to use the divisibility tricks for 2, 3, and 5 that we learned back in Section 1.4.

Is 91 divisible by 2? No, because 91 is odd.

Is 91 divisible by 3? No, because $9 + 1 = 10$ and this sum is not divisible by 3.

Is 91 divisible by 5? No, because 91 does not end in 0 or 5.

Is 91 divisible by 7? There isn't a good divisibility trick for 7, so we use long division.

$$
\begin{array}{r}
13 \\
7{\overline{)91}} \\
-7 \\
\hline
21 \\
-21 \\
\hline
0
\end{array}
$$
   Yes, 91 is divisible by 7.

Answer: Because we found a number other than 1 and 91 that divides 91 with no remainder, we conclude that 91 is a composite number.

Now let's determine whether 127 is prime or composite. Again, we divide 127 by each of the prime numbers.

Is 127 divisible by 2? No, because 127 is odd.

Is 127 divisible by 3? No, because $1 + 2 + 7 = 10$ and this sum is not divisible by 3.

Is 127 divisible by 5? No, because 127 does not end in 0 or 5.

Is 127 divisible by 7? There isn't a good trick for 7 so we just divide.

$$
\begin{array}{r}
18 \\
7{\overline{)127}} \\
-7 \\
\hline
57 \\
-56 \\
\hline
1
\end{array}
$$
   No, 127 is not divisible by 7.

Is 127 divisible by 11?

$$
\begin{array}{r}
11 \\
11{\overline{)127}} \\
-11 \\
\hline
17 \\
-11 \\
\hline
6
\end{array}
$$
   No, 127 is not divisible by 11.

How far do we need to go? Actually, we can stop testing now and conclude that 127 is a prime number. We can stop because the quotient is equal to the divisor (both 11). If we get to a point where the quotient is *equal to* or *less than* the prime divisor, then we can stop testing and conclude that the given number is a prime because if we test larger prime divisors, the quotients will continue to get smaller and actually be in the realm of numbers we've already tested. It is therefore pointless to continue.

From what we've learned in testing 91 and 127, we can write the following procedure for determining if a given number is prime or composite.

PROCEDURE *To determine if a given number is prime or composite, divide the given number by the primes on the list of prime numbers and consider the results.*
1. If the given number is divisible by the prime, then stop and conclude that the given number is a composite number.
2. If the given number is not divisible by the prime, then consider the quotient.
   a. If the quotient is greater than the current prime divisor, then repeat the process with the next prime on the list of prime numbers.
   b. If the quotient is equal to or less than the current prime divisor, then stop and conclude that the given number is itself a prime number.

Example 1 Determine if 157 is prime or composite.

Solution: Divide by the list of primes.

Is 157 divisible by 2? No, because 157 is odd.

Is 157 divisible by 3? No, because $1 + 5 + 7 = 13$ and this sum is not divisible by 3.

Is 157 divisible by 5? No, because 157 does not end in 5 or 0.

Is 157 divisible by 7?

$$\begin{array}{r} 22 \\ 7\overline{)157} \\ -14 \\ \hline 17 \\ -14 \\ \hline 3 \end{array}$$

◀ No, 157 is not divisible by 7. Notice that the quotient, 22, is greater than the divisor, 7. This means we need to go on to the next prime, which is 11.

Is 157 divisible by 11?

$$\begin{array}{r} 14 \\ 11\overline{)157} \\ -11 \\ \hline 47 \\ -44 \\ \hline 3 \end{array}$$

◀ No, 157 is not divisible by 11. Because the quotient, 14, is greater than the divisor, 11, we must go on to the next prime, which is 13.

Is 157 divisible by 13?

$$\begin{array}{r} 12 \\ 13\overline{)157} \\ -13 \\ \hline 27 \\ -26 \\ \hline 1 \end{array}$$

◀ No, 157 is not divisible by 13. This time the quotient, 12, is less than the divisor, 13. This means that we can stop testing. We can conclude that 157 is a prime number.

Answer: 157 is a prime number.

Do Your Turn 1 ▷

Your Turn 1

Determine if the given number is prime, composite, or neither.

a. 39

b. 119

c. 1

d. 107

Answers to Your Turn 1
a. composite b. composite
c. neither d. prime

OBJECTIVE 2 Find the prime factorization of a given number.

We have learned that every composite number is divisible by at least one prime number. It turns out that we can write any composite number as a product of only prime factors. We call this expression a **prime factorization.**

DEFINITION Prime factorization: A number written as a product of only prime factors.

For example, the prime factorization of 20 is $2 \cdot 2 \cdot 5$, which we can write in exponential form as $2^2 \cdot 5$.

There are several popular methods for finding a number's prime factorization. The factor tree method is one of the most flexible. We will use 20 to demonstrate. First, draw two branches below the given number.

At the end of these branches we place two factors whose product is the given number. For 20, we could use 2 and 10, or 4 and 5. It is helpful to circle the prime factors as they appear.

We then keep breaking down all the composite factors until we only have prime factors.

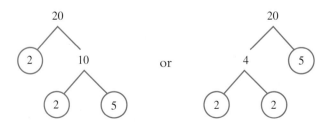

All the circled prime numbers in the factor tree form the prime factorization, so $20 = 2 \cdot 2 \cdot 5 = 2^2 \cdot 5$. Notice that it didn't matter whether we used 2 and 10 or 4 and 5 to begin the factor tree. Either way we ended up with two 2's and one 5 circled.

PROCEDURE *To find the prime factorization of a number, use a factor tree.*

1. Draw two branches below the number.
2. Place two factors that multiply to equal the given number at the end of the two branches.
3. Repeat steps 1 and 2 for every composite factor.
4. Place all the prime factors together in a multiplication sentence.

Example 2 Find the prime factorization. Write the answer in exponential form.

a. 84

Solution: Use a factor tree.

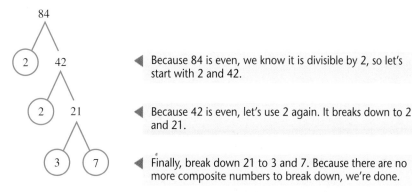

◄ Because 84 is even, we know it is divisible by 2, so let's start with 2 and 42.

◄ Because 42 is even, let's use 2 again. It breaks down to 2 and 21.

◄ Finally, break down 21 to 3 and 7. Because there are no more composite numbers to break down, we're done.

Answer: $2^2 \cdot 3 \cdot 7$

We could have started the factor tree for 84 many different ways and ended up with the same result. For example, we could have started with 4 and 21, or 7 and 12, or 3 and 28.

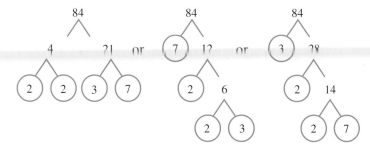

No matter how we start, we end up with the same result each time. The number 84 has two 2's, a 3, and a 7 as its prime factors.

In fact, 84 is the only number that has two 2's, a single 3, and a single 7. The product of two 2's, a single 3, and a single 7 will always equal 84.

Conclusion: No two numbers have the exact same prime factorization. In math language, we say that prime factorizations are unique.

◀ Your Turn 2

Find the prime factorization of each given number. Write your answer in exponential form.

a. 64

b. 105

c. 560

d. 910

b. 3500

Solution: Use a factor tree. Since 3500 is divisible by 10, let's start by dividing out the tens. Each division by 10 removes a zero.

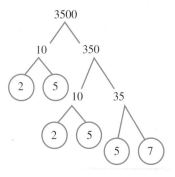

Note: We can check the prime factorization by multiplying the prime factors together. The product should be the given number.

Answer: $2^2 \cdot 5^3 \cdot 7$

◀ **Do Your Turn 2**

OBJECTIVE 3 Find all possible factors of a given number.

In mathematics, we often need to consider the factors or divisors of a number. Recall that factors of a given number are numbers that divide the given number with no remainder. Let's consider how to find all possible factors of a given number.

Your Turn 3

List all possible factors of each given number.

a. 36

b. 60

c. 84

Example 3 List all possible factors of 24.

Solution: To list all the factors, we'll divide by the natural numbers 1, 2, 3, and so on until we have them all. We can start with 1 and 24, then go to 2 and 12 and so forth.

$$1 \cdot 24$$
$$2 \cdot 12$$
$$3 \cdot 8$$
$$4 \cdot 6$$

Explanation: Once we established 2 and 12, all remaining factors must be between 2 and 12. When we established 3 and 8, all remaining factors must be between 3 and 8. When we established 4 and 6, any remaining factors must be between 4 and 6. The only natural number in between 4 and 6 is 5. Notice that 24 is not divisible by 5. This means we have all possible factors.

◀ **Do Your Turn 3**

OBJECTIVE 4 Find the greatest common factor of a given set of numbers by listing.

Suppose a fencing company has a job to enclose three sides of a yard. The left side is 32 feet long, the center is 40 feet, and the right side is 24 feet. The company is to use prefabricated sections. What is the longest section that can be used so that no partial sections are needed?

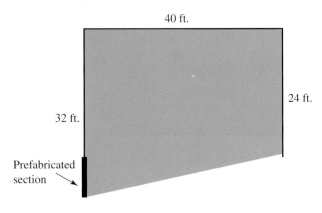

Answers to Your Turn 2
a. 2^6 b. $3 \cdot 5 \cdot 7$ c. $2^4 \cdot 5 \cdot 7$
d. $2 \cdot 5 \cdot 7 \cdot 13$

Answers to Your Turn 3
a. 1, 2, 3, 4, 6, 9, 12, 18, 36
b. 1, 2, 3, 4, 5, 6, 10, 12, 15, 20, 30, 60
c. 1, 2, 3, 4, 6, 7, 12, 14, 21, 28, 42, 84

Notice that several sizes would work. They could use 2-foot sections, 4-foot sections, and 8-foot sections because 32, 40, and 24 are all divisible by those numbers. But, the longest possible section would be the 8-foot section. The number 8 is the greatest number that divides 32, 40, and 24 with no remainder and is therefore called the **greatest common factor.**

DEFINITION　**Greatest common factor:** The largest number that divides all given numbers with no remainder.

Note: The greatest common factor is sometimes called the greatest common divisor.

We abbreviate greatest common factor as GCF. One way to find the GCF for a set of numbers is by listing factors. We list all possible factors for each given number and then search the lists for the greatest factor common to all lists.

PROCEDURE　*To find the greatest common factor by listing:*
1. List all possible factors for each given number.
2. Search the lists for the greatest factor common to all lists.

Example 4　Find the GCF of 24 and 60.

Solution: All possible factors of 24 were listed in Example 3, and you listed all possible factors of 60 in *Your Turn 3b*.

For 24 we have: 1, 2, 3, 4, 6, 8, 12, 24

For 60 we have: 1, 2, 3, 4, 5, 6, 10, 12, 15, 20, 30, 60

Notice that 12 is the greatest factor common to both lists.

Answer: GCF = 12

We can check the greatest common factor by dividing it into the given numbers.

These quotients share no common factors except the number 1.

$$
\begin{array}{r} 2 \\ 12\overline{)24} \\ -24 \\ \hline 0 \end{array}
\qquad
\begin{array}{r} 5 \\ 12\overline{)60} \\ -60 \\ \hline 0 \end{array}
$$

Notice that 12 divides 24 and 60 with no remainder, so it is a common factor. We can tell 12 is the *greatest* common factor because 1 is the only common factor of the quotients.

Conclusion: If you divide each given number by its GCF, each remainder will be 0, and the only common factor of the quotients will be 1.

Suppose we had mistakenly concluded that 6 was the GCF of 24 and 60. When we check 6 by dividing it into 24 and 60, we see that it is a common factor because the remainders are 0, but not the GCF because the quotients have a common factor other than 1.

These quotients are both divisible by 2. This means that 6 is not the greatest common factor.

$$
\begin{array}{r} 4 \\ 6\overline{)24} \\ -24 \\ \hline 0 \end{array}
\qquad
\begin{array}{r} 10 \\ 6\overline{)60} \\ -60 \\ \hline 0 \end{array}
$$

Your Turn 4

Find the GCF by listing.

a. 36 and 54

b. 32 and 45

c. 28, 32, and 40

This check technique also offers a very easy way to fix a mistake. Multiplying the incorrect answer by the greatest common factor of the quotients in the check produces the correct GCF. For example, multiplying our incorrect answer, 6, by the greatest common factor of the quotients, which is 2, produces the GCF: $6 \cdot 2 = 12$.

◀ **Do Your Turn 4**

OBJECTIVE 5 Find the greatest common factor of a given set of numbers using prime factorization.

For smaller numbers, listing is a good method, but for larger numbers, there is a more efficient method we can use. It turns out we can use prime factorization to generate the GCF of a given set of numbers. Consider the prime factorizations for the numbers from Example 4.

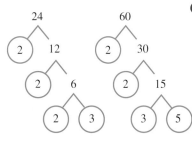

$$24 = 2 \cdot 2 \cdot 2 \cdot 3 = 2^3 \cdot 3$$
$$60 = 2 \cdot 2 \cdot 3 \cdot 5 = 2^2 \cdot 3 \cdot 5$$

CONNECTION You may wonder why we learn several methods for solving the same type of problem. Think of methods as if they were tools. A good analogy is a saw. For small jobs where you only need to cut through a few pieces of wood, a hand saw is a great tool and simple to use. However, for a big job involving lots of cutting or large pieces of wood, a power saw, although it takes more effort to set up, is much more efficient to use. The listing method for GCF, like the hand saw, is a good method for simpler problems involving small numbers. The prime factorization method is like the power saw. It's a good method for the bigger problems, involving large numbers or many numbers.

We already know the answer is 12 from our listing method in Example 4. Lets look at the prime factorization for 12 to see if we can discover a rule for finding the primes to get the GCF.

$$12 = 2 \cdot 2 \cdot 3 = 2^2 \cdot 3$$

Notice that 12 contains two factors of 2 and one factor of 3, which are the factors that are common to 24's and 60's factorizations. This suggests that we only use primes that are common to all factorizations involved. The factor 5 is not common to both 24's and 60's factorizations, so it is not included in the GCF.

$$24 = \boxed{2} \cdot \boxed{2} \cdot 2 \cdot \boxed{3} = 2^3 \cdot 3$$
$$60 = \boxed{2} \cdot \boxed{2} \cdot \boxed{3} \cdot 5 = 2^2 \cdot 3 \cdot 5$$
$$\text{GCF} = 2 \cdot 2 \cdot 3 \quad = 2^2 \cdot 3 = 12$$

Answers to Your Turn 4
a. 18 b. 1 c. 4

We could also compare exponents for a particular common prime factor to determine how many of that factor should be included in the GCF. For example, 2^2 is included instead of 2^3 because 2^2 has the lesser exponent. This suggests the following.

> **PROCEDURE** *To find the greatest common factor of a given set of numbers:*
> 1. Write the prime factorizations of each number in exponential form.
> 2. Create a factorization for the GCF that contains only those prime factors common to all the factorizations, each raised to its least exponent in all the factorizations.
> 3. Multiply.
>
> **Note:** If there are no common primes, then the GCF is 1.

Example 5 Find the GCF of 3024 and 2520.

Solution: First, we write the prime factorization of 3024 and 2520 in exponential form.

$$3024 = 2^4 \cdot 3^3 \cdot 7$$
$$2520 = 2^3 \cdot 3^2 \cdot 5 \cdot 7$$

The common prime factors are 2, 3, and 7. Now we compare the exponents of each of these common factors. For 2, the least exponent is 3. For 3, the least exponent is 2. For 7, the least exponent is 1. The GCF will be the product of 2^3, 3^2, and 7.

$$GCF = 2^3 \cdot 3^2 \cdot 7$$
$$= 8 \cdot 9 \cdot 7$$
$$= 504$$

Let's check by dividing 504 into 3024 and 2520. We should find that 504 divides both with no remainder and that 1 is the only factor common to both quotients.

The only factor common to both quotients is 1.

$$
\begin{array}{r}
6 \\
504\overline{)3024} \\
-3024 \\
\hline
0
\end{array}
\qquad
\begin{array}{r}
5 \\
504\overline{)2520} \\
-2520 \\
\hline
0
\end{array}
$$

0 ←—— No remainder ——→ 0

504 checks. It is the GCF.

Do Your Turn 5 ▶

Your Turn 5

Find the GCF using primes.

a. 84 and 48

b. 28 and 140

c. 60 and 77

d. 42, 63, and 105

Your Turn 6

A rectangular kitchen floor measures 18 feet long by 16 feet wide. What is the largest-size square tile that can be used to cover the floor without cutting or overlapping the tiles?

Example 6 The floor of a 42-foot-by-30-foot room is to be covered with colored squares like a checkerboard. All the squares must be of equal size with whole-number dimensions and may not be cut or overlapped to fit inside the room. Find the dimensions of the largest possible square that can be used.

Solution: Because we want the largest possible square we must find the GCF of 42 and 30.

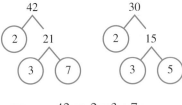

$$42 = 2 \cdot 3 \cdot 7$$
$$30 = 2 \cdot 3 \cdot 5$$
$$\text{GCF} = 2 \cdot 3 = 6$$

Answer: The largest square that can be used is 6 ft. by 6 ft.

◀ **Do Your Turn 6**

OBJECTIVE 6 Find the greatest common factor of a set of monomials.

Now let's find the greatest common factor of a set of monomials. We use the same prime factorization method that we have learned and treat variables as prime factors.

Example 7 Find the GCF of the monomials.

a. $18x^4$ and $12x^3$

Solution: Find the prime factorization of each monomial. We treat the variables like prime factors.

$$18x^4 = 2 \cdot 3^2 \cdot x^4 \qquad \text{Write the prime factorization of } 18x^4 \text{ in exponential form.}$$
$$12x^3 = 2^2 \cdot 3 \cdot x^3 \qquad \text{Write the prime factorization of } 12x^3 \text{ in exponential form.}$$

The GCF contains the common prime factors raised to their least exponent. The common prime factors are 2, 3, and x. The least exponent for 2 is 1, so we have a single 2 factor in the GCF. The least exponent for 3 is also 1. The least exponent for x is 3.

$$\text{GCF} = 2 \cdot 3 \cdot x^3 = 6x^3$$

CONNECTION In Section 3.7, we will learn how to factor a monomial out of the terms in a given polynomial. The monomial we factor out will be the greatest common factor of the terms in the polynomial.

b. $24n^2$ and $40n^5$

Solution: $24n^2 = 2^3 \cdot 3 \cdot n^2 \qquad \text{Write the prime factorization of } 24n^2 \text{ in exponential form.}$
$40n^5 = 2^3 \cdot 5 \cdot n^5 \qquad \text{Write the prime factorization of } 40n^5 \text{ in exponential form.}$

The common prime factors are 2 and n. The least exponent for 2 is 3 and the least exponent for n is 2.

$$\text{GCF} = 2^3 \cdot n^2 = 8n^2$$

Answer to Your Turn 6
2 ft. by 2 ft.

c. $14t^4$ and 45

Solution: $14t^4 = 2 \cdot 7 \cdot t^2$ Write the prime factorization of $14t^4$ in exponential form.

$45 = 3^3 \cdot 5$ Write the prime factorization of 45 in exponential form.

There are no common prime factors, so the GCF is 1.

d. $18y^4$, $27y^3$, and $54y^2$

Solution: $18y^4 = 2 \cdot 3^2 \cdot y^4$ Write the prime factorization of $18y^4$ in exponential form.

$27y^3 = 3^3 \cdot y^3$ Write the prime factorization of $27y^3$ in exponential form.

$54y^2 = 2 \cdot 3^3 \cdot y^2$ Write the prime factorization of $54y^2$ in exponential form.

The common prime factors are 3 and y. The least exponent for 3 is 2 and the least exponent for y is 2. Although 2 appears in two of the factorizations, it is not in all three, so it is not included in the GCF.

$$\text{GCF} = 3^2 \cdot y^2 = 9y^2$$

Do Your Turn 7 ▷

Your Turn 7

Find the GCF.

a. $20x$ and $32x^3$

b. $36m^4$ and $12m^2$

c. $16x^3$ and 30

d. $24h^5$, $16h^4$, and $48h^3$

Answers to Your Turn 7
a. $4x$ **b.** $12m^2$ **c.** 2 **d.** $8h^3$

3.6 Exercises

For Extra Help

 Videotape DVT

 Addison-Wesley Tutor Center

 Math XL

 MyMathLab

 Student Solutions Manual

1. What is a prime number? Give three examples.

2. What is a composite number? Give three examples.

3. What is the greatest common factor of a set of numbers?

4. Explain how to use prime factorization to find the GCF of a set of numbers.

For Exercises 5–20, determine whether the given number is prime, composite, or neither.

5. 49	**6.** 93	**7.** 89	**8.** 71
9. 0	**10.** 153	**11.** 179	**12.** 187
13. 1	**14.** 253	**15.** 247	**16.** 311
17. 377	**18.** 389	**19.** 409	**20.** 323

21. Are all prime numbers odd? Explain.

22. Are all odd numbers prime? Explain.

For Exercises 23–38, find the prime factorization.

23. 80	**24.** 72	**25.** 156	**26.** 120
27. 268	**28.** 180	**29.** 200	**30.** 324
31. 343	**32.** 220	**33.** 975	**34.** 462
35. 378	**36.** 875	**37.** 952	**38.** 1404

For Exercises 39–44, list all possible factors of the given number.

39. 60	**40.** 40
41. 81	**42.** 80
43. 120	**44.** 54

For Exercises 45–50, find the GCF by listing.

45. 24 and 60	**46.** 40 and 32
47. 81 and 65	**48.** 80 and 100
49. 72 and 120	**50.** 54 and 63

For Exercises 51–62, find the GCF by prime factorization.

51. 140 and 196

52. 240 and 150

53. 130 and 78

54. 324 and 270

55. 336 and 504

56. 270 and 675

57. 99 and 140

58. 252 and 143

59. 60, 120 and 140

60. 120, 300, and 420

61. 64, 160, and 224

62. 315, 441, and 945

For Exercises 63–66, solve.

63. A company produces plastic storage containers. For shipping purposes, the containers must be 60 inches wide by 70 inches long. The containers are sectioned into squares. What is the largest-size square that can be used without cutting or overlapping?

64. The ceiling of a 45-foot-long by 36-foot-wide lobby of a new hotel is to be sectioned into a grid of square regions by wooden beams. What is the largest square region possible so that all regions are of equal size, and no overlapping or cutting of any region takes place?

65. A landscaper is planning a sprinkler system that will require precut PVC pipe to be connected and placed in the ground. The trenches he will dig are to be 40 feet, 32 feet, and 24 feet in length. What is the longest length of PVC pipe he can use that will exactly fit the length of each trench and not require any further cutting? How many of the precut pipes will be used in each trench?

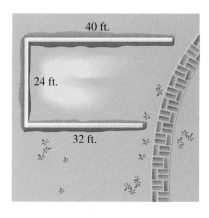

40 ft.

24 ft.

32 ft.

66. A professor has three classes. The first class has 30 people, the second class has 42 people, and the third class has 24 people. She wants to break the classes into project groups so that all groups are the same size no matter which class they are in. What is the largest-size group she can make? How many groups will be in each class?

For Exercises 67–80, find the GCF.

67. $20x$ and 30

68. 18 and $9t$

69. $8x^2$ and $14x^5$

70. $12a^5$ and $15a$

71. $24h^3$ and $35h^4$

72. $15m^5$ and $36m^4$

73. $27n^4$ and 49

74. 15 and $42r^3$

75. $56x^9$ and $72x^7$

76. $48t^5$ and $60t^8$

77. $18n^2$, $24n^3$, and $30n^4$

78. $20y^3$, $30y^4$, and $40y^6$

79. $42x^4$, $35x^6$, and $28x^3$

80. $36m^5$, $48m^4$, and $60m^6$

Review Exercises

[1.1, 1.3] **1.** Write expanded form for 5,784,209.

[1.4] **2.** Divide. $34{,}328 \div 17$

[1.3] **3.** Evaluate. 2^7

[1.3] **4.** A college has 265 employees. The college wishes to assign an ID to each employee. The proposed ID would contain one of the digits 0–9 and one of the letters A–Z. Is the proposed ID system sufficient?

[1.4] **5.** A company must produce square plastic panels that have an area of 36 square feet. What must be the dimensions of the squares?

3.7 Introduction to Factoring

OBJECTIVE 1 Divide monomials.

OBJECTIVES
1. Divide monomials.
2. Divide a polynomial by a monomial.
3. Find a missing factor.
4. Factor the GCF out of a polynomial.

In Section 3.5, we learned how to multiply exponential forms. We concluded that when we multiply exponential forms that have the same base, we can add the exponents and keep the same base. For example,

$$2^3 \cdot 2^4 = 2^{3+4} = 2^7$$

Now let's consider how to divide exponential forms. Because multiplication and division are inverse operations, we can write a related division statement for the preceding multiplication statement by dividing the product by one of the factors.

$$2^3 \cdot 2^4 = 2^7 \qquad \text{implies} \qquad 2^7 \div 2^4 = 2^3$$

Since the exponents were added in the multiplication problem, it follows that they should be subtracted in the division problem.

$$2^7 \div 2^4 = 2^{7-4} = 2^3$$

Note that we subtracted the divisor's exponent from the dividend's exponent. This order is important because subtraction is not commutative.

> **RULE** *When dividing exponential forms that have the same base, we can subtract the divisor's exponents from the dividend's exponent and keep the same base.*
>
> **In math language:** $n^a \div n^b = \dfrac{n^a}{n^b} = n^{a-b}$, when $n \neq 0$
>
> **Note:** $n \neq 0$ because if n were replaced with 0, we would have $\dfrac{0^a}{0^b}$. In Section 1.4, we concluded that when the dividend and divisor are both 0, the problem is indeterminate.

> **CONNECTION** We could simplify $2^7 \div 2^4$ using the order of operations.
>
> $2^7 \div 2^4 = 128 \div 16 = 8$
>
> The prime factorization of 8 is 2^3.
>
> $2^7 \div 2^4 = 128 \div 16 = 8 = 2^3$

> **Note:** Remember we can write division in fraction form.
>
> $$2^7 \div 2^4 = \frac{2^7}{2^4}$$

Example 1 Divide. $x^6 \div x^2$

Solution: $x^6 \div x^2 = x^{6-2} = x^4$ or $\dfrac{x^6}{x^2} = x^{6-2} = x^4$

Explanation: Because the bases were the same, we subtracted the exponents and kept the same base.

Check: Multiplying the answer, x^4, by the divisor, x^2, should equal the dividend, x^6.

$$x^4 \cdot x^2 = x^{4+2} = x^6$$

Do Your Turn 1 ▷

Now let's look at exponential forms that have the same base and the same exponent, as in $\dfrac{2^3}{2^3}$. Using the rule that we developed, we subtract the divisor's exponent from the dividend's exponent.

$$\frac{2^3}{2^3} = 2^{3-3} = 2^0$$

> **Your Turn 1**
>
> *Divide.*
>
> a. $3^9 \div 3^2$
>
> b. $\dfrac{a^{12}}{a^3}$
>
> c. $y^6 \div y^5$
>
> **Answers to Your Turn 1**
> a. 3^7 b. a^9 c. y

What does it mean to have 0 as an exponent? To answer this question we could evaluate $\frac{2^3}{2^3}$ using the order of operations. The result must be equal to 2^0.

$$\frac{2^3}{2^3} = \frac{8}{8} = 1$$

Because 2^0 and 1 are correct results from the same problem, they must be equal.

$$2^0 = 1$$

Conclusions: When dividing exponential forms that have the same base and the same exponent, the result is 1; and any nonzero base with an exponent of 0 simplifies to the number 1.

Note that the situation 0^0 would arise from a division situation where two exponential forms have 0 as the base and both have the same exponent.

$$\frac{0^n}{0^n} = 0^{n-n} = 0^0$$

In Section 1.4, we concluded that $\frac{0}{0}$ is indeterminate, therefore 0^0 is indeterminate.

Conclusion: 0^0 is indeterminate.

> **RULE** *Any base other than 0 raised to the 0 power simplifies to the number 1.*
>
> **In math language:** $n^0 = 1$, when $n \neq 0$
>
> **Note:** 0^0 is indeterminate.

Your Turn 2

Simplify.

a. $\dfrac{x^6}{x^6}$

b. 5^0

c. $(-71)^0$

Example 2 Divide. $\dfrac{n^3}{n^3}$ (Assume $n \neq 0$.)

Solution: Because we are to divide exponential forms that have the same base, we can subtract the exponents and keep the same base.

$$\frac{n^3}{n^3} = n^{3-3} = n^0 = 1$$

To assume that $n \neq 0$ is important because if we let $n = 0$ we would have $\frac{0^3}{0^3}$, which is indeterminate. If we replace n in $\frac{n^3}{n^3}$ with any value other than 0, the result is 1.

In all future division problems involving variables, assume that divisor variables do not equal 0.

◁ **Do Your Turn 2**

Now let's divide monomials that have coefficients, such as $\dfrac{20m^2}{4m}$. When we multiplied monomials, we multiplied the coefficients and added the exponents of the like bases. Because multiplication and division are inverse operations, it follows that to divide monomials we divide the coefficients and subtract the exponents of the like bases.

$$\frac{20m^2}{4m} = \frac{20}{4}m^{2-1} = 5m$$

Answers to Your Turn 2
a. 1   b. 1   c. 1

Note that we can check by multiplying the quotient by the divisor to equal the dividend.

$$5m \cdot 4m = 5 \cdot 4m^{1+1} = 20m^2$$

PROCEDURE *To divide monomials:*
1. Divide the coefficients.
2. For like bases, subtract the exponent of the divisor base from the exponent of the dividend base and keep the base. If the bases have the same exponent, then they divide out, becoming 1.
3. Bases in the dividend that have no like base in the divisor are written unchanged in the quotient.

Your Turn 3

Divide.

Example 3 Divide.

a. $8a^7 \div (-4a)$

a. $10u^7 \div 2u^3$

Solution: $8a^7 \div (-4a) = 8 \div (-4) \cdot a^{7-1}$ Divide coefficients and subtract exponents of the like bases.

$$= -2a^6$$

b. $\dfrac{28x^3}{-4x}$

Check: $-2a^6 \cdot (-4a) = (-2)(-4)a^{6+1} = 8a^7$

b. $\dfrac{16m^2}{8m^2}$

Solution: $\dfrac{16m^2}{8m^2} = \dfrac{16}{8}m^{2-2}$ Divide the coefficients and subtract the exponents of the like bases.

$\qquad\quad = 2m^0$ Simplify the subtraction.

$\qquad\quad = 2(1)$ Replace m^0 with 1.

$\qquad\quad = 2$ Multiply.

◀ **Note:** Since the quotient of m^2 and m^2 is 1, no m appears in the result. Consequently we say m^2 "divides out."

Check: $2 \cdot 8m^2 = 16m^2$

Do Your Turn 3 ▶

OBJECTIVE 2 Divide a polynomial by a monomial.

Recall that every multiplication operation can be checked using division. Consider the following multiplication:

$(3x + 4) \cdot 2 = 3x \cdot 2 + 4 \cdot 2$ ◀ **Note:** We used the distributive property to multiply each term in $3x + 4$ by 2.

$\qquad\qquad\quad = 6x + 8$

To check our result, we would divide $6x + 8$ by 2 and the result *must* be $3x + 4$.

$(6x + 8) \div 2 = 6x \div 2 + 8 \div 2$ ◀ **Note:** To get $3x + 4$, we must divide each term in $6x + 8$ by 2.

$\qquad\qquad\quad = 3x + 4$

We can also write the division using a fraction bar.

$$\frac{6x + 8}{2} = \frac{6x}{2} + \frac{8}{2}$$

$$= 3x + 4$$

Answers to Your Turn 3
a. $5u^4$ b. $-7x^2$

To divide a polynomial by a monomial:

1. Divide each term in the polynomial dividend by the monomial divisor.
2. Simplify.

Example 4 Divide.

a. $(14x^2 + 16x) \div (2x)$

Solution: $(14x^2 + 16x) \div (2x)$

$= 14x^2 \div (2x) + 16x \div (2x)$ **Divide each term in the polynomial by the monomial.**

$= 14 \div (2) \cdot x^{2-1} + 16 \div (2) \cdot x^{1-1}$ **Divide the coefficients and simplify variables using the rule for exponents.**

$= 7x + 8$

Check: $(7x + 8) \cdot 2x = 7x \cdot 2x + 8 \cdot 2x = 14x^2 + 16$

b. $\dfrac{15x^3 - 10x^2}{5x^2}$

Solution: $\dfrac{15x^3 - 10x^2}{5x^2} = \dfrac{15x^3}{5x^2} - \dfrac{10x^2}{5x^2}$ **Divide each term of the polynomial by $5x^2$.**

$= \dfrac{15}{5}x^{3-2} - \dfrac{10}{5}x^{2-2}$

$= 3x - 2$

Check: $(3x - 2) \cdot 5x^2 = 3x \cdot 5x^2 - 2 \cdot 5x^2$
$= 15x^3 - 10x^2$

c. $(18x^4 - 12x^3 + 6x^2) \div (-6x)$

Solution: $(18x^4 - 12x^3 + 6x^2) \div (-6x)$

$= 18x^4 \div (-6x) - 12x^3 \div (-6x) + 6x^2 \div (-6x)$ **Divide each term by $-6x$.**

$= 18 \div (-6) \cdot x^{4-1} - 12 \div (-6) \cdot x^{3-1} + 6 \div (-6) \cdot x^{2-1}$

$= -3x^3 + 2x^2 - x$

Note: Dividing by a monomial with a negative coefficient causes the signs in the resulting polynomial to be opposite the signs in the dividend polynomial.

Note: The coefficient of $-12x^3$ is -12, so when we divide that term by $-6x$, we get positive $2x^2$, which is written as $+2x^2$ in the resulting polynomial.

Check:

$(-3x^3 + 2x^2 - x) \cdot (-6x) = -3x^3 \cdot (-6x) + 2x^2 \cdot (-6x) - x \cdot (-6x)$
$= -3 \cdot (-6) \cdot x^{3+1} + 2 \cdot (-6) \cdot x^{2+1} - 1 \cdot (-6) \cdot x^{1+1}$
$= 18x^4 - 12x^3 + 6x^2$

d. $\dfrac{24x^9 - 20x^7 + 4x^6}{4x^6}$

Solution:

$\dfrac{24x^9 - 20x^7 + 4x^6}{4x^6} = \dfrac{24x^9}{4x^6} - \dfrac{20x^7}{4x^6} + \dfrac{4x^6}{4x^6}$ **Divide each term by $4x^6$.**

$= \dfrac{24}{4}x^{9-6} - \dfrac{20}{4}x^{7-6} + \dfrac{4}{4}x^{6-6}$

$= 6x^3 - 5x + 1$

Check: $(6x^3 - 5x + 1) \cdot 4x^6 = 6x^3 \cdot 4x^6 - 5x \cdot 4x^6 + 1 \cdot 4x^6$

$$= 6 \cdot 4x^{3+6} - 5 \cdot 4x^{1+6} + 1 \cdot 4x^6$$

$$= 24x^9 - 20x^7 + 4x^6$$

Do Your Turn 4 ▷

WARNING Remember that although multiplication is commutative, division is not.

$$2 \cdot (3x + 4) \text{ is equivalent to } (3x + 4) \cdot 2.$$

$$(6x + 8) \div 2 \text{ is not equivalent to } 2 \div (6x + 8).$$

We will learn how to divide by polynomials later in algebra.

OBJECTIVE 3 Find a missing factor.

Now let's use what we've learned about dividing monomials to find missing factors. Remember that we find a missing factor by dividing the product by the known factor.

Example 5 Find the missing factor. $6x^3 \cdot (?) = 24x^5$

Solution: We solve for a missing factor by writing a related division sentence.

$(?) = \dfrac{24x^5}{6x^3}$ **Divide the product, 24x^5, by the known factor, 6x^3.**

$(?) = \dfrac{24}{6}x^{5-3}$ **Divide the coefficients and subtract the exponents of the like bases.**

$(?) = 4x^2$ **Simplify.**

Check: Verify that multiplying $6x^3$ by $4x^2$ gives $24x^5$.

$$6x^3 \cdot 4x^2 = 6 \cdot 4x^{3+2} = 24x^5$$

Do Your Turn 5 ▷

Now let's find missing factors that are multiple-term polynomials.

Example 6 Find the missing factor.

a. $3 \cdot (?) = 6x + 15$

Solution: To find a missing factor, we divide the product by the known factor.

$(?) = \dfrac{6x + 15}{3}$ **Divide the product, 6x + 15, by the known factor, 3.**

$(?) = \dfrac{6x}{3} + \dfrac{15}{3}$ **Divide each term in the polynomial by the monomial.**

$(?) = 2x + 5$ **Simplify.**

Check: Verify that multiplying $2x + 5$ by 3 gives $6x + 15$.

$3 \cdot (2x + 5) = 3 \cdot 2x + 3 \cdot 5$ **Use the distributive property.**

$= 6x + 15$ **It checks.**

Your Turn 4

Divide.

a. $(9x + 21) \div 3$

b. $\dfrac{52y^4 - 39y^2}{13y^2}$

c. $(18x^4 - 12x^3 + 6x^2) \div (-3x)$

d. $\dfrac{-28b^{13} + 49b^8 - 63b^5}{-7b^5}$

Your Turn 5

Find the missing factor.

a. $7m^2 \cdot (?) = 21m^6$

b. $-8y \cdot (?) = 40y^2$

Answers to Your Turn 4
a. $3x + 7$
b. $4y^2 - 3$
c. $-6x^3 + 4x^2 - 2x$
d. $4b^8 - 7b^3 + 9$

Answers to Your Turn 5
a. $3m^4$ **b.** $-5y$

Your Turn 6

Find the missing factor.

a. $8x - 10 = 2 \cdot (?)$

b. $12h^2 + 18h = (?) \cdot 6h$

c. $9y^5 - 7y^4 + 2y^3 = (?)y^3$

d. $24n^5 - 30n^3 - 12n^2 = 6n^2 (?)$

b. $16x^4 - 12x^3 + 8x^2 = 4x^2 \cdot (?)$

Solution: $\dfrac{16x^4 - 12x^3 + 8x^2}{4x^2} = (?)$ Divide the product by the known factor.

$\dfrac{16x^4}{4x^2} - \dfrac{12x^3}{4x^2} + \dfrac{8x^2}{4x^2} = (?)$ Divide each term in the polynomial by the monomial.

$4x^2 - 3x + 2 = (?)$ Simplify.

Check: Verify that multiplying $4x^2 - 3x + 2$ by $4x^2$ gives $16x^4 - 12x^3 + 8x^2$.

$4x^2 \cdot (4x^2 - 3x + 2) = 4x^2 \cdot 4x^2 + 4x^2 \cdot (-3x) + 4x^2 \cdot 2$ Use the distributive property.

$= 16x^4 - 12x^3 + 8x^2$

◄ **Do Your Turn 6**

OBJECTIVE **4** Factor the GCF out of a polynomial.

All of the work we've done to this point has been to develop the concept of factoring. When we rewrite a number or an expression as a product of factors, we say we are writing the number or expression in **factored form.**

DEFINITION   Factored form: A number or expression written as a product of factors.

For example, $3(2x + 5)$ is factored form for the polynomial $6x + 15$. We can check that it is its factored form by multiplying.

$$3(2x + 5) = 3 \cdot 2x + 3 \cdot 5 \quad \text{Use the distributive property.}$$
$$= 6x + 15$$

When we factor, we begin with the product $6x + 15$ and try to find the factored form, $3(2x + 5)$. First, notice that the monomial factor, 3, is the GCF of the terms in $6x + 15$. Our first step in factoring will always be to find the GCF of the terms in the polynomial. That GCF will be one of two factors. Suppose we didn't already know the other factor in our example is $2x + 5$. We could write the following missing factor statement:

$$6x + 15 = 3 \cdot (?)$$

Notice that the preceding statement is the same as in Example 6a. Look back at how we found the missing factor. We said $(?) = \dfrac{6x + 15}{3}$, so in our factored form we can replace the missing factor with $\dfrac{6x + 15}{3}$. Finding that quotient reveals the missing factor to be $2x + 5$.

$6x + 15 = 3 \cdot \left(\dfrac{6x + 15}{3} \right)$ Replace the missing factor with $\dfrac{6x + 15}{3}$. ◄ **Note:** The missing factor is the quotient of the given polynomial and the GCF.

$= 3 \cdot \left(\dfrac{6x}{3} + \dfrac{15}{3} \right)$ Divide each term by 3.

$= 3(2x + 5)$ Divide to complete the factored form.

Answers to Your Turn 6
a. $4x - 5$
b. $2h + 3$
c. $9y^2 - 7y + 2$
d. $4n^3 - 5n - 2$

Our example suggests the following procedure.

PROCEDURE *To factor a monomial GCF out of a given polynomial:*

1. Find the GCF of the terms that make up the polynomial.
2. Rewrite the polynomial as a product of the GCF and parentheses that contain the quotient of the given polynomial and the GCF.

$$\text{given polynomial} = \text{GCF}\left(\frac{\text{given polynomial}}{\text{GCF}}\right)$$

3. Simplify in the parentheses.

CONNECTION Keep in mind that when we factor, we are simply writing the original expression in a different form, called *factored form*. The factored form expression is equal to the original expression. If we were given a value for the variable(s), it wouldn't matter which version of the expression we used to evaluate that value. Both factored form and the original polynomial form will give the same result.

For example, let's evaluate $6x + 15$ and $3(2x + 5)$ when $x = 4$ (any number would do).

$$
\begin{array}{ll}
6x + 15 & 3(2x + 5) \\
6 \cdot 4 + 15 & 3(2 \cdot 4 + 5) \\
= 24 + 15 & = 3(8 + 5) \\
 & = 3(13) \\
= 39 \longleftarrow \text{Same result} \longrightarrow = 39
\end{array}
$$

> **Example 7** Factor.

a. $12y - 8$

Solution: First, we find the GCF of $12y$ and 8, which is 4.

$$12y - 8 = 4\left(\frac{12y - 8}{4}\right)$$ Rewrite the polynomial as a product of the GCF and parentheses containing the quotient of the polynomial and the GCF.

$$= 4\left(\frac{12y}{4} - \frac{8}{4}\right)$$ Divide each term in the polynomial by the GCF.

$$= 4(3y - 2)$$ Simplify in the parentheses to complete the factored form.

Check: Verify that the factored form multiplies to equal the original polynomial.

$$4(3y - 2) = 4 \cdot 3y - 4 \cdot 2$$ Distribute 4.

$$= 12y - 8$$ Simplify.

b. $18n^2 + 45n$

Solution: The GCF of $18n^2$ and $45n$ is $9n$.

$$18n^2 + 45n = 9n\left(\frac{18n^2 - 45n}{9n}\right)$$ Rewrite the polynomial as a product of the GCF and parentheses containing the quotient of the polynomial and the GCF.

$$= 9n\left(\frac{18n^2}{9n} - \frac{45n}{9n}\right)$$ Divide each term in the polynomial by the GCF.

$$= 9n(2n - 5)$$ Simplify in the parentheses to complete the factored form.

CONNECTION In Section 3.6, we learned that the GCF contains common prime factors each raised to its least exponent. In the case of $18n^2$ and $45n$, the common primes are 3 and n. The least exponent for 3 is 2 and for n is 1, so the GCF $= 3^2 \cdot n = 9n$.

Your Turn 7

Factor.

a. $12y - 9$

b. $10x^3 + 15x$

c. $20a^4 + 30a^3 - 40a^2$

We can check by multiplying the factored form to verify that the product is the original polynomial. We will leave the check to the reader.

c. $24x^6 + 18x^4 - 30x^3$

Solution: The GCF of $24x^6$, $18x^4$, and $30x^3$ is $6x^3$.

$$24x^6 + 18x^4 - 30x^3 = 6x^3\left(\frac{24x^6 + 18x^4 - 30x^3}{6x^3}\right)$$

Rewrite the polynomial as a product of the GCF and parentheses containing the quotient of the polynomial and the GCF.

$$= 6x^3\left(\frac{24x^6}{6x^3} + \frac{18x^4}{6x^3} - \frac{30x^3}{6x^3}\right)$$

Divide each term in the polynomial by the GCF.

$$= 6x^3(4x^3 + 3x - 5)$$

Simplify in the parentheses to complete the factored form.

We will leave the check to the reader.

◀ **Do Your Turn 7**

Answers to Your Turn 7
a. $3(4y - 3)$
b. $5x(2x^2 + 3)$
c. $10a^2(2a^2 + 3a - 4)$

3.7 Exercises

For Extra Help

Videotape DVT

 Addison-Wesley Tutor Center

 Math XL

 MyMathLab

Student Solutions Manual

1. When dividing two exponential forms that have the same base, we _____ the exponents and keep the same base.

2. Explain how to divide monomials.

3. Explain how to divide a multi-term polynomial by a monomial.

4. Explain how to factor a monomial out of the terms of a polynomial.

For Exercises 5–22, divide.

5. $x^9 \div x^2$

6. $a^{11} \div a^5$

7. $\dfrac{m^7}{m}$

8. $\dfrac{y^5}{y}$

9. $u^4 \div u^4$

10. $k^6 \div k^6$

11. $12a^7 \div (3a^4)$

12. $18w^5 \div (9w^2)$

13. $\dfrac{-20t^6}{4t^2}$

14. $\dfrac{18x^{10}}{-2x^4}$

15. $-40n^9 \div (-2n)$

16. $-33h^{12} \div (-11h)$

17. $12x^5 \div (-4x)$

18. $38m^7 \div (-2m^4)$

19. $\dfrac{-36b^7}{9b^5}$

20. $\dfrac{54x^5}{-27x^4}$

21. $14b^4 \div (-b^4)$

22. $-25h^7 \div (h^7)$

For Exercises 23–40, divide.

23. $(9a + 6) \div 3$

24. $(4x + 10) \div 2$

25. $\dfrac{14c - 8}{-2}$

26. $\dfrac{9t - 18}{-9}$

27. $(12x^2 + 8x) \div (4x)$

28. $(20y^2 + 10y) \div (5y)$

29. $\dfrac{-63d^2 + 49d}{7d}$

30. $\dfrac{-15R^2 + 25R}{5R}$

31. $(16a^5 + 24a^3) \div (4a^2)$

32. $(21b^6 - 36b^4) \div (3b^3)$

33. $\dfrac{16x^4 - 8x^3 + 4x^2}{4x}$

34. $\dfrac{20y^5 - 15y^3 - 25y}{5y}$

35. $(6a^5 - 9a^4 + 12a^3) \div (-3a^2)$

36. $(25c^6 - 30c^5 + 10c^3) \div (-5c^3)$

37. $\dfrac{30a^4 - 24a^3}{6a^3}$

38. $\dfrac{55p^5 - 66p^7}{11p^5}$

39. $\dfrac{32c^2 + 16c^5 - 40c^9}{-4c^2}$

40. $\dfrac{24n^8 - 12n^5 + 30n^3}{-6n^2}$

For Exercises 41–58, find the missing factor.

41. $5a \cdot (?) = 10a^3$

42. $7m^2 \cdot (?) = 21m^5$

43. $(?) \cdot (-6x^3) = 42x^7$

44. $(?) \cdot 10y^4 = -40y^9$

45. $-36u^5 = (?) \cdot (-9u^4)$

46. $-48x^4 = -6x \cdot (?)$

47. $8a^9 = 8a^4 (?)$

48. $25t^7 = -5t^2 (?)$

49. $8x + 12 = 4 \cdot (?)$

50. $12y - 6 = 6 \cdot (?)$

51. $5a^2 - a = a (?)$

52. $10m^4 + m^3 = (?) \ m^3$

53. $18x^3 - 30x^4 = 6x^3 (?)$

54. $44n^3 + 33n^6 = 11n^3 (?)$

55. $6t^5 - 9t^4 + 12t^2 = 3t^2 (?)$

56. $25k^7 - 30k^5 - 20k^2 = 5k^2 (?)$

57. $45n^4 - 36n^3 - 18n^2 = 9n^2(?)$

58. $28y^5 - 56y^3 + 35y^2 = 7y^2(?)$

For Exercises 59–62, find the missing side length.

59. $A = 63m^2$

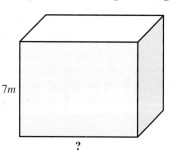

7m

?

60. $A = 24x^5$

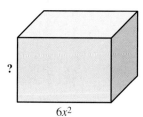

?

$6x^2$

61. $V = 60t^5$

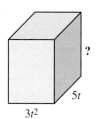

?

5t

$3t^2$

62. $V = 120a^7$

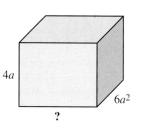

4a

$6a^2$

?

For Exercises 63–82, factor.

63. $8x - 4$

64. $14y + 7$

65. $10a + 20$

66. $8m - 20$

67. $2n^2 + 6n$

68. $5x^3 - 10x^2$

69. $7x^3 - 3x^2$

70. $12t^2 + 5t$

71. $20r^5 - 24r^3$

72. $32b^8 + 24b^4$

73. $6y^3 + 3y^2$

74. $5a^4 - 10a^2$

75. $12x^3 + 20x^2 + 32x$

76. $28y^4 - 42y^3 + 35y^2$

77. $9a^7 - 12a^5 + 18a^3$

78. $20x^5 - 15x^3 - 25x$

79. $14m^8 + 28m^6 + 7m^5$

80. $15a^2b^5 + 25a^2b^3 - 30a^2b^2$

81. $10x^9 - 20x^5 - 40x^3$

82. $45u^9 - 18u^6 + 27u^3$

Review Exercises

[1.2] **1.** Find the perimeter of the shape.

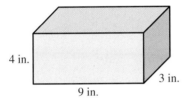

[1.6] **2.** Find the area of the parallelogram.

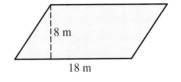

[1.6] **3.** Find the volume of the box.

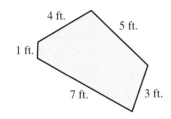

[3.4] **4.** Subtract. $(6x^3 - 9x^2 + x - 12) - (4x^3 + x + 5)$

[3.5] **5.** Multiply. $(x + 4)(x - 6)$

3.8 Applications and Problem Solving

OBJECTIVE 1 Solve polynomial problems involving perimeter, area, and volume.

We can use mathematics to describe situations. If we can write an expression that describes the situation, then we can experiment with numbers without actually having to do real experiments. To illustrate this process, let's write expressions for perimeter, area, and volume, then evaluate those expressions using given numbers.

Example 1 Write an expression in simplest form for the perimeter of the shape shown. What would the perimeter be if x is 7 feet?

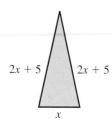

Understand: We are to find an expression for the perimeter of a triangle given expressions for its side lengths. We must then find the perimeter if x is 7 feet.

Plan: Since perimeter means to add the side lengths, we will add the expressions for the side lengths and then we will evaluate the perimeter expression using $x = 7$ feet.

Execute: perimeter $= x + 2x + 5 + 2x + 5$
$$= 5x + 10$$

Now evaluate $5x + 10$ when x is 7 feet.

$$5(7) + 10 \quad \text{Replace } x \text{ with 7.}$$
$$= 35 + 10 \quad \text{Multiply.}$$
$$= 45 \quad \text{Add.}$$

Answer: The perimeter expression is $5x + 10$; when x is 7 ft., the perimeter is 45 ft.

CONNECTION Simplifying an expression makes it much easier to evaluate. If we didn't simplify $x + 2x + 5 + 2x + 5$ and instead replaced all the x's with 7, we'd still get 45, but it is much easier to use $5x + 10$.

Evaluating the unsimplified expression	Evaluating the simplified expression
$x + 2x + 5 + 2x + 5$	$5x + 10$
$7 + 2(7) + 5 + 2(7) + 5$	$5(7) + 10$
$= 7 + 14 + 5 + 14 + 5$	$= 35 + 10$
$= 45$	$= 45$

Check: To check, let's replace x in the expressions for the side lengths with 7 so that we get numeric amounts for each side length. We can then verify that the perimeter is 45 feet by adding the side length values.

Evaluate $2x + 5$ when $x = 7$.

$$2(7) + 5$$
$$= 14 + 5$$
$$= 19 \text{ feet}$$

Note: We can draw a picture with the values of the side lengths when $x = 7$. ▶

We can now find the perimeter by adding the three side lengths together.

$P = 7 + 19 + 19$

$P = 45$ feet   **We get the same result.**

Do Your Turn 1 ▷

OBJECTIVE 2 Solve surface area problems.

We've considered the volume of a box, but there is another measurement that we can consider with boxes or any object. We can find an object's **surface area.** When we speak of surfaces, we mean the skin or shell of an object. Because area is a measure of the total number of square units, we can say a surface area is the total number of square units that completely cover the skin or shell of an object.

> **DEFINITION**   **Surface area:** The total number of square units that completely cover the outer shell of an object.

In order to develop a formula for surface area, it is helpful to break apart the object into all its surfaces. For a box, each surface is a rectangle. If we were to cut the box along the seams and fold it out flat, we would have the shape shown to the right below.

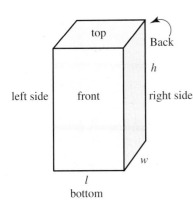
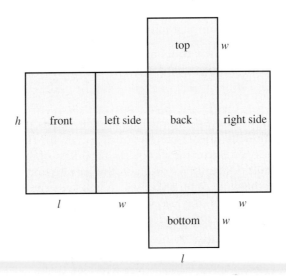

Notice that there are six surfaces: top, bottom, front, back, left side, and right side. To find the total surface area, we can add the areas of the six surfaces.

surface area = top + bottom + front + back + left side + right side
of a box   area   area   area   area   area   area

$SA = lw + lw + lh + lh + wh + wh$

We can simplify the polynomial that describes the surface area of a box by combining like terms.

$$SA = 2lw + 2lh + 2wh$$

We could also factor out a common factor, 2.

$$SA = 2(lw + lh + wh)$$

Your Turn 1

Solve.

a. Write an expression in simplest form for the perimeter of the shape. What would the perimeter be if $d = 9$ meters?

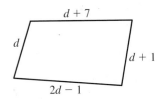

b. Write an expression in simplest form for the area of the shape. What would the area be if $w = 4$ inches?

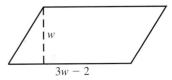

c. Write an expression in simplest form for the volume of the box. What would the volume be if $h = 2$ centimeters?

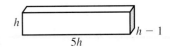

CONNECTION Remember, like terms have the same variables raised to the same exponents. The two lw terms are like, as are the two lh terms and the two wh terms.

Answers to Your Turn 1
a. $5d + 7$; 52 m
b. $3w^2 - 2w$; 40 in.²
c. $5h^3 - 5h^2$; 20 cm³

Your Turn 2

A carpenter builds wooden chests that are 2 feet wide by 3 feet long by 2 feet high. He finishes the outside of the chests with stain. What is the surface area of the chest? What total area must he plan to cover if he intends to put three coats of stain on the box?

Example 2 A company produces metal boxes that are 2 feet long by 3 feet wide by 1 foot high. How many square feet of metal are needed to produce one box?

Understand: We must calculate how many square feet of metal are needed to produce a box. Because the metal will become the shell or skin of the box, we are dealing with surface area. The formula for the surface area of a box is $SA = 2(lw + lh + wh)$.

Plan: Replace l with 2, w with 3, and h with 1 in the formula $SA = 2(lw + lh + wh)$.

Execute:

$SA = 2[(2)(3) + (2)(1) + (3)(1)]$	Replace the variables with the given values.
$SA = 2[6 + 2 + 3]$	Multiply in the brackets.
$SA = 2[11]$	Add in the brackets.
$SA = 22$	Multiply.

Answer: Each box will require 22 ft.^{2} of metal.

Check: An interesting way to check would be to use another version of the formula for surface area such as $SA = 2lw + 2lh + 2wh$ and make sure we get the same result. We will leave this to the reader.

◁ **Do Your Turn 2**

OBJECTIVE 3 Solve problems involving a falling object.

The polynomial $-16t^2 + h_0$ describes the height in feet of a falling object at any time during the fall. The variable h_0 represents the initial height of the object in feet from the ground. The variable t represents the number of seconds after the object is dropped from that initial height.

If we choose h to represent the height of the object after falling for t seconds, we can write the following formula:

$$h = -16t^2 + h_0$$

In Section 2.4, we mentioned that Galileo Galilei discovered that all objects fall at the same rate. Therefore, our formula for the height of an object after falling for t seconds is true for all objects, regardless of their weight. Note, however, that the formula does not take into account air resistance. If we drop a feather and a lead weight at the same time, air resistance will keep the feather from reaching the ground at the same time as the lead weight. However, if we perform the same experiment in a vacuum, the feather and lead weight would reach the ground at the same time.

OF INTEREST

Born near Pisa, Italy, on February 18, 1564, Galileo Galilei was to revolutionize scientific thought. His father sent him to study medicine at the University of Pisa in 1581. However, by chance he was led to study geometry. Galileo was so successful in mathematics that he was able to persuade his father to allow him to give up medicine for mathematics and science.

In 1589, Galileo became a professor of mathematics at the University of Pisa, and in 1592, he became professor of mathematics at the University of Padua. At Padua he discovered the laws of falling bodies and projectiles and conducted research on the motion of pendulums.

In 1609, Galileo constructed his own telescope, and his observations confirmed his belief in the Copernican theory that Earth and the planets revolve around the Sun. At the time, the prevailing view was that Earth was the center of the universe so this radical theory was rejected. In 1616, the Vatican arrested Galileo on charges of heresy.

In 1632, after years of silence on the subject, he repeated his belief in the Copernican system and was promptly summoned to Rome. He was brought to trial before the Inquisition in April 1633 and given a choice: either be tortured on the rack or renounce his defense of the Copernican system. He chose to renounce the Copernican system and was sentenced to house arrest for the rest of his life. He died in 1642. In 1992, the Vatican admitted its error in the Galileo case.

Answer to Your Turn 2

32 ft.^{2}; 96 ft.^{2}

Example 3 A skydiver jumps from a plane at an altitude of 2400 feet and deploys her parachute after 8 seconds of free fall. What was her altitude upon deploying the parachute?

Understand: We are given the initial altitude, or height, of the skydiver and the time for the free fall and we must find her altitude after 8 seconds of free fall.

Plan: Using the formula $h = -16t^2 + h_0$, replace t with 8 and h_0 with 2400, then calculate.

Execute:
$$h = -16(8)^2 + 2400$$
$$h = -16(64) + 2400 \qquad \text{Square 8.}$$
$$h = -1024 + 2400 \qquad \text{Multiply. } -16(64) = -1024$$
$$h = 1376 \qquad \text{Add.}$$

Answer: The skydiver deployed her parachute at an altitude of 1376 ft.

Check: We can check the calculations by reversing each step.

$$1376 - 2400 \overset{?}{=} -1024 \qquad -1024 \div 64 \overset{?}{=} -16 \qquad \sqrt{64} = 8$$
$$-1024 = -1024 \qquad\qquad -16 = -16$$

Do Your Turn 3 ▷

Your Turn 3

A marble is dropped from a 100-foot tower. How far is the marble from the ground after 2 seconds?

OBJECTIVE **4** Solve net-profit problems.

In Section 2.3, we developed the formula for net profit or loss given revenue and cost.

$$N = R - C$$

Let's consider the situation where we are given expressions for revenue and cost. We can write an expression for net by subtracting the cost expression from the revenue expression.

Example 4 The expression $7b + 9d + 1265$ describes the revenue for a toy manufacturer where b represents the number of toy bears sold and d represents the number of toy dogs sold. The expression $8b + 3d + 742$ describes the cost of producing the toy bears and toy dogs.

a. Write an expression in simplest form for the net.

Understand: We are to find net, given expressions for revenue and cost. The formula for net is $N = R - C$.

Plan: Using $N = R - C$, replace R with $7b + 9d + 1265$ and C with $8b + 3d + 742$, then subtract.

Execute: $N = R - C$
$$N = (7b + 9d + 1265) - (8b + 3d + 742)$$
$$N = (7b + 9d + 1265) + (-8b - 3d - 742) \qquad \text{Write as an equivalent addition.}$$
$$N = -b + 6d + 523 \qquad \text{Combine like terms.}$$

Note: Use parentheses when replacing a variable with a value or an expression.

DISCUSSION Does $-b$ in the expression indicate that the net is a total loss? Can profit still be made?

Answer to Your Turn 3
36 ft.

Answer: The expression that describes net is $-b + 6d + 523$.

A business sells two types of small motors. The revenue from the sales of motor A and motor B is described by $85A + 105B + 215$. The total cost of producing the two motors is described by $45A + 78B + 345$.

a. Write an expression in simplest form that describes the net.

b. If the business sells 124 of motor A and 119 of motor B in one month, what is the net for the month?

Check: We can reverse the process. If we add the expression for net to the expression for cost, we should get the expression for revenue.

$$R = N + C$$
$$R = (-b + 6d + 523) + (8b + 3d + 742)$$
$$R = 7b + 9d + 1265$$

b. If the company sold 2345 bears and 3687 dogs in one month, what was the net?

Understand: We are to find the net, given the number of toy bears and the number of toy dogs sold.

Plan: Using $-b + 6d + 523$, replace b with **2345** and d with **3687**, then calculate.

Execute: $-b + 6d + 523$
$$-(2345) + 6(3687) + 523$$
$$= -2345 + 22{,}122 + 523 \qquad \text{Evaluate the additive inverse of 2345 and multiply } 6(3687).$$

$$= 20{,}300 \qquad \text{Add from left to right.}$$

Answer: The business made a profit of $20,300.

Check: An interesting way to check is to calculate the revenue and cost using the original expressions for revenue and cost, then calculate net. Because this procedure is the same as the check for perimeter in Example 1, we will leave this to the reader.

CONNECTION Think about the terms in the expressions for revenue, cost, and net. In the revenue expression, $7b + 9d + 1265$, the $7b$ term represents the revenue from selling toy bears and $9d$ represents the revenue from selling toy dogs. The coefficient, 7, means each toy bear sells for $7 and the 9 means each toy dog sells for $9. The constant, 1265, is an income that does not depend on the number of toy bears or dogs sold. It could be interest or fees of some kind.

In the cost expression, $8b + 3d + 742$, the $8b$ term represents the cost of producing the toy bears and $3d$ represents the cost of producing the toy dogs. The coefficient, 8, means each bear costs $8 to produce and 3 means each dog costs $3 to produce. The constant, 742, is a cost that does not depend on the number of toy bears or dogs produced. It could be utility and operation costs for the plant.

When we developed the expression for net by subtracting the cost expression from the revenue expression, we combined like terms. In combining like terms, we subtracted the coefficients. Subtracting $8b$ from $7b$ equals $-b$. Because the $8 cost of producing each bear is more than the $7 revenue from the sale of each bear, the resulting -1 coefficient means the company has a net loss of $1 for each bear sold. Subtracting $3d$ from $9d$ equals $6d$. Because the $3 cost is less than the $9 revenue, the company has a net profit of $6 for each toy dog sold. Subtracting 742 from 1265 equals 523. Because the constant cost of $742 is less than the constant revenue of $1265, the $523 is a profit.

Putting it all together:

$$(7b + 9d + 1265) - (8b + 3d + 742) = -b \quad + \quad 6d \quad + \quad 523$$

$-b$	$6d$	523
net loss of $1 for each toy bear sold	net profit of $6 for each toy dog sold	net profit of $523 from constant revenue and costs

◀ **Do Your Turn 4**

Answers to Your Turn 4
a. $40A + 27B - 130$
b. $8043

3.8 Exercises

For
Extra
Help

 Videotape
DVT

 Tutor Center
Addison-Wesley
Tutor Center

 Math XL

 MyMathLab

 Student Solutions
Manual

1. What is the surface area of an object?

2. What is the formula for the surface area of a box?

3. What is the formula $h = -16t^2 + h_0$ used for, and what does each variable represent?

4. Given polynomials for revenue and cost, how is net profit or loss determined?

For Exercises 5–10, write an expression in simplest form for the perimeter of the shape.

5.

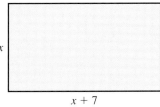

6.

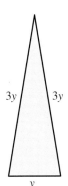

7.

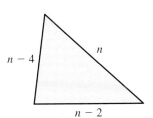

8.

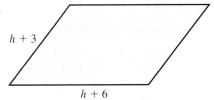

9.

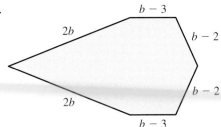

10.

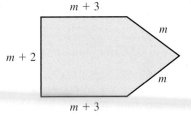

11. Calculate the perimeter of the rectangle in Exercise 5 if x is
   a. 19 inches
   b. 27 centimeters

12. Calculate the perimeter of the triangle in Exercise 6 if y is
   a. 15 miles
   b. 22 meters

13. Calculate the perimeter of the triangle in Exercise 7 if n is
   a. 12 feet
   b. 8 yards

14. Calculate the perimeter of the parallelogram in Exercise 8 if h is
   a. 6 kilometers
   b. 15 centimeters

15. Calculate the perimeter of the shape in Exercise 9 if b is
   a. 10 inches
   b. 20 feet

16. Calculate the perimeter of the shape in Exercise 10 if m is
   a. 14 miles
   b. 32 meters

For Exercises 17 and 18, write an expression in simplest form for the area of the shape.

17.

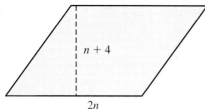

$n + 4$

$2n$

18.

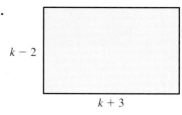

$k - 2$

$k + 3$

19. Calculate the area for the parallelogram in Exercise 17 if *n* is
   a. 6 kilometers
   b. 12 feet

20. Calculate the area of the rectangle in Exercise 18 if *k* is
   a. 8 inches
   b. 13 centimeters

For Exercises 21 and 22, write an expression in simplest form for the volume of the shape.

21.

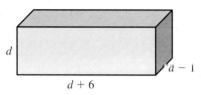

d

$d - 1$

$d + 6$

22.

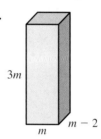

$3m$

m

$m - 2$

23. Calculate the volume of the box in Exercise 21 if *d* is
   a. 3 inches
   b. 5 feet

24. Calculate the volume of the box in Exercise 22 if *m* is
   a. 7 miles
   b. 14 inches

For Exercises 25–29, solve.

25. A company produces cardboard boxes. Each box is 2 feet long by 2 feet wide by 3 feet high. What is the area of the cardboard that is required for each box?

26. A cube that is 5 centimeters along each side is to be covered with paper. What is the total area that will be covered by the paper?

27. Therese designs and builds metal sculptures. One particular sculpture is to be a large cube that will rest on one of its corners. She will cover the cube with panels that are 1 square foot in size. If the cube is to be 16 feet on each side, how many panels will be needed to cover the cube?

28. The main body of a certain satellite is a box that measures 3 feet wide by 5 feet long by 6 feet high. The box is covered with a skin of a thin metal alloy. What is the area covered by the metal alloy?

29. The leaning tower of Pisa is 180 feet tall. When Galileo dropped a cannonball from the top of the leaning tower of Pisa, what was the height of the cannonball after 2 seconds? After 3 seconds?

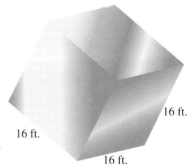

16 ft.

16 ft.

16 ft.

OF INTEREST

Construction on the tower at Pisa began in 1173 and continued for more than 200 years due to continuous structural problems and soft, unstable soil. Because of the soft soil, one side of the tower began sinking before the first three stories were completed, creating a noticeable lean. The lean is estimated to be increasing at a rate of 1 millimeter per year.

30. Victoria Falls is a waterfall on the Zambezi river in south central Africa that is 355 feet high. If a coin were dropped from the top of the falls, how high would the coin be after 3 seconds? 4 seconds?

31. A skydiver jumps from a plane at 12,000 feet and deploys her parachute after 22 seconds of free fall. What is her altitude at the time she deploys her parachute?

32. Two skydivers jump from a plane at an altitude of 16,000 feet. One skydiver deploys his parachute after 27 seconds and the other 2 seconds later. What were their respective altitudes at the time each deployed his parachute?

33. A company produces two different tables, a round top and a square-top. If r represents the number of round-top tables and s represents the number of square-top tables, then $145r + 215s + 100$ describes the revenue from the sales of the two types of table. The polynomial $110r + 140s + 345$ describes the cost of producing the two types of table.
   a. Write an expression for net in simplest form.
   b. If the company sells 120 round-top tables and 106 square-top tables in one month, what would be the net profit or loss?

34. Calvin installs tile. The expression $9lw + 45$ describes the revenue in dollars he receives for tiling a rectangular room, where l represents the length of the room and w represents the width. The expression $3lw + t$ describes his cost in dollars for tiling a rectangular room, where t represents the number of tiles required.
   a. Write an expression for net in simplest form.
   b. Calvin gets a job to tile a 12-foot-wide by 14-foot-long room. He uses 672 tiles. Calculate his profit.

35. Candice makes reed baskets in three sizes, small, medium, and large. She sells the small baskets for $5, medium for $9, and large for $15. The small baskets cost her $2 each to make, medium $4 each, and large $7 each.
   a. Write a polynomial that describes the revenue she receives from the sale of all three baskets.
   b. Write a polynomial that describes her cost to produce all three baskets.

   c. Write an expression for her net in simplest form.
   d. If in one day she sells 6 small, 9 medium, and 3 large baskets, what would be her net profit or loss?

36. Devon paints portraits in three sizes, small, medium, and large. He receives $50 for a small, $75 for a medium, and $100 for a large portrait. It costs him $22 for each small, $33 for each medium, and $47 for each large portrait.
   a. Write a polynomial that describes the revenue he receives from the sale of all three portrait sizes.
   b. Write a polynomial that describes his cost for all three sizes.

   c. Write an expression for his net in simplest form.
   d. In one month he sells 4 small, 6 medium, and 2 large portraits. What is his net profit or loss?

PUZZLE PROBLEM How many cubic feet of dirt are in a hole 4 feet wide, 6 feet long, and 3 feet deep?

Review Exercises

[3.1] **1.** Evaluate $\sqrt{2a + 7} - b^3$, when $a = 21$ and $b = -2$.

[3.2] **2.** What is the coefficient of $-x^3$?

[3.2] **3.** What is the degree of 5?

[3.3] **4.** Combine like terms. $8x^3 - 9x^2 + 11 - 12x^2 + 2x - 9x^3 + 3 - 2x$

[3.7] **5.** Evaluate. 6^0

[3.5] **6.** Multiply. $-4(3a + 7)$

Defined Terms

Review the following terms, and for any you do not know, study its definition on the page number next to it.

Section 3.1
Equation *(p. 150)*
Expression *(p. 150)*

Section 3.2
Monomial *(p. 155)*
Coefficient *(p. 156)*
Term *(p. 155)*
Degree of a monomial
  (p. 156)

Like terms *(p. 157)*
Polynomial *(p. 157)*
Simplest form *(p. 158)*
Polynomial in one
  variable *(p. 158)*
Multivariable polynomial
  (p. 158)
Degree of a multiple-term
  polynomial *(p. 159)*

Section 3.5
Conjugates *(p. 180)*

Section 3.6
Prime number *(p. 187)*
Composite number
  (p. 187)
Prime factorization
  (p. 190)

Greatest common factor
  (p. 193)

Section 3.7
Factored form *(p. 206)*

Section 3.8
Surface area *(p. 213)*

Two actions that can be performed with an expression:

1. Evaluate (replace the variables with numbers and calculate)

2. Rewrite (simplify, distribute, factor)

Procedures, Rules, and Key Examples

Procedures/Rules	*Key Example(s)*

Section 3.1 Evaluating Expressions
To evaluate an expression:
1. Replace the variables with the corresponding given values.
2. Calculate using the order of operations agreement.

Example 1: Evaluate $-2x + 8$ when $x = 3$.
$-2(3) + 8$ **Replace x with 3.**
$= -6 + 8$ **Multiply.**
$= 2$ **Add.**

Section 3.2 Introduction to Polynomials
To write a polynomial in descending order of degree, write the term with the greatest degree first, then the term with the next greatest degree, and so on.

Example 1: Write $6x^3 + 9x^7 + 8 - 4x^5$ in descending order of degree.
Answer: $9x^7 - 4x^5 + 6x^3 + 8$

Section 3.3 Simplifying the Polynomials
To combine like terms:
1. Add or subtract the coefficients.
2. Keep the variables and their exponents the same.

Example 1: Combine like terms.
a. $5x + 8x = 13x$
b. $9a^4 - 10a^4 = -1a^4 = -a^4$

Section 3.4 Adding and Subtracting Polynomials
To add polynomials, combine like terms.

Example 1: Add.
$(3x^2 - 5x + 1) + (7x^2 - 4x - 6)$
$= 3x^2 + 7x^2 - 5x - 4x + 1 - 6$
$= 10x^2 - 9x - 5$

To subtract polynomials:
1. Write the subtraction statement as an equivalent addition statement.
  a. Change the operation symbol from a minus sign to a plus sign.
  b. Change the subtrahend to its additive inverse. To get the additive inverse we change *all* the signs in the polynomial.
2. Combine like terms.

Example 2: Subtract.
$(5a^3 - 9a + 2) - (a^3 - 6a - 7)$
$= (5a^3 - 9a + 2) + (-a^3 + 6a + 7)$
$= 5a^3 - a^3 - 9a + 6a + 2 + 7$
$= 4a^3 - 3a + 9$

Procedures/Rules	Key Example(s)

Section 3.5 Multiplying Polynomials

When multiplying exponential forms that have the same base, we can add the exponents and keep the same base.

$$n^a \cdot n^b = n^{a+b}$$

To multiply monomials:

1. Multiply coefficients.
2. Add the exponents of the like bases.

When an exponential form is raised to a power, we can multiply the exponents and keep the same base.

$$\left(n^a\right)^b = n^{a \cdot b}$$

To simplify a monomial raised to a power:

1. Evaluate the coefficient raised to that power.
2. Multiply each variable's exponent by the power.

To multiply a polynomial by a monomial:

Use the distributive law and multiply each term in the polynomial by the monomial.

To multiply two polynomials:

1. Multiply every term in the first polynomial by every term in the second polynomial.
2. Combine like terms.

The product of two conjugates is a difference of two squares.

$$(a + b)(a - b) = a^2 - b^2$$

Example 1: Multiply.

a. $t^5 \cdot t^4 = t^{5+4} = t^9$

b. $-6a^4bc^2 \cdot 5a^3b^2 = -6 \cdot 5a^{4+3}b^{1+2}c^2$
$$= -30a^7b^3c^2$$

Example 2: Simplify.

a. $\left(x^2\right)^4 = x^{2 \cdot 4} = x^8$

b. $\left(-4hk^4\right)^3 = (-4)^3 h^{1 \cdot 3} k^{4 \cdot 3}$
$$= -64h^3k^{12}$$

Example 3: Multiply.

a. $2m\left(m^4 - 5m^2n + 3n^2\right)$
$$= 2m \cdot m^4 - 2m \cdot 5m^2n$$
$$+ 2m \cdot 3n^2$$
$$= 2m^5 - 10m^3n + 6mn^2$$

b. $(2x + 3)(5x - 1)$
$$= 2x \cdot 5x + 2x \cdot (-1) +$$
$$3 \cdot 5x + 3(-1)$$
$$= 10x^2 - 2x + 15x - 3$$
$$= 10x^2 + 13x - 3$$

c. $(x + 8)(x - 8) = x^2 - 64$

Section 3.6 Prime Numbers and GCF

To determine if a given number is prime or composite, divide it by the primes on the list of prime numbers and consider the results.

1. If the prime number divides the given number with no remainder, stop and conclude that the given number is a composite number.
2. If the prime does not divide the given number with no remainder, consider the quotient.
   a. If the quotient is larger than the current prime divisor, repeat the process with the next prime on the list of prime numbers.
   b. If the quotient is equal to or less than the current prime divisor, stop and conclude that the given number is itself a prime number.

Note: 0 and 1 are neither prime nor composite.

Example 1: Determine if 37 is prime or composite.

Divide by the list of primes.

Is 37 divisible by 2?	No, because it is odd.
Is 37 divisible by 3?	No, because $3 + 7 = 10$ and this sum is not divisible by 3.
Is 37 divisible by 5?	No, because it does not end with 5 or 0.

Is 37 divisible by 7?

$$\begin{array}{r} 5 \\ 7\overline{)37} \\ -35 \\ \hline 2 \end{array}$$

37 is not divisible by 7. The quotient is smaller than the divisor. Stop test! 37 is prime.

To find the prime factorization of a number using a factor tree.

1. Draw two branches below the number.
2. Place two factors that multiply to equal the given number at the end of the two branches.
3. Repeat steps 1 and 2 for every composite factor.
4. Place all the prime factors together in a multiplication sentence.

Example 2: Find the prime factorization.

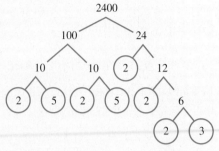

$$2400 = 2 \cdot 2 \cdot 2 \cdot 2 \cdot 2 \cdot 3 \cdot 5 \cdot 5$$
$$= 2^5 \cdot 3 \cdot 5^2$$

To find the greatest common factor by listing:

1. List all possible factors for each given number.
2. Search the lists for the largest factor common to all lists.

Example 3: Find the GCF of 32 and 40 by listing.

Factors of 32: 1, 2, 4, 8, 16, 32
Factors of 40: 1, 2, 4, 5, 8, 10, 20, 40

GCF = 8

To find the greatest common factor of a given set of numbers:

1. Write the prime factorization of each given number in exponential form.
2. Create a factorization for the GCF that contains only those prime factors common to all the factorizations, each raised to its least exponent in all the factorizations.
3. Multiply.

    Note: If there are no common primes, then the GCF is 1.

Example 4: Find the GCF.

a. 84 and 120

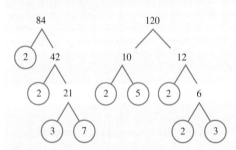

$$84 = 2^2 \cdot 3 \cdot 7$$
$$120 = 2^3 \cdot 3 \cdot 5$$
$$\text{GCF} = 2^2 \cdot 3 = 2 \cdot 2 \cdot 3 = 12$$

b. $36m^3$ and $54m^2$

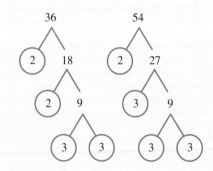

$$36m^3 = 2^2 \cdot 3^2 \cdot m^3$$
$$54m^2 = 2 \cdot 3^3 \cdot m^2$$
$$\text{GCF} = 2 \cdot 3^2 \cdot m^2 = 18m^2$$

Section 3.7 Introduction to Factoring

When dividing exponential forms that have the same base we can subtract the divisor's exponent from the dividend's exponent and keep the same base.

$$n^a \div n^b = \frac{n^a}{n^b} = n^{a-b}, \qquad \text{when } n \neq 0$$

Note: $n \neq 0$ because if n were replaced with 0, we would have $0^a \div 0^b$. In Section 1.4, we concluded that when the dividend and divisor are both 0, the problem is indeterminate.

Any base other than 0 raised to the 0 power simplifies to the number 1.

$$n^0 = 1, \qquad \text{when } n \neq 0$$

Note: 0^0 is indeterminate.

To divide monomials:

1. Divide the coefficients.
2. For like bases, subtract the exponent of the divisor base from the exponent of the dividend base and keep the base. If the bases have the same exponent, then they divide out, becoming 1.
3. Bases in the dividend that have no like base in the divisor are written unchanged in the quotient.

To divide a polynomial by a monomial:

1. Divide each term of the polynomial by the monomial.
2. Simplify.

Example 1: Divide.

a. $t^9 \div t^2 = t^{9-2} = t^7$

b. $\dfrac{y^6}{y^2} = y^{6-2} = y^4$

Example 2: Evaluate.

a. $245^0 = 1$

b. $(-5)^0 = 1$

Example 3: Divide.

a. $-18m^3 \div 2m = -18 \div 2 \cdot m^{3-1}$

$$= -9m^2$$

b. $\dfrac{30x^4}{5x^4} = \dfrac{30}{5} x^{4-4}$

$$= 6x^0$$
$$= 6(1)$$
$$= 6$$

Example 4: Divide.

a. $(20x^3 + 28x^2) \div (4x^2)$
$$= 20x^3 \div (4x^2) + 28x^2 \div (4x^2)$$
$$= (20 \div 4)x^{3-2} + (28 \div 4)x^{2-2}$$
$$= 5x + 7$$

b. $\dfrac{14y^4 - 28y^3 + 35y^2}{7y^2}$

$$= \frac{14y^4}{7y^2} - \frac{28y^3}{7y^2} + \frac{35y^2}{7y^2}$$

$$= \frac{14}{7}y^{4-2} - \frac{28}{7}y^{3-2} + \frac{35}{7}y^{2-2}$$

$$= 2y^2 - 4y + 5$$

To factor a monomial out of a given polynomial:
1. Find the GCF of the terms that make up the polynomial.
2. Rewrite the polynomial as a product of the GCF and parentheses that contain the quotient of the given polynomial and the GCF.

$$\text{given polynomial} = \text{GCF}\left(\frac{\text{given polynomial}}{\text{GCF}}\right)$$

3. Simplify in the parentheses.

Example 5: Factor.

a. $24x - 15 = 3\left(\dfrac{24x - 15}{3}\right)$

$$= 3\left(\dfrac{24x}{3} - \dfrac{15}{3}\right)$$

$$= 3(8x - 5)$$

b. $9u^7 + 27u^5 - 18u^3$

$$= 9u^3\left(\dfrac{9u^7 + 27u^5 - 18u^3}{9u^3}\right)$$

$$= 9u^3\left(\dfrac{9u^7}{9u^3} + \dfrac{27u^5}{9u^3} - \dfrac{18u^3}{9u^3}\right)$$

$$= 9u^3(u^4 + 3u^2 - 2)$$

Formulas

Area of a parallelogram:	$A = bh$
Volume of a box:	$V = lwh$
Surface area of a box:	$SA = 2lw + 2lh + 2wh$ or $SA = 2(lw + lh + wh)$
Height of a falling object:	$h = -16t^2 + h_0$
Net profit or loss:	$N = R - C$

For Exercises 1–6, answer true or false.

[3.1] **1.** $4x - 8y + 9$ is an equation.

[3.1] **2.** $3x^2 - 8x = 10$ is an equation.

[3.2] **3.** 27 is a monomial.

[3.2] **4.** The degree of 7 is 1.

[3.7] **5.** $(-16)^0 = 0$

[3.2] **6.** $14x^2$ and $12x$ are like terms.

For Exercises 7–10, complete the rule.

[3.3] **7.** When combining like terms, we add or subtract the _____ and keep the _____ the same.

[3.5] **8.** When multiplying exponential forms that have the same base, we _____ the exponents and keep the same ____.

[3.7] **9.** When dividing exponential forms that have the same base, we _____ the _____ exponent from the _____ exponent and keep the bases the same.

[3.5] **10.** When raising an exponential form to a power, we _____ the exponents and keep the bases the same.

[3.1] *For Exercises 11–14, evaluate the expression using the given values.*

11. $5n^2 - 9n + 2$; $n = -2$

12. $\sqrt{3x + 4} - 2y$; $x = 7$, $y = 9$

13. $4|t^2 - 11|$; $t = 3$

14. $\dfrac{-3ab + 4}{a - b^2}$; $a = 2$, $b = -4$

[3.2] **15.** Is $9x^3yz$ a monomial? Explain.

[3.2] **16.** Is $4x - 5$ a monomial? Explain.

[3.2] *For Exercises 17–20, identify the coefficient and degree of each monomial.*

17. $18x$

18. y^3

19. -9

20. $-3m^5n$

[3.2] **21.** Explain why $4y^3$ and $9y^3$ are like terms.

[3.2] **22.** Explain why $7a^2$ and $7a$ are not like terms.

[3.2] *For Exercises 23–26, tell whether the expression is a monomial, binomial, trinomial, or none of these.*

23. $9y - 5$

24. $4x^3 + 9x^2 - x + 7$

25. $-3a$

26. $x^2 + 5x - 4$

[3.2] **27.** What is the degree of $7a^4 - 9a + 13 + 5a^6 - a^2$?

[3.2] **28.** Write $7a^4 - 9a + 13 + 5a^6 - a^2$ in descending order of degree.

[3.3] *For Exercises 29–31, combine like terms and write the resulting polynomial in descending order of degree.*

29. $3a^2 + 2a - 4a^2 - 1$

30. $5m^3 + 9m - 12 + 2m^3 + 5$

31. $3x^2 - 5x + x^7 + 13 + 5x - 6x^7$

[3.4] *For Exercises 32 and 33, add.*

32. $(x - 4) + (2x + 9)$

33. $(y^4 + 2y^3 - 8y + 5) + (3y^4 - 2y - 9)$

[3.4] *For Exercises 34 and 35, subtract.*

34. $(a^2 - a) - (-2a^2 + 4a)$

35. $(19h^3 - 4h^2 + 2h - 1) - (6h^3 + h^2 + 7h + 2)$

[3.5] *For Exercises 36–39, multiply.*

36. $m^3 \cdot m^4$

37. $2x \cdot (-5x^4)$

38. $-4y \cdot 3y^4$

39. $-6t^3 \cdot (-t^5)$

[3.5] *For Exercises 40–43, simplify.*

40. $(5x^4)^3$

41. $(-2y^3)^2$

42. $(4t^3)^3$

43. $(-2a^4)^3$

[3.5] *For Exercises 44–52, multiply and simplify.*

44. $2(3x - 4)$

45. $-4y(4y - 5)$

46. $3n(5n^2 - n + 7)$

47. $(a + 5)(a - 7)$

48. $(2x - 3)(x + 1)$

49. $(2y - 1)(5y - 8)$

50. $(3t + 4)(3t - 4)$

51. $(u + 2)(u^2 - 5u + 3)$

52. $-p^2(3p + 1)(p - 2)$

[3.5] *For Exercises 53 and 54, find the conjugate.*

53. $3y - 4$

54. $-7x + 2$

[3.6] *For Exercises 55 and 56, determine whether the given number is prime, composite, or neither.*

55. 119

56. 97

[3.6] *For Exercises 57 and 58, find the prime factorization of the given number.*

57. 360

58. 4200

[3.6] *For Exercises 59–62, find the GCF.*

59. 140 and 196

60. 45 and 28

61. $48x^5$ and $36x^6$

62. $18a^2$, $30a^4$, and $24a^3$

[3.7] *For Exercises 63–65, divide.*

63. $r^8 \div r^2$

64. $\dfrac{20x^5}{-5x^2}$

65. $\dfrac{-30x^6 + 24x^4 - 12x^2}{-6x^2}$

[3.7] *Evaluate.*

66. 15^0

[3.7] *For Exercises 67–69, find the missing factor.*

67. $6m^3 \cdot (?) = -54m^8$

68. $9x^4 + 12x^6 = 3x^4 \cdot (?)$

69. $40y^6 - 30y^4 - 20y^2 = 10y^2 (?)$

70. Find an expression for the missing length in the rectangle shown if its area is $42b^6$.

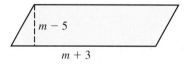

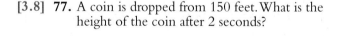

$3b^5$

?

[3.7] *For Exercises 71–73, factor.*

71. $30x + 12$

72. $9n^4 - 15n^3$

73. $18x^3 + 24x^2 - 36x$

[3.8] **74. a.** Write an expression in simplest form for the perimeter of the rectangle shown.

  b. Find the perimeter if x is 9 meters.

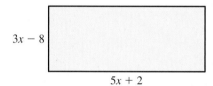

$3x - 8$

$5x + 2$

[3.8] **75. a.** Write an expression in simplest form for the area of the parallelogram shown.

  b. Find the area of the parallelogram if m is 8 inches.

$m - 5$

$m + 3$

[3.8] **76. a.** Write an expression for the volume of the box shown.

  b. Find the volume of the box if n is 4 feet.

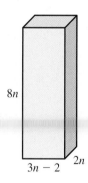

$8n$

$3n - 2$ $2n$

[3.8] **77.** A coin is dropped from 150 feet. What is the height of the coin after 2 seconds?

[3.8] **78.** A company produces two different tennis racquets, a normal size and an oversize version. If n represents the number of normal racquets and b represents the number of oversize racquets, then $65n + 85b + 345$ describes the revenue from the sales of the two racquets. The polynomial $22n + 42b + 450$ describes the cost of producing the two racquets.

  a. Write an expression in simplest form for net.

  b. If the company sells 487 normal-size racquets and 246 oversize racquets in one week, what would be the net?

1. _____   [3.1]   **1.** Evaluate $2n^3 - 3p^2$ when $n = -2$ and $p = -3$.

2. _____   [3.2]   **2.** Is $4x^3 - 9$ a monomial, binomial, trinomial, or none of these?

3. _____   [3.2]   **3.** What are the coefficient and degree of $-y^4$?

4. _____   [3.2]   **4.** What is the degree of $x^2 - 9x + 5x^4 - x^3$?

5. _____   [3.3]   **5.** Combine like terms and write the resulting polynomial in descending order of degree. $10a^4 + 3a^2 - 5a + 2a^2 + 5a - 11a^4$

6. _____   [3.4]   **6.** Add. $(2y^3 - 8y + 5) + (3y^3 - 2y - 9)$

7. _____   [3.4]   **7.** Subtract. $(12y^5 + 5y^3 - 7y - 3) - (6y^5 + y^2 - 7y + 1)$

8. _____   [3.5]   **8.** Multiply. $(-4u^5) \cdot (-5u^2)$

9. _____   [3.5]   **9.** Simplify. $(-2a^3)^5$

[3.5] *For 10–12, multiply.*

10. _____   **10.** $-2t(t^3 - 3t - 7)$

11. _____   **11.** $(x - 6)(x + 6)$

12. _____   **12.** $-2a^2(a + 1)(2a - 1)$

13. _____   [3.5]   **13.** What is the conjugate of $6x - 5$?

14. _____   [3.6]   **14.** Determine whether 91 is prime, composite, or neither.

15. _____   [3.6]   **15.** Find the prime factorization of 340.

For 16 and 17, find the GCF.

[3.6] **16.** 180 and 396

16. _____

[3.7] **17.** $60h^5$ and $48h^7$

17. _____

[3.7] **18.** Evaluate. 38^0

18. _____

[3.7] *For 19 and 20, divide.*

19. $m^6 \div m^4$

19. _____

20. $\dfrac{16x^5 - 18x^3 + 22x^2}{-2x}$

20. _____

21. Find the missing factor. $-8x^4 \cdot (?) = -40x^7$

21. _____

[3.7] *For 22 and 23, factor.*

22. $12x - 20$

22. _____

23. $10y^4 - 18y^3 + 14y^2$

23. _____

[3.8] **24. a.** Write an expression in simplest form for the area of the parallelogram.

24. a._____

b. Calculate the area if n is 7 centimeters.

b._____

$n - 4$

$2n + 1$

[3.8] **25.** A company makes two different styles of frames for glasses. One style allows for rounded lenses, and the other is for rectangular lenses. If a represents the number of rounded-style and b represents the number of rectangular-style frames, then $145a + 176b$ describes the revenue from the sales of the frames. The polynomial $61a + 85b$ describes the cost of producing the frames.

a. Write an expression in simplest form for the net.

25. a._____

b. If the company sold 128 rounded frames and 115 rectangular frames in one month, what would be the net profit or loss?

b._____

For Exercises 1–6, answer true or false.

[1.1] **1.** 0 is a natural number.   [1.1] **2.** $245 < 387$ [2.1] **3.** $|8| = -8$

[2.2] **4.** $9 + (-12) = -12 + 9$ [3.2] **5.** $15xy^2$ is a monomial.   [3.2] **6.** The degree of 18 is 1.

$\begin{bmatrix} 1.5 \\ 2.5 \end{bmatrix}$ **7.** List the order of operations agreement.   [2.2] **8.** The sum of two negative numbers is a _____ number.

       1.

       2.

       3.

       4.

[2.4] **9.** The product of a negative number and a positive number is a _____ number.   [3.5] **10.** Explain how to multiply two binomials.

[1.1] **11.** Write the word name for 2,480,045.   [1.3] **12.** Write 7×10^8 in standard form.

[2.1] **13.** Graph -9 on a number line. $\begin{bmatrix} 1.1 \\ 2.1 \end{bmatrix}$ **14.** Round $-105{,}612$ to the nearest ten thousand.

[1.3] **15.** Estimate $682 \cdot 246$ by rounding to the nearest hundred.   [3.2] **16.** What is the coefficient of $-x^2$?

[3.2] **17.** What is the degree of $5x$?   [3.2] **18.** What is the degree of $7t^2 - 9t + 4t^5 + 12$?

For Exercises 19–29, simplify.

[2.3] **19.** $|14 - 20|$ [2.1] **20.** $-(-(-9))$ [2.3] **21.** $58 - 184 - 32 + 14$

[2.3] **22.** $-63 - (-25)$ [2.4] **23.** $-1(-9)(-3)(-4)$ [2.4] **24.** $(-5)^3$

[2.4] **25.** -3^4 [2.4] **26.** $1442 \div (-14)$ [1.4] **27.** $\sqrt{16 \cdot 25}$

[2.5] **28.** $2[9 + (3 - 16)] - 35 \div 5$ [2.5] **29.** $4^3 - \{[6 + (2)8] - [12 + 9(6)]\}$

[3.1] **30.** Evaluate $\sqrt{x - y}$ when $x = 100$ and $y = 36$.   [3.3] **31.** Combine like terms.
$$5x^2 - 9x + 11x^2 - 7 - x^3 + 8x - 5$$

[3.4] **32.** Subtract. $(4y^3 - 6y^2 + 9y - 8) -$
$(3y^2 - 12y - 2)$

[3.5] **33.** Multiply. $(9x^3)(7x)$

[3.5] **34.** Multiply. $(b - 8)(2b + 3)$

[3.6] **35.** Find the prime factorization of 360.

[3.7] **36.** Find the GCF of $40x^2$ and $30x^3$.

[3.7] **37.** Divide. $\dfrac{18n^5}{3n^2}$

[3.7] **38.** Divide. $\dfrac{4x^3 - 8x^2 + 12x}{-2x}$

[3.7] **39.** Factor. $12m^4 - 18m^3 + 24m^2$

For Exercises 40 and 41, solve and check.

[1.2] **40.** $x + 19 = 25$

[2.4] **41.** $-16x = -128$

[1.6] **42.** Calculate the area of the figure.

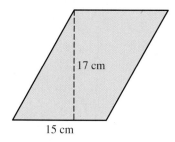

17 cm

15 cm

[3.8] **43.** Write an expression in simplest form for the perimeter.

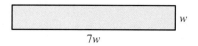

w

$7w$

[3.8] **44.** Write an expression in simplest form for the volume.

$y + 1$

$5y$

y

For Exercises 45–51, solve.

[1.3] **45.** Andre has a monthly mortgage payment of $785. If he makes 360 payments (30 years), then what would be the total he pays over the 30 years?

[1.3] **46.** An entrance lock requires a person to enter three things. First a digit, 0–9, then a letter of the alphabet, A–H, and finally a day of the week. How many possible combinations are there?

[1.4] **47.** A patient is to receive 600 milliliters of 5% D/W (dextrose in water) solution in an IV drip over the course of an hour. How many milliliters should the patient receive each minute?

[2.2] **48.** Carla has a balance of −$353 in a credit card account. If she makes a payment of $150, then makes two purchases at $38 each, what is her new balance?

[3.8] **49.** A 10-foot-wide by 12-foot-long by 7-foot-high box is to be made out of plastic. What is the surface area of the box?

[3.8] **50.** A ride at an amusement park is a ship that rotates to an inverted position 80 feet above the ground. While inverted, loose items such as sunglasses, coins, etc., tend to fall out. How far from the ground will the falling items be after 2 seconds?

[3.8] **51.** A company produces two different car alarm systems, basic and advanced. If b represents the number of basic alarms and a represents the number of advanced alarms, then $225b + 345a + 200$ describes the revenue from the sales of the two types of alarm systems. The polynomial $112b + 187a + 545$ describes the cost of producing the two alarm systems.

   a. Write an expression in simplest form for net.

   b. If the company sells 88 basic and 64 advanced alarm systems in one month, what would be the net profit or loss?

CHAPTER 4

Equations

> " *Spectacular achievement is always preceded by spectacular preparation.* "
>
> —ROBERT H. SCHULLER,
> RELIGIOUS LEADER

> " *What I do is prepare myself until I know I can do what I have to do.* "
>
> —JOE NAMATH,
> QUARTERBACK

Preparing for a Test

If you have followed the day-by-day routine of practice and study as discussed in the *To the Student* section, the Chapter 3 opener, and the Learning Strategy boxes, then most of your test preparation is already done. However, some additional preparation in the days leading up to a test can make it even more likely that you will perform well.

A few days before the test, read through all your notes and the summary at the end of the chapter. Once you know the rules and procedures, work through the review exercises that follow the chapter summary, then take a break.

When you return, take a practice test. If your instructor's tests closely resemble the practice tests in the text, then those practice tests are an obvious choice. However, if your instructor constructs tests that are very different from the practice tests in the text, then construct your own practice test (see the *To the Student* section of the text).

Keep repeating this review process until you can correctly and effortlessly solve every type of problem you are likely to see. By beginning this review process a few days before a test, you give yourself enough time to solidify your understanding of the material.

4.1 Equations and Their Solutions

OBJECTIVES

1. Differentiate between an expression and an equation.

2. Check a given number to see if it is a solution for a given equation.

OBJECTIVE 1 Differentiate between an expression and an equation.

In Section 3.1, we considered the difference between an expression and an equation. We learned that equations have an equal sign whereas expressions do not.

Expressions: $3x + 5$ Equations: $3x + 5 = 9$
$$7x^2 - 8x + 1 \qquad\qquad 7x^2 - 8x + 1 = 5x - 2$$

Think of expressions as phrases and equations as complete sentences. The expression $3x + 5$ is read "three x plus five," which is not a complete sentence. The equation $3x + 5 = 9$ is read "three x plus five is equal to nine" or simply "three x plus five is nine," which are complete sentences. Notice that the equal sign translates to the verb "is."

$$3x + 5 = 9$$
$$\updownarrow$$
"Three x plus five **is** nine."

Your Turn 1

Determine whether each of the following is an expression or an equation.

a. $5x - 8$

b. $7y + 9 = 15$

c. $P = 2l + 2w$

d. $4^2 + 3(9 - 1) + \sqrt{25}$

Example 1 Determine whether each of the following is an expression or an equation.

a. $7x^2 + 9x + 5$

Answer: Expression, because it has no equal sign.

b. $9x - 10 = 3x + 1$

Answer: Equation, because it has an equal sign.

◄ Do Your Turn 1

OBJECTIVE 2 Check a given number to see if it is a solution for a given equation.

In Chapter 3, we learned that we can perform two actions with expressions.

1. Evaluate (replace the variables with given numbers and perform the arithmetic).

2. Rewrite (simplify, distribute, or factor).

Our objective with an equation is to **solve** it. To solve an equation, we must find its **solution** or solutions.

DEFINITIONS   **Solve:** To find the solution or solutions to an equation.

       Solution: A number that makes an equation true when it replaces the variable in the equation.

For example, 5 is a solution for the equation $x + 3 = 8$ because the equation is true when 5 replaces x.

$$x + 3 = 8$$
$$\downarrow$$
$$5 + 3 = 8 \qquad \text{True.}$$

Answers to Your Turn 1
a. expression b. equation
c. equation d. expression

Showing that $x + 3 = 8$ is true when x is replaced by 5 serves as a *check* for the solution.

To check to see if a value is a solution to an equation:

1. Replace the variable(s) with the value.
2. Simplify both sides of the equation as needed. If the resulting equation is true, then the value is a solution.

Example 2 Is 3 a solution for $2x - 5 = 1$?

Solution:
$$2x - 5 = 1$$
$$2(3) - 5 \stackrel{?}{=} 1 \quad \text{Replace } x \text{ with 3.}$$
$$6 - 5 \stackrel{?}{=} 1 \quad \text{Simplify.}$$
$$1 = 1 \quad \text{True.}$$

◀ **Note:** The $\stackrel{?}{=}$ symbol means we are asking whether the left side of the equation equals the right side.

Answer: Since the resulting equation is true, 3 is a solution for $2x - 5 = 1$.

Do Your Turn 2 ▶

Your Turn 2

Is 2 a solution for
$12 = 9b - 5$?

Example 3 Is -2 a solution for $y^2 + 7 = 13$?

Solution:
$$y^2 + 7 = 13$$
$$(-2)^2 + 7 \stackrel{?}{=} 13 \quad \text{Replace } y \text{ with } -2.$$
$$4 + 7 \stackrel{?}{=} 13 \quad \text{Simplify.}$$
$$11 = 13 \quad \text{False.}$$

Answer: Since the resulting equation is false, -2 is not a solution for $y^2 + 7 = 13$.

Do Your Turn 3 ▶

Your Turn 3

Is -4 a solution for
$m^2 + 5 = 21$?

Example 4 Is 4 a solution for $-3n + 8 = 2n - 17$?

Solution:
$$-3n + 8 = 2n - 17$$
$$-3(4) + 8 \stackrel{?}{=} 2(4) - 17 \quad \text{Replace } n \text{ with 4.}$$
$$-12 + 8 \stackrel{?}{=} 8 - 17 \quad \text{Simplify.}$$
$$-4 = -9 \quad \text{False.}$$

Note: We simplify each side of the equation separately until we can determine if it is true or false.

Answer: 4 is not a solution for $-3n + 8 = 2n - 17$.

Do Your Turn 4 ▶

Your Turn 4

Is 2 a solution for
$5m - 29 = -8m - 3$?

Example 5 Is -8 a solution for $3x + 6 = 2(x - 1)$?

Solution:
$$3x + 6 = 2(x - 1)$$
$$3(-8) + 6 \stackrel{?}{=} 2((-8) - 1) \quad \text{Replace } x \text{ with } -8.$$
$$-24 + 6 \stackrel{?}{=} 2(-9) \quad \text{Simplify.}$$
$$-18 = -18 \quad \text{True.}$$

Answer: -8 is a solution for $3x + 6 = 2(x - 1)$.

Do Your Turn 5 ▶

Your Turn 5

Is -2 a solution for
$5x - 6 = 4(x + 2)$?

Answer to Your Turn 2
2 is not a solution.

Answer to Your Turn 3
-4 is a solution.

Answer to Your Turn 4
2 is a solution.

Answer to Your Turn 5
-2 is not a solution.

CONNECTION In previous chapters, we always placed results below the expression from which they came.

We wrote: $2(4) - 17$

$$= 8 - 17$$

$$= -9$$

By bringing results down we were preparing for this checking technique. To check, we bring down each result from the left expression on the left side of the equation and each result from the right expression on the right side of the equation, and see if the results equate.

Your Turn 6

Is 4 a solution for
$5x - 2(x + 3) = 7(x + 2)$?

Example 6   Is -5 a solution for $2(a + 6) = 6a - 4(3a + 7)$?

Solution: $2(a + 6) = 6a - 4(3a + 7)$

$$2(-5 + 6) \stackrel{?}{=} 6(-5) - 4(3(-5) + 7) \qquad \text{Replace } a \text{ with } -5.$$

$$2(1) \stackrel{?}{=} -30 - 4(-15 + 7) \qquad \text{Simplify.}$$

We keep bringing down 2 until we resolve the other side. ▶

$$2 \stackrel{?}{=} -30 - 4(-8)$$

$$2 \stackrel{?}{=} -30 - (-32)$$

$$2 \stackrel{?}{=} -30 + 32$$

$$2 = 2 \qquad \text{True.}$$

Answer: -5 is a solution for $2(a + 6) = 6a - 4(3a + 7)$.

◀ **Do Your Turn 6**

OF INTEREST

Pierre de Fermat is often referred to as the Prince of Amateurs. He was born in Beaumont-de-Lomagne, France, in 1601. He lived a very normal, quiet, honest life as king's council in the local parliament of Toulouse, France. Fermat is probably best known for a theorem that is an extension of the Pythagorean theorem, an equation relating the side lengths of right triangles.

Fermat had the habit of writing his theorems as commentary in the margins of math books or letters to mathematicians. Consequently, he did not supply proofs for his theorems. The proof of the theorem mentioned above eluded mathematicians for centuries. It became known as *Fermat's last theorem*. In 1993, Andrew Wiles of Princeton University submitted a potential proof of the theorem. Wiles' initial proof was found to have a subtle error that he was able to correct and his 150-page proof that Fermat's theorem is correct has now been accepted.

Source: E. T. Bell, *Men of Mathematics*

Answer to Your Turn 6
4 is not a solution.

1. What is the difference between an expression and an equation?

2. What is a solution to an equation?

3. What does it mean to solve an equation?

4. Explain how to check a value to see if it is a solution for an equation?

For Exercises 5–10, determine whether each of the following is an expression or an equation.

5. $9x + 7 = 5$

6. $7x - 8$

7. $2n + 8m^3$

8. $10x^2 = 12x + 1$

9. $14 + 2^5 - 2(3 + 7)$

10. $2y = 16$

For Exercises 11–30, check to see if the given number is a solution for the given equation.

11. $x + 9 = 14$; check $x = -5$

12. $c + 7 = -2$; check $c = -9$

13. $3t = -9$; check $t = -9$

14. $2p = -18$; check $p = -8$

15. $3y + 1 = 13$; check $y = 3$

16. $5d - 1 = 19$; check $d = 4$

17. $6x - 7 = 41$; check $x = 8$

18. $-5a + 19 = 9$; check $a = 2$

19. $3m + 10 = -2m$; check $m = 5$

20. $18 - 7n = 2n$; check $n = -2$

21. $x^2 - 15 = 2x$; check $x = -3$

22. $x^2 - 15 = 2x$; check $x = 5$

23. $b^2 = 5b + 6$; check $b = -3$

24. $b^2 = 5b + 6$; check $b = -1$

25. $-5y + 20 = 4y + 20$; check $y = 0$

26. $10n - 14 = -n + 19$; check $n = 1$

27. $-2(x - 3) - x = 4(x + 5) - 21$; check $x = 3$

28. $3(x - 5) - 1 = 11 - 2(x + 1)$; check $x = 5$

29. $a^3 - 5a + 6 = a^2 + 3(a - 1)$; check $a = -2$

30. $2t(t + 1) = t^3 - 5t$; check $t = -3$

PUZZLE PROBLEM A farmer has a fox, a goat, and a large head of cabbage to transport across a river. His boat is so small that it can hold only the farmer and the fox, goat, or cabbage at a time. If the farmer leaves the fox with the goat, the fox will eat the goat. If he leaves the goat with the cabbage, the goat will eat the cabbage. Somehow he transports them all across safely in the boat. Explain the process.

Review Exercises

[3.3] **1.** Combine like terms. $2x - 19 + 3x + 7$

[3.3] **2.** Combine like terms. $7y + 4 - 12 - 9y$

[3.5] **3.** Multiply. $5(m + 3)$

[3.5] **4.** Multiply. $-(3m - 4)$

[1.2] **5.** Solve and check. $x + 8 = 15$

4.2 The Addition/Subtraction Principle of Equality

OBJECTIVE 1 Determine whether a given equation is linear.

There are many types of equations. We will focus on **linear equations.**

DEFINITION **Linear equation:** An equation that is made of polynomials or monomials that are at most degree 1.

This means that linear equations can contain constants or variable terms with a single variable raised to an exponent of 1. All other equations are called *nonlinear equations.*

> **Example 1** Determine whether the given equation is linear.
>
> **a.** $5x + 9 = 24$
>
> **Answer:** $5x + 9 = 24$ is a linear equation because 1 is the greatest degree involved.
>
> **b.** $2x - 9x^2 = 15x + 7$
>
> **Answer:** $2x - 9x^2 = 15x + 7$ is not linear because there is a term with a degree greater than 1. $-9x^2$ is a degree-2 term.

If an equation has the same variable throughout, then we say it is an equation in one variable.

> $2x + 9 = 5x - 12$ is a linear equation in one variable, x.
>
> $y^2 - 2y + 5 = 8$ is a nonlinear equation in one variable, y.

Different variables can appear in the same equation. If an equation has two different variables, we say it is an equation in two variables. Such an equation is linear as long as the degree of every variable term is 1.

> $2x - 3y = 6$ is a linear equation in two variables, x and y.
>
> $y = x^3$ is a nonlinear equation in two variables.

Do Your Turn 1 ▷

OBJECTIVES

1 Determine whether a given equation is linear.

2 Solve linear equations in one variable using the addition/subtraction principle of equality.

3 Solve equations with variables on both sides of the equal sign.

4 Solve application problems.

Your Turn 1

Determine whether the given equation is linear.

a. $2x - 8 = 14$

b. $7y = 21$

c. $x^3 = 5x + 2$

d. $y = x^2 - 5$

e. $y = 2x + 1$

OBJECTIVE 2 Solve linear equations in one variable using the addition/subtraction principle of equality.

In Chapters 1 and 2, we solved simple equations like $x + 7 = 9$. Now we will build on what we learned about solving those simple equations and develop a technique for solving more complex equations. This technique for solving equations is called the **balance technique.** We can illustrate the nature of the technique by imagining an equation as scales with the equal sign acting as the pivot point on the scales.

Answers to Your Turn 1
a. yes **b.** yes **c.** no **d.** no **e.** yes

When weight is added or removed on one side of the balanced scales, the scales tip out of balance. In the figure, we add 4 to the left side of the scale, tipping it out of balance.

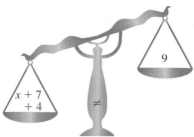

To maintain balance, the same weight must be added or removed on *both* sides of the scale. Adding 4 to the right side of our scale, balances it.

Though we illustrated addition on both sides, we could have used any operation, as long as we placed the same operation on both sides. In the language of mathematics, this technique can be summarized by two principles: the *addition/subtraction principle of equality* and the *multiplication/division principle of equality*. In this section, we will use only the addition/subtraction principle. We will use the multiplication/division principle in Section 4.3.

RULE *The addition/subtraction principle of equality*

The same amount can be added to or subtracted from both sides of an equation without affecting its solution(s).

Consider our equation that we used with the scales, $x + 7 = 9$. Notice that its solution is 2 because $2 + 7 = 9$. The addition principle tells us that adding a number on both sides will not change the fact that 2 is its solution. So, when we added 4 to both sides, which balanced the equation, 2 should still be the solution, and it is.

$$
\begin{array}{rl}
x + 7 = & 9 \\
\underline{+4 \quad} & \underline{+4} \\
x + 11 = & 13
\end{array}
$$

◀ **Note:** After adding 4 to both sides, the solution to the resulting equation is still 2 because $2 + 11 = 13$.

The goal for solving equations using the balance technique is to rewrite the equation in a simpler form, $x = c$, where the variable is isolated on one side of the equation and the solution, c, appears on the other side. In the case of $x + 7 = 9$, we would use the addition principle so that the equation becomes $x = 2$. Notice that adding -7 to both sides (or subtracting 7 from both sides) achieves this goal.

$$
\begin{array}{rl}
x + 7 = & 9 \\
\underline{-7 \quad} & \underline{-7} \\
x + 0 = & 2 \\
x = & 2
\end{array}
$$

◀ **Note:** Since 7 and -7 are additive inverses, their sum is 0, thereby isolating x on the left side and revealing the solution, 2, on the right side.

Our example suggests the following procedure.

PROCEDURE *To use the addition/subtraction principle of equality to clear a term, add the additive inverse of that term to both sides of the equation (that is, add or subtract appropriately so that the term you want to clear becomes 0.)*

LEARNING STRATEGY

If you are a visual learner, consider writing the amount added or subtracted on both sides of an equation in a different color than the rest of the equation. Once you are comfortable seeing the amount added or subtracted, you may not need the color any longer.

CONNECTION In Chapters 1 and 2, we solved a missing addend equation by writing a related subtraction statement. By recognizing that the inverse operation is used to solve for the missing addend, we laid the foundation for the addition/subtraction principle and the balance technique for solving equations.

Related subtraction:

$$x + 2 = 5$$

$$x = 5 - 2$$

$$x = 3$$

Balance technique:

$$x + 2 = 5$$
$$\underline{-2 \quad -2}$$
$$x + 0 = 3$$
$$x = 3$$

Note that both methods subtract 2 from 5 to arrive at the solution, 3.

Example 2 Solve and check.

a. $x - 8 = -15$

Solution: To isolate x, we must clear -8 from the left side, so we add 8. To keep the equation balanced, we must add 8 on both sides.

To isolate x we add $+8$ so that $-8 + 8 = 0$. ▶

$$x - 8 = -15$$
$$\underline{+8 \qquad +8}$$
$$x + 0 = \quad -7$$
$$x = \quad -7$$

◀ Because we added $+8$ on the left side we must add $+8$ on the right side as well.

Check: In Section 4.1, we learned that to check that -7 is the solution for $x - 8 = -15$, we replace x with -7 and see if the resulting equation is true.

$$x - 8 = -15$$
$$-7 - 8 \overset{?}{=} -15 \qquad \text{Replace } x \text{ with } -7.$$
$$-15 = -15 \qquad \text{True.}$$

The equation is true, so -7 is the solution.

b. $14 = m + 5$

Solution: To isolate m, we add -5 to both sides (or subtract 5 from both sides).

$$14 = m + 5$$
$$\underline{-5 \qquad -5}$$
$$9 = m + 0$$
$$9 = m$$

◀ **Note:** It is not necessary to write the resulting 0 because when we add 0, it doesn't affect anything. However, we will continue writing this step in the text to remind you of what is happening.

Check: $14 = m + 5$
$$14 \overset{?}{=} 9 + 5 \qquad \text{Replace } m \text{ with } 9.$$
$$14 = 14 \qquad \text{True.}$$

The equation is true, so 9 is the solution.

Do Your Turn 2 ▷

Your Turn 2

Solve and check.

a. $16 = x - 5$

b. $n + 13 = 20$

c. $-10 = y - 7$

d. $21 = -9 + k$

Answers to Your Turn 2
a. $x = 21$ b. $n = 7$
c. $y = 3$ d. $k = 30$

Some equations contain expressions that can be simplified. If like terms are on the same side of an equation, we combine the like terms before isolating the variable.

Your Turn 3

Solve and check.

a. $4x + 5 - 3x = 12 - 13$

b. $25 - 18 = 6c + 2 - 5c$

Note: It may be tempting to check a solution using an equation that occurs in one of the solution steps instead of the original equation. For example, in Example 3, it would be easier to check 9 using $n - 14 = 25$ instead of using $5n - 14 - 4n = -12 + 7$. However, remember that we are solving the original equation, so we always check using the original equation.

Example 3 Solve and check. $5n - 14 - 4n = -12 + 7$

Solution: On the left side of the equation, we have the like terms $5n$ and $-4n$. On the right side, we have -12 and 7. We will combine those like terms first, then isolate n.

$$5n - 14 - 4n = -12 + 7$$

Combine $5n$ and $-4n$ to get n. $\quad n - 14 = -5$ Combine -12 and 7 to get -5.

Add 14 on both sides to isolate n.
$$\underline{+14 \quad +14}$$
$$n + 0 = \quad 9$$
$$n = 9$$

◄ To keep balance, if we add $+14$ to the left side, we must add $+14$ to the right side as well.

Check:
$$5n - 14 - 4n = -12 + 7$$
$$5(9) - 14 - 4(9) \stackrel{?}{=} -12 + 7 \quad \text{Replace } n \text{ with 9.}$$
$$45 - 14 - 36 \stackrel{?}{=} -5 \quad \text{Simplify each side.}$$
$$31 - 36 \stackrel{?}{=} -5$$
$$-5 = -5$$

The equation is true, so 9 is the solution.

◄ **Do Your Turn 3**

If an equation contains parentheses, we can use the distributive property to clear them. After using the distributive property, we will usually have to simplify further by combining like terms that are on the same side of the equation.

Your Turn 4

Solve and check.

a. $14 - 17 = 4(u - 3) - 3u$

b. $7h - 2(3h + 4) = 15 - 12$

Example 4 Solve and check. $3(y - 5) - 2y = -20 + 2$

Solution: Use the distributive property to simplify the expressions, and then isolate y.

$$3(y - 5) - 2y = -20 + 2$$

Distribute to clear parentheses. $\quad 3y - 15 - 2y = -18$ Combine -20 and 2 to get -18.

Combine $3y$ and $-2y$ to get y. $\quad y - 15 = -18$

Add 15 on both sides to isolate y.
$$\underline{+15 \quad +15}$$
$$y + 0 = \quad -3$$
$$y = -3$$

Check:
$$3(y - 5) - 2y = -20 + 2$$
$$3((-3) - 5) - 2(-3) \stackrel{?}{=} -20 + 2 \quad \text{Replace } y \text{ with } -3.$$
$$3(-8) - 2(-3) \stackrel{?}{=} -18 \quad \text{Simplify.}$$
$$-24 - (-6) \stackrel{?}{=} -18$$
$$-24 + 6 \stackrel{?}{=} -18$$
$$-18 = -18$$

The equation is true, so -3 is the solution.

◄ **Do Your Turn 4**

Answers to Your Turn 3
a. $x = -6$ b. $c = 5$

Answers to Your Turn 4
a. $u = 9$ b. $h = 11$

OBJECTIVE 3 Solve equations with variables on both sides of the equal sign.

In Examples 3 and 4, the like terms were together on the same side of the equation. If like terms appear on opposite sides of an equation, as in $7x - 9 = 6x - 11$, we use the addition/subtraction principle of equality to get the like terms together on the same side so that they can be combined.

Example 5 Solve and check. $7x - 9 = 6x - 11$

Solution: Because like terms appear on opposite sides of the equal sign, we use the addition/subtraction principle of equality to get them together on the same side of the equal sign.

Note: By subtracting $6x$ from both sides, we clear the $6x$ term from the right side and combine it with the $7x$ on the left side. It does not matter which term you move first (see explanation below).

$$
\begin{array}{rcl}
7x - 9 &=& 6x - 11 \\
- 6x & & - 6x \\
\hline
x - 9 &=& 0 - 11
\end{array}
$$
Subtract $6x$ from both sides.

$$
\begin{array}{rcl}
x - 9 &=& -11 \\
+ 9 & & + 9 \\
\hline
x + 0 &=& -2
\end{array}
$$
Add 9 to both sides to isolate x.

$$x = -2$$

Check:
$$7x - 9 = 6x - 11$$
$$7(-2) - 9 \stackrel{?}{=} 6(-2) - 11 \quad \text{Replace } x \text{ with } -2.$$
$$-14 - 9 \stackrel{?}{=} -12 - 11 \quad \text{Simplify.}$$
$$-23 = -23$$

The equation is true, so -2 is the solution.

When variable terms appear on both sides of the equal sign, it does not matter which term you clear first, although some choices make the process easier. Suppose in Example 5, we clear $7x$ first:

Note: Clearing the $7x$ term first gives $-x$, whereas when we cleared the $6x$ term first, we got x. Most people prefer working with positive coefficients.

$$
\begin{array}{rcl}
7x - 9 &=& 6x - 11 \\
- 7x & & - 7x \\
\hline
0 - 9 &=& -x - 11
\end{array}
$$

$$
\begin{array}{rcl}
-9 &=& -x - 11 \\
+ 11 & & + 11 \\
\hline
2 &=& -x + 0
\end{array}
$$

$$2 = -x$$

$$-2 = x$$

Note: The equation $2 = -x$ means we must find a number whose additive inverse is 2, so x must be equal to -2. In effect, we simply changed the signs of both sides. In Section 4.3, we will see another way to interpret $-x = 2$.

Notice that the term we choose to clear first does not affect the solution, but it does affect our approach to the rest of the problem.

Conclusion: When solving an equation that has variable terms on both sides, clearing the variable term with the lesser coefficient will avoid negative coefficients.

Do Your Turn 5 ▷

Your Turn 5

Solve and check.

a. $8x - 5 = 7x + 2$

b. $3b - 2 = 4b + 3$

c. $1 + 5m = 7 + 4m$

Answers to Your Turn 5
a. $x = 7$ b. $b = -5$ c. $m = 6$

Solve and check.

a. $5n - (n - 7) =$
$\quad 2(n + 6) + n$

Example 6 Solve and check. $y - (2y + 9) = 6(y - 2) - 8y$

Solution: We must simplify the expressions, get the y terms together on the same side of the equal sign, and then isolate y.

Note: When we distribute a minus sign in an expression like $-(2y + 9)$, we can think of the minus sign as -1, so that we have:
$$-(2y + 9) = -1(2y + 9)$$
$$= -1(2y) + (-1)(9)$$
$$= -2y - 9$$

▼

$$y - (2y + 9) = 6(y - 2) - 8y$$

Distribute -1. $\qquad y - 2y - 9 = 6y - 12 - 8y$ Distribute 6.

Combine y and $-2y$. $\qquad -y - 9 = -2y - 12$ Combine $6y$ and $-8y$.

$$-y - 9 = -2y - 12$$

Note: The coefficient of $-2y$ is less than the coefficient of $-y$ (which means $-1y$), so we chose to clear the $-2y$ term by adding $2y$.

$$\underline{+ 2y \qquad\qquad + 2y}$$
$$y - 9 = 0 - 12$$
$$y - 9 = -12$$

Add 9 to both sides to isolate y.
$$\underline{+ 9 \qquad\qquad + 9}$$
$$y + 0 = \quad -3$$

$$y = -3$$

Check: $\qquad y - (2y + 9) = 6(y - 2) - 8y$

$$-3 - (2(-3) + 9) \overset{?}{=} 6(-3 - 2) - 8(-3)$$ Replace y with -3.

$$-3 - (-6 + 9) \overset{?}{=} 6(-5) - 8(-3)$$ Simplify.

$$-3 - 3 \overset{?}{=} -30 - (-24)$$

$$-6 \overset{?}{=} -30 + 24$$

$$-6 = -6$$

b. $10 + 3(5x - 2) =$
$\quad 7(2x - 1) + 9$

The equation is true, so -3 is the solution.

Let's put together an outline based on everything we've learned so far.

PROCEDURE *To solve equations:*

1. Simplify both sides of the equation as needed.
   a. Distribute to clear parentheses.
   b. Combine like terms.
2. Use the addition/subtraction principle so that all variable terms are on one side of the equation and all constants are on the other side. Then combine like terms.

Note: Clearing the variable term with the lesser coefficient will avoid negative coefficients.

c. $9 - 2(3c + 5) =$
$\quad -3 - 5(c - 1)$

◀ **Do Your Turn 6**

OBJECTIVE 4 Solve application problems.

Let's put what we've learned about using the addition/subtraction principle of equality into the context of some situations, the most common being the missing addend problem. You may recall that the way to recognize a missing addend problem is to look for key words such as "How much more is needed?"

Answers to Your Turn 6
a. $n = 5$ b. $x = -2$ c. $c = -3$

Example 7   Laura wants to buy a car stereo that costs $275. She currently has $142. How much more does she need?

Understand: We are given the total required and the amount she currently has, and we must find how much she needs. This is a missing addend situation. We must add the needed amount to 142 and end up with 275.

Plan: Let x represent the amount Laura needs. We will translate to a missing addend equation, then solve.

Execute: current amount + needed amount = 275

$$142 \quad + \quad x \quad\quad = 275$$
$$142 + x = 275$$
$$\underline{-\ 142 \qquad -\ 142} \qquad \text{Subtract 142 from both sides.}$$
$$0 + x = 133$$
$$x = 133$$

Answer: Laura needs $133 to buy the stereo.

Check: Does $142 plus the additional $133 equal $275?

$$142 + 133 \overset{?}{=} 275$$
$$275 = 275 \qquad \text{It checks.}$$

Do Your Turn 7 ▶

OF INTEREST

The word *algebra* first appeared in a book by Arab mathematician Mohammed ibn Musa al-Khowarizmi (about A.D. 825), entitled *al-jabr w'al-muqabalah*. The word *al-jabr* translates to *the reunification*. *Al-jabr* went through many spelling changes and finally emerged as our modern word, *algebra*. When the Moors brought the word to Spain, it came to mean the reunification of broken bones, an *algebrista* being one who sets broken bones.

Source: D. E. Smith, *History of Mathematics*, 1953

Your Turn 7

Solve.

a. A patient must receive 350 cubic centimeters of a medication in three injections. He has received two injections of 110 cubic centimeters each. How much should the third injection measure?

b. Daryl has a balance of −$568 on a credit card. How much must he pay to bring his balance to −$480?

Answers to Your Turn 7
a. 130 cc b. $88

For Extra Help

Videotape DVT

Addison-Wesley Tutor Center

Math*XL*
Math XL

MyMathLab

Student Solutions Manual

1. What is a linear equation?

2. Give an example of a linear equation and an example of a nonlinear equation.

3. What does the addition/subtraction principle of equality say?

4. How do you use the addition/subtraction principle to clear a positive term?

5. How do you use the addition/subtraction principle of equality to clear a negative term?

6. If an equation has variable terms on both sides of the equation, what should you do first?

For Exercises 7–18, determine whether the given equation is linear.

7. $9x - 7 = 4x + 3$

linear

8. $7y = 14$

linear

9. $t^2 - 5 = 20$

nonlinear

10. $2n - 9 = n^3 + 6n - 1$

nonlinear

11. $5u - 17 = u^2 + 1$

nonlinear

12. $14n - 9n = 5n + 3$

linear

13. $2x + y = 5$

linear

14. $y = x^2 - 5$

nonlinear

15. $t = u^3 + 2$

nonlinear

16. $x = -6$

linear

17. $y = 3x + 2$

linear

18. $(y - 5) = 3(x - 1)$

linear

For Exercises 19–38, solve and check.

19. $n + 14 = 20$

20. $y + 19 = -6$

21. $-7 = x - 15$

22. $5 = t - 11$

23. $10 = 5x - 2 - 4x$

24. $3r + 13 - 2r = 9$

25. $9y - 8y + 11 = 12 - 2$

26. $2x - 15 - x = 9 - 17$

27. $4x - 2 = 3x + 5$

28. $6b + 7 = 5b + 3$

29. $m - 1 = 2m - 5$

30. $7t + 2 = 8t - 9$

31. $3n - 5 + 4n = 6n - 12$

32. $8u + 13 - 2u = 2u + 3u$

33. $7y + 1 - 3y = 2y - 10 + 3y$

34. $-11 + x + 2 + 5x = 9x - 4 - 2x$

35. $3(n + 2) = 4 + 2(n - 1)$

36. $9 - 5(b - 1) = -4(b - 2)$

37. $6t - 2(t - 5) = 9t - (6t + 2)$

38. $8 + 3(h - 5) = 2(5h + 1) - 6h$

For Exercises 39–48, translate to an equation, then solve.

39. Yolanda is to close on her new house in three weeks. The amount that she must have at closing is $4768. She currently has $3295. How much more does she need?

40. Nikki has a balance of −$457 on her credit card. What payment should she make to get the balance to −$325?

41. A patient is to receive 450 cubic centimeters of a medication in three injections. The first injection is to be 200 cubic centimeters and the second injection 180 cubic centimeters. How much must the last injection be?

42. An entry in a chemist's notebook is smudged. The entry is the initial temperature measurement of a chemical in an experiment. In the experiment, a substance was introduced to the chemical, decreasing its temperature by 19°C. The final temperature is listed as 165°C. What was the initial temperature?

43. Jerry sells medical equipment. The company quota is set at $10,500 each month. The spreadsheet shows Jerry's sales as of the end of the second week of July. How much more does Jerry need to sell to make the July quota? Do you think Jerry will make the quota? Why?

Date	Item No.	Quantity	Price	Total Sale
7/3	45079	2	$ 800	$1600
7/8	47002	1	$4500	$4500
7/13	39077	3	$ 645	$1935

44. Marc is a waiter at a restaurant. His rent of $675 is due at the end of the week. Currently he has $487 in the bank. At the end of his shift he finds that he made $85. How much more does he need in order to pay his rent? If he has two more shifts before rent is due, do you think he will make enough to pay the rent? Explain.

45. Connie is considering a new apartment. What is the length of the dining area? Is this area large enough to accommodate a 4-foot by 4-foot square table and four chairs?

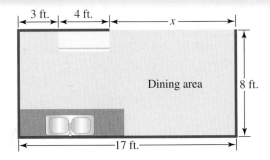

46. At an archaeological dig, the archaeologists suspect there may be a secret passage or small chamber between two chambers of a tomb. They have mapped and documented distances. What is the distance between the two chambers?

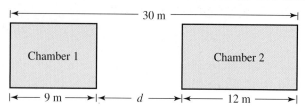

47. The perimeter of the triangle shown is 84 centimeters. Find the length of the missing side.

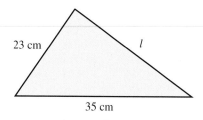

48. To take more than the standard deduction on taxes, itemized deductions must exceed the standard deduction. Because the Colemans' filing status is married filing jointly, their standard deduction is $9500 in 2003. If the Colemans have the following itemized deductions already, how much are they lacking to reach the level of the standard deduction?

> Medical and dental expenses: $242
> Taxes paid: $2447
> Charitable donations: $225

DISCUSSION What else might the Colemans deduct?

Review Exercises

For Exercises 1–4, evaluate.

[2.4] **1.** $5(-8)$

[2.4] **2.** $24 \div (-3)$

[2.5] **3.** $12 + 3(-5 - 2)$

[2.5] **4.** $\dfrac{16 + 2(3 - 5)}{-2(5) + 4}$

[1.4] **5.** Solve and check. $5x = 15$

4.3 The Multiplication/Division Principle of Equality

OBJECTIVE 1 Solve equations using the multiplication/division principle of equality.

OBJECTIVES

1 Solve equations using the multiplication/division principle of equality.

2 Solve equations using both the addition/subtraction and the multiplication/division principles of equality.

3 Solve application problems.

In Section 4.2, when we isolated the variable, the coefficient of that variable was always 1 or -1. What if we ended up with a coefficient other than 1 or -1? Consider the equation $5n - 9 = 3n + 1$. Let's follow our plan from Section 4.2 and get the variable terms together first, then isolate n.

$$
\begin{array}{r}
5n - 9 = 3n + 1 \\
\underline{-3n \qquad -3n} \\
2n - 9 = 0 + 1 \\
2n - 9 = 1 \\
\underline{+9 \quad +9} \\
2n + 0 = 10 \\
2n = 10
\end{array}
$$

Note: We have a coefficient other than 1 or -1 here. ▶

How do we solve $2n = 10$? Remember it is a missing factor statement because the 2 coefficient multiplies n. We learned in Chapter 1 that we solve for a missing factor by dividing. Connecting what we learned in Chapter 1 with the balance technique means we divide on both sides to keep the equation balanced. We can write:

$$2n \div 2 = 10 \div 2 \qquad \text{or the more popular form} \qquad \frac{2n}{2} = \frac{10}{2}$$

Notice that the coefficient, 2, divides out.

Because $2 \div 2 = 1$, we are left with $1n$, which can be simplified to n. ▶

$$
\frac{2n}{2} = \frac{10}{2}
$$
$$
1n = 5
$$
$$
n = 5
$$

If we manipulate an equation using the addition/subtraction principle of equality and we still have a coefficient other than 1, we can clear that coefficient by dividing. This is the main purpose of the multiplication/division principle of equality.

RULE *The Multiplication/Division Principle of Equality*
We can multiply or divide both sides of an equation by the same nonzero amount without affecting its solution(s).

PROCEDURE *To use the multiplication/division principle of equality to clear a coefficient, divide both sides by that coefficient.*

Example 1 Solve and check.

a. $7x = 21$

Solution: To isolate x we must clear the coefficient 7, so we use the multiplication/division principle of equality.

$$
\frac{7x}{7} = \frac{21}{7} \qquad \text{Divide both sides by 7 to clear the 7 coefficient.}
$$
$$
1x = 3
$$
$$
x = 3
$$

Solve and check.

a. $5y = 45$

b. $24 = 6n$

c. $-48 = -8a$

d. $-2x = 16$

Check: $7x = 21$

$7(3) \overset{?}{=} 21$ Replace x with 3.

$21 = 21$ Simplify.

The resulting equation is true, so 3 is the solution.

b. $-6m = 30$

Solution:

Note: It is not necessary to write the resulting 1 because multiplying by 1 has no effect on the variable. However, we will continue to write the 1 in the text to remind you of what is happening.

▶ $\dfrac{-6m}{-6} = \dfrac{30}{-6}$ Divide both sides by −6 to clear the −6 coefficient.

$1m = -5$

$m = -5$

Check: $-6m = 30$

$-6(-5) \overset{?}{=} 30$ Replace m with −5.

$30 = 30$ Simplify.

True, so -5 is the solution.

◀ **Do Your Turn 1**

OBJECTIVE 2 Solve equations using both the addition/subtraction and the multiplication/division principles of equality.

Now let's put the multiplication/division principle of equality together with the addition/subtraction principle of equality. We'll follow the same plan we developed in Section 4.2, except now we will have an extra step: to clear a coefficient at the end.

PROCEDURE *To solve equations:*

1. Simplify both sides of the equation as needed.
   a. Distribute to clear parentheses.
   b. Combine like terms.
2. Use the addition/subtraction principle of equality so that all variable terms are on one side of the equation and all constants are on the other side. (Clear the variable term with the lesser coefficient.) Then combine like terms.
3. Use the multiplication/division principle of equality to clear the remaining coefficient.

Example 2   Solve and check.

a. $-2x - 9 = 7$

Solution: To isolate x we first use the addition/subtraction principle of equality and add 9 to both sides, which isolates the $-2x$ term. We then use the multiplication/division principle of equality to clear the -2 coefficient.

CONNECTION After adding 9 to both sides, the equation becomes $-2x = 16$, which is the equation in Your Turn 1d.

▶

$$-2x - 9 = 7$$ Add 9 to both sides to isolate the −2x term.

$$\underline{+9 \quad +9}$$

$$-2x + 0 = 16$$

$$-2x = 16$$

$$\dfrac{-2x}{-2} = \dfrac{16}{-2}$$ Divide both sides by −2 to clear the −2 coefficient.

$$1x = -8$$

$$x = -8$$

Answers to Your Turn 1
a. $y = 9$ b. $n = 4$
c. $a = 6$ d. $x = -8$

Check: $\quad -2x - 9 = 7$

$\qquad -2(-8) - 9 \overset{?}{=} 7 \qquad$ Replace *x* with −8.

$\qquad\qquad 16 - 9 \overset{?}{=} 7 \qquad$ Simplify.

$\qquad\qquad\qquad 7 = 7$

True, so −8 is the solution.

b. $12 = 26 - 7x$

Solution: $\qquad 12 = \quad 26 - 7x$

$\qquad\qquad \dfrac{-26 \quad -26}{-14 = \quad 0 - 7x} \qquad$ Subtract 26 from both sides to isolate the −7*x* term.

$\qquad\qquad \dfrac{-14}{-7} = \dfrac{-7x}{-7} \qquad$ Divide both sides by −7 to clear the −7 coefficient.

$\qquad\qquad\quad 2 = 1x$

$\qquad\qquad\quad 2 = x$

Check: $12 = 26 - 7x$

$\qquad 12 \overset{?}{=} 26 - 7(2) \qquad$ Replace *x* with 2.

$\qquad 12 \overset{?}{=} 26 - 14 \qquad$ Simplify.

$\qquad 12 = 12$

True, so −2 is the solution.

Do Your Turn 2 ▶

Your Turn 2

Solve and check.

a. $4x - 1 = -21$

b. $-7b + 1 = 22$

c. $12 = 17 + 5x$

d. $-4 = 32 - 9y$

Example 3   Solve and check. $7x - 1 = 3x - 21$

Solution: We must get the variable terms together on one side of the equation and constants on the other side.

Note: We chose to clear 3*x* because its coefficient, 3, is less ▶ than the coefficient of 7*x*.

$$7x - 1 = 3x - 21$$
$$\underline{-3x \qquad -3x} \qquad \text{Subtract } 3x \text{ from both sides.}$$
$$4x - 1 = \quad 0 - 21$$

CONNECTION After moving ▶ the 3*x* term, we have the same equation as in Your Turn 2a.

$$4x - 1 = -21$$
$$\underline{+1 \quad \cdot \quad +1} \qquad \text{Add 1 to both sides to isolate } 4x.$$
$$4x + 0 = -20$$

$$\dfrac{4x}{4} = \dfrac{-20}{4} \qquad \text{Divide both sides by 4 to clear the 4 coefficient.}$$

$$1x = -5$$

$$x = -5$$

Check: $\qquad 7x - 1 = 3x - 21$

$\qquad 7(-5) - 1 \overset{?}{=} 3(-5) - 21 \qquad$ Replace *x* with −5.

$\qquad -35 - 1 \overset{?}{=} -15 - 21 \qquad$ Simplify.

$\qquad\qquad -36 = -36$

True, so −5 is the solution.

Do Your Turn 3 ▶

Your Turn 3

Solve and check.

a. $9b + 2 = 5b + 26$

b. $n - 11 = 8n + 3$

Answers to Your Turn 2
a. $x = -5$ **b.** $b = -3$
c. $x = -1$ **d.** $y = 4$

Answers to Your Turn 3
a. $b = 6$ **b.** $n = -2$

Example 4 Solve and check. $9n + 6 - 2n = 16 + 10n - 4$

Solution: Because we have like terms that are already on the same sides of the equation, we should simplify these like terms.

$$9n + 6 - 2n = 16 + 10n - 4 \qquad \text{Combine like terms.}$$

CONNECTION After simplifying, we have the same equation as in Your Turn 3c.

$$7n + 6 = 12 + 10n$$

$$\begin{array}{l} 7n + 6 = 12 + 10n \\ \underline{-7n \qquad\qquad -7n} \\ 0 - 6 = 12 + 3n \end{array} \qquad \begin{array}{l}\text{Subtract } 7n \text{ from} \\ \text{both sides.}\end{array}$$

$$\begin{array}{l} \underline{-12 \cdot -12} \\ -6 = 0 + 3n \end{array} \qquad \begin{array}{l}\text{Subtract 12 from both} \\ \text{sides to isolate } 3n.\end{array}$$

$$\frac{-6}{3} = \frac{3n}{3} \qquad \begin{array}{l}\text{Divide both sides by 3} \\ \text{to clear the 3 coefficient.}\end{array}$$

$$-2 = 1n$$

$$-2 = n$$

b. $15 - 2y + 6 = 12y + 30 - 5y$

Check:
$$9n + 6 - 2n = 16 + 10n - 4$$
$$9(-2) + 6 - 2(-2) \stackrel{?}{=} 16 + 10(-2) - 4 \qquad \text{Replace } n \text{ with } -2.$$
$$-18 + 6 - (-4) \stackrel{?}{=} 16 + (-20) - 4$$
$$-12 - (-4) \stackrel{?}{=} -4 - 4$$
$$-12 + 4 \stackrel{?}{=} -8$$
$$-8 = -8$$

True, so -2 is the solution.

◀ **Do Your Turn 4**

Example 5 Solve and check. $15 - 2(y - 3) = 6(2y + 5) - 5y$

Solution: We must simplify each side first.

$$15 - 2(y - 3) = 6(2y + 5) - 5y$$

CONNECTION After distributing, we have the same equation as in Your Turn 4b.

$$15 - 2y + 6 = 12y + 30 - 5y \qquad \text{Distribute } -2 \text{ and 6.}$$

$$21 - 2y = 7y + 30 \qquad \text{Combine like terms.}$$

$$\begin{array}{l} 21 - 2y = 7y + 30 \\ \underline{+2y \quad +2y} \\ 21 + 0 = 9y + 30 \end{array} \qquad \text{Add } 2y \text{ to both sides.}$$

$$\begin{array}{l} \underline{-30 \qquad\qquad -30} \\ -9 = 9y + 0 \end{array} \qquad \begin{array}{l}\text{Subtract 30 from both} \\ \text{sides to isolate } 9y.\end{array}$$

$$\frac{-9}{9} = \frac{9y}{9} \qquad \begin{array}{l}\text{Divide both sides by} \\ \text{9 to clear the 9} \\ \text{coefficient.}\end{array}$$

$$-1 = 1y$$

$$-1 = y$$

Answers to Your Turn 4
a. $t = 11$ b. $y = -1$

Check:

$$15 - 2(y - 3) = 6(2y + 5) - 5y$$
$$15 - 2((-1) - 3) \stackrel{?}{=} 6(2(-1) + 5) - 5(-1) \quad \text{Replace } y \text{ with } -1.$$
$$15 - 2(-4) \stackrel{?}{=} 6(-2 + 5) - 5(-1)$$
$$15 - (-8) \stackrel{?}{=} 6(3) - 5(-1)$$
$$15 + 8 \stackrel{?}{=} 18 - (-5)$$
$$23 \stackrel{?}{=} 18 + 5$$
$$23 = 23$$

True, so -1 is the solution.

Do Your Turn 5 ▶

OBJECTIVE 3 Solve application problems.

In Section 4.2, we revisited many of the applications that we solved earlier in the text and put those applications in the context of using the addition/subtraction principle of equality. Now we will do the same for the multiplication/division principle of equality. The formulas we will use involve multiplying variables, such as

> **Area of a parallelogram:** $A = bh$
> **Volume of a box:** $V = lwh$
> **Force:** $F = ma$
> **Distance:** $d = rt$
> **Voltage:** $V = ir$

Notice that in all of these formulas we multiply numbers to achieve the result. In fact, the area of a parallelogram, force, distance, and voltage are all mathematically identical because they are found by multiplying two numbers together. The only difference with volume is that we multiply three numbers.

Example 6 Juanita is driving at a rate of 59 feet per second (about 40 miles per hour). A large piece of debris lies in the road 177 feet ahead. How much time does she have to react?

Understand: We are given a rate and a distance and must find the time. Based on the given information we can use the formula $d = rt$.

Plan: Use $d = rt$. Replace d with **177** and r with **59**, then solve for t.

Execute:
$$d = rt$$
$$177 = 59t$$
$$\frac{177}{59} = \frac{59t}{59} \quad \text{Divide both sides by 59 to isolate } t.$$
$$3 = 1t$$
$$3 = t$$

Answer: Juanita has 3 sec. to react to the debris in the road ahead. (The time is in units of seconds because the rate was in terms of feet per *second*.)

Check: Verify that an object traveling at 59 feet per second for 3 seconds will travel a distance of 177 feet.

$$177 \stackrel{?}{=} 59(3)$$
$$177 = 177 \quad \text{It checks.}$$

Do Your Turn 6 ▶

Your Turn 5

Solve and check.

a. $3(3t - 5) + 13 = 4(t - 3)$

b. $9 - 5(x + 4) = 7x - (6x + 5)$

Your Turn 6

Solve.

A code in a certain region requires that upstairs windows have a minimum area of 864 square inches. Doris wants an upstairs window that is 18 inches wide. What must the length be?

Answers to Your Turn 5
a. $t = -2$ **b.** $x = -1$

Answer to Your Turn 6
48 in.

Some problems involve using both the addition/subtraction and multiplication/division principles of equality. If a formula has multiplication *and* addition or subtraction, then we will likely have to use both principles to solve for the unknown amount. Some formulas where this would apply are

Perimeter of a rectangle: $P = 2l + 2w$
Surface area of a box: $SA = 2lw + 2lh + 2wh$

Your Turn 7

Solve.

a. The length of a rectangle is 207 centimeters. Find the width of the rectangle if the perimeter is 640 centimeters.

b. The formula $C = 8f + 50$ describes the total cost of a fence with a gate, where C is the total cost and f is the number of feet to be enclosed. If Mario and Rosa have a budget of $1810, then how many feet of fencing can they afford?

Note: In the formula $C = 8f + 50$, the coefficient 8 represents the price per foot of fencing. It costs $8 for each foot of fencing. The 50 represents the cost of the gate that was mentioned. Therefore, the total cost is $8 times the number of feet of fencing plus the $50 cost of the gate.

Example 7   The total material allotted for the construction of a metal box is 3950 square inches. The length is to be 25 inches and the width is to be 40 inches. Find the height.

Understand: We are given the surface area, length, and width of a box and must find the height.

Plan: Using the formula $SA = 2lw + 2lh + 2wh$, replace SA with **3950**, l with **25**, w with **40**, and solve for h.

Execute:

$$3950 = 2(25)(40) + 2(25)h + 2(40)h$$

$3950 = 2000$	$+$	$50h +$	$80h$	Simplify.
$3950 = 2000$	$+$	$130h$		Combine like terms.
$-2000\ -2000$				Subtract 2000 from both sides.

$$1950 = \quad 0 + 130h$$

$$\frac{1950}{130} = \frac{130h}{130} \qquad \text{Divide both sides by 130.}$$

$$15 = 1h$$

$$15 = h$$

Answer: The height must be 15 in.

Check: Verify that the surface area of the box is 3950 square inches when the length is 25 inches, width is 40 inches, and height is 15 inches.

$$3950 \stackrel{?}{=} 2(25)(40) + 2(25)(15) + 2(40)(15)$$
$$3950 \stackrel{?}{=} 2000 \qquad + 750 \qquad + 1200$$
$$3950 = 3950 \qquad \text{It checks.}$$

◀ **Do Your Turn 7**

Answers to Your Turn 7
a. 113 cm b. 220 ft.

4.3 Exercises

For Extra Help

 Videotape DVT
 Addison-Wesley Tutor Center
Math XL
 MyMathLab
 Student Solutions Manual

1. What does the multiplication/division principle of equality say?

2. Explain how to use the multiplication/division principle to clear an integer coefficient.

For Exercises 3–12, solve using the multiplication/division principle, then check.

3. $3x = 21$

4. $5t = 20$

5. $-4a = 36$

6. $-8m = 48$

7. $-7b = 77$

8. $-9x = 108$

9. $-14a = -154$

10. $-11c = -132$

11. $-12y = -72$

12. $-15n = 45$

For Exercises 13–32, solve using the multiplication/division and addition/subtraction principles of equality, then check.

13. $2x + 9 = 23$

14. $5k - 4 = 31$

15. $29 = -6b - 13$

16. $45 = 15 - 10h$

17. $9u - 17 = -53$

18. $-7t + 11 = -73$

19. $7x - 16 = 5x + 6$

20. $3b + 13 = 7b + 17$

21. $9k + 19 = -3k - 17$

22. $-4y + 15 = -9y - 20$

23. $-2h + 17 - 8h = 15 - 7h - 10$

24. $5 - 8x - 24 = 7x + 11 - 5x$

25. $13 - 9h - 15 = 7h + 22 - 8h$

26. $-4m - 7 + 13m = 12 + 2m - 5$

27. $3(t - 4) = 5(t + 1) - 1$

28. $-2(x + 7) - 9 = 5x + 12$

29. $9 - 2(3x + 5) = 4x + 3(x - 9)$

30. $11n - 4(2n - 3) = 18 + 5(n + 2)$

31. $15 + 3(3u - 1) = 16 - (2u - 7)$

32. $8y - (3y + 7) = 14 - 4(y - 6)$

For Exercises 33–36, the check for each equation indicates a mistake was made. Find and correct the mistake.

33.

$$5x - 11 = 7x - 9$$
$$\underline{-5x \qquad\quad -5x}$$
$$11 = 2x - 9$$
$$\underline{+9 \qquad\quad +9}$$
$$20 = 2x$$
$$\frac{20}{2} = \frac{2x}{2}$$
$$10 = x$$

Check:

$$5x - 11 = 7x - 9$$
$$5(10) - 11 \overset{?}{=} 7(10) - 9$$
$$50 - 11 \overset{?}{=} 70 - 9$$
$$39 \neq 61$$

34.

$$6y + 11 = -3y + 38$$
$$\underline{-11 \qquad\qquad -11}$$
$$6y + 0 = -3y + 27$$
$$6y = -3y + 27$$
$$\underline{+3y \qquad +3y}$$
$$9y = \quad 0 + 27$$
$$\frac{9y}{9} = \frac{27}{9}$$
$$1y = 3$$
$$y = 3$$

Check:

$$6y + 11 = -3y + 38$$
$$6(3) + 11 \overset{?}{=} -3(3) + 38$$
$$18 + 11 \overset{?}{=} 9 + 38$$
$$29 \neq 47$$

35.

$$6 - 2(x + 5) = 3x - (x - 8)$$
$$6 - 2x - 10 = 3x - x - 8$$
$$-4 - 2x = 2x - 8$$
$$\underline{+2x \quad +2x}$$
$$-4 + 0 = 4x - 8$$
$$-4 = 4x - 8$$
$$\underline{+8 \qquad\quad +8}$$
$$4 = 4x + 0$$
$$\frac{4}{4} = \frac{4x}{4}$$
$$1 = 1x$$
$$1 = x$$

Check:

$$6 - 2(x + 5) = 3x - (x - 8)$$
$$6 - 2(1 + 5) \overset{?}{=} 3(1) - (1 - 8)$$
$$6 - 2(6) \overset{?}{=} 3 - (-7)$$
$$6 - 12 \overset{?}{=} 3 + 7$$
$$-6 \neq 10$$

36.

$$2x + 3(x - 3) = 10x - (3x - 11)$$
$$2x + 3x - 9 = 10x - 3x + 11$$
$$5x - 9 = 7x + 11$$
$$\underline{-7x \qquad -7x}$$
$$-2x - 9 = 0 + 11$$
$$-2x - 9 = 11$$
$$\underline{+9 \quad +9}$$
$$-2x + 0 = 20$$
$$\frac{-2x}{2} = \frac{20}{2}$$
$$x = 10$$

Check:

$$2x + 3(x - 3) = 10x - (3x - 11)$$
$$2(10) + 3(10 - 3) \overset{?}{=} 10(10) - (3(10) - 11)$$
$$20 + 3(7) \overset{?}{=} 100 - (30 - 11)$$
$$20 + 21 \overset{?}{=} 100 - 19$$
$$41 \neq 81$$

For Exercises 37–48, solve for the missing amount.

37. The area of a computer screen is to be 180 square inches. Find the length if the width must be 10 inches. (Use the formula $A = lw$.)

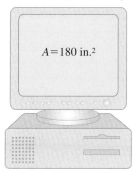

$A = 180$ in.²

38. A fish tank is to have a volume of 4320 cubic inches. The length is to be 20 inches and the width 12 inches. Find the height. (Use $V = lwh$.)

39. The Jones family is planning a 455-mile trip. If they travel at an average speed of 65 miles per hour, what will be their travel time? (Use $d = rt$.)

40. Jacob weighs 160 pounds. If the acceleration due to gravity is about 32 feet per second per second, what is his mass? (Use $F = ma$.)

41. The surface area of a large cardboard box is to be 11,232 square inches. The length is to be 36 inches and the width is to be 28 inches. Find the height. (Use $SA = 2lw + 2lh + 2wh$.)

42. A stage is designed in the shape of a rectangle with a perimeter of 122 feet. If the width is 26 feet, find the length. (Use $P = 2l + 2w$.)

43. The formula $C = 28h + 40$ describes the total cost for a plumber to visit a home, where C is the total cost and h is the number of hours on site. If Li has $152, how long can he afford for the plumber to work?

DISCUSSION What do you think 28 and 40 describe?

44. The formula $C = 6h + 32$ describes the total cost of renting a tiller, where C is the total cost and h is the number of hours rented. How many hours can the tiller be rented for $50?

DISCUSSION What do you think 6 and 32 describe?

45. The formula $v = v_i + at$ describes the final velocity (speed) of an object after being accelerated, where v_i is the initial velocity, a is the acceleration, t is the number of seconds that the object is accelerated, and v is the final velocity. Suppose the initial velocity of a car is 30 miles per hour and the car accelerates at a rate of 2 miles per hour per second. How long will it take the car to reach a speed of 40 miles per hour?

46. A sports car is said to be able to go from 0 to 60 miles per hour in 4 seconds. What acceleration is required? (Use $v = v_i + at$.)

47. The expression $54b + 1215$ describes the cost of materials for a certain computer chip, where b is the number of chips produced. The expression $25b + 4200$ describes the labor costs. How many chips can be produced with a total cost budget of $54,000?

48. Two different circuits operate within a television. The equation $V = 7i + 8 - 4i$ describes the output voltage of the first circuit. The equation $V = 3(3i - 1) - 13$ describes the output voltage of the second circuit. The variable i represents the current. The design calls for both circuits to have the same output voltage. What must be the current in amperes?

PUZZLE PROBLEM The letters below correspond to digits. Find the digit that corresponds to each letter.

$$
\begin{array}{r}
\text{F O R T Y} \\
\text{T E N} \\
+ \quad \text{T E N} \\
\hline
\text{S I X T Y}
\end{array}
$$

Review Exercises

[1.1] **1.** Write the word name for 6,784,209.

[3.7] **2.** Evaluate 7^0.

[3.1] **3.** Evaluate $x^2 - 9x + 7$ when $x = -3$.

[3.6] **4.** Find the prime factorization of 240.

[3.7] **5.** Factor. $24x^5 - 30x^4 + 18x^3$

4.4 Translating Word Sentences to Equations

In this section, we will further develop our problem-solving strategies by focusing on translating key words to equations.

The following tables contain some basic phrases and their translations. Since the equations we have learned to solve contain only addition, subtraction, or multiplication, we will only consider phrases that translate to those operations.

OBJECTIVE

1 Translate sentences to equations using key words, then solve.

Note: Notice that *a number* means *an unknown number*, so we use a variable. We can select any variable we wish.

Addition	Translation	Subtraction	Translation
the sum of x and 3	$x + 3$	the difference of x and 3	$x - 3$
h plus k	$h + k$	h minus k	$h - k$
7 added to t	$t + 7$	7 subtracted from t	$t - 7$
3 more than a number	$n + 3$	3 less than a number	$n - 3$
y increased by 2	$y + 2$	y decreased by 2	$y - 2$

Note: Because addition is commutative, it does not matter in what order we write the translation.

For "the sum of x and 3" we can write $x + 3$ or $3 + x$.

Note: Subtraction is not commutative, therefore the way we write the translation matters. We must translate each key phrase exactly as it was presented above. Particularly, note how we translate *less than* or *subtracted from*.

Multiplication	Translation	Multiplication	Translation
the product of 3 and x	$3x$	twice a number	$2n$
4 times k	$4k$	triple the number	$3n$

Note: Like addition, multiplication is commutative. This means we can write the translation order any way we wish.

So, 4 times k can be $4k$ or $k \cdot 4$. However, remember that coefficients are written to the left of a variable so $4k$ is more appropriate.

The key words *sum, difference,* and *product* are the answer words for their respective operations. Notice that they all involve the word *and*. In the translation, the word *and* becomes the operation symbol indicated by the key word *sum, difference,* or *product*.

$$\text{sum of } x \text{ and } 3 \qquad \text{difference of } x \text{ and } 3$$
$$\downarrow \qquad\qquad\qquad\qquad \downarrow$$
$$x + 3 \qquad\qquad\qquad x - 3$$

$$\text{product of } x \text{ and } 3$$
$$\downarrow$$
$$x \cdot 3$$

If we are going to translate sentences to equations, we must also know the key words that indicate an equal sign.

KEY WORDS FOR AN EQUAL SIGN

is equal to	produces
is the same as	yields
is	results in

Translate to an equation, then solve.

a. 15 more than a number is 40.

b. The length increased by 9 is 25.

Example 1 The sum of seventeen and a number is equal to fifteen. Translate to an equation, then find the number.

Understand: The key word *sum* indicates addition, *is equal to* indicates an equal sign, and *a number* indicates a variable.

Plan: Translate the key words to an equation, then solve the equation. We'll use n as the variable.

Execute: Translate: The **sum** of seventeen **and a number is equal to** fifteen.

$$17 + n = 15$$

Solve: $17 + n = 15$
$\underline{-17 \qquad -17}$ **Subtract 17 from both sides.**
$0 + n = -2$
$n = -2$

Answer: The unknown number is -2.

Check: Verify that the sum of 17 and -2 is equal to 15.

$$17 + (-2) \stackrel{?}{=} 15$$
$$15 = 15$$

Yes, the sum of 17 and -2 is equal to 15, so -2 is correct.

◀ **Do Your Turn 1**

Translate to an equation, then solve.

a. Twelve subtracted from a number is negative eight.

b. The difference of a number and twenty-one is negative six.

Example 2 Twelve less than a number is ten. Translate to an equation, then solve.

Understand: The key words *less than* indicate subtraction. Further, we must be careful how we translate because subtraction is not commutative. The key word *is* means an equal sign.

Plan: Translate the key words to an equation, then solve. We'll use n as the variable.

Execute: Translate: Twelve **less than a number is** ten.

$$n - 12 = 10$$

Solve: $n - 12 = 10$
$\underline{+12 \quad +12}$ **Add 12 to both sides to isolate n.**
$n + 0 = 22$
$n = 22$

Note: The key words *less than* require careful translation. In the sentence, the word *twelve* comes before the words *less than*, and the words *a number* come after. This order is reversed in the translation.

Answer: The unknown number is 22.

Check: Verify that 12 less than 22 is 10.

$$22 - 12 \stackrel{?}{=} 10$$
$$10 = 10$$

Yes, 12 less than 22 is 10, so 22 is correct.

◀ **Do Your Turn 2**

Answers to Your Turn 1
a. $n + 15 = 40$; $n = 25$
b. $l + 9 = 25$; $l = 16$

Answers to Your Turn 2
a. $n - 12 = -8$; $n = 4$
b. $m - 21 = -6$; $m = 15$

Example 3 The product of seven and y is negative thirty-five. Translate to an equation, then solve.

Understand: *Product* means multiply and *is* means an equal sign.

Plan: Translate to an equation, then solve.

Execute: Translate: The **product** of seven **and** y **is** negative thirty-five.

$$7 \cdot y = -35$$

Solve: $\dfrac{7y}{7} = \dfrac{-35}{7}$ **Divide both sides by 7 to isolate y.**

$$1y = -5$$

Answer: $y = -5$

Check: Verify that the product of 7 and -5 is -35.

$$7 \cdot (-5) \overset{?}{=} -35$$
$$-35 = -35$$

Yes, the product of 7 and -5 is -35, so -5 is correct.

Do Your Turn 3 ▷

Your Turn 3

Translate to an equation, then solve.

a. The product of negative six and a number is negative forty-two.

b. Twelve times a number is negative seventy-two.

Example 4 Seven more than the product of three and r is equal to nineteen. Translate to an equation, then solve.

Understand: *More than* indicates addition, *product* indicates multiplication, and *is equal to* indicates an equal sign.

Plan: Translate to an equation, then solve.

Execute: Translate: Seven **more than** the **product** of three **and** r **is equal to** nineteen.

$$7 \quad + \quad\quad\quad\quad 3 \quad \cdot \quad r \quad = \quad 19$$

Solve: $7 + 3r = 19$

$$\underline{-7 \qquad\quad -7} \quad \text{\textbf{Subtract 7 from both sides to isolate 3r.}}$$
$$0 + 3r = 12$$
$$\dfrac{3r}{3} = \dfrac{12}{3} \quad \text{\textbf{Divide both sides by 3 to isolate r.}}$$
$$1r = 4$$

Answer: $r = 4$

Check: Verify that 7 more than the product of 3 and 4 is equal to 19.

$$7 + 3 \cdot 4 \overset{?}{=} 19$$
$$7 + 12 \overset{?}{=} 19$$
$$19 = 19$$

Yes, 7 more than the product of 3 and 4 is equal to 19, so 4 is correct.

Do Your Turn 4 ▷

Your Turn 4

Translate to an equation, then solve.

a. Eight less than the product of three and x is equal to thirteen.

b. Negative five times n plus eighteen is twenty-eight.

Answers to Your Turn 3
a. $-6n = -42$; $n = 7$
b. $12n = -72$; $n = -6$

Answers to Your Turn 4
a. $3x - 8 = 13$; $x = 7$
b. $-5n + 18 = 28$; $n = -2$

Translate to an equation, then solve.

a. −6 times the sum of y and 3 is equal to 5 times y minus 40.

b. The difference of 8 times b and 9 is the same as triple the sum of b and 2.

Example 5 Twice the difference of five and n is the same as seventeen subtracted from seven times n. Translate to an equation, then solve.

Understand: *Twice* means to multiply by 2, *difference* is the result of subtraction, *is the same as* means an equal sign, and *subtracted from* indicates 17 is the subtrahend. Finally, *times* indicates multiplication.

Plan: Translate the key words to an equation, then solve.

Execute: Translate:

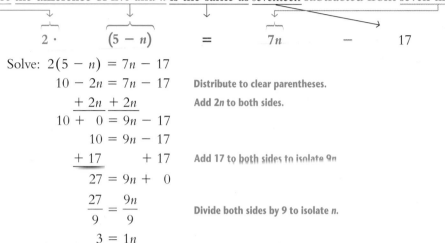

Twice **the difference of five and n is the same as** seventeen **subtracted from seven times n**.

$$2 \cdot \qquad (5 - n) \qquad = \qquad 7n \qquad - \qquad 17$$

Solve: $2(5 - n) = 7n - 17$

$\ 10 - 2n = 7n - 17$ **Distribute to clear parentheses.**

$\underline{+\, 2n \quad +\, 2n}$ **Add 2n to both sides.**

$\ 10 + \ \ 0 = 9n - 17$

$\ 10 = 9n - 17$

$\ \underline{+\, 17 \qquad\quad +\, 17}$ **Add 17 to both sides to isolate 9n**

$\ 27 = 9n + \ \ 0$

$$\frac{27}{9} = \frac{9n}{9}$$ **Divide both sides by 9 to isolate n.**

$\ 3 = 1n$

Answer: $\qquad\qquad 3 = n$

Check: Verify that twice the difference of 5 and 3 is the same as 17 subtracted from 7 times 3.

$$2(5 - 3) \stackrel{?}{=} 7(3) - 17$$
$$2(2) \stackrel{?}{=} 21 - 17$$
$$4 = 4$$

Yes, twice the difference of 5 and 3 is the same as 17 subtracted from 7 times 3, so 3 is correct.

◄ **Do Your Turn 5**

Answers to Your Turn 5
a. $-6(y + 3) = 5y - 40$; $y = 2$
b. $8b - 9 = 3(b + 2)$; $b = 3$

4.4 Exercises

For
Extra
Help

 Videotape DVT

 Addison-Wesley Tutor Center

 Math XL

 MyMathLab

 Student Solutions Manual

1. List three key words that indicate addition.

2. List three key words that indicate subtraction.

3. List three key words that indicate multiplication.

4. List three key words that indicate an equal sign.

For Exercises 5–28, translate to an equation, then solve.

5. Five more than a number is equal to negative seven.

6. The difference of a number and nine is equal to four.

7. Six less than a number is fifteen.

8. The sum of a number and twelve is negative twenty-seven.

9. A number increased by seventeen is negative eight.

10. A number decreased by twenty-four is negative seven.

11. The product of negative three and a number is twenty-one.

12. Negative eight times a number is equal to forty.

13. A number multiplied by nine is negative thirty-six.

14. Negative seven times a number is equal to negative forty-two.

15. Four more than the product of five and *x* yields fourteen.

16. Eighteen subtracted from negative seven times *y* is equal to three.

17. The difference of negative six times *m* and sixteen is fourteen.

18. Twenty minus eight times *b* is negative four.

19. Thirty-nine minus five times *x* is equal to the product of eight and *x*.

20. Forty less than the product of three and *y* is equal to seven times *y*.

21. The sum of seventeen and four times *t* is the same as the difference of six times *t* and nine.

22. The difference of ten times *n* and seven is equal to twenty-five less than four times *n*.

23. Two times the difference of *b* and eight is equal to five plus nine times *b*.

24. Nine more than the product of eight and *m* is the same as three times the sum of *m* and thirteen.

25. Six times *x* plus five times the difference of *x* and seven is equal to nineteen minus the sum of *x* and six.

26. Two times *r* subtracted from seven times the sum of *r* and one is equal to three times the difference of *r* and five.

27. Fourteen less than negative eight times the difference of y and 3 is the same as the difference of y and 5 subtracted from the product of negative two and y.

28. The sum of n and three subtracted from twelve times n is the same as negative eleven plus the product of 2 and the difference of n and five.

For Exercises 29–34, explain the mistake in the translation. Then write the correct translation.

29. Seven less than a number is fifteen.
Translation: $7 - n = 15$

30. Nineteen subtracted from a number is eleven.
Translation: $19 - n = 11$

31. Two times the sum of x and thirteen is equal to negative nine.
Translation: $2x + 13 = -9$

32. Five times x minus twelve is the same as three times the difference of x and four.
Translation: $5x - 12 = 3x - 4$

33. Sixteen decreased by the product of six and n is the same as twice the difference of n and four.
Translation: $6n - 16 = 2(n - 4)$

34. Negative three times the sum of y and five is equal to twice y minus the difference of y and one.
Translation: $-3(y + 5) = 2y - (1 - y)$

Review Exercises

[3.7] **1.** Find the GCF of $30x^3y^5$ and $24xy^7z$.

[3.7] **2.** Factor the polynomial. $18b^5 - 27b^3 + 54b^2$

[3.8] **3.** Write an expression in simplest form for the perimeter of the rectangle.

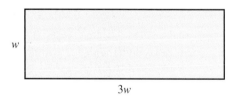

[3.1] **4.** Evaluate the perimeter expression from Review Exercise 3 when w is 6 feet.

[1.7] **5.** Jasmine sells two sizes of holly shrub, large and small. One day, she sold 7 large holly shrubs at $7 each and some small holly shrubs at $5 each. If she sold a total of 16 holly shrubs, how many of the small shrubs did she sell? How much money did she make from the sale of the holly shrubs altogether?

4.5 Applications and Problem Solving

OBJECTIVE 1 Solve problems involving two unknown amounts.

Let's extend what we learned about translating sentences to problems that have two or more unknown amounts. In general, if there are two unknowns, there will be two relationships. We will use the following procedure as a guide.

PROCEDURE *To solve problems with two or more unknowns:*

1. Determine which unknown will be represented by a variable.
   Tip: Let the unknown that is acted on be represented by the variable.
2. Use one of the relationships to describe the other unknown(s) in terms of the variable.
3. Use the other relationship to write an equation.
4. Solve the equation.

Example 1 One number is six more than another number. The sum of the numbers is forty. Find the two numbers.

Understand: We are to find two unknown numbers given two relationships.

Plan: Select a variable for one of the unknowns, translate to an equation, then solve the equation.

Execute: Use the first relationship to determine which unknown will be represented by the variable.

Relationship 1: One number is six more than another number.

$$\text{one number} = 6 + \textbf{another number}$$
$$= 6 + n$$

Note: Since "another number" is the unknown that is acted on (six is added to it), we let it be represented by n.

Now we can use the second relationship to write an equation.

Relationship 2: The sum of the numbers is forty.

$$\begin{aligned}
\text{one number} + \text{another number} &= 40 \\
6 + n \quad + \quad n \quad &= 40
\end{aligned}$$

$$6 + 2n = 40 \quad \text{Combine like terms.}$$
$$\underline{-6 \qquad\quad -6} \quad \text{Subtract 6 from both sides.}$$
$$0 + 2n = 34$$
$$\frac{2n}{2} = \frac{34}{2} \quad \text{Divide both sides by 2.}$$
$$1n = 17$$
$$n = 17$$

Answer: The second number, "another number," is 17. To find the first number, we use relationship 1.

$$\begin{aligned}
\text{one number} &= 6 + n \\
&= 6 + 17 \quad \text{In } 6 + n, \text{ replace } n \text{ with 17.} \\
&= 23
\end{aligned}$$

Check: Verify that 17 and 23 satisfy both relationships: 23 is 6 more than 17 and the sum of 23 and 17 is 40.

Do Your Turn 1 ▶

OBJECTIVES

1 Solve problems involving two unknown amounts.
2 Use a table in solving problems with two unknown amounts.

Your Turn 1

Solve.

a. One number is five more than another number. The sum of the numbers is forty-seven. Find the numbers.

b. One number is twice another. The sum of the numbers is thirty. Find the numbers.

Answers to Your Turn 1
a. 21 and 26 b. 10 and 20

Your Turn 2

A TV is to have a rectangular screen with a length that is 5 inches less than twice the width and a perimeter of 44 inches. What are the dimensions of the screen?

Sometimes, problems do not give all the needed relationships in an obvious manner. This often occurs in geometry problems in which the definition of a geometry term provides the needed relationship. For example, *perimeter* indicates to add the lengths of the sides of the shape.

Example 2 A carpenter is asked to make a rectangular frame out of an 8-foot strip of wood. The length of the frame must be three times the width. What must the dimensions be?

Understand: Draw a picture.

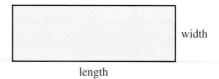

List the relationships:

Relationship 1: The length is three times the width.

Relationship 2: The total frame material is 8 feet. (This means that the perimeter must be 8 feet.)

There are two amounts missing, the length and the width. Because we only know how to solve linear equations in one variable, we must choose the length or width to be the variable and write our equation in terms of that variable.

Plan: Translate to an equation using the key words, then solve the equation.

Execute: Remember that there are two amounts missing, length and width, and we'll have to select one of those amounts to be our variable.

Relationship 1: Length must be three times the width.

$$\text{length} = 3 \cdot \textbf{width}$$
$$= 3 \cdot w$$

◄ **Note:** Since width is multiplied by 3, we will let width be represented by w.

Now we use the second relationship to write an equation to solve.

Relationship 2: The perimeter must be 8 feet.

Note: Recall *perimeter* indicates to add the lengths of all the sides.

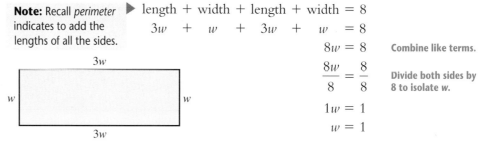

$$\text{length} + \text{width} + \text{length} + \text{width} = 8$$
$$3w + w + 3w + w = 8$$
$$8w = 8 \quad \text{Combine like terms.}$$
$$\frac{8w}{8} = \frac{8}{8} \quad \begin{array}{l}\text{Divide both sides by}\\ \text{8 to isolate } w.\end{array}$$
$$1w = 1$$
$$w = 1$$

Answer: The width, w, is 1 ft. To determine the length, we use relationship 1.

$$\text{length} = 3w$$
$$= 3(1) \quad \text{In } 3w, \text{ replace } w \text{ with 1.}$$
$$= 3$$

The frame must have a length of 3 ft. and a width of 1 ft.

Check: Verify that the length is three times the width and the perimeter is 8 feet. A length of 3 feet is three times a width of 1 foot. The perimeter is $3 + 1 + 3 + 1 = 8$ feet.

Answer to Your Turn 2
9 in. wide by 13 in. long

◄ **Do Your Turn 2**

Let's consider problems involving **equilateral** and **isosceles** triangles.

DEFINITION **Equilateral triangle:** A triangle with all three sides of equal length.

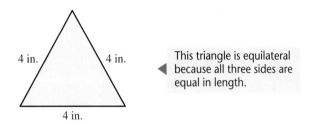

4 in.   4 in.

4 in.

This triangle is equilateral because all three sides are equal in length.

DEFINITION **Isosceles triangle:** A triangle with two sides of equal length.

6 cm   6 cm

3 cm

This triangle is isosceles because two of its sides are equal in length.

Example 3   Suppose each of the equal-length sides of an isosceles triangle is twice the length of the other side. If the perimeter is 75 centimeters, what are the lengths of the sides?

Understand: Isosceles triangles have two equal-length sides.

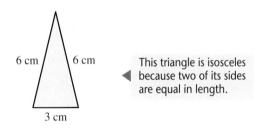

Note: These two sides are equal in length. We are told they are twice the length of the other side (bottom side).

We are also told that the perimeter is 75 centimeters.

Plan: Translate the relationships, write an equation, then solve.

Execute:

Relationship 1: The equal-length sides are twice the length of the other side.

equal-length sides = 2 · **length of the other side**

= 2L   **Note:** Because the length of the other side is acted on (multiplied by 2), we will use the variable L to represent its length.

Suppose each of the equal-length sides of an isosceles triangle is 3 inches more than the base. The perimeter is 30 inches. What are the lengths of the sides?

Now we use the second relationship to write an equation to solve.

Relationship 2: The perimeter is 75 centimeters.

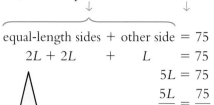

$$\text{equal-length sides} + \text{other side} = 75$$
$$2L + 2L \quad + \quad L \quad = 75$$
$$5L = 75 \qquad \text{Combine like terms.}$$
$$\frac{5L}{5} = \frac{75}{5} \qquad \text{Divide both sides by 5 to isolate } L.$$
$$1L = 15$$
$$L = 15$$

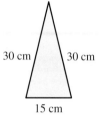

Answer: The other side is 15 cm. To determine the equal-length sides, we use relationship 1:

$$\text{equal-length sides} = 2L$$
$$= 2(15) \qquad \text{In } 2L, \text{ replace } L \text{ with 15.}$$
$$= 30$$

The equal-length sides are 30 cm and the other side is 15 cm.

Check: Because 30 centimeters is twice 15 centimeters, the equal-length sides are in fact twice the length of the other side which verifies relationship 1. The perimeter is $15 + 30 + 30 = 75$ centimeters, which verifies relationship 2.

◁ **Do Your Turn 3**

Let's consider problems involving angles. Angles are formed when two lines or line segments intersect or touch. Lines segments AB and BC form an angle in the following figure. The measure of the angle is 40°.

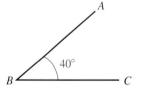

Note: The symbol ∠ means *angle*. To name this angle, we can write all three letters with *B* in the center like ∠*ABC* or ∠*CBA*. Or, we can simply write ∠*B*. To indicate the measure of the angle, we write ∠*B* = 40°.

Problems involving angle measurements may indicate that two unknown angles are **congruent, complementary,** or **supplementary.**

DEFINITION **Congruent angles:** Angles that have the same measurement.

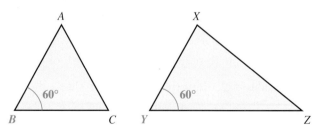

In the two triangles, ∠*B* is congruent to ∠*Y* because they have the same measurement of 60°. The symbol for congruent is ≅.

$$\angle B \cong \angle Y$$

Answer to Your Turn 3
8 in., 11 in., and 11 in.

DEFINITION   **Complementary angles:** Two angles whose sum is 90°.

In the following figure, angles $\angle ABD$ and $\angle DBC$ are complementary because $32° + 58° = 90°$.

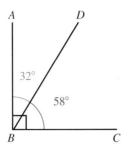

DEFINITION   **Supplementary angles:** Two angles whose sum is 180°.

For example, a straight line forms an angle that measures 180°. Any line that intersects a straight line will cut that 180° angle into two angles that are supplementary. For example, angles $\angle ABD$ and $\angle DBC$ are supplementary angles because $20° + 160° = 180°$.

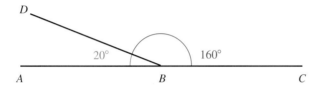

Let's look at a problem containing one of these terms.

Example 4   In designing a roof frame, the architect wants the angled beam to meet the horizontal truss so that the outer angle measurement is 30° more than twice the inner angle measurement. What are the angle measurements?

Understand: We must find the inner and outer angle measurements. The sketch for this situation is shown (notice it is similar to the preceding figure).

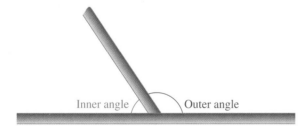

Relationship 1: The outer angle measurement is 30° more than twice the inner angle measurement.

Relationship 2: Looking at the picture, we see that the inner and outer angles are supplementary, which means the sum of the measurements is 180°.

Plan: Translate the relationships to an equation, then solve.

Execute:

Relationship 1: The outer angle measurement is 30° more than twice the inner angle measurement.

> outer angle $= 30 + 2 \cdot$ **inner angle**
>
> $\quad\quad\quad\;\; = 30 + 2a$

Note: Because the inner angle measurement is multiplied by 2, we let the variable represent the inner angle. Let's use the letter a.

Solve.

a. Two angles are to be constructed from metal beams so that when they are joined they form a 90° angle. One of the angles is to be 6° less than three times the other angle. What are the angles?

b. Two electric doors are designed to swing open simultaneously when activated. The design calls for the doors to open so that the angle made between each door and the wall is the same. Because the doors are slightly different in weight, the equation $15i - 1$ describes the angle for one of the doors and $14i + 5$ describes the angle for the other door, where i represents the current supplied to the controlling motors. What current must be supplied so that the door angles are the same?

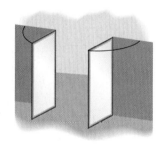

Answers to Your Turn 4
a. 24° and 66° **b.** 6 A

Now, we use relationship 2 to write an equation that we can solve.

Relationship 2: inner angle + outer angle = 180

$$a + 30 + 2a = 180$$
$$3a + 30 = 180 \qquad \text{Combine like terms.}$$
$$\underline{-30 \quad -30} \qquad \text{Subtract 30 from both sides to isolate } 3a.$$
$$3a + 0 = 150$$
$$\frac{3a}{3} = \frac{150}{3} \qquad \text{Divide both sides by 3 to isolate } a.$$
$$1a = 50$$
$$a = 50$$

Answer: Because *a* represents the inner angle, we can say the inner angle measures 50°. To find the outer angle measurement, we use relationship 1.

$$\text{outer angle} = 30 + 2a$$
$$= 30 + 2(50) \qquad \text{In 30 + 2a, replace } a \text{ with 50.}$$
$$= 30 + 100$$
$$= 130$$

The inner angle is 50° and outer angle is 130°.

Check: Verify the relationships. The outer angle measuring 130° is 30° more than twice the inner angle measuring 50°. The sum of the inner and outer angles is 50° + 130° = 180°, which means they are in fact supplementary.

◁ **Do Your Turn 4**

OBJECTIVE 2 Use a table in solving problems with two unknown amounts.

We have seen how drawing a picture, making a list, and using key words have helped us to understand a problem. Let's now consider problems in which a table is helpful in organizing information. These problems will describe two types of items. We will be given the value of each type of item and asked to find the number of items that combine to make a given total amount. Our tables will be four-column tables that look like this:

Categories	Value	Number	Amount

The first column, categories, will contain descriptions of the two types of items. The second column, value, will list the given values of each category. The third column, number, will contain the number of items in each category. The amount column will be found by the relationship:

$$\text{value} \cdot \text{number} = \text{amount}$$

For the number column, you will either be given a relationship about the number of items or a total number of items to split. Let's consider an example to see how it all fits together.

Example 5 Salvador is an artist. A company produces prints of two of his paintings. The first print sells for $45 and the second print for $75. He is told that during the first day of sales, the company sold six more of the $75 prints than the $45 prints, with a total income of $1410. How many of each print were sold?

Understand: We are given that the total income is $1410. From this we can say:

$$\boxed{\text{income from first print}} + \boxed{\text{income from second print}} = 1410$$

How can we describe the income from each print? If we knew the number of prints sold, we could multiply that number times the selling price to get the income from that particular print.

$$\text{selling price} \cdot \text{number of prints} = \text{income from the print}$$

Because of all the information we have, it is helpful to use a table. The value column is the selling price of each print. The amount column is the income from each print.

Categories	Selling price	Number of prints	Income
First print	45	n	$45n$
Second print	75	$n + 6$	$75(n + 6)$

▲ We were given these.

▲ We selected n to represent the number of the first print, then translated: "six more of the $75 print than the $45 print."

▲ We multiplied straight across because:

$$\left(\begin{array}{c}\text{selling}\\\text{price}\end{array}\right) \cdot \left(\begin{array}{c}\text{number}\\\text{of prints}\end{array}\right) = \left(\begin{array}{c}\text{income from}\\\text{each print}\end{array}\right)$$

The expressions in the last column in the table describe the income from the sale of each print.

Plan: Translate the information in the table into an equation, then solve.

Execute: Now we can use our initial relationship:

$$\boxed{\text{income from first print}} + \boxed{\text{income from second print}} = 1410$$

$$45n + 75(n + 6) = 1410$$
$$45n + 75n + 450 = 1410 \qquad \text{Distribute.}$$
$$120n + 450 = 1410 \qquad \text{Combine like terms.}$$
$$120n + 450 = 1410$$
$$\underline{\qquad -450 \quad -450} \qquad \text{Subtract 450 from both sides.}$$
$$120n + 0 = 960$$
$$\frac{120n}{120} = \frac{960}{120} \qquad \text{Divide both sides by 120 to isolate } n.$$
$$1n = 8$$
$$n = 8$$

a. A marketing manager wants to research the sale of two different sizes of perfume. The small bottle sells for $35 and the large bottle for $50. The report indicates that the number of large bottles sold was 24 less than the number of small bottles and the total sales were $4070. How many of each size bottle was sold?

b. A bottling company produces 12-ounce and 16-ounce bottles. In one day the company produces three times as many 12-ounce bottles as 16-ounce bottles. If the company produces a total of 7280 ounces of the beverage, how many 12-ounce and 16-ounce bottles were filled?

Answer: Use the relationships from the table about the number of each print. Since n represents the number of $45 prints, 8 of the $45 prints were sold. Since $n + 6$ represents the number of $75 prints, $8 + 6 = 14$ of the $75 prints were sold.

Check: The number of $75 prints sold, 14, is six more than the number of $45 prints sold, 6. Also, 8 of the $45 prints and 14 of the $75 prints make a total income of $1410.

$$\text{Total income} = 8(45) + 14(75)$$
$$= 360 + 1050$$
$$= 1410 \qquad \textbf{It checks.}$$

◀ **Do Your Turn 5**

Example 6 Jasmine sells two sizes of holly shrub. The larger size sells for $7 and the smaller size for $5. She knows she sold 16 shrubs for a total of $94, but forgot how many of each size. How many of each size did she sell?

Understand: We are given the price or value of two different-sized shrubs. We are also given a total amount of income from the sale of the two different shrubs and a total number of shrubs to split up. We can say:

income from small $+$ income from large $=$ total income

Because we have two categories of shrubs along with a value and a number of shrubs, this problem lends itself to a table setup. We are given the selling prices: $5 and $7. The tricky part is filling in the number column.

We are told that Jasmine sold a total of 16 shrubs. We can say:

number of small $+$ number of large $=$ 16

Or, if we knew one of the numbers, we could find the other by subtracting from 16. We can write a related subtraction this way:

number of small $=$ 16 $-$ number of large ◀

Note: We could also say:

$$\text{number of large} = 16 - \text{number of small}$$

It works either way.

Choosing L to represent the number of large shrubs, we can complete our table.

Categories	Selling price	Number	Income
Small	5	$16 - L$	$5(16 - L)$
Large	7	L	$7L$

We were given these.

In general, if given a total number of items, select one of the categories to be the variable. The other will be:

total number − variable

Since selling price · number = income, we multiply straight across the columns to get the expressions in the income column.

Answers to Your Turn 5
a. 62 small bottles, 38 large bottles
b. 420 12-oz. bottles, 140 16-oz. bottles

Plan: Translate to an equation, then solve.

Execute: We can now use our initial relationship:

income from small + income from large = total income

$$5(16 - L) + \quad 7L \quad = 94$$
$$80 - 5L + \quad 7L \quad = 94 \qquad \textbf{Distribute.}$$
$$80 + 2L = 94 \qquad \textbf{Combine like terms.}$$
$$80 + 2L = 94$$
$$\underline{-80 \qquad\qquad -80} \qquad \textbf{Subtract 80 from both sides to isolate } 2L.$$
$$0 + 2L = 14$$
$$\frac{2L}{2} = \frac{14}{2} \qquad \textbf{Divide both sides by 2 to isolate } L.$$
$$1L = 7$$
$$L = 7$$

Answer: Number of large size shrubs sold: $L = 7$

Number of small size shrubs sold: $16 - L = 16 - 7 = 9$

Check: Verify that Jasmine sold 16 shrubs for $94. Since she sold 7 large-size shrubs and 9 small-size shrubs, she did indeed sell 16 shrubs. Now, verify that 7 large-size shrubs at $7 each and 9 small-size shrubs at $5 each is a total income of $94.

$$\text{total income} = 7(7) + 9(5)$$
$$= 49 + 45$$
$$= 94 \qquad \textbf{It checks.}$$

Do Your Turn 6 ▶

Tips for Table Problems

You could be given a relationship about the number of items in each category, as in Example 5, or a total number of items to split up, as in Example 6. If you are given a relationship, use key words to translate to expressions. If you are given a total number of items to split up, choose one of the categories to be the variable and the other will be: total number − variable.

Your Turn 6

Complete a four-column table, write an equation, then solve.

a. A farmer has a total of 17 pigs and chickens. The combined number of legs is 58. How many pigs and how many chickens are there?

b. A tire factory produces two different-size tires. Tire A costs $37 to produce while tire B costs $42. At the end of the day, management gets the following report:

tires produced = 355

total cost = $13,860

How many of each tire was produced?

Answers to Your Turn 6
a. 12 pigs, 5 chickens
b. 210 of tire A, 145 of tire B

1. What is an equilateral triangle?

2. What is an isosceles triangle?

3. What are supplementary angles?

4. What are complementary angles?

5. Suppose a problem gives the following information: "the length of a rectangle is five more than the width." Which is easier, representing the length or the width with a variable? What expression describes the other amount?

6. Suppose a problem gives the following information: "one angle is 20 less than a second angle." Which is easier, representing the first angle or the second angle with a variable? What expression describes the other amount?

For Exercises 7–28, translate to an equation, then solve.

7. One number is four more than a second number. The sum of the numbers is thirty-four. Find the numbers.

8. One number is eight less than another number. The sum of the numbers is seventy-two. Find the numbers.

9. One number is three times another number. The sum of the numbers is thirty-two. Find the numbers.

10. One number is four times another number. The sum of the numbers is thirty-five. Find the numbers.

11. An architect feels that the optimal design for a new building would be a rectangular shape where the length is four times the width. Budget restrictions force the building perimeter to be 300 feet. What will be the dimensions of the building?

12. 84 feet of border strip was used to go around a rectangular room. The width of the room is 4 feet less than the length. What are the dimensions?

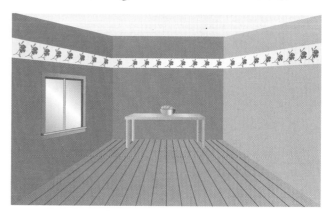

13. The Johnsons can afford 220 feet of fencing material. They want to fence in a rectangular area with a length 20 feet more than twice the width. What will the dimensions be?

14. After completing the trim on some kitchen cabinets, Candice has 48 inches of trim wood left. She decides to make a picture frame out of the wood. She wants to make the frame so that the length is 4 inches less than three times the width. What must the dimensions be?

15. A frame section for the roof of a building is an isosceles triangle. The equal sides of the triangle are 5 meters less than twice the base. The perimeter is 70 meters. What are the lengths of the base and the sides?

16. A section of a public park is in the shape of an isosceles triangle. Each equal side of the triangle is three times the length of the base. If the perimeter is 210 feet, find the lengths of the base and sides.

17. A corporate logo has an equilateral triangle sitting atop a rectangle. Each side of the equilateral triangle is the same distance as the length of the rectangle. The width of the rectangle is 10 feet less than three times the length. The perimeter is the same for both shapes. What are the dimensions of the shapes?

18. A refrigerator magnet is to have the shape of an isosceles triangle sitting atop a rectangle. The base of the triangle and width of the rectangle are equal. The other two sides of the triangle are 2 centimeters more than the base. The length of the rectangle is twice the width. If the perimeter of the magnet is 39 centimeters, what are all the dimensions?

19. A security laser is placed 3 feet above the floor in the corner of a rectangular room. It is aimed so that its beam contacts a detector 3 feet above the floor in the opposite corner. The beam forms two angles in the corner as shown. The angle on one side of the beam is four times the angle on the other side. Find the angle measurements.

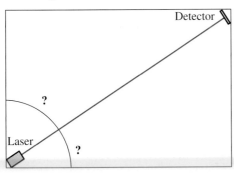

20. A security camera is placed in the corner of a room in a museum. The camera is set so that the angle between it and one of the walls is 16° less than the angle made with the other wall. Find the angle measurements. (Assume the angle formed by the walls is 90°.)

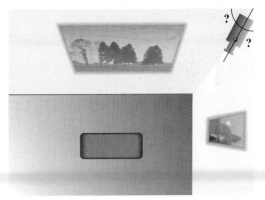

21. In an experiment, a laser contacts a flat detector so that one angle between the laser beam and detector is 15° more than four times the other angle. What are the two angle measurements?

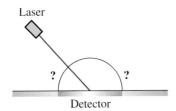

22. A suspension bridge has a steel beam that connects to the flat surface of the bridge forming two angles. The measure of the smaller of the two angles is 38° less than the measure of the other angle. What are the two angles?

23. A laser beam strikes a flat mirror forming an angle with the mirror. The beam is deflected at the same angle. The angle in between the entering and exiting beams is 10° more than three times the angle formed between the beam and the mirror. What are the three angles?

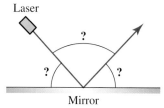

Laser

Mirror

24. A billiard ball strikes the bumper at an angle. Because the ball has side spin, it is deflected at an angle that is 6° less than the initial angle. The angle formed between the initial and deflected lines of travel is 10° more than twice the initial angle. What are the three angles?

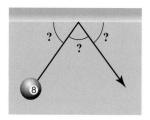

25. The sum of the angles in any triangle is 180°. One of the angles is 43°, the second is two more than twice the third. Find the other two angles.

26. The sum of the angles in any triangle is 180°. Suppose one of the angles of a triangle is 5° more than three times another. The third angle is 115°. Find the two angles.

27. A decorative shelf is to be in the shape of a triangle with the second angle measuring 10° more than the first and the third angle 7° less than the first. What are the three angle measurements?

28. A mounting apparatus for a telescope is to be in the shape of a triangle. The second angle in the triangle is to be twice the first. The third angle is to be 12° less than the first. What are the three angle measurements?

For Exercises 29–36, complete a four-column table, write an equation, then solve.

29. Monique sells makeup products. One of the products is a lotion that comes in two different-size bottles. The smaller-size bottle sells for $8 and the larger-size bottle for $12. She remembers that she sold 5 more smaller bottles than larger bottles that day, but doesn't remember exactly how many of each. She also remembers the total sales of the lotion was $260. How many of each size did she sell?

Categories	Value	Number	Amount

30. Byron is a wedding photographer. He develops two different-size prints for a newlywed couple. He sells the larger print for $7 each and the smaller print for $3 each. He develops twice as many smaller prints as larger prints at a total cost of $78. How many of each print did he sell?

Categories	Value	Number	Amount

31. Ian has some $5 bills and $10 bills in his wallet. If he has 19 bills worth a total of $125, how many of each bill is in his wallet?

Categories	Value	Number	Amount

32. A paper mill produces two different sizes of paper stock. It costs $5 to produce a box of the smaller stock and $7 to produce a box of the larger stock. Management gets a report that 455 boxes were produced at a total cost of $2561 in one day. How many boxes of each stock were produced?

Categories	Value	Number	Amount

33. Nina has two different types of foreign bills to exchange for American currency. She is told the more valuable bill is worth three times that of the other bill. She has 12 of the more valuable bills and 18 of the other bills and receives a total of $108. What is the value of each foreign bill?

34. Cal has two different stocks. One of the stocks is more valuable than the other. He notices in the newspaper that one of his stocks is $3 more valuable than the other. He has 24 shares of the more valuable stock and 22 shares of the other stock. His total assets in the stocks are $348. How much is each stock worth?

35. Jada has her friend take a certain number of play bills. The friend says she has 13 of one bill and 17 of another. The two different bills together are worth $16. The total she has in her hand is $244. How much is each bill worth?

36. Andre sells furniture. Two different chairs go with a dining set, a chair with arms or a chair without. One armchair and one chair without arms sell together for $184. One day he sold 5 armchairs and 17 chairs without arms for a total of $1868. How much does each chair sell for?

PUZZLE PROBLEM At 3:00:05 P.M. what are the approximate angles made between the second hand and the minute and hour hands?

Review Exercises

[4.1] **1.** Is -3 a solution for $4x - 8 = 5x - 6$?

[4.2] **2.** Is $x^2 + 5 = 9$ a linear equation?

For Exercises 3–5, solve and check.

$\begin{bmatrix} 2.3 \\ 4.2 \end{bmatrix}$ **3.** $m + 19 = 12$

$\begin{bmatrix} 2.4 \\ 4.3 \end{bmatrix}$ **4.** $-15n = 75$

[4.3] **5.** $9x - 7(x + 2) = -3x + 1$

Defined Terms

Review the following terms, and for those you do not know, study its definition on the page number next to it.

Section 4.1
Solve *(p. 234)*
Solution *(p. 234)*

Section 4.2
Linear equation *(p. 239)*

Section 4.5
Equilateral triangle
 (p. 267)
Isosceles triangle *(p. 267)*
Congruent angles *(p. 268)*

Complementary angles
 (p. 269)
Supplementary angles
 (p. 269)

LEARNING STRATEGY

Your study sheet (described in the *To the Student* section of the text) should resemble the summaries at the end of each chapter. To see how well you know the rules and procedures, try duplicating your study sheet from memory. This is sometimes called a *memory dump*.

Procedures, Rules, and Key Examples

Procedures/Rules

Key Example(s)

Section 4.1 Equations and Their Solutions

To check to see if a value is a solution for a given equation:

1. Replace the variable(s) with the value.
2. Simplify both sides of the equation as needed. If the resulting equation is true, then the value is a solution.

Example 1: Is -2 a solution for $x^2 - 5x = 14$?

$$x^2 - 5x = 14$$
$$(-2)^2 - 5(-2) \overset{?}{=} 14 \quad \text{Replace } x \text{ with } -2.$$
$$4 - (-10) \overset{?}{=} 14$$
$$4 + 10 \overset{?}{=} 14$$
$$14 = 14$$

Since the equation is true, -2 is a solution.

Example 2: Is 4 a solution for
$2y - 9 = 6y - 23$?

$$2y - 9 = 6y - 23$$
$$2(4) - 9 \overset{?}{=} 6(4) - 23 \quad \text{Replace } y \text{ with 4.}$$
$$8 - 9 \overset{?}{=} 24 - 23$$
$$-1 \neq 1$$

Since the equation is false, 4 is not a solution.

Section 4.2 The Addition/Subtraction Principle of Equality

The Addition/Subtraction Principle of Equality

The same amount can be added to or subtracted from both sides of an equation without affecting its solution(s).

To use the addition/subtraction principle of equality to clear a term, add the additive inverse of that term to both sides of the equation (that is, add or subtract appropriately so that the term you want to clear becomes 0.)

Example 1: Solve and check. $t + 15 = 7$

$$t + 15 = \quad 7$$
$$\underline{-15 \quad -15} \quad \text{Subtract 15 from}$$
$$t + 0 = -8 \quad \text{both sides.}$$
$$t = -8$$

Check: $\quad t + 15 = 7$
$$(-8) + 15 \overset{?}{=} 7$$
$$7 = 7$$

The equation is true, so -8 is the solution.

Section 4.3 The Multiplication/Division Principle of Equality

The Multiplication/Division Principle of Equality

We can multiply or divide both sides of an equation by the same nonzero amount without affecting its solution(s).

To use the multiplication/division principle of equality to clear a coefficient from a term, divide both sides by that coefficient.

Example 1: Solve and check. $-7x = 42$

$$\frac{-7x}{-7} = \frac{42}{-7} \qquad \text{Divide both sides by } -7.$$

$$1x = -6$$

$$x = -6$$

Check: $\quad -7x = 42$

$$-7(-6) \stackrel{?}{=} 42$$

$$42 = 42$$

The equation is true, so -6 is the solution.

To solve equations:

1. Simplify both sides of the equation as needed.
   a. Distribute to clear parentheses.
   b. Combine like terms.
2. Use the addition/subtraction principle of equality so that all variable terms are on one side of the equation and all constants are on the other side. (Clear the variable term with the lesser coefficient.) Then combine like terms.
3. Use the multiplication/division principle of equality to clear the remaining coefficient.

Example 2: Solve and check.

$$12 - (2x + 7) = 13x - 5(x - 7)$$

$$12 - 2x - 7 = 13x - 5x + 35$$

$$5 - 2x = 8x + 35$$

$$\underline{+ 2x \qquad + 2x}$$

$$5 + 0 = 10x + 35$$

$$5 = 10x + 35$$

$$\underline{- 35 \qquad\qquad - 35}$$

$$-30 = 10x + \ 0$$

$$\frac{-30}{10} = \frac{10x}{10}$$

$$-3 = 1x$$

$$-3 = x$$

Check:

$$12 - (2x + 7) = 13x - 5(x - 7)$$

$$12 - (2(-3) + 7) \stackrel{?}{=} 13(-3) - 5(-3 - 7)$$

$$12 - (-6 + 7) \stackrel{?}{=} 13(-3) - 5(-10)$$

$$12 - 1 \stackrel{?}{=} -39 - (-50)$$

$$11 \stackrel{?}{=} -39 + 50$$

$$11 = 11$$

The equation is true, so -3 is the solution.

Section 4.4 Translating Word Sentences to Equations
Translating

Addition	Translation	Subtraction	Translation	Multiplication	Translation
the sum of x and 3	$x + 3$	the difference of x and 3	$x - 3$	the product of x and 3	$3x$
h plus k	$h + k$	h minus k	$h - k$	h times k	hk
7 added to t	$t + 7$	7 subtracted from t	$t - 7$	twice a number	$2n$
3 more than a number	$n + 3$	3 less than a number	$n - 3$	triple the number	$3n$
y increased by 2	$y + 2$	y decreased by 2	$y - 2$		

Note: Because addition is a commutative operation, it does not matter what order we write the translation.

For "the sum of x and 3" we can write $x + 3$ or $3 + x$.

Note: Subtraction is not a commutative operation; therefore the way we write the translation matters. We must translate each key phrase exactly as it was presented above. Notice that when we translate *less than* or *subtracted from*, the translation is in reverse order from what we read.

Note: Like addition, multiplication is a commutative operation. This means we can write the translation order any way we wish.

h times k can be hk or kh.

Key words for an equal sign: is equal to, is the same as, is, produces, yields, results in

Section 4.5 Applications and Problem Solving

To solve problems with two or more unknowns,
1. Determine which unknown will be represented by the variable.
   Tip: Let the unknown that is acted on be represented by the variable.
2. Use one of the relationships to describe the other unknown(s) in terms of the variable.
3. Use the other relationship to write an equation.
4. Solve the equation.

Example 1: The length of a rectangular yard is 15 feet more than the width. If the perimeter of the yard is 150 feet, find the length and the width.

Solution:

Relationship 1: The length is 15 feet more than the width.

Translation: length $= 15 + w$
Note that w represents the width.

Relationship 2: The perimeter is 144 feet.

Translation:
length $+$ width $+$ length $+$ width $= 150$

$15 + w + \quad w \quad + 15 + w + \quad w \quad = 150$

$$30 + 4w = 150 \quad \text{Combine like terms.}$$
$$\underline{-\ 30 \qquad\qquad -\ 30} \quad \text{Subtract 30 from both sides.}$$
$$0 + 4w = 120$$
$$\frac{4w}{4} = \frac{120}{4} \quad \text{Divide both sides by 4.}$$
$$w = 30$$

Answer: The width is 30 ft.
The length is $15 + 30 = 45$ ft.

Formulas

Perimeter of a rectangle:	$P = 2l + 2w$	Volume of a box:	$V = lwh$
Area of a rectangle:	$A = lw$	Force:	$F = ma$
Area of a parallelogram:	$A = bh$	Distance:	$d = rt$
Surface area of a box:	$SA = 2lw + 2lh + 2wh$	Voltage:	$V = ir$

For Exercises 1–6, answer true or false.

[4.1]   **1.** $4x - 9 + 11x$ is an equation.

[4.1]   **2.** $9y - 6 = 4y - 1$ is an equation.

[4.2]   **3.** $t^2 - 9 = 16$ is a linear equation.

[4.2]   **4.** $4(m - 2) = 6m + 16$ is a linear equation.

[4.1]   **5.** -6 is a solution for $x - 9 = -15$.

[4.4]   **6.** The phrase "7 less than a number" translates to $7 - n$.

For Exercises 7–9, fill in the blank.

[4.2]   **7.** The addition/subtraction principle of equality says that we can _____ or _____ the _____ amount on both sides of an equation without affecting its solution(s).

[4.3]   **8.** The multiplication/division principle of equality says that we can _____ or _____ both sides of an equation by the _____ amount without affecting its solution(s).

[4.3]   **9.** To solve equations:

   1. _____ both sides of the equation as needed.

   a. _____ to clear parentheses.

   b. _____ like terms.

   2. Use the _____ principle of equality so that all variable terms are on one side of the equation and all constants are on the other side.

   3. Use the _____ principle of equality to clear any remaining coefficient.

[4.1]   **10.** Explain in your own words how to check a potential solution.

[4.1]   *For Exercises 11–18, check to see if the given number is a solution for the given equation.*

11. $u - 7 = 12$; check $u = 5$

12. $3a = -12$; check $a = -4$

13. $-7n + 12 = 5$; check $n = 1$

14. $14 - 4y = 3y$; check $y = -2$

15. $b^2 - 15 = 2b$; check $b = -3$

16. $9r - 17 = -r + 23$; check $r = 5$

17. $5(x - 2) - 3 = 10 - 4(x - 1)$; check $x = 3$

18. $u^3 - 7 = u^2 + 6u$; check $u = -1$

[4.2]   *For Exercises 19–30, solve and check.*

19. $n + 19 = 27$

20. $y - 6 = -8$

21. $-4k = 36$

22. $-8m = -24$

23. $2x + 11 = -3$

24. $-3h - 8 = 19$

25. $9t - 14 = 3t + 4$

26. $20 - 5y = 34 + 2y$

27. $10m - 17 + m = 2 + 12m - 20$

28. $-16v - 18 + 7v = 24 - v + 6$

29. $4x - 3(x + 5) = 16 - 7(x + 1)$

30. $9y - (2y + 3) = 4(y - 5) + 2$

[4.2] *For Exercises 31 and 32, write a linear equation to describe the situation, then solve.*

31. The temperature at 5 A.M. was reported to be $-15°$F. By 11 A.M., the temperature rose to $28°$F. How much did the temperature rise?

32. Kari has a balance of $-\$547$ on a credit card. How much should she pay on the account to get a balance of $-\$350$?

[4.3] *For Exercises 33 and 34, use the formula and solve for the missing amount.*

33. A swimming pool is to be designed to have a volume of 5000 cubic feet. The length of the pool is to be 50 feet and the width is to be 25 feet. Find the depth (height). (Use $V = lwh$.)

34. A box company shapes flat pieces of cardboard into boxes. The flat pieces have an area of 136 square feet. If the length is 3 feet and height is 4 feet, find the width. (Use $SA = 2lw + 2lh + 2wh$.)

$\begin{bmatrix} 4.4 \\ 4.5 \end{bmatrix}$ *For Exercises 35–46, translate to an equation, then solve.*

35. Fifteen minus seven times a number is twenty-two.

36. The difference of five times x and four is the same as the product of three and x.

37. Twice the sum of n and twelve is equal to eight less than negative six times n.

38. Twelve minus three times the difference of x and seven is the same as three less than the product of six and x.

39. One number is twelve more than another number. The sum of the numbers is forty-two. Find the numbers.

40. One number is five times another number. The sum of the numbers is thirty-six. Find the numbers.

41. A field is developed so that the length is 2 meters less than three times the width. The perimeter is 188 meters. Find the dimensions of the field.

42. A window over the entrance to a library is to be an isosceles triangle. The equal sides are to be 38 inches longer than the base. If the perimeter must be 256 inches, what will be the dimensions of the triangle?

43. The supports to the basketball goals in a park are to be angled for safety. The angle made on the side facing away from the goal is to be 15° less than twice the angle made on the goal side. Find the angle measurements.

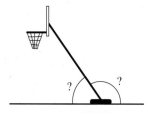

44. A security laser is placed in the corner of a room near the floor and aimed to a detector across the room. The angle made with the wall on one side of the beam is 16° more than the angle made with the other wall. What are the angles?

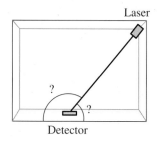

45. Karen has nine more ten-dollar bills than twenty-dollar bills. If the total amount of money is $330, how many of each bill does she have?

46. A store sells two different-size bags of mulch. The larger bag costs $6 and the smaller bag costs $4. In one day the store sold 27 bags of mulch for a total of $146. How many of each size were sold?

LEARNING STRATEGY

Treat a practice test as a dress rehearsal. Have ready the same materials you will use during the actual test. Set a timer for the same amount of time as you will have for the actual test. When you finish, calculate your score and resolve any mistakes you made. Take the practice test over and over until you feel confident with your performance.

[4.1] **1.** Is $8y - 17 = 7$ an expression or equation? Why?

1. _____

[4.2] **2.** Is $9x + 7 = x^2 - 4$ linear or nonlinear? Why?

2. _____

[4.1] **3.** Check to see if -4 is a solution for $-9 = 3x + 5$.

3. _____

[4.1] **4.** Check to see if 2 is a solution for $3x - 11 = 2(x - 2) - 5$.

4. _____

$\begin{bmatrix} 4.2 \\ 4.3 \end{bmatrix}$ *For Problems 5–12, solve and check.*

5. $n - 15 = -7$

5. _____

6. $-9m = 54$

6. _____

7. $-6y + 3 = -21$

7. _____

8. $9k + 5 = 17 - 3k$

8. _____

9. $-13x + 26 = 11 - 8x$

9. _____

10. $4t - 13 + t = 11 + 6t - 3$

10. _____

11. $7(u - 2) + 12 = 4(u - 2)$

11. _____

12. $6 + 3(k + 4) = 5k + (k - 9)$

12. _____

For Problems 13–20, translate to an equation and solve.

[4.2] **13.** Daryl owes $458 in taxes. He currently has $375. How much more does he need?

13. _____

[4.3] **14.** A farmer needs a fenced pasture with an area of 96,800 square yards. The width must be 242 yards. Find the length. (Use $A = lw$.)

14. _____

[4.4] **15.** Nine subtracted from four times a number is equal to twenty-three. Find the number.

15. _____

16. _____

[4.4] **16.** Three times the difference of x and five is the same as nine less than four times x. Find x.

17. _____

[4.5] **17.** One number is three times another number. The sum of the numbers is forty-four. Find the numbers.

18. _____

[4.5] **18.** An entrance to a new restaurant is to be in the shape of an isosceles triangle with the equal sides 9 inches more than twice the base. The perimeter is to be 258 inches. Find the dimensions of the triangle.

19. _____

[4.5] **19.** A microphone is placed on a stand on the floor of a stage so that the angle made by the microphone and the floor on one side is 30° less than the angle made on the other side of the microphone. Find the two angle measurements.

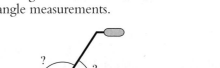

20. _____

[4.5] **20.** A music store sells two models of the same guitar. Model A costs $450 and model B costs $675. In one week the store sold 12 of those guitars for a total of $6300. How many of each model did the store sell? (Use a four-column table.)

For Exercises 1–6, answer true or false.

[1.2] **1.** $(5 + 4) + 2 = 5 + (4 + 2)$

[2.1] **2.** $-65 < -68$

[3.6] **3.** 91 is a prime number.

[3.7] **4.** $x^0 = 1$, where x is any integer.

[4.2] **5.** $4y - 9 = 3(y + 5)$ is a linear equation.

[4.1] **6.** -6 is a solution for $5x - 8 = -24$

For Exercises 7–10, fill in the blank.

[2.4] **7.** The quotient of two negative numbers is a _____ number.

[3.5] **8.** When multiplying two exponential forms that have the same base, we can _____ the exponents and keep the same base.

[3.7] **9.** When dividing two exponential forms that have the same base, we can _____ the exponents and keep the same base.

[3.1] **10.** Explain in your own words how to evaluate an expression.

$\begin{bmatrix} 1.1 \\ 1.3 \end{bmatrix}$ **11.** Write expanded form for 36,097.

[2.1] **12.** Graph 7 on a number line.

[1.1] **13.** Round 29,512 to the nearest thousand.

[1.4] **14.** Estimate $5826 \div 224$ by rounding so that there is only one nonzero digit in each number.

[3.2] **15.** What is the coefficient of $-8x^3$?

[3.2] **16.** What is the degree of 51?

[3.2] **17.** What is the degree of $4x^2 - 8x + 9x^5 - 6$?

[2.5] *For Exercises 18–24, simplify.*

18. $|2 - 3(7)|$

19. $45 - (-52)$

20. $19 - 5(9)$

21. $3 + (-4)^3$

22. -2^4

23. $2(-9) - 20 \div (5)$

24. $7^2 + [8 - 5(6)][6 + 2(7)]$

[3.1] **25.** Evaluate $2x - 5\sqrt{x + y}$ when $x = 9$ and $y = 16$.

[3.3] **26.** Combine like terms and write your answer in descending order.
$7t^3 - 10t + 14t^2 - 7 + t^3 - 15t$

[3.4] **27.** Subtract. $(10x^3 - 7x^2 - 15) - (4x^3 + x - 8)$

[3.5] **28.** Multiply. $(-6a^4)(9a^2)$

[3.5] 29. Multiply. $(3x - 5)(3x + 5)$

[3.6] 30. Find the prime factorization of 420.

[3.7] 31. Find the GCF of $36m^2n$ and $72m^5$.

[3.7] 32. Divide. $\dfrac{28k^6}{-4k^5}$

[3.7] 33. Factor. $20n^5 + 15mn^3 - 10n^2$

$\begin{bmatrix} 4.2 \\ 4.3 \end{bmatrix}$ *For Exercises 34–38, solve and check.*

34. $x - 11 = 28$ **35.** $-7t = 42$ **36.** $30 = 2b - 14$

37. $9y - 13 = 4y + 7$ **38.** $6(n + 2) = 3n - 9$

For Exercises 39–50, solve.

[4.2] 39. In training for a new job, Carlos must log 500 hours of training. He has completed 278 hours. How many more hours must he log to complete the training?

[1.6] 40. The voltage in a circuit measures -60 volts. The resistance measures 15 ohms. What is the current? (Use $V = ir$.)

[2.2] 41. The following table lists the assets and debts for the Goodman family. Calculate their net worth.

Assets	Debts
Savings = $985	Credit-card balance = $2345
Checking = $1862	Mortgage = $75,189
Furniture = $12,006	Automobile = $4500

[2.3] 42. In one month, a company's total revenue was $96,408. If the profit was $45,698, find the company's costs.

[2.6] 43. Find the area of the shape.

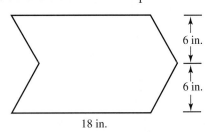

[3.8] 44. Write an expression in simplest form for the area.

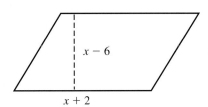

[4.3] 45. A small storage building is a box that has a length of 10 feet and a width of 9 feet. If the volume of the building is 630 cubic feet, what is the height?

[3.8] 46. A wooden jewelry box has a lid that is 8 inches wide by 9 inches long. The surface area of the box is 314 square inches. What is the height of the box?

[2.6] 47. Adrian drives for 45 minutes, stops for a break, then drives for 1 hour and 15 minutes to reach his destination. If the total trip was 124 miles, what was his average speed in miles per hour? (Use $d = rt$.)

[4.5] 48. One number is twice another. Their sum is ninety-six. Find the numbers.

[4.5] 49. The length of a rectangle is 4 feet more than the width. If the perimeter is 44 feet, find the length and width.

[4.5] 50. Kedra sells two sizes of handmade blankets. The large size sells for $75 and the small size for $45. If she sold 16 blankets at a crafts fair and made a total of $930, how many of each size did she sell? (Use a four-column table.)

Fractions and Rational Expressions

> " *You play the way you practice.* "
>
> —POP WARNER,
> FOOTBALL COACH

> " *When you're prepared you're more confident. When you have a strategy you're more comfortable.* "
>
> —FRED COUPLES,
> PROFESSIONAL GOLFER

Taking a Test

If you've prepared as suggested in the *To the Student* section of the text, then the key to success on the test is to relax and trust in your preparation. When taking the test, do the same things that you practiced. When you get the test, do not look at it. Instead, duplicate your study sheet on your scratch paper, just as you've practiced. This way, you don't have to worry about forgetting rules as you take the test.

As you work through the test, work all the problems that you are sure about first. Show all your steps and work neatly so that your instructor can see everything you did in solving the problems. Instructors that give partial credit are much more likely to do so if they can follow the steps. After you've worked your way through the entire test, go back and work any problems that you weren't sure about. Always write something down for every problem, even if you have to guess.

Finally, use all the time you are given. If you finish early, take a break from looking at the test and then go through the entire test again. You may find it helpful to repeat your work on separate paper, then compare your solutions. Keep checking until time is up. Change your answer only if you are *certain* that it is wrong.

5.1 Fractions, Mixed Numbers, and Rational Expressions

OBJECTIVE 1 Name the fraction represented by a shaded region.

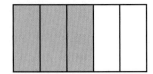

A commercial farmer has one lot of land divided into 5 fields of equal size. Three of the fields are planted while the other 2 fields are left unplanted. In the diagram, the 3 shaded regions represent the 3 planted fields and the other 2 regions represent the fields that are not planted.

We can make the following statements about the fields:

3 out of the 5 fields are planted.

2 out of the 5 fields are not planted.

We can express the part of the lot that is planted or not planted as a **fraction.**

DEFINITION   **Fraction:** A number that describes a part of a whole.

In fraction notation, the 3 fields planted out of the 5 total fields is written as follows:

$$\frac{3 \leftarrow \textbf{Numerator}}{5 \leftarrow \textbf{Denominator}}$$

◀ $\frac{3}{5}$ is read "three-fifths" and it means 3 out of 5.

We can say that $\frac{3}{5}$ of the lot is planted.

DEFINITIONS   **Numerator:** The number written in the top position in a fraction.

Denominator: The number written in the bottom position in a fraction.

The denominator describes the total number of equal-sized divisions in the whole. The numerator describes the number of those divisions that we are interested in. If the numerator and denominator are integers, then we say the number is a **rational number.**

DEFINITION   **Rational number:** A number that can be expressed as a ratio of integers.

In math language: A number that can be expressed in the form $\frac{a}{b}$, where a and b are integers and $b \neq 0$.

Example 1   Name the fraction represented by the shaded region.

a.

Answer: $\dfrac{7}{15}$

Explanation: There are 15 equal-size divisions in the whole region and 7 are shaded.

---

b.

Answer: $\dfrac{3}{8}$

Explanation: There are 8 equal-size divisions in the whole region and 3 are shaded.

Do Your Turn 1 ▶

Example 2 In a group of 35 people at a conference, 17 are wearing glasses. What fraction of the people at the conference are wearing glasses? What fraction are not wearing glasses?

Answer: The fraction of the people at the conference that are wearing glasses is $\frac{17}{35}$. The fraction of the people at the conference that are not wearing glasses is $\frac{18}{35}$.

Explanation: Because a person can either wear glasses, or not wear glasses, there are only two possible categories. This means that if 17 out of the 35 people are wearing glasses, then everyone else is not. We can subtract 17 from 35 to get the number of people that are not wearing glasses. There are $35 - 17 = 18$ people who are not wearing glasses out of the 35 total people; therefore, the fraction is $\frac{18}{35}$.

Do Your Turn 2 ▶

OBJECTIVE 2 Graph fractions on a number line.

To graph a fraction on a number line, we draw marks so that the space between each integer has the number of equal-sized divisions indicated by the denominator and then draw a dot on the mark indicated by the numerator.

CONNECTION The fractional marks on a number line are like the marks on a ruler. A typical ruler is marked so that halves, quarters, eighths, and sixteenths of an inch are shown.

Example 3 Graph the fraction on a number line.

a. $\dfrac{3}{4}$

Solution:

Explanation: Because the denominator is 4, we draw marks to divide the distance between 0 and 1 into 4 equal spaces. Then go to the third mark to the right of 0.

DISCUSSION Where are $\frac{2}{4}$ and $\frac{4}{4}$ located on a number line?

b. $-\dfrac{5}{8}$

Solution:

Explanation: Because the denominator is 8, we draw marks to divide the distance between 0 and -1 into 8 equal spaces and go to the fifth mark to the left of 0.

DISCUSSION Where are $-\frac{2}{8}$, $-\frac{4}{8}$, $-\frac{6}{8}$, and $-\frac{8}{8}$ located on the number line?

Your Turn 1

Name the fraction represented by the shaded region.

a.

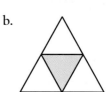

b.

Your Turn 2

Solve.

a. According to a pamphlet, 1 in 5 teenagers smokes cigarettes. What fraction of teenagers smoke?

b. A football quarterback throws 28 passes during a game and 19 of those passes are completed. What fraction of the quarterback's passes were completed? What fraction were not completed?

c. Margaret spends 8 hours each day at her office. Her company allows 1 of those hours to be a lunch break. What fraction of her time at the office is spent at lunch? What fraction is spent working?

Answers to Your Turn 1
a. $\dfrac{11}{12}$ b. $\dfrac{1}{4}$

Answers to Your Turn 2
a. $\dfrac{1}{5}$ b. $\dfrac{19}{28}$, $\dfrac{9}{28}$ c. $\dfrac{1}{8}$, $\dfrac{7}{8}$

Your Turn 3

Graph each fraction on a number line.

a. $\dfrac{1}{4}$

b. $\dfrac{5}{6}$

c. $-\dfrac{3}{8}$

d. $\dfrac{-4}{5}$

We can write negative fractions three ways: the negative sign can be placed to the left of the fraction bar, to the left of the numerator, or to the left of the denominator.

$$-\frac{5}{8} = \frac{-5}{8} = \frac{5}{-8}$$

Placing the negative sign to the left of the fraction bar or to the left of the numerator are the preferred positions.

> **CONNECTION** Recall that a fraction bar also indicates division. When dividing two numbers that have different signs, the quotient is *always* negative.
>
> $$\frac{-6}{2} = -3 \quad \text{and} \quad \frac{6}{-2} = -3$$

◀ **Do Your Turn 3**

OBJECTIVE 3 Simplify fractions.

In the discussion questions in Example 3a, you were asked to consider where fractions like $\frac{4}{4}$ and $\frac{2}{4}$ would be on a number line. The purpose in those questions was to get you thinking about how we might **simplify** fractions to **simplest form.**

DEFINITIONS **Simplify:** Write an equivalent expression with fewer symbols or smaller numbers.

Simplest form: An equivalent expression written with the fewest symbols and the smallest numbers possible.

For example, we can simplify a fraction that has a denominator of 1.

Your Turn 4

Simplify:

a. $\dfrac{257}{1}$

b. $\dfrac{-9}{1}$

> **Example 4** Simplify. $\dfrac{5}{1}$
>
> **Answer:** 5
>
> **Explanation:** $\frac{5}{1}$ is read as "five ones." Five ones can be interpreted as five wholes. On a ruler, it would mean 5 inches.

> **CONNECTION** Remember that the fraction line also means divide. So $\frac{5}{1}$ means $5 \div 1$, which is 5.

> **RULE** *If the denominator of a fraction is 1, then the fraction can be simplified to the numerator.*
>
> **In math language:** $\dfrac{n}{1} = n$, when n is any number.

Answers to Your Turn 3

a.
$0 \quad \frac{1}{4} \qquad 1$

b.
$0 \qquad \frac{5}{6} \ 1$

c.
$-1 \qquad -\frac{3}{8} \quad 0$

d.
$-1 \ \frac{-4}{5} \qquad 0$

Answers to Your Turn 4
a. 257 b. -9

◀ **Do Your Turn 4**

We can also simplify a fraction that has a numerator of 0 and a denominator that is not 0.

> **Example 5** Simplify. $\dfrac{0}{9}$
>
> **Answer:** 0
>
> **Explanation:** Suppose we divided a picture into 9 equal-size pieces, then didn't shade any pieces. Because the fraction represents the shaded region and there is nothing shaded, the fractional amount is $\frac{0}{9}$. This fraction is equal to 0.

RULE *If the numerator of a fraction is 0 and the denominator is any number other than 0, then the fraction can be simplified to 0.*

In math language: $\dfrac{0}{n} = 0$, when $n \neq 0$.

CONNECTION This same rule appeared in Section 1.4 when we learned about dividing whole numbers. We said that 0 divided by any number other than 0 is 0.

In math language: $0 \div n = \dfrac{0}{n} = 0$, when $n \neq 0$.

Do Your Turn 5 ▷

What if the denominator is 0?

Example 6 Simplify. $\dfrac{5}{0}$

Answer: $\dfrac{5}{0}$ is undefined.

Explanation: Because $\frac{5}{0}$ means $5 \div 0$, we should be able to check the answer by multiplying the quotient by the divisor, in this case 0, and get the dividend, 5.

Check: quotient · divisor = dividend

$$? \cdot 0 = 5$$

The product of 0 and any number is 0 so there's no way to get 5 by multiplying a number by 0. Therefore, there is no quotient that will satisfy $\frac{5}{0}$, so $\frac{5}{0}$ is undefined.

RULE *If the denominator is 0 and the numerator is any number other than 0, we say the fraction is undefined.*

In math language: $\dfrac{n}{0}$ is undefined, when $n \neq 0$.

CONNECTION We developed this rule in Section 1.4. We said that any number other than 0 divided by 0 is undefined.

In math language: $n \div 0 = \dfrac{n}{0}$ is undefined, when $n \neq 0$.

Do Your Turn 6 ▷

What if the numerator and denominator contain the same number?

Example 7 Simplify. $\dfrac{4}{4}$

CONNECTION Remember that the fraction bar also means divide. So $\dfrac{4}{4} = 4 \div 4 = 1$.

Answer: 1

Explanation: On a number line, $\frac{4}{4}$ is the same as 1, which answers the discussion question in Example 3a.

Or, think about money. Four-fourths of a dollar, or 4 quarters is a whole dollar. Using a picture, if we divide a whole into 4 equal parts, then shade all 4, we have shaded the whole picture.

Your Turn 5

Simplify.

a. $\dfrac{0}{19}$

b. $\dfrac{0}{-8}$

c. $-\dfrac{0}{11}$

Your Turn 6

Simplify.

a. $\dfrac{97}{0}$

b. $\dfrac{-13}{0}$

c. $-\dfrac{49}{0}$

Answers to Your Turn 5
a. 0 b. 0 c. 0

Answers to Your Turn 6
a. undefined b. undefined
c. undefined

CONNECTION The same rules appeared in Section 1.4 when we learned about dividing whole numbers.

Your Turn 7

Simplify.

a. $\dfrac{59}{59}$

b. $\dfrac{-7}{-7}$

◀ **Do Your Turn 7**

OBJECTIVE 4 Write equivalent fractions.

Let's return to the discussion questions from Example 3a. The fractions $\frac{2}{4}$ and $\frac{4}{8}$ are both halfway between 0 and 1. In other words, they're both equivalent to the fraction $\frac{1}{2}$.

$$\frac{4}{8} = \frac{2}{4} = \frac{1}{2}$$

When two fractions name the same number, we say they are **equivalent fractions.**

DEFINITION Equivalent fractions: Fractions that name the same number.

We can see other equivalent fractions using rulers like those to the right.

The rulers reaffirm our rule that fractions with the same nonzero denominator and numerator are equal to 1.

$$1 = \frac{2}{2} = \frac{4}{4} = \frac{8}{8}$$

Look at how we can express $\frac{1}{2}$ in various ways.

$$\frac{1}{2} = \frac{2}{4} = \frac{4}{8}$$

We can express the quarters in terms of eighths.

$$\frac{1}{4} = \frac{2}{8} \quad \text{and} \quad \frac{3}{4} = \frac{6}{8}$$

DISCUSSION How many sixths would be the same as a half?

How can we write a fraction as one of its equivalent fraction? For example, how do we write $\frac{1}{2}$ as $\frac{2}{4}$ or $\frac{4}{8}$? Notice, multiplying the numerator and denominator of $\frac{1}{2}$ by 2 gives the equivalent fraction $\frac{2}{4}$.

$$\frac{1 \cdot 2}{2 \cdot 2} = \frac{2}{4}$$

Similarly, multiplying both the numerator and denominator of $\frac{1}{2}$ by 4 gives the equivalent fraction $\frac{4}{8}$.

Answers to Your Turn 7
a. 1 b. 1

$$\frac{1 \cdot 4}{2 \cdot 4} = \frac{4}{8}$$

Conclusion: Multiplying both the numerator and denominator by the same nonzero number creates an equivalent fraction. We will call this process *upscaling* a fraction.

We can also divide both the numerator and denominator by the same nonzero number to create an equivalent fraction. This process is called *reducing* a fraction. Or, because the numerator and denominator are smaller values, we say we are *simplifying* the fraction.

For example, we can simplify $\frac{4}{8}$ by dividing its numerator and denominator by 2 to get $\frac{2}{4}$, or by 4 to get $\frac{1}{2}$.

$$\frac{4 \div 2}{8 \div 2} = \frac{2}{4} \quad \text{or} \quad \frac{4 \div 4}{8 \div 4} = \frac{1}{2}$$

PROCEDURE *To write an equivalent fraction, multiply or divide both the numerator and denominator by the same nonzero number.*

Example 8 Fill in the blank so that the fractions are equivalent.

$$\frac{3}{8} = \frac{?}{16}$$

Solution: $\dfrac{3}{8} = \dfrac{3 \cdot 2}{8 \cdot 2} = \dfrac{6}{16}$ Upscale $\frac{3}{8}$ by multiplying the numerator and denominator by 2.

Do Your Turn 8 ▶

Your Turn 8

Fill in the blank so that the fractions are equivalent.

a. $\dfrac{5}{9} = \dfrac{?}{36}$

b. $-\dfrac{3}{4} = \dfrac{?}{12}$

Example 9 Fill in the blank so that the fractions are equivalent.

$$\frac{20}{24} = \frac{5}{?}$$

Solution: $\dfrac{20}{24} = \dfrac{20 \div 4}{24 \div 4} = \dfrac{5}{6}$ Simplify $\frac{20}{24}$ by dividing the numerator and denominator by 4.

Do Your Turn 9 ▶

Your Turn 9

Fill in the blank so that the fractions are equivalent.

a. $\dfrac{28}{35} = \dfrac{4}{?}$

b. $\dfrac{-24}{42} = \dfrac{-4}{?}$

OBJECTIVE 5 Use <, >, or = to make a true statement.

We can easily compare fractions that have the same denominator by looking at their numerators. For example, $\frac{2}{5}$ is less than $\frac{3}{5}$ because 2 is less than 3. In symbols we write

$$\frac{2}{5} < \frac{3}{5}$$

What if the fractions do not have the same denominator?

Consider $\frac{1}{2}$ and $\frac{1}{3}$. We could draw pictures to compare.

$\frac{1}{2} =$

$\frac{1}{3} =$

From the pictures we see that $\frac{1}{2}$ is greater than $\frac{1}{3}$. In symbols, we write

$$\frac{1}{2} > \frac{1}{3}$$

Answers to Your Turn 8
a. 20 b. -9

Answers to Your Turn 9
a. 5 b. 7

Alternatively, we could rewrite $\frac{1}{2}$ and $\frac{1}{3}$ so that they have a common denominator, then compare their numerators. Because we multiply when we upscale, the common denominator must have both 2 and 3 as factors. That is, the common denominator must be a number that is divisible by both original denominators. When a number is divisible by a given number we say it is a **multiple** of the given number.

DEFINITION   Multiple: A number that is evenly divisible by a given number.

Multiples of 2 would be 2, 4, 6, 8, 10, . . .
Multiples of 3 would be 3, 6, 9, 12, 15, . . .

Notice that a common multiple for 2 and 3 is 6, which we can find by multiplying 2 times 3. To upscale $\frac{1}{2}$, we multiply its numerator and denominator by 3. To upscale $\frac{1}{3}$, we multiply its numerator and denominator by 2.

$$\frac{1}{2} = \frac{1 \cdot 3}{2 \cdot 3} = \frac{3}{6} \quad \text{and} \quad \frac{1}{3} = \frac{1 \cdot 2}{3 \cdot 2} = \frac{2}{6}$$

Comparing the numerators in the rewritten fractions, since 3 is greater than 2 we can say, $\frac{1}{2} > \frac{1}{3}$.

PROCEDURE   *To compare two fractions:*
1. Write equivalent fractions that have a common denominator.
2. Compare the numerators in the rewritten fractions.

We will discuss common multiples in more detail later, but for now, a simple way to create a common denominator is to multiply the denominators as we did with 2 and 3 to get 6.

Example 10   Use $<$, $>$, or $=$ to write a true statement.

a. $\dfrac{5}{8}$? $\dfrac{4}{7}$

Solution: Write equivalent fractions that have a common denominator, then compare numerators. A common multiple for 8 and 7 is 56, so our common denominator is 56.

$$\frac{5}{8} = \frac{5 \cdot 7}{8 \cdot 7} = \frac{35}{56} \qquad \text{Upscale } \tfrac{5}{8} \text{ by multiplying its numerator and denominator by 7.}$$

$$\frac{4}{7} = \frac{4 \cdot 8}{7 \cdot 8} = \frac{32}{56} \qquad \text{Upscale } \tfrac{4}{7} \text{ by multiplying its numerator and denominator by 8.}$$

Comparing numerators, we see that 35 is greater than 32, so $\frac{5}{8}$ is greater than $\frac{4}{7}$.

Answer: $\dfrac{5}{8} > \dfrac{4}{7}$

CONNECTION In our method of comparing fractions, we find a common denominator by multiplying the denominators. As a result, when we upscale, each new numerator is the product of the original numerator and the other fraction's denominator. We can find those numerators quickly using a process called *cross multiplying*. Consider Example 10a again.

$$\frac{35}{56} = \frac{5 \cdot 7}{8 \cdot 7} = \frac{5}{8} \times \frac{4}{7} = \frac{4 \cdot 8}{7 \cdot 8} = \frac{32}{36}$$

Notice that the cross products, 35 and 32, are the numerators of the fractions equivalent to $\frac{5}{8}$ and $\frac{4}{7}$ with 56 as their denominator.

Since 35 is the greater cross product and it is the numerator of the fraction equivalent to $\frac{5}{8}$, we can conclude that $\frac{5}{8}$ is the greater fraction.

Conclusion: To compare two fractions, we can compare their cross products. The greater cross product indicates the greater fraction. If the cross products are equal, so are the fractions.

b. $\dfrac{-7}{8}$? $\dfrac{-9}{11}$

Solution: Write equivalent fractions that have a common denominator, then compare numerators. Multiplying the denominators gives us a common denominator of 88.

$$\frac{-7}{8} = \frac{-7 \cdot 11}{8 \cdot 11} = \frac{-77}{88}$$
$$\frac{-9}{11} = \frac{-9 \cdot 8}{11 \cdot 8} = \frac{-72}{88}$$

Note: Remember that when comparing two negative numbers, the greater number has the smaller absolute value.

Because -72 is closer to zero than -77, -72 is the greater of the two numerators. Therefore, $\frac{-72}{88}$ or $\frac{-9}{11}$ is the greater fraction.

Answer: $\dfrac{-7}{8} < \dfrac{-9}{11}$

Using the cross products method, we get the same result:

$$\frac{-77}{88} = \frac{-7 \cdot 11}{8 \cdot 11} = \frac{-7}{8} \times \frac{-9}{11} = \frac{-9 \cdot 8}{11 \cdot 8} = \frac{-72}{88}$$

Comparing cross products, -72 is greater than -77, so $\frac{-9}{11}$ is the greater fraction.

Do Your Turn 10 ▷

Your Turn 10

Use $<$, $>$, *or* $=$ *to write a true statement.*

a. $\dfrac{3}{7}$? $\dfrac{4}{9}$

b. $\dfrac{5}{11}$? $\dfrac{4}{9}$

c. $-\dfrac{2}{7}$? $-\dfrac{12}{42}$

d. $\dfrac{-9}{10}$? $-\dfrac{8}{9}$

OBJECTIVE 6 Write improper fractions as mixed numbers.

If the absolute value of the numerator of a fraction is greater than or equal to the absolute value of the denominator, we call the fraction an **improper fraction.**

DEFINITION Improper fraction: A fraction in which the absolute value of the numerator is greater than or equal to the absolute value of the denominator.

DISCUSSION Why do we have to say "absolute value?" Why can't we just say an improper fraction is a fraction with a numerator greater than the denominator?

Answers to Your Turn 10

a. $\dfrac{3}{7} < \dfrac{4}{9}$ b. $\dfrac{5}{11} > \dfrac{4}{9}$

c. $-\dfrac{2}{7} = -\dfrac{12}{42}$ d. $\dfrac{-9}{10} < -\dfrac{8}{9}$

$\frac{9}{4}$ is an improper fraction because $|9| = 9$ is greater than $|4| = 4$.

$\frac{-7}{2}$ is an improper fraction because $|-7| = 7$ is greater than $|2| = 2$.

$\frac{8}{8}$ is improper because $|8| = 8$ is equal to $|8| = 8$.

What does $\frac{9}{4}$ mean?

Think about money. One-fourth of a dollar is a quarter. If we had 9 quarters and every 4 quarters is a dollar, then we have 2 dollars plus 1 quarter left over. Let's translate this into mathematical terms:

$$9 \text{ quarters} = 2 \text{ whole dollars} + 1 \text{ quarter}$$
$$\frac{9}{4} = 2 + \frac{1}{4}$$

The customary notation for this result is to leave out the plus sign and write $2\frac{1}{4}$, which is called a **mixed number.**

DEFINITION **Mixed number:** An integer combined with a fraction.

When we say combined, we literally mean *added*.

$$2\frac{1}{4} = 2 + \frac{1}{4}$$

◀ **Note:** $2\frac{1}{4}$ is read as "two and one-fourth" and means 2 wholes plus $\frac{1}{4}$ of another whole.

How does this apply to negative mixed numbers?

$$-3\frac{5}{8} = -\left(3\frac{5}{8}\right) = -\left(3 + \frac{5}{8}\right) = -3 - \frac{5}{8}$$

◀ **Note:** The negative sign applies to both the integer and the fraction. In terms of a number line (shown below), we are saying go left 3 and then left another $\frac{5}{8}$.

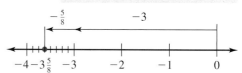

How do we express an improper fraction as a mixed number? Think about what we did when we expressed $\frac{9}{4}$ as a mixed number. We noted how many groups of 4 were in 9, which is division.

PROCEDURE *To write an improper fraction as a mixed number:*

1. Divide the denominator into the numerator.
2. Write the results in the following form:

$$\text{quotient } \frac{remainder}{original\ denominator}$$

Example 11 Write the improper fraction as a mixed number.

a. $\dfrac{20}{3}$

Solution: Divide 20 by 3.

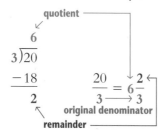

CONNECTION $\frac{20}{3}$ is the same as $20 \div 3$. Mixed numbers are another way to write quotients that have remainders.

Instead of 6 r2 we can now write $6\frac{2}{3}$.

b. $-\dfrac{41}{7}$

Solution: Divide 41 by 7 and write the negative sign with the integer in the mixed number.

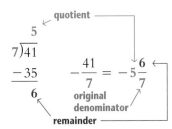

Do Your Turn 11 ▷

As we learned in the Connection box in Example 11a, we can write the result of a division as a mixed number. Remember, the divisor is the denominator.

Example 12 Divide $-2035 \div (-19)$ and write the quotient as a mixed number.

Solution: Because dividing two numbers with the same sign equals a positive number, we can divide and write the quotient as a positive mixed number.

$$
\begin{array}{r}
107 \leftarrow \text{quotient} \\
19\overline{)2035} \\
\underline{-19} \\
13 \\
\underline{-0} \\
135 \\
\underline{-133} \\
2 \leftarrow \text{remainder}
\end{array}
\qquad 107\dfrac{2}{19} \; \overset{\text{divisor}}{}
$$

Do Your Turn 12 ▷

OBJECTIVE ▇**7** Write mixed numbers as improper fractions.

In Example 11a, we wrote $\dfrac{20}{3}$ as $6\dfrac{2}{3}$. How can we verify that $6\dfrac{2}{3}$ is correct? Because we divided 20 by 3 to get $6\dfrac{2}{3}$, we can check the division by multiplying.

$$
\begin{array}{ccccccc}
\text{quotient} & \cdot & \text{divisor} & + & \text{remainder} & = & \text{dividend} \\
6 & \cdot & 3 & + & 2 & \overset{?}{=} & 20 \\
 & 18 & & + & 2 & \overset{?}{=} & 20 \\
 & & 20 & & & = & 20 \quad \textbf{Yes, it checks.}
\end{array}
$$

In mixed-number format, it looks like this:

1. Multiply. $3 \cdot 6 = 18$ 2. Add the product to numerator. $18 + 2 = 20$

$$
6\dfrac{2}{3} = \dfrac{3 \cdot 6 + 2}{3} = \dfrac{20}{3}
$$

3. Keep the same denominator.

Your Turn 11

Write each improper fraction as a mixed number.

a. $\dfrac{13}{3}$

b. $\dfrac{32}{5}$

c. $-\dfrac{37}{4}$

d. $\dfrac{-79}{10}$

Your Turn 12

Divide and write the quotient as a mixed number.

a. $49 \div 2$

b. $139 \div 13$

c. $-920 \div 9$

d. $-3249 \div -16$

Answers to Your Turn 11

a. $4\dfrac{1}{3}$ b. $6\dfrac{2}{5}$

c. $-9\dfrac{1}{4}$ d. $-7\dfrac{9}{10}$

Answers to Your Turn 12

a. $24\dfrac{1}{2}$ b. $10\dfrac{9}{13}$

c. $-102\dfrac{2}{9}$ d. $203\dfrac{1}{16}$

To write a mixed number as an improper fraction:

1. Multiply the denominator by the integer.
2. Add the resulting product to the numerator to find the numerator of the improper fraction.
3. Keep the same denominator.

Your Turn 13

Write as an improper fraction.

a. $10\frac{5}{6}$

b. $-11\frac{2}{9}$

c. 19

Example 13 Write the mixed number as an improper fraction.

a. $5\frac{7}{8}$

Solution: $5\frac{7}{8} = \frac{8 \cdot 5 + 7}{8} = \frac{40 + 7}{8} = \frac{47}{8}$

◀ **Note:** We have shown the steps. However, many people perform the multiplication and addition mentally.

Explanation: We multiplied 8 by 5 to get 40, then added 7 to get 47 as the numerator of the improper fraction. We kept the denominator the same.

b. $-9\frac{3}{7}$

Solution: $-9\frac{3}{7} = -\frac{7 \cdot 9 + 3}{7} = -\frac{63 + 3}{7} = -\frac{66}{7}$

Explanation: Remember, the negative sign just means that the amount is to the left of 0 on a number line. We simply followed the same process and wrote the negative sign with the fraction. We multiplied 7 by 9 to get 63, then added 3 to get 66 for the numerator.

WARNING Do not make the mistake of thinking that we multiply 7 times -9 to get -63, then add 3 to get -60. Think of the negative sign as applying to all of the mixed number.

$$-9\frac{3}{7} = -\left(9\frac{3}{7}\right) = -\frac{66}{7}$$

You may also think of $-9\frac{3}{7}$ as the "additive inverse of $9\frac{3}{7}$." So we can write $9\frac{3}{7}$ as an improper fraction and then find its additive inverse.

◀ **Do Your Turn 13**

Answers to Your Turn 13

a. $\frac{65}{6}$ b. $-\frac{101}{9}$ c. $\frac{19}{1}$

For
Extra
Help

Videotape
DVT

Tutor Center
Addison-Wesley
Tutor Center

Math XL
Math XL

MyMathLab

Student Solutions
Manual

1. What is a rational number?

2. If the numerator of a fraction is 0 and the denominator is a number other than 0, then the fraction simplifies to _____.

3. If the denominator of a fraction is 0 and the numerator is a number other than 0, then we say the fraction is _____.

4. What does the property of equivalent fractions say?

5. Explain how to write an improper fraction as a mixed number.

6. Explain how to write a mixed number as an improper fraction.

For Exercises 7–12, name the fraction represented by a shaded region.

7.

8.

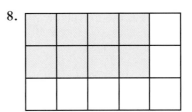

9.

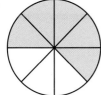

10.

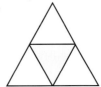

11.

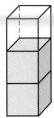

12.

For Exercises 13–18, write the fraction for each situation.

13. a. In an experiment, a subject is blindfolded and her nose pinned so that she cannot smell. She could accurately identify 5 out of 16 foods by taste. What fraction did she accurately identify?

  b. Still blindfolded but without her nose pinned, she could then identify 15 out of 16 foods. What fraction did she accurately identify with just the blindfold?

  c. What can we conclude from the experiment?

  d. Are there any flaws in the experiment?

14. a. In a memory experiment, a subject is shown 20 words in sequence. Each word is shown for 3 seconds. The subject is asked to write down all the words he can remember in any order. The subject was able to write down 11 words. What fraction of the 20 words was he able to write down?

  b. The experiment is repeated with new words and the subject is asked to write the words in the order in which they appeared. The subject was able to write 5 in order. What fraction did he get correct?

c. What can you conclude from the experiment?

d. Are there any flaws in this experiment?

15. A certain fish lays 800 eggs out of which only 17 survive to adulthood. What fraction describes the portion that survives to adulthood? What fraction describes the portion that does not survive to adulthood?

16. At a certain university, of the 179 entering students that declared premed as their major, only 15 completed the premed program. What fraction completed the program? What fraction did not? What can you conclude about this program?

17. Of the 258 students that have ever taken math with Mrs. Jones, 249 passed. What fraction passed? What fraction did not pass? What can you conclude about taking math with Mrs. Jones as the instructor?

18. An ad claims that a certain medicine is safe according to the FDA. You read an article that claims the medicine caused cancer in laboratory rats. As you read, you see that 9 out of 50 rats got cancer. What fraction of the rats got cancer? What can you conclude about the claims of the FDA vs. the claims of the ad?

For Exercises 19–26, graph the fraction on a number line.

19. $\dfrac{1}{4}$

20. $\dfrac{3}{4}$

21. $\dfrac{5}{6}$

22. $\dfrac{2}{3}$

23. $-\dfrac{5}{8}$

24. $-\dfrac{1}{2}$

25. $\dfrac{-3}{16}$

26. $\dfrac{9}{-10}$

For Exercises 27–34, simplify.

27. $\dfrac{23}{1}$

28. $\dfrac{19}{19}$

29. $\dfrac{0}{16}$

30. $\dfrac{12}{0}$

31. $\dfrac{-6}{-6}$

32. $\dfrac{-45}{1}$

33. $\dfrac{-2}{0}$

34. $\dfrac{0}{-18}$

For Exercises 35–42, fill in the blank so that the fractions are equivalent.

35. $\dfrac{5}{9} = \dfrac{?}{27}$

36. $\dfrac{3}{5} = \dfrac{?}{35}$

37. $\dfrac{21}{36} = \dfrac{?}{12}$

38. $\dfrac{12}{42} = \dfrac{?}{7}$

39. $-\dfrac{9}{15} = -\dfrac{18}{?}$

40. $-\dfrac{28}{36} = -\dfrac{7}{?}$

41. $\dfrac{-6}{16} = \dfrac{?}{80}$

42. $-\dfrac{24}{60} = \dfrac{2}{?}$

For Exercises 43–50, use $<$, $>$, or $=$ to make a true statement.

43. $\dfrac{4}{9}$? $\dfrac{2}{5}$

44. $\dfrac{25}{30}$? $\dfrac{5}{6}$

45. $\dfrac{12}{18}$? $\dfrac{9}{16}$

46. $\dfrac{5}{7}$? $\dfrac{8}{11}$

47. $-\dfrac{4}{15}$? $-\dfrac{6}{17}$

48. $\dfrac{-2}{9}$? $\dfrac{-5}{23}$

49. $-\dfrac{9}{12}$? $-\dfrac{15}{20}$

50. $\dfrac{-7}{8}$? $\dfrac{9}{-10}$

For Exercises 51–58, write the improper fraction as a mixed number.

51. $\dfrac{30}{7}$

52. $\dfrac{52}{9}$

53. $\dfrac{85}{4}$

54. $\dfrac{19}{2}$

55. $\dfrac{-64}{5}$

56. $-\dfrac{97}{6}$

57. $\dfrac{103}{-8}$

58. $\dfrac{-111}{20}$

For Exercises 59–66, write as an improper fraction.

59. $5\dfrac{1}{6}$

60. $4\dfrac{3}{8}$

61. 11

62. $13\dfrac{1}{2}$

63. $-9\dfrac{7}{8}$

64. -24

65. $-1\dfrac{9}{20}$

66. $-15\dfrac{3}{4}$

Review Exercises

[2.4]   **1.** Divide. $-48 \div (-12)$

[3.6]   **2.** Find the prime factorization of 840.

[3.6]   **3.** Find the GCF of 24 and 60.

[3.7]   **4.** Find the GCF of $40x^5$ and $56x^2$.

[3.7]   **5.** Factor. $40x^5 - 56x^2$

5.2 Simplifying Fractions and Rational Expressions

OBJECTIVES

1 Simplify fractions to lowest terms.

2 Simplify improper fractions or fractions within mixed numbers.

3 Simplify rational expressions.

OBJECTIVE 1 Simplify fractions to lowest terms.

In Section 5.1, we learned how to write equivalent fractions. In this section, we will learn to write equivalent fractions that are in **lowest terms.**

> **DEFINITION** **Lowest terms:** A fraction is in lowest terms when the greatest common factor of its numerator and denominator is 1.

For example, the fraction $\frac{2}{3}$ is in lowest terms because the greatest common factor of 2 and 3 is 1. The fraction $\frac{8}{12}$ is not in lowest terms because the greatest common factor of 8 and 12 is 4 (not 1).

Recall from Section 5.1 that we can simplify (or reduce) a fraction by dividing both its numerator and denominator by any common factor. So to simplify $\frac{8}{12}$, we could divide by either 2 or 4 because those are factors common to 8 and 12. However, as we'll see, dividing by 4, which is the greatest common factor, takes us directly to the equivalent fraction in lowest terms.

$$\frac{8}{12} = \frac{8 \div 2}{12 \div 2} = \frac{4}{6} \leftarrow \text{Though equivalent, } \tfrac{4}{6} \text{ is not in lowest terms.}$$

$$\frac{8}{12} = \frac{8 \div 4}{12 \div 4} = \frac{2}{3} \leftarrow \text{This is the equivalent fraction in lowest terms.}$$

> **CONNECTION** In Section 3.6, we learned that the greatest common factor is the greatest number that divides all the given numbers evenly.

Notice, we could continue simplifying $\frac{4}{6}$ by dividing by the common factor 2 to get lowest terms.

$$\frac{8}{12} = \frac{8 \div 2}{12 \div 2} = \frac{4}{6} = \frac{4 \div 2}{6 \div 2} = \frac{2}{3}$$

> **Note:** We divided by two factors of 2. The product of those two factors of 2 is 4, which is the greatest common factor of 8 and 12. No matter how we approach reducing to lowest terms, we will always be dividing out the greatest common factor of the numerator and denominator.

Our example suggests the following procedure.

> **PROCEDURE** *To simplify a fraction to lowest terms, divide the numerator and denominator by their greatest common factor.*

Your Turn 1

Reduce to lowest terms.

a. $\dfrac{30}{40}$

b. $-\dfrac{24}{54}$

Example 1 Reduce $\frac{18}{24}$ to lowest terms.

Solution: The greatest common factor for 18 and 24 is 6, so we divide both 18 and 24 by 6.

$$\frac{18}{24} = \frac{18 \div 6}{24 \div 6} = \frac{3}{4}$$

Notice that $\frac{3}{4}$ is in lowest terms because the greatest common factor for 3 and 4 is 1.

◁ **Do Your Turn 1**

Answers to Your Turn 1

a. $\dfrac{3}{4}$ b. $-\dfrac{4}{9}$

Another way that we can divide out the GCF involves primes. In Section 3.6, we learned that the GCF of two numbers contains all the primes that are common in their prime factorizations. Therefore, dividing out all the common prime factors in the prime factorizations of the numerator and denominator of a fraction would be dividing out the GCF of the numerator and denominator. Consider $\frac{18}{24}$ again. Replacing 18 and 24 with their prime factorizations gives:

$$\frac{18}{24} = \frac{2 \cdot 3 \cdot 3}{2 \cdot 2 \cdot 2 \cdot 3}$$ ◀ **Note:** We have a common factor of 2 and a common factor of 3, which, when multiplied, equal the GCF, 6.

Dividing out those common factors divides out the GCF. Multiplying the remaining factors gives lowest terms.

$$\frac{18}{24} = \frac{\overset{1}{2} \cdot 3 \cdot \overset{1}{3}}{\underset{1}{2} \cdot 2 \cdot 2 \cdot \underset{1}{3}} = \frac{1 \cdot 3 \cdot 1}{1 \cdot 2 \cdot 2 \cdot 1} = \frac{3}{4}$$

PROCEDURE *To simplify a fraction to lowest terms using primes:*

1. Replace the numerator and denominator with their prime factorizations.
2. Divide out all the common prime factors.
3. Multiply the remaining factors.

This method is especially useful for simplifying fractions with large numerators or denominators.

Example 2 Simplify $-\frac{72}{90}$ to lowest terms.

Solution: Since 72 and 90 are large numbers, we will use the prime factorization method. We need the prime factorization of 72 and 90.

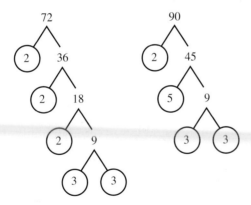

Now replace 72 and 90 with their prime factorizations, divide out all the common primes, and then multiply the remaining prime factors.

$$-\frac{72}{90} = -\frac{\overset{1}{2} \cdot 2 \cdot 2 \cdot \overset{1}{3} \cdot \overset{1}{3}}{\underset{1}{2} \cdot \underset{1}{3} \cdot \underset{1}{3} \cdot 5} = -\frac{4}{5}$$

Do Your Turn 2 ▶

a. In a survey of 300 people, 252 responded that they believed there is life elsewhere in the universe. Write in lowest terms the fraction of people in the survey who believe life exists elsewhere in the universe.

DISCUSSION Do you think the result could be skewed?

b. According to the *CIA World Fact Book*, in 2003 about 60,000,000 out of 300,000,000 people living in the United States were between 0 and 14 years of age. Write in lowest terms the fraction of people living in the United States who are between 0 and 14 years of age.

Example 3 According to a survey of 182 people, 112 responded that they had seen a particular movie. Write a fraction in lowest terms of people in the survey who had not seen the movie. Interpret the results.

Understand: We must write a fraction in lowest terms for the number of people who did not see the movie, then interpret the result.

Plan: Because 112 saw the movie, we subtract 112 from 182 to find the number of people who had not seen the movie. We can then write the fraction and simplify.

Execute: $182 - 112 = 70$ people did not see the movie. So the fraction of people who did not see it is $\frac{70}{182}$.

Now we need to simplify to lowest terms. We find the prime factorization of 70 and 182, divide out the common primes, then multiply the remaining factors.

$$\frac{70}{182} = \frac{\overset{1}{2} \cdot 5 \cdot \overset{1}{7}}{\underset{1}{2} \cdot \underset{1}{7} \cdot 13} = \frac{5}{13}$$

$\frac{5}{13}$ is in lowest terms because the GCF of 5 and 13 is 1.

Answer: $\frac{5}{13}$ of the people surveyed had not seen the movie.

Check: Verify that $\frac{5}{13}$ is equivalent to $\frac{70}{182}$. We will use the cross-products method.

$$\frac{910}{2366} = \frac{5 \cdot 182}{13 \cdot 182} \overset{\longleftarrow}{=} \frac{5}{13} \times \frac{70}{182} \overset{\longrightarrow}{=} \frac{70 \cdot 13}{182 \cdot 13} = \frac{910}{2366}$$

The cross products are equal, so $\frac{5}{13}$ is equivalent to $\frac{70}{182}$.

Interpret: $\frac{5}{13}$ means that 5 out of any 13 people in the survey did *not* see the movie. If the survey was given to people of differing ethnic groups, genders, and ages, then we should be able to ask any 13 people of those surveyed if they had seen the movie and we would expect 5 people to say no.

Note: If a survey is not accurately conducted, we say the survey's results are *skewed*.

◀ Do Your Turn 3

OBJECTIVE 2 Simplify improper fractions or fractions within mixed numbers.

What if an improper fraction is not in lowest terms or a mixed number contains a fraction that can be simplified?

Example 4 Write $\frac{26}{12}$ as a mixed number.

Solution: Since we are always expected to write fractions in lowest terms, let's simplify first and then write the mixed number.

$$\frac{26}{12} = \frac{2 \cdot \overset{1}{13}}{\underset{1}{2} \cdot 2 \cdot 3} = \frac{13}{6} = 2\frac{1}{6}$$

Equivalent fraction in lowest terms.

Mixed number with its fraction in lowest terms.

a. $\frac{21}{25}$ b. $\frac{1}{5}$

Alternately, we can write the mixed number first and then simplify its fraction.

$$\frac{26}{12} = 2\frac{2}{12} = 2\frac{\overset{1}{\cancel{2}}}{\underset{1}{\cancel{2} \cdot 2 \cdot 3}} = 2\frac{1}{6}$$

Either approach produces the same results.

Do Your Turn 4 ▶

OBJECTIVE 3 Simplify rational expressions.

Let's apply what we've learned about simplifying to a special class of expressions called **rational expressions.**

DEFINITION Rational expression: A fraction that is a ratio of monomials or polynomials.

Some examples of rational expressions are

$$\frac{4x^3}{8x} \qquad \frac{5x}{7x + 1} \qquad \frac{2x^2 - 3x + 4}{8x - 5}$$

CONNECTION A rational number is a number that can be expressed as a ratio of integers, whereas a rational expression is a ratio of polynomials.

The rational expressions we will consider will only contain monomials, as in $\frac{4x^3}{8x}$. We simplify rational expressions containing monomials in the same way that we simplify fractions.

Example 5 Simplify $\dfrac{4x^3}{8x}$ to lowest terms.

Solution: Replace the numerator and denominator with their prime factorizations (treat variables like primes), divide out all the common prime factors, then multiply the remaining factors.

$$\frac{4x^3}{8x} = \frac{2 \cdot 2 \cdot x \cdot x \cdot x}{2 \cdot 2 \cdot 2 \cdot x}$$

Write the numerator and denominator in factored form.

$$= \frac{\overset{1}{\cancel{2}} \cdot \overset{1}{\cancel{2}} \cdot x \cdot x \cdot \overset{1}{\cancel{x}}}{\underset{1}{\cancel{2}} \cdot \underset{1}{\cancel{2}} \cdot 2 \cdot \underset{1}{\cancel{x}}}$$

Divide out two 2's and one x.

$$= \frac{x^2}{2}$$

Multiply the remaining factors.

CONNECTION x^3 means there are three x factors. If we divide out one x in the numerator and denominator, we have two x factors left. This goes back to the rule of exponents. When we divide exponential forms that have the same base, we can subtract the exponents.

Do Your Turn 5 ▶

Your Turn 4

Write each fraction as a mixed number and reduce.

a. $\dfrac{28}{6}$

b. $\dfrac{80}{15}$

c. $-\dfrac{54}{24}$

d. $\dfrac{-180}{24}$

Your Turn 5

Simplify to lowest terms.

a. $\dfrac{9x^4}{24x^3}$

b. $\dfrac{10a^5}{20a}$

c. $-\dfrac{4x^3y}{30x}$

d. $\dfrac{-15m^4}{20m^2n}$

Answers to Your Turn 4

a. $4\dfrac{2}{3}$ b. $5\dfrac{1}{3}$ c. $-2\dfrac{1}{4}$ d. $-7\dfrac{1}{2}$

Answers to Your Turn 5

a. $\dfrac{3x}{8}$ b. $\dfrac{a^4}{2}$ c. $-\dfrac{2x^2y}{15}$

d. $\dfrac{-3m^2}{4n}$

$$\frac{2^3}{2} = 2^3 \div 2 = 2^{3-1} = 2^2$$

If we write all the factors and divide out the common 2, we have: $\frac{2^3}{2} = \frac{2 \cdot 2 \cdot \overset{1}{\cancel{2}}}{\underset{1}{\cancel{2}}} = 2^2$.

Conclusion: Subtracting exponents corresponds to dividing out common factors.

We can also get a better understanding of what happens when the exponents are the same.

$$\frac{3^2}{3^2} = 3^2 \div 3^2 = 3^{2-2} = 3^0$$

If we write all the factors and divide out the 3's we have: $\frac{3^2}{3^2} = \frac{\overset{1}{\cancel{3}} \cdot \overset{1}{\cancel{3}}}{\underset{1}{\cancel{3}} \cdot \underset{1}{\cancel{3}}} = 1$.

Conclusion: Because $\frac{3^2}{3^2}$ equals both 3^0 and 1, we can conclude that $3^0 = 1$.

What if the variables in the denominator have greater exponents than the like variables in the numerator?

Your Turn 6

Reduce to lowest terms.

a. $\dfrac{14x^4y}{18x^6}$

b. $\dfrac{20m^2}{40m^5}$

c. $-\dfrac{13a^6b}{26a^4b^4}$

d. $\dfrac{-10hk^2}{5k^7}$

Example 6 Simplify $\dfrac{-16p^3r^2s}{28p^5r^2}$ to lowest terms.

Solution: Write the numerator and denominator in factored form, divide out the common factors, then multiply the remaining factors.

$$\frac{-16p^3r^2s}{28p^5r^2} = \frac{-2 \cdot 2 \cdot 2 \cdot 2 \cdot p \cdot p \cdot p \cdot r \cdot r \cdot s}{2 \cdot 2 \cdot 7 \cdot p \cdot p \cdot p \cdot p \cdot p \cdot r \cdot r}$$

Write the numerator and denominator in factored form.

$$= \frac{-\overset{1}{\cancel{2}} \cdot \overset{1}{\cancel{2}} \cdot 2 \cdot 2 \cdot \overset{1}{\cancel{p}} \cdot \overset{1}{\cancel{p}} \cdot \overset{1}{\cancel{p}} \cdot \overset{1}{\cancel{r}} \cdot \overset{1}{\cancel{r}} \cdot s}{\underset{1}{\cancel{2}} \cdot \underset{1}{\cancel{2}} \cdot 7 \cdot \underset{1}{\cancel{p}} \cdot \underset{1}{\cancel{p}} \cdot \underset{1}{\cancel{p}} \cdot p \cdot p \cdot \underset{1}{\cancel{r}} \cdot \underset{1}{\cancel{r}}}$$

Divide out two 2's, three p's, and two r's.

$$= \frac{-4s}{7p^2}$$

Multiply the remaining factors.

Note: We ended up with p^2 in the denominator because there were more factors of p in the denominator then in the numerator. p^5 meant that there were five p's in the denominator, while p^3 meant that there were only three p's in the numerator. Therefore, dividing out the three common p's leaves two p's in the denominator.

◀ **Do Your Turn 6**

Answers to Your Turn 6

a. $\dfrac{7y}{9x^2}$ b. $\dfrac{1}{2m^3}$

c. $-\dfrac{a^2}{2b^3}$ d. $\dfrac{-2h}{k^5}$

5.2 **Exercises**

For
Extra
Help

Videotape
DVT

Addison-Wesley
Tutor Center

Math XL

MyMathLab

Student Solutions
Manual

1. A fraction is in lowest terms if the _____ _____ for the numerator and denominator is _____.

2. Explain how to simplify a fraction to lowest terms.

For Exercises 3–22, reduce to lowest terms.

3. $\dfrac{25}{30}$

4. $\dfrac{12}{28}$

5. $\dfrac{-14}{35}$

6. $\dfrac{-16}{24}$

7. $\dfrac{26}{52}$

8. $\dfrac{15}{45}$

9. $-\dfrac{24}{40}$

10. $-\dfrac{20}{45}$

11. $\dfrac{66}{88}$

12. $\dfrac{32}{80}$

13. $\dfrac{-57}{76}$

14. $\dfrac{-51}{85}$

15. $\dfrac{120}{140}$

16. $\dfrac{196}{210}$

17. $\dfrac{182}{234}$

18. $\dfrac{221}{357}$

19. $\dfrac{-121}{187}$

20. $-\dfrac{270}{900}$

21. $\dfrac{-360}{480}$

22. $\dfrac{-210}{294}$

For Exercises 23–26, find the requested fraction in lowest terms.

23. 25 minutes is what fraction of an hour?

24. 9 hours is what fraction of a day?

25. A survey is conducted in front of a department store. The person conducting the survey asks 248 women if they used a certain cosmetic product and 96 respond that they do use the product. What is the lowest terms fraction of the women surveyed that said they used the product? Interpret the results. Do you see any flaws in the survey?

26. A marketing firm develops two different television commercials for a product. To determine which is more effective, the firm conducts an experiment. Using the phone directory, they randomly select 400 people and pay 200 to watch one of the commercials and the other 200 to watch the other. They then ask those viewers whether they would buy the product. 180 said yes in the first group, while 148 said yes in the second group. Write the lowest terms fractions for each group. Which commercial should be used? Do you see any flaws in the experiment?

Use the table below to answer Exercises 27–29. The table lists the number of miles driven by a company's sales representatives for various types of trips that occurred in 2004.

Type of Trip by Destination	Miles
Same state	448 million
Different state, same division	212 million
Different state, different division, but same region	133 million
Different state, different division, and different region	208 million

27. Find the lowest terms fraction of total miles driven traveling to a different state within the same division.

28. Find the lowest terms fraction of total miles driven traveling to a different state within the same region.

29. Find the lowest terms fraction of total miles driven traveling to a different state, different region, and different division of the country.

30. A company produces 540 computer chips on a particular day. Quality-control inspectors discover 21 defective chips. What lowest terms fraction of the chips produced is defective? What lowest terms fraction is not defective?

For Exercises 31–38, write as a mixed number.

31. $\dfrac{30}{8}$

32. $\dfrac{69}{9}$

33. $\dfrac{-50}{15}$

34. $\dfrac{-84}{16}$

35. $\dfrac{116}{28}$

36. $\dfrac{168}{105}$

37. $\dfrac{-186}{36}$

38. $\dfrac{486}{-63}$

For Exercises 39–50, reduce to lowest terms.

39. $\dfrac{10x}{32}$

40. $\dfrac{15y^2}{27}$

41. $\dfrac{-x^3}{xy}$

42. $\dfrac{-a^4b}{a^3b^5}$

43. $-\dfrac{6m^4n}{15m^7}$

44. $-\dfrac{30h^2k}{18h^4k}$

45. $\dfrac{9t^2u}{36t^5u^2}$

46. $\dfrac{7x^3y}{35xy^2}$

47. $\dfrac{21a^6bc}{35a^5b^2}$

48. $\dfrac{10m^7n}{30m^9n^4p}$

49. $\dfrac{-14a^4b^5c^2}{28a^{10}bc^7}$

50. $\dfrac{8x^2yz^3}{-32xy^4z^3}$

Review Exercises

[2.4] **1.** Multiply. $165 \cdot (-91)$

[2.4] **2.** Multiply. $(-12)(-6)$

[3.5] **3.** Multiply. $(5x^2)(7x^3)$

[3.5] **4.** Simplify. $(-2x^3)^5$

[1.6] **5.** Calculate the area of the figure.

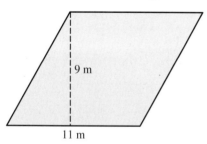

9 m

11 m

5.3 Multiplying Fractions, Mixed Numbers, and Rational Expressions

OBJECTIVE 1 Multiply fractions.

Let's see what we can discover about how to multiply fractions. Suppose a recipe calls for $\frac{1}{4}$ of a cup of oil but we only want to make half of the recipe. We must find half of $\frac{1}{4}$. Consider the picture of $\frac{1}{4}$:

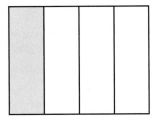

If we want half of $\frac{1}{4}$, we need to cut the single fourth into two pieces then shade one of those two pieces.

In order to name the fraction represented by half of the fourth, we need equal-size divisions. Extending the line across makes equal-size divisions.

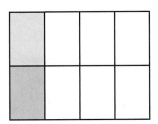

Notice that the darker shaded region is half of the fourth that we originally shaded and is $\frac{1}{8}$ of the whole picture. So we can now say

$$\text{Half of } \frac{1}{4} \text{ is } \frac{1}{8}.$$

We can translate this statement to an equation. The word *of* with a fraction in front of it is a key word for multiplication and *is* indicates an equal sign, so we have

$$\text{Half of } \frac{1}{4} \text{ is } \frac{1}{8}.$$
$$\frac{1}{2} \cdot \frac{1}{4} = \frac{1}{8}$$

Our example suggests that to multiply fractions, we multiply numerator by numerator and denominator by denominator.

Multiply.

a. $\dfrac{3}{4} \cdot \dfrac{7}{8}$

b. $\dfrac{1}{5} \cdot -\dfrac{4}{9}$

c. $\dfrac{-5}{6} \cdot -\dfrac{1}{6}$

d. $\dfrac{1}{-4} \cdot \dfrac{-1}{5}$

Example 1 Multiply.

a. $\dfrac{3}{4} \cdot \dfrac{5}{8}$

Solution: $\dfrac{3}{4} \cdot \dfrac{5}{8} = \dfrac{3 \cdot 5}{4 \cdot 8} = \dfrac{15}{32}$

Explanation: We multiplied numerator by numerator and denominator by denominator. Notice that $\frac{15}{32}$ is in lowest terms.

b. $-\dfrac{2}{3} \cdot \dfrac{1}{5}$

Note: Remember that the product of two numbers that have different signs is always a negative number.

Solution: $-\dfrac{2}{3} \cdot \dfrac{1}{5} = -\dfrac{2 \cdot 1}{3 \cdot 5} = -\dfrac{2}{15}$

Explanation: We multiplied numerator by numerator and denominator by denominator. Notice that $-\frac{2}{15}$ is in lowest terms.

c. $\dfrac{-5}{7} \cdot -\dfrac{9}{11}$

Solution: $\dfrac{-5}{7} \cdot \dfrac{9}{-11} = \dfrac{5 \cdot 9}{7 \cdot 11} = \dfrac{45}{77}$

Note: Remember that we can write $\dfrac{-5}{7}$ as $-\dfrac{5}{7}$ and $\dfrac{9}{-11}$ as $-\dfrac{9}{11}$, so both fractions are negative. The product of two numbers that have the same sign is always a positive number.

Explanation: We multiplied numerator by numerator and denominator by denominator. Notice that $\frac{45}{77}$ is in lowest terms.

◀ **Do Your Turn 1**

OBJECTIVE 2 Multiply and simplify fractions.

Consider $\dfrac{5}{6} \cdot \dfrac{3}{4}$.

$$\dfrac{5}{6} \cdot \dfrac{3}{4} = \dfrac{15}{24}$$

Notice that $\frac{15}{24}$ can be simplified because we can divide out the common factor 3.

$$\dfrac{5}{6} \cdot \dfrac{3}{4} = \dfrac{15}{24} = \dfrac{\overset{1}{\cancel{3}} \cdot 5}{2 \cdot 2 \cdot 2 \cdot \underset{1}{\cancel{3}}} = \dfrac{5}{8}$$

◀ **Note:** This is now in lowest terms.

Alternatively, because the original numerators are factors of the resulting numerator and the original denominators are factors of the resulting denominator, we can actually divide out common factors *before* multiplying. If we replace each numerator and denominator with its prime factorization, we can divide out the same factor of 3 as above and then multiply the remaining factors to get the same product as above.

$$\dfrac{5}{6} \cdot \dfrac{3}{4} = \dfrac{5}{2 \cdot \underset{1}{\cancel{3}}} \cdot \dfrac{\overset{1}{\cancel{3}}}{2 \cdot 2} = \dfrac{5 \cdot 1}{2 \cdot 1 \cdot 2 \cdot 2} = \dfrac{5}{8}$$

Answers to Your Turn 1
a. $\dfrac{21}{32}$ b. $-\dfrac{4}{45}$ c. $\dfrac{5}{36}$ d. $\dfrac{1}{20}$

Or, some prefer to divide out the common factor(s) this way:

$$\frac{5}{\overset{2}{\cancel{6}}} \cdot \frac{\overset{1}{\cancel{3}}}{4} = \frac{5}{8}$$

◄ **Note:** For now, we will write the numerators and denominators in prime factored form so that the factors are easier to see. In Sections 5.7 and 5.8, we'll leave out the prime factored form.

The following procedure summarizes our methods for multiplying.

PROCEDURE *To multiply fractions:*

Method 1:

1. Multiply the numerators and multiply the denominators.
2. Simplify the product.

Method 2:

1. Divide out any numerator factor with any like denominator factor.
2. Multiply the remaining factors.
3. Simplify as needed so that the result is in lowest terms. (This would be necessary if you miss common factors in the first step.)

Example 2 Multiply $\frac{6}{8} \cdot \frac{10}{15}$. Write the product in lowest terms.

Solution:

Method 1: Multiply straight across and then simplify:

$$\frac{6}{8} \cdot \frac{10}{15} = \frac{60}{120}$$ Multiply numerators and multiply denominators.

$$= \frac{\overset{1}{\cancel{2}} \cdot \overset{1}{\cancel{2}} \cdot \overset{1}{\cancel{3}} \cdot \overset{1}{\cancel{5}}}{\underset{1}{\cancel{2}} \cdot \underset{1}{\cancel{2}} \cdot 2 \cdot \underset{1}{\cancel{3}} \cdot \underset{1}{\cancel{5}}}$$ Divide out the common factors.

$$= \frac{1}{2}$$ Multiply the remaining factors.

Method 2: Divide out the common factors in the numerators and denominators before multiplying.

$$\frac{6}{8} \cdot \frac{10}{15} = \frac{2 \cdot 3}{2 \cdot 2 \cdot 2} \cdot \frac{2 \cdot 5}{3 \cdot 5}$$ Replace each numerator and denominator with its prime factorization.

$$= \frac{\overset{1}{\cancel{2}} \cdot \overset{1}{\cancel{3}}}{\underset{1}{\cancel{2}} \cdot 2 \cdot \underset{1}{\cancel{2}}} \cdot \frac{\overset{1}{\cancel{2}} \cdot \overset{1}{\cancel{5}}}{\underset{1}{\cancel{3}} \cdot \underset{1}{\cancel{5}}}$$ Divide out the common factors.

$$= \frac{1}{2}$$ Multiply the remaining factors.

CONNECTION No matter which method we use, we divide out the same common factors; in this case, two 2's, a 3 and a 5.

Do Your Turn 2 ▶

OBJECTIVE 3 Multiply mixed numbers.

How do we multiply mixed numbers? In Section 5.1, we learned how to write mixed numbers as improper fractions and we have just learned how to multiply fractions. The logical conclusion would be to write the mixed numbers as improper fractions, then multiply.

Your Turn 2

Multiply and write the product in lowest terms.

a. $\frac{5}{9} \cdot \frac{3}{10}$

b. $\frac{6}{8} \cdot \frac{12}{30}$

c. $-\frac{10}{15} \cdot \frac{6}{16}$

d. $\frac{-18}{20} \cdot \frac{30}{-32}$

Calculator Tip

On a scientific calculator, use the $\boxed{A^{b/c}}$ key to enter fractions. We would work Example 2 this way:

$6 \boxed{A^{b/c}} 8 \boxed{\times} 10 \boxed{A^{b/c}}$

$15 \boxed{\overset{\text{ENTER}}{=}}$

Answers to Your Turn 2

a. $\frac{1}{6}$ b. $\frac{3}{10}$ c. $-\frac{1}{4}$ d. $\frac{27}{32}$

Example 3 Estimate the product, then multiply. Write the answer as a mixed number in simplest form.

a. $3\dfrac{1}{6} \cdot 10\dfrac{1}{2}$

Solution: To estimate, we must round the numbers to comfortable amounts, then multiply.

Note: By *comfortable*, we mean numbers that are simple enough to work with in your head—whole numbers or simple fractions.

$$3\dfrac{1}{6} \cdot 10\dfrac{1}{2}$$

$3\frac{1}{6}$ rounds to 3 because $\frac{1}{6}$ is below the halfway point between 3 and 4.

$10\frac{1}{2}$ rounds to 11 because $\frac{1}{2}$ is at the halfway point between 10 and 11, and we must round up.

$$3 \cdot 11 = 33$$

Remember, the purpose of the estimate is to get an idea of what the actual product should be. We know that $3\dfrac{1}{6} \cdot 10\dfrac{1}{2}$ should be around 33. Now let's find the actual product.

$$3\dfrac{1}{6} \cdot 10\dfrac{1}{2} = \dfrac{19}{6} \cdot \dfrac{21}{2} \qquad \text{Write as improper fractions.}$$

$$= \dfrac{19}{2 \cdot \overset{1}{\cancel{3}}} \cdot \dfrac{\overset{1}{\cancel{3}} \cdot 7}{2} \qquad \text{Divide out a common factor of 3.}$$

$$= \dfrac{133}{4} \qquad \text{Multiply.}$$

$$= 33\dfrac{1}{4} \qquad \text{Write as a mixed number.}$$

Notice that our estimate is very close to the actual product.

b. $20 \cdot \dfrac{3}{8}$

Solution: To estimate, we round the fraction to a comfortable amount, then multiply.

$$20 \cdot \dfrac{3}{8}$$

No rounding necessary.

$\dfrac{3}{8}$ rounds to $\dfrac{1}{2}$.

Note: If we were to round $\frac{3}{8}$ to the nearest whole number, we would get $20 \cdot 0 = 0$. We can get a better estimate by rounding $\frac{3}{8}$ to $\frac{1}{2}$.

$$20 \cdot \dfrac{1}{2} = 10$$

▦ Calculator Tip

We would enter Example 3 this way:

| 10 | Aᵇ/c | 1 | Aᵇ/c | 2 |

| × | 3 | Aᵇ/c | 1 | Aᵇ/c |

| 6 | ENTER = |

On the screen you will see:

10⌊1⌊2 and 3⌊1⌊4

The answer will look like:

33⌊1/4, which represents $33\frac{1}{4}$.

Now let's find the actual product.

$$20 \cdot \frac{3}{8} = \frac{20}{1} \cdot \frac{3}{8}$$ **Write as improper fractions.**

$$= \frac{\overset{1}{2} \cdot \overset{1}{2} \cdot 5}{1} \cdot \frac{3}{2 \cdot 2 \cdot 2}$$ **Divide out the common factors, which are two 2's.**

$$= \frac{15}{2}$$ **Multiply.**

$$= 7\frac{1}{2}$$ **Write the improper fraction as a mixed number.**

Notice that our estimate is close to the actual product.

Do Your Turn 3 ▶

Your Turn 3

Estimate the product, then find the actual product.

a. $3\frac{1}{3} \cdot 1\frac{4}{5}$

b. $5\frac{2}{3} \cdot 2\frac{1}{7}$

c. $\frac{2}{5} \cdot 12$

d. $20 \cdot 3\frac{3}{4}$

Example 4 Multiply.

a. $-3\frac{1}{5} \cdot 2\frac{2}{9}$

Solution: We write the mixed numbers as improper fractions, then multiply.

$$-3\frac{1}{5} \cdot 2\frac{2}{9} = -\frac{16}{5} \cdot \frac{20}{9}$$ **Write as improper fractions.**

$$= -\frac{16}{\underset{1}{\cancel{5}}} \cdot \frac{2 \cdot 2 \cdot \overset{1}{\cancel{5}}}{9}$$ **Divide out the common factor 5.**

$$= -\frac{64}{9}$$ **Multiply.**

$$= -7\frac{1}{9}$$ **Write the improper fraction as a mixed number.**

CONNECTION In Chapter 2, we learned that the product of two numbers with different signs is negative.

b. $2\frac{3}{8} \cdot (-12)$

Solution: $2\frac{3}{8} \cdot (-12) = \frac{19}{8} \cdot \left(-\frac{12}{1}\right)$ **Write as improper fractions.**

$$= \frac{19}{\underset{1}{2} \cdot \underset{1}{2} \cdot 2} \cdot \left(-\frac{\overset{1}{2} \cdot \overset{1}{2} \cdot 3}{1}\right)$$ **Divide out the common factors of two 2's.**

$$= -\frac{57}{2}$$ **Multiply.**

$$= -28\frac{1}{2}$$ **Write the improper fraction as a mixed number.**

Do Your Turn 4 ▶

Your Turn 4

Multiply and simplify.

a. $-4\frac{1}{8} \cdot 5\frac{1}{2}$

b. $\left(-2\frac{3}{5}\right)\left(-7\frac{1}{4}\right)$

c. $\left(2\frac{4}{7}\right)\left(-11\frac{2}{5}\right)$

d. $\left(\frac{-4}{9}\right)(-21)$

Answers to Your Turn 3

a. 6; 6   b. 12; $12\frac{1}{7}$ c. 6; $4\frac{4}{5}$

d. 80; 75

Answers to Your Turn 4

a. $-22\frac{11}{16}$ b. $18\frac{17}{20}$

c. $-29\frac{11}{35}$ d. $9\frac{1}{3}$

OBJECTIVE 4 Multiply rational expressions.

We multiply rational expressions the same way that we multiply fractions. Although we can use either of the two methods we have learned, it is usually easiest to divide out the common factors first and then multiply (method 2).

Your Turn 5

Multiply.

a. $\dfrac{x}{6} \cdot \dfrac{9}{x^2}$

b. $\dfrac{10a^3}{3} \cdot \dfrac{12}{a}$

c. $\dfrac{18tu}{5} \cdot \left(-\dfrac{15}{27u^4}\right)$

d. $\dfrac{-6}{13hk^2} \cdot \left(-\dfrac{26h}{9k}\right)$

Example 5 Multiply. $-\dfrac{3a}{5b} \cdot \left(-\dfrac{10ab^2}{9}\right)$

Solution:

$$-\dfrac{3a}{5b} \cdot \left(-\dfrac{10ab^2}{9}\right) = -\dfrac{3 \cdot a}{5 \cdot b} \cdot \left(-\dfrac{2 \cdot 5 \cdot a \cdot b \cdot b}{3 \cdot 3}\right)$$

Write the prime factorization of each numerator and denominator.

$$= -\dfrac{\overset{1}{\cancel{3}} \cdot a}{\underset{1}{\cancel{5}} \cdot \underset{1}{\cancel{b}}} \cdot \left(-\dfrac{2 \cdot \overset{1}{\cancel{5}} \cdot a \cdot \overset{1}{\cancel{b}} \cdot b}{\underset{1}{\cancel{3}} \cdot 3}\right)$$

Divide out a common 3, a common 5, and a common b factor.

$$= \dfrac{2a^2 b}{3}$$

Multiply the remaining numerator and denominator factors.

CONNECTION In Chapter 2, we learned that the product of two numbers that have the same sign is positive.

◄ **Do Your Turn 5**

Calculator Tip

We would enter Example 6 this way:

(2 Ab/c 3)

^ 4 ENTER/=

If your calculator gives a decimal answer, then use the F ◄►D function to change the decimal back to a fraction. The F stands for fraction and the D stands for decimal. The double arrow in between indicates that pressing that key changes fractions to decimals or decimals to fractions.

On most scientific calculators the F ◄►D function is a secondary function for an existing key, which means you will have to first press the 2nd or shift key to use the function.

OBJECTIVE 5 Simplify fractions raised to a power.

Remember that an exponent indicates the number of times to use the base as a factor.

Example 6 Evaluate. $\left(\dfrac{2}{3}\right)^4$

Solution: We write the base $\frac{2}{3}$ as a factor four times, then multiply.

$$\left(\dfrac{2}{3}\right)^4 = \dfrac{2}{3} \cdot \dfrac{2}{3} \cdot \dfrac{2}{3} \cdot \dfrac{2}{3} = \dfrac{16}{81}$$

Explanation: There are four 2's that multiply together to make 16 in the numerator and four 3's that multiply to make 81 in the denominator. We can say

$$\left(\dfrac{2}{3}\right)^4 = \dfrac{2^4}{3^4}$$

Example 6 suggests the following rule.

RULE

When a fraction is raised to a power, we evaluate both the numerator and denominator raised to that power.

In math language: $\left(\dfrac{a}{b}\right)^n = \dfrac{a^n}{b^n}$,   when $b \neq 0$

Answers to Your Turn 5

a. $\dfrac{3}{2x}$ b. $40a^2$ c. $-\dfrac{2t}{u^3}$ d. $\dfrac{4}{3k^3}$

WARNING

$\dfrac{2^4}{3}$ and $\left(\dfrac{2}{3}\right)^4$ are not the same.

$$\dfrac{2^4}{3} = \dfrac{2 \cdot 2 \cdot 2 \cdot 2}{3} = \dfrac{16}{3} \text{ while } \left(\dfrac{2}{3}\right)^4 = \dfrac{2 \cdot 2 \cdot 2 \cdot 2}{3 \cdot 3 \cdot 3 \cdot 3} = \dfrac{16}{81}.$$

Do Your Turn 6 ▷

Example 7 Simplify. $\left(\dfrac{-2x^2}{5y}\right)^3$

Solution:

$$\left(\dfrac{-2x^2}{5y}\right)^3 = \dfrac{(-2x^2)^3}{(5y)^3}$$

Write both the numerator and denominator raised to the third power.

$$= \dfrac{(-2x^2) \cdot (-2x^2) \cdot (-2x^2)}{5y \cdot 5y \cdot 5y}$$

Write $-2x^2$ as a factor three times and $5y$ as a factor three times.

$$= \dfrac{-8x^6}{125y^3}$$

Multiply coefficients and add exponents for the like bases.

We could have also simplified this expression using our procedure for simplifying a monomial raised to a power.

$$\left(\dfrac{-2x^2}{5y}\right)^3 = \dfrac{(-2x^2)^3}{(5y)^3}$$

Write both the numerator and denominator raised to the 3rd power.

$$= \dfrac{(-2)^3 x^{2 \cdot 3}}{5^3 y^{1 \cdot 3}}$$

Write the coefficient to the 3rd power and multiply each variables, exponent by the power.

$$= \dfrac{-8x^6}{125y^3}$$

Simplify.

CONNECTION In Section 3.5, we learned how to multiply monomials and how to simplify a monomial raised to a power.

Do Your Turn 7 ▷

OBJECTIVE 6 Solve applications involving multiplying fractions.

One of the difficulties in translating application problems that involve fractions is identifying the arithmetic operation. To identify multiplication, look for language that indicates we are to find a fraction *of* an amount. Whenever the word *of* is directly preceded by a fraction and followed by an amount, we multiply the amount by the fraction. Consider the sentence:

$$\dfrac{3}{4} \text{ of the 28 people in the class got an A.}$$

Since *of* is preceded by the fraction $\frac{3}{4}$ and followed by the amount 28, to calculate the number of people in the class who got an A, we multiply 28 by $\frac{3}{4}$.

Your Turn 6

Evaluate.

a. $\left(\dfrac{2}{5}\right)^3$

b. $\left(\dfrac{1}{2}\right)^6$

c. $\left(-\dfrac{3}{4}\right)^3$

d. $\left(\dfrac{-5}{6}\right)^2$

Your Turn 7

Simplify.

a. $\left(\dfrac{7x^3}{9}\right)^2$

b. $\left(\dfrac{3x^2}{4z}\right)^3$

c. $\left(-\dfrac{2a^4}{3c^3}\right)^4$

d. $\left(\dfrac{-m^3}{2p}\right)^5$

Answers to Your Turn 6

a. $\dfrac{8}{125}$ b. $\dfrac{1}{64}$ c. $-\dfrac{27}{64}$ d. $\dfrac{25}{36}$

Answers to Your Turn 7

a. $\dfrac{49x^6}{81}$ b. $\dfrac{27x^6}{64z^3}$

c. $\dfrac{16a^{16}}{81c^{12}}$ d. $\dfrac{-m^{15}}{32p^5}$

The word *of* does not always mean multiply. Sometimes *of* indicates a fraction. Consider the sentence:

21 out of 28 people in the class got an A.   or   21 of 28 people in the class got an A.

In *21 of 28*, the word *of* indicates the fraction $\frac{21}{28}$. Whenever the word *of* is preceded by a whole number, it indicates a fraction.

Conclusion: The word *of* indicates multiplication when preceded by a fraction and indicates a fraction when preceded by a whole number.

The two uses of the word *of* are related.

$\frac{3}{4}$ of the 28 people in the class got an A.

$\frac{3}{4} \cdot 28 = $ the number of people that got an A

$$\frac{3}{\cancel{4}} \cdot \frac{\overset{7}{\cancel{28}}}{1} = 21$$

21 of 28 people in the class got an A.

$\frac{21}{28} = $ the fraction of the people in the class that got an A

$$\frac{21}{28} = \frac{3}{4}$$

Your Turn 8

Solve.

a. A doctor says that 8 out of 10 patients receiving a particular treatment have no side effects. If a hospital gives 600 patients this treatment, how many can be expected to have no side effects? How many can be expected to have side effects?

b. A population study finds that there are $2\frac{1}{3}$ children for each household in a given area. If there are 6252 households in the area, how many children are there?

DISCUSSION   How can there be $2\frac{1}{3}$ children per household?

Example 8   An advertisement claims that 4 out of 5 dentists choose Crest toothpaste. If you were to visit a dental conference that has 345 dentists in attendance, how many would you expect to choose Crest toothpaste based on the ad's claim?

Understand: 4 out of 5 is the fraction $\frac{4}{5}$. If the claim of the ad is accurate, then we can say:

$\frac{4}{5}$ of the 345 dentists at the conference should choose Crest.

Plan: We can translate our sentence to an equation and solve.

Execute: $\frac{4}{5}$ of the 345 dentists at the conference should choose Crest.

$$\frac{4}{5} \cdot 345 = n$$

Note: The variable *n* represents the number of dentists out of the 345 dentists who should choose Crest.

$$\frac{4}{5} \cdot \frac{345}{1} = n$$

$$\frac{4}{\cancel{5}} \cdot \frac{3 \cdot \overset{1}{\cancel{5}} \cdot 23}{1} = n$$

$$276 = n$$

Answer: If we were to survey the dentists at the conference, we should find that out of the 345 dentists, 276 would choose Crest.

DISCUSSION   How many of the 345 dentists would you expect to choose a brand of toothpaste other than Crest? What if you conducted the survey and found an amount other than 276?

Check: 276 out of 345 dentists should be the same ratio as 4 out of 5. In other words, $\frac{276}{345}$ should be the same fraction as $\frac{4}{5}$. We can verify by reducing $\frac{276}{345}$ to $\frac{4}{5}$.

◀ **Do Your Turn 8**

Answers to Your Turn 8
a. 480; 120 b. 14,588

Example 9 $\frac{3}{4}$ of the students taking a particular history course passed the course. Of these, $\frac{1}{3}$ got an A. What fraction of all students taking the course got an A?

Understand: The students that got an A are a fraction of the group that passed. We can say:

$$\frac{1}{3} \text{ of the } \frac{3}{4} \text{ that passed got an A.}$$

Plan: We can translate the sentence to an equation, then solve.

Execute: Let *a* represent the number of students that received an A.

$$\frac{1}{3} \underset{\big\downarrow}{\text{ of }} \text{ the } \frac{3}{4} \text{ that passed got an A.}$$

$$\frac{1}{3} \cdot \frac{3}{4} = a$$

◄ **Note:** The variable *a* represents the fraction of the students in the course that received an A.

$$\frac{1}{\cancel{3}} \cdot \frac{\cancel{3}}{4} = a$$

$$\frac{1}{4} = a$$

Answer: $\frac{1}{4}$ of all students in the class not only passed, but got an A.

Check: We can use a picture to check.

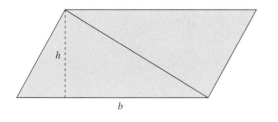

$\frac{3}{4}$ that passed

$\frac{1}{3}$ of the $\frac{3}{4}$ ⟶

◄ **Note:** The fact that there are 12 divisions in the picture does not necessarily mean that there are 12 students total in the class.

The darker shaded region is $\frac{3}{12}$ of the whole picture, which simplifies to $\frac{1}{4}$.

Do Your Turn 9 ▷

Your Turn 9

Solve.

a. A study estimates that $\frac{2}{3}$ of all Americans own a car and $\frac{1}{4}$ of these cars are blue. What fraction of the population drives a blue car?

b. A survey of an area indicates that $\frac{5}{8}$ of the people responding to it said they watch more than 2 hours of television each day. Of these, $\frac{7}{10}$ were female. What fraction of all those surveyed were females who watch more than 2 hours of television each day?

DISCUSSION Errors were made in conducting this survey. What factors might have skewed the results?

OBJECTIVE 7 Calculate the area of a triangle.

In Section 1.6, we learned that the area of a parallelogram is $A = bh$. We can use this fact to develop the formula for the area of a triangle. By joining any two identical triangles as shown, we can create a parallelogram.

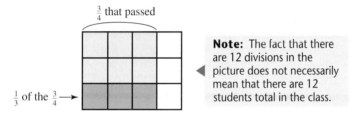

◄ **Note:** Because the triangles are identical, their angle measurements are identical. Rotating one triangle and joining it to the other along their corresponding side forms a four-sided figure with opposing angles that have the same measurement and with opposing sides that are parallel. Therefore, the figure is a parallelogram.

Answers to Your Turn 9
a. $\frac{1}{6}$ b. $\frac{7}{16}$

Since the triangles are identical, one of them occupies half of the area of the parallelogram.

area of a triangle = half of the area of a parallelogram with the same base and height

$$A = \frac{1}{2} \cdot bh$$

Your Turn 10

Calculate the area of each triangle.

a.

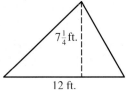

7¼ ft.

12 ft.

b.

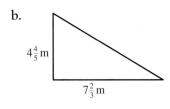

$4\frac{4}{5}$ m

$7\frac{2}{3}$ m

Example 10 A recycling company produces triangular stickers that are $5\frac{1}{2}$ inches along the base and $4\frac{3}{4}$ inches high. Find the area.

Solution: Use the formula $A = \frac{1}{2}bh$.

$4\frac{3}{4}$in.

$5\frac{1}{2}$in.

$A = \frac{1}{2}\left(5\frac{1}{2}\right)\left(4\frac{3}{4}\right)$ **Replace b with $5\frac{1}{2}$ and h with $4\frac{3}{4}$.**

$A = \frac{1}{2}\left(\frac{11}{2}\right)\left(\frac{19}{4}\right)$ **Write the mixed numbers as improper fractions**

$A = \frac{209}{16}$ **Multiply.**

$A = 13\frac{1}{16}$ **Write the result as a mixed number.**

Answer: The area is $13\frac{1}{16}$ in.²

◀ **Do Your Turn 10**

OBJECTIVE 8 Calculate the radius and diameter of a circle.

Now let's learn about **circles.** The distance from the center to any point on a circle is the same for all points on the circle. This distance is the called **radius** (plural *radii*). The distance across a circle along a straight line that passes through the center is the **diameter.**

DEFINITIONS   **Circle:** A collection of points that are equally distant from a central point, called the *center.*

Radius: The distance from the center to any point on the circle.

Diameter: The distance across a circle along a straight line through the center.

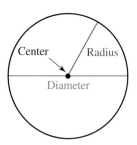

Center   Radius

Diameter

Answers to Your Turn 10

a. $43\frac{1}{2}$ ft.²   **b.** $18\frac{2}{5}$ m²

We can state the following relationships for diameter and radius.

The diameter is twice the radius.
The radius is half of the diameter.

We can translate these relationships to equations.

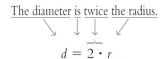

$$d = 2 \cdot r$$

$$r = \frac{1}{2} \cdot d$$

Example 11 Calculate the diameter of the circle shown.

Solution: The radius is $3\frac{5}{8}$ centimeters. Diameter is twice the radius, so we can use $d = 2r$.

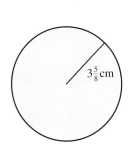

$d = 2 \cdot 3\frac{5}{8}$ In $d = 2r$, replace r with $3\frac{5}{8}$.

$d = \frac{\overset{1}{2}}{1} \cdot \frac{29}{\underset{4}{8}}$ Write the mixed number as an improper fraction and divide out the common factor 2.

$d = \frac{29}{4}$ Multiply.

$d = 7\frac{1}{4}$ Write the result as a mixed number.

Answer: The diameter is $7\frac{1}{4}$ cm.

Do Your Turn 11 ▷

Your Turn 11

Solve.

a. Calculate the diameter of a circle with a radius of $2\frac{3}{5}$ feet.

b. Calculate the diameter of a circle with a radius of $\frac{7}{8}$ inches.

Example 12 Calculate the radius of the circle shown.

Solution: The diameter is 17 feet. Radius is half of the diameter so we can use $r = \frac{1}{2}d$.

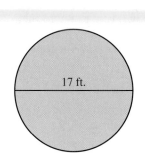

$r = \frac{1}{2} \cdot 17$ In $r = \frac{1}{2}d$, replace d with 17.

$r = \frac{1}{2} \cdot \frac{17}{1}$ Write 17 as an improper fraction.

$r = \frac{17}{2}$ Multiply.

$r = 8\frac{1}{2}$ Write the result as a mixed number.

Answer: The radius is $8\frac{1}{2}$ ft.

Do Your Turn 12 ▷

Your Turn 12

Solve.

a. Calculate the radius of a circle with a diameter of 25 feet.

b. Calculate the radius of a circle with a diameter of $9\frac{3}{8}$ meters.

Answers to Your Turn 11

a. $5\frac{1}{5}$ ft.   **b.** $1\frac{3}{4}$ in.

Answers to Your Turn 12

a. $12\frac{1}{2}$ ft.   **b.** $4\frac{11}{16}$ m

OBJECTIVE 🔟 Calculate the circumference of a circle.

Now that we have learned how diameter and radius are related, we can learn about **circumference.**

DEFINITION   Circumference: The distance around a circle.

When ancient mathematicians began experimenting with circles, they found an interesting property. When they measured the circumference and diameters of different-size circles and expressed the ratio of the circumference to diameter, they found the fractions were very close no matter how large or small the circles. The fractions were so close that mathematicians concluded the ratio must actually be the same for all circles and the slight differences could be accounted for in their inability to accurately measure the circumference and diameter. It turns out that the value of this fraction cannot be expressed exactly, only closely approximated. We call a number like this an **irrational number.**

DEFINITION   Irrational number: A number that cannot be expressed exactly as a fraction.

In math language: A number that cannot be expressed in the form $\frac{a}{b}$, where a and b are integers and $b \neq 0$.

Because the ratio of the circumference to the diameter is an irrational number, mathematicians decided to use a symbol to represent the value. They chose the Greek letter π (pronounced *pi*).

DEFINITION π: An irrational number that is the ratio of the circumference of a circle to its diameter.

In math language: $\pi = \dfrac{C}{d}$, where C represents the circumference and d represents the diameter of any circle.

A circle with a diameter of 7 inches will have a circumference of about 22 inches. This means that the value of π is approximately $\frac{22}{7}$. Symbolically, we write $\pi \approx \frac{22}{7}$. The symbol $\approx$ is read, "is approximately."

Remember, $\frac{22}{7}$ is only an approximation for the value of π. If we were to try to measure the circumference of a circle with diameter 7 inches, we would find the distance to be close to 22 inches but not exactly. As we wrap the measuring tape around the circle, we would find the circumference mark would never line up with an exact mark on the measuring tape, no matter how precisely the tape could measure.

We can now use this approximate value of π to calculate circumference, given the diameter or radius of a circle. We can rearrange the relationship and say

$$C = \pi d$$

Or, since $d = 2r$, we can say

$$C = 2\pi r$$

Example 13 Calculate the circumference of the circle shown.

Solution: We can use the formula $C = \pi d$ or $C = 2\pi r$. Because we were given radius, let's use $C = 2\pi r$.

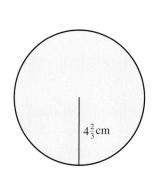

$4\frac{2}{3}$ cm

$$C \approx 2 \cdot \frac{22}{7} \cdot 4\frac{2}{3}$$

In $C = 2\pi r$, replace π with $\frac{22}{7}$ and r with $4\frac{2}{3}$.

$$C \approx \frac{2}{1} \cdot \frac{22}{\cancel{7}_1} \cdot \frac{\cancel{14}^2}{3}$$

Write as improper fractions and divide out a common factor of 7.

$$C \approx \frac{88}{3}$$

Multiply.

$$C \approx 29\frac{1}{3}$$

Write as a mixed number.

Answer: The circumference is $29\frac{1}{3}$ cm.

Note: When we replace π with $\frac{22}{7}$, the calculation becomes an approximation, so we use the $\approx$ symbol instead of an equal sign.

Do Your Turn 13 ▶

Your Turn 13

Solve.

a. Calculate the circumference of a circle with a radius of $1\frac{1}{6}$ feet.

b. Calculate the circumference of a circle with a diameter of $5\frac{5}{6}$ meters.

Answers to Your Turn 13

a. $7\frac{1}{3}$ ft.   b. $18\frac{1}{3}$ m

1. Explain how to multiply two fractions.

2. Explain how to multiply two mixed numbers.

3. What is the formula $A = \dfrac{1}{2}bh$ used for, and what does each variable represent?

4. What are the radius and diameter of a circle?

5. What is the circumference of a circle?

6. What is an irrational number?

7. What is π?

8. What are the two formulas for calculating the circumference of a circle?

For Exercises 9–16, multiply.

9. $\dfrac{2}{5} \cdot \dfrac{4}{7}$

10. $\dfrac{5}{8} \cdot \dfrac{1}{4}$

11. $-\dfrac{1}{6} \cdot \dfrac{1}{9}$

12. $-\dfrac{1}{4} \cdot \dfrac{1}{10}$

13. $-\dfrac{3}{4} \cdot \left(-\dfrac{5}{7}\right)$

14. $-\dfrac{9}{10} \cdot \left(-\dfrac{3}{10}\right)$

15. $\dfrac{7}{100} \cdot \dfrac{7}{10}$

16. $\dfrac{6}{11} \cdot \dfrac{10}{19}$

For Exercises 17–30, multiply and express the product in lowest terms.

17. $\dfrac{6}{7} \cdot \dfrac{14}{15}$

18. $\dfrac{5}{9} \cdot \dfrac{12}{25}$

19. $-\dfrac{8}{12} \cdot \dfrac{10}{20}$

20. $-\dfrac{14}{18} \cdot \dfrac{9}{21}$

21. $\dfrac{24}{32} \cdot \dfrac{26}{30}$

22. $\dfrac{15}{33} \cdot \dfrac{22}{27}$

23. $\dfrac{16}{38} \cdot \left(\dfrac{-57}{80}\right)$

24. $\dfrac{36}{42} \cdot \left(-\dfrac{28}{45}\right)$

25. $\dfrac{19}{20} \cdot \dfrac{16}{38}$

26. $\dfrac{25}{34} \cdot \dfrac{17}{30}$

27. $-\dfrac{18}{40} \cdot \left(-\dfrac{28}{30}\right)$

28. $\dfrac{-22}{35} \cdot \left(\dfrac{15}{-33}\right)$

29. $\dfrac{36}{-40} \cdot \left(\dfrac{-30}{54}\right)$

30. $\dfrac{60}{-81} \cdot \dfrac{45}{72}$

For Exercises 31–42, estimate the product, then multiply and write the product as a mixed number in simplest form.

31. $3\frac{1}{5} \cdot 1\frac{3}{4}$

32. $4\frac{1}{6} \cdot 3\frac{2}{5}$

33. $\frac{5}{6}(28)$

34. $12 \cdot \frac{7}{8}$

35. $4\frac{5}{8} \cdot 16$

36. $27 \cdot 2\frac{4}{9}$

37. $-6\frac{1}{8} \cdot \frac{4}{7}$

38. $2\frac{7}{10} \cdot \left(-8\frac{1}{3}\right)$

39. $-5\frac{2}{3} \cdot \left(-7\frac{1}{5}\right)$

40. $-3\frac{9}{16} \cdot \left(-6\frac{1}{8}\right)$

41. $\frac{-3}{10}(45)$

42. $-18 \cdot \frac{7}{24}$

For Exercises 43–54, multiply and express the product in lowest terms.

43. $\frac{x^2}{5} \cdot \frac{2}{3}$

44. $\frac{3y}{7} \cdot \frac{2}{5}$

45. $\frac{4x}{9} \cdot \frac{3x}{8}$

46. $\frac{2}{15a} \cdot \frac{5}{6a^2}$

47. $\frac{xy}{10} \cdot \frac{4y^3}{9}$

48. $\frac{3m^2}{14} \cdot \frac{7mn}{12}$

49. $-\frac{5hk^3}{9} \cdot \frac{3}{4h}$

50. $\frac{9u}{20t^2} \cdot \left(-\frac{4tu^2}{15}\right)$

51. $-\frac{10x^4y}{11z} \cdot \left(-\frac{22z}{14x^2}\right)$

52. $\frac{-a^4b^3}{5c^3} \cdot \frac{10c^5}{9b}$

53. $\frac{9m^3}{25n^4} \cdot \left(\frac{-15n}{18m^2}\right)$

54. $-\frac{2t^6}{28t} \cdot \left(\frac{-14u^3}{18uv}\right)$

For Exercises 55–66, simplify.

55. $\left(\frac{5}{6}\right)^2$

56. $\left(\frac{4}{9}\right)^2$

57. $\left(-\frac{3}{4}\right)^3$

58. $\left(-\frac{1}{5}\right)^3$

59. $\left(\frac{-1}{2}\right)^6$

60. $\left(\frac{-2}{3}\right)^4$

61. $\left(\frac{x}{2}\right)^3$

62. $\left(\frac{3}{y}\right)^4$

63. $\left(\frac{2x^2}{3}\right)^3$

64. $\left(\frac{t^4}{4y}\right)^3$

65. $\left(-\frac{m^3n}{3p^2}\right)^4$

66. $\left(\frac{-2xy^4}{z^3}\right)^5$

For Exercises 67–76, solve. Write all answers in lowest terms.

67. A shipping company estimates that $\frac{1}{8}$ of the produce shipped will go bad during shipment. If a single truck carries about 12,480 pieces of fruit, how many can be expected to go bad during shipment?

68. A single share of a certain computer company's stock is listed at $6\frac{3}{8}$ per share. What would be the total value of 500 shares of the stock?

69. $\frac{5}{6}$ of a company's employees live within a 15-mile radius of the company. Of these, $\frac{3}{4}$ live within a 10-mile radius. What fraction of all employees live within a 10-mile radius?

70. $\frac{3}{4}$ of a lot is to be landscaped. $\frac{2}{3}$ of the landscaped area is to be covered in sod. What fraction of the lot is to be covered in sod?

71. On Tanya's phone bill, 24 out of the 30 long-distance calls were in state. $\frac{5}{6}$ of the in-state calls were to her parents. What fraction of all her long-distance calls were to her parents? How many calls were to her parents?

72. 16 out of the 24 people enrolled in a CPR course plan to go into a health-related field. $\frac{5}{8}$ of the people who intend to pursue a health field plan to go into nursing. What fraction of all people enrolled plan to become nurses? How many people is this?

73. On a standard-size guitar, the length of the strings between the bridge and nut is $25\frac{1}{2}$ inches. Guitar makers must place the 12th fret at exactly half of the length of the string. How far from the saddle or nut should one measure to place the 12th fret?

74. A recipe calls for $1\frac{1}{3}$ cups of sugar. How much sugar should be used in making half of the recipe?

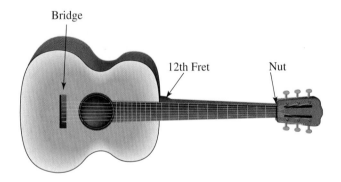

Bridge

12th Fret   Nut

75. A hand weight has a mass of $15\frac{1}{10}$ kilograms. What is its weight if the acceleration due to gravity is $-9\frac{4}{5}$ meters per second squared (m/sec.^2)?

CONNECTION Exercises 75 and 76 are similar to exercises in Section 2.6.

CONNECTION We treat units like rational expressions:

$$\text{kg} \cdot \text{m/sec.}^2 = \frac{\text{kg}}{1} \cdot \frac{\text{m}}{\text{sec.}^2} = \frac{\text{kg} \cdot \text{m}}{\text{sec.}^2}$$

$\dfrac{\text{kg} \cdot \text{m}}{\text{sec.}^2}$ is the unit for force. It was renamed the newton (N) in recognition of Isaac Newton's development of the mathematics of force.

76. The Liberty Bell has a mass of $64\frac{3}{5}$ slugs. How much does it weigh if the acceleration due to gravity is $-32\frac{1}{5}$ feet per second squared (ft./sec.^2)?

Note: Remember that the slug is the American unit for mass.

CONNECTION When we used $F = ma$ in the past, we said the values for the acceleration due to gravity were -10 meters per second squared (m/sec.^2) for metric and -32 feet per second squared (ft./sec.^2) for American. Now that we have developed fractions, we can use more accurate values.

For Exercises 77–78, find the area of each triangle.

77.

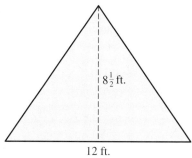

$8\frac{1}{2}$ ft.

12 ft.

78.

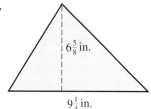

$6\frac{5}{8}$ in.

$9\frac{1}{4}$ in.

For Exercises 79–82, solve. Use $\frac{22}{7}$ for π.

79. A giant sequoia tree has a diameter of 9 meters.
  a. What is its radius?
  b. What is the circumference?

80. Big Ben is the name of the clock in the tower of London's Westminster Palace. The face of the clock is a circle with a diameter of 23 feet.
  a. What is the radius?
  b. What is the circumference?

81. The Fermi National Accelerator Laboratory is a circular tunnel that is used to accelerate elementary particles. The radius of the tunnel is $\frac{273}{440}$ miles.
  a. What is the diameter?
  b. What is the circumference?

82. The radius of the bottom of the Liberty Bell is $22\frac{87}{88}$ inches.
  a. What is the diameter?
  b. What is the circumference?

PUZZLE PROBLEM The Milky Way galaxy in which our solar system exists is a disk shape with spiral arms. The diameter of the galaxy is approximately 100,000 light-years. If our solar system is about $\frac{2}{3}$ of the distance from the center to the edge of the galaxy, then how far is our solar system from the center?

Review Exercises

[2.4]   **1.** Evaluate. $1836 \div (-9)$

$\begin{bmatrix} 1.4 \\ 2.4 \end{bmatrix}$ **2.** Evaluate. $\sqrt{256}$

[3.7]   **3.** Factor. $9x^2 - 12x$

[2.4]   **4.** Solve and check. $-6x = 30$

[1.4]   **5.** A 12-foot board is to be cut into 6 pieces. How long will each piece be?

5.4 Dividing Fractions, Mixed Numbers, and Rational Expressions

OBJECTIVE 1 Divide fractions.

Suppose we want change for a $20 bill in $5 bills. We can determine the number of $5 bills we receive using division. Since $20 \div 5 = 4$, we would receive four $5 bills.

Now suppose we want the change in quarters. To determine the number of quarters in $20, we can use a division statement similar to the one we used for finding the number of $5 bills in $20. Remember, a quarter is $\frac{1}{4}$ of a dollar, so this division statement is:

$$\text{number of quarters in } \$20 = 20 \div \frac{1}{4}$$

Alternatively, since there are 4 quarters in each dollar, we can find the number of quarters in $20 by multiplying 20 by 4.

$$\text{number of quarters in } \$20 = 20 \cdot 4 = 80 \text{ quarters}$$

This means $20 \div \frac{1}{4}$ and $20 \cdot 4$ are equivalent.

$$\text{number of quarters in } \$20 = 20 \div \frac{1}{4} = 20 \cdot 4 = 80 \text{ quarters}$$

In our example, $\frac{1}{4}$ and 4 are **multiplicative inverses** or **reciprocals.**

> **DEFINITION** **Reciprocals:** Two numbers whose product is 1.

Notice that the product of $\frac{1}{4}$ and 4 is 1.

$$\frac{1}{4} \cdot 4 = \frac{1}{\cancel{4}} \cdot \frac{\cancel{4}^{1}}{1} = 1$$

CONNECTION Remember that we use additive inverses to write subtraction statements as equivalent addition statements. Similarly, we use *multiplicative inverses* to write division statements as equivalent multiplication statements. Also, additive inverses *undo* one another in that their sum is 0. Similarly, multiplicative inverses *undo* one another in that their product is 1.

Example 1 Find the reciprocal.

a. $\dfrac{3}{4}$

Answer: $\dfrac{4}{3}$

Explanation: The product of $\frac{3}{4}$ and $\frac{4}{3}$ is 1.

b. $\dfrac{1}{5}$

Answer: 5

Explanation: The product of $\frac{1}{5}$ and 5 is 1.

c. 6

Answer: $\dfrac{1}{6}$

Explanation: The product of 6 and $\frac{1}{6}$ is 1.

Note: To find the reciprocal, *invert* the numerator and denominator. In other words, write the numerator in the denominator and the denominator in the numerator.

$$\frac{\cancel{3}^{1}}{\cancel{4}_{1}} \cdot \frac{\cancel{4}^{1}}{\cancel{3}_{1}} = 1$$

$$\frac{1}{5} \cdot 5 = \frac{1}{\cancel{5}} \cdot \frac{\cancel{5}^{1}}{1} = 1$$

$$6 \cdot \frac{1}{6} = \frac{\cancel{6}^{1}}{1} \cdot \frac{1}{\cancel{6}} = 1$$

d. $-\dfrac{5}{7}$

Answer: $-\dfrac{7}{5}$ ◀ **Note:** The sign stayed the same. The reciprocal of a number has the same sign as the number.

Explanation: The product of $-\frac{5}{7}$ and $-\frac{7}{5}$ is 1.
$$-\dfrac{\overset{1}{\cancel{5}}}{\underset{1}{\cancel{7}}}\left(-\dfrac{\overset{1}{\cancel{7}}}{\underset{1}{\cancel{5}}}\right) = 1$$

e. 0

Answer: 0 has no reciprocal.

Explanation: The product of 0 and any number is 0, hence the product can never equal 1.

Do Your Turn 1 ▶

We use reciprocals in our procedure for dividing fractions.

PROCEDURE *To divide fractions:*
1. Change the operation symbol from division to multiplication and change the divisor to its reciprocal.
2. Divide out any numerator factor with any like denominator factor.
3. Multiply numerator by numerator and denominator by denominator.
4. Simplify as needed.

Example 2 Divide.

a. $\dfrac{5}{8} \div \dfrac{3}{4}$

Solution: $\dfrac{5}{8} \div \dfrac{3}{4} = \dfrac{5}{8} \cdot \dfrac{4}{3}$ **Write an equivalent multiplication statement.**

$$= \dfrac{5}{\underset{1}{2} \cdot \underset{1}{2} \cdot 2} \cdot \dfrac{\overset{1}{2} \cdot \overset{1}{2}}{3}$$ **Divide out the common factors.**

$$= \dfrac{5}{6}$$ **Multiply.**

b. $\dfrac{-4}{9} \div (-12)$

Solution: $\dfrac{-4}{9} \div (-12) = \dfrac{-4}{9} \cdot \left(-\dfrac{1}{12}\right)$ **Write an equivalent multiplication statement.**

$$= \dfrac{-\overset{1}{2} \cdot \overset{1}{2}}{9} \cdot \left(-\dfrac{1}{\underset{1}{2} \cdot \underset{1}{2} \cdot 3}\right)$$ **Divide out the common factors.**

$$= \dfrac{1}{27}$$ **Multiply.**

WARNING It may be tempting to divide out common factors in the division statement, but we can divide out common factors *only* when multiplying.

Your Turn 1

Find the reciprocal.

a. $\dfrac{3}{8}$

b. $\dfrac{1}{4}$

c. -2

d. $-\dfrac{5}{11}$

CONNECTION The quotient is always positive when dividing two numbers that have the same sign.

Answers to Your Turn 1

a. $\dfrac{8}{3}$ b. 4 c. $-\dfrac{1}{2}$ d. $-\dfrac{11}{5}$

Your Turn 2

Divide.

a. $\dfrac{3}{16} \div \dfrac{5}{12}$

b. $\dfrac{\frac{1}{14}}{\frac{1}{7}}$

c. $\dfrac{5}{18} \div \left(-\dfrac{2}{9}\right)$

d. $\dfrac{-1}{5} \div (-4)$

Recall that we can write division using a fraction line. For example, we can write $18 \div 3 = 6$ as $\frac{18}{3} = 6$. If we express the division of fractions using fraction form, we call the expression a **complex fraction.**

DEFINITION   Complex fraction: An expression that is a fraction with fractions in the numerator and/or denominator.

For example, $\dfrac{\frac{5}{8}}{\frac{3}{4}}$ is a complex fraction equivalent to $\dfrac{5}{8} \div \dfrac{3}{4}$.

Because most people prefer the regular division form, if we are given a complex fraction, we rewrite it in that form.

◀ **Do Your Turn 2**

OBJECTIVE 2 Divide mixed numbers.

We learned in Section 5.3 that when we multiply mixed numbers, we first write them as improper fractions. The same process applies to division.

PROCEDURE   *To divide mixed numbers:*
1. Write the mixed numbers as improper fractions.
2. Write the division statement as an equivalent multiplication statement using the reciprocal of the divisor.
3. Divide out any numerator factor with any like denominator factor.
4. Multiply.
5. Simplify as needed.

Example 3   Estimate, then find the quotient. Write the answer as a mixed number in simplest form.

a. $8\dfrac{2}{5} \div 2\dfrac{1}{4}$

Solution: To estimate, we round the fraction to a comfortable amount, then divide.

$$8\dfrac{2}{5} \div 2\dfrac{1}{4}$$

$8\frac{2}{5}$ rounds to 8 because $\frac{2}{5}$ is below the halfway point between 8 and 9.

$2\frac{1}{4}$ rounds to 2 because $\frac{1}{4}$ is below the halfway point between 2 and 3.

$$8 \div 2 = 4$$

Now let's find the actual product.

$$8\dfrac{2}{5} \div 2\dfrac{1}{4} = \dfrac{42}{5} \div \dfrac{9}{4}$$ Write the mixed numbers as improper fractions.

$$= \dfrac{42}{5} \cdot \dfrac{4}{9}$$ Write as equivalent multiplication.

$$= \dfrac{2 \cdot \cancel{3} \cdot 7}{5} \cdot \dfrac{4}{\cancel{3} \cdot 3}$$ Divide out the common factor 3.

$$= \dfrac{56}{15}$$ Multiply.

$$= 3\dfrac{11}{15}$$ Write the improper fraction as a mixed number.

Notice that our estimate is close to the actual quotient.

Answers to Your Turn 2

a. $\dfrac{9}{20}$ b. $\dfrac{1}{2}$ c. $-1\dfrac{1}{4}$ d. $\dfrac{1}{20}$

b. $\dfrac{-5\frac{3}{7}}{3\frac{1}{3}}$

Solution: Write an equivalent division statement. To estimate, we round the fraction to a comfortable amount, then divide.

$$\dfrac{-5\frac{3}{7}}{3\frac{1}{3}} = -5\frac{3}{7} \div 3\frac{1}{3}$$

$-5\frac{3}{7}$ rounds to -5. $3\frac{1}{3}$ rounds to 3.

$$-5 \div 3 = -\frac{5}{3} = -1\frac{2}{3}$$

Now let's find the actual product.

$$-5\frac{3}{7} \div 3\frac{1}{3} = -\frac{38}{7} \div \frac{10}{3}$$ 　Write the mixed numbers as improper fractions.

$$= -\frac{38}{7} \cdot \frac{3}{10}$$ 　Write as equivalent multiplication.

$$= -\frac{\overset{1}{2} \cdot 19}{7} \cdot \frac{3}{\underset{1}{2 \cdot 5}}$$ 　Divide out the common factor 2.

$$= -\frac{57}{35}$$ 　Multiply.

$$= -1\frac{22}{35}$$ 　Write the improper fraction as a mixed number.

Notice that our estimate is close to the actual quotient.

c. $4\dfrac{3}{5} \div \dfrac{3}{10}$

Solution: To estimate, we round the fraction to a comfortable amount, then divide.

$$4\frac{3}{5} \div \frac{3}{10}$$

$4\frac{3}{5}$ rounds to 5. $\frac{3}{10}$ rounds to $\frac{1}{3}$.

$$5 \div \frac{1}{3} = 5 \cdot 3 = 15$$

Note: If we were to round $\frac{3}{10}$ to the nearest whole number, we would get $5 \div 0$, which is undefined. We can get a better estimate by rounding $\frac{3}{10}$ to $\frac{1}{3}$.

Now let's find the actual product.

$$4\frac{3}{5} \div \frac{3}{10} = \frac{23}{5} \div \frac{3}{10}$$ 　Write the mixed numbers as improper fractions.

$$= \frac{23}{5} \cdot \frac{10}{3}$$ 　Write as equivalent multiplication.

$$= \frac{23}{\underset{1}{5}} \cdot \frac{2 \cdot \overset{1}{5}}{3}$$ 　Divide out the common factor 5.

$$= \frac{46}{3}$$ 　Multiply.

$$= 15\frac{1}{3}$$ 　Write the improper fraction as a mixed number.

Notice that our estimate is close to the actual quotient.

Do Your Turn 3 ▶

Your Turn 3

Estimate each quotient, then find the actual quotient.

a. $5\dfrac{1}{4} \div 2\dfrac{1}{3}$

b. $\dfrac{7\frac{2}{7}}{7\frac{1}{2}}$

c. $-13\dfrac{1}{5} \div 5\dfrac{1}{10}$

d. $-4\dfrac{1}{2} \div \left(-11\dfrac{2}{5}\right)$

Answers to Your Turn 3

a. $2\dfrac{1}{2}; 2\dfrac{1}{4}$ b. $\dfrac{7}{8}; \dfrac{34}{35}$

c. $-2\dfrac{3}{5}; -2\dfrac{10}{17}$ d. $\dfrac{5}{11}; \dfrac{15}{38}$

To divide rational expressions we follow the same procedure as with numeric fractions.

Your Turn 4

Divide.

a. $\dfrac{8}{13m} \div \dfrac{12m}{5}$

b. $\dfrac{10x^3}{14} \div \dfrac{15x}{9}$

c. $-\dfrac{11n^3}{12p^4} \div \dfrac{33n^3}{18p}$

d. $-\dfrac{21c}{30a^3b} \div \left(-\dfrac{14c^3b}{20a^4}\right)$

Example 4 Divide. $-\dfrac{6a^3b}{35c^4} \div \dfrac{15a}{14c^2}$

Solution: $-\dfrac{6a^3b}{35c^4} \div \dfrac{15a}{14c^2} = -\dfrac{6a^3b}{35c^4} \cdot \dfrac{14c^2}{15a}$ Write an equivalent multiplication statement.

$= -\dfrac{2 \cdot \overset{1}{3} \cdot \overset{1}{4} \cdot a \cdot a \cdot b}{5 \cdot 7 \cdot \underset{1}{c} \cdot \underset{1}{c} \cdot c \cdot c} \cdot \dfrac{2 \cdot \overset{1}{7} \cdot \overset{1}{c} \cdot \overset{1}{c}}{\underset{1}{3} \cdot 5 \cdot \underset{1}{a}}$ Write in factored form and divide out common factors.

$= -\dfrac{4a^2b}{25c^2}$ Multiply the remaining factors.

◄ **Do Your Turn 4**

OBJECTIVE 4 Find the square root of a fraction.

Remember that a square root is a number that, when squared, equals the given number. To find a square root we must find the base that can be multiplied by itself to equal the given amount.

For example, $\sqrt{\dfrac{9}{16}} = \dfrac{3}{4}$ because $\dfrac{3}{4} \cdot \dfrac{3}{4} = \dfrac{9}{16}$.

Our example suggests that we can find the square root of a fraction by finding the square root of the numerator and denominator separately.

$$\sqrt{\dfrac{9}{16}} = \dfrac{\sqrt{9}}{\sqrt{16}} = \dfrac{3}{4}$$

If a numerator or denominator is not a perfect square, then we try simplifying first to see if the simplified number is a perfect square. Consider $\sqrt{\dfrac{6}{24}}$.

$$\sqrt{\dfrac{6}{24}} = \sqrt{\dfrac{1}{4}} = \dfrac{\sqrt{1}}{\sqrt{4}} = \dfrac{1}{2}$$

Our examples suggest the following procedure.

Note: Some expressions such as $\sqrt{\frac{5}{7}}$ cannot be simplified using either technique. You will learn how to simplify those expressions in other courses.

PROCEDURE *To find the square root of a fraction, try the following:*
Find the square root of the numerator and denominator separately, then simplify.
or
Simplify the fraction, then find the square root of the quotient.

Example 5 Simplify.

a. $\sqrt{\dfrac{25}{36}}$

Solution: We find the square root of the numerator and denominator separately.

$$\sqrt{\dfrac{25}{36}} = \dfrac{\sqrt{25}}{\sqrt{36}} = \dfrac{5}{6}$$

Answers to Your Turn 4
a. $\dfrac{10}{39m^2}$ b. $\dfrac{3x^2}{7}$ c. $\dfrac{-1}{2p^3}$ d. $\dfrac{a}{b^2c^2}$

b. $\sqrt{\dfrac{45}{5}}$

Solution: Since 45 and 5 are not perfect squares, finding the square root of the numerator and denominator is not a good approach. However, $\frac{45}{5}$ simplifies to 9, which is a perfect square.

$$\sqrt{\dfrac{45}{5}} = \sqrt{9} = 3$$

Do Your Turn 5 ▷

Your Turn 5

Simplify.

a. $\sqrt{\dfrac{4}{9}}$

b. $\sqrt{\dfrac{49}{100}}$

c. $\sqrt{\dfrac{125}{5}}$

d. $\sqrt{\dfrac{392}{8}}$

OBJECTIVE 5 Solve equations involving fractions.

In Chapter 4, we learned about two principles that we use to solve equations. 1. The addition/subtraction principle of equality allowed us to add or subtract the same amount on both sides of an equation without affecting its solution(s). 2. The multiplication/division principle of equality allowed us to multiply or divide both sides of an equation by the same nonzero amount without affecting its solution(s).

Let's use the multiplication/division principle of equality to solve equations that contain fractions or mixed numbers. Remember from Section 4.3 that to use the multiplication/division principle of equality to clear a coefficient, we divide both sides of the equation by the coefficient.

Example 6 Solve and check. $\dfrac{3}{4}x = 2\dfrac{5}{8}$

Solution: First, we write the mixed number as an improper fraction so the equation is $\frac{3}{4}x = \frac{21}{8}$. To isolate x we must clear the coefficient $\frac{3}{4}$. There are two ways that we can write the steps. In method 1, we divide both sides of the equation by $\frac{3}{4}$, writing the division in complex fraction form. In method 2, we multiply both sides of the equation by the reciprocal of $\frac{3}{4}$, which is $\frac{4}{3}$. The two methods are equivalent.

Method 1:

$$\dfrac{\frac{3}{4}x}{\frac{3}{4}} = \dfrac{\frac{21}{8}}{\frac{3}{4}}$$ Divide both sides by $\frac{3}{4}$.

$$1x = \dfrac{21}{8} \div \dfrac{3}{4}$$

$$x = \dfrac{21}{8} \cdot \dfrac{4}{3}$$ Write the right side as multiplication.

$$x = \dfrac{\overset{1}{\cancel{3}} \cdot 7}{2 \cdot 2 \cdot \underset{1}{\cancel{2}}} \cdot \dfrac{\overset{1}{\cancel{2}} \cdot \overset{1}{\cancel{2}}}{\underset{1}{\cancel{3}}}$$ Divide out the common factors.

$$x = \dfrac{7}{2}$$ Multiply.

$$x = \dfrac{7}{2} \text{ or } 3\dfrac{1}{2}$$

Method 2:

$$\dfrac{4}{3} \cdot \dfrac{3}{4}x = \dfrac{21}{8} \cdot \dfrac{4}{3}$$ Multiply both sides by $\frac{4}{3}$.

$$\dfrac{\overset{1}{\cancel{4}}}{3} \cdot \dfrac{\overset{1}{\cancel{3}}}{\underset{1}{\cancel{4}}}x = \dfrac{\overset{1}{\cancel{3}} \cdot 7}{2 \cdot 2 \cdot \underset{1}{\cancel{2}}} \cdot \dfrac{\overset{1}{\cancel{2}} \cdot \overset{1}{\cancel{2}}}{\underset{1}{\cancel{3}}}$$ Divide out the common factors.

$$1x = \dfrac{7}{2}$$ Multiply.

$$x = \dfrac{7}{2} \text{ or } 3\dfrac{1}{2}$$

Note: Most people prefer using method 2 to avoid complex fractions. We will use method 2 hereafter.

Answers to Your Turn 5
a. $\dfrac{2}{3}$ **b.** $\dfrac{7}{10}$ **c.** 5 **d.** 7

Your Turn 6

Solve and check.

a. $\dfrac{3}{4}a = \dfrac{2}{3}$

b. $\dfrac{5}{6}y = \dfrac{3}{10}$

c. $\dfrac{4}{9}k = -\dfrac{8}{15}$

d. $\dfrac{-4}{15} = -\dfrac{1}{6}n$

Your Turn 7

Solve and check.

a. $\dfrac{x}{10} = 6$

b. $\dfrac{y}{7} = \dfrac{5}{14}$

c. $\dfrac{-3}{4} = \dfrac{n}{8}$

d. $\dfrac{h}{-9} = -2\dfrac{1}{3}$

Answers to Your Turn 6

a. $a = \dfrac{8}{9}$ b. $y = \dfrac{9}{25}$

c. $k = -1\dfrac{1}{5}$ d. $n = 1\dfrac{3}{5}$

Answers to Your Turn 7

a. $x = 60$ b. $y = 2\dfrac{1}{2}$

c. $n = -6$ d. $h = 21$

Check: Replace x in $\frac{3}{4}x = 2\frac{5}{8}$ with $\frac{7}{2}$ $\left(\text{or } 3\frac{1}{2}\right)$, and verify that the equation is true.

$$\frac{3}{4}x = 2\frac{5}{8}$$

$$\frac{3}{4} \cdot \frac{7}{2} \stackrel{?}{=} 2\frac{5}{8}$$

$$\frac{21}{8} \stackrel{?}{=} 2\frac{5}{8}$$

$$2\frac{5}{8} = 2\frac{5}{8} \qquad \text{It checks.}$$

◄ **Do Your Turn 6**

Example 7 Solve and check. $\dfrac{m}{5} = -3\dfrac{1}{10}$

Solution: Since $\frac{m}{5}$ is the same as $\frac{1m}{5}$ or $\frac{1}{5}m$, the coefficient is $\frac{1}{5}$. To isolate m, we multiply both sides of the equation by the reciprocal of $\frac{1}{5}$, which is $\frac{5}{1}$.

$$\frac{m}{5} = -\frac{31}{10} \qquad \text{Write } -3\tfrac{1}{10} \text{ as an improper fraction } -\tfrac{31}{10}.$$

$$\frac{5}{1} \cdot \frac{m}{5} = -\frac{31}{10} \cdot \frac{5}{1} \qquad \text{Multiply both sides by } \tfrac{5}{10}.$$

$$\frac{\overset{1}{\cancel{5}}}{1} \cdot \frac{m}{\underset{1}{\cancel{5}}} = -\frac{31}{2 \cdot \underset{1}{\cancel{5}}} \cdot \frac{\overset{1}{\cancel{5}}}{1} \qquad \text{Divide out the common factors.}$$

$$1m = -\frac{31}{2} \qquad \text{Multiply.}$$

$$m = -\frac{31}{2} \text{ or } -15\frac{1}{2}$$

Check: Replace m in $\frac{m}{5} = -3\frac{1}{10}$ with $-15\frac{1}{2}$ and verify that the equation is true.

$$\frac{m}{5} = -3\frac{1}{10}$$

$$\frac{-15\frac{1}{2}}{5} \stackrel{?}{=} -3\frac{1}{10}$$

$$-15\frac{1}{2} \div 5 \stackrel{?}{=} -3\frac{1}{10}$$

$$-\frac{31}{2} \div 5 \stackrel{?}{=} -3\frac{1}{10}$$

$$-\frac{31}{2} \cdot \frac{1}{5} \stackrel{?}{=} -3\frac{1}{10}$$

$$-\frac{31}{10} \stackrel{?}{=} -3\frac{1}{10}$$

$$-3\frac{1}{10} = -3\frac{1}{10} \qquad \text{It checks.}$$

Note: We get a complex fraction when we replace m with $-15\frac{1}{2}$. Remember, the complex fraction means we divide: $-15\frac{1}{2} \div 5$.

◄ **Do Your Turn 7**

OBJECTIVE 6 Solve applications involving division of fractions.

Let's return to the problem about changing a $20 bill into quarters. At the beginning of this section, we translated the problem to division: $20 \div \frac{1}{4}$. You should note that $20 is the amount that we split up, and the size of each division is a quarter, $\frac{1}{4}$. We can also translate to a missing factor statement:

$$\text{value of the currency} \cdot \text{number of coins/bills} = \text{total money}$$

$$\frac{1}{4} \qquad \cdot \qquad n \qquad = \qquad \$20$$

$$\frac{1}{4}n = 20$$

Note: We end up multiplying 20 by 4 just as we did earlier in this section. ▶ $\dfrac{\overset{1}{\cancel{4}}}{1} \cdot \dfrac{1}{\underset{1}{\cancel{4}}}n = \dfrac{20}{1} \cdot \dfrac{4}{1}$ **Multiply both sides by 4.**

$$n = 80$$

Conclusion: Whenever we are given a total amount and either the size of the parts or the number of parts, we can write a missing factor statement. We can say

$$\text{size of each part} \cdot \text{number of parts} = \text{whole amount}$$

Of course, if you recognize the situation for what it is, then go directly to the related division statement.

$$\text{size of each part} = \text{whole amount} \div \text{number of parts}$$

or

$$\text{number of parts} = \text{whole amount} \div \text{size of each part}$$

> **Example 8** A board is $35\frac{3}{4}$ inches long. The board is to be cut into pieces that are $6\frac{1}{2}$ inches each. How many pieces can be cut?

Understand: We are given the total length of a board and the size of each part, and we want to find the number of parts. This is a missing factor/division problem.

Plan: We will write a missing factor statement, then solve.

Execute: Let n represent the number of parts:

$$\text{size of each part} \cdot \text{number of parts} = \text{whole amount}$$

$$6\frac{1}{2} \qquad \cdot \qquad n \qquad = \qquad 35\frac{3}{4}$$

$$\frac{13}{2}n = \frac{143}{4} \qquad \text{Write the mixed numbers as improper fractions.}$$

$$\frac{2}{13} \cdot \frac{13}{2}n = \frac{143}{4} \cdot \frac{2}{13} \qquad \text{Multiply both sides by } \tfrac{2}{13}.$$

$$\frac{1}{\cancel{4}} \cdot \frac{\overset{1}{\cancel{4}}}{1}n = \frac{11 \cdot \overset{1}{\cancel{13}}}{2 \cdot 2} \cdot \frac{\overset{1}{\cancel{2}}}{\underset{1}{\cancel{13}}} \qquad \text{Divide out the common factors.}$$

$$1n = \frac{11}{2} \qquad \text{Multiply.}$$

$$n = 5\frac{1}{2} \qquad \text{Write as a mixed number.}$$

Solve.

a. How many $4\frac{1}{2}$-ounce servings are there in $30\frac{1}{4}$ ounces of cereal?

b. A circle has a circumference of $25\frac{41}{56}$ inches. What is the diameter?

Answer: The board can be cut into $5\frac{1}{2}$ pieces that are each $6\frac{1}{2}$ inches long. Note that this means that we could cut 5 whole pieces that are each $6\frac{1}{2}$ inches long and would be left with $\frac{1}{2}$ of another $6\frac{1}{2}$-inch piece.

Check: If we multiply the length of each part by the number of parts, we should get the total length of the original board.

$$\text{size of each part} \cdot \text{number of parts} = \text{whole amount}$$

$$6\frac{1}{2} \cdot 5\frac{1}{2} \stackrel{?}{=} 35\frac{3}{4}$$

$$\frac{13}{2} \cdot \frac{11}{2} \stackrel{?}{=} 35\frac{3}{4}$$

$$\frac{143}{4} \stackrel{?}{=} 35\frac{3}{4}$$

$$35\frac{3}{4} = 35\frac{3}{4} \quad \text{It checks.}$$

◀ **Do Your Turn 8**

Answers to Your Turn 8

a. $6\frac{13}{18}$ servings b. $8\frac{3}{16}$ in.

5.4 Exercises

For
Extra
Help

 Videotape DVT

 Tutor Center
Addison-Wesley
Tutor Center

 Math XL
Math XL

 MyMathLab

 Student Solutions
Manual

1. What are reciprocals? Give an example of a pair of reciprocals.

2. Explain how to divide fractions.

3. Explain how to divide mixed numbers.

4. What are two approaches for finding the square root of a fraction?

For Exercises 5–12, write the reciprocal.

5. $\dfrac{2}{3}$

6. $\dfrac{3}{8}$

7. $\dfrac{1}{6}$

8. $-\dfrac{1}{9}$

9. -15

10. 7

11. $-\dfrac{x}{4}$

12. $\dfrac{-9y}{7}$

For Exercises 13–24, divide. Express the quotient in lowest terms.

13. $\dfrac{1}{2} \div \dfrac{1}{6}$

14. $\dfrac{5}{8} \div \dfrac{2}{3}$

15. $\dfrac{3}{10} \div \dfrac{5}{6}$

16. $\dfrac{9}{16} \div \dfrac{3}{4}$

17. $\dfrac{14}{15} \div \left(-\dfrac{7}{12}\right)$

18. $-\dfrac{10}{13} \div \dfrac{5}{12}$

19. $\dfrac{-7}{12} \div (-14)$

20. $-6 \div \left(-\dfrac{3}{4}\right)$

21. $\dfrac{\dfrac{2}{5}}{\dfrac{4}{15}}$

22. $\dfrac{\dfrac{9}{10}}{\dfrac{21}{10}}$

23. $\dfrac{12}{\dfrac{-2}{3}}$

24. $\dfrac{\dfrac{-3}{5}}{18}$

For Exercises 25–32, estimate, then find the actual quotient. Write the quotient as a mixed number in simplest form.

25. $5\dfrac{1}{6} \div \dfrac{4}{9}$

26. $7\dfrac{1}{3} \div \dfrac{3}{8}$

27. $9\dfrac{1}{2} \div 3\dfrac{3}{4}$

28. $2\dfrac{7}{12} \div 2\dfrac{1}{6}$

29. $-2\dfrac{9}{16} \div 10\dfrac{1}{4}$

30. $4\dfrac{2}{9} \div \left(-8\dfrac{1}{3}\right)$

31. $\dfrac{-12\dfrac{1}{2}}{-6\dfrac{3}{4}}$

32. $\dfrac{-10\dfrac{2}{3}}{-2\dfrac{5}{6}}$

For Exercises 33–38, divide. Express the quotient in lowest terms.

33. $\dfrac{7}{15x} \div \dfrac{1}{6x^3}$

34. $\dfrac{ab}{10} \div \dfrac{4b^3}{25}$

35. $\dfrac{9m^5}{20n} \div \dfrac{12m^2n}{25}$

36. $\dfrac{t^2u^4}{18v^3} \div \left(\dfrac{-10t^5}{12v}\right)$

37. $\dfrac{14x^6}{28y^4z} \div \left(\dfrac{-8x^2y}{18z}\right)$

38. $-\dfrac{13h^6}{20hk} \div \left(\dfrac{-65h^3k}{25k^2}\right)$

For Exercises 39–46, simplify.

39. $\sqrt{\dfrac{64}{81}}$

40. $\sqrt{\dfrac{49}{100}}$

41. $\sqrt{\dfrac{121}{36}}$

42. $\sqrt{\dfrac{169}{100}}$

43. $\sqrt{\dfrac{180}{5}}$

44. $\sqrt{\dfrac{960}{15}}$

45. $\sqrt{\dfrac{1053}{13}}$

46. $\sqrt{\dfrac{3400}{34}}$

For Exercises 47–54, solve and check.

47. $\dfrac{3}{4}x = 12$

48. $\dfrac{2}{5}a = 10$

49. $\dfrac{5}{8} = \dfrac{-3}{16}y$

50. $\dfrac{7}{10} = \dfrac{-1}{5}x$

51. $\dfrac{n}{12} = -\dfrac{3}{20}$

52. $-\dfrac{1}{16} = \dfrac{k}{14}$

53. $\dfrac{-5m}{6} = \dfrac{-5}{21}$

54. $-\dfrac{9}{40} = -\dfrac{3u}{10}$

For Exercises 55–66, solve.

55. A child is to receive $1\frac{1}{3}$ teaspoons of amoxicillin (an antibiotic). How many doses can be given if the bottle contains 40 teaspoons? If the instructions are to give the dosage twice a day for 14 days, is the amount of amoxicillin in the bottle correct? If it is not correct, why would the pharmacist put more or less in the bottle than the amount prescribed?

56. A cereal box contains $13\frac{3}{10}$ ounces of cereal. The label indicates a serving size is $1\frac{4}{5}$ ounces. How many servings are in a box of the cereal? The label indicates that there are about 8 servings per container. Is this accurate?

57. A baker only had enough ingredients to make $10\frac{1}{2}$ cups of batter for muffins. If he is to make a dozen muffins of equal size, how much of the batter should he place in each cup of the muffin pan?

58. Ellyn makes curtains. She wants to make 4 curtains of equal length from 78 inches of material. How long should each piece be?

59. A rectangular field has an area of $2805\frac{1}{2}$ square feet and has a width of $30\frac{1}{2}$ feet. What is the length?

60. The area of a photograph is $18\frac{1}{3}$ square inches and has a length of $5\frac{1}{2}$ inches. What is its width?

61. The desired area for a triangular garden is to be $14\frac{1}{2}$ square feet. If the base is to be $6\frac{1}{2}$ feet, what must the height be?

62. A triangular company logo is to be $1\frac{3}{4}$ square inches. If it must be $2\frac{1}{3}$ inches high, what must be its base?

63. The largest radio telescope in the world is the Arecibo dish located in Puerto Rico. It has a circumference of $3111\frac{2}{5}$ feet. What is the diameter?

64. A circular track has a circumference of $\frac{1}{4}$ miles. What is the radius?

65. Ron uses the odometer in his car to measure the distance from his house to where he works. He finds the distance to be $18\frac{3}{10}$ miles. It takes him $\frac{2}{5}$ hours to get there. What was his average rate?

66. How much time does it take an F-16 fighter plane traveling at Mach 2 to travel 20 miles? Mach 2 is about $\frac{2}{5}$ (miles per second), which is twice the speed of sound.

Review Exercises

[3.6] **1.** Find the prime factorization of 378.

[3.5] **2.** Simplify. $2^2 \cdot 3 \cdot 5 \cdot x^3 \cdot y$

[5.1] **3.** Use $<$, $>$, or $=$ to make a true statement.
$$\dfrac{5}{6} \; ? \; \dfrac{10}{12}$$

[3.3] **4.** Combine like terms.
$$5t^3 - 8t + 9t^2 + 4t^3 - 16 - 5t$$

[3.7] **5.** Factor. $32x^2 + 16x$

5.5 Least Common Multiple

OBJECTIVE 1 Find the least common multiple (LCM) by listing.

In Section 5.1, we defined a *multiple* to be a number that is divisible by a given number. Below we list multiples of 6 and multiples of 4.

Multiples of 6: 6, 12, 18, 24, 30, 36, 42, 48, . . .
Multiples of 4: 4, 8, 12, 16, 20, 24, 28, 32, 36, 40, . . .

Notice that the numbers 6 and 4 have several multiples in common. In our lists we can see 12, 24, and 36 are all common multiples for 6 and 4. We will find it useful to consider the **least common multiple.**

DEFINITION   Least common multiple (LCM): The smallest natural number that is divisible by all the given numbers.

Looking again at our list of multiples for 6 and 4 we see that the least common multiple is 12. Our example suggests that to find the LCM for a set of numbers, we list multiples of each number until we find a number common to all the lists. However, we can quicken that process by listing only multiples of the greatest given number until we find a multiple that is divisible by all the other given numbers.

PROCEDURE   *To find the LCM by listing, list multiples of the greatest given number until you find a multiple that is divisible by all the other given numbers.*

Example 1   Find the LCM of 36 and 120 by listing.

Solution: The greatest given number is 120, so we will list multiples of 120 until we find a number that is also divisible by 36. Multiples of 120 are 120, 240, 360, etc.

    120   240   360

36 does not divide 120 evenly so we go to the next multiple of 120.	36 does not divide 240 evenly so we go to the next multiple of 120.	36 divides 360 evenly, 10 times. This is the LCM.

Because 360 is the smallest multiple of 120 that is also divisible by 36, it is the LCM. In math language, we write LCM(36, 120) = 360.

We can check the LCM by dividing it by the given numbers.

$$
\begin{array}{r}
10 \\
36\overline{)360} \\
-36 \\
\hline
00 \\
-0 \\
\hline
0
\end{array}
\qquad\qquad
\begin{array}{r}
3 \\
120\overline{)360} \\
-360 \\
\hline
0
\end{array}
$$

OBJECTIVES

1. Find the least common multiple (LCM) by listing.
2. Find the LCM using prime factorization.
3. Find the LCM of a set of monomials.
4. Write fractions as equivalent fractions with the least common denominator (LCD).
5. Write rational expressions as equivalent expressions with the LCD.

CONNECTION In Section 5.6, we will learn that to add or subtract fractions they must have a common denominator. If we are to add or subtract fractions that do not have a common denominator, we will rewrite them so that they do. We will use the LCM of the denominators as their common denominator.

Your Turn 1

Find the LCM of the given numbers by listing.

a. 18 and 24

b. 21 and 35

c. 3, 4, and 10

d. 4, 6, and 9

Look at the quotients 10 and 3. Notice that the only common factor for 10 and 3 is 1. This means that 360 is in fact the LCM for 36 and 120. If any factor other than 1 divides all the quotients evenly, it means our result is a common multiple but not the least common multiple. For example, suppose we mistakenly get 720 as our answer. Look at the check:

$$
\begin{array}{r}
20 \\
36\overline{)720} \\
-72 \\
\hline
00 \\
-0 \\
\hline
0
\end{array}
\qquad
\boxed{\text{Notice that the quotients, 20 and 6, are both divisible by 2. This means we overshot the LCM by a factor of 2.}}
\qquad
\begin{array}{r}
6 \\
120\overline{)720} \\
-720 \\
\hline
0
\end{array}
$$

To correct the mistake, we can divide our incorrect answer by the factor common to the quotients. In this case, that factor is 2, so we divide 720 by 2, which gives the correct answer 360.

◀ **Do Your Turn 1**

OBJECTIVE 2 Find the LCM using prime factorization.

We can use prime factorization to find the LCM of a given set of numbers. In Example 1, we found the LCM of 36 and 120 to be 360. Let's look at the prime factorizations of 36 and 120, then compare their factorizations with the prime factorization of 360 to see what we can learn.

WARNING Don't confuse LCM with GCF. When we use primes to find a GCF, we include only those primes common to all the factorizations. For an LCM, we include each prime factor.

$$36 = 2 \cdot 2 \cdot 3 \cdot 3 = 2^2 \cdot 3^2$$
$$120 = 2 \cdot 2 \cdot 2 \cdot 3 \cdot 5 = 2^3 \cdot 3 \cdot 5$$
$$\text{LCM}(36, 120) = 360 = 2 \cdot 2 \cdot 2 \cdot 3 \cdot 3 \cdot 5 = 2^3 \cdot 3^2 \cdot 5$$

Notice that the factorization for 360 contains each prime factor that appeared in the factorizations of 36 and 120. Further, each of those prime factors is included in the LCM the greatest number of times it appears in the factorizations of 36 and 120. That is, since 2 occurs most in 120, three times, it is included three times in the LCM. Since 3 occurs most in 36, twice, two 3's are included in the LCM. The 5 occurs most in 120, once, so one 5 is included in the LCM.

Writing the factorizations in exponential form can be helpful because we can then simply compare the exponents of each prime factor and include each prime raised to the greatest exponent that occurred in the factorizations. That is, the 2^3 in 120 has a greater exponent than the 2^2 in 36, so the LCM contains 2^3. Similarly, the 3^2 in 36 has a greater exponent than the 3 in 120, so the LCM contains 3^2. Last, the lone 5 in 120 has the greatest exponent of the 5's in the factorizations, so the LCM contains one 5. Our example suggests the following procedure.

PROCEDURE *To find the LCM using prime factorization:*
1. Find the prime factorization of each given number.
2. Write a factorization that contains each prime factor the greatest number of times it occurs in the factorizations. Or, if you prefer exponents, the factorization contains each prime factor raised to the greatest exponent that occurs in the factorizations.
3. Multiply to get the LCM.

Answers to Your Turn 1
a. 72 b. 105 c. 60 d. 36

Example 2   Find the LCM of 24, 90, and 70 using prime factorization.

Solution: We find the prime factorization of 24, 90, and 70, then use the greatest exponent of each prime factor.

$$24 = 2^3 \cdot 3$$

$$90 = 2 \cdot 3^2 \cdot 5$$

$$70 = 2 \cdot 5 \cdot 7$$

$$\text{LCM}(24, 90, 70) = 2^3 \cdot 3^2 \cdot 5 \cdot 7$$

◀ 2^3 has the greatest exponent for 2, 3^2 has the greatest exponent for 3, 5 has the greatest exponent for 5, and 7 has the greatest exponent for 7.

$$= 8 \cdot 9 \cdot 5 \cdot 7$$

$$= 2520$$

Do Your Turn 2 ▶

Your Turn 2

Find the LCM using prime factorization.

a. 36 and 80

b. 30 and 75

c. 20, 35, and 40

d. 26, 40, and 65

The listing method is simple, direct, and good when the given numbers are small. The prime factorization method has a more complex setup but is good when given large numbers. Of course, we can use either method to find the LCM, but the idea is to select the most effective method.

OBJECTIVE 3 Find the LCM of a set of monomials.

To find the LCM of a group of monomials, we follow the same procedure and treat the variables as prime factors.

Example 3   Find the LCM of $18x^3y$ and $24xz^2$.

Solution: We use the prime factorization method.

$$18x^3y = 2 \cdot 3^2 \cdot x^3 \cdot y$$
$$24xz^2 = 2^3 \cdot 3 \cdot x \cdot z^2$$
$$\text{LCM}(18x^3y, 24xz^2) = 2^3 \cdot 3^2 \cdot x^3 \cdot y \cdot z^2$$
$$= 72x^3yz^2$$

◀ 2^3 has the greatest exponent for 2, 3^2 has the greatest exponent for 3, x^3 has the greatest exponent for x, y has the greatest exponent for y, and z^2 has the greatest exponent for z.

Do Your Turn 3 ▶

Your Turn 3

Find the LCM of each pair of monomials.

a. $6a^3$ and $12a$

b. $9m^2n$ and $12m^4$

c. $10hk^5$ and $8h^2k$

OBJECTIVE 4 Work fractions as equivalent fractions with the least common denominator (LCD).

In Section 5.1, we learned that multiplying both the numerator and denominator of a fraction by the same number gives an equivalent fraction. We used this rule to rewrite two fractions so that they have a common denominator, which allowed us to easily compare the fractions. Now, let's rewrite fractions so that they have the least common multiple of their denominators as their common denominator. When we use the LCM as a common denominator, we say we are using the **least common denominator (LCD).**

DEFINITION   **Least common denominator (LCD):** The least common multiple of the denominators.

Answers to Your Turn 2
a. 720 b. 150 c. 280 d. 520

Answers to Your Turn 3
a. $12a^3$ b. $36m^4n$ c. $40h^2k^5$

For example, the LCD of $\frac{3}{4}$ and $\frac{5}{6}$ is 12 because 12 is the least common multiple of the denominators, 4 and 6. Now, let's rewrite the fractions so that 12 is their denominator. We will need to determine what to multiply each fraction by so that it becomes an equivalent fraction with 12 as its denominator.

$$\frac{3}{4} = \frac{3 \cdot ?}{4 \cdot ?} = \frac{}{12} \qquad\qquad \frac{5}{6} = \frac{5 \cdot ?}{6 \cdot ?} = \frac{}{12}$$

Looking at the denominator, we have a missing factor statement, $4 \cdot ? = 12$. Notice that the missing factor must be 3, so to get a fraction equivalent to $\frac{3}{4}$ with a denominator of 12, we must multiply its numerator and denominator by 3.

Looking at the denominator, we must multiply 6 by 2 to get 12, so to get a fraction equivalent to $\frac{5}{6}$ with a denominator of 12, we must multiply its numerator and denominator by 2.

$$\frac{3}{4} = \frac{3 \cdot 3}{4 \cdot 3} = \frac{9}{12} \qquad\qquad \frac{5}{6} = \frac{5 \cdot 2}{6 \cdot 2} = \frac{10}{12}$$

Your Turn 4

Rewrite the fractions as equivalent fractions with the LCD.

a. $\frac{2}{3}$ and $\frac{1}{4}$

b. $\frac{5}{8}$ and $\frac{7}{12}$

c. $\frac{2}{5}$ and $\frac{3}{10}$

d. $\frac{9}{20}$ and $\frac{8}{15}$

Example 4 Write $\frac{7}{12}$ and $\frac{2}{15}$ as equivalent fractions with the least common denominator.

Solution: The least common denominator (LCD) is the LCM of the denominators, so we need to find the LCM of 12 and 15. We will use the listing method, listing multiples of 15 until we find a number that is also a multiple of 12.

15	30	45	60

12 does not divide 15 evenly, so we go to the next multiple of 15.

12 does not divide 30 evenly, so we go to the next multiple of 15.

12 does not divide 45 evenly, so we go to the next multiple of 15.

12 divides 60 evenly, so 60 is the LCM.

Now we write $\frac{7}{12}$ and $\frac{2}{15}$ as equivalent fractions with 60 as their denominator.

$$\frac{7}{12} = \frac{7 \cdot ?}{12 \cdot ?} = \frac{}{60} \qquad\qquad \frac{2}{15} = \frac{2 \cdot ?}{15 \cdot ?} = \frac{}{60}$$

We think, "What factor will multiply by 12 to equal 60?" Notice that to find this missing factor, we can divide.

$$60 \div 12 = 5$$

So we need to multiply the numerator and denominator by 5.

We think, "What factor will multiply by 15 to equal 60?" Notice that to find this missing factor, we can divide.

$$60 \div 15 = 4$$

So we need to multiply the numerator and denominator by 4.

$$\frac{7}{12} = \frac{7 \cdot 5}{12 \cdot 5} = \frac{35}{60} \qquad\qquad \frac{2}{15} = \frac{2 \cdot 4}{15 \cdot 4} = \frac{8}{60}$$

◁ **Do Your Turn 4**

Answers to Your Turn 4

a. $\frac{8}{12}$ and $\frac{3}{12}$

b. $\frac{15}{24}$ and $\frac{14}{24}$

c. $\frac{4}{10}$ and $\frac{3}{10}$

d. $\frac{27}{60}$ and $\frac{32}{60}$

Rational expressions are rewritten in the same way as numeric fractions.

Example 5 Write $\dfrac{3}{8x}$ and $\dfrac{5}{6x^2}$ as equivalent rational expressions with the LCD.

Solution: The LCD is the LCM of the denominators, so we need to find the LCM of $8x$ and $6x^2$.

The LCM for 8 and 6 is 24. For the variables, x^2 has the greater exponent, so the LCD is $24x^2$. Now we rewrite $\frac{3}{8x}$ and $\frac{5}{6x^2}$ with $24x^2$ as the denominator.

$$\frac{3}{8x} = \frac{3 \cdot ?}{8x \cdot ?} = \frac{}{24x^2} \qquad\qquad \frac{5}{6x^2} = \frac{5 \cdot ?}{6x^2 \cdot ?} = \frac{}{24x^2}$$

▲

We think, "What factor will multiply by $8x$ to equal $24x^2$?" Notice that to find this missing factor, we can divide.

$$24x^2 \div 8x = 3x$$

So we need to multiply the numerator and denominator by $3x$.

▼

$$\frac{3}{8x} = \frac{3 \cdot 3x}{8x \cdot 3x} = \frac{9x}{24x^2}$$

We think, "What factor will multiply by $6x^2$ to equal $24x^2$?" Notice that to find this missing factor, we can divide.

$$24x^2 \div 6x^2 = 4$$

So we need to multiply the numerator and denominator by 4.

▼

$$\frac{5}{6x^2} = \frac{5 \cdot 4}{6x^2 \cdot 4} = \frac{20}{24x^2}$$

Do Your Turn 5 ▷

Your Turn 5

Rewrite the rational expressions as equivalent expressions with the LCD.

a. $\dfrac{5}{9}$ and $\dfrac{2}{3a}$

b. $\dfrac{3}{4x}$ and $\dfrac{7}{6x^2}$

c. $-\dfrac{m}{n^4}$ and $\dfrac{3}{mn}$

d. $\dfrac{11}{12t^2}$ and $-\dfrac{7u}{9tv}$

Answers to Your Turn 5

a. $\dfrac{5a}{9a}$ and $\dfrac{6}{9a}$

b. $\dfrac{9x}{12x^2}$ and $\dfrac{14}{12x^2}$

c. $-\dfrac{m^2}{mn^4}$ and $\dfrac{3n^3}{mn^4}$

d. $\dfrac{33v}{36t^2v}$ and $-\dfrac{28tu}{36t^2v}$

5.5 Exercises

For Extra Help

 Videotape DVT

Tutor Center Addison-Wesley Tutor Center

 Math XL Math XL

 MyMathLab

 Student Solutions Manual

1. What is the least common multiple (LCM) of a set of numbers?

2. Explain how to find the least common multiple by listing.

3. Explain how to find the least common multiple using prime factorization.

4. Given two fractions with different denominators, what is the least common denominator?

5. Suppose the LCD of two fractions is 24. If one of the denominators is 6, what factor do you multiply that fraction's numerator and denominator by in order to write an equivalent fraction with the LCD?

6. Suppose the LCD of two rational expressions is $18x$. If one of the denominators is 3, what factor do you multiply that rational expression's numerator and denominator by to write an equivalent rational expression with the LCD?

For Exercises 7–14, find the LCM by listing.

7. 10 and 6

8. 15 and 9

9. 12 and 36

10. 8 and 32

11. 20 and 30

12. 16 and 20

13. 6, 9, and 15

14. 6, 10, and 15

For Exercises 15–26, find the LCM using prime factorization.

15. 18 and 24

16. 24 and 50

17. 63 and 28

18. 42 and 56

19. 52 and 28

20. 68 and 56

21. 180 and 200

22. 210 and 420

23. 28, 32, and 60

24. 26, 30, and 39

25. 42, 56, and 80

26. 36, 49, and 72

For Exercises 27–34, find the LCM.

27. $12x$ and $8y$

28. $9a$ and $6b$

29. $10y^3$ and $6y$

30. $14t^2$ and $5t$

31. $16mn$ and $8m$

32. $20h^5k^3$ and $15h^2k$

33. $18x^2y$ and $12xy^3$

34. $24a^4b$ and $30ab^2$

For Exercises 35–42, rewrite the fractions as equivalent fractions with the LCD.

35. $\dfrac{3}{10}$ and $\dfrac{5}{6}$

36. $\dfrac{2}{15}$ and $\dfrac{4}{9}$

37. $\dfrac{7}{12}$ and $\dfrac{11}{36}$

38. $\dfrac{1}{8}$ and $\dfrac{15}{32}$

39. $\dfrac{1}{20}$ and $\dfrac{17}{30}$

40. $\dfrac{3}{16}$ and $\dfrac{9}{20}$

41. $\dfrac{3}{4}, \dfrac{1}{6},$ and $\dfrac{7}{9}$

42. $\dfrac{2}{5}, \dfrac{5}{6},$ and $\dfrac{13}{15}$

For Exercises 43–50, rewrite the rational expressions as equivalent expressions with the LCD.

43. $\dfrac{7}{12x}$ and $\dfrac{3}{4}$

44. $\dfrac{5}{9b}$ and $\dfrac{2}{3b}$

45. $\dfrac{9}{16mn}$ and $\dfrac{3n}{8m}$

46. $\dfrac{1}{14t^2}$ and $\dfrac{2}{5}$

47. $\dfrac{7}{10y^3z}$ and $\dfrac{5z}{6y}$

48. $\dfrac{-13}{20h^5k^3}$ and $\dfrac{4}{15h^2k}$

49. $\dfrac{z}{18x^2y}$ and $\dfrac{-5}{12xy^3}$

50. $\dfrac{5}{24a^4b}$ and $\dfrac{c}{30ab^2}$

PUZZLE PROBLEM Planetary conjunctions occur when two or more planets form a rough line with the Sun. Using rough estimates of the orbital periods of the following planets, calculate the time that will elapse between alignments of these planets.

Planet	Orbit Period
Earth	1 year
Mars	2 years
Jupiter	12 years
Saturn	30 years
Uranus	84 years

OF INTEREST

Because the planets do not all orbit in the same plane, it is virtually impossible for more than two planets to form a straight line with the Sun. However, if we were to view the solar system from above, the planets can form an apparent line (conjunction) with the Sun. Theoretically, this type of alignment for all nine planets could occur every 334,689,064 years. When conjunction dates are calculated, we relax the criterion that the planets be in a straight line and choose an acceptable angle of dispersion.

The conjunction that occurred in May of 2000 included Mercury, Venus, Earth, the Moon, Mars, Jupiter, and Saturn with an angle of dispersion of about 15 degrees in the sky.

Review Exercises

For Exercises 1 and 2, simplify.

[2.2] **1.** $16 + (-28)$

[2.3] **2.** $-8 - (-19)$

[3.3] **3.** Combine like terms. $14x^2 - 16x + 25x^2 - 12 + 4x^3 + 9$

For Exercises 4–6, solve and check.

[4.3] **4.** $m + 14 = 6$

[4.2] **5.** $x - 37 = 16$

[4.3] **6.** $-3a = \dfrac{12}{5}$

Adding and Subtracting Fractions, Mixed Numbers, and Rational Expressions

OBJECTIVE **1** Add and subtract fractions with the same denominator.

To see how we add fractions that have the same denominator, consider the fact that when we add two halves we get a whole.

$$\frac{1}{2} + \frac{1}{2} = 1$$

Notice that if we add the numerators and keep the same denominator, we'll have $\frac{2}{2}$, which simplifies to 1.

$$\frac{1}{2} + \frac{1}{2} = \frac{1+1}{2} = \frac{2}{2} = 1$$

WARNING It is tempting to treat addition like multiplication and add numerator plus numerator and denominator plus denominator. But when we follow this approach, we have a half plus a half equals a half.

$$\frac{1}{2} + \frac{1}{2} = \frac{1+1}{2+2} = \frac{2}{4} = \frac{1}{2} \quad \text{This can't be true.}$$

The only situation where we can add an amount to a number and end up with the same number is $n + 0 = n$.

Our example suggests the following procedure.

PROCEDURE *To add or subtract fractions that have the same denominator:*

1. Add or subtract the numerators.
2. Keep the same denominator.
3. Simplify.

Your Turn 1

Add or subtract.

a. $\dfrac{5}{7} + \dfrac{1}{7}$

b. $\dfrac{3}{16} + \dfrac{1}{16}$

c. $\dfrac{9}{11} - \dfrac{3}{11}$

d. $\dfrac{7}{8} - \dfrac{5}{8}$

Answers to Your Turn 1

a. $\dfrac{6}{7}$ b. $\dfrac{1}{4}$ c. $\dfrac{6}{11}$ d. $\dfrac{1}{4}$

Example 1 Add or subtract.

a. $\dfrac{1}{8} + \dfrac{3}{8}$

Solution: Because the denominators are the same add the numerators, then simplify.

$$\frac{1}{8} + \frac{3}{8} = \frac{1+3}{8} = \frac{4}{8} = \frac{1}{2}$$

b. $\dfrac{8}{9} - \dfrac{2}{9}$

Solution: Because the denominators are the same, subtract the numerators, then simplify.

$$\frac{8}{9} - \frac{2}{9} = \frac{8-2}{9} = \frac{6}{9} = \frac{2}{3}$$

◀ **Do Your Turn 1**

Example 2 Add.

a. $-\dfrac{3}{10} + \left(-\dfrac{1}{10}\right)$

Solution: Because the denominators are the same, add the numerators, then simplify.

> **Note:** Since these two numbers have the same sign, we add their absolute values and keep the same sign.

$$-\dfrac{3}{10} + \left(-\dfrac{1}{10}\right) = \dfrac{-3 + (-1)}{10} = \dfrac{-4}{10} = -\dfrac{2}{5}$$

> **Note:** Remember, to indicate a negative fraction, we can place the minus sign to the left of the numerator, denominator, or fraction bar. In an answer, we will place the sign to the left of the fraction bar.

b. $\dfrac{-7}{12} + \dfrac{5}{12}$

Solution: Because the denominators are the same, add the numerators, then simplify.

> **Note:** Since these two numbers have different signs, we subtract their absolute values and keep the sign of the number with the greater absolute value.

$$\dfrac{-7}{12} + \dfrac{5}{12} = \dfrac{-7 + 5}{12} = \dfrac{-2}{12} = -\dfrac{1}{6}$$

Do Your Turn 2 ▶

Example 3 Subtract.

a. $\dfrac{5}{9} - \dfrac{7}{9}$

Solution: Because the denominators are the same, subtract the numerators, then simplify.

> **Note:** To write a subtraction statement as an equivalent addition statement, we change the subtraction sign to an addition sign and change the subtrahend to its additive inverse.

$$\dfrac{5}{9} - \dfrac{7}{9} = \dfrac{5 - 7}{9}$$

$$= \dfrac{5 + (-7)}{9} \quad \text{Write an equivalent addition statement.}$$

$$= -\dfrac{2}{9} \quad \text{Add.}$$

b. $\dfrac{-3}{16} - \left(\dfrac{-5}{16}\right)$

Solution: Because the denominators are the same, subtract the numerators, then simplify.

$$\dfrac{-3}{16} - \left(\dfrac{-5}{16}\right) = \dfrac{-3 - (-5)}{16}$$

$$= \dfrac{-3 + 5}{16} \quad \text{Write an equivalent addition statement.}$$

$$= \dfrac{2}{16} \quad \text{Add.}$$

$$= \dfrac{1}{8} \quad \text{Simplify.}$$

Do Your Turn 3 ▶

OBJECTIVE 2 Add and subtract rational expressions with the same denominator.

To add or subtract rational expressions, we follow the same procedure as for numeric fractions. Keep in mind that we can add or subtract only like terms. Remember, like terms have the same variable(s) raised to the same exponent(s). If the terms are not like, we must indicate the addition or subtraction as a polynomial expression.

Your Turn 4

Add.

a. $\dfrac{3}{8m} + \dfrac{7}{8m}$

b. $\dfrac{5t}{u} + \dfrac{2}{u}$

c. $\dfrac{x+7}{9} + \dfrac{2-x}{9}$

d. $\dfrac{3t^2+8}{5} + \dfrac{4t^2-t-10}{5}$

Example 4   Add.

a. $\dfrac{3x}{10} + \dfrac{x}{10}$

Solution: Because the rational expressions have the same denominator, we add numerators and keep the same denominator.

$$\frac{3x}{10} + \frac{x}{10} = \frac{3x + x}{10} = \frac{4x}{10} = \frac{2x}{5}$$

> **Note:** Because $3x$ and x are like terms, they can be combined to equal $4x$.

b. $\dfrac{x}{3} + \dfrac{2}{3}$

Solution: $\dfrac{x}{3} + \dfrac{2}{3} = \dfrac{x+2}{3}$

> **Note:** We cannot combine x and 2 because they are not like terms. We must express the addition as a polynomial.

c. $\dfrac{x^2 - 3x + 1}{7} + \dfrac{x+4}{7}$

Solution: $\dfrac{x^2 - 3x + 1}{7} + \dfrac{x+4}{7} = \dfrac{(x^2 - 3x + 1) + (x + 4)}{7}$

$$= \frac{x^2 - 2x + 5}{7}$$

> **Note:** To add polynomials, we combine like terms. Since x^2 has no like terms, we write it as-is in the sum. Combining $-3x$ with x gives $-2x$. Combining 1 with 4 gives 5.

◀ **Do Your Turn 4**

Now let's consider subtraction of rational expressions that have the same denominator. We can still write subtraction as an equivalent addition. Remember, if we do not have like terms, we must express the subtraction as a polynomial.

Example 5   Subtract.

a. $\dfrac{4n}{9} - \left(\dfrac{-5}{9}\right)$

Solution: Because the denominators are the same, we subtract numerators and keep the same denominator.

> **Note:** We cannot combine $4n$ and 5 because they are not like terms. Because $4n + 5$ cannot be factored, the expression cannot be simplified.

$$\frac{4n}{9} - \left(\frac{-5}{9}\right) = \frac{4n - (-5)}{9}$$

$$= \frac{4n + 5}{9}$$

Write as an equivalent addition statement.

Answers to Your Turn 4

a. $\dfrac{5}{4m}$ b. $\dfrac{5t+2}{u}$ c. 1

d. $\dfrac{7t^2 - t - 2}{5}$

b. $\dfrac{8x^2 - 5x + 3}{y} - \dfrac{2x - 7}{y}$

Solution:

$$\dfrac{8x^2 - 5x + 3}{y} - \dfrac{2x - 7}{y} = \dfrac{(8x^2 - 5x + 3) - (2x - 7)}{y}$$

Note: To write the equivalent addition, we must find the additive inverse of the subtrahend, $2x - 7$. To write the additive inverse of a polynomial, we change all the signs in the polynomial.

▶ $$= \dfrac{(8x^2 - 5x + 3) + (-2x + 7)}{y}$$

Write as an equivalent addition statement.

$$= \dfrac{8x^2 - 7x + 10}{y}$$ ◀ **Note:** $8x^2$ had no like terms. $-5x$ and $-2x$ combine to equal $-7x$ and 3 and 7 combine to equal 10.

Do Your Turn 5 ▶

OBJECTIVE 3 Add and subtract fractions with different denominators.

To add or subtract fractions that have different denominators, we need to find a common denominator. We will apply what we learned in Section 5.5 about rewriting fractions as equivalent fractions with the LCD.

PROCEDURE *To add or subtract fractions with different denominators:*

1. Write the fractions as equivalent fractions with a common denominator.
   Note: Any common multiple of the denominators will do, but using the LCD makes the numbers more manageable.
2. Add or subtract the numerators and keep the common denominator.
3. Simplify.

Example 6 Add. $\dfrac{7}{12} + \dfrac{2}{15}$

CONNECTION These are the same fractions we used in Example 4 in Section 5.5.

Solution: We must rewrite the fractions with a common denominator, then add the numerators and keep the common denominator. The LCD for 12 and 15 is 60. (See Example 4, Section 5.5.)

Note: To write an equivalent fraction with 60 as the denominator, we rewrite $\frac{7}{12}$ by multiplying its numerator and denominator by 5.

▶ $$= \dfrac{7(5)}{12(5)} + \dfrac{2(4)}{15(4)}$$ ◀ **Note:** To write an equivalent fraction with 60 as the denominator, we rewrite $\frac{2}{15}$ by multiplying its numerator and denominator by 4.

$$= \dfrac{35}{60} + \dfrac{8}{60}$$

$$= \dfrac{35 + 8}{60}$$

$$= \dfrac{43}{60}$$

Do Your Turn 6 ▶

Your Turn 5

Subtract.

a. $\dfrac{7}{15x} - \dfrac{2}{15x}$

b. $\dfrac{9n}{5} - \left(\dfrac{-n}{5}\right)$

c. $\dfrac{y}{z} - \dfrac{4}{z}$

d. $\dfrac{3 + b}{7a} - \dfrac{9 - b}{7a}$

Your Turn 6

Add or subtract.

a. $\dfrac{3}{8} + \dfrac{1}{6}$

b. $\dfrac{1}{6} + \left(\dfrac{-2}{3}\right)$

c. $-\dfrac{5}{9} - \dfrac{2}{15}$

d. $\dfrac{-7}{18} - \left(\dfrac{-5}{24}\right)$

Answers to Your Turn 5

a. $\dfrac{1}{3x}$ b. $2n$

c. $\dfrac{y - 4}{z}$

d. $\dfrac{-6 + 2b}{7a}$

Answers to Your Turn 6

a. $\dfrac{13}{24}$ b. $-\dfrac{1}{2}$ c. $-\dfrac{31}{45}$ d. $-\dfrac{13}{72}$

OBJECTIVE 4 Add and subtract rational expressions with different denominators.

We add and subtract rational expressions using the same procedure as for adding and subtracting numeric fractions.

Your Turn 7

Add or subtract.

a. $\dfrac{3m}{4} + \dfrac{2}{3m}$

b. $\dfrac{7}{3x} - \dfrac{9}{6x^3}$

c. $\dfrac{5}{6a} + \dfrac{2}{4b}$

d. $\dfrac{4}{5xy^2} - \dfrac{1}{3y}$

CONNECTION These are the same expressions as in Example 5 of Section 5.5.

Example 7 Add.

a. $\dfrac{2x}{5} + \dfrac{3}{x}$

Solution: We rewrite the rational expressions with a common denominator, then add the numerators and keep the common denominators. Keep in mind that if the numerators are not like terms, we will have to express the sum as a polynomial.

For 5 and 1 the LCD is 5. For the x's, x^1 has the greater exponent, so the LCD is $5x^1$ or $5x$.

Note: To write an equivalent fraction with $5x$ as the denominator, we rewrite $\frac{2x}{5}$ by multiplying its numerator and denominator by x.

Note: To write an equivalent fraction with $5x$ as the denominator, we rewrite $\frac{3}{x}$ by multiplying its numerator and denominator by 5.

$$= \dfrac{2x(x)}{5(x)} + \dfrac{3(5)}{x(5)}$$

$$= \dfrac{2x^2}{5x} + \dfrac{15}{5x}$$

$$= \dfrac{2x^2 + 15}{5x}$$

Note: Because $2x^2$ and 15 are not like terms, we had to express the sum as a polynomial. Because we cannot factor $2x^2 + 15$, we cannot reduce the rational expression.

b. $\dfrac{3}{8x} + \dfrac{5}{6x^2}$

Solution: The LCM for 8 and 6 is 24. For the x's, x^2 has the greater exponent, so the LCD is $24x^2$. (See Example 5, Section 5.5.)

Note: To write an equivalent fraction with $24x^2$ as the denominator, we rewrite $\frac{3}{8x}$ by multiplying its numerator and denominator by $3x$.

Note: To write an equivalent fraction with $24x^2$ as the denominator, we rewrite $\frac{5}{6x^2}$ by multiplying its numerator and denominator by 4.

$$= \dfrac{3(3x)}{8x(3x)} + \dfrac{5(4)}{6x^2(4)}$$

$$= \dfrac{9x}{24x^2} + \dfrac{20}{24x^2}$$

$$= \dfrac{9x + 20}{24x^2}$$

Note: Because $9x$ and 20 are not like terms, we had to express the sum as a polynomial. Because we cannot factor $9x + 20$, we cannot reduce the rational expression.

◀ **Do Your Turn 7**

OBJECTIVE 5 Add mixed numbers.

There are two methods for adding mixed numbers. 1. We can write the mixed numbers as improper fractions, then add. Or, 2. we can leave them as mixed numbers and add the integers and fractions separately. For example, consider $2\dfrac{1}{5} + 3\dfrac{2}{5}$.

Method 1: Write them as improper fractions.

$$2\dfrac{1}{5} + 3\dfrac{2}{5} = \dfrac{11}{5} + \dfrac{17}{5} = \dfrac{28}{5} = 5\dfrac{3}{5}$$

Answers to Your Turn 7

a. $\dfrac{9m^2 + 8}{12m}$ b. $\dfrac{14x^2 - 9}{6x^3}$

c. $\dfrac{10b + 6a}{12ab}$ d. $\dfrac{12 - 5xy}{15xy^2}$

Method 2: Because mixed numbers are, by definition, the result of addition, we can use the commutative property of addition and add the integers and fractions separately.

$$2\frac{1}{5} + 3\frac{2}{5}$$

$$= \left(2 + \frac{1}{5}\right) + \left(3 + \frac{2}{5}\right) \quad \text{Use the definition of mixed numbers.} \blacktriangleleft$$

Note: We are showing all these steps only to illustrate why we can add the integers and fractions separately. From here on, we will no longer show all these steps.

$$= 2 + 3 + \frac{1}{5} + \frac{2}{5} \quad \text{Use the commutative property of addition.}$$

$$= 5 + \frac{3}{5} \quad \text{Add the integers and fractions separately.}$$

$$= 5\frac{3}{5} \quad \text{Write the sum as a mixed number.}$$

We can summarize the methods as follows.

PROCEDURE *To add mixed numbers:*

Method 1: Write as improper fractions, then follow the procedure for adding fractions.

Method 2: Add the integer parts and fraction parts separately.

Example 8 Add. $5\frac{3}{4} + 4\frac{2}{3}$

Solution:

Method 1: Write the mixed numbers as improper fractions, then add.

$$5\frac{3}{4} + 4\frac{2}{3} = \frac{23}{4} + \frac{14}{3} \quad \text{Write as improper fractions. Notice the LCD is 12.}$$

Note: To rewrite $\frac{23}{4}$ with a denominator of 12, we multiply the numerator and denominator by 3.

$$\blacktriangleright \quad = \frac{23(3)}{4(3)} + \frac{14(4)}{3(4)} \quad \blacktriangleleft$$

Note: To rewrite $\frac{14}{3}$ with a denominator of 12, we multiply the numerator and denominator by 4.

$$= \frac{69}{12} + \frac{56}{12}$$

$$= \frac{125}{12} \quad \text{Add the numerators.}$$

$$= 10\frac{5}{12} \quad \text{Write the result as a mixed number.}$$

Method 2: Add the integers and fractions separately.

$$5\frac{3}{4} + 4\frac{2}{3} = 5\frac{3(3)}{4(3)} + 4\frac{2(4)}{3(4)} \quad \text{Write equivalent fractions with the LCD, 12.}$$

Note: An improper fraction in a mixed number is not considered simplest form. Since $\frac{17}{12} = 1\frac{5}{12}$, we add the 1 to the 9 so that we have $10\frac{5}{12}$. Writing all the steps we have: $9\frac{17}{12} = 9 + \frac{17}{12} = 9 + 1\frac{5}{12} = 10\frac{5}{12}$

$$= 5\frac{9}{12} + 4\frac{8}{12}$$

$$\blacktriangleright \quad = 9\frac{17}{12} \quad \text{Add integers and fractions separately.}$$

$$= 10\frac{5}{12} \quad \text{Simplify the improper fraction within the mixed number.}$$

Do Your Turn 8 ▷

Your Turn 8

Add.

a. $4 + 9\frac{2}{7}$

b. $6\frac{1}{8} + 3\frac{5}{8}$

c. $5\frac{5}{6} + 7\frac{1}{3}$

d. $3\frac{3}{4} + 8\frac{4}{5}$

Answers to Your Turn 8

a. $13\frac{2}{7}$ b. $9\frac{3}{4}$ c. $13\frac{1}{6}$ d. $12\frac{11}{20}$

OBJECTIVE 6 Subtract mixed numbers.

The same two methods we used to add mixed numbers apply to subtracting mixed numbers.

PROCEDURE *To subtract mixed numbers:*

Method 1: Write as improper fractions, then follow the procedure for adding/subtracting fractions.

Method 2: Subtract the integer parts and fraction parts separately.

Example 9 Subtract.

a. $6\dfrac{4}{9} - 2\dfrac{1}{9}$

Method 1: Write the mixed numbers as improper fractions, then subtract.

$$6\frac{4}{9} - 2\frac{1}{9} = \frac{58}{9} - \frac{19}{9} = \frac{39}{9} = 4\frac{3}{9} = 4\frac{1}{3}$$

Method 2: Leave the mixed numbers and subtract the integers and fractions separately. We will show why here, but leave out these "extra" steps in the future.

$$6\frac{4}{9} - 2\frac{1}{9}$$

Note: We are including these extra steps in part a to show why we can subtract the integers and fractions separately.

$$= \left(6 + \frac{4}{9}\right) - \left(2 + \frac{1}{9}\right) \quad \text{Use the definition of mixed numbers.}$$

$$= \left(6 + \frac{4}{9}\right) + \left(-2 - \frac{1}{9}\right) \quad \text{Change the subtraction to addition.}$$

$$= 6 - 2 + \frac{4}{9} - \frac{1}{9} \quad \text{Use the commutative property of addition to separate the integers and fractions.}$$

$$= 4 + \frac{3}{9} \quad \text{Subtract the integers and fractions separately.}$$

$$= 4\frac{1}{3} \quad \text{Write the mixed number and simplify the fraction.}$$

b. $7\dfrac{1}{4} - 2\dfrac{5}{6}$

Solution:

Method 1: Write the mixed numbers as improper fractions, then subtract.

$$7\frac{1}{4} - 2\frac{5}{6} = \frac{29}{4} - \frac{17}{6} \quad \text{Write as improper fractions.}$$

$$= \frac{29(3)}{4(3)} - \frac{17(2)}{6(2)} \quad \text{Write equivalent fractions with the LCD, 12.}$$

$$= \frac{87}{12} - \frac{34}{12}$$

$$= \frac{53}{12} \quad \text{Subtract numerators and keep the LCD.}$$

$$= 4\frac{5}{12} \quad \text{Write as a mixed number.}$$

Method 2: Subtract the integers and fractions separately.

$$7\frac{1}{4} - 2\frac{5}{6} = 7\frac{1(3)}{4(3)} - 2\frac{5(2)}{6(2)} \qquad \text{Rewrite the fractions with the LCD, 12.}$$

$$= 7\frac{3}{12} - 2\frac{10}{12}$$

We have a problem now. Although we can subtract the 2 from 7, if we subtract $\frac{10}{12}$ from $\frac{3}{12}$, we will get a negative number. We need to rewrite $7\frac{3}{12}$ so that its fraction is greater than $\frac{10}{12}$. The steps look like this:

Rewrite 7 as 6 + 1.   Add $\frac{12}{12}$ and $\frac{3}{12}$.

$$7\frac{3}{12} = 6 + 1 + \frac{3}{12} = 6 + \frac{12}{12} + \frac{3}{12} = 6\frac{15}{12}$$

Rewrite 1 as $\frac{12}{12}$.

Note: These steps can be performed mentally by subtracting 1 from the integer, then adding the denominator to the original numerator to get the new numerator.
$$7\frac{3}{12} = 6\frac{12 + 3}{12} = 6\frac{15}{12}$$

Now we can complete the subtraction.

$$= 6\frac{15}{12} - 2\frac{10}{12} \qquad \text{Take 1 from 7 and add it to } \tfrac{3}{12}.$$

$$= 4\frac{5}{12} \qquad \text{Subtract the integers and fractions separately.}$$

Do Your Turn 9 ▷

Your Turn 9

Subtract.

a. $9\frac{7}{8} - 4\frac{5}{8}$

b. $7\frac{9}{10} - 1\frac{1}{5}$

c. $10\frac{1}{5} - 3\frac{3}{4}$

d. $8 - 5\frac{1}{4}$

OBJECTIVE **7** Add and subtract signed mixed numbers.

We have seen that using the second method to add mixed numbers sometimes leads to a mixed number sum that must be simplified because it contains an improper fraction. Similarly, when we subtract using the second method, we sometimes have to rewrite one of the mixed numbers so that the fraction parts can be subtracted. But using the first method avoids these extra steps. Since these extra steps can be distracting when we have signed mixed numbers, when adding or subtracting signed mixed numbers, we will only use the first method.

Your Turn 10

Add or subtract.

a. $5\frac{3}{4} + \left(-5\frac{1}{8}\right)$

b. $7\frac{1}{2} - 10\frac{1}{3}$

c. $-5\frac{7}{10} - 2\frac{1}{4}$

d. $8\frac{2}{9} - \left(\frac{5}{6}\right)$

Example 10 Subtract. $-9\frac{1}{6} - \left(-2\frac{1}{3}\right)$

Solution: First, we will write an equivalent addition. Also, to avoid potential extra steps, we will write the mixed numbers as improper fractions.

$$-9\frac{1}{6} - \left(-2\frac{1}{3}\right) = -9\frac{1}{6} + 2\frac{1}{3} \qquad \begin{array}{l}\text{Write an equivalent addition using the}\\\text{additive inverse of the subtrahend.}\end{array}$$

$$= -\frac{55}{6} + \frac{7}{3} \qquad \begin{array}{l}\text{Write the mixed numbers as improper}\\\text{fractions. Note that the LCD is 6.}\end{array}$$

Note: $-\frac{55}{6}$ already has 6 as its denominator, so we do not need to rewrite it. ▶

$$= -\frac{55}{6} + \frac{7(2)}{3(2)}$$

◀ **Note:** To write $\frac{7}{3}$ as an equivalent fraction with 6 as the denominator, we multiply the numerator and denominator by 2.

$$= -\frac{55}{6} + \frac{14}{6}$$

$$= -\frac{41}{6} \qquad \text{Add.}$$

$$= -6\frac{5}{6} \qquad \text{Write the sum as a mixed number.}$$

Do Your Turn 10 ▷

Answers to Your Turn 9

a. $5\frac{1}{4}$ b. $6\frac{7}{10}$ c. $6\frac{9}{20}$ d. $2\frac{3}{4}$

Answers to Your Turn 10

a. $\frac{5}{8}$ b. $-2\frac{5}{6}$

c. $-7\frac{19}{20}$ d. $-7\frac{7}{18}$

Your Turn 11

Solve and check.

a. $\dfrac{1}{3} + y = \dfrac{3}{4}$

b. $a - \dfrac{1}{5} = \dfrac{5}{8}$

c. $m + 3\dfrac{1}{2} = -6\dfrac{1}{4}$

d. $-4\dfrac{1}{7} = t - \dfrac{2}{5}$

OBJECTIVE 8 Solve equations.

In Section 5.4, we revisited solving equations. We used the multiplication/division principle of equality to clear fraction coefficients. Now we will use the addition/subtraction principle of equality to clear fraction terms in an equation. In Section 4.2, we learned that to clear a term using the addition/subtraction principle of equality, we add its additive inverse to both sides of the equation. That is, we add or subtract appropriately so that the undesired term is eliminated.

Example 11 Solve and check. $x + \dfrac{3}{8} = \dfrac{5}{6}$

Solution: We want to isolate x, so we subtract $\dfrac{3}{8}$ from both sides.

$$x + \dfrac{3}{8} = \dfrac{5}{6}$$

Note: $\frac{3}{8} - \frac{3}{8} = 0$, which isolates x on the left side. ▶

$$x + \dfrac{3}{8} - \dfrac{3}{8} = \dfrac{5}{6} - \dfrac{3}{8}$$

$$x + 0 = \dfrac{5(4)}{6(4)} - \dfrac{3(3)}{8(3)}$$

◀ **Note:** To complete the subtraction on the right side, we must find a common denominator. The LCD is 24.

$$x = \dfrac{20}{24} - \dfrac{9}{24}$$

$$x = \dfrac{11}{24}$$

Check: In the original equation, replace x with $\frac{11}{24}$ and verify the equation is true.

$$x + \dfrac{3}{8} = \dfrac{5}{6}$$

$$\dfrac{11}{24} + \dfrac{3}{8} \overset{?}{=} \dfrac{5}{6} \qquad \text{Replace } x \text{ with } \tfrac{11}{24}.$$

$$\dfrac{11}{24} + \dfrac{3(3)}{8(3)} \overset{?}{=} \dfrac{5}{6} \qquad \text{Write equivalent fractions with the LCD, 24.}$$

$$\dfrac{11}{24} + \dfrac{9}{24} \overset{?}{=} \dfrac{5}{6}$$

$$\dfrac{20}{24} \overset{?}{=} \dfrac{5}{6} \qquad \text{Add the fractions.}$$

$$\dfrac{5}{6} = \dfrac{5}{6} \qquad \text{Simplify. The equation is true, so } \tfrac{11}{24} \text{ is the solution.}$$

◀ **Do Your Turn 11**

OBJECTIVE 9 Solve applications.

What clues can we look for that indicate that we need to add or subtract? With addition we look for words that indicate we must find a total. For subtraction, we look for clues that indicate we must find a difference. Another clue for addition/subtraction situations is that the fractions represent distinct categories.

Answers to Your Turn 11

a. $y = \dfrac{5}{12}$ b. $a = \dfrac{33}{40}$

c. $m = -9\dfrac{3}{4}$ d. $t = -3\dfrac{26}{35}$

Example 12 In a poll, $\frac{1}{4}$ of the respondents said they agreed with the passing of a certain bill, $\frac{1}{5}$ said they disagreed, and the rest said they had no opinion. What fraction of the respondents had no opinion?

Understand: There are three distinct ways to respond to the poll: one can agree, disagree, or have no opinion. Because the three categories together make up the whole group that responded, we can say this:

$$\boxed{\text{respondents who agreed}} + \boxed{\text{respondents who disagreed}} + \boxed{\text{respondents who had no opinion}} = \boxed{\text{all respondents}}$$

Because we are dealing with fractions, *all respondents* means a whole amount, which is the number 1. The number 1 always represents the whole amount.

Plan: Use the preceding formula to write an equation, then solve.

Execute: Let *n* represent the number of respondents that had no opinion.

$$\boxed{\text{respondents who agreed}} + \boxed{\text{respondents who disagreed}} + \boxed{\text{respondents who had no opinion}} = \boxed{\text{all respondents}}$$

$$\frac{1}{4} \quad + \quad \frac{1}{5} \quad + \quad n \quad = \quad 1$$

$$\frac{1(5)}{4(5)} + \frac{1(4)}{5(4)} + n = \frac{1(20)}{1(20)} \qquad \text{Write equivalent fractions with the LCD, 20.}$$

$$\frac{5}{20} + \frac{4}{20} + n = \frac{20}{20}$$

$$\frac{9}{20} + n = \frac{20}{20} \qquad \text{Combine fractions on the left side.}$$

$$\frac{9}{20} - \frac{9}{20} + n = \frac{20}{20} - \frac{9}{20} \qquad \text{Subtract } \tfrac{9}{20} \text{ from both sides to isolate } n.$$

$$n = \frac{11}{20}$$

Answer: $\frac{11}{20}$ had no opinion.

Check: Add the three fractions together. The whole group should be the sum, which is represented by the number 1.

$$\frac{1}{4} + \frac{1}{5} + n = 1$$

$$\frac{1}{4} + \frac{1}{5} + \frac{11}{20} \stackrel{?}{=} 1 \qquad \text{Replace } n \text{ with } \tfrac{11}{20}.$$

$$\frac{1(5)}{4(5)} + \frac{1(4)}{5(4)} + \frac{11}{20} \stackrel{?}{=} 1 \qquad \text{Write equivalent fractions with the LCD, 20.}$$

$$\frac{5}{20} + \frac{4}{20} + \frac{11}{20} \stackrel{?}{=} 1$$

$$\frac{20}{20} \stackrel{?}{=} 1 \qquad \text{Add.}$$

$$1 = 1 \qquad \text{Simplify. It checks.}$$

Do Your Turn 12 ▶

Your Turn 12

Solve.

a. A report claims that $\frac{3}{8}$ of all cars are blue, $\frac{1}{4}$ are red, and the rest are all other colors. What fraction of all cars are other colors?

b. At the opening of the stock market, a company's stock is worth $\$5\frac{3}{8}$ per share. At the close of the stock market, the same stock is worth $\$6\frac{1}{4}$. How much did the stock increase in value?

Tip If you are uncertain how to approach a problem containing fractions, try replacing the fractions with whole numbers and read the problem again. Suppose Example 12 read like this:

In a poll, 25 of the respondents said they agreed with the passing of a certain bill, 20 said they disagreed, and the rest said they had no opinion. How many of the respondents had no opinion?

If we knew the total number of people in the poll, to solve we could add the number that agreed plus the number that disagreed, then subtract from the total number of people polled. With fractions we do not need the total number because it is always represented by 1. Therefore, to solve our equation, we added the fraction that agreed plus the fraction that disagreed, then subtracted the sum from all respondents, which is represented by 1.

Answers to Your Turn 12

a. $\frac{3}{8}$ b. $\frac{7}{8}$

5.6 **Exercises**

For
Extra
Help

Videotape
DVT

Addison-Wesley
Tutor Center

Math XL
Math XL

MyMathLab

Student Solutions
Manual

1. Explain how to add or subtract fractions that have the same denominator.

2. Explain how to add or subtract fractions that have different denominators.

3. Suppose you are given the problem $\frac{3}{4} + \frac{5}{6}$ and you know the LCD is 12. Explain how to write the equivalent fractions so that they have the LCD.

4. Explain how to add or subtract mixed numbers.

For Exercises 5–22, add or subtract.

5. $\frac{2}{7} + \frac{3}{7}$

6. $\frac{1}{13} + \frac{6}{13}$

7. $\frac{-4}{9} + \left(\frac{-2}{9}\right)$

8. $-\frac{2}{15} + \left(-\frac{8}{15}\right)$

9. $\frac{10}{17} - \frac{4}{17}$

10. $\frac{9}{11} - \frac{1}{11}$

11. $\frac{6}{35} - \frac{13}{35}$

12. $\frac{1}{30} - \frac{4}{30}$

13. $\frac{9}{x} + \frac{3}{x}$

14. $\frac{4}{ab} + \frac{16}{ab}$

15. $\frac{8x^2}{9} - \frac{2x^2}{9}$

16. $\frac{x}{6} + \frac{1}{6}$

17. $\frac{4m}{n} - \frac{6m}{n}$

18. $\frac{4}{5h} - \frac{k}{5h}$

19. $\frac{3x^2 + 4x}{7y} + \frac{x^2 - 7x + 1}{7y}$

20. $\frac{9t - 12}{11u^2} + \frac{6t + 5}{11u^2}$

21. $\frac{7n^2}{5m} - \frac{2n^2 + 3}{5m}$

22. $\frac{h^2 - 3h + 5}{k} - \frac{h^2 + 4h - 1}{k}$

For Exercises 23–38, add or subtract.

23. $\frac{3}{10} + \frac{5}{6}$

24. $\frac{2}{15} - \frac{4}{9}$

25. $\frac{7}{12} - \frac{11}{36}$

26. $\frac{1}{8} + \frac{15}{32}$

27. $\frac{1}{20} + \frac{17}{30}$

28. $\frac{3}{16} - \frac{9}{20}$

29. $\frac{3}{4} + \frac{1}{6} + \frac{7}{9}$

30. $\frac{2}{5} + \frac{5}{6} + \frac{13}{15}$

31. $\frac{7x}{12} + \frac{3x}{8}$

32. $\frac{a}{9} + \frac{2a}{3}$

33. $\frac{9}{16m} - \frac{3}{8m}$

34. $\frac{7}{10x} - \frac{5}{6x}$

35. $\frac{13}{20} + \frac{4}{5h}$

36. $\frac{7}{4x} + \frac{5}{12}$

37. $\frac{2}{3n^2} - \frac{7}{9n}$

38. $\frac{1}{2t} - \frac{3}{5t^2}$

For Exercises 39–46, add.

39. $3\frac{4}{9} + 7$

40. $4\frac{2}{5} + \frac{1}{5}$

41. $2\frac{1}{4} + 5\frac{1}{4}$

42. $6\frac{1}{7} + 1\frac{5}{7}$

43. $5\frac{5}{6} + 1\frac{2}{3}$

44. $3\frac{4}{5} + 7\frac{1}{4}$

45. $6\frac{5}{8} + 3\frac{7}{12}$

46. $9\frac{1}{2} + 10\frac{4}{9}$

For Exercises 47–54, subtract.

47. $9\frac{7}{8} - \frac{1}{8}$

48. $4\frac{5}{9} - 3$

49. $11\frac{7}{12} - 2\frac{5}{12}$

50. $8\frac{3}{7} - 2\frac{1}{7}$

51. $5\frac{1}{8} - 1\frac{5}{6}$

52. $6\frac{2}{5} - 2\frac{3}{4}$

53. $8\frac{1}{3} - 7\frac{5}{6}$

54. $10\frac{5}{12} - 6\frac{5}{8}$

For Exercises 55–62, add or subtract.

55. $6\frac{3}{4} + \left(-2\frac{1}{3}\right)$

56. $-5\frac{1}{2} + 2\frac{1}{8}$

57. $-7\frac{5}{6} - 2\frac{2}{3}$

58. $-\frac{4}{5} - 3\frac{1}{2}$

59. $\frac{5}{8} - 4\frac{1}{4}$

60. $6\frac{7}{8} - \left(-3\frac{1}{4}\right)$

61. $-7\frac{3}{4} - \left(-1\frac{1}{8}\right)$

62. $-2\frac{1}{6} - \left(-3\frac{1}{2}\right)$

For Exercises 63–70, solve and check.

63. $x + \frac{3}{5} = \frac{7}{10}$

64. $\frac{1}{4} + y = \frac{5}{8}$

65. $n - \frac{4}{5} = \frac{1}{4}$

66. $\frac{5}{8} = k - \frac{1}{6}$

67. $-\frac{1}{6} = b + \frac{3}{4}$

68. $-\frac{2}{3} = h - \frac{1}{8}$

69. $-3\frac{1}{2} + t = -4\frac{1}{5}$

70. $-2\frac{1}{6} = m - \frac{3}{4}$

For Exercises 71–78, solve.

71. A tabletop is $\frac{3}{4}$ inches thick. During sanding, $\frac{1}{16}$ inch is removed off the top. How thick is the tabletop after sanding?

72. Amanda wants to hang a picture frame so that the bottom of the frame is $54\frac{1}{2}$ inches from the floor. The hanger on the back of the picture is $11\frac{5}{8}$ inches from the bottom of the frame. How high should she place the nail?

73. A poll is taken to assess the president's approval rating. Respondents can answer four ways: excellent, good, fair, or poor. $\frac{1}{8}$ said "excellent," $\frac{3}{5}$ said "good," and $\frac{1}{6}$ said "fair."

   a. What fraction of the respondents said "excellent" or "good?"

   b. What fraction said "poor?"

74. In Mrs. Robinson's English class, students can earn a grade of A, B, C, or F (no D's are given). $\frac{1}{3}$ of the students got an A, $\frac{3}{8}$ got a B, and $\frac{1}{6}$ got a C.

   a. What fraction of the students received an A or B?

   b. What fraction of the students received an F?

75. A wooden frame is to be $20\frac{3}{4}$ inches long by $13\frac{1}{2}$ inches wide. What total length of wood is needed to make the frame?

76. A game preserve is in the shape of a triangle. What is the total distance around the park?

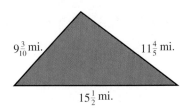

77. What is the width, w, of the hallway in the following floor plan?

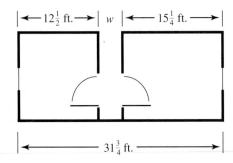

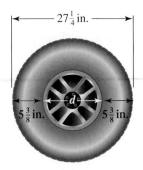

78. What is the diameter, d, of the hubcap on the tire shown?

Review Exercises

[3.1]   **1.** Evaluate $x^2 - 2x + \sqrt{x}$ when $x = 4$.

[3.1]   **2.** Evaluate $y^2 + 2y + 3$ when $y = -2$.

[2.5]   **3.** Simplify. $3^2 - 5[2 + (18 \div (-2))] + \sqrt{49}$

[3.4]   **4.** Add. $(4x^2 + 5x - 6) + (6x^2 - 2x + 1)$

[3.4]   **5.** Subtract. $(8y^3 - 4y + 2) - (9y^3 + y^2 + 7)$

[3.5]   **6.** Multiply. $(-6b^3)(9b)$

[3.5]   **7.** Multiply. $(x + 3)(2x - 5)$

[3.7]   **8.** Divide. $\dfrac{-16x^7y^6}{4x^5y}$

[3.8]   **9.** Write an expression in simplest form for the volume of the box shown.

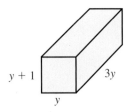

Order of Operations and Evaluating and Simplifying Expressions

OBJECTIVE 1 Use the order of operations agreement to simplify expressions containing fractions and mixed numbers.

OBJECTIVES

1 Use the order of operations agreement to simplify expressions containing fractions and mixed numbers.

2 Evaluate expressions.

3 Find the area of a trapezoid.

4 Find the area of a circle.

5 Simplify polynomials containing fractions.

Example 1 Simplify. $\frac{3}{4} + 2\frac{4}{5} \cdot \left(-\frac{5}{6}\right)$

Solution: Follow the order of operations agreement. We multiply first.

$$\frac{3}{4} + 2\frac{4}{5} \cdot \left(-\frac{5}{6}\right)$$

$$= \frac{3}{4} + \frac{14}{5} \cdot \left(-\frac{5}{6}\right) \qquad \text{Write } 2\frac{4}{5} \text{ as an improper fraction so that we can multiply.}$$

$$= \frac{3}{4} + \frac{\overset{7}{\cancel{14}}}{\cancel{5}} \cdot \left(-\frac{\overset{1}{\cancel{5}}}{\cancel{6}}\right) \qquad \text{Divide out a common factor of 2 in 14 and 6, then divide out the common 5's.}$$

$$= \frac{3}{4} + \left(-\frac{7}{3}\right) \qquad \text{Multiply the remaining factors.}$$

$$= \frac{3(3)}{4(3)} + \left(-\frac{7(4)}{3(4)}\right) \qquad \text{Write equivalent fractions with the LCD, 12.}$$

$$= \frac{9}{12} + \left(-\frac{28}{12}\right)$$

$$= -\frac{19}{12} \qquad \text{Add.}$$

$$= -1\frac{7}{12} \qquad \text{Write as a mixed number.}$$

Do Your Turn 1 ▷

Your Turn 1

Simplify.

a. $\frac{5}{8} - \frac{2}{9} \cdot \frac{3}{4}$

b. $3\frac{1}{2} + \frac{4}{5} \div \frac{3}{10}$

c. $\frac{1}{6} + \left(-4\frac{1}{3}\right) \cdot 2\frac{1}{2}$

d. $-5\frac{3}{8} - \frac{7}{10} \div \left(-\frac{7}{30}\right)$

Example 2 Simplify. $\left(\frac{1}{2}\right)^3 - 6\left(\frac{3}{5} + \frac{2}{3}\right)$

Solution: First, we find the sum in the parentheses. The LCD for $\frac{3}{5}$ and $\frac{2}{3}$ is 15.

$$\left(\frac{1}{2}\right)^3 - 6\left(\frac{3}{5} + \frac{2}{3}\right) = \left(\frac{1}{2}\right)^3 - 6\left(\frac{3(3)}{5(3)} + \frac{2(5)}{3(5)}\right) \qquad \text{Write equivalent fractions with the LCD, 15.}$$

$$= \left(\frac{1}{2}\right)^3 - 6\left(\frac{9}{15} + \frac{10}{15}\right)$$

$$= \left(\frac{1}{2}\right)^3 - 6 \cdot \frac{19}{15} \qquad \text{Add.}$$

$$= \frac{1}{8} - 6 \cdot \frac{19}{15} \qquad \text{Evaluate the expression with the exponent. } \frac{1}{2} \cdot \frac{1}{2} \cdot \frac{1}{2} = \frac{1}{8}$$

$$= \frac{1}{8} - \frac{\overset{2}{\cancel{6}}}{1} \cdot \frac{19}{\underset{5}{\cancel{15}}} \qquad \text{Divide out the common factors.}$$

Answers to Your Turn 1

a. $\frac{11}{24}$ b. $6\frac{1}{6}$ c. $-10\frac{2}{3}$ d. $-2\frac{3}{8}$

Simplify.

a. $\left(\dfrac{1}{3}\right)^2 + 16\left(\dfrac{1}{4} + \dfrac{3}{8}\right)$

b. $\left(\dfrac{1}{25} - \dfrac{4}{5}\right) - \left(\dfrac{2}{5}\right)^2$

c. $\left(1\dfrac{2}{3}\right)^2 + 2\dfrac{1}{2} \div \dfrac{5}{8}$

d. $\left(\dfrac{1}{2} + \dfrac{1}{3}\right) \div \left(\dfrac{3}{4} - \dfrac{1}{8}\right)$

$$= \dfrac{1}{8} - \dfrac{38}{5} \qquad \text{Multiply.}$$

$$= \dfrac{1(5)}{8(5)} - \dfrac{38(8)}{5(8)} \qquad \text{Write equivalent fractions with the LCD, 40.}$$

$$= \dfrac{5}{40} - \dfrac{304}{40}$$

$$= -\dfrac{299}{40} \qquad \text{Subtract.}$$

$$= -7\dfrac{19}{40} \qquad \text{Write the answer as a mixed number.}$$

◀ **Do Your Turn 2**

OBJECTIVE 2 Evaluate expressions.

Now let's evaluate algebraic expressions when we are given variable values that are fractions. In Section 3.1, we learned that to evaluate an expression, we replace the variables with the corresponding given values and then use the order of operations agreement to simplify the numeric expression.

Evaluate each expression using the given values.

a. mv^2; $m = 3\dfrac{1}{4}$, $v = 10$

b. $\dfrac{1}{2}at^2$; $a = -9\dfrac{4}{5}$, $t = 2$

c. $a^2 - 5b$; $a = \dfrac{3}{4}$, $b = \dfrac{7}{20}$

d. $2m(n + 3)$; $m = \dfrac{5}{8}$, $n = 1\dfrac{1}{3}$

Example 3 Evaluate $4xy^2 + z$ when $x = \frac{3}{4}$, $y = \frac{2}{3}$, and $z = 3\frac{1}{2}$.

Solution:

$$4\left(\dfrac{3}{4}\right)\left(\dfrac{2}{3}\right)^2 + 3\dfrac{1}{2} \qquad \text{Replace } x \text{ with } \tfrac{3}{4}, y \text{ with } \tfrac{2}{3}, \text{ and } z \text{ with } 3\tfrac{1}{2}.$$

$$= \dfrac{4}{1}\left(\dfrac{3}{4}\right)\dfrac{4}{9} + 3\dfrac{1}{2} \qquad \text{Evaluate the expression with the exponent and write 4 as } \tfrac{4}{1}.$$

$$= \dfrac{\overset{1}{\cancel{4}}}{1}\left(\dfrac{\overset{1}{\cancel{3}}}{\underset{1}{\cancel{4}}}\right)\dfrac{4}{\underset{3}{\cancel{9}}} + 3\dfrac{1}{2} \qquad \text{Divide out the common factors.}$$

$$= \dfrac{4}{3} + 3\dfrac{1}{2} \qquad \text{Multiply.}$$

$$= 1\dfrac{1}{3} + 3\dfrac{1}{2} \qquad \text{Change } \quad \text{to a mixed number in order to add.}$$

$$= 1\dfrac{1(2)}{3(2)} + 3\dfrac{1(3)}{2(3)} \qquad \text{Write equivalent fractions with the LCD, 6.}$$

$$= 1\dfrac{2}{6} + 3\dfrac{3}{6}$$

$$= 4\dfrac{5}{6} \qquad \text{Add.}$$

Note: We could have written $3\frac{1}{2}$ as $\frac{7}{2}$ and then added the improper fractions. Choose the method that seems simplest to you.

◀ **Do Your Turn 3**

Answers to Your Turn 2

a. $10\dfrac{1}{9}$ b. $-\dfrac{23}{25}$ c. $6\dfrac{7}{9}$ d. $1\dfrac{1}{3}$

Answers to Your Turn 3

a. 325 b. $-19\dfrac{3}{5}$

c. $-1\dfrac{3}{16}$ d. $5\dfrac{5}{12}$

OBJECTIVE 3 Find the area of a trapezoid.

We have learned formulas for finding the area of a parallelogram ($A = bh$) and a triangle ($A = \frac{1}{2}bh$). Now let's develop a formula for finding the area of a **trapezoid**.

DEFINITION Trapezoid: A four-sided figure with one pair of parallel sides.

The following figures are examples of trapezoids:

In each figure, notice that the top and bottom sides are parallel but the other sides are not. Because the top and bottom sides are different lengths, we will label them a and b.

To develop the formula for the area of a trapezoid, we use the same process we used for finding the formula for the area of a triangle. First, we make two identical trapezoids. By inverting one of the trapezoids, then joining the corresponding sides, we create a parallelogram, as shown here.

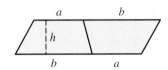

The area of this parallelogram is found by multiplying the base and height. Notice that the base is $a + b$ and the height is h, so the area of the parallelogram is $h(a + b)$. However, we really want the area of one of the trapezoids, which is half the area of the parallelogram.

Conclusion: The formula for the area of a trapezoid is $A = \dfrac{1}{2}h(a + b)$.

Example 4 Find the area of the trapezoid shown.

Solution: Use the formula $A = \dfrac{1}{2}h(a + b)$.

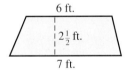

$$A = \frac{1}{2}h(a + b)$$

$$A = \frac{1}{2}\left(2\frac{1}{2}\right)(6 + 7) \qquad \text{Replace } h \text{ with } 2\frac{1}{2}, a \text{ with 6, and } b \text{ with 7.}$$

$$A = \frac{1}{2}\left(2\frac{1}{2}\right)(13) \qquad \text{Add 6 and 7 inside the parentheses.}$$

$$A = \frac{1}{2}\left(\frac{5}{2}\right)\left(\frac{13}{1}\right) \qquad \text{Write as improper fractions.}$$

$$A = \frac{65}{4} \qquad \text{Multiply.}$$

$$A = 16\frac{1}{4}$$

Note: Remember, area is always in square units. Because the distance units are in terms of feet, the area unit is in terms of square feet.

Answer: The area is $16\frac{1}{4}$ ft.^{2}

Do Your Turn 4 ▷

Your Turn 4

Calculate the area of the trapezoid shown.

a.

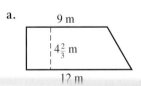

b.

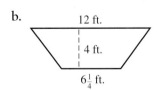

Answers to Your Turn 4

a. 49 m^2 b. $36\frac{1}{2}$ ft.^{2}

OBJECTIVE 4 Find the area of a circle.

In Section 5.3, we developed the circumference formula. Now let's develop a formula for the area of a circle. As we did for triangles and trapezoids, to develop the formula for the area of a circle, we will transform a circle into a parallelogram.

To make a circle look like a parallelogram, we cut it up like a pizza. First, we cut the top half into six pizza slices and fan those slices out. Then we do the same with the bottom half of the circle.

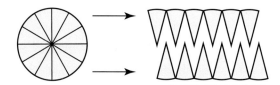

Next, the slices are fit together like teeth. The resulting figure closely resembles a parallelogram. This parallelogram has the same area as the circle.

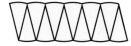 **Note:** If we were to cut more slices, the *slight curves* along the top and bottom would be less prominent. In fact, if we were to use extremely slender slices, the curves would be almost imperceptible.

Because the area of a parallelogram is found by multiplying base and height, we need to relate the base and height of the parallelogram back to the circle. If our circle is a pizza, then the height goes from the center to crust, so the height corresponds to the radius of the circle. The base distance corresponds to the crust of the bottom half of the pizza, so the base is half of the circumference of the circle.

The height corresponds to the radius.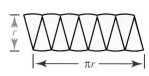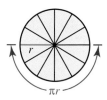

The base corresponds to half of the circumference. The full circumference is found by $2\pi r$, so half circumference is πr.

Because the area of a parallelogram is found by multiplying base and height, we can say

$$A = \text{base} \cdot \text{height}$$
$$A = \pi r \cdot r$$
$$A = \pi r^2$$

Conclusion: The formula for the area of a circle with radius r is $A = \pi r^2$.

Example 5 Calculate the area of a circle with a diameter of 5 meters.

Solution: Use the formula $A = \pi r^2$. Because we were given the diameter, we must first find the radius.

$$r = \frac{1}{2}d$$

$$r = \frac{1}{2} \cdot 5 \qquad \text{Replace } d \text{ with 5.}$$

$$r = \frac{1}{2} \cdot \frac{5}{1}$$

$$r = \frac{5}{2} \qquad$$ **Note:** Because we'll square the radius to find area, we will leave the fraction as an improper fraction.

The radius is $\frac{5}{2}$ meters. Now we can find the area using $A = \pi r^2$.

Note: We approximate π as $\frac{22}{7}$. When we replace π with $\frac{22}{7}$, the entire calculation becomes an approximation, so we use $\approx$ instead of an equal sign.

▶ $A = \pi r^2$

$A \approx \frac{22}{7} \cdot \left(\frac{5}{2}\right)^2$ **Replace π with $\frac{22}{7}$ and r with $\frac{5}{2}$.**

$A \approx \frac{22}{7} \cdot \frac{25}{4}$ **Square $\frac{5}{2}$ to get $\frac{25}{4}$.**

$A \approx \frac{\overset{11}{\cancel{22}}}{7} \cdot \frac{25}{\underset{2}{\cancel{4}}}$ **Divide out the common factor of 2 in 22 and 4.**

$A \approx \frac{275}{14}$ **Multiply.**

$A \approx 19\frac{9}{14}$ **Write as a mixed number.**

Answer: The area is $19\frac{9}{14}$ m^2. ◀ **Note:** Area is always in square units. Because the distance unit for the radius is meters, the area unit is square meters.

Do Your Turn 5 ▷

Your Turn 5

Solve.

a. Calculate the area of a circle with a radius of 6 yards.

b. Calculate the area of a circle with a diameter of 7 inches.

OBJECTIVE 5 Simplify polynomials containing fractions.

We learned in Section 3.3 that we can simplify expressions by combining like terms. Like terms have the same variables raised to the same exponents. We combine like terms by adding or subtracting the coefficients.

Example 6 Combine like terms. $\frac{1}{4}a^3 - \frac{3}{5}a + \frac{1}{6}a^3 + \frac{1}{2}a$

Solution: Add or subtract coefficients for the like terms. Because we cannot combine an a^3 term with an a term, we only need to get common denominators for the fractions that we will actually combine.

$\frac{1}{4}a^3 - \frac{3}{5}a + \frac{1}{6}a^3 + \frac{1}{2}a$

$= \frac{1}{4}a^3 + \frac{1}{6}a^3 - \frac{3}{5}a + \frac{1}{2}a$ **Use the commutative property to group the like terms together. Remember, this is optional.**

$= \frac{1(3)}{4(3)}a^3 + \frac{1(2)}{6(2)}a^3 - \frac{3(2)}{5(2)}a + \frac{1(5)}{2(5)}a$ **Write equivalent fractions. The LCD for the a^3 coefficients is 12. The LCD for the a coefficients is 10.**

$= \frac{3}{12}a^3 + \frac{2}{12}a^3 - \frac{6}{10}a + \frac{5}{10}a$

$= \frac{5}{12}a^3 - \frac{1}{10}a$ **Add or subtract coefficients.**

Note: Since a^3 and a are not like terms, we cannot combine further. The expression is in simplest form.

Do Your Turn 6 ▷

Your Turn 6

Combine like terms.

a. $\frac{5}{8}x + \frac{1}{2} - \frac{3}{4}x^2 + \frac{2}{3} - \frac{1}{4}x$

b. $\frac{1}{6}m^2 - 3m + \frac{2}{3}m^2 + \frac{1}{7} - \frac{1}{2}m$

Answers to Your Turn 5
a. $113\frac{1}{7}$ yd.^{2} b. $38\frac{1}{2}$ in.^{2}

Answers to Your Turn 6
a. $-\frac{3}{4}x^2 + \frac{3}{8}x + \frac{7}{6}$
b. $\frac{5}{6}m^2 - \frac{7}{2}m + \frac{1}{7}$

In Section 3.4, we learned how to add and subtract polynomials. To add polynomials, we combine like terms. To subtract, we write an equivalent addition statement using the additive inverse of the subtrahend, which is the second polynomial.

Your Turn 7

Simplify.

a.

$$\left(\frac{3}{8}n - \frac{2}{3}\right) + \left(\frac{1}{4}n - \frac{1}{6}\right)$$

b.

$$\left(\frac{2}{5}x^2 + \frac{5}{16}\right) - \left(\frac{1}{10}x^2 - \frac{3}{4}\right)$$

Example 7 Simplify. $\left(\frac{3}{4}h - \frac{5}{6}\right) - \left(\frac{1}{3}h - \frac{2}{5}\right)$

Solution: Write an equivalent addition by changing the operation to addition and the subtrahend to its additive inverse. To write the additive inverse of a polynomial, we change the sign of each term in the polynomial.

$$\left(\frac{3}{4}h - \frac{5}{6}\right) - \left(\frac{1}{3}h - \frac{2}{5}\right)$$

$$= \left(\frac{3}{4}h - \frac{5}{6}\right) + \left(-\frac{1}{3}h + \frac{2}{5}\right) \quad \text{Write an equivalent addition statement.}$$

$$= \frac{3}{4}h - \frac{1}{3}h - \frac{5}{6} + \frac{2}{5} \quad \text{Collect like terms.}$$

$$= \frac{3(3)}{4(3)}h - \frac{1(4)}{3(4)}h - \frac{5(5)}{6(5)} + \frac{2(6)}{5(6)} \quad \text{Write equivalent fractions with their LCD.}$$

$$= \frac{9}{12}h - \frac{4}{12}h - \frac{25}{30} + \frac{12}{30}$$

$$= \frac{5}{12}h - \frac{13}{30} \quad \text{Combine the like terms.}$$

◀ **Do Your Turn 7**

We can also multiply monomials and polynomials with fractional coefficients. Consider multiplying monomials first. We learned in Section 3.5 that to multiply monomials, we multiply the coefficients and add the exponents of the like bases.

Your Turn 8

Multiply.

a. $\left(\frac{1}{4}a\right)\left(\frac{2}{7}a^5\right)$

b. $\left(\frac{-5}{12}h^4\right)\left(\frac{6}{-11}h^3\right)$

Example 8 Multiply. $\left(-\frac{3}{10}m^4\right)\left(\frac{2}{9}m\right)$

Solution: Multiply the coefficients and add the exponents of the like bases.

$$\left(-\frac{3}{10}m^4\right)\left(\frac{2}{9}m\right) = -\frac{\overset{1}{\cancel{3}}}{\underset{5}{\cancel{10}}} \cdot \frac{\overset{1}{2}}{\underset{3}{9}}m^{4+1} \quad \begin{array}{l}\text{Divide out the common factors in}\\ \text{the coefficients and add the exponents}\\ \text{of the bases.}\end{array}$$

$$= -\frac{1}{15}m^5 \quad \begin{array}{l}\text{Multiply the coefficients and simplify the}\\ \text{exponents.}\end{array}$$

◀ **Do Your Turn 8**

Next, as we learned in Section 3.5, when we multiply a polynomial by a monomial, we use the distributive property and multiply each term inside the polynomial by the monomial.

Answers to Your Turn 7

a. $\frac{5}{8}n - \frac{5}{6}$ b. $\frac{3}{10}x^2 + \frac{17}{16}$

Answers to Your Turn 8

a. $\frac{1}{14}a^6$ b. $\frac{5}{22}h^7$

Example 9 Multiply. $\dfrac{3}{5}\left(\dfrac{2}{3}x - 10\right)$

Solution: We use the distributive property and multiply each term in $\frac{2}{3}x - 10$ by $\frac{3}{5}$.

$$\dfrac{3}{5}\left(\dfrac{2}{3}x - 10\right) = \dfrac{3}{5} \cdot \dfrac{2}{3}x - \dfrac{3}{5} \cdot \dfrac{10}{1} \qquad \text{Distribute } \tfrac{3}{5}.$$

$$= \dfrac{\overset{1}{\cancel{3}}}{5} \cdot \dfrac{2}{\underset{1}{\cancel{3}}}x - \dfrac{3}{\underset{1}{\cancel{5}}} \cdot \dfrac{\overset{2}{\cancel{10}}}{1} \qquad \text{Divide out the common factors.}$$

$$= \dfrac{2}{5}x - 6 \qquad \text{Multiply.}$$

Do Your Turn 9 ▷

Your Turn 9

Multiply.

a. $\dfrac{2}{3}\left(\dfrac{3}{7}a^3 - 9a^2\right)$

b. $-\dfrac{5}{12}\left(6m^2 + \dfrac{2}{5}mn\right)$

Now let's multiply polynomials. In Section 3.5, we learned that when we multiply polynomials, we multiply every term in the second polynomial by every term in the first polynomial.

Example 10 Multiply. $\left(\dfrac{3}{4}x - 5\right)\left(\dfrac{1}{2}x + 8\right)$

Solution: We multiply every term in the second polynomial by every term in the first polynomial.

$$\left(\dfrac{3}{4}x - 5\right)\left(\dfrac{1}{2}x + 8\right)$$

$$= \dfrac{3}{4}x \cdot \dfrac{1}{2}x + \dfrac{3}{4}x \cdot 8 - 5 \cdot \dfrac{1}{2}x - 5 \cdot 8 \qquad \begin{array}{l}\text{Multiply every term in the second}\\\text{polynomial by every term in the first}\\\text{polynomial.}\end{array}$$

$$= \dfrac{3}{4}x \cdot \dfrac{1}{2}x + \dfrac{3}{\underset{1}{\cancel{4}}}x \cdot \dfrac{\overset{2}{\cancel{8}}}{1} - \dfrac{5}{1} \cdot \dfrac{1}{2}x - 5 \cdot 8 \qquad \text{Divide out the common 4 in 4 and 8.}$$

$$= \dfrac{3}{8}x^2 + 6x - \dfrac{5}{2}x - 40 \qquad \text{Multiply.}$$

$$= \dfrac{3}{8}x^2 + \dfrac{6(2)}{1(2)}x - \dfrac{5}{2}x - 40 \qquad \begin{array}{l}\text{Write the coefficients of the like}\\\text{terms with their LCD, 2.}\end{array}$$

$$= \dfrac{3}{8}x^2 + \dfrac{12}{2}x - \dfrac{5}{2}x - 40$$

$$= \dfrac{3}{8}x^2 + \dfrac{7}{2}x - 40 \qquad \begin{array}{l}\text{Subtract coefficients of the like}\\\text{terms.}\end{array}$$

Do Your Turn 10 ▷

Your Turn 10

Multiply.

a. $\left(3a - \dfrac{1}{8}\right)\left(2a - \dfrac{1}{4}\right)$

b. $\left(\dfrac{1}{5}m - 2\right)\left(\dfrac{1}{5}m + 2\right)$

Answers to Your Turn 9

a. $\dfrac{2}{7}a^3 - 6a^2$

b. $-\dfrac{5}{2}m^2 - \dfrac{1}{6}mn$

Answers to Your Turn 10

a. $6a^2 - a + \dfrac{1}{32}$

b. $\dfrac{1}{25}m^2 - 4$

5.7 **Exercises**

For Extra Help

Videotape DVT

Addison-Wesley Tutor Center

Math XL

MyMathLab

Student Solutions Manual

1. What is a trapezoid?

2. What is the formula for the area of a trapezoid?

3. What do the variables a and b represent in the formula for the area of a trapezoid?

4. What is the formula $A = \pi r^2$ used for, and what does each symbol in it represent?

For Exercises 5–18, simplify.

5. $\dfrac{1}{4} + 2 \cdot \dfrac{5}{8}$

6. $\dfrac{1}{6} - 5 \cdot \dfrac{7}{10}$

7. $4\dfrac{3}{5} - \dfrac{1}{2} \div \dfrac{5}{6}$

8. $5\dfrac{1}{3} + \dfrac{3}{4} \div \dfrac{5}{12}$

9. $2\dfrac{1}{2} - \left(\dfrac{3}{4}\right)^2$

10. $\left(\dfrac{4}{5}\right)^2 + 5\dfrac{2}{3}$

11. $\dfrac{1}{4} - \dfrac{2}{3}\left(6 - 1\dfrac{1}{2}\right)$

12. $\dfrac{2}{5} + 9\left(\dfrac{1}{3} + \dfrac{1}{6}\right)$

13. $5\dfrac{3}{4} - 2\sqrt{\dfrac{80}{5}}$

14. $7\dfrac{1}{4} + \dfrac{3}{4}\sqrt{\dfrac{4}{9}}$

15. $\left(\dfrac{1}{2}\right)^3 + 2\dfrac{1}{4} - 5\left(\dfrac{3}{10} + \dfrac{1}{5}\right)$

16. $\left(\dfrac{2}{3}\right)^4 \div \dfrac{1}{3}\left(1\dfrac{1}{3} + \dfrac{2}{9}\right)$

17. $5\left(\dfrac{1}{2} - 3\dfrac{4}{5}\right) - 2\left(\dfrac{1}{6} + 4\right)$

18. $1\dfrac{1}{4} - \left(2\dfrac{1}{4}\right)\left(\dfrac{1}{3} + \dfrac{1}{2}\right)^2$

For Exercises 19–26, evaluate the expression using the given values.

19. $x + vt$; $x = 2\dfrac{1}{2}$, $v = 30$, $t = \dfrac{1}{4}$

20. $x + vt$; $x = 6\dfrac{4}{5}$, $v = 40$, $t = \dfrac{1}{6}$

21. mv^2; $m = 1\dfrac{1}{4}$, $v = \dfrac{3}{4}$

22. mv^2; $m = 6\dfrac{1}{2}$, $v = -\dfrac{2}{5}$

23. $\dfrac{1}{2}at^2$; $a = -9\dfrac{4}{5}$, $t = \dfrac{1}{2}$

24. $\dfrac{1}{2}at^2$; $a = -32\dfrac{1}{5}$, $t = 2\dfrac{1}{2}$

25. $2xyz$; $x = \dfrac{1}{6}$, $y = -1\dfrac{3}{4}$, $z = -\dfrac{8}{7}$

26. $2xyz$; $x = \dfrac{-1}{5}$, $y = -2\dfrac{1}{3}$, $z = \dfrac{-10}{7}$

For Exercises 27–30, find the area.

27.
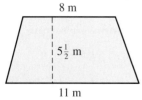
8 m
$5\dfrac{1}{2}$ m
11 m

28.
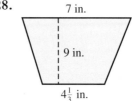
7 in.
9 in.
$4\dfrac{1}{3}$ in.

29.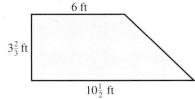
6 ft

$3\frac{2}{3}$ ft

$10\frac{1}{2}$ ft

30.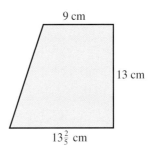
9 cm

13 cm

$13\frac{2}{5}$ cm

For Exercises 31–34, use $\frac{22}{7}$ for π.

31. Find the area of a circle with a radius of 14 inches.

32. Find the area of a circle with a radius of 3 feet.

33. Find the area of a circle with a diameter of $2\frac{1}{2}$ meters.

34. Find the area of a circle with a diameter of $10\frac{1}{3}$ centimeters.

35. Find the area of the shape shown.

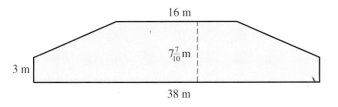
16 m

$7\frac{7}{10}$ m

3 m

38 m

36. Find the area of the shaded region. The diameter of the circle is $4\frac{1}{2}$ inches.

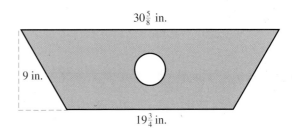
$30\frac{5}{8}$ in.

9 in.

$19\frac{3}{4}$ in.

For Exercises 37 and 38, combine like terms.

37. $\dfrac{1}{2}x^2 - \dfrac{3}{4}x - \dfrac{1}{3}x^2 - 6x$

38. $\dfrac{3}{8}n^3 - 3n^3 - \dfrac{1}{4} - 2$

For Exercises 39–42, add or subtract.

39. $\left(5y^3 - \dfrac{4}{5}y^2 + y - \dfrac{1}{6}\right) + \left(y^3 + 3y^2 - \dfrac{2}{3}\right)$

40. $\left(a^4 + \dfrac{1}{4}a^2 - a - \dfrac{1}{6}\right) + \left(\dfrac{3}{10}a^3 + a^2 - \dfrac{2}{3}a + 4\right)$

41. $\left(\dfrac{4}{5}t^3 + \dfrac{2}{3}t^2 - \dfrac{1}{6}\right) - \left(\dfrac{7}{10}t^3 + \dfrac{1}{4}t^2 + \dfrac{1}{2}\right)$

42. $\left(2x^4 + \dfrac{3}{5}x^2 - \dfrac{1}{8}x + 1\right) - \left(6x^4 + \dfrac{1}{4}x^2 - \dfrac{1}{2}x - \dfrac{1}{3}\right)$

For Exercises 43–48, multiply.

43. $\left(-\dfrac{1}{6}m\right)\left(\dfrac{3}{5}m^3\right)$

44. $\left(\dfrac{-7}{15}x^4\right)\left(\dfrac{5}{-21}x^2\right)$

45. $\dfrac{5}{8}\left(\dfrac{4}{5}t^2 - \dfrac{2}{3}t\right)$

46. $-\dfrac{1}{15}\left(3h^2 - \dfrac{3}{4}k^2\right)$

47. $\left(x - \dfrac{1}{2}\right)\left(x + \dfrac{1}{4}\right)$

48. $\left(\dfrac{1}{4}u - 1\right)\left(\dfrac{1}{2}u + 2\right)$

PUZZLE PROBLEM The original size of Khutu's pyramid in Egypt was 754 feet. along the base with a start height of 481 feet. It was originally covered with blocks cut so that the faces were all smooth. (Most of these blocks are now gone.) What was the total surface area of the four triangular faces of Khufu's pyramid?

Review Exercises

For Exercises 1 and 2, solve and check.

[4.3] **1.** $4w - 9 = 11$

[4.3] **2.** $5(n - 3) = 3n + 7$

[4.5] **3.** Six times the sum of n and 2 is equal to 3 less than n. Translate to an equation, then solve.

[2.6] **4.** Florence studies a map and finds that her trip will be 240 miles. If she averages 60 miles per hour, how long will it take her to get to her destination?

[4.5] **5.** David has $105 all in tens and fives. If he has 15 bills, how many tens and how many fives does he have?

5.8 Solving Equations

OBJECTIVE 1 Use the LCD to eliminate fractions from equations.

In Section 5.4, we reviewed solving equations using the multiplication/division principle of equality. In Section 5.6, we reviewed solving equations using the addition/subtraction principle of equality. Now let's consider equations that require both principles. We will also develop a way to use the multiplication/division principle of equality to simplify equations that contain fractions.

OBJECTIVES

1. Use the LCD to eliminate fractions from equations.
2. Translate sentences to equations, then solve.
3. Solve applications involving one unknown.
4. Solve applications involving two unknowns.

Example 1 Solve and check. $\dfrac{3}{4}x - \dfrac{2}{3} = \dfrac{1}{6}$

Solution: First, we use the addition/subtraction principle to isolate the variable term. Then we use the multiplication/division principle to clear the remaining coefficient.

$$\frac{3}{4}x - \frac{2}{3} = \frac{1}{6}$$

$$\frac{3}{4}x - \frac{2}{3} + \frac{2}{3} = \frac{1}{6} + \frac{2}{3} \qquad \text{Add } \tfrac{2}{3} \text{ to both sides.}$$

Note: $-\frac{2}{3}$ and $\frac{2}{3}$ and are additive inverses, so their sum is 0.

$$\frac{3}{4}x + 0 = \frac{1}{6} + \frac{2(2)}{3(2)} \qquad \text{Write equivalent fractions with the LCD, 6.}$$

$$\frac{3}{4}x = \frac{1}{6} + \frac{4}{6}$$

$$\frac{3}{4}x = \frac{5}{6} \qquad \text{Add the fractions.}$$

$$\frac{\cancel{4}^{1}}{\cancel{3}_{1}} \cdot \frac{\cancel{3}^{1}}{\cancel{4}_{1}}x = \frac{5}{\cancel{6}_{3}} \cdot \frac{\cancel{4}^{2}}{3} \qquad \text{Multiply both sides by } \tfrac{4}{3} \text{ to isolate } x. \text{ Divide out the common factors.}$$

Note: $\frac{3}{4}$ and $\frac{4}{3}$ are multiplicative inverses/reciprocals, so their product is 1.

$$1x = \frac{10}{9} \qquad \text{Multiply.}$$

$$x = 1\frac{1}{9} \qquad \text{Write the solution as a mixed number.}$$

Check: Replace x in the original equation with $1\frac{1}{9}$ and verify that the equation is true.

$$\frac{3}{4}x - \frac{2}{3} = \frac{1}{6}$$

$$\frac{3}{4}\left(1\frac{1}{9}\right) - \frac{2}{3} \stackrel{?}{=} \frac{1}{6} \qquad \text{Replace } x \text{ with } 1\tfrac{1}{9}.$$

$$\frac{3}{4} \cdot \frac{10}{9} - \frac{2}{3} \stackrel{?}{=} \frac{1}{6} \qquad \text{Write } 1\tfrac{1}{9} \text{ as an improper fraction.}$$

$$\frac{\cancel{3}^{1}}{\cancel{4}_{2}} \cdot \frac{\cancel{10}^{5}}{\cancel{9}_{3}} - \frac{2}{3} \stackrel{?}{=} \frac{1}{6} \qquad \text{Divide out common factors.}$$

$$\frac{5}{6} - \frac{2}{3} \stackrel{?}{=} \frac{1}{6} \qquad \text{Multiply.}$$

$$\frac{5}{6} - \frac{2(2)}{3(2)} \stackrel{?}{=} \frac{1}{6} \qquad \text{Write equivalents fractions with the LCD.}$$

$$\frac{5}{6} - \frac{4}{6} \stackrel{?}{=} \frac{1}{6}$$

$$\frac{1}{6} = \frac{1}{6} \qquad \text{Subtract. It checks.}$$

Example 1 illustrates the approach to solving equations that we learned in Chapter 4. But working with fractions can be tedious. We can use the multiplication/division principle of equality to eliminate the fractions, leaving us with integers, which simplifies the process considerably.

Remember, the multiplication/division principle of equality simply says that we can multiply or divide both sides of an equation by the same amount without affecting its solution(s). The principle does not say that we have to wait until the last step of the process to use it. If we multiply both sides by a multiple of all the denominators, we can divide out all the denominators. To keep the numbers as small as possible, it is best to use the least common multiple of the denominators, which is the LCD (least common denominator).

Consider Example 1 again using this technique. We will eliminate the fractions by multiplying both sides by the LCD, which is 12.

$$\frac{3}{4}x - \frac{2}{3} = \frac{1}{6}$$

$$12\left(\frac{3}{4}x - \frac{2}{3}\right) = \left(\frac{1}{6}\right)12 \qquad \text{Multiply both sides by 12.}$$

$$\frac{12}{1} \cdot \frac{3}{4}x - \frac{12}{1} \cdot \frac{2}{3} = \frac{1}{6} \cdot \frac{12}{1} \qquad \text{Distribute.}$$

$$\frac{\overset{3}{\cancel{12}}}{1} \cdot \frac{3}{\underset{1}{\cancel{4}}}x - \frac{\overset{4}{\cancel{12}}}{1} \cdot \frac{2}{\underset{1}{\cancel{3}}} = \frac{1}{\underset{1}{\cancel{6}}} \cdot \frac{\overset{2}{\cancel{12}}}{1}$$

Note: After dividing out the common factors, we are left with an equation that contains only integers, which is easier to work with for most people.

$$9x - 8 = 2 \blacktriangleleft$$

$$\underline{+8 \quad +8} \qquad \text{Add 8 to both sides to isolate } 9x.$$

$$9x + 0 = 10$$

$$\frac{9x}{9} = \frac{10}{9} \qquad \text{Divide both sides by 9 to isolate } x.$$

$$1x = \frac{10}{9} \qquad \text{Simplify.}$$

Note: Our alternative method leads to the same solution. ▶

$$x = 1\frac{1}{9} \qquad \text{Write the solution as a mixed number.}$$

Though we used a different method to solve the equation, the check is the same as in Example 1.

WARNING When using the LCD to eliminate fractions, it is tempting to check the solution in the rewritten equation (or another later step). In Example 1, it would certainly be easier to check using $9x - 8 = 2$ instead of $\frac{3}{4}x - \frac{2}{3} = \frac{1}{6}$ and, as long as we made no mistakes, the solution would check in $9x - 8 = 2$. However, suppose we made a mistake in getting to $9x - 8 = 2$ and then made no mistakes thereafter. Our incorrect answer would check in $9x - 8 = 2$, but it would *not* check in the original equation. Therefore, always use the original equation to check.

We can modify the procedure for solving equations that we developed in Chapter 4 to include this procedure for eliminating fractions.

To solve equations:

1. Simplify both sides of the equation as needed.
   a. Distribute to clear parentheses.
   b. Eliminate fractions by multiplying both sides by the LCD of all the fractions. (Optional)
   c. Combine like terms.
2. Use the addition/subtraction principle of equality so that all variable terms are on one side of the equation and all constants are on the other side. (Clear the variable term that has the lesser coefficient. This will avoid negative coefficients.) Then combine like terms.
3. Use the multiplication/division principle of equality to clear any remaining coefficients.

Your Turn 1

Solve and check.

a. $\dfrac{5}{8}y - 3 = \dfrac{1}{2}$

Do Your Turn 1 ▷

b. $\dfrac{1}{4} + \dfrac{1}{8}m = \dfrac{5}{16}$

Example 2 Solve and check. $\dfrac{2}{3}(x - 3) = \dfrac{1}{2}x - \dfrac{2}{5}$

Solution: We will start by distributing the $\frac{2}{3}$ to clear the parentheses. We will then clear the fractions by multiplying by their LCD. The LCD for 3, 2, and 5 is 30.

Note: We could have multiplied by the LCD as the first step, but this can be tricky with the parentheses, so we chose to clear the parentheses first. We will show how to multiply this equation by the LCD in the first step next.

▶ $\dfrac{2}{3}(x - 3) = \dfrac{1}{2}x - \dfrac{2}{5}$

$\dfrac{2}{3}x - \dfrac{2}{\overset{1}{3}} \cdot \dfrac{\overset{1}{3}}{1} = \dfrac{1}{2}x - \dfrac{2}{5}$ Distribute $\frac{2}{3}$ to clear the parentheses and divide out common factors.

$\dfrac{2}{3}x - 2 = \dfrac{1}{2}x - \dfrac{2}{5}$ Multiply.

$30\left(\dfrac{2}{3}x - 2\right) = \left(\dfrac{1}{2}x - \dfrac{2}{5}\right)30$ Eliminate the fractions by multiplying both sides by the LCD, 30.

$\dfrac{\overset{10}{30}}{1} \cdot \dfrac{2}{\overset{3}{1}}x - 30 \cdot 2 = \dfrac{\overset{15}{30}}{1} \cdot \dfrac{1}{\overset{2}{1}}x - \dfrac{\overset{6}{30}}{1} \cdot \dfrac{2}{\overset{5}{1}}$ Distribute 30 and divide out the common factors.

$20x - 60 = 15x - 12$ Multiply.

$\underline{-15x \qquad\quad -15x}$ Subtract 15x from both sides.

$5x - 60 = \quad 0 - 12$

$5x - 60 = -12$

$\underline{\quad +60 \qquad +60}$ Add 60 to both sides to isolate the 5x term.

$5x - 0 = \quad 48$

$\dfrac{5x}{5} = \dfrac{48}{5}$ Divide both sides by 5 to isolate x.

$1x = \dfrac{48}{5}$ Simplify.

$x = 9\dfrac{3}{5}$ Write the solution as a mixed number.

c. $-\dfrac{7}{9} = \dfrac{2}{3}n + \dfrac{1}{6}$

d. $\dfrac{-1}{7} = -2 - \dfrac{3}{5}k$

Answers to Your Turn 1

a. $y = 5\dfrac{3}{5}$ b. $m = \dfrac{1}{2}$

c. $n = -1\dfrac{5}{12}$ d. $k = -3\dfrac{2}{21}$

Solve and check.

a. $\frac{1}{4}y + 5 = \frac{1}{6}y + 2$

b. $\frac{2}{3} - \frac{4}{5}a = \frac{1}{5} + a$

c. $\frac{1}{4}(m + 2) = \frac{3}{4}m - \frac{2}{3}$

d. $\frac{3}{5}(x - 10) = \frac{1}{2}(x + 6) - 4$

Eliminating fractions that are multiplying parentheses can be tricky, which is why we chose to multiply by the LCD at a later step in our solution above. Below, we multiply our equation by the LCD in the first step.

$$\frac{2}{3}(x - 3) = \frac{1}{2}x - \frac{2}{5}$$

Note: Because $\frac{2}{3}(x - 3)$ is a product, when we multiply it by 30, we multiply the first two factors, 30 and $\frac{2}{3}$, to get 20, then multiply the third factor, $x - 3$, by that result.

$$30 \cdot \frac{2}{3}(x - 3) = 30 \cdot \left(\frac{1}{2}x - \frac{2}{5}\right)$$

Multiply both sides by 30 to eliminate the fractions.

$$\frac{\overset{10}{\cancel{30}}}{1} \cdot \frac{2}{\underset{1}{\cancel{3}}}(x - 3) = \frac{\overset{15}{\cancel{30}}}{1} \cdot \frac{1}{\underset{1}{\cancel{2}}}x - \frac{\overset{6}{\cancel{30}}}{1} \cdot \frac{2}{\underset{1}{\cancel{5}}}$$

Note: Because $\frac{1}{2}x - \frac{2}{5}$ is a difference, we distribute the 30 to each term.

$$20(x - 3) = 15x - 12$$

Multiply.

$$2x - 60 = 15x - 12$$

Distribute 20.

Note: This equation is the same as the one in the 6th step of our previous solution. The remaining steps would match our previous solution.

Check: Replace x in the original equation with $9\frac{3}{5}$ and verify the equation is true. We will leave this check to the reader.

◁ **Do Your Turn 2**

OBJECTIVE 2 Translate sentences to equations, then solve.

We have learned some new key words that are associated with fractions. In Section 5.3, we learned that the key word *of* indicates multiplication when preceded by a fraction. Also, the word *reciprocal* indicates to invert the number.

$$\frac{2}{5} \textbf{ of } \text{a number} \quad \text{translates to} \quad \frac{2}{5}n$$

$$\text{The } \textbf{reciprocal} \text{ of } n \quad \text{translates to} \quad \frac{1}{n}$$

The key words we've learned in the past, such as *sum, difference, less than,* and *product* translate the same regardless of the types of numbers involved.

Example 3 The sum of $2\frac{2}{3}$ and $\frac{3}{5}$ of n is the same as 2 less than $\frac{1}{3}$ of n. Solve for n.

Understand: We must translate using the key words and solve for n. The word *sum* indicates addition. *Less than* indicates subtraction in reverse order of the sentence. $\frac{3}{5}$ of n and $\frac{1}{3}$ of n indicate multiplication because *of* is preceded by the fraction in both cases.

Note: Remember, when used with the key words *sum, difference, product,* and *quotient,* the word *and* becomes the operation symbol.

Answers to Your Turn 2

a. -36 b. $\frac{7}{27}$ c. $2\frac{1}{3}$ d. 50

Plan: Translate to an equation, then solve.

Execute: The sum of $2\frac{2}{3}$ and $\frac{3}{5}$ of n is the same as 2 less than $\frac{1}{3}$ of n.

$$2\frac{2}{3} + \frac{3}{5}n = \frac{1}{3}n - 2$$

$$15\left(\frac{8}{3} + \frac{3}{5}n\right) = \left(\frac{1}{3}n - 2\right)15 \qquad \text{Eliminate the fractions by multiplying both sides by the LCD, 15.}$$

$$\frac{\overset{5}{\cancel{15}}}{1} \cdot \frac{8}{\underset{1}{\cancel{3}}} + \frac{\overset{3}{\cancel{15}}}{1} \cdot \frac{3}{\underset{1}{\cancel{5}}}n = \frac{\overset{5}{\cancel{15}}}{1} \cdot \frac{1}{\underset{1}{\cancel{3}}}n - \frac{15}{1} \cdot 2 \qquad \text{Distribute 15 and divide out the common factors.}$$

$$40 + 9n = 5n - 30 \qquad \text{Multiply.}$$

$$\underline{\quad -5n \quad -5n \quad} \qquad \text{Subtract 5n from both sides.}$$

$$40 + 4n = 0 - 30 \qquad \text{Subtract 40 from both sides.}$$

$$\underline{-40 \qquad\qquad -40\quad}$$

$$0 + 4n = \qquad -70$$

$$\frac{4n}{4} = \frac{-70}{4} \qquad \text{Divide both sides by 4.}$$

$$1n = -\frac{35}{2} \qquad \text{Simplify.}$$

$$n = -17\frac{1}{2} \qquad \text{Write the solution as a mixed number.}$$

Check: Replace n with $-17\frac{1}{2}$ to show that the answer is a solution for the equation. We will leave this check to the reader.

Do Your Turn 3 ▶

Your Turn 3

Translate to an equation, then solve.

a. $\frac{2}{3}$ of a number is $3\frac{5}{6}$.

b. The difference of x and $\frac{4}{9}$ is equal to $-3\frac{1}{6}$.

c. The sum of 3 and $\frac{3}{4}$ of k is the same as the reciprocal of 8.

d. $\frac{1}{2}$ of the difference of n and 4 is equal to $\frac{2}{5}$ of n.

OBJECTIVE 3 Solve applications involving one unknown.

Example 4 Find the length a in the trapezoid shown if the area is 9 square feet.

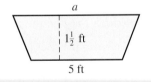

Understand: We must solve for a. Because we are given the area of the trapezoid, we can use the formula $A = \frac{1}{2}h(a + b)$.

Plan: Replace the variables in $A = \frac{1}{2}h(a + b)$ with the known amounts and solve for a.

Execute:

$$9 = \frac{1}{2}\left(1\frac{1}{2}\right)(a + 5) \qquad \text{Replace } A \text{ with 9, } h \text{ with } 1\frac{1}{2} \text{ and } b \text{ with 5.}$$

$$9 = \frac{1}{2}\left(\frac{3}{2}\right)(a + 5) \qquad \text{Write the mixed number as an improper fraction.}$$

$$9 = \frac{3}{4}(a + 5) \qquad \text{Multiply } \frac{1}{2} \text{ and } \frac{3}{2}.$$

Answers to Your Turn 3

a. $\frac{2}{3}n = 3\frac{5}{6}$; $n = 5\frac{3}{4}$

b. $x - \frac{4}{9} = -3\frac{1}{6}$; $x = -2\frac{13}{18}$

c. $3 + \frac{3}{4}k = \frac{1}{8}$; $k = -3\frac{5}{6}$

d. $\frac{1}{2}(n - 4) = \frac{2}{5}n$; $n = 20$

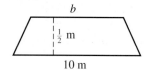

Find the missing side if the area of the trapezoid is $4\frac{1}{2}$ square meters.

b

$1\frac{1}{2}$ m

10 m

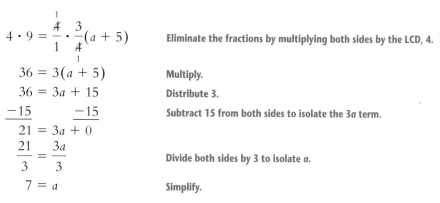

$$4 \cdot 9 = \frac{\overset{1}{\cancel{4}}}{1} \cdot \frac{3}{\cancel{4}}(a + 5)$$ Eliminate the fractions by multiplying both sides by the LCD, 4.

$$36 = 3(a + 5)$$ Multiply.

$$36 = 3a + 15$$ Distribute 3.

$$\underline{-15 \qquad\qquad -15}$$ Subtract 15 from both sides to isolate the 3*a* term.

$$21 = 3a + 0$$

$$\frac{21}{3} = \frac{3a}{3}$$ Divide both sides by 3 to isolate *a*.

$$7 = a$$ Simplify.

Answer: The length *a* is 7 feet.

Check: Verify that the area is 9 square feet when *a* is 7 feet, *b* is 5 feet, and *h* is $1\frac{1}{2}$ feet. We will leave this check to the reader.

◀ **Do Your Turn 4**

OBJECTIVE 4 Solve applications involving two unknowns.

In Chapter 4, we developed techniques for solving problems that have two un-knowns. In general, these problems have two relationships that we translate to an equation.

Example 5 A small company has two salespersons. In one particular week, Benjamin's total sales are half of the amount sold by Florence. If the total amount of sales for the week is $12,000, how much did each person sell?

Understand: There are two sentences that describe relationships.

Relationship 1: Benjamin's total sales are half of the amount sold by Florence.

Relationship 2: . . . the total amount of sales for the week is $12,000. . . .

We are to find the amount that Benjamin sold and the amount that Florence sold.

Plan: Translate the relationships to an equation and solve.

Execute: Let *f* be the amount sold by Florence.

Relationship 1: Benjamin's total sales are half of the amount sold by Florence.

$$\text{Benjamin's sales} = \frac{1}{2} \cdot f$$

Relationship 2: . . . the total amount of sales for the week is $12,000. . . .

$$\text{Benjamin's sales} + \text{Florence's sales} = \$12,000$$

$$\frac{1}{2}f \quad + \quad f \quad = 12,000$$

$$2\left(\frac{1}{2}f \quad + \quad f\right) \quad = (12,000)2$$ Multiply both sides by 2 to clear the fraction.

$$2 \cdot \frac{1}{2}f \quad + \quad 2 \cdot f \quad = 12,000 \cdot 2$$ Distribute the 2.

Answer to Your Turn 4
$b = 8$ m

$$f + 2f = 24{,}000 \qquad \textbf{Multiply.}$$

$$3f = 24{,}000 \qquad \textbf{Combine like terms.}$$

$$\frac{3f}{3} = \frac{24{,}000}{3} \qquad \begin{array}{l}\textbf{Divide both sides by 3 to clear}\\ \textbf{the 3 coefficient.}\end{array}$$

$$f = 8000$$

Answer: Because f represents Florence's sales, we can say her sales amounted to $8000. If Benjamin's sales were $\frac{1}{2}$ of Florence's then his totaled only $4000.

Check: Notice that the sum of their sales is in fact $12,000.

$$\$4000 + \$8000 = \$12{,}000$$

Do Your Turn 5 ▶

In Chapter 4, we solved problems with two unknowns involving geometry terms such as perimeter, complementary angles, and supplementary angles. Let's look again at these problems.

Example 6 Figure $ABCD$ is a rectangle. If $\angle DBC$ is $14\frac{1}{3}°$ less than $\angle ABD$, then what are the angle measurements?

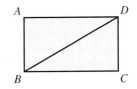

Understand: We must find the angle measurements. It seems we are given only one relationship:

$$\angle DBC \text{ is } 14\tfrac{1}{3}° \text{ less than } \angle ABD.$$

The other relationship is in the shape. Because the shape is a rectangle, the corner angles measure $90°$. Therefore, $\angle DBC$ and $\angle ABD$ are complementary, which means the sum of their measurements is $90°$.

Plan: Translate the relationships to an equation, then solve.

Execute: Let a represent the measurement of $\angle ABD$.

Relationship 1: $\angle DBC$ is $14\frac{1}{3}°$ less than $\angle ABD$.

$$\angle DBC = a - 14\tfrac{1}{3}$$

Relationship 2: The sum of the angle measurements is 90.

$$\angle DBC + \angle ABD = 90$$

$$\overbrace{a - 14\tfrac{1}{3}}\ +\quad a\ \ = 90$$

$$2a - 14\tfrac{1}{3} = 90 \qquad \textbf{Combine like terms.}$$

$$2a - 14\tfrac{1}{3} + 14\tfrac{1}{3} = 90 + 14\tfrac{1}{3} \qquad \textbf{Add } 14\tfrac{1}{3} \textbf{ to both sides.}$$

$$2a + 0 = 104\tfrac{1}{3} \qquad \textbf{Add.}$$

$$\frac{1}{2} \cdot 2a = \frac{313}{3} \cdot \frac{1}{2} \qquad \textbf{Multiply both sides by } \tfrac{1}{2} \textbf{ to isolate } a.$$

$$1a = \frac{313}{6} \qquad \textbf{Simplify.}$$

$$a = 52\tfrac{1}{6} \qquad \textbf{Write the result as a mixed number.}$$

Your Turn 5

Two boards are joined. One board is $\frac{2}{3}$ the length of the other. The two boards combine to be $7\frac{1}{2}$ feet. What are the lengths of both boards?

Answer to Your Turn 5

$4\frac{1}{2}$ ft., 3 ft.

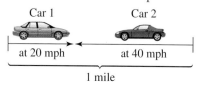

Your Turn 6

a. The length of a rectangular frame is $3\frac{1}{2}$ inches longer than the width. If the perimeter is 42 inches, what are the dimensions of the frame?

b. A piece of steel is welded to a horizontal truss forming two angles with the truss. The larger of the two angles is $2\frac{1}{2}°$ more than twice the smaller angle. Find the angle measurements.

c. A rectangle has a width that is half of the length. The perimeter is $37\frac{1}{2}$ meters. What are the dimensions of the rectangle?

Answer: Because $\angle ABD$ is represented by the variable a, we can say $\angle ABD = 52\frac{1}{6}°$. Because $\angle DBC = a - 14\frac{1}{3}$, we can find its value replacing a with $52\frac{1}{6}$ and then subtracting:

$$\angle DBC = 52\frac{1}{6} - 14\frac{1}{3}$$

$$\angle DBC = 52\frac{1}{6} - 14\frac{1(2)}{3(2)} \qquad \text{Write equivalent fractions.}$$

$$\angle DBC = 52\frac{1}{6} - 14\frac{2}{6}$$

$$\angle DBC = 51\frac{7}{6} - 14\frac{2}{6} \qquad \text{Rewrite } 52\frac{1}{6} \text{ so that we can subtract.}$$

$$\angle DBC = 37\frac{5}{6} \qquad \text{Subtract.}$$

$\angle ABD$ is $52\frac{1}{6}°$ and $\angle DBC$ is $37\frac{5}{6}°$.

Check: Verify that the sum of the angle measurements is 90°.

$$\angle DBC + \angle ABD = 90$$

$$37\frac{5}{6} + 52\frac{1}{6} \overset{?}{=} 90$$

Note: $\frac{6}{6}$ is 1, which adds to 89 to get 90. ▶

$$89\frac{6}{6} \overset{?}{=} 90$$

$$90 = 90$$

◀ **Do Your Turn 6**

In Section 4.5, we developed the use of tables with currency problems to find the number of bills of two different types of currency. Now we will use a similar table to help organize information involving distance, rate, and time. Recall the formula for finding the distance given the rate and time is $d = rt$.

The problems we will consider involve two people or objects traveling in opposite directions. It won't matter whether the two people or objects are traveling toward one another or away from one another, the math will be the same. The tables we will use will look like the one on the right.

Categories	Rate	Time	Distance

Example 7 In a crash test, two remote-guided cars are 1 mile apart and driven toward each other to collide somewhere in between. One car is traveling at a speed of 20 miles per hour, while the other is traveling at 40 miles per hour. How long will it take them to collide?

Understand: Let's draw a picture of the situation:

Answers to Your Turn 6

a. $8\frac{3}{4}$ in. by $12\frac{1}{4}$ in.

b. $59\frac{1}{6}°$, $120\frac{5}{6}°$

c. $12\frac{1}{2}$ m by $6\frac{1}{4}$ m

Car 1   Car 2

at 20 mph   at 40 mph

1 mile

Note: Car 2 will make up more of the distance that separates the cars because it is traveling faster.

From our picture, we can say that the sum of the individual distances traveled will be 1 mile.

distance traveled by car 1 + distance traveled by car 2 = total distance apart

We will use a table to organize the information and determine the individual distances.

Categories	Rate	Time	Distance
Car 1	20 mph	t	$20t$
Car 2	40 mph	t	$40t$

Note: If we used a stopwatch to time the crash test, we would start the watch when the cars are separated by 1 mile, and stop the watch when they collide. Both cars travel for the same amount of time, so the time is t for both.

Note: We determine the distance traveled by multiplying the rate and the time.

We now have expressions for the individual distances and know that the sum of the individual distances will be 1 mile.

Plan: Use the information to write an equation, then solve.

Execute: distance traveled by car 1 + distance traveled by car 2 = total distance apart

$$20t \quad + \quad 40t \quad = 1$$

$$60t = 1 \qquad \text{Combine like terms.}$$

$$\frac{60t}{60} = \frac{1}{60} \qquad \text{Divide both sides by 60.}$$

$$1t = \frac{1}{60} \qquad \text{Simplify.}$$

DISCUSSION How many minutes is $\frac{1}{60}$ of an hour?

$$t = \frac{1}{60}$$

Answer: The cars will collide in $\frac{1}{60}$ of an hour.

Check: Verify that in $\frac{1}{60}$ of an hour the cars will travel a combined distance of 1 mile.

Car 1 individual distance: $d = 20\left(\frac{1}{60}\right) = \frac{20}{60} = \frac{1}{3}$ mi.

Car 2 individual distance: $d = 40\left(\frac{1}{60}\right) = \frac{40}{60} = \frac{2}{3}$ mi.

Note: Car 2 travels twice as far because it is going twice as fast.

Combined distance: car 1 + car 2 = total distance

$$\frac{1}{3} + \frac{2}{3} \stackrel{?}{=} 1$$

$$\frac{3}{3} \stackrel{?}{=} 1$$

$$1 = 1$$

Do Your Turn 7 ▶

Your Turn 7

Solve.

a. Andrea is bicycling east at 8 miles per hour along a trail. Carla is running west on the same trail at $4\frac{1}{2}$ miles per hour. If they are $\frac{1}{4}$ mile apart, how long will it be until they meet?

b. Jennifer and Phil are traveling in cars going toward one another on an interstate highway. Jennifer is traveling at 70 miles per hour and Phil at 65 miles per hour. From the time they eventually meet and pass each other, how long until they are 27 miles apart?

Note: When two objects travel toward one another or away from one another, we can describe their closing or separation speed. When we combined like terms, we combined the numbers representing the speeds of the cars.

20 mph + 40 mph = 60 mph

60 miles per hour is the closing speed for the two cars. It is the same as if we kept one car still and drove the other car at 60 miles per hour for 1 mile. It would take the same amount of time.

The reasoning is the same if the two cars are going away from each other. If they are going 40 miles per hour and 20 miles per hour, then they are separating at a rate of 60 miles per hour. This is why the math is the same, whether the cars are going toward each other or away from each other.

Answers to Your Turn 7

a. $\frac{1}{50}$ hr. b. $\frac{1}{5}$ hr.

For
Extra
Help

Videotape
DVT

Tutor Center
Addison-Wesley
Tutor Center

Math XL
Math XL

MyMathLab
MyMathLab

Student Solutions
Manual

1. Explain how to use the multiplication/division principle of equality to eliminate fractions from an equation.

2. Suppose two objects begin moving toward each other on a collision course. What can we conclude about the amount of time it takes them to meet?

3. Suppose two objects begin "back to back" and travel in opposite directions. Object 1 travels $5t$ miles whereas object 2 travels $10t$ miles. In the following figure, which arrow represents which object. Explain.

4. Suppose the two objects described in Exercise 3 are separated by 45 miles after traveling t hours, write the equation that you can use to find t.

Start

For Exercises 5–24, use the LCD to simplify the equation, then solve and check.

5. $x + \dfrac{3}{4} = \dfrac{1}{2}$

6. $y + \dfrac{4}{9} = \dfrac{2}{5}$

7. $p - \dfrac{1}{6} = \dfrac{1}{3}$

8. $q - \dfrac{2}{3} = \dfrac{3}{7}$

9. $\dfrac{5}{6} + c = \dfrac{3}{5}$

10. $\dfrac{1}{3} + d = \dfrac{3}{4}$

11. $\dfrac{3}{4}a = \dfrac{1}{2}$

12. $\dfrac{2}{5}b = \dfrac{1}{10}$

13. $\dfrac{6}{35}f = \dfrac{8}{15}$

14. $\dfrac{14}{15}g = \dfrac{21}{20}$

15. $3b - \dfrac{1}{4} = \dfrac{1}{2}$

16. $\dfrac{2}{3}y - 8 = \dfrac{1}{6}$

17. $\dfrac{3}{4}x + \dfrac{1}{6} = \dfrac{1}{2}$

18. $\dfrac{1}{5}p + \dfrac{2}{3} = 2$

19. $\dfrac{4}{5} - \dfrac{a}{2} = \dfrac{3}{4} - 1$

20. $\dfrac{1}{3} - \dfrac{b}{4} = \dfrac{1}{2} - 2$

21. $\dfrac{1}{8} + n = \dfrac{5}{6}n - \dfrac{2}{3}$

22. $\dfrac{1}{6} + m = \dfrac{3}{4}m + \dfrac{1}{3}$

23. $\dfrac{1}{2}(x - 6) = \dfrac{1}{4}x - \dfrac{2}{5}$

24. $\dfrac{5}{6}(k - 8) = \dfrac{1}{5}k - \dfrac{1}{3}$

For Exercises 25–32, translate to an equation, then solve.

25. $\frac{3}{8}$ of a number is $4\frac{5}{6}$.

26. The product of $2\frac{1}{3}$ and a number is $-4\frac{1}{2}$.

27. $8\frac{7}{10}$ less than y is $-2\frac{1}{2}$.

28. $6\frac{2}{3}$ more than m is $5\frac{1}{12}$.

29. $3\frac{1}{4}$ more than twice n is $-\frac{1}{6}$.

30. $\frac{5}{8}$ less than $3\frac{4}{5}$ times k is $\frac{9}{16}$.

31. $\frac{3}{4}$ of the sum of b and 10 is equal to $1\frac{1}{6}$ added to b.

32. $\frac{2}{5}$ of the difference of 1 and h is the same as $1\frac{1}{4}$ times h.

For Exercises 33–46, solve.

33. Find the length of the missing side if the area is $7\frac{41}{50}$ square centimeters.

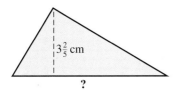

$3\frac{2}{5}$ cm

?

34. Find the length of the missing side if the area is $122\frac{1}{2}$ square inches.

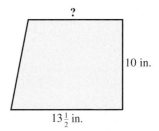

?

10 in.

$13\frac{1}{2}$ in.

35. Mary is part of a team of social workers trying to raise funds to build an outreach center for a neighborhood. Mary alone raises $2\frac{1}{2}$ times what the rest of the team does. The total amount raised was $31,542. How much did Mary raise?

36. Rosa discovers that her salary is $\frac{5}{6}$ of Rick's salary, yet they have the same job. The difference in their salaries is $4700. What are each of their salaries?

37. Two chemical storage tanks hold the same amount. Tank A is $\frac{3}{4}$ full while tank B is $\frac{2}{3}$ full. The two tanks currently have a combined total of 1700 gallons of CO_2. How much do the tanks hold?

38. A computer has two equal-size hard drives. Drive C is $\frac{4}{5}$ full while drive D is only $\frac{1}{4}$ full. If the two drives have a combined total of $6\frac{3}{10}$ gigabytes stored, how much does each drive hold?

39. In a survey, $\frac{3}{4}$ of the people contacted agreed to answer a single yes-or-no question. Of these, $\frac{1}{5}$ answered yes and the rest answered no. What fraction of all people contacted said no?

40. In a political poll, $\frac{2}{5}$ of the respondents said they would vote for a certain candidate. Of these, $\frac{1}{3}$ were men. What fraction of all respondents that voted for the candidate were women?

41. Two boards come together as shown. The smaller angle measurement is $\frac{1}{3}$ the measurement of the larger angle. Find the angle measurements.

42. At exactly 3:00:10, find the angles made between the second hand and the minute and hour hands of a clock.

43. An F-16 and MiG-29 are flying directly toward one another. The F-16 is flying at Mach 2, which is $1483\frac{3}{5}$ miles per hour. The MiG-29 is flying at Mach 1, which is $741\frac{4}{5}$ miles per hour. If the two planes are 20 miles apart, how long until they meet?

OF INTEREST

Taken from the last name of Ernst Mach (1838–1916), an Austrian physicist, *Mach* is a term used to describe how many times the speed of sound an object is traveling. The speed of sound at 32°F at sea level is $741\frac{4}{5}$ miles per hour. Mach 2 means twice the speed of sound.

44. A jury is listening to testimony in a case of a head-on collision. As the defendant swerved to avoid a construction crew, he moved slightly into the lane of on-coming traffic. After passing the construction crew, he hit an oncoming car. He claims he did not have time to react to the oncoming car because of a bend in the road ahead. It was determined that the plaintiff was at the bend in the road at the same time that the defendant was passing the construction crew. From the site where the construction crew was working to the bend was measured to be 200 feet. The defendant was traveling at 44 feet per second. The plaintiff was traveling at $51\frac{1}{3}$ feet per second. How much time did the defendant have to react? Do you think it was sufficient time?

45. Greg and Debbie start back-to-back and walk away from each other until they are $\frac{1}{2}$ mile apart. If Greg walks at 2 miles per hour and Debbie at $2\frac{1}{2}$ miles per hour, how long will this take?

46. Laura and Thomas are swimming in opposite directions. Laura is swimming 2 meters per second and Thomas is swimming $2\frac{3}{5}$ meters per second. If we begin a stopwatch at the time they pass one another, how long will it take them to get 40 meters apart?

PUZZLE PROBLEM Two people are in a desert together. They have half of a gallon (8 cups) of water in a canteen to split between them. They decide to divide the water so that they each get 4 cups. The problem is that one person has a container that holds 3 cups while the other person has a container that holds 5 cups. How can they use their containers and the canteen to get exactly 4 cups each?

Review Exercises

[5.1] **1.** Use $<$, $>$, or $=$ to make a true sentence.

$$\frac{6}{17} \; ? \; \frac{7}{19}$$

[5.2] **2.** Simplify to lowest terms. $\dfrac{9x}{24x^2}$

[5.3] **3.** Multiply. $\dfrac{-4}{9} \cdot \dfrac{21}{30}$

[5.4] **4.** Divide. $\dfrac{14a}{15ab^4} \div \dfrac{21}{25ab}$

[5.6] **5.** Add. $\dfrac{7}{8} + 5\dfrac{2}{3}$

Defined Terms

Review the following terms and for those you do not know, study its definition on the page number next to it.

Section 5.1
Fraction *(p. 290)*
Numerator *(p. 290)*
Denominator *(p. 290)*
Rational number *(p. 290)*
Simplify *(p. 292)*
Simplest form *(p. 292)*
Equivalent fractions
  (p. 294)
Multiple *(p. 296)*

Improper fraction *(p. 297)*
Mixed number *(p. 298)*

Section 5.2
Lowest terms *(p. 304)*
Rational expression
  (p. 307)

Section 5.3
Circle *(p. 320)*
Radius *(p. 320)*

Diameter *(p. 320)*
Circumference *(p. 322)*
Irrational number *(p. 322)*
Pi (π) *(p. 322)*

Section 5.4
Reciprocals *(p. 328)*
Complex fraction *(p.330)*

Section 5.5
Least common muliple
  (LCM) *(p. 339)*
Least common
  denominator (LCD)
  (p. 341)

Section 5.7
Trapezoid *(p. 361)*

Procedures, Rules, and Key Examples

Procedures/Rules	Key Example(s)

Section 5.1 Fractions, Mixed Numbers, and Rational Expressions

If the denominator of a fraction is 1, the fraction can be simplified to the numerator.

$$\frac{n}{1} = n, \text{ when } n \text{ is any number.}$$

If the numerator of a fraction is 0 and the denominator is any number other than 0, the fraction can be simplified to 0.

$$\frac{0}{n} = 0, \text{ when } n \neq 0.$$

If the denominator is 0 with any number other than 0 in the numerator, we say the fraction is undefined.

$$\frac{n}{0} \text{ is undefined, when } n \neq 0.$$

A fraction with the same numerator and denominator (other than zero) can be simplified to 1.

$$\frac{n}{n} = 1, \text{ when } n \neq 0.$$

Zero over zero is indeterminate.

$$\frac{0}{0} \text{ is indeterminate.}$$

Example 1: Simplify.

a. $\dfrac{8}{1} = 8$

b. $\dfrac{0}{11} = 0$

c. $\dfrac{14}{0}$ is undefined

d. $\dfrac{60}{60} = 1$

e. $\dfrac{0}{0}$ is indeterminate

To write an equivalent fraction multiply or divide both the numerator and denominator by the same non zero number.

Example 2: Fill in the blank so that the fractions are equivalent.

a. $\dfrac{4}{5} = \dfrac{?}{15}$ b. $\dfrac{12}{15} = \dfrac{?}{5}$

Solution: Solution:

$\dfrac{4}{5} = \dfrac{4 \cdot 3}{5 \cdot 3} = \dfrac{12}{15}$ $\dfrac{12}{15} = \dfrac{12 \div 3}{15 \div 3} = \dfrac{4}{5}$

To compare two fractions:
1. Write equivalent fractions that have a common denominator.
2. Compare the numerators in the rewritten fractions.

Tip: We can also use cross products. If the cross products are equal, so are the fractions. The greater cross product will indicate the greater fraction.

Example 3: Use $<$, $>$, or $=$ to make a true statement.

a. $\dfrac{6}{9}$? $\dfrac{12}{18}$

Simplify each:

$$\dfrac{6}{9} = \dfrac{6 \div 3}{9 \div 3} = \dfrac{2}{3} \qquad \dfrac{12}{18} = \dfrac{12 \div 6}{18 \div 6} = \dfrac{2}{3}$$

They are equal: $\dfrac{6}{9} = \dfrac{12}{18}$

b. $\dfrac{5}{8}$? $\dfrac{4}{7}$

Upscale each (using cross products here):

$$\dfrac{35}{56} = \dfrac{5 \cdot 7}{8 \cdot 7} = \dfrac{5}{8} \diagdown \dfrac{4}{7} = \dfrac{4 \cdot 8}{7 \cdot 8} = \dfrac{32}{56}$$

$\dfrac{35}{56} > \dfrac{32}{56}$, therefore $\dfrac{5}{8} > \dfrac{4}{7}$

To write an improper fraction as a mixed number:
1. Divide the denominator into the numerator.
2. Write the results in the form:

$$\text{quotient} \; \dfrac{\text{remainder}}{\text{original denominator}}$$

Example 4: Write $\frac{19}{8}$ as an improper fraction.

$$\begin{array}{r} 2 \\ 8\overline{)19} \\ -16 \\ \hline 3 \end{array} \qquad 2\dfrac{3}{8}$$

To write a mixed number as an improper fraction:
1. Multiply the denominator by the integer.
2. Add the resulting product to the numerator to find the numerator of the improper fraction.
3. Keep the same denominator.

Example 5: Write $5\frac{7}{8}$ as an improper fraction.

$$5\dfrac{7}{8} = \dfrac{8 \cdot 5 + 7}{8} = \dfrac{40 + 7}{8} = \dfrac{47}{8}$$

Many people do not write these steps, performing the calculations mentally.

Section 5.2 Simplifying Fractions and Rational Expressions
To simplify a fraction to lowest terms, divide the numerator and denominator by their greatest common factor.

Example 1: Simplify $\frac{70}{182}$ to lowest terms.

$$\dfrac{70}{182} = \dfrac{70 \div 14}{182 \div 14} = \dfrac{5}{13} \quad \text{or}$$

$$\dfrac{70}{182} = \dfrac{\overset{1}{2} \cdot 5 \cdot \overset{1}{7}}{2 \cdot 7 \cdot 13} = \dfrac{5}{13}$$

To simplify a fraction to lowest terms using prime:
1. Replace the numerator and denominator with their prime factorizations.
2. Divide out all the common prime factors.
3. Multiply the remaining factors.

Example 2: Simplify $-\dfrac{18x^3}{24x^2}$ to lowest terms.

$$-\dfrac{18x^3}{24x^2} = -\dfrac{\overset{1}{2} \cdot \overset{1}{3} \cdot 3 \cdot \overset{1}{x} \cdot \overset{1}{x} \cdot x}{2 \cdot 2 \cdot 2 \cdot \underset{1}{3} \cdot \underset{1}{x} \cdot \underset{1}{x}} = -\dfrac{3x}{4}$$

Section 5.3 Multiplying Fractions, Mixed Numbers, and Rational Expressions

To multiply fractions:

Method 1:

1. Multiply the numerators and multiply the denominators.
3. Simplify the products.

Method 2:

1. Divide out any numerator factor with any like denominator factor.
2. Multiply the remaining factors.
3. Simplify as needed so that the result is in lowest terms. (This would be necessary if you miss common factors in the first step.)

Example 1: Multiply. $-\dfrac{14}{15} \cdot \dfrac{9}{16}$

$$-\frac{14}{15} \cdot \frac{9}{16} = -\frac{\overset{1}{2} \cdot 7}{\underset{1}{3} \cdot 5} \cdot \frac{\overset{1}{3} \cdot 3}{2 \cdot 2 \cdot \underset{1}{2} \cdot 2} = -\frac{21}{40}$$

Example 2: Multiply. $\dfrac{9a}{10b} \cdot \dfrac{-16}{21a^2}$

$$\frac{9a}{10b} \cdot \frac{-16}{21a^2} = \frac{\overset{1}{3} \cdot 3 \cdot \overset{1}{a}}{2 \cdot 5 \cdot b} \cdot \frac{-\overset{1}{2} \cdot 2 \cdot 2 \cdot 2}{\underset{1}{3} \cdot 7 \cdot \underset{1}{a} \cdot a}$$

$$= -\frac{24}{35ab}$$

To multiply mixed numbers:

1. Write the mixed numbers as improper fractions.
2. Divide out any numerator factor with any like denominator factor.
3. Multiply numerator by numerator and denominator by denominator.
4. Write the product as a mixed number in simplest form.

Example 3: Multiply. $-6\dfrac{3}{8} \cdot \left(-1\dfrac{1}{3}\right)$

$$-6\frac{3}{8} \cdot \left(-1\frac{1}{3}\right) = -\frac{51}{8} \cdot \left(-\frac{4}{3}\right)$$

$$= -\frac{\overset{1}{3} \cdot 17}{2 \cdot 2 \cdot \underset{1}{2}} \cdot \left(-\frac{\overset{1}{2} \cdot \overset{1}{2}}{\underset{1}{3}}\right)$$

$$= \frac{17}{2}$$

$$= 8\frac{1}{2}$$

When a fraction is raised to a power, we evaluate both the numerator and denominator raised to that power.

$$\left(\frac{a}{b}\right)^n = \frac{a^n}{b^n}, \quad \text{when } b \neq 0.$$

Example 4: Simplify. $\left(\dfrac{2}{3}\right)^4$

$$\left(\frac{2}{3}\right)^4 = \frac{2^4}{3^4} = \frac{2}{3} \cdot \frac{2}{3} \cdot \frac{2}{3} \cdot \frac{2}{3} = \frac{16}{81}$$

Example 5: Simplify. $\left(\dfrac{-2x^2}{5}\right)^3$

$$\left(\frac{-2x^2}{5}\right)^3 = \frac{(-2x^2)^3}{5^3}$$

$$= \frac{(-2x^2)(-2x^2)(-2x^2)}{5 \cdot 5 \cdot 5}$$

$$= \frac{-8x^6}{125}$$

Section 5.4 Dividing Fractions, Mixed Numbers, and Rational Expressions

To divide fractions:

1. Change the operation symbol from division to multiplication and change the divisor to its reciprocal.
2. Divide out any numerator factor with any like denominator factor.
3. Multiply numerator by numerator and denominator by denominator.
4. Simplify as needed.

Example 1: Divide. $\dfrac{9}{16} \div \dfrac{3}{10}$

$$\frac{9}{16} \div \frac{3}{10} = \frac{9}{16} \cdot \frac{10}{3}$$

$$= \frac{\overset{1}{\cancel{3}} \cdot 3}{\underset{1}{2 \cdot 2 \cdot 2} \cdot 2} \cdot \frac{\overset{1}{\cancel{2}} \cdot 5}{\underset{1}{\cancel{3}}}$$

$$= \frac{15}{8}$$

$$= 1\frac{7}{8}$$

Example 2: Divide. $-\dfrac{5k}{16m^2} \div \dfrac{25k}{2m}$

$$-\frac{5k}{16m^2} \div \frac{25k}{2m} = -\frac{5k}{16m^2} \cdot \frac{2m}{25k}$$

$$= -\frac{\overset{1}{\cancel{5}} \cdot k}{\underset{1}{2 \cdot 2 \cdot 2 \cdot 2 \cdot \underset{1}{\cancel{m}} \cdot m}} \cdot \frac{\overset{1}{\cancel{2}} \cdot \overset{1}{\cancel{m}}}{\underset{1}{\cancel{5}} \cdot 5 \cdot \underset{1}{\cancel{k}}}$$

$$= -\frac{1}{40m}$$

To divide mixed numbers:

1. Write the mixed numbers as improper fractions.
2. Write the division statement as an equivalent multiplication statement using the reciprocal of the divisor.
3. Divide out any numerator factor with any like denominator factor.
4. Multiply.
5. Simplify as needed.

Example 3: Divide. $-2\dfrac{2}{5} \div 5\dfrac{1}{4}$

$$-2\frac{2}{5} \div 5\frac{1}{4} = -\frac{12}{5} \div \frac{21}{4}$$

$$= -\frac{12}{5} \cdot \frac{4}{21}$$

$$= -\frac{2 \cdot 2 \cdot \overset{1}{\cancel{3}}}{5} \cdot \frac{2 \cdot 2}{\underset{1}{\cancel{3}} \cdot 7}$$

$$= -\frac{16}{35}$$

To find the square root of a fraction, try the following:

Find the square root of the numerator and denominator separately, then simplify.

or

Simplify the fraction, then find the square root of the quotient.

Example 4: Simplify.

a. $\sqrt{\dfrac{49}{100}} = \dfrac{\sqrt{49}}{\sqrt{100}} = \dfrac{7}{10}$

b. $\sqrt{\dfrac{75}{3}} = \sqrt{25} = 5$

Procedures/Rules	Key Example(s)

Section 5.5 Least Common Multiple

To find the LCM by listing:

List multiples of the greatest given number until you find a multiple that is divisible by all the other given numbers.

To find the LCM using prime factorization:

1. Find the prime factorization of each given number.
2. Write a factorization that contains each prime factor the greatest number of times it occurs in the factorizations. Or, if you prefer exponents, the factorization contains each prime factor raised to the greatest exponent that occurs in the factorizations.
3. Multiply to get the LCM.

Example 1: Find the LCM of 18 and 24.
Listing method:

24	48	72
is not divisible by 18.	is not divisible by 18.	is divisible by 18.

$\text{LCM}(18, 24) = 72$

Prime factorization method:

$$18 = 2 \cdot 3^2$$
$$24 = 2^3 \cdot 3$$
$$\text{LCM}(18, 24) = 2^3 \cdot 3^2$$
$$= 2 \cdot 2 \cdot 2 \cdot 3 \cdot 3$$
$$= 72$$

Example 2: Find the LCM of $18x^3y$ and $24xz^2$.

$$\text{LCM}(18x^3y, 24xz^2) = 2^3 \cdot 3^2 x^3 yz^2$$
$$= 8 \cdot 9x^3 yz^2$$
$$= 72x^3 yz^2$$

Section 5.6 Adding and Subtracting Fractions, Mixed Numbers, and Rational Expressions

To add or subtract fractions that have the same denominator:

1. Add or subtract numerators.
2. Keep the same denominator.
3. Simplify.

Note: With rational expressions we can only combine numerators that have like terms. For unlike terms, we must express the sum or difference as a polynomial.

Example 1: Add or subtract.

a. $\dfrac{2}{9} + \dfrac{1}{9} = \dfrac{3}{9} = \dfrac{1}{3}$

b. $\dfrac{5h}{2} - \dfrac{1}{2} = \dfrac{5h - 1}{2}$

c. $\dfrac{2x - 9}{5y} - \dfrac{x + 1}{5y}$

$= \dfrac{(2x - 9) - (x + 1)}{5y}$

$= \dfrac{(2x - 9) + (-x - 1)}{5y}$

$= \dfrac{x - 10}{5y}$

To add or subtract fractions with different denominators:

1. Write the fractions as equivalent fractions with a common denominator.

Note: Any common multiple of the denominators will do, but using the LCD will make the numbers more manageable.

2. Add or subtract the numerators and keep the common denominator.
3. Simplify.

Example 2: Add or subtract.

a. $\dfrac{1}{6} + \dfrac{3}{4} = \dfrac{1(2)}{6(2)} + \dfrac{3(3)}{4(3)}$

$= \dfrac{2}{12} + \dfrac{9}{12}$

$= \dfrac{11}{12}$

b. $\dfrac{1}{2} - \dfrac{5}{8} = \dfrac{1(4)}{2(4)} - \dfrac{5(1)}{8(1)}$

$= \dfrac{4}{8} - \dfrac{5}{8}$

$= -\dfrac{1}{8}$

c. $\dfrac{3}{5x^2} - \dfrac{9}{10xy} = \dfrac{3(2y)}{5x^2(2y)} - \dfrac{9(x)}{10xy(x)}$

$$= \dfrac{6y}{10x^2y} - \dfrac{9x}{10x^2y}$$

$$= \dfrac{6y - 9x}{10x^2y}$$

To add mixed numbers:

Method 1: Write as improper fractions, then follow the procedure for adding fractions.

Example 3: Add. $2\dfrac{3}{5} + 5\dfrac{3}{4}$

$$2\dfrac{3}{5} + 5\dfrac{3}{4} = \dfrac{13}{5} + \dfrac{23}{4}$$

$$= \dfrac{13(4)}{5(4)} + \dfrac{23(5)}{4(5)}$$

$$= \dfrac{52}{20} + \dfrac{115}{20}$$

$$= \dfrac{167}{20}$$

$$= 8\dfrac{7}{20}$$

Method 2: Add the integer parts and fraction parts separately.

$$2\dfrac{3}{5} + 5\dfrac{3}{4} = 2\dfrac{3(4)}{5(4)} + 5\dfrac{3(5)}{4(5)}$$

$$= 2\dfrac{12}{20} + 5\dfrac{15}{20}$$

$$= 7\dfrac{27}{20}$$

$$= 8\dfrac{7}{20}$$

To subtract mixed numbers:

Method 1: Write as improper fractions, then follow the procedure for adding/subtracting fractions.

Example 4: Subtract. $8\dfrac{1}{4} - 5\dfrac{2}{3}$

$$8\dfrac{1}{4} - 5\dfrac{2}{3} = \dfrac{33}{4} - \dfrac{17}{3}$$

$$= \dfrac{33(3)}{4(3)} - \dfrac{17(4)}{3(4)}$$

$$= \dfrac{99}{12} - \dfrac{68}{12}$$

$$= \dfrac{31}{12}$$

$$= 2\dfrac{7}{12}$$

Procedures/Rules	Key Example(s)

Method 2: Subtract the integer parts and fraction parts separately.

$$8\frac{1}{4} - 5\frac{2}{3} = 8\frac{1(3)}{4(3)} - 5\frac{2(4)}{3(4)}$$

$$= 8\frac{3}{12} - 5\frac{8}{12}$$

$$= 7\frac{15}{12} - 5\frac{8}{12}$$

$$= 2\frac{7}{12}$$

Section 5.7 Order of Operations and Evaluating and Simplifying Expressions

Order of Operations

Perform operations in the following order:

1. Grouping symbols: parentheses ( ), brackets [ ], braces { }, absolute value | |, radicals $\sqrt{}$, and fraction bars —
2. Exponents
3. Multiplication or division from left to right, in order as they occur.
4. Addition or subtraction from left to right, in order as they occur.

Example 1: Simplify. $\left(\dfrac{3}{4}\right)^2 - \dfrac{1}{6} \div \dfrac{2}{3}$

$$= \frac{9}{16} - \frac{1}{6} \div \frac{2}{3} \qquad \text{Simplify the exponential form.}$$

$$= \frac{9}{16} - \frac{1}{6} \cdot \frac{3}{2} \qquad \text{Rewrite the division as multiplication.}$$

$$= \frac{9}{16} - \frac{1}{\overset{}{\underset{2}{6}}} \cdot \frac{\overset{1}{3}}{2} \qquad \text{Divide out the common factor.}$$

$$= \frac{9}{16} - \frac{1}{4} \qquad \text{Multiply.}$$

$$= \frac{9}{16} - \frac{1(4)}{4(4)} \qquad \text{Rewrite with the LCD, 16.}$$

$$= \frac{9}{16} - \frac{4}{16}$$

$$= \frac{5}{16} \qquad \text{Subtract.}$$

To evaluate an algebraic expression:

1. Replace the variables with the corresponding given numbers.
2. Calculate following the order of operations agreement.

Example 2: Evaluate $\dfrac{2}{3}x - 5xy$ when $x = -6$ and $y = \dfrac{1}{10}$.

$$\frac{2}{3}(-6) - 5(-6)\left(\frac{1}{10}\right) \qquad \substack{\text{Replace vari-}\\ \text{ables with}\\ \text{values.}}$$

$$= \frac{2}{\overset{}{\underset{1}{3}}}\left(-\frac{\overset{2}{6}}{1}\right) - \frac{\overset{1}{5}}{1}\left(-\frac{\overset{3}{6}}{1}\right)\left(\frac{1}{\underset{1}{10}}\right) \qquad \substack{\text{Write as im-}\\ \text{proper fractions}\\ \text{and divide out}\\ \text{common}\\ \text{factors.}}$$

$$= -4 + 3 \qquad \text{Multiply.}$$

$$= -1 \qquad \text{Add.}$$

To combine like terms:

1. Add or subtract the coefficients.
2. Keep the variables and their exponents the same.

Example 3: Combine like terms.

$$\frac{7}{12}x^2 + \frac{1}{4}x - \frac{2}{3}x^2 + \frac{1}{2}x$$

$$= \frac{7}{12}x^2 - \frac{2}{3}x^2 + \frac{1}{4}x + \frac{1}{2}x \qquad \text{Collect like terms.}$$

$$= \frac{7}{12}x^2 - \frac{2(4)}{3(4)}x^2 + \frac{1}{4}x + \frac{1(2)}{2(2)}x$$

Rewrite the fractions with their LCD.

$$= \frac{7}{12}x^2 - \frac{8}{12}x^2 + \frac{1}{4}x + \frac{2}{4}x$$

$$= -\frac{1}{12}x^2 + \frac{3}{4}x \qquad \text{Add and subtract appropriately.}$$

To add polynomials, combine like terms.

Example 4: Add or subtract.

a. $\left(\frac{1}{6}x + \frac{4}{5}\right) + \left(\frac{1}{6}x - \frac{1}{5}\right)$

$$= \frac{1}{6}x + \frac{1}{6}x + \frac{4}{5} - \frac{1}{5} \qquad \text{Collect the like terms.}$$

$$= \frac{2}{6}x + \frac{3}{5} \qquad \text{Add and subtract appropriately.}$$

$$= \frac{1}{3}x + \frac{3}{5} \qquad \text{Simplify.}$$

To subtract two polynomials:

1. Write the subtraction statement as an equivalent addition statement.
   a. Change the operation sign from minus to plus.
   b. Change the subtrahend to its additive inverse. To get the additive inverse, we change all the signs in the polynomial.
2. Combine like terms.

b. $\left(\frac{5}{9}x - \frac{3}{4}\right) - \left(\frac{1}{9}x - \frac{1}{12}\right)$

$$= \left(\frac{5}{9}x - \frac{3}{4}\right) + \left(-\frac{1}{9}x + \frac{1}{12}\right) \qquad \text{Rewrite as addition.}$$

$$= \frac{5}{9}x - \frac{1}{9}x - \frac{3}{4} + \frac{1}{12} \qquad \text{Collect the like terms.}$$

$$= \frac{5}{9}x - \frac{1}{9}x - \frac{3(3)}{4(3)} + \frac{1}{12} \qquad \text{Rewrite the fractions with their LCD.}$$

$$= \frac{5}{9}x - \frac{1}{9}x - \frac{9}{12} + \frac{1}{12}$$

$$= \frac{4}{9}x - \frac{8}{12} \qquad \text{Add and subtract appropriately.}$$

$$= \frac{4}{9}x - \frac{2}{3} \qquad \text{Simplify.}$$

Section 5.8 Solving equations.

To solve equations:

1. Simplify both sides of the equation as needed.
   a. Distribute to clear parentheses.
   b. Eliminate fractions by multiplying both sides by the LCD of all fractions (optional).
   c. Combine like terms.

2. Use the addition/subtraction principle of equality so that all variable terms are on one side of the equation and all constants are on the other side. (Clear the variable term that has the lesser coefficient. This will avoid negative coefficients.) Then combine like terms.

3. Use the multiplication/division principle of equality to clear any remaining coefficients.

Example 1: Solve.

$$\frac{2}{3}(x - 1) = \frac{1}{2}x - \frac{2}{5}$$

$$30\left(\frac{2}{3}x - \frac{2}{3}\right) = \left(\frac{1}{2}x - \frac{2}{5}\right)30$$

Eliminate the fractions by multiplying both sides by the LCD, 30.

$$\overset{10}{\frac{30}{1}} \cdot \frac{2}{\underset{1}{3}}x - \overset{10}{\frac{30}{1}} \cdot \frac{2}{\underset{1}{3}} = \overset{15}{\frac{30}{1}} \cdot \frac{1}{\underset{1}{2}}x - \overset{6}{\frac{30}{1}} \cdot \frac{2}{\underset{1}{5}}$$

$$20x - 20 = 15x - 12$$
Subtract 15x from both sides.

$$\underline{-15x \qquad\quad -15x}$$

$$5x - 20 = \quad 0 - 12$$

$$5x - 20 = -12$$

$$\underline{+20 \qquad +20}$$
Add 20 to both sides.

$$5x - \quad 0 = \quad 8$$

$$\frac{5x}{5} = \frac{8}{5}$$
Divide both sides by 5.

$$1x = \frac{8}{5}$$

$$x = 1\frac{3}{5}$$

Formulas

Area of a parallelogram: $A = bh$

Area of a triangle: $A = \frac{1}{2}bh$

Area of a trapezoid: $A = \frac{1}{2}h(a + b)$

Diameter of a circle: $d = 2r$

Radius of a circle: $r = \frac{1}{2}d$

Circumference of a circle: $C = \pi d$ or $C = 2\pi r$

Area of a circle: $A = \pi r^2$

For Exercises 1–6, answer true or false.

[5.2]   **1.** $\frac{3}{5}$ is in lowest terms.

[5.1]   **2.** Every integer can be expressed in fraction form.

[5.1]   **3.** $\frac{54}{0} = 0$

[5.1]   **4.** A fraction with the same numerator and denominator (other than 0) can be simplified to the number 1.

[5.5]   **5.** When finding the LCM using primes, the rule is to use only those prime factors that are common to all the factorizations.

[5.3]   **6.** When multiplying mixed numbers, we can multiply the integer parts and fraction parts separately.

[5.5]   **7.** We can write an equivalent fraction by _____ or _____ both the numerator and denominator by the same nonzero number.

[5.2]   **8.** Explain in your own words how to reduce a fraction to lowest terms.

[5.4]   **9.** Explain in your own words how to divide fractions.

[5.6]   **10.** Explain in your own words how to add fractions.

[5.1]   **11.** Name the lowest terms fraction represented by the shaded region.

[5.1]   **12.** Graph each number on a number line.

[5.1]   **a.**

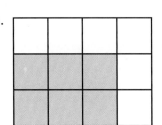

b.

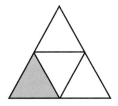

a. $4\frac{1}{5}$

b. $-2\frac{5}{8}$

[5.1]   **13.** Write each improper fraction as a mixed number.

a. $\frac{40}{9}$

b. $-\frac{29}{4}$

[5.1]   **14.** Write each mixed number as an improper fraction.

a. $6\frac{2}{3}$

b. $-5\frac{1}{2}$

[5.1] **15.** Simplify.

    a. $\dfrac{18}{1}$

    b. $\dfrac{-14}{0}$

    c. $\dfrac{0}{19}$

    d. $\dfrac{-2}{-2}$

[5.2] **16.** Simplify to lowest terms.

    a. $\dfrac{15}{35}$

    b. $-\dfrac{84}{105}$

[5.2] **17.** Simplify to lowest terms.

    a. $-\dfrac{8m^3n}{26n^2}$

    b. $\dfrac{6xy^2}{20y^5}$

$\begin{bmatrix}5.1\\5.2\end{bmatrix}$ **18.** Use $<$, $>$, or $=$ to write a true statement.

    a. $\dfrac{5}{9}$? $\dfrac{7}{13}$

    b. $\dfrac{3}{16}$? $\dfrac{5}{24}$

    c. $\dfrac{10}{16}$? $\dfrac{25}{40}$

[5.3] **19.** Multiply.

    a. $\dfrac{4}{9} \cdot \dfrac{12}{20}$

    b. $-4\dfrac{1}{2} \cdot 2\dfrac{2}{3}$

[5.3] **20.** Multiply.

    a. $\dfrac{10n}{6p^3} \cdot \dfrac{14np}{8n}$

    b. $\dfrac{h^3}{4k} \cdot \dfrac{-12k}{15h}$

[5.4] **21.** Divide. Write your answer as a mixed number where appropriate.

    a. $-\dfrac{5}{6} \div \dfrac{-15}{28}$

    b. $4\dfrac{1}{5} \div \dfrac{3}{5}$

[5.4] **22.** Divide.

    a. $\dfrac{20a}{12ab^3} \div \dfrac{10}{8b}$

    b. $-\dfrac{8x^6y}{25x^3} \div \dfrac{2xy}{5z^2}$

[5.4] **23.** Evaluate the square root.

    a. $\sqrt{\dfrac{100}{36}}$

    b. $\sqrt{\dfrac{50}{2}}$

[5.5] **24.** Find the LCM.

    a. 28 and 24

    b. $15x$ and $20x^2y$

[5.6] 25. Add or subtract.

    a. $\dfrac{5}{6} + \dfrac{1}{5}$

    b. $4\dfrac{5}{8} + 6\dfrac{2}{3}$

    c. $\dfrac{9}{15} - \dfrac{4}{5}$

    d. $-5\dfrac{1}{6} - \left(-2\dfrac{1}{3}\right)$

[5.6] 26. Add or subtract.

    a. $\dfrac{3n}{8} + \dfrac{n}{8}$

    b. $-\dfrac{2}{9x} + \left(-\dfrac{4}{9x}\right)$

    c. $\dfrac{7}{6h} - \dfrac{3}{4h}$

    d. $\dfrac{5}{8a} - \dfrac{7}{12}$

[5.7] 27. Simplify.

    a. $3\dfrac{1}{2} + \dfrac{1}{4}\left(\dfrac{2}{3} - \dfrac{1}{6}\right)$

    b. $\left(4 + \dfrac{1}{12}\right) - 8 \div \dfrac{3}{4}$

[5.3] 28. Evaluate the expressions.

    a. $\left(\dfrac{1}{2}\right)^{6}$

    b. $\left(-\dfrac{3}{5}xy^{3}\right)^{2}$

[5.7] 29. Simplify.

    a. $\dfrac{3}{5}x^{2} + 9x - \dfrac{1}{2}x^{2} - 2 - 11x$

    b. $\left(\dfrac{1}{4}n^{2} - \dfrac{2}{3}n - 3\right) + \left(\dfrac{3}{8}n^{2} + 1\right)$

    c. $\left(12y^{3} - \dfrac{5}{6}y + \dfrac{2}{5}\right) - \left(\dfrac{1}{2}y^{3} + 2y^{2} + \dfrac{1}{3}\right)$

    d. $\left(\dfrac{2}{7}b^{3}\right)\left(-\dfrac{7}{8}b\right)$

    e. $\left(\dfrac{1}{4}x + 2\right)\left(5x - \dfrac{3}{2}\right)$

[5.8] 30. Solve and check.

    a. $y - 3\dfrac{4}{5} = \dfrac{1}{3}$

    b. $-\dfrac{3}{8}n = \dfrac{4}{3}$

    c. $\dfrac{1}{2}m - 5 = \dfrac{3}{4}$

    d. $\dfrac{1}{2}\left(n - \dfrac{2}{3}\right) = \dfrac{3}{4}n + 2$

[5.3] 31. Find the area of the triangle.

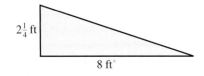

[5.7] 32. Find the area of the trapezoid.

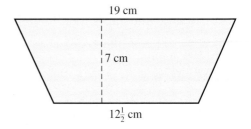

[5.3] **33.** The circular top of a can of beans has a diameter of $3\frac{1}{2}$ inches. Find the radius.

[5.3] **34.** A wheel on a piece of exercise equipment has a radius of $14\frac{1}{5}$ centimeters. Find the diameter.

[5.3] **35.** A tree has a diameter of $2\frac{1}{2}$ feet. What is the circumference? Use $\frac{22}{7}$ for π.

[5.7] **36.** Find the surface area of a circular tabletop with a diameter of 4 feet. Use $\frac{22}{7}$ for π.

[5.7] **37.** Find the area of the shaded region.

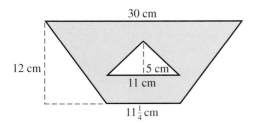

30 cm

12 cm 5 cm 11 cm $11\frac{1}{4}$ cm

[5.3] **38.** At a clothing store, $\frac{5}{9}$ of the merchandise is discounted. $\frac{3}{10}$ of these discounted items are shirts. What fraction of all items in the store are discounted shirts?

[5.4] **39.** A box of rice contains $4\frac{1}{2}$ cups of uncooked rice. The label indicates that a single serving is $\frac{1}{4}$ cup of uncooked rice. How many servings are in the box?

[5.6] **40.** Respondents to a survey can answer in three ways: agree, disagree, or no opinion. $\frac{2}{5}$ said they agreed, while $\frac{1}{4}$ said they disagreed. What fraction had no opinion?

[5.8] **41.** $\frac{1}{4}$ of the sum of n and 2 is the same as $\frac{3}{5}$ less than $\frac{3}{4}$ of n. Translate to an equation, then solve.

[5.7] **42.** The area of the trapezoid shown is 19 square meters. Find the length of the missing side.

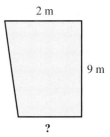

2 m

9 m

?

[5.7] **43.** A section of PVC pipe is $2\frac{1}{3}$ times the length of a second section. The two pipes together have a length of 35 feet. What are the lengths of the two sections?

For Exercises 44 and 45, use a four-column table.

[5.7] **44.** Emma and Bill pass each other going in opposite directions. Emma is walking at $2\frac{1}{2}$ miles per hour and Bill is jogging at 5 miles per hour. How long until they are $\frac{7}{10}$ miles apart?

[5.7] **45.** Two asteroids are separated by 2000 miles and headed directly at one another. Asteroid A is traveling at 50 miles per second and asteroid B is traveling at 120 miles per second. How long until they collide?

[5.1] **1.** Name the fraction represented by the shaded region.

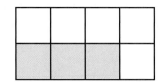

1. _____

[5.1] **2.** Graph $-3\frac{1}{4}$ on a number line.

2. _____

[5.1]
[5.2] **3.** Use $<$, $>$, or $=$ to write a true statement.

a. $\dfrac{3}{4}$? $\dfrac{9}{12}$ b. $\dfrac{14}{15}$? $\dfrac{5}{6}$

3. a. _____

b. _____

[5.1] **4.** Write $\frac{37}{6}$ as a mixed number.

4. _____

[5.1] **5.** Write $-4\frac{5}{8}$ as an improper fraction.

5. _____

[5.1] **6.** Simplify.

a. $\dfrac{-16}{1}$ b. $\dfrac{3}{0}$ c. $\dfrac{0}{14}$ d. $\dfrac{27}{27}$

6. a. _____
   b. _____
   c. _____
   d. _____

[5.2] **7.** Reduce to lowest terms.

a. $\dfrac{24}{40}$ b. $-\dfrac{9x^5y}{30x^3}$

7. a. _____

b. _____

[5.3] **8.** Multiply.

a. $-2\dfrac{1}{4} \cdot 5\dfrac{1}{6}$ b. $\dfrac{2a}{7b^3} \cdot \dfrac{14b}{3a}$

8. a. _____

b. _____

[5.4] **9.** Divide.

a. $3\dfrac{1}{8} \div \dfrac{5}{4}$ b. $\dfrac{-2m^4n}{9n^2} \div \dfrac{5m}{12n^2}$

9. a. _____

b. _____

[5.5] **10.** Find the LCM of $18t^3$ and $12t^2u$.

10. _____

[5.6] **11.** Add or subtract.

a. $\dfrac{3}{10} + \dfrac{1}{4}$ b. $-3\dfrac{1}{8} - \left(-1\dfrac{1}{2}\right)$

11. a. _____

b. _____

[5.6] **12.** Add or subtract.

a. $-\dfrac{4x}{15} + \left(-\dfrac{x}{15}\right)$ b. $\dfrac{7}{4a} - \dfrac{5}{6a}$

12. a. _____

b. _____

[5.3] 13. Evaluate.

    a. $\left(\dfrac{2}{5}\right)^2$ **b.** $\left(\dfrac{1}{4}\right)^3$

[5.7] 14. Calculate. $2\dfrac{1}{4} + \dfrac{5}{8}\left(\dfrac{4}{5} - \dfrac{3}{10}\right)$

[5.7] 15. Simplify.

    a. $\left(\dfrac{1}{4}n^2 - \dfrac{1}{2}n + 2\right) - \left(\dfrac{1}{4}n^2 + n + 5\right)$ **b.** $\left(m + \dfrac{3}{4}\right)\left(2m - \dfrac{1}{2}\right)$

[5.8] 16. Solve and check. $\dfrac{1}{4}m - \dfrac{2}{3} = \dfrac{5}{6}$

[5.3] 17. Find the area of the triangle.

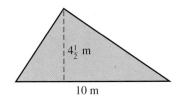

    $4\frac{1}{2}$ m

    10 m

[5.7] 18. Find the area of the trapezoid.

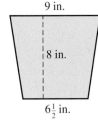

    9 in.

    8 in.

    $6\frac{1}{2}$ in.

[5.3] 19. A compact disc has a diameter of $4\frac{3}{4}$ inches. What is the circumference? Use $\frac{22}{7}$ for π.

[5.7] 20. Calculate the area of a circle with a diameter of 5 inches. Use $\frac{22}{7}$ for π.

[5.3] 21. In a new housing development, $\frac{3}{4}$ of the lots were sold in the first year. $\frac{5}{6}$ of these lots had finished houses on them by the end of the first year. What fraction of all lots in the development had finished houses on them by the end of the first year?

[5.4] 22. Jason needs pieces of wire that are $4\frac{3}{8}$ inches long. How many pieces can he make out of 35 inches of wire?

[5.3] 23. A testing company examines the results of a certain multiple-choice test question that has four choices, A, B, C, or D. $\frac{5}{8}$ of those who took the test chose A, while $\frac{1}{6}$ chose B, and nobody chose D. What fraction chose C?

[5.8] 24. $\frac{2}{3}$ of the sum of n and 5 is the same as $\frac{3}{4}$ less than $\frac{1}{2}$ of n. Translate to an equation, then solve.

[5.8] 25. Ken and Keisha are traveling toward each other on the same path. Ken is bicycling at 8 miles per hour and Keisha is jogging at $5\frac{1}{2}$ miles per hour. If they are $\frac{3}{4}$ of a mile apart, how long until they meet?

13. a. _____

    b. _____

14. _____

15. a. _____

    b. _____

16. _____

17. _____

18. _____

19. _____

20. _____

21. _____

22. _____

23. _____

24. _____

25. _____

For Exercises 1–6, answer true or false.

$\begin{bmatrix} 1.1 \\ 1.2 \end{bmatrix}$ **1.** -1 is a whole number.

[2.2] **2.** The sum of two numbers that have the same sign is always positive.

[4.2] **3.** $2x = 5x^2 + 8$ is a linear equation.

[4.1] **4.** -7 is a solution for $3a + 9 = -30$.

[5.3] **5.** $\left(\dfrac{3}{4}\right)^2 = \dfrac{3^2}{4^2}$

[5.1] **6.** $\dfrac{3}{8} > \dfrac{5}{16}$

[3.5] **7.** When raising an exponential form to a power, we can _____ the exponents and keep the same base.

[3.7] **8.** Explain in your own words how to find the GCF/GCD.

[4.3] **9.** Explain the mistake.

$$4x - 9 = 10x + 3$$
$$4x - 9 = 10x + 3$$
$$\underline{-4x \qquad\quad -4x}$$
$$0 - 9 = 6x + 3$$
$$-9 = 6x + 3$$
$$\underline{-3 \qquad\quad -3}$$
$$-6 = 6x + 0$$
$$\dfrac{-6}{6} = \dfrac{6x}{6}$$
$$-1 = x$$

[5.6] **10.** Explain in your own words how to add or subtract fractions.

[1.1] **11.** Write the word name for 4,582,601.

[2.1] **12.** Graph -9 on a number line.

[1.1] **13.** Round 851,412 to the nearest ten thousand.

[1.3] **14.** Estimate $461 \cdot 72$ by rounding so that there is only one nonzero digit in each number.

[3.2] **15.** What is the degree of $-x$?

[3.6] **16.** Find the prime factorization of 3600.

[3.6] **17.** Find the GCF of $54x^3y$ and $45x$.

For Exercises 18–27, simplify.

[2.1] **18.** $-[-(-15)]$

[2.2] **19.** $5 - 12 + (-17) + 9$

[2.5] **20.** $2^3 - 7(-4)$

[2.5] **21.** $-3^4 - 5\sqrt{100 - 64}$

[2.5] **22.** $18 - 5(6) \div [9 + 3(-7)]$

[5.2] **23.** $\dfrac{36}{40}$

[5.4] **24.** $4\dfrac{1}{3} \div 5\dfrac{1}{9}$

[5.3] **25.** $-\dfrac{5x^3}{9} \cdot \dfrac{27}{40x}$

[5.6] **26.** $7\dfrac{1}{6} - 3\dfrac{1}{2}$

[5.6] **27.** $\dfrac{4}{5x} + \dfrac{2}{3}$

[5.7] **28.** Evaluate $mn - 2n^3$ when $m = \frac{5}{6}$ and $n = -4$.

[5.7] **29.** Combine like terms and write your answer in descending order of degree.

$$12b^2 - \frac{2}{3}b + 3b^2 - b^3 - \frac{1}{4}b$$

[3.4] **30.** Subtract. $(x^4 - 5x^3 - 10x + 18) - (x^4 + 2x^3 - 3x - 8)$

[3.5] **31.** Multiply. $(-8x^3)(-4x^2)$

[3.5] **32.** Multiply. $(x - 7)(x + 7)$

[3.7] **33.** Divide. $\dfrac{t^9}{t^5}$

[3.7] **34.** Factor. $18m^3 + 24m^2 - 30m$

For Exercises 35–38, solve and check.

[4.2] **35.** $x + 24 = 8$

[5.4] **36.** $\dfrac{2}{3}a = -18$

[5.6] **37.** $\dfrac{3}{4}n - 1 = -\dfrac{2}{5}$

[4.3] **38.** $3(y - 4) - 8 = 7y - 16$

For Exercises 39–50, solve.

[1.3] **39.** A combination lock consists of 4 rollers with the numerals 0–9 on each roller. How many different combinations are possible?

[2.3] **40.** The following table lists the assets and debts for the Krueger family. Calculate their net worth.

Assets	Debts
Savings = $1260	Credit-card balance = $872
Checking = $945	Mortgage = $57,189
Furniture = $13,190	Automobile 1 = $3782
	Automobile 2 = $12,498

[3.8] **41.** An acorn falls from a height of 92 feet. What is the height of the acorn after 2 seconds? (Use the formula $h = -16t^2 + h_0$.)

[3.8] **42.** **a.** Write an expression in simplest form for the perimeter of the rectangle shown.

b. Find the perimeter if y is 8 feet.

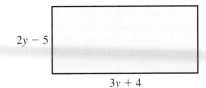

$2y - 5$

$3y + 4$

[3.8] **43.** Write an expression in simplest form for the area.

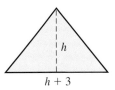

h

$h + 3$

[5.3]
[5.4] **44.** A tree has a circumference of 44 inches. Calculate the diameter.

[4.3] **45.** A storage chest has a volume of 24 cubic feet. If the length is 4 feet and the width is 3 feet, find the height?

[4.5] 46. A support beam is to be angled so that the angle made between the beam and the ground on one side of the beam is 35° less than twice the adjacent angle on the other side of the beam. What are the angle measurements?

[4.5] 47. The length of a rectangle is three times the width. If the perimeter is 48 feet, find the length and width.

[4.5] 48. The sum of two numbers is 25. One number is one more than twice the other. Find the two numbers.

[4.5] 49. A movie theater sells two different-size boxes of popcorn. The large box costs $3 and the small box costs $2. In one day the theater sold 657 boxes of popcorn for a total of $1612. How many of each size were sold? (Use a four-column table.)

[5.2] 50. Shelly and Ryan pass each other going in opposite directions. Shelly is running at $5\frac{1}{2}$ miles per hour and Ryan is jogging at 3 miles per hour. How long until they are $3\frac{2}{5}$ miles apart?

Decimals

> *"In the middle of difficulty lies opportunity"*
>
> —ALBERT EINSTEIN, PHYSICIST

> *"No matter what accomplishments you make, someone helped you."*
>
> —ALTHEA GIBSON, TENNIS PLAYER

Getting Help

Remember that to learn, you must be challenged. There will be times in your education that you will encounter difficulties. The key to success is learning to respond to those difficulties in a positive way.

When you encounter difficulties that you cannot resolve, seek help immediately. Even if your question or problem seems small, it can often make a difference in understanding future topics. As soon as possible, talk with your instructor. Most instructors have office hours or are available by appointment to discuss your problems.

In addition, most colleges have tutorial centers with flexible hours that offer a variety of services. The centers will likely have tutors as well as the videotapes and computer software that are keyed to the text. Some schools may have these resources available in their library as well. Another tutorial resource you may use is the Addison-Wesley Math Tutor Center. If you received a PIN with your text, you can call, e-mail, or fax your question, and an experienced math teacher will respond.

Another great source of help is your classmates. Form a study group and exchange phone numbers and email addresses so that you can discuss problems and encourage and support each other. Above all, do not let questions go unanswered. Take advantage of all the resources available to you to get the help you need to succeed.

6.1 Decimals and Rational Numbers

OBJECTIVES

1 Write decimals as fractions or mixed numbers.

2 Write a word name for a decimal number.

3 Graph decimals on a number line.

4 Use < or > to make a true statement.

5 Round decimal numbers to a specified place.

OBJECTIVE 1 Write decimals as fractions or mixed numbers.

In this chapter, we will use a different notation for writing fractions, **decimal notation.**

DEFINITION Decimal notation: A base-10 notation for expressing fractions.

By *base-10 notation* we mean that each place value is a power of 10. Whole numbers are expressed to the left of the decimal point and have familiar place values such as ones, tens, hundreds, etc. To the right of the decimal point are the fractional place values: tenths, hundredths, thousandths, etc. For each fractional place, the power of 10 is in the denominator of a fraction.

Decimal Point

Whole numbers			Fractions			
Hundreds	Tens	Ones	Tenths	Hundredths	Thousandths	Ten thousandths

In Chapter 5, we defined **rational numbers.** Let's recall that definition:

DEFINITION Rational number: number that can be expressed as a ratio of integers.

In math language: rational number is a number that can be expressed in the form $\frac{a}{b}$, where a and b are integers and $b \neq 0$.

CONNECTION Notice that the fraction names follow the same pattern as the whole-number side of the decimal point. We simply place *th* in each whole place name. *Tens* becomes *tenths, hundreds* becomes *hundredths,* and so on.

DISCUSSION Why is there no oneths place?

Because all integers can be expressed as fractions by placing the integer over 1, all integers are rational numbers. All mixed numbers are rational numbers because we can write them as fractions. If we can write decimal numbers as fractions, then they, too, are rational numbers. But can we write every decimal number as a fraction? We need to explore further to answer this question.

To see the connection between decimals and fractions, let's have a look at a decimal number written in expanded form. Consider the number 23.791.

Decimal Point

Whole numbers		Fractions		
Tens	Ones	Tenths	Hundredths	Thousandths
2	3	7	9	1

Expanded form: $23.791 = 2 \cdot 10 + 3 \cdot 1 + 7 \cdot \dfrac{1}{10} + 9 \cdot \dfrac{1}{100} + 1 \cdot \dfrac{1}{1000}$

We can determine the equivalent fraction by multiplying and adding.

$$23.791 = 2 \cdot 10 + 3 \cdot 1 + 7 \cdot \frac{1}{10} + 9 \cdot \frac{1}{100} + 1 \cdot \frac{1}{1000}$$

$$= 20 + 3 + \frac{7}{1} \cdot \frac{1}{10} + \frac{9}{1} \cdot \frac{1}{100} + \frac{1}{1} \cdot \frac{1}{1000}$$

$$= 23 + \frac{7}{10} + \frac{9}{100} + \frac{1}{1000}$$

$$= 23 + \frac{7(100)}{10(100)} + \frac{9(10)}{100(10)} + \frac{1}{1000}$$

Note: The whole-number part becomes the integer part of a mixed number.

$$= 23 + \frac{700}{1000} + \frac{90}{1000} + \frac{1}{1000}$$

$$= 23 + \frac{791}{1000}$$

◀ **Note:** Recall that to add the fractions we need a common denominator. Notice that the last place value determines the common denominator. In this case, the LCD is 1000.

$$= 23\frac{791}{1000}$$

◀ **Note:** The numerator of the fraction contains all the decimal digits and the denominator is the last place value.

CONNECTION We read $23\frac{791}{1000}$ as "twenty-three and seven hundred ninety-one thousandths," so 23.791 is read exactly the same way.

Our example suggests a fast way to write a decimal number as a fraction or mixed number:

PROCEDURE *To write a decimal number as a reduced fraction or mixed number:*

1. Write all digits to the left of the decimal (whole digits) as the integer in a mixed number.
2. Write all digits to the right of the decimal (fraction digits) in the numerator of a fraction.
3. Write the last place value in the denominator.
4. Simplify to lowest terms.

Example 1 Write the number as a fraction or mixed number in simplest form.

a. 0.84

Solution: Write all the decimal digits in the numerator and write the last place value in the denominator. Then simplify to lowest terms.

$$0.84 = \frac{84}{100} = \frac{84 \div 4}{100 \div 4} = \frac{21}{25}$$

hundredths

Because $0.84 = \frac{84}{100}$, we read 0.84 as "eighty-four hundredths."

b. −217.5

Solution: We write the integer part of the decimal number as the integer part of a mixed number, then write the decimal digits in the numerator and the last place value in the denominator.

$$-217.5 = -217\frac{5}{10} = -217\frac{5 \div 5}{10 \div 5} = -217\frac{1}{2}$$

tenths

We read −217.5 as "negative two hundred seventeen and five tenths" or, in simplest form, "negative two hundred seventeen and one-half." Note that it is also common to say "negative two hundred seventeen point five."

Do Your Turn 1 ▷

Your Turn 1

Write as a fraction or mixed number in lowest terms.

a. 0.79

b. −0.08

c. 2.6

d. −14.675

CONNECTION

Because the denominators will always be powers of 10, and the primes that divide powers of 10 are 2 and 5, these fractions will only reduce if the numerators are divisible by 2 or 5.

Answers to Your Turn 1

a. $\frac{79}{100}$ **b.** $-\frac{2}{25}$ **c.** $2\frac{3}{5}$

d. $-14\frac{27}{40}$

OBJECTIVE 2 Write a word name for a decimal number.

In Example 1, we learned to read decimal numbers using their equivalent fraction or mixed number. Similarly, when we write a word name for a decimal number, we write the name of its equivalent fraction or mixed number.

$0.39 = \frac{39}{100}$ and is written: thirty-nine hundredths

$19.8571 = 19\frac{8571}{10,000}$ and is written: nineteen and eight thousand five hundred seventy-one ten-thousandths.

Notice, the word *and* separates the integer and fraction parts of both mixed numbers and decimal numbers, so *and* takes the place of the decimal point in the word name for a decimal number that has an integer amount.

> *To write the word name for decimal digits (fraction part of a decimal number):*
> 1. Write the word name for digits as if they represented a whole number.
> 2. Write the name of the last place value.
>
> *To write a word name for a decimal number with both integer and fraction parts:*
> 1. Write the name of the integer number part.
> 2. Write the word *and* for the decimal point.
> 3. Write the name of the fractional part.

Your Turn 2

Write the word name.

a. 0.91

b. −0.602

c. 2.7

d. 124.90017

Example 2 Write the word name.

a. 0.0000489

Answer: four hundred eighty-nine ten-millionths

b. 98.10479

Answer: ninety-eight and ten thousand four hundred seventy-nine hundred-thousandths

◁ **Do Your Turn 2**

OBJECTIVE 3 Graph decimals on a number line.

Decimal numbers can be graphed on a number line. The denominator of the equivalent fraction describes the number of divisions between integers. The numerator of the equivalent fraction describes how many of those divisions we are interested in.

Example 3 Graph on a number line.

a. 3.6

Solution: Because 3.6 means $3\frac{6}{10}$, we divide the distance between 3 and 4 into 10 equal-size divisions and draw a dot on the 6th division mark.

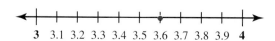

3   3.1 3.2 3.3 3.4 3.5 3.6 3.7 3.8 3.9   4

b. 14.517

Solution: Because 14.517 means $14\frac{517}{1000}$, we should divide the space between 14 and 15 into 1000 divisions and count to the 517th mark. Of course, this is rather tedious, so we'll gradually zoom in on smaller and smaller sections of the number line.

Answers to Your Turn 2
a. ninety-one hundredths
b. negative six hundred two thousandths
c. two and seven tenths
d. one hundred twenty-four and ninety thousand seventeen hundred-thousandths.

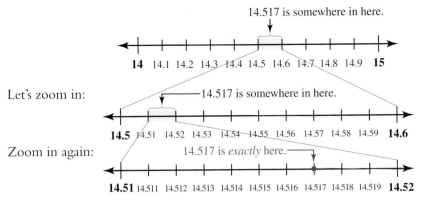

14.517 is somewhere in here.

14 14.1 14.2 14.3 14.4 14.5 14.6 14.7 14.8 14.9 15

Let's zoom in:

14.517 is somewhere in here.

14.5 14.51 14.52 14.53 14.54 14.55 14.56 14.57 14.58 14.59 14.6

Zoom in again:

14.517 is *exactly* here.

14.51 14.511 14.512 14.513 14.514 14.515 14.516 14.517 14.518 14.519 14.52

Note: 14.0 and 14 are the same just as 14.5 and 14.50 are the same. Zero digits are understood to continue on forever beyond the last nonzero digit in a decimal number. Writing those zeros is optional, but it is simpler not to write them.

Rather than going through all the *zoom* stages, we could just go directly to the last number line. Notice that the endpoints on the last number line are determined by the next-to-last place, in this case the hundredths place. We then draw 9 marks to divide the space between the endpoints into 10 equal divisions and count to the 7th mark.

c. -2.75

Solution: Let's zoom in on tenths in order to graph the hundredths. Because the number is negative, the values of the numbers decrease as you move left. However, the absolute value of the numbers will increase toward the left.

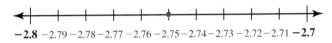

−2.8 −2.79 −2.78 −2.77 −2.76 −2.75 −2.74 −2.73 −2.72 −2.71 −2.7

Do Your Turn 3 ▷

OBJECTIVE 4 Use $<$ or $>$ to make a true statement.

Remember that we can use a number line to compare two numbers. Let's compare 13.7 and 13.2 by graphing them on a number line.

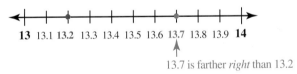

13 13.1 **13.2** 13.3 13.4 13.5 13.6 **13.7** 13.8 13.9 **14**

13.7 is farther *right* than 13.2

Since 13.7 is farther right than 13.2, we can say $13.7 > 13.2$.

We can also compare the numbers without graphing them. Look at the digits in 13.7 and 13.2 from left to right. They match until we get to the tenths place, and it is this place with its different digits that tells us which number is greater. The greater digit, 7, is in the greater number.

> **PROCEDURE** *To determine which of two positive decimal numbers is greater:*
> 1. Compare the digits in the numbers from left to right until you find two different digits in the same place value.
> 2. The greater of those two digits is in the greater number.
>
> **Note:** If both numbers are negative, the lesser digit is in the greater number because the number is closer to zero.

Your Turn 3

Graph on a number line.

a. 0.7

b. 12.45

c. -6.21

d. -19.106

Answers to Your Turn 3

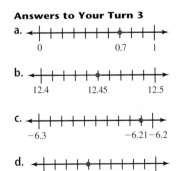

a.
0   0.7   1

b.
12.4   12.45   12.5

c.
−6.3   −6.21 −6.2

d.
−19.11   −19.106   −19.1

Your Turn 4

Use < or > to write a true statement.

a. 0.61 ? 0.65

b. 45.192 ? 45.092

c. 0.002 ? 0.00056

d. −5.71 ? −5.8

e. −0.04701 ? −0.0471

Example 4 Use < or > to make a true statement.

a. 35.619 ? 35.625

Answer: 35.619 < 35.625

Explanation: Compare the digits in the numbers from left to right.

35.619

↑↓

35.625

◄ Each digit matches until we get to the hundredths place.

In the hundredths places, because 2 is greater than 1, we can say 35.625 is greater than 35.619. On a number line, 35.625 would be farther to the right.

b. −0.00981 ? −0.009806

Answer: −0.00981 < −0.009806

Explanation: When we compare the digits from left to right, we see the hundred-thousandths place is the first place to have different digits.

−0.00981

↑↓

−0.009806

◄ Hundred-thousandths digits do not match.

Because both numbers are negative, the number closest to zero will be the greater number. This means the lesser of the two different digits is in the greater number. Because 0 is less than 1, −0.009806 is the greater number.

◄ **Do Your Turn 4**

OBJECTIVE 5 Round decimal numbers to a specified place.

Now that we have learned how to graph and compare decimal numbers, we can study how to round them. The idea is to round a given number to the nearest number in a specified place. When we graph on a number line, the endpoints are determined by the place prior to the last nonzero digit. This place will, therefore, determine how we round. Suppose we want to round 14.517 to the nearest hundredth. We graphed this number in Example 3b. Let's recall that graph:

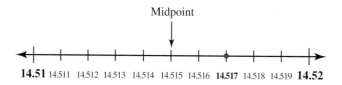

Notice that the nearest hundredths to 14.517 are 14.51 and 14.52. Since 14.517 is closer to 14.52, we say it rounds up to 14.52.

Rounding to a different place value would require a different scale. For example, to round 14.517 to the nearest tenth, we would use a number line with the nearest tenths, 14.5 and 14.6, as endpoints.

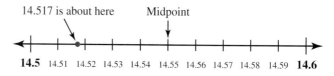

Because 14.517 is closer to 14.5, it rounds down to 14.5.

Answers to Your Turn 4
a. 0.61 < 0.65
b. 45.192 > 45.092
c. 0.002 > 0.00056
d. −5.71 > −5.8
e. −0.04701 > −0.0471

If we had to round to the nearest whole (ones place), we would use 14 and 15 as our endpoints and the tenths place would determine whether we round up or down.

14.517 is about here.

14 14.1 14.2 14.3 14.4 14.5 14.6 14.7 14.8 14.9 **15**

Because 14.517 is slightly higher than the midpoint on this scale, it rounds up to 15. Even if it were exactly halfway, we agree to round up.

PROCEDURE *To round a number to a given place value, consider the digit to the right of the desired place value.*
If this digit is 5 or higher, round up.
If this digit is 4 or less, round down.

Example 5 Round 917.28053 to the specified place.

a. Tenths

Solution: The nearest tenths are 917.2 and 917.3. Rounding to the nearest tenth means we consider the digit in the hundredths place, which is 8. Since 8 is greater than 5, we round up.

Answer: 917.3

b. Thousandths

Solution: The nearest thousandths are 917.280 and 917.281. Looking at the ten-thousandths place, we see a 5, so we round up to 917.281.

Answer: 917.281

c. Whole number

Solution: The nearest whole numbers are 917 and 918. The 2 in the tenths place means we round down to 917.

Answer: 917

Do Your Turn 5 ▶

Your Turn 5

Round 59.61538 to the specified place.

a. Tenths

b. Hundredths

c. Ten-thousandths

d. Whole number

Answers to Your Turn 5
a. 59.6 b. 59.62
c. 59.6154 d. 60

6.1 Exercises

For Extra Help

 Videotape DVT

 Addison-Wesley Tutor Center

 Math XL

 MyMathLab

 Student Solutions Manual

1. Complete the table of place values.

Decimal Point

Whole numbers				Fractions				

2. Explain how to write a decimal number as a fraction or mixed number.

3. When writing the word name for a decimal number greater than 1, as in 2.56, what word takes the place of the decimal point?

4. Explain how to determine which of two positive decimal numbers is the greater number.

For Exercises 5–24, write as a fraction or mixed number in lowest terms.

5. 0.2   **6.** 0.6   **7.** 0.25   **8.** 0.75

9. 0.375   **10.** 0.125   **11.** 0.24   **12.** 0.48

13. 1.5   **14.** 4.2   **15.** 18.75   **16.** 6.4

17. 9.625   **18.** 14.375   **19.** 7.36   **20.** 9.54

21. −0.008   **22.** −0.0005   **23.** −13.012   **24.** −219.105

For Exercises 25–38, write the word name.

25. 0.097   **26.** 0.415

27. 0.002015   **28.** 0.030915

29. 31.98   **30.** 15.73

31. 521.608   **32.** 343.406

33. 4159.6   **34.** 2109.003

35. −107.99   **36.** −2106.1

37. −0.50092   **38.** −0.000982

For Exercises 39–50, graph on a number line.

39. 0.8

40. 0.4

41. 1.3

42. 6.9

43. 4.25

44. 13.75

45. 8.06

46. 11.05

47. −3.21

48. −6.45

49. −19.017

50. −15.002

For Exercises 51–62, use < or > to make a true statement.

51. 0.81 ? 0.8

52. 0.15 ? 0.5

53. 2.891 ? 2.8909

54. 153.77 ? 153.707

55. 0.001983 ? 0.001985

56. 0.090194 ? 0.091094

57. 0.0245 ? 0.00963

58. 0.007 ? 0.0602

59. −1.01981 ? −1.10981

60. −18.1095 ? −18.1085

61. −145.7183 ? −14.57183

62. −26.71803 ? −267.1803

For Exercises 63–68, round 610.28315 to the specified place.

63. Tenths

64. Hundredths

65. Ten-thousandths

66. Whole number

67. Tens

68. Hundreds

For Exercises 69–74, round 0.95106 to the specified place.

69. Whole number

70. Tens

71. Tenths

72. Hundredths

73. Thousandths

74. Ten-thousandths

For Exercises 75–80, round −408.06259 to the specified place.

75. Tenths

76. Hundredths

77. Ten-thousandths

78. Thousandths

79. Tens

80. Whole number

Review Exercises

[5.6] **1.** Add. $\dfrac{37}{100} + \dfrac{4}{100}$

[2.3] **2.** Subtract. $-29 - (-16)$

For Exercises 3 and 4, simplify.

[3.3] **3.** $7a^2 - a + 15 - 10a^2 + a - 7a^3$

[3.4] **4.** $(5x^2 + 9x - 18) - (x^2 - 12x - 19)$

[4.2] **5.** Solve and check. $x - 19 = -36$

[5.6] **6.** Find the perimeter of a $12\frac{1}{2}$-foot-wide by 16-foot-long room.

6.2 Adding and Subtracting Decimal Numbers

OBJECTIVE 1 Add decimal numbers.

We could add or subtract decimal numbers by rewriting them as mixed numbers or fractions. Consider $2.82 + 43.01$.

$$2.82 + 43.01 = 2\frac{82}{100} + 43\frac{1}{100} \qquad \text{Write the decimal numbers as mixed numbers.}$$

$$= 45\frac{83}{100} \qquad \text{Add the whole numbers and fractions separately.}$$

$$= 45.83 \qquad \text{Write the result as a decimal number.}$$

However, there is an easier method. Notice that adding the digits in the corresponding place values, just as we did with whole numbers and integers, gives the same sum.

$$\begin{array}{r} 2.82 \\ + 43.01 \\ \hline 45.83 \end{array}$$

CONNECTION When we add mixed numbers, we add the integer parts and fraction parts separately. Because the decimal point separates the integer part from the fractional part in a decimal number, aligning the decimal points helps us add the integer and fraction parts separately when adding two decimal numbers.

PROCEDURE *To add decimal numbers:*

1. Stack the numbers so that the place values align. (Line up the decimal points.)
2. Add the digits in the corresponding place values.
3. Place the decimal point in the sum so that it aligns with the decimal points in the addends.

Example 1 Add.

a. $51.092 + 23.64$

Solution: We stack the numbers by place value, then add. We can align the corresponding place values by aligning the decimal points.

$$\begin{array}{r} \overset{1}{}51.092 \\ + 23.640 \\ \hline 74.732 \end{array}$$

Note: Some people choose to write the understood 0's beyond the last nonzero digit instead of having blank spaces.

Answer: 74.732

b. $32.09 + 4.103 + 19.6423$

Solution:

$$\begin{array}{r} \overset{1}{}\overset{1}{}32.0900 \\ 4.1030 \\ + 19.6423 \\ \hline 55.8353 \end{array}$$

Answer: 55.8353

c. $76 + 0.0189 + 3.99$

Solution:

$$
\begin{array}{r}
\overset{1\ 1\ \ 1}{76.0000} \\
0.0189 \\
+\ 3.9900 \\
\hline
80.0089
\end{array}
$$

Note: Integers have an understood decimal point to the right of the ones place. Once again, we can write the understood 0's as desired.

Answer: 80.0089

Do Your Turn 1 ▶

Your Turn 1

Add.

a. $14.103 + 7.035$

b. $0.1183 + 0.094$

c. $6.981 + 0.57 + 23.1$

d. $0.46 + 781 + 34$

OBJECTIVE **2** Subtract decimal numbers.

We subtract decimal numbers the same way we add, except that we subtract the digits in the corresponding place values.

PROCEDURE *To subtract decimal numbers:*

1. Stack the number with the greater absolute value on top so that the place values align. (Line up the decimal points.)
2. Subtract the digits in the corresponding place values.
3. Place the decimal point in the difference so that it aligns with the decimal points in the minuend and subtrahend.

Example 2 Subtract.

a. $58.941 - 2.54$

Solution: Stack the number with the greater absolute value on top.

$$
\begin{array}{r}
58.941 \\
-\ 2.540 \\
\hline
56.401
\end{array}
$$

Answer: 56.401

b. $179 - 48.165$

Solution:

$$
\begin{array}{r}
\overset{8\ \ 9\ 9\ 10}{179.000} \\
-\ 48.165 \\
\hline
130.835
\end{array}
$$

Note: Because 179 is an integer, the decimal point is understood to be to the right of the 9 digit. We can write as many 0's to the right of the decimal point as desired.

Answer: 130.835

c. $98.036 - 9.482$

Solution:

$$
\begin{array}{r}
\overset{\ \ \ \ \ 17}{\overset{8\ 7\ \ 9\ 13}{98.036}} \\
-\ \ \ 9.482 \\
\hline
88.554
\end{array}
$$

Answer: 88.554

Do Your Turn 2 ▶

Your Turn 2

Subtract.

a. $146.79 - 35.14$

b. $809 - 162.648$

c. $0.80075 - 0.06791$

d. $10.302 - 9.8457$

Answers to Your Turn 1
a. 21.138 b. 0.2123
c. 30.651 d. 815.46

Answers to Your Turn 2
a. 111.65 b. 646.352
c. 0.73284 d. 0.4563

OBJECTIVE 3 Add and subtract signed decimals.

To add or subtract signed decimal numbers, we follow the same rules that we developed in Chapter 2 for integers.

Your Turn 3

Add.

a. $90.56 + (-59.8)$

b. $-104 + (-54.88)$

c. $-5.14 + 0.75 + (-4.50) + 2.95$

d. $54.6 + (-14.50) + (-6.78) + (-1.3)$

Example 3 Add.

a. $-15.79 + 8.4$

Solution: Because the two numbers have different signs, we subtract their absolute values and keep the sign of the number with the greater absolute value.

> **Note:** To subtract, stack the number with the greater absolute value on top and align the decimal points.

$$\begin{array}{r} {}^{0\,1\,5}\!\!\not{1}\not{5}.79 \\ -\ 8.40 \\ \hline 7.39 \end{array}$$

> **Note:** To avoid confusion in the calculation, we left out the negative signs for -15.79 and -7.39. Although it is okay to leave signs out of a calculation, we must be careful to write them properly in our answer statement.

Answer: -7.39 Because -15.79 has the greater absolute value, our sum is negative.

Think: We have a debt of $15.79 and make a payment of $8.40. Because our payment is not enough to pay off the debt, we still have a debt of $7.39.

b. $-68.95 + (-14.50)$

Solution: Because we must add two numbers that have the same sign, we add and keep the same sign.

$$\begin{array}{r} {}^{1\ 1}68.95 \\ +\ 14.50 \\ \hline 83.45 \end{array}$$

> **Note:** Again, to avoid confusion, we left the negative signs out of the calculation, then put them back in the answer statement.

Answer: -83.45 Because both addends are negative, the sum is negative.

Think: We have a debt of $68.95 and add another debt of $14.50, so we are increasing our debt to $83.45.

◁ **Do Your Turn 3**

Recall from Chapter 2 that to subtract signed numbers, we can write an equivalent addition statement. To write an equivalent addition, we change the subtraction sign to an addition sign and change the subtrahend to its additive inverse.

Example 4 Subtract.

a. $0.08 - 5$

Solution: Write an equivalent addition statement.

Change the minus sign to a plus sign. $0.08 - 5$ Change the subtrahend to its additive inverse.

$$= 0.08 + (-5)$$

Because we are adding numbers with different signs, we subtract and keep the sign of the number with the greater absolute value.

$$\begin{array}{r} {}^{4\ \ 9\,10}\!\!\not{5}.\not{0}\not{0} \\ -0.08 \\ \hline 4.92 \end{array}$$ Subtract absolute values.

Answer: -4.92 In the equivalent addition, -5 has the greater absolute value, so the result is negative.

Answers to Your Turn 3
a. 30.76 b. -158.88
c. -5.94 d. 32.02

Think: We have $0.08 in a bank account (asset) and write a check for $5. This means we are overdrawn by $4.92.

b. $-16.2 - (-9.07)$

Solution: $-16.2 - (-9.07) = -16.2 + 9.07$ **Write the equivalent addition statement.**

Because we are adding two numbers with different signs, we subtract their absolute values and keep the sign of the number with the greater absolute value.

$$\begin{array}{r} {\scriptstyle 0\,16\ \ 1\,10} \\ \cancel{16}.\cancel{20} \\ -\ 9.07 \\ \hline 7.13 \end{array}$$ **Subtract absolute values.**

Answer: -7.13 **In the equivalent addition, -16.2 has the greater absolute value, so the result is negative.**

Do Your Turn 4 ▷

Your Turn 4

Subtract.

a. $1.987 - 10.002$

b. $4.1 - (-0.951)$

c. $-213.4 - 90.9$

d. $-0.017 - (-0.056)$

OBJECTIVE 4 Combine like terms.

In Section 3.3, we learned that we can simplify expressions by combining like terms. Recall that like terms have the same variables raised to the same exponents. To combine like terms, add or subtract the coefficients and keep the variables the same.

Your Turn 5

Simplify.

a. $5.2x^2 - 9.87x + 15x^2 - 2.35x - 10x^3 + 1.6$

b. $4.2a^4 - 2.67a + 1 + 15.7a^2 + a^4 - 5.1 - 3.9a$

Example 5   Simplify. $4.2n^2 - 6.1n + 9 + 2.35n - 10n^2 + 2.09$

Solution: Combine like terms.

$$4.2n^2 - 6.1n + 9 + 2.35n - 10n^2 + 2.09$$
$$= \underbrace{4.2n^2 - 10n^2}_{} \;\; \underbrace{-\,6.1n + 2.35n}_{} \;\; \underbrace{+\,9 + 2.09}_{}$$ **Collect like terms.**

$$= \qquad -5.8n^2 \qquad\quad -\,3.75n \qquad\quad +\,11.09$$ **Combine like terms.**

Note: Remember, collecting the like terms is an optional step. ◀

Do Your Turn 5 ▷

Your Turn 6

Add.

a. $(t^3 - 4.15t^2 + 7.3t - 16) + (0.4t^3 + 3.9t^2 + 1.73)$

b. $(8.1m^4 + 2.78m^2 - 5.15m + 10.3n) + (1.7m^4 - 5.2m^2 - 3.1m - 15n)$

OBJECTIVE 5 Add polynomials.

We learned in Section 3.4 that we add polynomials by combining like terms.

Example 6   Add. $(8.2y^3 - 6.1y^2 + 10.65) + (y^3 - y - 12)$

Solution: Combine like terms.

$$(8.2y^3 - 6.1y^2 + 10.65) + (y^3 - y - 12)$$
$$= \underbrace{8.2y^3 + y^3}_{} - 6.1y^2 - y + \underbrace{10.65 - 12}_{}$$ **Collect like terms.**

$$= \quad 9.2y^3 \quad\; -\,6.1y^2 - y \quad\; -\,1.35$$ **Combine like terms.**

Note: Remember, if a term has no visible coefficient, as in y^3, then the coefficient is understood to be 1, so $8.2y^3 + 1y^3 = 9.2y^3$. Also, notice that we simply brought down $-6.1y^2$ and $-y$ because they had no like terms to combine with.

Do Your Turn 6 ▷

Answers to Your Turn 4
a. -8.015 **b.** 5.051
c. -304.3 **d.** 0.039

Answers to Your Turn 5
a. $-10x^3 + 20.2x^2 - 12.22x + 1.6$
b. $5.2a^4 + 15.7a^2 - 6.57a - 4.1$

Answers to Your Turn 6
a. $1.4t^3 - 0.25t^2 + 7.3t - 14.27$
b. $9.8m^4 - 2.42m^2 - 8.25m - 4.7n$

OBJECTIVE 6 Subtract polynomials.

Also in Section 3.4 we learned how to subtract polynomials. To subtract polynomials, we write an equivalent addition statement using the additive inverse of the subtrahend and then combine like terms.

Example 7   Subtract. $(19.15x^2 + 8x - 7) - (10.1x^2 + x - 9.1)$

Solution: Write the subtraction as an equivalent addition. Then combine like terms.

$$(19.15x^2 + 8x - 7) - (10.1x^2 + x - 9.1)$$
$$= (19.15x^2 + 8x - 7) + (-10.1x^2 - x + 9.1) \quad \text{Write an equivalent addition.}$$
$$= 19.15x^2 - 10.1x^2 + 8x - x - 7 + 9.1 \quad \text{Collect like terms.}$$
$$= \quad 9.05x^2 \qquad\qquad + 7x \qquad + 2.1 \quad \text{Combine like terms.}$$

Note: Remember, to get the additive inverse of a polynomial, change the sign of each term in the polynomial, so the additive inverse of $10.1x^2 + x - 9.1$ is $-10.1x^2 - x + 9.1$.

◀ **Do Your Turn 7**

Your Turn 7

Subtract.

a. $(15.7n^2 + 0.2n - 1.8) - (2.17n^2 + 0.97)$

b. $(0.14x^2 + 2.9x + 4.8) - (x^2 - 9.35x + 12.8)$

OBJECTIVE 7 Solve equations using the addition/subtraction principle.

In Section 4.2 we learned to solve equations using the addition/subtraction principle of equality. The addition/subtraction principle states that the same number may be added or subtracted on both sides of an equation without affecting its solution(s).

Example 8   Solve and check. $n + 4.78 = -5.62$

Solution: To isolate n we subtract 4.78 from both sides.

$$n + 4.78 = -5.62$$
$$\underline{-4.78 \qquad -4.78} \quad \text{Subtract 4.78 from both sides.}$$
$$n + \quad 0 = -10.40$$
$$n = -10.4$$

Note: Because $+4.78$ and -4.78 are additive inverses, their sum is 0. This isolates n.

Check: Replace n in the original equation with -10.4, and verify that the equation is true.

$$n + 4.78 = -5.62$$
$$-10.4 + 4.78 \overset{?}{=} -5.62 \quad \text{Replace } n \text{ with } -10.4.$$
$$-5.62 = -5.62 \quad \text{True, so } -10.4 \text{ is the solution.}$$

◀ **Do Your Turn 8**

Your Turn 8

Solve and check.

a. $-10.15 = x - 2.8$

b. $-14.6 = y - 2.99$

OBJECTIVE 8 Solve applications.

Many applications of adding and subtracting decimals can be found in the field of accounting. We developed some principles of accounting in Chapters 1 and 2. Let's extend those principles to decimal numbers.

Example 9   Following is Reed's register from his checkbook. Find his final balance.

Date	No.	Transaction Description	Debits		Credits		Balance	
3 - 30		Deposit			1254	88	1380	20
3 - 30	2341	National Mortgage	756	48				
3 - 30	2342	Electric & Gas	125	56				
3 - 30	2343	Bell South	43	70				
3 - 30	2344	Visa	150	00				
3 - 30	2345	Domino's	12	50				
4 - 01		Transfer from savings			200	00		

Answers to Your Turn 7
a. $13.53n^2 + 0.2n - 2.77$
b. $-0.86x^2 + 12.25x - 8$

Answers to Your Turn 8
a. $x = 12.95$ b. $y = -11.61$

Understand: We must find Reed's final balance. Check amounts are debits, so they are subtracted, whereas a transfer into the account is a credit, which adds to the account balance.

Plan: Subtract the sum of all the debits (check amounts) from the initial balance and then add the $200 credit (transfer in).

Execute: balance = initial balance − sum of all debits + credit

$B = 1380.20 - (756.48 + 125.56 + 43.70 + 150.00 + 12.50) + 200.00$

$B = 1380.20 - 1088.24 + 200.00$

$B = 291.96 + 200.00$

$B = 491.96$

Answer: Reed's final balance is $491.96.

Check: We can check by reversing the process. We start with $491.96, deduct the transfer, and add in all the check amounts. The result should be the initial balance of $1380.20.

$491.96 - 200 + 756.48 + 125.56 + 43.70 + 150 + 12.50$
$= 1380.20$ **It checks.**

Do Your Turn 9 ▶

Date	No.	Transaction Description	Debits		Credits		Balance	
12-28	1320	Publix	45	28			580	20
12-30		Deposit			640	50		
12-30	1321	Electric & Gas	155	48				
12-30	1322	Bell South	89	45				
12-30	1323	Mastercard	100	00				
12-30	1324	Riverside Apartments	565	00				

Meyer, Lewis, Callahan, Garrity and Pederson, Esq.
One Financial Center
Boston, Massachusetts 02111

700 Philadelphia Avenue
Washington, D.C. 20004
Stephen L. Meyer

Direct Dial Number
617.555.0000

October 23, 2004

Gross amount from borrower: (Debts)

Amounts paid by the borrower: (Credits)

Deposit Money   1,000.00

Contract Sales Price   89,980.00   Principal Amount of
Settlement Charges   1,025.12   New Loan(s)   85,300.00
                                          Portion of Appraisal   125.10
                                          County Taxes   188.36

Your Turn 9

Solve.

a. To the left is Carol's checkbook register. Find her final balance.

b. To the left is a settlement sheet for the Jacksons, who will be closing on their new house in a few days. How much should they bring to closing?

Answers to Your Turn 9
a. $310.77 b. $4391.66

6.2 Exercises

For Extra Help

Videotape DVT

Addison-Wesley Tutor Center

Math XL

MyMathLab

Student Solutions Manual

1. Explain how to add decimal numbers.

2. Explain how to subtract decimal numbers.

3. Place the decimal point correctly in the sum shown.

$$
\begin{array}{r}
24.5 \\
+\ 2.81 \\
\hline
27\ 31
\end{array}
$$

4. Place the decimal point correctly in the difference shown.

$$
\begin{array}{r}
497.6 \\
-\ 34.2 \\
\hline
463\ 4
\end{array}
$$

For Exercises 5–10, add.

5. $34.51 + 125.2$

6. $9.167 + 12.32$

7. $58.915 + 0.8361$

8. $0.5082 + 1.946$

9. $312.98 + 6.337 + 14$

10. $0.183 + 75 + 245.9$

For Exercises 11–16, subtract.

11. $864.55 - 23.1$

12. $1964.86 - 213.5$

13. $809.06 - 24.78$

14. $6.002 - 4.135$

15. $0.1005 - 0.08261$

16. $5.0002 - 0.19873$

For Exercises 17–28, add or subtract.

17. $25 + (-14.89)$

18. $-20 + 8.007$

19. $-0.16 + (-4.157)$

20. $-7.09 + (-10.218)$

21. $0.0015 - 1$

22. $0.0209 - 1$

23. $-30.75 - 159.27$

24. $-980.4 - 19.62$

25. $-90 - (-16.75)$

26. $-100 - (-63.912)$

27. $-0.008 - (-1.0032)$

28. $-0.005 - (-2.0031)$

For Exercises 29–32, simplify.

29. $5y - 2.81y^2 + 7.1 - 10y^2 - 8.03$

30. $0.8x^3 - 7x + 2.99 - 1.2x - 9 + 0.91x^3$

31. $a^3 - 2.5a + 9a^2 + a - 12 + 3.5a^3 + 1.4$

32. $4.2m^2 + m^3 - 9m + 0.6 - 1.02m - m^3 + 0.5$

For Exercises 33–36, add.

33. $(x^2 + 5.2x + 3.4) + (9.2x^2 - 6.1x + 5)$ **34.** $(0.4y^3 - 8.2y + 2.1) + (3.4y^3 - y^2 - 3)$

35. $(3.1a^3 - a^2 + 7.5a - 0.01) + (6.91a^3 - 3.91a^2 + 6)$

36. $(20.5m^4 + 16m^2 - 18.3m - 0.77) + (0.9m^4 - 2.08m^2 + 1)$

For Exercises 37–44, subtract.

37. $(1.4n^2 + 8.3n - 0.6) - (n^2 - 2.5n + 3)$

38. $(3.1y^3 - y^2 + 0.7) - (2.3y^3 + 1.7y^2 - 4)$

39. $(5k^3 + 2.5k^2 - 6.2k - 0.44) - (2.2k^3 - 6.2k - 0.5)$

40. $(1.2n^3 - 7.1n + 0.4) - (3.1n^3 - 6n^2 + 8.9n + 1.3)$

41. $(0.08t^4 + 1.9t^2 - 0.2t + 0.45) - (1.5t^4 + 2.4t^3 + 0.95t^2 - 1)$

42. $(2.12s^5 - 4s^3 - 2.85s + 1.29) - (1.73s^4 + 0.31s^2 - 1.93s - 3)$

43. $(12.91x^3 - 16.2x^2 - x + 7) - (10x^3 - 5.1x^2 - 3.6x + 11.45)$

44. $(6a^4 - 3.1a^2 + 1.2a - 1) - (5.4a^4 + 1.7a^2 - 2.9a - 1.32)$

For Exercises 45–50, solve and check

45. $m + 3.67 = 14.5$ **46.** $x - 5.8 = 4.11$ **47.** $2.19 = 8 + y$

48. $6.09 = n + 7.1$ **49.** $k - 4.8 = -8.02$ **50.** $-16.7 = h - 7.59$

For Exercises 51–60, solve.

51. Following is Adam's checkbook register. Find his final balance.

Date	No.	Transaction Description	Debits		Credits		Balance	
2-24		Deposit			428	45	682	20
2-25	952	Food Lion	34	58				
2-25	953	Cash	30	00				
2-25	954	Green Hills Apartments	560	00				
2-25	955	Electric & Gas	58	85				
2-26	956	American Express	45	60				

52. Following is Jada's checkbook register. Find her balance.

Date	No.	Transaction Description	Debits		Credits		Balance	
10-26	1011	Sears	36	70			250	20
10-27	1012	Frank's Costumes	25	00				
10-30	1013	Credit Union Mortgage	858	48				
10-30	1014	Electric Co.	119	45				
10-30	1015	Water Co.	17	50				
10-30	1016	Food Mart	84	95				
10-31		Deposit			1355	98		

53. Tim runs a landscaping company. Following is a balance sheet that shows the charges for a client. What is the client's final balance?

Charges		Payments Received	$200
Trencher rental	$75.00		
PVC pipe	$15.50		
Sprinkler heads	$485.65		
Back flow prevention	$45.79		

54. Following is a settlement sheet for the Smiths, who will be closing on their new house in a few days. How much should they bring to closing?

Meyer, Lewis, Callahan, Garrity and Pederson, Esq.
One Financial Center
Boston, Massachusetts 02111

700 Philadelphia Avenue   Direct Dial Number
Washington, D.C. 20004   617.555.0000
Stephen L. Meyer

November 13, 2004

Gross amount from borrower: (Debts)   Amounts paid by the borrower: (Credits)

		Earnest Money	1,500.00
Contract Sales Price	124,480.00	Principal Amount of	
Settlement Charges	1,265.08	New Loan(s)	118,300.00
Construction Overage	345.00	Portion of Appraisal	145.05
		County Taxes	218.24

55. Following is Catherine's receipt from the grocery store. If she gave the cashier $20, how much would she get back in change?

Milk	$1.59
Yogurt	$2.49
Bread	$1.79
Soup	$0.79
Cereal	$3.45
Tax	$0.51

56. Following is Selina's receipt from a clothing store. She has a $50 bill; is this enough? If not, how much more does she need?

Tank T	$12.95
Jeans	$35.95
Tax	$ 2.45

57. Maria gets paid once a month. Following is a list of deductions from her paycheck. Find Maria's net monthly pay.

CURRENT GROSS:	2904.17		
Taxes		Other Deductions	
FICA	174.25	MED SPEND	45.00
MEDICARE	39.64	DENT	12.74
FED W/H	235.62	BLUE CROSS	138.15
STATE W/H	131.56	401K	200.00
		STD LIFE	7.32

58. Tony gets paid twice each month. Following is a list of deductions from his paycheck. Find Tony's net pay.

CURRENT GROSS:	1067.08		
Taxes		Other Deductions	
FICA	62.27	HMO BLUE	210.96
MEDICARE	14.57	DENT	7.54
FED W/H	128.05	STATE RET	64.02
STATE W/H	53.35	STATE OP LIFE	6.45

59. Find the missing side length if the perimeter is 96.1 meters.

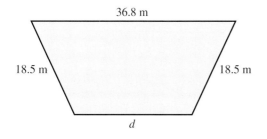

60. Find the missing side length if the perimeter is 5.9 miles.

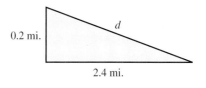

PUZZLE PROBLEM At a bookstore, a calculator and pen cost a total of $15.50. The calculator costs $15 more than the pen. What is the cost of each item?

Review Exercises

For Exercises 1 and 2, multiply.

[2.4] **1.** $-12 \cdot 25$

[5.3] **2.** $\dfrac{3}{10} \cdot \dfrac{7}{10}$

For Exercises 3 and 4, divide.

[1.4] **3.** $10{,}404 \div 34$

[5.4] **4.** $9\dfrac{1}{6} \div 3\dfrac{1}{4}$

[1.4] **5.** Evaluate. $\sqrt{121}$

For Exercises 6 and 7, multiply.

[3.5] **6.** $-2x^2 \cdot 9x^5$

[3.5] **7.** $(y + 3)(y - 5)$

$\begin{bmatrix} 2.4 \\ 4.3 \end{bmatrix}$ **8.** Solve and check. $-8x = 72$

Multiplying Decimal Numbers; Exponents with Decimal Bases

OBJECTIVE **1** Multiply decimal numbers.

Because decimal numbers are base-10 numbers, we should be able to multiply them in the same way that we multiplied whole numbers. But, we must be careful in placing the decimal point in the product.

Consider $(0.3)(0.7)$. Since we know how to multiply fractions, let's write these decimals as equivalent fractions to see what we can discover about the decimal point placement in the product.

$$(0.3)(0.7) = \frac{3}{10} \cdot \frac{7}{10}$$

Note: 0.3 is equivalent to $\frac{3}{10}$ and 0.7 is equivalent to $\frac{7}{10}$.

$$= \frac{21}{100}$$

$$= 0.21$$

Note: There are two decimal places in the product because we multiplied tenths times tenths to create hundredths.

What if the problem were $0.03 \cdot 0.7$?

$$0.03 \cdot 0.7 = \frac{3}{100} \cdot \frac{7}{10}$$

Note: 0.03 is equivalent to $\frac{3}{100}$.

$$= \frac{21}{1000}$$

$$= 0.021$$

Note: There are now three decimal places in the product because we multiplied hundredths times tenths to create thousandths.

Our examples suggest the following conclusion and procedure.

Conclusion: Since the numerators of the fraction equivalents are whole numbers, when we multiply decimal numbers, we can multiply the decimal digits just as we would multiply whole numbers. Also, because we multiply the denominators, we can say that the number of places in the product will be equal to the total number of decimal places in the decimal factors.

PROCEDURE *To multiply decimal numbers:*

1. Stack the numbers. (It is not necessary to align the decimal points in the factors.)
2. Multiply as if the numbers were whole numbers.
3. Place the decimal in the product so that it has the same number of decimal places as the total number of decimal places in the factors.

Note: When multiplying decimal numbers, we do not need to align the decimal points because the process does not depend on decimal alignment.

Example 1

a. Multiply. $(2.31)(1.7)$

Solution: We multiply the numbers as if they were whole numbers. Because 2.31 has two decimal places and 1.7 has one decimal place, the product will have three decimal places.

$$
\begin{array}{r}
\overset{2}{2.31} \leftarrow \text{ 2 decimal places} \\
\times \ 1.7 \leftarrow \text{ +1 decimal place} \\
\hline
1617 \\
+ \ 231 \\
\hline
3.927 \leftarrow \text{ 3 decimal places}
\end{array}
$$

Note: We can check our decimal placement by estimating the product. Rounding the factors to the nearest whole, we have:

$$
\begin{array}{cc}
(2.31) & (1.7) \\
\downarrow & \downarrow \\
(2) & (2) = 4
\end{array}
$$

Our actual product, 3.927, and estimate, 4, agree. Notice that other decimal placements like 39.27 or 0.3927 disagree with the estimate.

Answer: 3.927

CONNECTION $(2.31)(1.7)$ is the same as:

$$(2.31)(1.7) = (2\tfrac{31}{100})(1\tfrac{7}{10})$$
$$= \tfrac{231}{100} \cdot \tfrac{17}{10}$$
$$= \tfrac{3927}{1000}$$
$$= 3\tfrac{927}{1000}$$
$$= 3.927$$

Adding the total number of places in the factors corresponds to adding exponents of the powers of 10 in the denominators. We add the exponents because we are multiplying exponential forms that have the same base.

$$\frac{231}{100} \cdot \frac{17}{10} = \frac{231}{10^2} \cdot \frac{17}{10^1} = \frac{3927}{10^{2+1}} = \frac{3927}{10^3} = 3.927$$

↑ Number of decimal places in 2.31 ↑ Number of decimal places in 1.7 ↑ Number of decimal places in the product, 3.927

Your Turn 1

Multiply.

a. $2.04 \cdot 0.09$

b. $1.005 \cdot 34.6$

c. $6.8 \cdot 12.5$

d. $15 \cdot 0.007$

b. $(3.04)(0.35)$

Solution: Because 3.04 has two decimal places and 0.35 has two decimal places, the product will have four places.

$$
\begin{array}{r}
\overset{\overset{1}{2}}{3.04} \leftarrow \textbf{2 decimal places} \\
\times\ 0.35 \leftarrow +\textbf{2 decimal places} \\
\hline
1520 \\
+\ 912 \\
\hline
1.0640 \leftarrow \textbf{4 decimal places}
\end{array}
$$

Answer: 1.064

◀ **Note:** After placing the decimal point, 0 digits that are to the right of the last nonzero decimal digit can be deleted. We can see why using fractions.

$$(3.04)(0.35) = \frac{304}{100} \cdot \frac{35}{100} - \frac{10{,}640}{10{,}000} = \frac{1064}{1000} = 1.064$$

↑ Dividing out the common factor of 10 deletes the last 0 in the numerator and denominator.

Do Your Turn 1 ▷

OBJECTIVE 2 Multiply signed decimal numbers.

Recall that when multiplying two numbers that have the same sign, the result is positive. When multiplying two numbers that have different signs, the result is negative.

Your Turn 2

Multiply.

a. $-1.5 \cdot 2.91$

b. $(-6.45)(-8)$

Example 2 Multiply. $(0.09)(-0.051)$

Solution:

Note: To avoid confusion, we ▶ took out the negative signs while calculating but put them back in the answer statement.

$$
\begin{array}{r}
0.051 \leftarrow \textbf{3 decimal places} \\
\times\ 0.09 \leftarrow +\textbf{2 decimal places} \\
\hline
0459 \\
0000 \\
+\ 0000 \\
\hline
0.00459 \leftarrow \textbf{5 decimal places}
\end{array}
$$

Answer: -0.00459 Because the two factors have different signs, the product is negative.

Do Your Turn 2 ▷

Answers to Your Turn 1
a. 0.1836 b. 34.773 c. 85
d. 0.105

Answers to Your Turn 2
a. -4.365 b. 51.6

OBJECTIVE 3 Evaluate exponential forms with decimal bases.

Because an exponent indicates to repeatedly multiply the base, we must be careful how we place the decimal in the result.

Your Turn 3

Evaluate.

a. $(0.008)^2$

b. $(0.32)^0$

c. $(-0.12)^2$

d. $(-0.6)^3$

Example 3 Evaluate. $(-0.04)^3$

Solution: Use -0.04 as a factor three times.

$(-0.04)^3 = (-0.04)(-0.04)(-0.04)$

$= -0.000064$

Note: The product is negative because we have an odd number of negative factors.

Note: We can quickly tell how many places should be in the product by multiplying the number of decimal places in the base by the exponent. $(-0.\underline{04})^3 = -0.\underline{000064}$

number of · exponent = number of
places in   places in
the base   the product

2 · 3 = 6

Answer: -0.000064

◁ **Do Your Turn 3**

OBJECTIVE 4 Write a number in scientific notation in standard form.

Scientists, mathematicians, and engineers often work with very large numbers. When these very large numbers have lots of 0 digits, it is convenient to express these numbers in **scientific notation.**

> **DEFINITION**   **Scientific notation:** A notation composed of a decimal number whose absolute value is greater than or equal to 1 but less than 10, multiplied by 10 raised to an integer exponent.

The number 2.37×10^4 is in scientific notation because the absolute value of 2.37 is 2.37, which is greater than or equal to 1 but less than 10, and the power of 10 is 4, which is an integer. The number 42.5×10^3 is not in scientific notation because the absolute value of 42.5 is greater than 10.

What does 2.37×10^4 mean?

$$2.37 \times 10^4 = 2.37 \times 10 \times 10 \times 10 \times 10$$

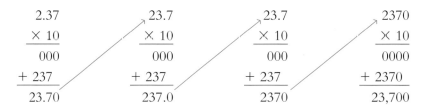

Notice that each multiplication by 10 moves the decimal point 1 place to the right. Because we multiply by four factors of 10, the decimal point moves a total of 4 places to the right from where it started in 2.37.

Conclusion: The number of 10's is determined by the exponent, so the exponent actually determines the number of places that the decimal point moves.

> **PROCEDURE**
>
> *To change from scientific notation with a positive integer exponent to standard notation for a given number, move the decimal point to the right the number of places indicated by the exponent.*

Answers to Your Turn 3
a. 0.000064 b. 1 c. 0.0144
d. -0.216

Example 4 Write the number in standard form, then write its word name.

a. 2.37×10^6

Solution: Multiplying 2.37 by 10^6 causes the decimal point to move 6 places to the right from its original position in 2.37.

$$2.37 \times 10^6 = 2\underbrace{3\,7\,0\,0\,0\,0}_{1\ 2\ 3\ 4\ 5\ 6}$$

$$= 2{,}370{,}000$$

Note: Moving two of those 6 places puts the decimal after the 37, so we must place four 0's after the 7 digit to account for all 6 places.

CONNECTION Because 10^6 is 1,000,000, the number 2.37×10^6 is often read as 2.37 million.

Word name: Two million, three hundred seventy thousand

b. -5.1782×10^9

Solution: Multiplying -5.1782 by 10^9 causes the decimal point to move 9 places to the right from its original position in -5.1782.

$$-5.1782 \times 10^9 = -5\underbrace{1\,7\,8\,2\,0\,0\,0\,0\,0}_{1\ 2\ 3\ 4\ 5\ 6\ 7\ 8\ 9}$$

$$= -5{,}178{,}200{,}000$$

Note: Moving 4 of the 9 places puts the decimal point after the 1782, so we must place five 0's after the 2 digit to account for all 9 places.

CONNECTION Because 10^9 is another way to represent 1,000,000,000, it is popular to read -5.1782×10^9 as -5.1782 billion.

Word name: Negative five billion, one hundred seventy-eight million, two hundred thousand

Do Your Turn 4 ▶

OBJECTIVE 5 Write standard form numbers in scientific notation.

Suppose we want to write 62,910,000 in scientific notation. First, we must determine the decimal position.

$$62{,}910{,}000$$
↑

The decimal must go between the 6 and 2 to express a number whose absolute value is greater than or equal to 1 but less than 10.

Next, we determine the power of 10.

$$62{,}910{,}000 = 6.2910000 \times 10^7$$

Since there are 7 places between the new decimal position and the original decimal position, the power of 10 is 7.

Finally, we can simplify by eliminating unnecessary 0's.

$$62{,}910{,}000 = 6.2910000 \times 10^7 = 6.291 \times 10^7$$

Since 0's to the right of the last nonzero decimal digit are not needed, we can delete these four 0's.

Our example suggests the following procedure:

PROCEDURE *To write a standard-form number whose absolute value is greater than 1 in scientific notation:*

1. Move the decimal point so that the number's absolute value is greater than or equal to 1 but less than 10. (Tip: Place the decimal point to the right of the first nonzero digit.)
2. Write the decimal number multiplied by 10^n, where n is the number of places between the new decimal position and the original decimal position.
3. Delete 0's to the right of the last nonzero digit.

Your Turn 4

Write each number in standard form, then write the word name.

a. 7.8×10^3

b. 9.102×10^9

c. 4.1518×10^{12}

d. 5.4×10^7

Answers to Your Turn 4
a. 7800; Seven thousand, eight hundred.
b. 9,102,000,000; Nine billion, one hundred two million
c. 4,151,800,000,000; Four trillion, one hundred fifty-one billion, eight hundred million.
d. 54,000,000; Fifty-four million

◀ Do Your Turn 5

Write the number in scientific notation.

a. 91,000

b. 108,300,000

c. −6,390,000

d. −40,913,000,000,000

Example 5 Write the number in scientific notation.

a. 417,000,000,000

Solution:

$$417,000,000,000 = 4.17 \times 10^{11}$$

Move the decimal here to express a number whose absolute value is greater than or equal to 1 but less than 10.

There are 11 places between the new decimal position and the original decimal position. After placing the decimal point, the nine 0's can be deleted.

b. −2,917,300,000

Solution:

$$-2,917,300,000 = -2.9173 \times 10^{9}$$

Move the decimal here to express a number whose absolute value is greater than or equal to 1 but less than 10.

There are 9 places between the new decimal position and the original decimal position. After placing the decimal point, the five 0's can be deleted.

◀ **Do Your Turn 5**

OBJECTIVE 6 Multiply monomials.

We learned how to multiply monomials in Section 3.5. Recall that we multiply the coefficients, add exponents for the like bases, and write any unlike bases unchanged in the product.

Example 6 Multiply.

a. $(2.1y^7)(3.43y^8)$

Solution:

$$(2.1y^7)(3.43y^8) = 2.1 \cdot 3.43 \cdot y^{7+8}$$ **Multiply the coefficients and add the exponents of the like bases.**

$$= 7.203y^{15}$$

Your Turn 6

Multiply.

a. $(1.7n^3)(6n^2)$

b. $(-4.03a)(-2.1a^3)$

CONNECTION Multiplying monomials is like multiplying numbers expressed in scientific notation. For example, $(2.1y^7)(3.43y^8)$ corresponds to $(2.1 \times 10^7)(3.43 \times 10^8)$.

$(2.1y^7)(3.43y^8)$	$(2.1 \times 10^7)(3.43 \times 10^8)$
$= 2.1 \cdot 3.43 \cdot y^7$	$= 2.1 \times 3.43 \times 10^{7+8}$
$= 7.203y^{15}$	$= 7.203 \times 10^{15}$

b. $(6.13x^2)(-4.5x^4)$

Solution:

$$(6.13x^2)(-4.5x^4) = 6.13 \cdot (-4.5) \cdot x^{2+4}$$ **Multiply the coefficients, add exponents for the like bases, and write the unlike bases as they are.**

$$= -27.585x^6$$

◀ **Do Your Turn 6**

Answers to Your Turn 5
a. 9.1×10^4
b. 1.083×10^8
c. -6.39×10^6
d. -4.0913×10^{13}

Answers to Your Turn 6
a. $10.2n^5$ b. $8.463a^4$

In Section 3.5, we learned how to simplify monomials raised to a power. We evaluate the coefficient raised to the power and multiply each variable's exponent by the power.

Example 7 Simplify. $(0.2a^2)^3$

Solution:

$$(0.2a^2)^3 = (0.2)^3 a^{2 \cdot 3}$$ Evaluate 0.2 to the 3rd power and multiply each variable's exponent by 3.
$$= 0.008a^6$$

Do Your Turn 7 ▶

Your Turn 7

Evaluate the expression.

a. $(0.3y^4)^3$

b. $(-2.5m^2n^3)^2$

OBJECTIVE 7 Multiply polynomials.

In Section 3.5, we learned to multiply polynomials using the distributive property. First, let's revisit multiplying a polynomial by a monomial.

Example 8 Multiply. $4.2x^2(5x^3 + 9x - 7)$

Solution: Using the distributive property, multiply every term in the polynomial by the monomial.

$$4.2x^2(5x^3 + 9x - 7)$$

$$= 4.2x^2 \cdot 5x^3 + 4.2x^2 \cdot 9x - 4.2x^2 \cdot 7$$ Distribute $4.2x^2$.
$$= 21x^5 + 37.8x^3 - 29.4x^2$$ Simplify.

Do Your Turn 8 ▶

Your Turn 8

Multiply.

a. $5.3(1.4a^2 + 2.5a - 1)$

b. $-2.8y^2(3.5y^3 - 4y + 5.1)$

Now let's multiply two polynomials.

Example 9 Multiply.

a. $(x + 1.5)(x - 0.3)$

Solution: Using the distributive property, we multiply every term in the second polynomial by every term in the first polynomial, then simplify.

$$
\begin{array}{c}
x \cdot (-0.3) \\
x \cdot x \\
(x + 1.5)(x - 0.3) \\
1.5 \cdot x \\
1.5 \cdot (-0.3)
\end{array}
$$

$$= x \cdot x + x \cdot (-0.3) + 1.5 \cdot x + 1.5 \cdot (-0.3)$$
$$= x^2 - 0.3x + 1.5x - 0.45$$ Multiply.
$$= x^2 + 1.2x - 0.45$$ Combine like terms.

LEARNING STRATEGY

Recall that we can use FOIL to remember how to multiply the terms in two binomials. We multiply the *first* terms, then the *outer* terms, then the *inner* terms, and finally, the *last* terms.

Answers to Your Turn 7
a. $0.027y^{12}$ b. $6.25m^4n^6$

Answers to Your Turn 8
a. $7.42a^2 + 13.25a - 5.3$
b. $-9.8y^5 + 11.2y^3 - 14.28y^2$

Your Turn 9

Multiply.

a. $(7x - 4.1)(3.2x - 2)$

b. $(1.9y + 5)(1.9y - 5)$

b. $(3.1n - 2)(1.7n + 6.5)$

Solution:

$$
\begin{array}{c}
3.1n \cdot 6.5 \\
3.1n \cdot 1.7n \\
(3.1n - 2) \quad (1.7n + 6.5) \\
-2 \cdot 1.7n \\
-2 \cdot (6.5)
\end{array}
$$

$= 3.1n \cdot 1.7n + 3.1n \cdot 6.5 - 2 \cdot 1.7n - 2 \cdot 6.5$ **Use FOIL.**

$= 5.27n^2 + 20.15n - 3.4n - 13$ **Multiply.**

$= 5.27n^2 + 16.75n - 13$ **Combine like terms.**

◀ **Do Your Turn 9**

OBJECTIVE 8 Solve applications.

A common situation involving multiplication is when a total amount of money is calculated based on a **unit price**.

> **DEFINITION**   Unit price: The price for each unit of an item.

For example, suppose gasoline is sold at a unit price of $1.09 per gallon, which means we pay $1.09 for every 1 gallon of gasoline that we purchase. The unit price can be expressed as a fraction.

$$\$1.09 \text{ per gallon} = \$1.09/\text{gallon} = \frac{\$1.09}{1 \text{ gallon}}$$

▲

Note: The word *per* indicates division and translates to the symbol /, or a fraction line.

Because the unit price is $1.09 for every 1 gallon, if we buy more or less than 1 gallon, we can multiply the unit price by the quantity we buy to get the total price. For example, if we buy 10 gallons, we multiply 10 by 1.09 to get the total price.

total price $=$ quantity $\cdot$ unit price

$= 10 \text{ gal} \cdot \dfrac{\$1.09}{1 \text{ gal}}$

$= \dfrac{10 \text{ gal}}{1} \cdot \dfrac{\$1.09}{1 \text{ gal}}$

$= \$10.90$

CONNECTION We left the units in to illustrate how the units of measurement can be *divided out* just like numbers and variables. Because *gal* appears in the numerator and denominator, we can *divide out* this common factor. Notice that the only unit of measurement left is $. This unit analysis, as it is called, helps clarify how unit pricing works. We will discuss unit analysis in more detail when we develop rates more fully in Chapter 7.

Example 10   For county property taxes, a vehicle is assessed to have a value of $1010. The property taxes are $0.2954 for each dollar of the assessed value of the vehicle. How much are the taxes?

Understand: We must find the property taxes for the vehicle. We are given the assessed value of the vehicle, $1010, and the tax rate, $0.2954 for each dollar of the assessed value. Notice that $0.2954 is a unit price.

Answers to Your Turn 9

a. $22.4x^2 - 27.12x + 8.2$

b. $3.61y^2 - 25$

Plan: Multiply the value of the vehicle by the rate to find the tax, which we will represent with t.

Execute: $t = 1010 \cdot 0.2954$

$\qquad t = 298.354$

Answer: Because taxes are monetary charges and we can only pay to the nearest hundredth of a dollar (cent), we must round 298.354 to the nearest hundredth, so the taxes are $298.35.

Check: The best check would be to reverse the process and divide. Because we have not covered division of decimals yet, we will check by estimating the product and verifying that the estimate is reasonably close to the actual product.

To estimate $1010 \cdot 0.2954$, we will round the numbers, then multiply the rounded numbers. When we estimate we want to round so that we can perform the calculation mentally. In this case, we'll round 1010 to the nearest thousand and 0.2954 to the nearest tenth.

$$1010 \cdot 0.2954$$
$$\downarrow \qquad \downarrow$$
$$1000 \cdot 0.3 = 300$$

Our estimate of 300 is very close to the actual product of 298.354.

Do Your Turn 10 ▶

Your Turn 10

Solve.

a. The long-distance phone company that Janice uses charges $0.09 per minute. If she talks for 76 minutes, how much will the call cost?

b. A broker purchases 150 shares of a stock that costs $4.25 per share. What is the total cost of the purchase?

In earlier chapters we discussed force. Newton's formula for calculating force is

$$F = ma.$$

Because we were working with whole numbers in Chapter 2, we rounded the measurements for the acceleration due to gravity to integers. In American units, we said the acceleration is -32 feet per second per second, and in metric, -10 meters per second per second. The acceleration is negative because gravity accelerates objects downwards. In Chapter 5, we developed fractions and refined these values to $-32\frac{1}{5}$ feet per second per second and $-9\frac{4}{5}$ meters per second per second. We can now use decimal values for the acceleration due to gravity. In American units, the decimal value is about -32.2 feet per second per second, and in metric it is about -9.8 meters per second per second.

CONNECTION If we write -32.2 and -9.8 as equivalent mixed numbers, we have

$$-32.2 = -32\frac{2}{10} = -32\frac{1}{5}$$
$$-9.8 = -9\frac{8}{10} = -9\frac{4}{5}$$

These are the same mixed number values we used in Chapter 5. In Section 6.4, we will see how to write fractions and mixed numbers as equivalent decimal numbers.

Remember, in American units, the mass unit is the *slug* while in metric it is the *kilogram* (kg). In the American system, the unit of force is the *pound* (lb.) while in the metric system the unit of force is the *newton* (N).

Answers to Your Turn 10
a. $6.84 b. $637.50

Your Turn 11

Solve.

a. Find the weight of a 0.0025 kilogram capsule of ibuprofen.

b. Find the weight of a 7.5 slug football player.

Example 11   Find the weight of a person with a mass of 68.1 kilograms.

Understand: We must calculate the weight given the mass. Weight is force due to gravity and force is calculated using Newton's formula, $F = ma$. The metric value for the acceleration due to gravity on Earth is -9.8 meters per second per second.

Plan: Use the formula $F = ma$ with $m = 68.1$ and $a = -9.8$.

Execute:
$$F = ma$$
$$F = (68.1)(-9.8)$$
$$F = -667.38$$

Answer: The weight of the person is 667.38 N. The negative sign indicates that the force is downward.

Check: We will check by estimating the product. If we round the mass and acceleration to the tens place, we have

$$(68.1)(-9.8)$$
$$\downarrow \qquad \downarrow$$
$$(70)(-10) = -700$$

The -700 N estimate is reasonably close to the actual force of -667.38 N.

Keep in mind that estimates are not as strong a check as reversing the process.

◀ **Do Your Turn 11**

Answers to Your Turn 11
a. 0.0245 N b. 241.5 lb.

6.3 Exercises

For Extra Help

 Videotape DVT

 Addison-Wesley Tutor Center

 Math XL

 MyMathLab

 Student Solutions Manual

1. When multiplying two decimal numbers, how do you determine the position of the decimal point in the product?

2. Place the decimal correctly in the product shown.

$$\begin{array}{r} 5.3 \\ \times\ 0.4 \\ \hline 212 \end{array}$$

3. Is 24.7×10^2 scientific notation? Explain.

4. Suppose we are writing 456,000 in scientific notation. So far we have $4.56 \times 10^?$. What will be the unknown exponent? Why?

For Exercises 5–20, multiply.

5. $(0.6)(0.9)$

6. $(0.7)(0.4)$

7. $(2.9)(3.65)$

8. $(6.81)(4.2)$

9. $(9.81)(71.62)$

10. $(10.4)(12.31)$

11. $(0.01)(619.45)$

12. $145(0.1)$

13. $19.6(10)$

14. $(1000)5.413$

15. $(0.1508)(-100)$

16. $(-10,000)(0.052)$

17. $152(-0.001)$

18. $-0.01(6)$

19. $(-0.029)(-0.15)$

20. $(-0.18)(-0.234)$

For Exercises 21–28, evaluate the expression.

21. $(0.9)^2$

22. $(0.13)^2$

23. $(2.1)^3$

24. $(3.5)^3$

25. $(-0.03)^4$

26. $(-0.2)^5$

27. $(-0.4)^3$

28. $(-1.43)^2$

For Exercises 29–34, write the scientific notation in standard form, then write the word name.

29. The speed of light is approximately 3×10^8 meters per second.

30. A light-year (lt-yr) is the distance light travels in 1 year and is approximately 5.76×10^{12} miles.

31. The approximate distance to the Andromeda galaxy is 2.14×10^6 light years.

32. The national debt is about $\$7.133 \times 10^{12}$. (*Source:* Bureau of Public Debt, June 7, 2004)

33. There are approximately 2.93×10^8 people in the United States. (*Source:* U.S. Bureau of the Census, June 7, 2004)

34. Earth is about 9.292×10^7 miles from the sun.

For Exercises 35–40, write each number in scientific notation.

35. There are 2,300,000 blocks of stone in Khufu's pyramid.

36. There are approximately 191,000 words in this textbook.

37. The distance from the Sun to Pluto is approximately 3,540,000,000 miles.

38. The nearest star to ours (the Sun) is Proxima Centauri at a distance of about 24,700,000,000,000 miles.

39. Human DNA consists of 5,300,000,000 nucleotide pairs.

40. An FM radio station broadcasts at a frequency of 102,300,000 hertz.

For Exercises 41–46, multiply.

41. $(3.2x^5)(1.8x^3)$

42. $(9.15y^2)(7.2y^4)$

43. $(0.2a^2)(-0.08a)$

44. $(-1.67n^3)(2.4n^2)$

45. $(0.5t^3)^2$

46. $(-0.2z^5)^3$

For Exercises 47–58, multiply.

47. $0.5(1.8y^2 + 2.6y + 38)$

48. $0.2(6.8m^3 + 0.44m - 7)$

49. $-6.8(0.02a^3 - a + 1.9)$

50. $-0.14(0.9n^2 - 40n - 8.7)$

51. $(x + 1.12)(x - 0.2)$

52. $(p - 1.5)(p + 7.9)$

53. $(0.4x - 1.22)(16x + 3)$

54. $(6.1y + 2)(0.8y - 5)$

55. $(5.8a + 9)(5.8a - 9)$

56. $(0.1m - 0.41)(0.1m + 0.41)$

57. $(k - 0.1)(14.7k - 2.8)$

58. $(9t - 0.5)(2t - 6.01)$

For Exercises 59–60, solve.

59. A gas company charges \$1.3548 per therm used. If Ruth uses 76 therms during a particular month, what will be the cost?

60. An electric company charges \$0.1316 per kilowatt-hour. Margaret uses 698 kilowatt-hours one particular month. What will be the cost?

For Exercises 61 and 62, use the following table that shows the pricing schedule for a phone company's long-distance rates.

Time	Rate
8 A.M.–5 P.M.	\$0.12/minute
5 P.M.–10 P.M.	\$0.10/minute
10 P.M.–8 A.M.	\$0.09/minute

61. Joanne makes a long-distance call at 4:12 P.M. and talks for 32 minutes. How much does the call cost?

62. Joanne makes another long-distance call at 10:31 P.M. and talks for 21 minutes. How much does the call cost?

For Exercises 63–78, solve.

63. A van is assessed to have a value of \$6400. Property tax on vans in the owner's county is \$0.2855 per dollar of the assessed value. What will be the tax on the van?

64. One year after the purchase of a new sedan, it is assessed at a value of \$9850. Property tax for the sedan in the county in which the owner lives is \$0.0301 per dollar of assessed value. What is the property tax on the sedan?

65. Stan buys 4.75 pounds of onions at \$0.79 per pound. What is the total cost of the onions?

66. Nadine buys 5.31 pounds of chicken at \$1.19 per pound. What is the total cost of the chicken?

67. Find the weight of 0.003 kilograms of aspirin.

68. Find the weight of a 16.5-kilogram backpack.

69. Find the weight of a piece of road maintenance equipment with a mass of 21.8 slugs.

70. Find the weight of a dog with a mass of 2.1 slugs.

71. In a test, the speed of a car is measured to be 12.5 feet per second. How far will the car travel in 4.5 seconds?

72. The speed of a DC-10 aircraft is 608.7 miles per hour. How far will the aircraft fly in 3.5 hours?

73. A square top covers the components in an electronic device. The top is 4.8 centimeters on each side. What is the area of the top?

74. A computer processor is 3.4 centimeters by 3.6 centimeters. What is the area of the processor?

75. Find the area of the lot shown.

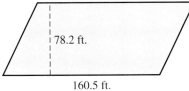

78.2 ft.

160.5 ft.

76. Find the area of the brace shown.

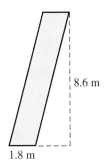

8.6 m

1.8 m

77. Find the volume.

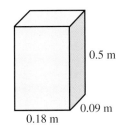

0.5 m

0.09 m

0.18 m

78. Find the volume.

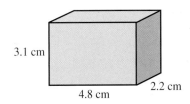

3.1 cm

4.8 cm

2.2 cm

Review Exercises

[1.4] *For Exercises 1 and 2, evaluate.*

1. $3280 \div 16$

2. $\sqrt{169}$

[5.3] **3.** Divide. $\dfrac{12x^7}{2x^3}$

[4.3] *For Exercises 4–6, solve and check.*

4. $9t = -81$

5. $3x + 5 = 32$

6. $-2a + 9 = 3a - 6$

[3.7] **7.** Find the radius and circumference of a circle with a 14-inch diameter. use 3.14 for π.

6.4 Dividing Decimal Numbers; Square Roots with Decimals

OBJECTIVE 1 Divide decimal numbers.

Suppose we must divide $4.8 \div 2$. We will see that we can use long division. However, to gain some insight into the process, we will divide by writing 4.8 as an equivalent mixed number.

$$4.8 \div 2 = 4\frac{8}{10} \div 2 \qquad \text{Write 4.8 as a mixed number.}$$

$$= \frac{48}{10} \div 2 \qquad \text{Write the mixed number as a fraction.}$$

$$= \frac{\overset{24}{\cancel{48}}}{10} \cdot \frac{1}{\underset{1}{\cancel{2}}} \qquad \blacktriangleleft \quad \textbf{Note:} \text{ After writing the equivalent multiplication we can divide 48 by 2. This suggests that we can divide the original digits as if they were whole numbers.}$$

$$= \frac{24}{10} \qquad \blacktriangleleft \quad \textbf{Note:} \text{ The denominator, 10, indicates that the quotient is in terms of tenths.}$$

$$= 2\frac{4}{10} \qquad \blacktriangleleft \quad \textbf{Note:} \text{ Although we could have reduced the fraction, we left it in terms of tenths so that writing the equivalent decimal number would be simple.}$$

$$= 2.4$$

Using this lengthy method, we can see that $4.8 \div 2 = 2.4$.

Notice that if we set up the problem in long division form and write the decimal in the quotient directly above its position in the dividend, we get the same result.

$$
\begin{array}{r}
2.4 \\
2\overline{)4.8} \\
\underline{-4} \\
08 \\
\underline{-8} \\
0
\end{array}
$$

Our example suggests the following procedure.

PROCEDURE *To divide a decimal number by a whole number using long division:*

1. Divide the divisor into the dividend as if both numbers were whole numbers. Be sure to align the digits in the quotient properly.
2. Write the decimal in the quotient directly above its position in the dividend.
3. Write extra 0 digits to the right of the last nonzero digit in the dividend as needed until you get a remainder of 0, or until a digit or group of digits repeats without end in the quotient.

Note: To indicate repeated digits, write a repeat bar over the digit or group of digits that repeat.

Example 1 Divide. $9 \div 2$

Solution: Set up a long division problem and divide. Write the decimal in the quotient directly above its position in 9.0.

Note: 9 can be expressed as 9.0000.... We can write as many 0's as needed to the right of the decimal point.

$$
\begin{array}{r}
4.5 \\
2\overline{)9.0} \\
-8 \\
\hline
10 \\
-10 \\
\hline
0
\end{array}
$$

CONNECTION We saw that we could express division using mixed numbers in Chapter 5. For $9 \div 2$, it looks like this:

$$
\begin{array}{r}
4 \\
2\overline{)9} \\
-8 \\
\hline
1
\end{array}
$$

Answer: $4\frac{1}{2}$

Our decimal answer 4.5 is the same as $4\frac{1}{2}$.

$$4.5 = 4\frac{5}{10} = 4\frac{1}{2}$$

Answer: 4.5

Do Your Turn 1 ▶

Your Turn 1

Divide.

a. $29 \div 5$

b. $211 \div 4$

c. $-12.8 \div 8$

d. $0.591 \div -2$

Example 2 Divide. $3.2 \div 3$

Solution: Set up a long division problem and divide. Write the decimal in the quotient directly above its position in 3.2.

$$
\begin{array}{r}
1.066... \\
3\overline{)3.200} \\
-3 \\
\hline
02 \\
-0 \\
\hline
20 \\
-18 \\
\hline
20
\end{array}
$$

◀ **Note:** The 6 digit will continue in an unending pattern. To indicate this repetitive pattern, we use a repeat bar over the 6.

Answer: $1.0\overline{6}$

WARNING The 0 digit in 1.066... is not part of the repeated pattern, so we do not write the repeat bar over the 0. Writing the bar over the 0 as well as the 6 indicates a different number.

$$1.\overline{06} = 1.060606...$$

◀ This is a different pattern than we see in our quotient.

Do Your Turn 2 ▶

Your Turn 2

Divide.

a. $10 \div 6$

b. $14 \div 9$

c. $-12.5 \div 3$

d. $-0.16 \div 15$

What if the divisor is a decimal number as in $1.68 \div 0.5$? We can rewrite the division statement with a whole-number divisor so that we can use the long division process we just learned. To see how this works, we will write the division in fraction form.

$$1.68 \div 0.5 = \frac{1.68}{0.5}$$

In Chapter 5, we learned that we can rewrite a fraction by multiplying its numerator and denominator by the same nonzero number. Notice that if we multiply the denominator by 10, the decimal moves to the right 1 place so that 0.5 becomes the whole number 5. This achieves our goal of writing the divisor as a whole number.

Answers to Your Turn 1
a. 5.8 b. 52.75 c. -1.6
d. -0.2955

Answers to Your Turn 2
a. $1.\overline{6}$ b. $1.\overline{5}$ c. $-4.1\overline{6}$
d. $-0.010\overline{6}$

But to keep the fractions equivalent, we must multiply the numerator by 10 as well so that 1.68 becomes 16.8. So we have

$$1.68 \div 0.5 = \frac{1.68}{0.5} = \frac{1.68 \cdot 10}{0.5 \cdot 10} = \frac{16.8}{5}$$

Now we can divide.

$$
\begin{array}{r}
3.36 \\
5\overline{)16.80} \\
-15 \\
\hline
18 \\
-15 \\
\hline
30 \\
-30 \\
\hline
0
\end{array}
$$

Note: Multiplying our dividend and divisor by 10 gives an equivalent division with a whole-number divisor. Also, notice that multiplying both numbers by the same power of 10 moves the decimal to the right the same number of places in the two numbers.

We can now amend our procedure for dividing decimal numbers.

PROCEDURE *To divide using long division when the divisor is a decimal number:*
1. Move the decimal point in the divisor to the right enough places to make the divisor an integer.
2. Move the decimal point in the dividend the same number of places.
3. Divide the divisor into the dividend as if both numbers were whole numbers. Be sure to align the digits in the quotient properly.
4. Write the decimal point in the quotient directly above its position in the dividend.

Your Turn 3

Divide.

a. $0.768 \div 0.6$

b. $13.49 \div 1.42$

c. $-605.3 \div 0.02$

d. $16 \div -0.03$

Example 3   Divide. $24.1 \div 0.04$

Solution: Because the divisor is a decimal number, we will write an equivalent division statement with a whole-number divisor. To do this, we multiply both the divisor and dividend by an appropriate power of 10. In this case, it's 100. This moves the decimal to the right 2 places in both divisor and dividend.

$$24.1 \div 0.04 = 2410 \div 4$$

Note: When we move the decimal point 2 places to the right in 24.1, we move past an understood 0 digit that is to the right of 1 so that the result is 2410.

Now divide.

$$
\begin{array}{r}
602.5 \\
4\overline{)2410.0} \\
-24 \\
\hline
01 \\
-0 \\
\hline
10 \\
-8 \\
\hline
20 \\
-20 \\
\hline
0
\end{array}
$$

Answer: 602.5

Answers to Your Turn 3
a. 1.28   b. 9.5
c. −30,265 d. −533.$\overline{3}$

◁ **Do Your Turn 3**

OBJECTIVE 2 Write fractions and mixed numbers as decimals.

Because a fraction is a notation for division, we can write fractions as decimals by dividing the denominator into the numerator. We follow the same procedure as for dividing decimal numbers when the divisor is a whole number.

> **PROCEDURE** *To write a fraction as a decimal number, divide the denominator into the numerator.*

Example 4 Write $\frac{5}{8}$ as a decimal number.

Solution: Divide the denominator into the numerator.

$$
\begin{array}{r}
0.625 \\
8\overline{)5.000} \\
-48 \\
\hline
20 \\
-16 \\
\hline
40 \\
-40 \\
\hline
0
\end{array}
$$

CONNECTION We can verify that $0.625 = \frac{5}{8}$ by writing 0.625 as a fraction in lowest terms.

$$0.625 = \frac{625}{1000} = \frac{625 \div 125}{1000 \div 125} = \frac{5}{8}$$

Answer: 0.625

Do Your Turn 4 ▷

Your Turn 4

Write as a decimal number.

a. $\frac{3}{4}$

b. $-\frac{5}{12}$

How can we write mixed numbers as decimal numbers? We saw in Section 6.1 that $23.791 = 23\frac{791}{1000}$. In reverse, if we want to write a mixed number as a decimal number, the integer part of the mixed number, in this case 23, is the integer part of the decimal number, and the fraction part, $\frac{791}{1000}$, becomes the decimal digits.

> **PROCEDURE** *To write a mixed number as a decimal:*
> 1. Write the integer part of the mixed number to the left of a decimal point.
> 2. Divide the denominator into the numerator to get the decimal digits.

LEARNING STRATEGY

If you are a visual learner, imagine the space between the integer and fraction part of a mixed number as the decimal point in its decimal number equivalent.

$$-17\frac{2}{11} = -17.\overline{18}$$

Example 5 Write $-17\frac{2}{11}$ as a decimal number.

Solution: The -17 part of the mixed number is written to the left of the decimal and the decimal equivalent of $\frac{2}{11}$ is written to the right of the decimal point. To find the decimal digits, we divide 11 into 2.

Note: The decimal digits repeat in an unending pattern, so we use a repeat bar over those digits that repeat.

$$
\begin{array}{r}
0.1818... \\
11\overline{)2.0000} \\
-11 \\
\hline
90 \\
-88 \\
\hline
20 \\
-11 \\
\hline
90
\end{array}
$$

CONNECTION Notice, $-17.\overline{18}$ is a rational number because it is equivalent to $-17\frac{2}{11}$, which we can express as the fraction $-\frac{189}{11}$. It turns out that all nonterminating decimal numbers with repeating digits can be expressed as fractions, so they are all rational numbers. But there are nonterminating decimal numbers without repeating digits, which means there are decimal numbers that are not rational numbers. We will learn about these numbers in the next objective.

Answer: $-17.\overline{18}$

Do Your Turn 5 ▷

Your Turn 5

Write as a decimal number.

a. $25\frac{1}{8}$

b. $-19\frac{5}{6}$

Answers to Your Turn 4
a. 0.75 b. $-0.41\overline{6}$

Answers to Your Turn 5
a. 25.125 b. $-19.8\overline{3}$

OBJECTIVE 3 Evaluate square roots.

In previous chapters, we discussed square roots of perfect squares, as in $\sqrt{64}$, and square roots of fractions with perfect-square numerators and denominators, as in $\sqrt{\frac{100}{121}}$. Recall that the radical sign indicates the principal square root, which is either 0, or a positive number whose square is the radicand.

$$\sqrt{64} = 8 \quad \text{because} \quad (8)^2 = 64$$

$$\sqrt{\frac{100}{121}} = \frac{10}{11} \quad \text{because} \quad \left(\frac{10}{11}\right)^2 = \frac{100}{121}$$

Note: The numbers 64, 100, and 121 are perfect squares because their square roots are whole numbers. (For a list of perfect squares, see Table 1.1.)

We now consider square roots of numbers that are not perfect squares, such as $\sqrt{14}$. To evaluate $\sqrt{14}$, we must find a positive number that can be squared to equal 14. Since 14 is between the perfect squares 9 and 16, we can conclude that it must be true that $\sqrt{14}$ is between $\sqrt{9}$ and $\sqrt{16}$. Therefore, $\sqrt{14}$ is between 3 and 4.

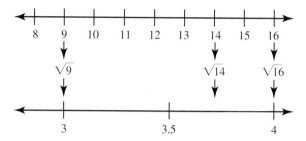

Notice that $\sqrt{14}$ is not an integer. Using our number line, we can approximate $\sqrt{14}$ to be about 3.7. We can test our approximation by squaring it.

$$(3.7)^2 = 13.69$$

Since 13.69 is less than 14, our approximation is too small. Let's try squaring 3.8.

$$(3.8)^2 = 14.44$$

So $\sqrt{14}$ is between 3.7 and 3.8. We need to go to hundredths to refine further.

We could continue guessing and refining indefinitely because $\sqrt{14}$ is a nonterminating decimal number that does not have a pattern of unending repeating digits. Because we cannot express the exact value of $\sqrt{14}$ as a decimal or fraction, we say it is an **irrational number.** In fact, the square root of any whole number that is not a perfect square is irrational.

CONNECTION We defined irrational numbers in Section 5.3 when we discussed the irrational number π.

DEFINITION **Irrational number:** A number that cannot be expressed in the form of $\frac{a}{b}$, where a and b are integers and $b \neq 0$.

The rational and irrational numbers together form the set of **real numbers.** If we were to travel the entire number line from end to end, every number we encounter would either be rational or irrational.

DEFINITION **Real numbers:** The set of all rational and irrational numbers.

 **Calculator Tips**

To find the decimal value of a square root on a calculator, use the $\sqrt{}$ function, which may require pressing $\boxed{2^{nd}}$ $\boxed{x^2}$. For example, to approximate $\sqrt{14}$, press $\boxed{2^{nd}}$ $\boxed{x^2}$ $\boxed{1}$ $\boxed{4}$ $\boxed{\text{ENTER} \atop =}$. Some calculators use a $\boxed{\sqrt{x}}$ key and require entering the radicand first, as in $\boxed{1}$ $\boxed{4}$ $\boxed{\sqrt{x}}$.

We express the exact value of the square root of a whole number that is not a perfect square using the radical sign. For example, $\sqrt{14}$ is the exact expression for the square root of 14. If the radical form is not desirable, then we can approximate an irrational square root with a rational number. To approximate we can guess, verify, and refine as we did earlier, or use a calculator.

Approximating $\sqrt{14}$ using a calculator gives

$$\sqrt{14} \approx 3.741657387$$

▲

Note: Remember this symbol means *is approximately*.

Example 6 Approximate $\sqrt{29}$ to the nearest hundredth.

Answer: ≈ 5.39

Do Your Turn 6 ▷

Your Turn 6

Approximate the square root to the nearest hundredth.

a. $\sqrt{50}$

b. $\sqrt{94}$

c. $\sqrt{128}$

d. $\sqrt{212}$

What if we must find the square root of a decimal number? Some decimal numbers have exact decimal square roots; others are irrational.

$$\sqrt{0.09} = 0.3 \quad \text{because} \quad (0.3)^2 = 0.09$$
$$\sqrt{1.21} = 1.1 \quad \text{because} \quad (1.1)^2 = 1.21$$
$$\sqrt{0.0064} = 0.08 \quad \text{because} \quad (0.08)^2 = 0.0064$$

Notice that each of the radicands is a perfect square if we take out the decimal point.

0.09 is similar to 9,

1.21 is similar to 121,

and 0.0064 is similar to 64.

Also notice that the square roots have the same digits as the square roots of the corresponding perfect squares. The tricky part is the decimal point placement. If we square a decimal number, the product will always have twice as many places as the original decimal number.

$$(0.3)^2 = 0.09 \qquad (0.03)^2 = 0.0009 \qquad (0.003)^2 = 0.000009$$

| 1 place | 2 places | 2 places | 4 places | 3 places | 6 places |

This means that when we find the square root of a decimal number that corresponds to a perfect square, the decimal number must have an *even number of decimal places* if it is to have an exact decimal square root. Otherwise, the root is irrational.

Conclusion: If the radicand has an even number of decimal places and corresponds to a perfect square, the square root will have half the number of decimal places and the same digits as the square root of the corresponding perfect square. If the radicand has an odd number of decimal digits, the square root is irrational.

$$\sqrt{0.0081} = 0.09 \quad \text{whereas} \quad \sqrt{0.081} \text{ is irrational}$$

| 4 places | 2 places |

◀ **Note:** Using a calculator we can approximate the irrational root:

$$\sqrt{0.081} \approx 0.284604989$$

RULE

The square root of a decimal number that corresponds to a perfect square and has an even number of decimal places will be a decimal number with half the number of places and the same digits as the square root of the corresponding perfect square.

Answers to Your Turn 6

a. ≈ 7.07 b. ≈ 9.70 c. ≈ 11.31

d. ≈ 14.56

Evaluate the square root.

a. $\sqrt{0.01}$

b. $\sqrt{0.0016}$

c. $\sqrt{0.25}$

d. $\sqrt{0.000169}$

Example 7 Find the square root. $\sqrt{0.000196}$

Solution: Because 196 is a perfect square and 0.000196 has an even number of decimal places, the square root will have half the number of places and the same digits as the square root of 196.

$$\sqrt{0.000196} = 0.014$$

◀ **Note:** Because 0.000196 has 6 places, its square root will have 3 places. Because $\sqrt{196} = 14$, we can say that $\sqrt{0.000196} = 0.014$.

◀ **Do Your Turn 7**

OBJECTIVE 4 Divide monomials with decimal coefficients.

We learned how to divide monomials in Section 3.7. Recall that the procedure is to divide the coefficients, subtract exponents for the like bases, and write any unlike bases as they are.

Example 8 Divide.

a. $8.2n^6 \div 2.5n^2$

Solution:

$$8.2n^6 \div 2.5n^2 = (8.2 \div 2.5)n^{6-2}$$

Divide the coefficients and subtract exponents for the like bases.

$$= 3.28n^4$$

b. $\dfrac{-3.85x^5yz}{0.5x^4y}$

Solution:

$$\frac{-3.85x^5yz}{0.5x^4y} = \frac{-3.85}{0.5}x^{5-4}y^{1-1}z$$

Note: Remember, y^0 simplifies to 1. ▶

$$= -7.7x^1y^0z$$
$$= -7.7xz$$

Divide the coefficients, subtract exponents for the like bases, and write the unlike bases as they are.

Your Turn 8

Divide.

a. $1.28y^5 \div 0.4y$

b. $\dfrac{-0.8tu^7}{-1.6u^3}$

CONNECTION Dividing monomials is like reducing rational expressions that have monomials in the numerator and denominator. Example 8a can be written as

$$8.2n^6 \div 2.5n^2 = \frac{8.2n^6}{2.5n^2} = \frac{8.2 \cdot \overset{1}{n} \cdot \overset{1}{n} \cdot n \cdot n \cdot n \cdot n}{2.5 \cdot \underset{1}{n} \cdot \underset{1}{n}} = 3.28n^4$$

We can use this same approach with part b.

$$\frac{-3.85x^5yz}{0.5x^4y} = \frac{-3.85 \cdot \overset{1}{x} \cdot \overset{1}{x} \cdot \overset{1}{x} \cdot \overset{1}{x} \cdot x \cdot \overset{1}{y} \cdot z}{0.5 \cdot \underset{1}{x} \cdot \underset{1}{x} \cdot \underset{1}{x} \cdot \underset{1}{x} \cdot \underset{1}{y}} = -7.7xz$$

◀ **Do Your Turn 8**

OBJECTIVE 5 Solve equations using the multiplication/division principle.

In Section 4.3, we learned to solve equations using the multiplication/division principle of equality, which says we can multiply or divide both sides of an equation by the same nonzero amount without affecting its solution(s). Let's apply the principle to equations that contain decimal numbers.

Example 9 Solve and check. $-14.2x = 39.05$

Solution: To isolate x, we must divide both sides by its coefficient, -14.2.

$$\frac{-14.2x}{-14.2} = \frac{39.05}{-14.2}$$ **Divide both sides by −14.2.**

$$1x = -2.75$$
$$x = -2.75$$

Check: Replace x in $-14.2x = 39.05$ with -2.75, and verify that the equation is true.

$$-14.2x = 39.05$$
$$-14.2(-2.75) \stackrel{?}{=} 39.05$$
$$39.05 = 39.05$$ **True, so −2.75 is the solution.**

Do Your Turn 9 ▷

Your Turn 9

Solve and check.

a. $37.994 = 3.14d$

b. $-9.8m = 158.76$

OBJECTIVE 6 Solve applications.

Example 10 Laura owes a hospital $1875.84. She arranges to pay the debt in monthly installments over 1 year. How much is each monthly payment?

Understand: Laura is to make monthly payments to pay off an $1875.84 debt in 1 year. Because there are 12 months in a year, she will make 12 payments.

Plan: To calculate the amount of each payment (P), we divide $1875.84 by 12.

Execute: $P = 1875.84 \div 12$
$$P = 156.32$$

Answer: Each payment should be $156.32.

Check: Verify that 12 payments of $156.32 is a total of $1875.84. We multiply $156.32 by 12.

$$156.32 \cdot 12 = 1875.84$$ **It checks.**

Do Your Turn 10 ▷

Your Turn 10

Solve.

a. To avoid interest charges, Mark takes a 90-day same-as-cash option to pay off a DVD player purchase. The cash amount would be $256.86. If he designs his budget to make 3 payments of equal amount, how much should each payment be?

b. A patient is to receive 1.5 liters of saline over 60 minutes. How much should the patient receive every minute?

In Chapter 5, we learned about circles. We learned that the circumference of a circle is the distance around its edge and is found using the formula $C = \pi d$ or $C = 2\pi r$.

Remember, π is an irrational number and in Chapter 5 we approximated its value with the fraction $\frac{22}{7}$. We can now find a decimal equivalent to $\frac{22}{7}$ by dividing 22 by 7.

$$
\begin{array}{r}
3.14... \\
7\overline{)22.00} \\
-21 \\
\hline
10 \\
-7 \\
\hline
30 \\
-28 \\
\hline
2
\end{array}
$$

Note: The π key on a calculator gives a more accurate decimal approximation, 3.141592654.

Answers to Your Turn 9
a. $d = 12.1$ b. $m = -16.2$

Answers to Your Turn 10
a. $85.62 b. 0.025 l

When solving problems involving π, if we are given no instructions about which approximation to use, we choose the approximation that seems best suited to the problem. Or, we can express the exact answer in terms of π (similar to using the radical sign to express the exact answer when a radicand is not a perfect square).

Your Turn 11

Solve.

a. The radius of an engine cylinder is 1.78 centimeters. What is the circumference? Use 3.14 for π.

b. The circumference of a crater is 16.8 kilometers. What is the diameter? Use 3.14 for π.

Example 11 The diameter of Earth is about 12,756 kilometers at the equator. Find the circumference of Earth at the equator. Use 3.14 for π.

Understand: We must find the circumference of a circle with a diameter of 12,756 kilometers.

Plan: Use the formula $C = \pi d$.

Execute: $C = \pi d$

$C = \pi(12{,}756)$	Replace d with 12,756.
$C = 12{,}756\pi$	This is the exact answer in terms of π.
$C \approx 12{,}756(3.14)$	Replace π with the approximate value 3.14.
$C \approx 40{,}053.84$	Multiply.

Answer: The circumference of Earth is approximately 40,053.84 km at the equator.

Check: We can check by dividing the product by one of the factors. We should get the other factor. Let's divide 40,053.84 by 3.14, and we should get the diameter.

$$40{,}053.84 \div 3.14 = 12{,}756 \qquad \text{It checks.}$$

CONNECTION Earlier in this section, we encountered other irrational numbers. We said that for square roots that are irrational, we express the exact irrational number by leaving the radical sign in the expression. Similarly, the only way we can express the *exact* value of a calculation involving π is to leave the symbol π in the expression. Calculating with 3.14 or $\frac{22}{7}$ is an approximation.

Exact	Approximation
$\sqrt{17}$	≈ 4.123105626
$12{,}756\pi$	$\approx 40{,}053.84$

◀ **Do Your Turn 11**

Answers to Your Turn 11
a. ≈ 11.1784 cm b. ≈ 5.35 km

6.4 Exercises

For Extra Help

Videotape DVT

Addison-Wesley Tutor Center

Math XL
Math XL

MyMathLab

Student Solutions Manual

1. Given the long division shown, place the decimal point correctly in the quotient.

$$
\begin{array}{r}
62 \\
4\overline{)\,24.8} \\
-24 \\
\overline{08} \\
-8 \\
\overline{0}
\end{array}
$$

2. To divide $65.76 \div 0.02$ using long division, how many places should the decimal point be moved in the dividend and divisor to complete the long division?

3. How do you write a fraction as a decimal number?

4. The square root of a decimal number that corresponds to a perfect square and has an even number of decimal places will be a decimal number with _____ the number of places and the same digits as the square root of the corresponding perfect square.

For Exercises 5–24, divide.

5. $286.2 \div 12$

6. $194.16 \div 24$

7. $0.48 \div 20$

8. $3.392 \div 32$

9. $5640 \div 9.4$

10. $93.96 \div 10.8$

11. $0.288 \div 0.06$

12. $0.7208 \div 0.08$

13. $-6.4 \div 0.16$

14. $28.5 \div (-0.15)$

15. $1 \div 0.008$

16. $0.1 \div 0.002$

17. $20.6 \div (-1000)$

18. $8.93 \div (-10,000)$

19. $-8.145 \div (-0.01)$

20. $-21.7 \div (-0.001)$

21. $19.6 \div 0.11$

22. $30.13 \div 0.99$

23. $-0.0901 \div (-0.085)$

24. $-339.2 \div (-0.106)$

For Exercises 25–40, write as a decimal.

25. $\dfrac{3}{5}$

26. $\dfrac{1}{8}$

27. $\dfrac{9}{20}$

28. $\dfrac{7}{25}$

29. $-\dfrac{7}{16}$

30. $-\dfrac{5}{32}$

31. $\dfrac{13}{30}$

32. $\dfrac{1}{6}$

33. $13\dfrac{1}{4}$

34. $25\dfrac{1}{2}$

35. $-17\dfrac{5}{8}$

36. $-5\dfrac{4}{5}$

37. $104\dfrac{2}{3}$

38. $76\dfrac{5}{6}$

39. $-216\dfrac{4}{7}$

40. $-99\dfrac{5}{13}$

For Exercises 41–52, evaluate the square root. If the root is irrational, approximate the square root to the nearest hundredth.

41. $\sqrt{0.0016}$

42. $\sqrt{2.25}$

43. $\sqrt{0.25}$

44. $\sqrt{0.0081}$

45. $\sqrt{24}$

46. $\sqrt{108}$

47. $\sqrt{200}$

48. $\sqrt{78}$

49. $\sqrt{1.69}$

50. $\sqrt{0.0256}$

51. $\sqrt{0.009}$

52. $\sqrt{2.5}$

For Exercises 53–58, divide.

53. $10.2x^6 \div 0.4x^4$

54. $5.44y^4 \div 1.7y$

55. $\dfrac{-0.96m^3n}{0.15m}$

56. $\dfrac{3.42hk^7}{-3.8k^2}$

57. $\dfrac{-3.03a^5bc}{-20.2a^4b}$

58. $\dfrac{-2.7t^4u^3v^2}{-0.75uv^2}$

For Exercises 59–66, solve and check.

59. $2.1b = 12.642$

60. $81.5y = 21.19$

61. $-0.88h = 1.408$

62. $4.066 = -0.38k$

63. $-2.28 = -3.8n$

64. $-2.89 = -8.5p$

65. $-28.7 = 8.2x$

66. $10.8t = -9.072$

For Exercises 67–74, solve.

67. Carla purchases a computer on a 90-day same-as-cash option. The purchase price is $1839.96. If she plans to make three equal payments to pay off the debt, how much should each payment be?

68. Brad has a student loan balance of $1875.96. If he agrees to make equal monthly payments over a 3-year period, how much will each payment be?

69. A patient is to receive radiation treatment for cancer. The treatment calls for a total of 324 rads administered in 8 bursts of focused radiation. How many rads should each burst be?

70. The FDA recommends that for a healthy diet a person should limit fat intake to about 65 grams of fat per day. If a person splits the recommended fat intake equally among three meals, how much fat is allowed in each meal? (*Source:* U.S. Food and Drug Administration, May 1999)

71. The voltage in a circuit is 12 volts. The current is measured to be 0.03 amps. What is the resistance? (Use $V = ir$.)

72. A pallet of merchandise weighs 485 pounds. If the acceleration due to gravity is -32.2 feet per second per second, then what is the mass? (Use $F = ma$.)

73. A plane is flying at 350 miles per hour. How long will it take the plane to reach a city that is 600 miles away? (Use $d = rt$.)

74. The area of a rectangular painting is 358.875 square inches. If the length is 16.5 inches, what is the width? (Use $A = bh$.)

For Exercises 75–80, use 3.14 to approximate π.

75. The radius of the Moon is about 1087.5 miles. Calculate the circumference of the Moon at its equator.

76. The solid-rocket boosters used to propel the space shuttle into orbit have O-rings that fit around the boosters at connections. If the diameter of one rocket booster is 12.17 feet, what is the circumference of the O-ring?

77. Haleakala crater on Maui island, Hawaii, is the largest inactive volcanic crater in the world and has a circumference of about 21 miles at its rim. What is the diameter at the rim?

78. A giant sequoia is measured to have a circumference of 26.4 meters. What is the diameter?

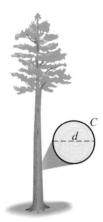

79. To manufacture a light fixture, a company bends 6.5 feet of metal tubing into a circle. What is the radius of this circle?

 80. The circumference of Earth along the equator is approximately 24,888.98 miles. What is the Earth's equatorial radius?

Review Exercises

For Exercises 1 and 2, simplify.

[2.5] **1.** $16 - 9[12 + (8 - 13)] \div 3$

[5.3] **2.** $\frac{3}{5} \cdot 30$

[5.7] **3.** Evaluate $\frac{1}{2}at^2$ when $a = -10$ and $t = 6$.

For Exercises 4 and 5, find the area.

[5.3] **4.**

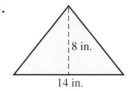

8 in.

14 in.

[5.7] **5.**

46 cm

20 cm

22 cm

Order of Operations and Applications in Geometry

OBJECTIVE 1 Simplify numerical expressions using the order of operations agreement.

Example 1 Simplify. $2.75 - 8.2(3.4 + 4)$

Solution:

$$2.75 - 8.2(3.4 + 4) = 2.75 - 8.2(7.4)$$

Add inside parentheses. **3.4 + 4 = 7.4**

$$= 2.75 - 60.68$$

Multiply. **8.2(7.4) = 60.68**

$$= -57.93$$

Subtract. **2.75 − 60.68 = −57.93**

◀ **Do Your Turn 1**

Example 2 Simplify. $12.6 \div 0.2 - 6.5(0.4)^2$

Solution:

$$12.6 \div 0.2 - 6.5(0.4)^2 = 12.6 \div 0.2 - 6.5(0.16)$$

Evaluate the expression with the exponent. **(0.4)² = 0.16**

$$= 63 - 1.04$$

Multiply or divide in order from left to right. **12.6 ÷ 0.2 = 63 and 6.5(0.16) = 1.04**

$$= 61.96$$

Subtract. **63 − 1.04 = 61.96**

◀ **Do Your Turn 2**

Your Turn 1

Simplify.

a. $6.5 - 1.2(14.4 + 0.6)$

b. $1.8 + 0.4(9.02 - 10.52)$

Your Turn 2

Simplify.

a. $-0.2 \div 1.6 - 20(0.2)^3$

b. $(1.5)^2 + 2.52 \div 0.4(-0.6)$

Example 3 Simplify. $[(1.2 + 0.3) \div 5] \div 2\sqrt{0.09}$

Solution:

$$[(1.2 + 0.3) \div 5] \div 2\sqrt{0.09} = [1.5 \div 5] \div 2\sqrt{0.09}$$

Calculate within the innermost parentheses. **1.2 + 0.3 = 1.5**

$$= 0.3 \div 2\sqrt{0.09}$$

Calculate in the brackets. **1.5 ÷ 5 = 0.3**

$$= 0.3 \div 2(0.3)$$

Calculate the square root. **√0.09 = 0.3**

$$= 0.15(0.3)$$

Multiply or divide from left to right. **0.3 ÷ 2 = 0.15**

$$= 0.045$$

Multiply. **0.15(0.3) = 0.045**

Do Your Turn 3 on Next Page ▶

Answers to Your Turn 1
a. −11.5 b. 1.2

Answers to Your Turn 2
a. −0.285 b. −1.53

OBJECTIVE **2** Simplify expressions containing fractions and decimals.

How do we handle problems that contain both decimals and fractions? Here are a few ways:

PROCEDURE *To simplify numerical expressions that contain both fractions and decimals:*

1. Write all decimals as fractions. (Method 1)   or
   Write all fractions as decimals. (Method 2)   or
   Write decimals over 1 in fraction form. (Method 3)
2. Follow the order of operations agreement.

Consider an example using all three methods.

Example 4   Simplify. $\dfrac{2}{3}(6.09) - \dfrac{3}{8}$

Method 1: Write all decimals as fractions.

$$\dfrac{2}{3}(6.09) - \dfrac{3}{8} = \dfrac{2}{3} \cdot \dfrac{609}{100} - \dfrac{3}{8}$$ Write all decimals as fractions. $6.09 = \frac{609}{100}$

$$= \dfrac{2}{\underset{1}{\cancel{3}}} \cdot \dfrac{\overset{203}{\cancel{609}}}{\underset{50}{\cancel{100}}} - \dfrac{3}{8}$$ Divide out the common factors.

$$= \dfrac{203}{50} - \dfrac{3}{8}$$ Multiply.

$$= \dfrac{203(4)}{50(4)} - \dfrac{3(25)}{8(25)}$$ The LCD for 50 and 8 is 200.

$$= \dfrac{812}{200} - \dfrac{75}{200}$$ Upscale each fraction to equivalent fractions with 200 as the denominator.

$$= \dfrac{737}{200}$$ Subtract numerators and keep the common denominator.

$$= 3\dfrac{137}{200} \quad \text{or} \quad 3.685$$

Working with fractions can be tedious, so let's explore two other methods.

Method 2: Write all fractions as decimals.

$$\dfrac{2}{3}(6.09) - \dfrac{3}{8} = 0.\overline{6}(6.09) - 0.375$$ Write $\frac{2}{3}$ as $0.\overline{6}$ and $\frac{3}{8}$ as 0.375.

Note: The repeated digits of $0.\overline{6}$ create a problem. We must round $0.\overline{6}$ in order to calculate further. This means our answer will not be exact.

▶ $\approx 0.667(6.09) - 0.375$ Round $0.\overline{6}$ to the nearest thousandth.

$\approx 4.06203 - 0.375$ Multiply.

≈ 3.68703 Subtract.

Answers to Your Turn 3
a. -0.49 b. 1.284

Because we used fractions in Method 1, we got an exact answer of 3.685. We can see from Method 2 that if we use a fraction's equivalent decimal number with nonterminating digits, after rounding that decimal number, the answer will be an approximation.

Method 3: Write decimals over 1 in fraction form.

Your Turn 4

Simplify.

a. $\dfrac{3}{4}(24.6)$

b. $(-5.052)\dfrac{5}{6}$

c. $\dfrac{1}{2}(7.9) - \dfrac{2}{5}$

d. $\dfrac{1}{4} + (-2.7) \div \dfrac{9}{2}$

Note: We can write the decimal number over 1 because any number divided by 1 is equal to that same number.

$$\dfrac{2}{3}(6.09) - \dfrac{3}{8} = \dfrac{2}{3} \cdot \dfrac{6.09}{1} - \dfrac{3}{8} \qquad \text{Write 6.09 over 1.}$$

$$= \dfrac{12.18}{3} - \dfrac{3}{8} \qquad \text{Multiply. } 2 \cdot 6.09 = 12.18 \text{ and } 3 \cdot 1 = 3$$

$$= 4.06 - 0.375 \qquad \text{Divide denominators into numerators to get decimals. } 12.18 \div 3 = 4.06 \text{ and } 3 \div 8 = 0.375$$

$$= 3.685 \qquad \text{Subtract.}$$

The answer from Method 3 is exact because dividing 12.18 by 3 gives a terminating decimal. Nonterminating decimal numbers result from division, so if we can avoid a division that leads to a nonterminating decimal, we can get an exact answer. The only way to always avoid nonterminating decimal numbers is to stick with fractions.

◀ **Do Your Turn 4**

OBJECTIVE 3 Evaluate expressions.

Remember, to evaluate a variable expression, replace the variables in the expression with the corresponding numbers and calculate following the order of operations agreement.

Your Turn 5

Evaluate the expression using the given values.

a. mv^2; $m = 10$, $v = 0.5$

b. $\dfrac{Mm}{d^2}$; $M = 3.5 \times 10^5$, $m = 4.8$, $d = 2$

c. $\left(1 + \dfrac{r}{n}\right)^{nt}$; $r = 0.06$, $n = 2$, $t = 0.5$

Example 5 Evaluate the expression using the given values.

a. $vt + \dfrac{1}{2}at^2$; $v = 30$, $t = 0.4$, $a = 12.8$

Solution:

$$(30)(0.4) + \dfrac{1}{2}(12.8)(0.4)^2 \qquad \text{Replace } v \text{ with 30, } t \text{ with 0.4, and } a \text{ with 12.8.}$$

$$= (30)(0.4) + 0.5(12.8)(0.4)^2 \qquad \text{Write } \tfrac{1}{2} \text{ as a 0.5.}$$

$$= (30)(0.4) + 0.5(12.8)(0.16) \qquad \text{Calculate. } (0.4)^2 = 0.16$$

$$= 12 + 1.024 \qquad \text{Multiply. } (30)(0.4) = 12 \text{ and } 0.5(12.8)(0.16) = 1.024$$

$$= 13.024 \qquad \text{Add. } 12 + 1.024 = 13.024$$

b. $\left(1 + \dfrac{r}{n}\right)^{nt}$; $r = 0.08$, $n = 4$, $t = 0.5$

Solution: $\left(1 + \dfrac{0.08}{4}\right)^{(4)(0.5)}$ Replace r with 0.08, n with 4, and t with 0.5.

$$= (1 + 0.02)^2 \qquad \text{Divide 0.08 by 4 in the parentheses and multiply (4)(0.5) to simplify the exponent.}$$

$$= (1.02)^2 \qquad \text{Add in the parentheses.}$$

$$= 1.0404 \qquad \text{Square 1.02.}$$

Note: When an exponent contains operations, we must simplify those operations before evaluating the exponential form.

◀ **Do Your Turn 5**

Answers to Your Turn 4
a. 18.45 b. −4.21 c. 3.55
d. −0.35

Answers to Your Turn 5
a. 2.5 b. 420,000 c. 1.03

OBJECTIVE 4 Find the area of a triangle.

Now that we have discussed how to handle fractions and decimals in the same problem, we can consider the area of a triangle with a decimal base or height. In Section 5.3, we learned the formula for the area of a triangle, $A = \frac{1}{2}bh$.

Example 6   Find the area.

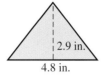

Solution: Use the formula $A = \frac{1}{2}bh$.

$A = \frac{1}{2}(4.8)(2.9)$ Replace b with 4.8 and h with 2.9.

$A = \frac{1}{2}(13.92)$ Multiply 4.8 and 2.9.

$A = \frac{1}{2} \cdot \frac{13.92}{1}$ Write 13.92 over 1.

$A = \frac{13.92}{2}$ Multiply.

$A = 6.96$ Divide 13.92 by 2.

Note: Although we chose to keep $\frac{1}{2}$ in fraction form, we could have written it as a decimal and then multiplied.
$A = 0.5(4.8)(2.9)$
$A = 2.4(2.9)$
$A = 6.96$

Answer: The area is 6.96 in.^{2}.

Do Your Turn 6 ▷

Your Turn 6

Find the area.

a.

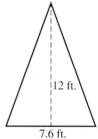

b.

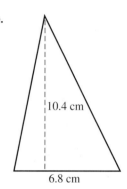

OBJECTIVE 5 Find the area of a trapezoid.

We can now calculate the area of a trapezoid with decimal lengths or height. In Section 5.7 we learned the formula $A = \frac{1}{2}h(a + b)$.

Example 7   Find the area.

Solution: Use the formula $A = \frac{1}{2}h(a + b)$.

$A = \frac{1}{2}(9)(4.86 + 5.34)$ Replace h with 9, a with 4.86, and b with 5.34.

$A = \frac{1}{2}(9)(10.2)$ Add inside the parentheses.

$A = \frac{1}{2}(91.8)$ Multiply 9 by 10.2.

$A = \frac{1}{2} \cdot \frac{91.0}{1}$ Write 91.8 over 1.

$A = \frac{91.8}{2}$ Multiply.

$A = 45.9$ Divide 91.8 by 2.

Answer: The area is 45.9 m^2.

Do Your Turn 7 ▷

Your Turn 7

Find the area.

a.

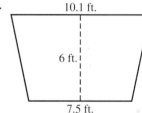

b.

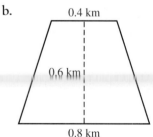

OBJECTIVE 6 Find the area of a circle.

In Chapter 5, we learned to find the area of a circle using $A = \pi r^2$. We saw, in Section 6.4, that the value of π can be approximated by the decimal number 3.14. Let's use this approximation to calculate the area of a circle.

Answers to Your Turn 6
a. 45.6 ft.^{2} b. 35.36 cm^2

Answers to Your Turn 7
a. 52.8 ft.^{2} b. 0.36 km^2

a.

0.5 m

b.

4.2 ft.

Example 8 Find the area of the circle shown. Use 3.14 for π.

5.6 cm

Solution: Use the formula $A = \pi r^2$. We are given the diameter of 5.6 centimeters but the formula requires the radius. Because radius is half of the diameter, the radius is 2.8 centimeters.

Note: Because π is irrational, 7.84π is the exact representation of the area. When we use an approximate value for π, the calculation becomes an approximation. This is why we use $\approx$.

$A = \pi(2.8)^2$ Replace *r* with 2.8.

$A = \pi(7.84)$ Square 2.8.

$A = 7.84\pi$ Express the exact answer with π.

$A \approx 7.84(3.14)$ Replace π with 3.14.

$A \approx 24.6176$ Multiply.

Answer: The area is approximately 24.6176 cm².

◀ **Do Your Turn 8**

OBJECTIVE 7 Find the volume of a cylinder.

To understand how to derive the formula for the volume of a cylinder, let's look more closely at the formula for the volume of a box, $V = lwh$. Notice that the base of a box is a rectangle, and we calculate the area of that base by multiplying *l* by *w*. So, the formula for the volume of a box is a result of multiplying the area of the base, *lw*, by the height of the box, *h*.

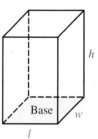
h
Base *w*
l

volume = **area of the base** · **height**

$V =$ lw ·   *h*

We can apply this same approach to deriving the volume of any three-dimensional object that is formed by repeatedly stacking a shape. To understand what we mean by repeatedly stacking a shape, think about a box as a stack of papers. A single sheet of paper is very thin when viewed along its edge but is a rectangle when viewed from the top or bottom. When we stack lots of these thin rectangles on top of each other, we form a box shape.

A cylinder is formed by repeatedly stacking a circular base. Imagine coins stacked on top of each other. Following the same approach that we used for a box, we find the volume of a cylinder by multiplying the area of its base, which is πr^2, by its height, *h*.

r
h
Base

volume = **area of the base** · **height**

$V =$ πr^2 ·   *h*

$V = \pi r^2 h$

Answers to Your Turn 8
a. 0.785 m² b. 13.8474 ft.²

Example 9 Find the volume of the cylinder. Use 3.14 for π.

Solution: Use the formula $V = \pi r^2 h$.

$$V = \pi(4.5)^2(12)$$ Replace π with 4.5 and h with 12.

$$V = \pi(20.25)(12)$$ Square 4.5.

Exact answer → $V = 243\pi$ Multiply 20.25 by 12.

$$V \approx 243(3.14)$$ Replace π with 3.14.

Approximation → $V \approx 763.02$ Multiply.

Note: Remember, volume is always expressed in cubic units. We write cubic inches as in.^{3}.

Answer: The volume is approximately 763.02 in.^{3}.

Do Your Turn 9 ▶

Your Turn 9

Find the volume.

a.   6.2 in.

1.2 in.

b.

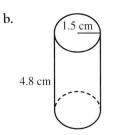

1.5 cm

4.8 cm

OBJECTIVE 8 Find the volume of a pyramid.

To help understand the formula for the volume of a four-sided pyramid, notice that one would fit inside a box that has a base of the same length and width. The pyramid's volume occupies $\frac{1}{3}$ of the volume of the box.

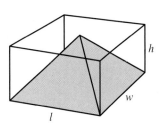

$$V = \frac{1}{3} \cdot \text{volume of a box with the same length, width, and height}$$

$$V = \frac{1}{3}lwh$$

Example 10 Find the volume of the pyramid.

Solution: Use the formula $V = \frac{1}{3}lwh$.

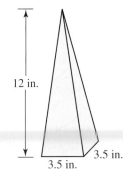

12 in.

3.5 in.

3.5 in.

$$V = \frac{1}{3}(3.5)(3.5)(12)$$ Replace l with 3.5, w with 3.5, and h with 12.

$$V = \frac{1}{3}(147)$$ Multiply 3.5, 3.5, and 12.

$$V = \frac{1}{3} \cdot \frac{147}{1}$$ Write 147 over 1. ◀

$$V = \frac{147}{3}$$ Multiply.

$$V = 49$$ Divide 147 by 3.

Note: We avoid the decimal equivalent of $\frac{1}{3}$ because it is a nonterminating decimal and its use would lead to an approximate answer.

Answer: The volume is 49 in.^{3}.

Do Your Turn 10 ▶

Your Turn 10

Find the volume.

a.

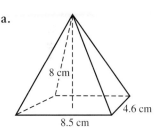

8 cm

8.5 cm

4.6 cm

b.

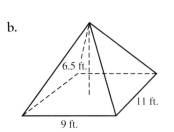

6.5 ft.

9 ft.

11 ft.

OBJECTIVE 9 Find the volume of a cone.

A cone is much like a pyramid except the base is a circle. Also, like a pyramid, which is $\frac{1}{3}$ the volume of a box with the same-size base and height, a cone is $\frac{1}{3}$ the volume of a cylinder with the same-size base and height.

Answers to Your Turn 9
a. ≈ 144.84 in.^{3}
b. ≈ 33.912 cm^3

Answers to Your Turn 10
a. $104.2\overline{6}$ cm^3 b. 214.5 ft.^{3}

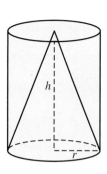

$$V = \frac{1}{3} \cdot \text{volume of a cylinder with the same-size base and height}$$

$$V = \frac{1}{3}\pi r^2 h$$

Your Turn 11

Find the volume. Use 3.14 for π.

a.

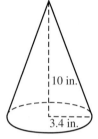

10 in.

3.4 in.

b.

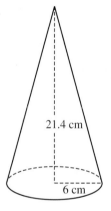

21.4 cm

6 cm

Example 11 Find the volume of the cone. Use 3.14 for π.

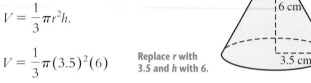

6 cm

3.5 cm

Solution: Use the formula $V = \frac{1}{3}\pi r^2 h$.

$$V = \frac{1}{3}\pi(3.5)^2(6) \qquad \text{Replace } r \text{ with } 3.5 \text{ and } h \text{ with } 6.$$

$$V = \frac{1}{3}\pi(12.25)(6) \qquad \text{Square 3.5.}$$

$$V = \frac{1}{3}(73.5)\pi \qquad \text{Multiply 12.25 by 6.}$$

$$V = \frac{1}{3} \cdot \frac{73.5}{1}\pi \qquad \text{Write 73.5 over 1.}$$

$$V = \frac{73.5}{3}\pi \qquad \text{Multiply.}$$

$$V = 24.5\pi \qquad \text{Divide 73.5 by 3.}$$

$$V \approx 24.5(3.14) \qquad \text{Replace } \pi \text{ with 3.14.}$$

$$V \approx 76.93 \qquad \text{Multiply.}$$

Answer: The volume is approximately 76.93 cm^3.

◀ **Do Your Turn 11**

OBJECTIVE 10 Find the volume of a sphere.

A ball is a sphere. Every point on the surface of a sphere is equally distant from the center. It turns out that the volume of a sphere is equal to four cones that have the same radius as the sphere and height equal to the radius.

$$\text{the volume of one such cone} = \frac{1}{3}\pi r^2 \cdot r = \frac{1}{3}\pi r^{2+1} = \frac{1}{3}\pi r^3$$

▲
Because the height of the cone is equal to the radius of the sphere, we use r in place of h.

Because the volume of a sphere is equal to four of these cones, we must multiply the expression we found above by 4.

$$\text{volume of a sphere} = 4 \cdot \frac{1}{3}\pi r^3 = \frac{4}{1} \cdot \frac{1}{3}\pi r^3 = \frac{4}{3}\pi r^3$$

The volume of a sphere can be found using the formula $V = \frac{4}{3}\pi r^3$.

Answers to Your Turn 11
a. ≈ 120.99 in.^{3}
b. ≈ 806.352 cm^3

Example 12 Find the volume of the sphere. Use 3.14 for π.

Solution: Use the formula $V = \frac{4}{3}\pi r^3$.

$V = \frac{4}{3}\pi(0.2)^3$ **Replace r with 0.2.**

$V = \frac{4}{3}\pi(0.008)$ **Cube 0.2.**

$V = \frac{4}{3}\left(\frac{0.008}{1}\right)\pi$ **Write 0.008 over 1.**

$V = \frac{0.032}{3}\pi$ **Multiply.**

$V = 0.010\overline{6}\pi$ **Divide 0.032 by 3.**

$V \approx 0.010\overline{6}(3.14)$ **Replace π with 3.14.**

$V \approx 0.033493$ **Multiply.**

Answer: The volume is approximately $0.03349\overline{3}$ m³.

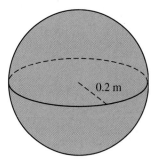

0.2 m

DISCUSSION   What is a reasonable number of decimal places in an answer?

Do Your Turn 12 ▶

Your Turn 12

Solve.

a. Find the volume of a sphere with a radius of 3 feet.

b. Find the volume of a sphere with a radius of 8 centimeters.

OBJECTIVE 11 Find the area or volume of a composite form.

Composite forms are geometric forms that combine several basic forms. The key to finding the area of a composite form is to determine whether two or more forms have been put together or whether a form or forms have been removed from a larger form.

Example 13 Find the area. Use 3.14 for π.

Understand: This shape is a combination of a half-circle and a triangle. The area of a full circle is found by $A = \pi r^2$, so for a half-circle, the area formula is $A = \frac{1}{2}\pi r^2$. Note that 1.6 meters is the diameter, so the radius is 0.8 meter. Also, the formula for the area of a triangle is $A = \frac{1}{2}bh$.

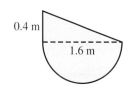

0.4 m

1.6 m

Plan: Add the area of the half-circle and triangle.

Execute: $A =$ area of half-circle $+$ area of triangle

$A = \frac{1}{2}\pi r^2 \qquad\quad + \quad \frac{1}{2}bh$

$A \approx \frac{1}{2}(3.14)(0.8)^2 \; + \; \frac{1}{2}(1.6)(0.4)$ ◀ 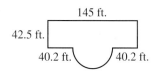 We use $\approx$ when we replace π with an approximate value.

$A \approx \frac{1}{2}(3.14)(0.64) \; + \; \frac{1}{2}(1.6)(0.4)$ **Square 0.8.**

$A \approx \frac{1}{2}(2.0096) \quad + \quad \frac{1}{2}(0.64)$ **Multiply.**

$A \approx \frac{1}{2}\cdot\frac{2.0096}{1} \; + \; \frac{1}{2}\cdot\frac{0.64}{1}$ **Write the decimal numbers over 1.**

$A \approx \frac{2.0096}{2} \qquad + \quad \frac{0.64}{2}$ **Multiply.**

$A \approx 1.0048 \qquad + \quad 0.32$ **Divide.**

$A \approx 1.3248$ **Add.**

Answer: The area of the shape is approximately 1.3248 m².

Check: We can check by reversing each calculation. We will leave the check to the reader.

Do Your Turn 13 ▶

Your Turn 13

Find the area.

145 ft.

42.5 ft.

40.2 ft.   40.2 ft.

Answers to Your Turn 12
a. ≈ 113.04 ft.³
b. ≈ 2143.57$\overline{3}$ cm³

Answer to Your Turn 13
≈ 7800.47 ft.²

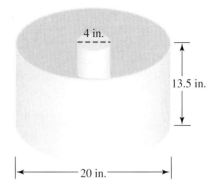

Your Turn 14

Find the amount of storage space under the counter-top. (Assume that the sink is half of a sphere.)

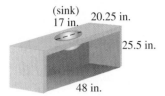

(sink)
17 in.   20.25 in.

25.5 in.

48 in.

Example 14   The tub inside a washing machine is a small cylinder within a larger cylinder. The space in between the cylinders is where the clothes are placed. Find the volume of the space for clothes inside the washing machine shown.

Understand: The volume of a cylinder is found by $V = \pi r^2 h$. The radius of the smaller cylinder is 2 inches. The radius of the larger cylinder is 10 inches. The height of both cylinders is 13.5 inches.

Plan: Subtract the volume of the inner cylinder from the volume of the outer cylinder.

Execute: $V =$ volume of outer cylinder $-$ volume of inner cylinder

$$V = \qquad \pi r^2 h \qquad - \qquad \pi r^2 h$$
$$V = \pi (10)^2 (13.5) - \pi (2)^2 (13.5)$$
$$V = \pi (100)(13.5) - \pi (4)(13.5)$$
$$V = 1350\pi - 54\pi$$
$$V = 1296\pi$$
$$V \approx 1296(3.14)$$
$$V \approx 4069.44$$

CONNECTION   Because the volume for both shapes involves π, we chose to leave π in symbolic form.

Because 1350π and -54π are both in terms of π, they are like terms. We can combine by subtracting the coefficients and keeping the result in terms of π.

Answer: The space for clothes inside the washing machine is about 4069.44 in.³

◁ **Do Your Turn 14**

OF INTEREST

In any list of the greatest mathematicians, the name Archimedes is usually at the top. Most scholars consider Archimedes, Isaac Newton, and Carl Friedrich Gauss the top three mathematicians to have ever lived.

Archimedes was born in 287 B.C. in the city of Syracuse in what is now Sicily. In his youth, he studied in Alexandria, Egypt, where he met and befriended Eratosthenes (famous for his sieve method for finding primes).

In pure mathematics, Archimedes developed methods for finding the areas of many shapes and objects, including the cylinder, cone, and sphere. He developed a method for calculating π. He placed the value as being between $3\frac{1}{7}$ and $3\frac{10}{71}$. He even developed some of the ideas involved in calculus, preceding Isaac Newton (the accepted inventor of calculus) by 1800 years.

Though he preferred mathematics, Archimedes also contributed to astronomy and mechanics. One popular invention was the Archimedian pump. This pump was an augerlike device within a cylinder. The pump worked by tilting the apparatus at an angle with one end in water. At the other end, the operator would turn the auger which would force water out the top of the cylinder. This pump remained in use for centuries.

Archimedes is also known for his discovery of principles of hydrostatics. The story goes that the king had given a smith a lump of pure gold to make a crown. After the crown was completed, the king became suspicious that the smith had substituted the gold for a lesser metal within the crown, keeping most of the gold for himself. The king asked Archimedes to figure a way to determine if the crown was pure gold without destroying the crown. Sometime thereafter Archimedes was taking a bath and as he lowered himself into the water, he realized that his body weighed less as a result of the water pushing upwards, and this force is equal to the volume displacement of the water. He realized that this principle could be applied to the problem of the crown. It is said that in his excitement over the discovery he jumped from the water and ran naked through the streets shouting, "Eureka, Eureka!" (I have found it, I have found it!)

Answer to Your Turn 14
$\approx 23,500.43$ in.³

1. To simplify $\frac{2}{3}(4.8) + \frac{1}{4}$, which is better, writing the decimals as fractions, or writing the fractions as decimals? Why?

2. What is the formula for the volume of a pyramid? How is it related to the volume of a box?

3. What are the formulas for the volume of a cylinder and a cone? How are they related?

4. What is the formula for the volume of a sphere? What do its symbols represent?

For Exercises 5–18, simplify.

5. $0.64 + 2.5(0.8)$

6. $9.28 - 0.56(12)$

7. $-5.1(3.4 - 5.7)$

8. $-0.25(0.7 + 5.2)$

9. $88.2 - 2.2(0.45 + 20.1)$

10. $9.5(2.6 - 0.99) - 4.33$

11. $(0.4)^2 - 2.8 \div 0.2(1.6)$

12. $16.1 \div 0.4(0.3) + (1.3)^2$

13. $10.7 - 18\sqrt{1.96} + (74.6 - 88.1)$

14. $7.5 + 2.2\sqrt{0.25} - 36.8 \div 8$

15. $\sqrt{0.0081} + 120.8 \div 4(2.5 - 6.4)$

16. $40.1 - 6.9 \div 4.6(1.4)^2 + \sqrt{0.36}$

17. $[-5.53 \div (0.68 + 0.9)] \div (0.6)^2$

18. $[(6.2 - 10) \div 8] - 2\sqrt{1.21}$

For Exercises 19–26, simplify.

19. $\frac{3}{5}(-0.85)$

20. $\frac{7}{10}(12.88 - 4.38)$

21. $\frac{1}{4} + \frac{2}{3}(0.06)$

22. $(-12.9)\frac{5}{6} + \frac{3}{4}$

23. $-2\frac{4}{9}(1.8) - \frac{3}{8}$

24. $3\frac{1}{8} - (0.45)\left(5\frac{1}{3}\right)$

25. $\frac{2}{5} \div (-0.8) + \left(\frac{1}{3}\right)^2$

26. $\frac{1}{6} - \frac{1}{24}(0.4 + 1.2)$

For Exercises 27–36, evaluate the expression using the given numbers.

27. $\frac{1}{2}mv^2$; $m = 1.25$, $v = -8$

OF INTEREST

$\frac{1}{2}mv^2$ is an expression used to calculate the energy of a moving object, where m is the mass of the object and v is the velocity. As a formula, we would write $E = \frac{1}{2}mv^2$.

28. $\frac{1}{2}mv^2$; $m = 84.6$, $v = 0.2$

29. $\frac{w}{h^2} \cdot 705$; $w = 150$, $h = 67$

OF INTEREST

$\frac{w}{h^2} \cdot 705$ is used to calculate a person's body mass index (BMI), where w is the person's weight in pounds and h is the person's height in inches. According to the National Institute of Health, people with a BMI of 25 or more have an increased risk for cardiovascular and other diseases.

30. $\frac{w}{h^2} \cdot 705$; $w = 185$, $h = 70$

31. mc^2; $m = 2.5 \times 10^6$, $c = 3 \times 10^8$

OF INTEREST

mc^2 is an expression developed by Albert Einstein to calculate the energy of a particle, where m is the mass of the particle and c is the speed of light. It appears in possibly the most famous formula of physics: $E = mc^2$.

32. mc^2; $m = 3.6 \times 10^4$, $c = 3 \times 10^8$

33. $vt + \frac{1}{2}at^2$; $v = 20$, $t = 0.6$, $a = -12.5$

OF INTEREST

$vt + \frac{1}{2}at^2$ is an expression used to calculate the distance an object travels that has an initial velocity, v, and acceleration or deceleration, a, over a period of time, t. As a formula, we would write $d = vt + \frac{1}{2}at^2$.

34. $vt + \frac{1}{2}at^2$; $v = 35$, $t = 1.2$, $a = 10$

35. $\left(1 + \frac{r}{n}\right)^{nt}$; $r = 0.12$, $n = 4$, $t = 0.5$

OF INTEREST

$\left(1 + \frac{r}{n}\right)^{nt}$ is a part of the formula used for calculating the balance in an account that has earned interest at an interest rate of r compounded n times in t years. We will study the formula for compound interest in Section 8.6.

36. $\left(1 + \frac{r}{n}\right)^{nt}$; $r = 0.04$, $n = 8$, $t = 0.25$

For Exercises 37–42, find the area. Use 3.14 for π.

37.
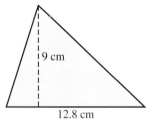

9 cm

12.8 cm

38.
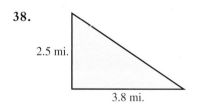

2.5 mi.

3.8 mi.

39.

1.28 m

1.08 m

1.12 m

40.
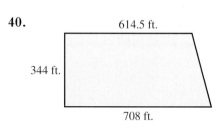

614.5 ft.

344 ft.

708 ft.

41.

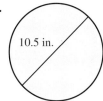

10.5 in.

42.

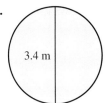

3.4 m

For Exercises 43–50, find the volume. Use 3.14 for π.

43. A quarter.

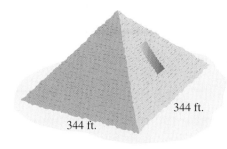

0.2 cm

2.4 cm

44. Canned drink.

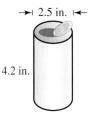

2.5 in.

4.2 in.

45. The Castillo at Chichen Itza in the Yucatan forest in Mexico is 90 feet tall.

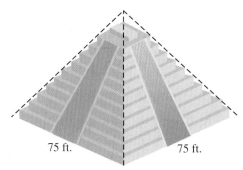

75 ft.   75 ft.

OF INTEREST

The Castillo at Chichen Itza is a pyramid temple built by the Mayans around A.D. 500. The pyramid is a calendar. There are four sets of steps, one set on each side of the pyramid, with 91 steps in each. At the top is a small temple with 1 step on all four sides for a total of 365 steps, one for each day of the year. At the bottom of each set of steps are serpent heads that stare out from either side of the steps. At the equinoxes, the sunlight hits the ridged sides of the pyramid at the perfect angle to cast a shadow that connects with the serpent head, giving the appearance of a serpent snaking its way down the pyramid.

46. Menkaure's pyramid at Giza, Egypt, is 203 feet tall.

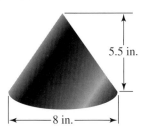

344 ft.

344 ft.

OF INTEREST

Menkaure was king of Egypt from 2490 to 2472 B.C. His pyramid at Giza is the smallest of the three pyramids. It is believed that Khufu, builder of the largest of the three pyramids, was Menkaure's grandfather. The original height of the pyramid was to be 215 feet, but it is only 203 feet because the smooth outer case of granite was never completed.

47. Onyx decorative cone.

5.5 in.

8 in.

48. Afterburner fire cone from a jet engine.

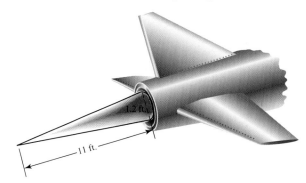

1.2 ft.

11 ft.

49. If Earth is a sphere with a radius of 3884.3 miles, find the volume of Earth.

50. If the Sun is a sphere with a radius of 423,858 miles, find the volume of the Sun.

DISCUSSION   Approximately how many Earths could fit inside the Sun?

For Exercises 51 and 52, find the area of the composite shape. Use 3.14 for π.

51.

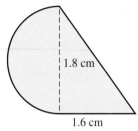

1.8 cm

1.6 cm

52.

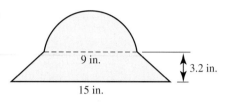

9 in.

3.2 in.

15 in.

For Exercises 53 and 54, find the area of the shaded region. Use 3.14 for π.

53.

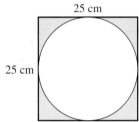

25 cm

25 cm

54.

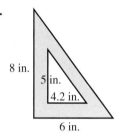

8 in.

5 in.

4.2 in.

6 in.

For Exercises 55 and 56, find the volume of the composite object. Use 3.14 for π.

55. Water tower.

64 ft.

45.5 ft.

20 ft.

56. Minisub.

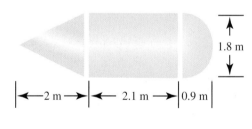

1.8 m

2 m   2.1 m   0.9 m

57. A pyramid built with cut blocks of stone has an inner chamber that is 5.8 meters long by 5.2 meters wide by 10.8 meters high. The height of the pyramid is 146.5 meters. Find the volume of stone used to built the pyramid.

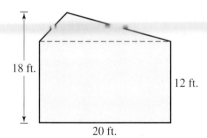

230 m

230 m

58. The diameter of a bowling ball is 10 inches. Each hole is a cylinder 3 inches deep with a 1-inch diameter. Find the volume of material used to make the ball. Use 3.14 for π.

59. a. A grain of salt is a cube that is 0.03 inch along each side. What is the volume of a grain of salt?

   b. About how many grains of salt would be in 1 cubic inch?

60. A grain of salt is a cube that is 0.03 inch along each side. Calculate the surface area of a grain of salt.

PUZZLE PROBLEM Estimate the volume of our galaxy. It is a spiral galaxy 100,000 light-years across and about 1000 light-years thick at the center.

Review Exercises

For Exercises 1 and 2, solve and check.

[4.3]   **1.** $4(x - 6) + 1 = 9x - 2 - 8x$ [5.8]   **2.** $\frac{1}{3}y - \frac{3}{4} = 6$

[5.7]   **3.** Find the area of the composite shape.   [5.7]   **4.** Find the area of the shaded region.

18 ft.

12 ft.

20 ft.

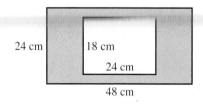

24 cm   18 cm

24 cm

48 cm

[5.8]   **5.** Translate to an equation, then solve. $\frac{1}{2}$ of the sum of x and 12 is equal to 9 less than $\frac{3}{4}$ of x.

[4.5]   **6.** Marla has $160 in tens and fives. If she has 4 more tens than fives, how many of each bill does she have?

6.6 Solving Equations and Problem Solving

OBJECTIVES

1 Solve equations using the addition/ subtraction and multiplication/division principles of equality.

2 Simplify decimal equations using the multiplication principle.

3 Solve application problems.

4 Solve problems involving one unknown.

5 Solve problems using the Pythagorean theorem.

6 Solve problems involving two unknowns.

OBJECTIVE 1 Solve equations using the addition/subtraction and multiplication/division principles of equality.

Example 1 Solve and check. $6 - 4.8t = 3.552$

Solution: To isolate t, we can start by using the addition/subtraction principle of equality. Subtract 6 from both sides to isolate $-4.8t$. Then clear the -4.8 coefficient by using the multiplication/division principle of equality and dividing by -4.8 on both sides.

$$6 - 4.8t = 3.552$$
$$\underline{-6 \qquad\qquad -6} \qquad \text{Subtract 6 from both sides to isolate } -4.8t.$$
$$0 - 4.8t = -2.448$$
$$\frac{-4.8t}{-4.8} = \frac{-2.448}{-4.8} \qquad \text{Divide both sides by } -4.8 \text{ to isolate } t.$$
$$t = 0.51$$

Check: Replace t in $6 - 4.8t = 3.552$ with 0.51, and verify that the equation is true.

$$6 - 4.8t = 3.552$$
$$6 - 4.8(0.51) \stackrel{?}{=} 3.552 \qquad \text{Replace } t \text{ with 0.51.}$$
$$6 - 2.448 \stackrel{?}{=} 3.552$$
$$3.552 = 3.552 \qquad \text{True, so 0.51 is the solution.}$$

◁ **Do Your Turn 1**

Your Turn 1

Solve and check.

a. $4.5y - 21.99 = 6$

b. $0.08n + 0.4 = 0.244$

OBJECTIVE 2 Simplify decimal equations using the multiplication principle.

In Chapter 5, we learned to use the multiplication principle of equality to clear fractions in an equation. The idea was to multiply both sides throughout by the LCD (lowest common denominator). Because decimal numbers are equivalent to fractions, we can use the same principle with equations that contain decimal numbers. When we write decimals as fractions, the denominators are always powers of 10. This means we can clear decimals by multiplying both sides by an appropriate power of 10.

We saw, in Section 6.3, that when we multiply a decimal number by a power of 10, the decimal point moves to the right. Multiplying a decimal number by a large enough power of 10 can yield a whole number. But how do we determine the power of 10? Consider $2.5x - 0.64 = 8$. If we multiply both sides by 10, the decimal point moves to the right 1 place in each number.

$$10(2.5x - 0.64) = 10 \cdot 8 \qquad \text{Multiply both sides by 10.}$$
$$10 \cdot 2.5x - 10 \cdot 0.64 = 10 \cdot 8 \qquad \text{Distribute 10.}$$
$$25x - 6.4 = 80$$

Notice that 6.4 still contains a decimal point, which means we should have used a larger power of 10. Because 0.64 has the most places, it determines the power of 10 we use. Since it has 2 places, we use 100.

Note: Multiplying both sides by 100 transforms the equation from a decimal equation to an equation that contains only integers.

$$100(2.5x - 0.64) = 100 \cdot 8 \qquad \text{Multiply both sides by 100.}$$
$$100 \cdot 2.5x - 100 \cdot 0.64 = 100 \cdot 8 \qquad \text{Distribute 100.}$$
$$250x - 64 = 800$$

Answers to Your Turn 1
a. $y = 6.22$ b. $n = -1.95$

We are still multiplying by the LCD in the same way that we did with fractions. We can see this by writing the original equation in terms of equivalent fractions.

$$\frac{25}{10}x - \frac{64}{100} = 8$$

100 is the LCD of 10 and 100.

With decimal numbers, the LCD will be a power of 10 with the same number of 0 digits as there are decimal places in the number with the most decimal places. We can amend our procedure for solving equations to include this technique.

PROCEDURE *To solve equations:*

1. Simplify both sides of the equation as needed.
   a. Distribute to clear parentheses.
   b. Clear fractions and/or decimals by multiplying both sides by the LCD of all denominators. In the case of decimals, the LCD is the power of 10 with the same number of 0 digits as decimal places in the number with the most decimal places.
   c. Combine like terms.
2. Use the addition/subtraction principle of equality so that all variable terms are on one side of the equation and all constants are on the other side.(Clear the variable term with the lesser coefficient.) Then combine like terms.
3. Use the multiplication/division principle of equality to clear any remaining coefficient.

Let's look at Example 1 again using this technique.

Example 2 Solve and check. $6 - 4.8t = 3.552$

Solution: To clear the decimals, we use the power of 10 with the same number of 0's as there are decimal places in the number with the most decimal places. In this case, 3.552 has the most decimal places so we multiply both sides by 1000.

$1000 \cdot (6 - 4.8t) = 1000 \cdot 3.552$	Multiply both sides by 1000 to clear the decimal numbers.
$1000 \cdot 6 - 1000 \cdot 4.8t = 1000 \cdot 3.552$	Distribute 1000.
$6000 - 4800t = 3552$	Multiply.
$\underline{-6000 \qquad\qquad -6000}$	Subtract 6000 from both sides to isolate $-4800t$.
$0 - 4800t = -2448$	
$\dfrac{-4800t}{-4800} = \dfrac{-2448}{-4800}$	Divide both sides by -4800 to isolate t.
$t = 0.51$	

Check: Replace t in $6 - 4.8t = 3.552$ with 0.51, and verify that the equation is true.

$$6 - 4.8t = 3.552$$
$$6 - 4.8(0.51) \stackrel{?}{=} 3.552 \qquad \text{Replace } t \text{ with 0.51.}$$
$$6 - 2.448 \stackrel{?}{=} 3.552$$
$$3.552 = 3.552 \qquad \text{True, so 0.51 is the solution.}$$

Do Your Turn 2 ▶

Your Turn 2

Solve and check.

a. $4.5y - 21.99 = 6$

b. $0.08n + 0.4 = 0.244$

Answers to Your Turn 2
a. $y = 6.22$ b. $n = -1.95$

Solve and check.

a. $18.6k - 1.25 = 9.4k + 1.326$

b. $11 - 2(5.6 + x) = 0.7x + 1.3$

Example 3 Solve and check. $0.3(x - 1.2) = 1.46 - (2.1x + 0.22)$

Solution: First, we will distribute to clear the parentheses. Then we will use the multiplication/division principle of equality to clear the decimal numbers. Because 1.46 and 0.22 have the most decimal places, we will multiply both sides throughout by 100 to clear the decimals.

$$0.3(x - 1.2) = 1.46 - (2.1x + 0.22)$$

$$0.3x - 0.36 = 1.46 - 2.1x - 0.22 \qquad \text{Distribute to clear parentheses.}$$

$$100 \cdot (0.3x - 0.36) = 100 \cdot (1.46 - 2.1x - 0.22) \qquad \text{Multiply both sides by 100 to clear the decimal numbers.}$$

$$100 \cdot 0.3x - 100 \cdot 0.36 = 100 \cdot 1.46 - 100 \cdot 2.1x - 100 \cdot 0.22 \qquad \text{Distribute.}$$

$$30x - 36 = 146 - 210x - 22 \qquad \text{Multiply.}$$

$$30x - 36 = 124 - 210x \qquad \text{Combine like terms.}$$

$$\underline{+\,210x \qquad\qquad\quad +\,210x} \qquad \text{Add } 210x \text{ to both sides so that the } x \text{ terms are on the same side of the equation.}$$

$$240x - 36 = 124 + 0$$

$$240x - 36 = 124$$

$$\underline{+\,36 \quad +\,36} \qquad \text{Add 36 to both sides to isolate } 240x.$$

$$240x + 0 = 160$$

$$\frac{240x}{240} = \frac{160}{240} \qquad \text{Divide both sides by 240 to isolate } x.$$

$$x = 0.\overline{6}$$

Check: Because $0.\overline{6}$ is a nonterminating decimal number, we will have to round if we intend to check with it. If we round, we must be aware that both sides will not match exactly. We should expect that they will be close, though.

The only way to get the exact same result on both sides is to use the fraction equivalent of $0.\overline{6}$, which is $\frac{2}{3}$. But the method for writing nonterminating decimals as fractions is beyond the scope of our discussion. Therefore, we will round the decimal and keep in mind that both sides will not match exactly, but should be reasonably close.

We will round $0.\overline{6}$ to the nearest thousandth, 0.667.

$$0.3(x - 1.2) = 1.46 - (2.1x + 0.22)$$

$$0.3(0.667 - 1.2) \stackrel{?}{=} 1.46 - (2.1[0.667] + 0.22) \qquad \text{Replace } x \text{ with 0.667.}$$

$$0.3(-0.533) \stackrel{?}{=} 1.46 - (1.4007 + 0.22)$$

$$-0.1599 \stackrel{?}{=} 1.46 - 1.6207$$

$$-0.1599 \approx -0.1607$$

▲

Note: As predicted, the values are not equal, but they are reasonably close. If we had rounded to ten-thousandths, or hundred-thousandths, the results would be even closer.

◁ **Do Your Turn 3**

Answers to Your Turn 3
a. $k = 0.28$ b. $x = -0.\overline{5}$

OBJECTIVE 3 Solve application problems.

Let's first consider more advanced problems similar to those we've discussed in previous sections.

Example 4 The table shows the pricing schedule for a phone company's long-distance rates.

Time	Rate
8 A.M.–5 P.M.	$0.12/minute
5 P.M.–10 P.M.	$0.10/minute
10 P.M.–8 A.M.	$0.09/minute

Below is a portion of Muffy's long-distance call summary. Find the total cost of the calls.

Time of Call		City	Number of Minutes
10:05	P.M.	Richmond, VA	13
9:52	A.M.	Columbus, OH	21
4:47	P.M.	Atlanta, GA	24

Understand: Because the unit cost depends on the time of the call, we must consider the time of each call.

The 13 minutes to Richmond will cost $0.09/minute because the call was made at 10:05 P.M.. The 21 minutes to Columbus will cost $0.12/minute because the call was made at 9:52 A.M.. Notice that the 24-minute call to Atlanta started at 4:47 P.M. and lasted until 5:11 P.M. This means the first 13 minutes of the call will cost $0.12/minute and the last 11 minutes will cost $0.10/minute because the rate changes at 5 P.M.

Plan: Multiply the number of minutes by their respective rates, then add to get the total cost.

Execute: total cost = $\dfrac{\text{cost of call to}}{\text{Richmond}}$ + $\dfrac{\text{cost of call}}{\text{to Columbus}}$ + $\dfrac{\text{cost of call}}{\text{to Atlanta}}$

$$
\begin{aligned}
C &= \;\; 0.09(13) \;\; + \;\; 0.12(21) \;\; + \;\; 0.12(13) + 0.10(11) \\
C &= \;\; 1.17 \;\;\;\;\;\; + \;\;\;\; 2.52 \;\;\;\; + \;\;\;\; 1.56 + 1.10 \\
C &= \;\; 6.35
\end{aligned}
$$

Answer: The total cost of the long-distance calls is $6.35.

Check: We can reverse the calculations. Start with 6.35 and subtract the last three addends and see if we get the first addend.

$$6.35 - 1.10 - 1.56 - 2.52 \overset{?}{=} 1.17$$
$$1.17 = 1.17$$

We can check the multiplication by dividing each product by one of the factors to see if we get the other factor.

$$1.17 \div 13 \overset{?}{=} 0.09 \quad 2.52 \div 21 \overset{?}{=} 0.12 \quad 1.56 \div 13 \overset{?}{=} 0.12 \quad 1.10 \div 11 \overset{?}{=} 0.10$$
$$0.09 = 0.09 \qquad\quad 0.12 = 0.12 \qquad\quad 0.12 = 0.12 \qquad\quad 0.10 = 0.10$$

Do Your Turn 4 ▶

Your Turn 4

Solve.

a. Use the rate schedule in Example 4 to find the total cost of the following long-distance calls.

Time of Call		City	Number of Minutes
2:35	P.M.	Los Angeles, CA	19
9:55	P.M.	Tulsa, OK	29
11:02	A.M.	Houston, TX	31

b. Angie has the following medical charges. She arranges with the hospital to make three equal payments to pay off the debt. How much is each payment?

Service	Cost
Lab	$125
Radiology	$287.25
Treatment	$482.56

Answers to Your Turn 4
a. $8.66 b. $298.27

Solve.

a. 18.6 less than y is the same as 1.25 times y plus 2.18. Translate to an equation, then solve for y.

b. 0.4 times the difference of m and 1.8 is equal to -0.8 times m minus 9.36. Translate to an equation, then solve for m.

OBJECTIVE 4 Solve problems involving one unknown.

Let's consider some problems where we translate key words directly to an equation.

Example 5 5.9 more than 0.2 times n is equal to 6.7 less than the product of 1.4 and n. Translate to an equation, then solve for n.

Understand: Focus on the key words.

5.9 <u>more than</u> 0.2 <u>times</u> n <u>is equal to</u> 6.7 <u>less than</u> the <u>product</u> of 1.4 and n.

    addition   multiplication   equal sign   subtraction   multiplication

Plan: Translate the key words to an equation, then solve.

Execute: 5.9 <u>more than</u> 0.2 <u>times</u> n <u>is equal to</u> 6.7 <u>less than</u> the <u>product</u> of 1.4 and n.

$$5.9 \quad + \quad 0.2n \quad = \quad 1.4n - 6.7$$
$$5.9 + 0.2n = 1.4n - 6.7$$
$$10(5.9 + 0.2n) = 10(1.4n - 6.7)$$
$$10 \cdot 5.9 + 10 \cdot 0.2n = 10 \cdot 1.4n - 10 \cdot 6.7$$
$$59 + 2n = 14n - 67$$
$$\underline{-2n \qquad -2n}$$
$$59 + 0 = 12n - 67$$
$$59 = 12n - 67$$
$$\underline{+67 \qquad\qquad +67}$$
$$126 = 12n + 0$$
$$\frac{126}{12} = \frac{12n}{12}$$

Answer: $\qquad\qquad\qquad 10.5 = n$

Note: When translating *less than* to subtraction, the subtrahend appears before the words *less than* and the minuend appears after.

Check: Verify that 10.5 satisfies the original equation.

$$5.9 + 0.2n = 1.4n - 6.7$$
$$5.9 + 0.2(10.5) \stackrel{?}{=} 1.4(10.5) - 6.7 \qquad \text{Replace } n \text{ with 10.5.}$$
$$5.9 + 2.1 \stackrel{?}{=} 14.7 - 6.7$$
$$8 = 8 \qquad\qquad \text{It checks.}$$

◀ Do Your Turn 5

Example 6 Karen uses an Internet provider that charges $14.95 per month; this allows for 200 minutes of online time. After 200 minutes, it costs $0.10 for each additional minute. If Karen's bill is $20.75, how many minutes did she spend online in all?

Understand: We must calculate Karen's total time online. The total charges are $20.75. To calculate the total charges, the company charged a flat fee of $14.95 for the first 200 minutes, then $0.10 for each additional minute.

flat fee + cost of additional minutes = total charges

Because $0.10 is a unit price, to get the cost of additional minutes, we multiply 0.10 times the number of additional minutes. If we let m be the number of additional minutes, then $0.10m$ describes the cost of those additional minutes.

Plan: We will write an equation and then solve.

Answers to Your Turn 5
a. $y - 18.6 = 1.25y + 2.18$; $y = -83.12$
b. $0.4(m - 1.8) = -0.8m - 9.36$; $m = -7.2$

Execute: flat fee $+$ cost of additional minutes $=$ total charges

$$14.95 + 0.10m = 20.75$$
$$100(14.95 + 0.10m) = 100(20.75)$$ Multiply both sides by 100 to eliminate decimals.

$$100 \cdot 14.95 + 100 \cdot 0.10m = 100 \cdot 20.75$$ Distribute 100.

$$1495 + 10m = 2075$$ Subtract 1495 from both sides.
$$\underline{-1495 \qquad\quad -1495}$$
$$0 + 10m = 580$$

$$\frac{10m}{10} = \frac{580}{10}$$ Divide both sides by 10.

$$m = 58$$

Answer: Karen spent 58 additional minutes on-line. This is in addition to the 200 minutes that she paid for with $14.95, so she spent a total of

$$200 + 58 = 258 \text{ minutes on-line}$$

Check: Verify that 258 minutes at $14.95 for the first 200 minutes and $0.10 per minute for the additional 58 minutes comes to a total of $20.75.

$$\text{total charges} = 14.95 + 0.10(58)$$
$$= 14.95 + 5.80$$
$$= 20.75$$ It checks.

Do Your Turn 6 ▶

Your Turn 6

Solve.

a. A plumber charges a flat fee of $45 plus $7.25 for every quarter of an hour spent working. Dina suspects that the $95.75 that he charged is too much. How long should he have worked?

b. A cell-phone company charges $19.95 for up to 30 minutes of use plus $0.50 for each additional minute. Ron's bill comes to a total of $33.45. How many minutes did he spend using his cell phone?

OBJECTIVE 5 Solve problems using the Pythagorean theorem.

One of the most popular theorems in mathematics is the Pythagorean theorem. The theorem is named after the Greek mathematician Pythagoras. The theorem describes a relationship that exists among the sides of all **right triangles.**

DEFINITION **Right triangle:** A triangle that has one right angle.

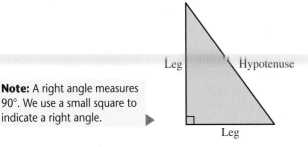

Note: A right angle measures 90°. We use a small square to indicate a right angle.

In a right triangle, the two sides that form the 90° angle are called **legs.** The side directly across from the 90° angle is called the **hypotenuse.**

DEFINITIONS **Legs:** The sides that form the 90° angle in a right triangle.

             Hypotenuse: The side directly across from the 90° angle in a right triangle.

Answers to Your Turn 6
a. 7 quarter hours, or 1.75 hr.
b. 57 min.

Pythagoras saw that the sum of the areas of the squares on the legs of a right triangle is the same as the area of the square on the hypotenuse. Consider a right triangle with side lengths of 3 feet, 4 feet, and 5 feet.

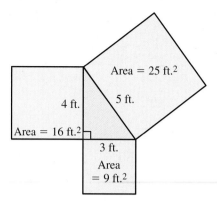

If we add the areas of the squares on the legs, 9 + 16, we get the same area as the square on the hypotenuse, 25.

$$16 \text{ ft.}^2 + 9 \text{ ft.}^2 = 25 \text{ ft.}^2$$

Pythagoras proved that this relationship is true for every right triangle, which is why it is called the Pythagorean theorem.

RULE **The Pythagorean Theorem**
Given a right triangle, where a and b represent the lengths of the legs and c represents the length of the hypotenuse, then $a^2 + b^2 = c^2$.

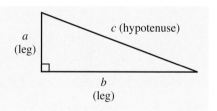

The Pythagorean theorem can be used to find a missing length in a right triangle if the other two lengths are known.

Example 7

a. To steady a telephone pole, a wire is to be attached 21 feet up on the pole and then attached to a stake 20 feet out from the base of the pole. What length of wire will be needed between the pole and the stake? (Ignore the wire needed for attachment.)

Understand: Draw a picture. We assume the pole makes a 90° angle with the ground. The wire connects to the pole to form a right triangle.

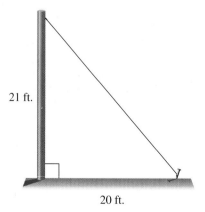

Plan: Use the Pythagorean theorem. The missing length is the hypotenuse, which is *c* in the formula.

Note: We have placed the units in the equation as a reminder that 441 ft.² and 400 ft.² are the areas of the squares on the legs. Their sum, 841 ft.², is the area of the square on the hypotenuse. To get the length from the area, we must use a square root.

▶
$$a^2 + b^2 = c^2$$
$$(21 \text{ ft.})^2 + (20 \text{ ft.})^2 = c^2 \quad \text{Replace } a \text{ with 21 and } b \text{ with 20.}$$
$$441 \text{ ft.}^2 + 400 \text{ ft.}^2 = c^2 \quad \text{Square 21 and 20.}$$
$$841 \text{ ft.}^2 = c^2 \quad \text{Add.}$$
$$\sqrt{841} \text{ ft.} = c$$
$$29 \text{ ft.} = c$$

◀ **Note:** The equation $841 = c^2$ means the square of the value *c* is 841, so the value of *c* must be the square root of 841.

Answer: The wire must be at least 29 ft. to connect the pole and the stake.

Check: Verify that $a^2 + b^2 = c^2$ is true when *a* is 21, *b* is 20, and *c* is 29.

$$(21)^2 + (20)^2 \overset{?}{=} (29)^2$$
$$441 + 400 \overset{?}{=} 841$$
$$841 = 841 \quad \text{It checks.}$$

b. How high up on the side of a building will a 20 foot ladder reach if the base is placed 8 feet from the base of the building?

Understand: Draw a picture of the situation.

When the ladder is placed against the building, it forms a right triangle. The 20-foot ladder is the hypotenuse and the 8-foot distance from the base of the building is one of the legs. We must find the vertical distance, which is the other leg.

Plan: Use the Pythagorean theorem.

Execute:
$$a^2 + b^2 = c^2$$
$$(8)^2 + b^2 = (20)^2 \quad \text{Replace } a \text{ with 8 and } c \text{ with 20.}$$
$$64 + b^2 = 400 \quad \text{Square 8 and 20.}$$
$$\underline{-64 \qquad\qquad -64} \quad \text{Subtract 64 from both sides to isolate } b^2.$$
$$0 + b^2 = 336$$

Note: This equation says the square of our unknown value is 336; so the unknown value, *b*, must be the square root of 336.

$$b^2 = 336$$
$$b = \sqrt{336}$$
$$b \approx 18.33$$

◀ Because $\sqrt{336}$ is irrational, this is the exact answer.

◀ This is the approximate answer rounded to the nearest hundredth.

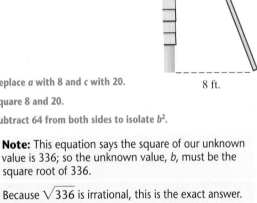
?   20 ft.

8 ft.

Your Turn 7

Solve.

a. In constructing a roof, three boards are used to form a right triangle frame. An 8-foot board and a 6-foot board are brought together to form a 90° angle. How long must the third board be?

b. A wire is attached to a telephone pole then pulled taut and attached to a stake in the ground 4 meters from the base of the pole. If the length of wire is 10 meters, how high up the pole is the wire attached?

Answer: The ladder would reach approximately 18.33 ft. up the side of the building.

Check: Verify that the sum of the squares of the legs is the same as the square of the hypotenuse.

$$(8)^2 + (18.33)^2 \overset{?}{=} (20)^2$$
$$64 + 335.9889 \overset{?}{=} 400$$
$$399.9889 \approx 400$$

Note: Because 18.33 is an approximation, we cannot expect the results to match exactly, but they should be close. 399.9889 is close enough to 400 for us to accept that 18.33 is a good approximation.

Do Your Turn 7 ▶

Answers to Your Turn 7
a. 10 ft. b. ≈ 9.17 m

Pythagoras was a Greek mathematician who lived in the town of Croton in what is now Italy. He is believed to have lived from 569 B.C. to 500 B.C. The Pythagorean theorem is named after Pythagoras not because he discovered the relationship, as commonly thought, but because he and his followers are the first to have proved that the relationship is true for all right triangles.

The Pythagoreans also believed in the mystical power of numbers. Pythagoras believed that everything in the universe was built on whole numbers. The very theorem that he and his followers proved wreaked havoc to their beliefs, however. It quickly became obvious that the legs could be measured by whole numbers as desired, but the hypotenuse would have to be a value that *couldn't* be expressed exactly. For example, if the legs are both 1 unit in length, we have

$$1^2 + 1^2 = c^2$$
$$1 + 1 = c^2$$
$$2 = c^2$$
$$\sqrt{2} = c$$

$\sqrt{2}$ is definitely not a whole amount and, it turns out, cannot be expressed exactly as a decimal or fraction. It is said that this distressed Pythagoras tremendously because he was so sure that the universe was based on whole numbers.

OBJECTIVE 6 Solve problems involving two unknowns.

In Section 4.5, we developed the use of a table to help organize information in problems involving two unknowns. The table we used had four columns:

Categories	Value	Number	Amount

The two unknowns are written in the categories column. The value column contains the given values of each category. The number column will describe the number of items in each category. The amount column will be found by the relationship:

$$\text{value} \cdot \text{number} = \text{amount}$$

Example 8 Betty tells her son that she has 15 coins in her change purse that total $2.85. If she keeps only quarters and dimes in her change purse, how many of each coin does she have?

Understand: There are two unknowns in the problem. We must find the number of quarters and the number of dimes. Since we only know how to solve equations involving one unknown, we must select a variable for either quarters or dimes.

Let's begin filling in the table. Let n represent the number of quarters. To see how can we represent the number of dimes, suppose she has 1 quarter. If there are 15 coins, then there must be 14 dimes. If she has 2 quarters, then there must be 13 dimes and so on. We can find the number of dimes by subtracting the number of quarters from 15. If she has n quarters, she must have $15 - n$ dimes.

Note: We let n represent the number of quarters because quarters have greater value than dimes. As we will see, this choice will avoid negative coefficients in the last steps of solving the equation.

Categories	Value	Number	Amount
Quarters	0.25	n	$0.25n$
Dimes	0.10	$15 - n$	$0.10(15 - n)$

Because value · number = amount, we multiply the expression in the Value column and in the Number column to get the expression in the Amount column.

We are also given the total amount of money, which is $2.85. This total is calculated by adding the amount in quarters plus the amount in dimes.

$$\text{amount in quarters} + \text{amount in dimes} = \text{total amount of money}$$

Plan: Write an equation, then solve.

Execute:

amount in quarters	+ amount in dimes	= total money
$0.25n$	$+\ 0.10(15 - n)$	$= 2.85$
$0.25n$	$+\ 1.5 - 0.1n$	$= 2.85$

Distribute 0.10 to clear parentheses.

$$100 \cdot 0.25n + 100 \cdot 1.5 - 100 \cdot 0.10n = 100 \cdot 2.85$$

Multiply both sides by 100 to clear decimals.

$$25n + 150 - 10n = 285$$

Multiply.

Note: When we combine $25n$ and $-10n$ we get $15n$, which has a positive coefficient. Choosing to let n represent the greater valued coin, quarters, leads to this positive coefficient. We'll see that if we let n be the number of dimes, we get a negative coefficient.

▶ $15n + 150 = 285$ Combine like terms.

$\underline{\quad\ -\ 150\ -\ 150\quad}$ Subtract 150 from both sides.

$15n + 0 = 135$

$\dfrac{15n}{15} = \dfrac{135}{15}$ Divide both sides by 15.

$n = 9$

Answer: Because n represents the number of quarters, there are 9 quarters. To find the number of dimes we use the expression $15 - n$ from the table: $15 - 9 = 6$ dimes.

Check: Verify that 9 quarters and 6 dimes are 15 coins that total $2.85.

$9 + 6 = 15$ coins	$0.25(9) + 0.10(6) \overset{?}{=} 2.85$
	$2.25 + 0.60 \overset{?}{=} 2.85$
	$2.85 = 2.85$ Everything checks.

We could have let n be the number of dimes. If so then $15 - n$ would describe the number of quarters. The problem would have proceeded this way:

Categories	Value	Number	Amount
Quarters	0.25	$15 - n$	$0.25(15 - n)$
Dimes	0.10	n	$0.10n$

amount in quarters	+ amount in dimes	= total
$0.25(15 - n)$	$+\quad 0.10n$	$= 2.85$
$3.75 - 0.25n$	$+\quad 0.10n$	$= 2.85$

◀ **Note:** We chose to work this version of the problem without clearing the decimals.

$3.75 \quad 0.15n \quad 2.85$

Note: Now when we combine $-0.25n$ and $0.10n$ the result, $-0.15n$, has a negative coefficient. Choosing to let n represent the number of the smaller-valued coin leads to this negative coefficient.

▶ $\underline{-\ 3.75\qquad\qquad\ -\ 3.75}$

$0 - 0.15n = -0.90$

$\dfrac{-0.15n}{-0.15} = \dfrac{-0.90}{-0.15}$

$n = 6$

There are 6 dimes and $15 - 6 = 9$ quarters.

Notice that we get the same final answer regardless of which number of coins we let n represent.

Do Your Turn 8 ▷

Your Turn 8

Solve.

a. Jack is a sales manager at a music supply store. He remembers selling packs of 6-string electric and 12-string acoustic strings to a customer. He also remembers the customer bought 20 packs but doesn't remember how many of each type. The price for a pack of 6-string electric strings is $6.95, and $10.95 for a pack of 12-string acoustic strings. If the total sale was $159.00, how many of each pack did the customer buy?

b. Kelly has only nickels and dimes in her pocket. If she has 12 more dimes than nickels and the total is $3.90, how many of each coin does she have?

Answers to Your Turn 8
a. 5 packs of 12-string, 15 packs of 6-string
b. 18 nickels, 30 dimes

For
Extra
Help

Videotape
DVT

Tutor
Center
Addison-Wesley
Tutor Center

Math XL
Math XL

MyMathLab

Student Solutions
Manual

1. Explain how to use the multiplication/division principle to eliminate decimal numbers from an equation.

2. To eliminate the decimal numbers in the equation $0.4x + 1.2 = 0.25x - 3$, what is the least power of ten that you could use?

3. What is a right triangle?

4. What are the legs of a right triangle?

5. What is the hypotenuse?

6. What is the formula for the Pythagorean theorem, and what do its variables represent?

For Exercises 7–24, solve and check.

7. $4.5n + 7 = 7.9$

8. $3x - 0.8 = 1.6$

9. $15.5y + 11.8 = 21.1$

10. $12.1x + 5.6 = 10.924$

11. $16.7 - 3.5t = 17.12$

12. $0.15 = 0.6 - 1.2n$

13. $0.62k - 12.01 = 0.17k - 14.8$

14. $0.72p - 1.21 = 1.37p - 3.29$

15. $0.8n + 1.22 = 0.408 - 0.6n$

16. $5.1m + 7.5 = 4.05 - 4.9m$

17. $4.1 - 1.96x = 4.2 - 1.99x$

18. $12.14 - 2.42g = 4.76 - 0.78g$

19. $0.4(8 + t) = 5t + 0.9$

20. $8k - 2.98 = 2.6(k - 0.8)$

21. $1.5(x + 8) = 3.2 + 0.7x$

22. $0.06(m - 11) = 22 - 0.05m$

23. $20(0.2n + 0.28) = 3.98 - (0.3 - 3.92n)$

24. $4(2.55 - x) - 5.8x = 12.2 - (8 + 11.4x)$

For Exercises 25 and 26, the following table shows the pricing schedule for a phone company's long-distance rates.

Time	Rate
8 A.M.–5 P.M.	$0.12/minute
5 P.M.–10 P.M.	$0.10/minute
10 P.M.–8 A.M.	$0.09/minute

25. Below is a portion of Vicki's long-distance call summary. Find the total cost of the calls.

Time of Call	City	Number of Minutes
1:09 P.M.	Denver, CO	15
12:32 A.M.	Austin, TX	38
7:57 A.M.	Phoenix, AZ	12

26. Below is a portion of Teshieka's long-distance call summary. Find the total cost of the calls.

Time of Call		City	Number of Minutes
5:01	P.M.	Boston, MA	22
9:07	P.M.	New York, NY	64
7:32	A.M.	Washington, DC	24

27. Below is a summary of charges from an electric company. Find the total charges if 1433 kilowatt-hours were used.

Transaction Summary	
Previous balance:	$50.94
Payment on 4/10:	$40.00
Late fee:	$ 0.55
Electric charge:	First 800 kWh @ $0.13162 per kWh
	Remaining kWh @ $0.11275 per kWh

28. Below is a summary of gas charges from an electric and gas company. Find the total charges if 38 therms were used in one month.

Transaction Summary	
Basic facilities charge:	$ 3.00
First 25 therms:	@ $ 1.32461 per therm
Remaining therms:	@ $ 1.24362 per therm

For Exercises 29 and 30, use the following table of long-distance phone charges for the Talk2Friends calling plan.

Charges Summary	
Monthly membership fee	$3.45
Calls to Canada	
First 30 minutes	Free
Additional minutes	$0.09
Calls to Europe	
First 20 minutes	Free
Additional minutes	$0.18
Calls to Asia	
First 15 minutes	Free
Additional minutes	$0.23

29. How much will Nina's long-distance bill be this month if she spent 119 minutes talking to her friend in Montreal, Canada, and 142 minutes talking to her mother in Europe?

30. Viktor's brother is traveling in Europe and Asia. Viktor called him and talked for 73 minutes when his brother was in Europe and 56 minutes when he was in Asia. How much will Viktor be charged this month on his long-distance bill?

For Exercises 31 and 32, use the following table of a company's reimbursements for travel on business.

Reimbursements for Travel on Business
$0.35 per mile when using own vehicle
$0.15 per mile when using company vehicle
$10 breakfast when traveling between 12 A.M. and 11 A.M.
$12 lunch when traveling between 11 A.M. and 3 P.M.
$15 dinner when traveling between 3 P.M. and 12 A.M.

31. Aimee uses her own vehicle on a business trip. She leaves on a Tuesday at 11:30 A.M. and notes that the odometer reads 75,618.4. She returns on Thursday and arrives at 6:45 P.M. At the conclusion of her trip, she finds that the odometer reads 76,264.1. How much should she be reimbursed for mileage and food from her company?

32. Miguel uses a company vehicle and leaves at 7:00 A.M. on a Monday. He notes that the odometer reads 45,981.6. He returns on Friday, arriving at 2:15 P.M. At the conclusion of the trip he notes that the odometer reads 46,610.8. How much should he be reimbursed for mileage and food?

33. Chan receives the following printout of charges that he plans to pay on a 90-day same-as-cash option. He wants to split the charges into three equal payments. How much is each payment?

Description	Amount
1 EPS keyboard	$1645.95
2 cables	$14.95 each
Sales Tax	$83.79

34. Shannon makes the following purchases for her new apartment using a special credit agreement with the store. The credit line is interest-free as long as she pays off the balance in 6 months. After 6 months, all 6 months' worth of interest will be added onto the remaining balance. To avoid the interest, she decides to split the charges into six equal payments. How much is each payment?

Description	Amount
1 table	$354.95
4 chairs	$74.95 each
Sales tax	$32.74

For Exercises 35–50, solve.

35. 7 more than 3.2 times p is equal to 9.56.

36. Twice m decreased by 9.9 is the product of m and -6.25.

37. 18.75 more than 3.5 times t is equal to 1.5 minus t.

38. The product of 3.2 and n is the same as the sum of n and 1.43.

39. 0.48 less than 0.2 times y is the same as 0.1 times the difference of 2.76 and y.

40. 0.6 times the sum of k and 1.5 is equal to 0.42 plus the product of 1.2 and k.

41. There is a huge crater on Mimas, one of Saturn's moons. The crater is circular and has an area of 7850 square kilometers. What is the diameter of the crater?

42. Crater Lake in southern Oregon is actually the top of an inactive volcano (Mount Mazama). The lake is circular and has an area of 20 square miles. What is the diameter of the lake?

43. Rebekah is a scientist studying plant life in the rain forests of South America. From her base camp, she hikes 4 miles straight south then heads east for 3 miles. Because it is getting late in the day, she decides to head straight back to camp from her present location rather than backtrack the way she came. What will be the distance back to camp?

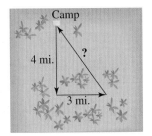

44. Jennifer, a long-distance swimmer, wants to estimate the distance between two docks on a lake. She measures from the north dock directly south along the shoreline, then makes a 90° turn and measures east until she reaches the southern dock. The measurements are shown in the diagram.

a. How far apart are the docks?

b. If there are 5280 feet in a mile, how many times must she swim the distance between the docks to swim 1 mile? (Ignore the lengths of the docks.)

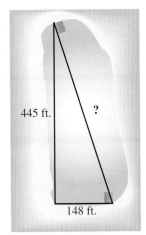

45. A 14.5-foot beam is to be attached to a vertical beam 3.5 feet above a horizontal support beam. How long must the horizontal support beam be to attach to the bottom of the 14.5-foot beam? (Ignore the width of the beams.)

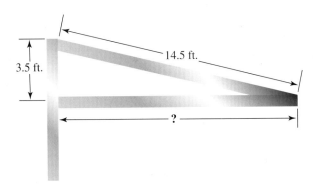

46. A steel support beam on a tower needs to be replaced. It is known that the connecting joint is 24 feet above the ground and the base of the support beam is 8 feet out from a point directly below the connecting joint. How long is the beam?

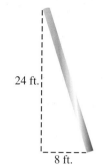

24 ft.

8 ft.

47. A fire truck has a ladder that can extend up to 60 feet. The bottom of the ladder sits atop the truck so that the bottom of the ladder is 8 feet above the ground.

    a. If the truck parks 30 feet away from the bottom of a building, what is the highest point the ladder can reach?

    b. If a person is in a seventh-story window awaiting rescue, can the ladder reach if each story is 10.5 feet?

48. A screen on a small notebook computer is 6 inches wide by 8 inches long. What is the distance along the diagonal?

49. A softball diamond is a square with bases at each corner. The distance between bases is 65 feet. What is the distance from home plate to second base?

50. An 8-foot wide by 15-foot long steel frame is to be fitted with a diagonal beam welded to the corners of the frame. How long must the diagonal beam be?

For Exercises 51–54, use a four-column table to solve.

51. Latisha has a change purse in which she keeps only quarters and half-dollars. She has 12 more quarters than half-dollars. If the total she has in the purse is $6.75, how many of each coin does she have?

52. Jose has 22 coins totaling $4.10. If all of the 22 coins are either nickels or quarters, how many of each coin does he have?

53. Bernice sells drinks at college football games and gets paid by the number of each size drink she sells. There are two drink sizes, 12 ounces and 16 ounces. The 12 ounce sells for $1.50 and the 16 ounce for $2.00. She knows she sold 65 drinks but cannot remember how many of each size. If her total sales are $109, how many of each size did she sell?

54. Arturo sells two different home security systems. He receives $225 commission for each A system and $275 commission for each B system that he sells. In one month, he sells 14 systems. If he receives a total commission of $3400, how many of each system did he sell?

> **PUZZLE PROBLEM** A cube has a surface area of 10.14 square feet. What are the dimensions of the cube?

Review Exercises

[1.1]   **1.** Write the word name for 24,915,000,204.

[1.1]   **2.** Round 46,256,019 to the nearest hundred-thousand.

[5.6]   **3.** Add. $4\dfrac{5}{8} + 8\dfrac{5}{6}$

[5.6]   **4.** Subtract. $12\dfrac{1}{4} - 7\dfrac{4}{5}$

[3.7]   **5.** Divide. $\dfrac{28x^9}{7x^3}$

[4.3]   **6.** Explain the mistake, then work the problem correctly.

$$
\begin{aligned}
-3x - 18 &= -24 \\
+18 \quad & +18 \\
\hline
-3x + 0 &= -6 \\
\frac{-3x}{3} &= \frac{-6}{3} \\
x &= -2
\end{aligned}
$$

Defined Terms

Review the following terms and for those you do not know, study its definition on the page number next to it.

Section 6.1
Decimal notation *(p. 400)*
Rational number *(p. 400)*

Section 6.3
Scientific notation *(p. 420)*
Unit price *(p. 424)*

Section 6.4
Irrational number *(p. 434)*
Real numbers *(p. 434)*

Section 6.6
Right triangle *(p. 461)*
Legs *(p. 461)*
Hypotenuse *(p. 461)*

The Real-Number System

Real Numbers

Irrational Numbers: Any real number that is not rational, such as: $-\sqrt{2}$, $-\sqrt{3}$, $\sqrt{0.8}$, and π.

Rational Numbers: Real numbers that can be expressed in the form $\frac{a}{b}$, where a and b are integers and $b \neq 0$, such as: $-4\frac{3}{4}$, $-\frac{2}{3}$, 0.018, $0.\overline{3}$, and $\frac{5}{8}$.

Integers: ... , -3, -2, -1, 0, 1, 2, 3, ...

Whole Numbers: 0, 1, 2, 3, ...

Natural Numbers: 1, 2, 3, ...

Our Place Value System

Decimal Point

Whole numbers				Fractions						
Hundreds	Tens	Ones	Tenths	Hundredths	Thousandths	Ten thousandths	Hundred thousandths	Millionths	Ten millionths	Hundred millionths

◄ ⋯   ⋯ ►

Procedures, Rules, and Key Examples

Procedures/Rules	Key Example(s)

Section 6.1 Decimals and Rational Numbers

To write a decimal number as a reduced fraction or mixed number:

1. Write all digits to the left of the decimal (whole digits) as the integer in a mixed number.
2. Write all digits to the right of the decimal (fraction digits) in the numerator of a fraction.
3. Write the last place value in the denominator.
4. Simplify to lowest terms.

Example 1: Write as a fraction or mixed number in simplest form.

a. $0.24 = \frac{24}{100} = \frac{6}{25}$

b. $16.082 = 16\frac{82}{1000} = 16\frac{41}{500}$

|

To write the word name for decimal digits (fraction part of a decimal number):

1. Write the word name for digits as if they represented a whole number.
2. Write the name of the last place value.

To write a word name for a decimal number with both integer and fraction parts:

1. Write the name of the integer number part.
2. Write the word *and* for the decimal point.
3. Write the name of the fractional part.

To determine which of two positive decimal numbers is greater:

1. Compare the digits in the numbers from left to right until you find two different digits in the same place value.
2. The greater of these two digits is in the greater number.

Note: If both numbers are negative, the lesser digit is in the greater number because the number is closer to zero.

To round a number to a given place value, consider the digit to the right of the desired place value.

If this digit is 5 or higher, round up.
If this digit is 4 or less, round down.

Example 2: Write the word name for the given number.

a. 0.0351

Answer: three hundred fifty-one ten-thousandths

b. 147.29

Answer: one hundred forty-seven and twenty-nine hundredths

Example 3: Use $<$, $>$, or $=$ to make a true sentence.

a. 0.517   ?   0.516

Answer: $0.517 > 0.516$

b. -0.4912 ? -0.4812

Answer: $-0.4912 < -0.4812$

Example 4: Round each number to the specified place.

a. 4.619 to the nearest tenth.

Answer: 4.6

b. 18.5141 to the nearest whole.

Answer: 19

c. 0.00127 to the nearest ten-thousandth.

Answer: 0.0013

Section 6.2 Adding and Subtracting Decimal Numbers

To add decimal numbers:

1. Stack the numbers so that the place values align. (Line up the decimal points.)
2. Add the digits in the corresponding place values.
3. Place the decimal point in the sum so that it aligns with the decimal points in the addends.

To subtract decimal numbers:

1. Stack the number with the greater absolute value on top so that the place values align. (Line up the decimal points.)
2. Subtract the digits in the corresponding place values.
3. Place the decimal point in the difference so that it aligns with the decimal points in the minuend and subtrahend.

To combine like terms, add or subtract the coefficients and keep the variables the same.

Example 1: Add. $2.035 + 15.98$

$$
\begin{array}{r}
{}^{1}{}^{1} \\
2.035 \\
+\ 15.980 \\
\hline
18.015
\end{array}
$$

Example 2: Subtract. $60.043 - 4.798$

$$
\begin{array}{r}
{}^{13} \\
{}^{5\,9}\ {}^{9\!\!/\,13} \\
6\!\!/0.0\!\!/4\!\!/3 \\
-\ 4.798 \\
\hline
55.245
\end{array}
$$

Example 3: Simplify.

$5.1x^3 + 2.91x - 7x^2 + 9.8 - 3.4x - 3.8x^2$

$= 5.1x^3 - 10.8x^2 - 0.49x + 9.8$

Section 6.3 Multiplying Decimal Numbers; Exponents with Decimal Bases

To multiply decimal numbers:

1. Stack the numbers. (It is not necessary to align the decimal points in the factors.)
2. Multiply as if the numbers were whole numbers.
3. Place the decimal in the product so that it has the same number of decimal places as the total number of decimal places in the factors.

To change from scientific notation with a positive integer exponent to standard notation for a given number, move the decimal point to the right the number of places indicated by the exponent.

To write a standard-form number whose absolute value is greater than 1 in scientific notation:

1. Move the decimal point so that the number's absolute value is greater than or equal to 1 but less than 10. (*Tip:* Place the decimal point to the right of the first nonzero digit.)
2. Write the decimal number multiplied by 10^n, where n is the number of places between the new decimal position and the original decimal position.
3. Delete 0's to the right of the last nonzero digit.

Example 1: Multiply 6.45×2.8.

$$
\begin{array}{r}
\overset{1\ 1}{6.45} \\
\times\ 2.8 \\
\hline
5\ 160 \\
12\ 90 \\
\hline
18.060
\end{array}
$$

Example 2: Write in standard form.

a. $3.59 \times 10^7 = 35{,}900{,}000$

b. $-4.2 \times 10^5 = -420{,}000$

Example 3: Write in scientific notation.

a. $4{,}791{,}000{,}000 = 4.791 \times 10^9$

b. $-87{,}050{,}000{,}000{,}000 = -8.705 \times 10^{13}$

Section 6.4 Dividing Decimal Numbers; Square Roots with Decimals

To divide a decimal number by a whole number using long division:

1. Divide the divisor into the dividend as if both numbers were whole numbers. Be sure to align the digits in the quotient properly.
2. Write the decimal in the quotient directly above its position in the dividend.
3. Write extra 0 digits to the right of the last nonzero digit of the dividend as needed until you get a remainder of 0, or until a digit or group of digits repeats without end in the quotient.

Note: To indicate repeated digits, write a repeat bar over the digit or group of digits that repeat.

Example 1: Divide.

a. $20.4 \div 6$

$$
\begin{array}{r}
3.4 \\
6\,\overline{)20.4} \\
-18 \\
\hline
24 \\
-24 \\
\hline
0
\end{array}
$$

Answer: 3.4

b. $3.68 \div 11$

$$
\begin{array}{r}
0.33454545\ldots \\
11\,\overline{)3.6800000} \\
-33 \\
\hline
38 \\
-33 \\
\hline
50 \\
-44 \\
\hline
60 \\
-55 \\
\hline
50
\end{array}
$$

Answer: $0.33\overline{45}$

To divide using long division when the divisor is a decimal number:

1. Move the decimal point in the divisor to the right enough places to make the divisor an integer.
2. Move the decimal point in the dividend the same number of places.
3. Divide the divisor into the dividend as if both numbers were whole numbers. Be sure to align the digits in the quotient properly.
4. Write the decimal point in the quotient directly above its position in the dividend.

c. $50.8 \div 0.02 = 5080 \div 2$

$$
\begin{array}{r}
2540 \\
2\,\overline{)5080} \\
-4 \\
\hline
10 \\
-10 \\
\hline
08 \\
-8 \\
\hline
0
\end{array}
$$

Answer: 2540

Procedures/Rules	Key Example(s)

d. $0.694 \div 0.22 = 69.4 \div 22$

$$
\begin{array}{r}
3.154545... \\
22\overline{)69.400000} \\
-66 \\ \hline
34 \\
-22 \\ \hline
120 \\
-110 \\ \hline
100 \\
-88 \\ \hline
120
\end{array}
$$

Answer: $3.1\overline{54}$

To write a fraction as a decimal,

Divide the denominator into the numerator.

Example 2: Write $\frac{5}{8}$ as a decimal.

$$
\begin{array}{r}
0.625 \\
8\overline{)5.000} \\
-48 \\ \hline
20 \\
-16 \\ \hline
40 \\
-40 \\ \hline
0
\end{array}
$$

Answer: $\frac{5}{8} = 0.625$

To write a mixed number as a decimal:

1. Write the integer part of the mixed number to the left of a decimal point.
2. Divide the denominator into the numerator to get the decimal digits.

Example 3: Write $14\frac{2}{3}$ as a decimal.

Answer: $14.\overline{6}$

$$
\begin{array}{r}
0.66... \\
3\overline{)2.00} \\
-18 \\ \hline
20 \\
-18 \\ \hline
20
\end{array}
$$

The square root of a decimal number that corresponds to a perfect square and has an even number of decimal places will be a decimal number with half the number of places and the same digits as the square root of the corresponding perfect square.

Example 4: Simplify.

$\sqrt{0.0049} = 0.07$

Section 6.5 Order of Operations and Applications in Geometry

To simplify numerical expressions that contain both fractions and decimals:

1. Write all decimals as fractions. (Method 1) or
   Write all fractions as decimals. (Method 2) or
   Write decimals over 1 in fraction form. (Method 3)
2. Follow the order of operations agreement.

Example 1: Simplify.

(Method 3 illustrated here)

$$
\begin{aligned}
\frac{2}{3}(6.09) - \frac{3}{8} &= \frac{2}{3} \cdot \frac{6.09}{1} - \frac{3}{8} \\
&= \frac{12.18}{3} - \frac{3}{8} \\
&= 4.06 - 0.375 \\
&= 3.685
\end{aligned}
$$

Section 6.6 Solving Equations and Problem Solving

To solve equations:

1. Simplify both sides of the equation as needed.
   a. Distribute to clear parentheses.
   b. Clear fractions and/or decimals by multiplying both sides by the LCD of all denominators. In the case of decimals, the LCD is the power of 10 with the same number of 0 digits as decimal places in the number with the most decimal places.
   c. Combine like terms.
2. Use the addition/subtraction principle of equality so that all variable terms are on one side of the equation and all constants are on the other side. (Clear the variable term with the lesser coefficient.) Then combine like terms.
3. Use the multiplication/division principle of equality to clear any remaining coefficient.

Example 1: Solve and check.

$$8.3 - (2.4x + 0.7) = 14.6x - 5(x + 6.16)$$
$$8.3 - 2.4x - 0.7 = 14.6x - 5x - 30.8$$
$$10(8.3 - 2.4x - 0.7) = (14.6x - 5x - 30.8)10$$
$$83 - 24x - 7 = 146x - 50x - 308$$
$$76 - 24x = 96x - 308$$

$$\underline{+\,24x \qquad +\,24x}$$
$$76 + 0 = 120x - 308$$
$$76 = 120x - 308$$
$$\underline{+\,308 \qquad\qquad +\,308}$$
$$384 = 120x + 0$$
$$\frac{384}{120} = \frac{120x}{120}$$
$$3.2 = 1x$$
$$3.2 = x$$

Check:
$$8.3 - (2.4x + 0.7) = 14.6x - 5(x + 6.16)$$
$$8.3 - [2.4(3.2) + 0.7] \overset{?}{=} 14.6(3.2) - 5(3.2 + 6.16)$$
$$8.3 - (7.68 + 0.7) \overset{?}{=} 46.72 - 5(9.36)$$
$$8.3 - 8.38 \overset{?}{=} 46.72 - 46.8$$
$$-0.08 = -0.08$$

Because both sides match, 3.2 is correct.

The Pythagorean Theorem

Given a right triangle, where a and b represent the lengths of the legs and c represents the length of the hypotenuse, then $a^2 + b^2 = c^2$.

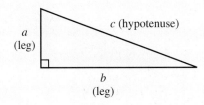

Example 2: A 15-foot ladder is leaning against a building with its base 9 feet from the base of the building. How high up the side of the building does the ladder reach?

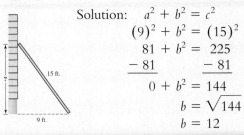

Solution:
$$a^2 + b^2 = c^2$$
$$(9)^2 + b^2 = (15)^2$$
$$81 + b^2 = 225$$
$$\underline{-\,81 \qquad\quad -\,81}$$
$$0 + b^2 = 144$$
$$b = \sqrt{144}$$
$$b = 12$$

Answer. The ladder is 12 ft. up the side of the building.

Formulas

Volume of a cylinder: $V = \pi r^2 h$

Volume of a cone: $V = \dfrac{1}{3}\pi r^2 h$

Volume of a pyramid: $V = \dfrac{1}{3}lwh$

Volume of sphere: $V = \dfrac{4}{3}\pi r^3$

For Exercises 1–6, answer true or false.

[6.1] 1. Every rational number can be expressed as a decimal number.

[6.4] 2. Every decimal number is a rational number.

[6.4] 3. 5.$\overline{16}$ means 5.16666 . . .

[6.2] 4. 0.11 − 0.095 equals a positive number.

[6.4] 5. $\sqrt{90} \approx 7.487$

[6.1] 6. 0.0851 is between 0.085 and 0.086.

[6.3] 7. Explain the mistake.

$$\begin{array}{r} 0.24 \\ \times\ 0.13 \\ \hline 72 \\ 24 \\ \hline 3.12 \end{array}$$

[6.2] 8. Explain in your own words how to add decimal numbers.

[6.4] 9. To write a fraction as a decimal number, divide the _____ into the _____ .

[6.1] 10. Explain in your own words how to write a decimal number as a fraction.

[6.1] 11. Write the word name.

   a. 24.39

   b. 581.459

   c. 0.2917

[6.1] 12. Write as a fraction in lowest terms.

   a. 0.8

   b. 0.024

   c. −2.65

[6.1] 13. Use <, >, or = to make a true statement.

   a. 2.001 ? 2.0009

   b. −0.016 ? −0.008

[6.1] 14. Round 31.8056 to the nearest:

   a. whole   b. tenth

      32   31.8

   c. hundredth   d. thousandth

      31.81   31.806

[6.2] 15. Add or subtract.

    a. $24.81 + 54.2 + 9.005$ **b.** $9.005 - 1.089$ **c.** $6.7 + (-12.9)$

    d. $(-20) - (-13.88)$ **e.** $-1.082 + 29.1 - 4.9$ **f.** $14.1 - (-6.39)$

[6.2] 16. Simplify.

    a. $9y^4 + 8y - 12.2y^2 + 2.1y^4 - 16y + 3.8$ **b.** $(0.3a^2 + 8.1a + 6) + (5.1a^2 - 2.9a + 4.33)$

    c. $(2.6x^3 - 98.1x^2 + 4) - (5.3x^3 - 2.1x + 8.2)$

[6.3] 17. Multiply.

    a. $(3.44)(5.1)$ **b.** $(1.5)(-9.67)$ **c.** $(1.2)^2$

    d. $(-0.2)^3$ **e.** $(2.5h)(0.01hk^3)$ **f.** $(-0.9x^3y)(4.2x^5y)$

    g. $(-0.3x^3)^3$ **h.** $(1.2y^3)^2$ **i.** $0.2(0.6n^3 - 1.4n^2 + n - 18)$

    j. $-1.1(2.8a - 4.9b)$ **k.** $(4.5a - 6.1)(2a - 0.5)$ **l.** $(1.3x + 4.2)(1.3x - 4.2)$

[6.4] 18. Divide.

    a. $21 \div 4$ **b.** $3.6 \div 9$ **c.** $17.304 \div 2.06$

    d. $-54 \div (-0.09)$ **e.** $\dfrac{3.18x^7}{0.4x^5}$ **f.** $\dfrac{-2.38a^4bc}{6.8ab}$

[6.4] 19. Find the square root.

    a. $\sqrt{0.64}$ **b.** $\sqrt{1.21}$

[6.4] 20. Approximate the square root to the nearest hundredth.

    a. $\sqrt{148}$ **b.** $\sqrt{1.6}$

[6.4] 21. Write as a decimal.

    a. $\dfrac{3}{5}$ **b.** $\dfrac{1}{6}$ **c.** $6\dfrac{1}{4}$ **d.** $-4\dfrac{2}{11}$

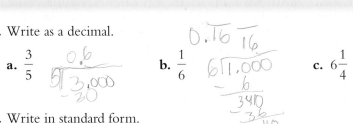

[6.3] 22. Write in standard form.

    a. 3.58×10^9 **b.** -4.2×10^5

[6.3] 23. Write in scientific notation.

    a. $9,200,000,000$ **b.** $-103,000$

[6.5] 24. Simplify:

a. $4.59 - 2.8 \cdot 6.7 + \sqrt{1.44}$

b. $3.6 + (2.5)^2 - 1.5(9 - 0.4)$

c. $\dfrac{3}{4}(0.86) - \dfrac{3}{8}$

d. $\dfrac{1}{6}(0.24) + \dfrac{2}{3}(-3.3)$

$\begin{bmatrix} 6.4 \\ 6.5 \end{bmatrix}$ **25.** Find the radius, circumference, and area of the circle. Use 3.14 for π.

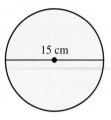

15 cm

[6.5] 26. Find the area. Use 3.14 for π.

a.

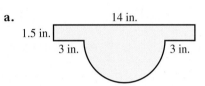

14 in.

1.5 in.

3 in.   3 in.

b.

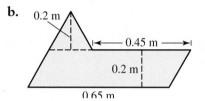

0.2 m

0.45 m

0.2 m

0.65 m

[6.5] 27. Find the area of the shaded region. Use 3.14 for π.

a.

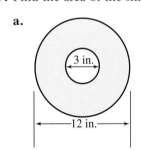

3 in.

12 in.

b.

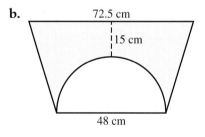

72.5 cm

15 cm

48 cm

[6.5] 28. Find the volume. Use 3.14 for π.

a.

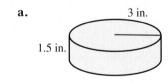

3 in.

1.5 in.

b.

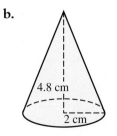

4.8 cm

2 cm

c.

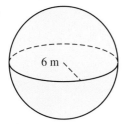

6 m

d.

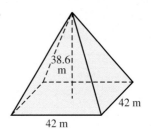

38.6 m

42 m

42 m

e.

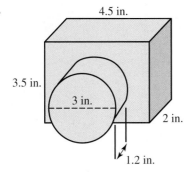

4.5 in.

3.5 in.

3 in.

2 in.

1.2 in.

[6.5] **29.** Find the volume of the shaded region. Use 3.14 for π.

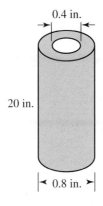

0.4 in.

20 in.

0.8 in.

[6.6] **30.** Solve and check.

a. $x + 0.56 = 2.1$

b. $y - 0.58 = -1.22$

c. $0.8y = 2.4$

d. $-1.6n = 0.032$

e. $-0.562b = 0$

f. $4.5k - 2.61 = 0.99$

g. $51.2 = 0.4a + 51.62$

h. $2.1x - 12.6 = 1.9x - 10.98$

i. $3.98 - 19.6x = 3.2(x - 4.1)$

j. $3.6n - (n - 13.2) = 10(0.4n + 0.62)$

For Exercises 31–36, solve.

[6.2] **31.** Shown is Sonja's checkbook register. Find her balance.

Date	No.	Transaction Description	Debits	Credits	Balance
3 - 21	1201	Papa John's	16 70		150 20
3 - 22	1202	Food Lion	84 61		
3 - 25	1203	Ford Motor Credit	245 16		
3 - 25	1204	Electric & Gas	67 54		
3 - 25	1205	Fleet Mortgage	584 95		
3 - 26		Deposit		1545 24	

[6.4] **32.** Naja has a student loan balance of $2854.82. If he agrees to make equal monthly payments over a 5-year period, how much will each payment be? How much is the last payment?

[6.5] **33.** Use $\text{BMI} = \dfrac{w}{h^2} \cdot 705$ to calculate a person's body mass index if the person weighs 210 pounds and is 72 inches tall.

[6.6] **34.** Below is a portion of Ngyen's long-distance call summary. Find the total cost of the calls using the rate table provided.

Time of Call		City	Number of Minutes
2:49	P.M.	Sacramento, CA	45
7:42	A.M.	Miami, FL	28
8:57	P.M.	Miami, FL	16

Time	Rate
8 A.M.–5 P.M.	$0.12/minute
5 P.M.–10 P.M.	$0.10/minute
10 P.M.–8 A.M.	$0.09/minute

[6.6] **35.** A 12-foot board is resting against a tree. The bottom of the board is 3 feet from the base of the tree. How far is the top of the board from the base of the tree? (Assume the tree forms a right angle with the ground.)

[6.6] **36.** An electrician charges $45 for all service calls, then $4.50 for every 15 minutes spent working. If Brittany has $72, how long can she afford for the electrician to work?

For Exercises 37 and 38, use the four-column table to solve.

[6.6] **37.** Elissa has a change purse that contains only nickels and dimes. She has 9 more dimes than nickels. If she has a total of $2.40 in the purse, how many of each coin is there?

[6.6] **38.** Conrad makes two different sizes of glass figurines. He sells the larger size for $14.50, and the smaller size sells for $9.50. If he sells a total of 24 glass figurines at a fair and makes a total of $273, how many figurines of each size did he sell?

[6.1]   **1.** Write the word name for 56.789.

1. _____

[6.1]   **2.** Write 0.68 as a fraction in lowest terms.

2. _____

[6.1]   **3.** Use $<$, $>$, or $=$ to make a true statement.
-0.0059 ? -0.0058

3. _____

[6.1]   **4.** Round 2.0915 to the nearest:
   a. whole   **b.** tenth   **c.** hundredth   **d.** thousandth

4. a._____
   b._____
   c._____
   d._____

For Exercises 5–10, evaluate.

[6.2]   **5.** $4.591 + 34.6 + 2.8$ [6.2]   **6.** $9.005 - 1.089$

5. _____

6. _____

[6.3]   **7.** $(4.5)(-8.61)$ [6.4]   **8.** $4.8 \div 16$

7. _____

8. _____

[6.4]   **9.** $(-55.384) \div (-9.2)$ [6.4]   **10.** $\sqrt{0.81}$

9. _____

10. _____

[6.4]   **11.** Write $3\frac{5}{6}$ as a decimal.

11. _____

$\begin{bmatrix} 6.3 \\ 6.4 \end{bmatrix}$ **12. a.** Write 2.97×10^6 in standard form.

12. a._____

   b. Write $-35,600,000$ in scientific notation.

   b._____

[6.5]   **13.** Simplify. $3.2 + (0.12)^2 - 0.6(2.4 - 5.6)$

13. _____

[6.5]   **14.** Evaluate $\frac{1}{2}mv^2$, when $m = 12.8$ and $v = 0.2$.

14. _____

[6.2]   **15.** Simplify. $(2.6x^3 - 98.1x^2 + 4) - (5.3x^3 - 2.1x + 8.2)$

15. _____

[6.3]   **16.** Multiply.
   a. $2.5x^3 \cdot 0.6x$ **b.** $(4.5a - 6.1)(2a - 0.5)$

16. a._____
   b._____

[6.6]   **17.** Solve and check. $6.5t - 12.8 = 13.85$

17. _____

[6.6]   **18.** Solve and check. $1.8(k - 4) = -3.5k - 4.02$

18. _____

19. a._____

b._____

$\begin{bmatrix} 6.4 \\ 6.5 \end{bmatrix}$ **19.** A circle has a radius of 12 feet. Use 3.14 for π.
  a. Find the circumference.
  b. Find the area.

20. a._____

b._____

[6.5] **20.** Find the volume. Use 3.14 for π.

a.

b.

2 cm

16.2 ft.   12 ft.

12 ft.

21. _____

[6.5] **21.** Find the area.

|← 35 ft. →|

12 ft.

22 ft.

|← 56.5 ft. →|

22. _____

[6.2] **22.** Following is a copy of Dedra's bill at a restaurant. She pays with a $20 bill. How much change should she receive?

Guest Check	
Club sandwich	7.95
Salad	2.95
Bev.	1.50
Tax	0.57

23. _____

[6.4] **23.** Bill receives the following printout of charges that he plans to pay on a 90-day same-as-cash option. He wants to split the charges into three equal payments. How much is each payment?

Description	Amount
1 TV	$385.95
2 speakers	$59.95 each
Sales tax	$30.35

24. _____

[6.6] **24.** Carlton has a 20-foot ladder that he places 5 feet from the base of his house. How high up the side of the house will the ladder reach?

25. _____

[6.6] **25.** Yolanda sells two different kinds of cakes, round-layered cakes and sheet cakes. The round-layered cakes sell for $14 and the sheet cakes sell for $12.50. In one month, she sold 12 cakes and made a total of $162. How many of each cake did she sell?

For Exercises 1–6, answer true or false.

[3.2] **1.** $4x^3 - 7$ is a binomial.

[3.6] **2.** 221 is a composite number.

[5.3] **3.** π is a rational number.

[3.5] **4.** The conjugate of $x - 9$ is $-x + 9$.

[4.2] **5.** The addition/subtraction principle of equality states that adding or subtracting the same amount on both sides of an equation will not affect its solution(s).

[6.1] **6.** $-0.118 < -0.119$

For Exercises 7 and 8, complete the rule.

[2.4] **7.** Any number other than 0 divided by 0 is _____ .

[3.3] **8.** When combining like terms we add or subtract the _____ and keep the _____ the same.

[3.5] **9.** Explain the mistake. $(x + 2)(x + 3) = x^2 + 6$

[5.4] **10.** Explain in your own words how to divide fractions.

[6.1] **11.** Write the word name for 29.6081.

[5.1] **12.** Graph $-2\frac{3}{4}$ on a number line.

[6.1] **13.** Round 2.0185 to the nearest hundredth.

[1.4] **14.** Estimate $2439 \div 49$ by rounding so that there is only one nonzero digit in each number.

[3.2] **15.** What is the degree of $3b^2 - b^4 + 9b + 4b^6 - 17$?

[3.6] **16.** Find the prime factorization of 840.

[5.5] **17.** Find the LCM of 56 and 42.

For Exercises 18–27, simplify.

[2.5] **18.** $\{9 - 4[6 + (-16)]\} \div 7$

[2.5] **19.** $(-2)^5 - 4\sqrt{16} + 9$

[5.2] **20.** $-\dfrac{48}{60}$

[5.2] **21.** $\dfrac{18x^2y}{24xy^4}$

[5.3] **22.** $-\dfrac{1}{9} \cdot 5\dfrac{1}{6}$

[6.3] **23.** $(0.5)^2$

[5.4] **24.** $-\dfrac{12n^2}{13m} \div \dfrac{-15n}{26m}$

[5.6] **25.** $4\dfrac{5}{8} + 7\dfrac{1}{2}$

[5.6] **26.** $\dfrac{x}{6} + \dfrac{3}{4}$

[6.5] **27.** $0.48 + 0.6\left(4\dfrac{1}{3}\right) + \sqrt{0.81}$

[6.5] **28.** Evaluate $4h^2 - k^3$, when $h = \dfrac{3}{4}$ and $k = -0.2$.

[6.2] 29. Combine like terms and write your answer in descending order.

$$2.6x^3 - \frac{1}{4}x^2 + 3.8 - x^3 + x^2 - \frac{1}{4}$$

[6.2] 30. Subtract.
$$\left(14m^3 - 5.7m^2 + 7\right) - \left(9.1m^3 + m^2 - 11.6\right)$$

[3.7] 31. Find the missing factor. $4x^3 \cdot \boxed{?} = -48x^5y$

[3.5] 32. Multiply. $(4y - 3)(y + 2)$

[6.3] 33. Write 7.68×10^8 in standard form, then write the word name.

[3.7] 34. Factor. $30n^5 + 15n^2 - 25n$

For Exercises 35–38, solve and check.

[6.6] 35. $y - 15.2 = 9.5$

[6.6] 36. $-0.06k = 0.408$

[5.8] 37. $\frac{2}{3}x + \frac{1}{4} = \frac{5}{6}$

[4.3] 38. $2n + 8 = 5(n - 3) + 2$

For Exercises 39–50, solve.

[2.3] 39. The temperature at sunset was reported to be 24°F. By midnight it is reported to be −8°F. What is the amount of the decrease?

[2.6] 40. The financial report for a business indicates that the total revenue for 2004 was $2,609,400 and the total costs were $1,208,500. What was the net? Did the business have a profit or loss?

[6.4] 41. Sherry has an annual salary of $23,272. What is her gross monthly salary?

[3.5] 42. a. Write an expression for the volume of the box shown.

b. Find the volume of the box if w is 6 centimeters.

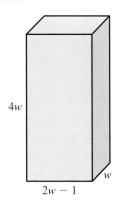

[6.5] 43. A box company shapes flat pieces of cardboard into boxes. The flat pieces have an area of 21.4 square feet. If the length is 2 feet and the width is 1.5 feet, find the height. (Use the formula $SA = 2lw + 2lh + 2wh$.)

[6.5] 44. Find the area of the shape. Use 3.14 for π.

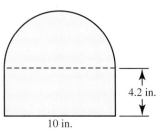

[6.4] **45.** Five minus twice the difference of x and 9 is the same as 7 more than the product of 6 and x.

[5.3] **46.** A study at a business indicates that $\frac{3}{4}$ of its employees have taken business supplies for personal use. Of those employees that have taken supplies, $\frac{2}{3}$ have been with the company more than 5 years. What fraction of all employees have taken supplies and been with the company for more than 5 years?

[5.6] **47.** A question on a survey allows for a response of yes, no, or no opinion. $\frac{1}{6}$ of the respondents said yes and $\frac{3}{5}$ said no. What fraction had no opinion?

[6.6] **48.** An entrance to an amusement park is to be an isosceles triangle. The equal sides are to be 12.5 feet longer than the base. If the perimeter must be 70 feet, then what will be the dimensions of the triangle?

[6.6] **49.** A 20-foot ladder is resting against the side of a building. The bottom of the ladder is 6 feet from the base of the building. How far is the top of the ladder from the base of the building?

[6.6] **50.** Cedrick makes two different sizes of clay bowls. He sells the larger size for $12.50, and the smaller size sells for $8.50. If he sells a total of 20 of the bowls at a fair and makes a total of $206, how many of each size did he sell? (Use a four-column table.)

Ratios, Proportions, and Measurement

> **" Alone we can do so little, together we can do so much."**
>
> —HELEN KELLER,
> ADVOCATE FOR THE BLIND

> **" Never doubt that a small group of thoughtful, committed citizens can change the world. Indeed, it is the only thing that ever has."**
>
> —MARGARET MEAD,
> ANTHROPOLOGIST

Working with a Group

One of the most rewarding ways to grow as a student is to form a study group. When you form a group, keep in mind that smaller is better, usually four people or fewer. By forming a group, you have another source for information. Group members can answer questions when you are unable to reach your instructor, or they can offer notes if you miss a class.

If you meet with your group to go over homework or prepare for a test, it is best to have completed as much of the homework or test preparation as possible on your own. Then you can go over problems together that you or other group members did not understand. You will find that explaining things to each other will strengthen your understanding of the material.

Keep in mind that your study group cannot be expected to carry you through a course. The responsibility for your success remains with you. However, the support, encouragement, and accountability you gain from a study group can be a valuable part of achieving that success.

7.1 Ratios, Probability, and Rates

OBJECTIVES

1 Write ratios in simplest form.

2 Calculate probabilities.

3 Calculate unit ratios.

4 Write rates in simplest form.

5 Use unit price to determine the better buy.

OBJECTIVE 1 Write ratios in simplest form.

In this section, we will solve problems involving **ratios.**

> **DEFINITION** **Ratio:** A comparison between two quantities using a quotient.

We usually write ratios as fractions in simplest form. When a ratio is written in English, the word *to* separates the numerator and denominator quantities. For example, if a committee consists of 12 males and 18 females, the ratio of males to females would be written as follows:

$$\text{ratio of } \textbf{males to females} = \frac{12}{18} = \frac{2}{3}$$

Note: The numerator appears to the left of the word *to* and the denominator appears to the right.

Our example suggests the following rule for translating problems involving a ratio.

> **RULE** *For two quantities, a and b, the ratio of a to b is* $\frac{a}{b}$.

Your Turn 1

Stacie measures 40 inches from her feet to her navel and 24 inches from her navel to the top of her head. Write the ratio of the smaller length to the larger length in simplest form.

Example 1 The pitch, or degree of slant, of a roof is to be 24 inches of vertical rise for every 40 inches of horizontal length. Write the ratio of vertical rise to horizontal length in simplest form.

Solution: $\text{ratio of } \textbf{vertical rise to horizontal length} = \frac{24}{40} = \frac{3}{5}$

This means that measuring out 40 inches then up 24 inches is equivalent to measuring out 5 inches then up 3 inches. The pitch of the roof is the same using either ratio.

> **CONNECTION** The ratio of the vertical rise to the horizontal length is known as the *slope* of an incline. Slope is often referred to as the ratio of *rise to run* or *rise over run.*

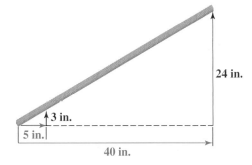

◁ **Do Your Turn 1**

A ratio can have fractions or decimals in its numerator or denominator, which can make simplifying more challenging.

Example 2 For every $2\frac{1}{2}$ rotations of the pedals on a bicycle, the back wheel rotates $3\frac{1}{4}$ times. Write the ratio of the number of pedal rotations to the number of back wheel rotations.

Answer to Your Turn 1

$\frac{3}{5}$

Solution: number of pedal rotations to number of back wheel rotations $= \dfrac{2\frac{1}{2}}{3\frac{1}{4}}$

OF INTEREST

Gears in machines operate on the principle of ratios. In an automobile, the number of revolutions of the driveshaft is converted by a gear to a number of tire revolutions. In a lower gear, the engine is turning at many more revolutions than the tire, while in a higher gear the opposite is true.

This is a complex fraction, which we learned to simplify in Chapter 5. To simplify, we divide the mixed numbers.

$$\dfrac{2\frac{1}{2}}{3\frac{1}{4}} = 2\frac{1}{2} \div 3\frac{1}{4} \qquad \text{Rewrite using a division sign.}$$

$$= \dfrac{5}{2} \div \dfrac{13}{4} \qquad \text{Write as improper fractions.}$$

$$= \dfrac{5}{\underset{1}{2}} \cdot \dfrac{\overset{2}{\cancel{4}}}{13} \qquad \text{Write as an equivalent multiplication and divide out the common factor 2.}$$

$$= \dfrac{10}{13} \qquad \text{Multiply.}$$

This means that every 10 rotations of the pedals result in 13 rotations of the back wheel.

Example 2 suggests the following procedure.

> **PROCEDURE** *To simplify a ratio of two fractions or mixed numbers, divide the fractions or mixed numbers.*

Do Your Turn 2 ▷

Your Turn 2

A recipe calls for $2\frac{1}{2}$ cups of milk and $\frac{3}{4}$ of a cup of sugar. Write the ratio of milk to sugar in simplest form.

Now let's try simplifying ratios of decimal numbers.

> **Example 3** The diameter of a piece of pipe is 0.42 meters. The length is 3.5 meters. Write the ratio of the diameter to the length in simplest form.

Solution: diameter to length $= \dfrac{0.42}{3.5}$

To write simplest form, we must clear the decimals so that the numerator and denominator are integers. In Chapter 6, we learned that we can clear decimals by multiplying by an appropriate power of 10. In Chapter 5, we learned that we write equivalent fractions by multiplying the numerator and denominator by the same number. Because 0.42 has the most decimal places, it determines the power of 10 that we use. Therefore, in this case, we multiply both the numerator and denominator by 100.

$$\dfrac{0.42}{3.5} = \dfrac{0.42(100)}{3.5(100)} = \dfrac{42}{350} = \dfrac{3}{25}$$

Example 3 suggests the following procedure.

Note: Because we multiply the numerator and denominator by the same power of 10, the decimal will move to the right the same number of places in the numerator and denominator.

Answer to Your Turn 2

$\dfrac{10}{3}$

◀ **Do Your Turn 3**

Your Turn 3

An ant weighing 0.004 grams can carry an object weighing up to 0.02 grams. Write the ratio of the object's weight to the ant's weight in simplest form.

OBJECTIVE 2 Calculate probabilities.

Ratios are used in calculating probabilities. Suppose a friend approached you with a coin and suggested the following game. If it lands heads up you win; if it lands tails up your friend wins. What is the likelihood that you will win? Because there are two possible outcomes, both of which are equally likely to occur, heads or tails, and you win with one of those two possibilities, we say you have a 1 in 2 chance of winning.

Note that we can write "1 in 2" as the fraction $\frac{1}{2}$. The fraction $\frac{1}{2}$ is the **theoretical probability** that you will win.

DEFINITION Theoretical probability of equally likely outcomes: The ratio of the number of favorable outcomes to the total number of possible outcomes.

$$\text{the probability of a coin landing heads up} = \frac{1}{2} \quad \begin{matrix} \leftarrow \text{1 side is heads} \\ \text{out of} \\ \leftarrow \text{2 possible outcomes} \end{matrix}$$

Notice that your friend has an equal chance of winning because there is 1 tail out of 2 possible outcomes.

$$\text{the probability of a coin landing tails up} = \frac{1}{2}$$

PROCEDURE *To write a probability:*
1. Write the number of favorable outcomes in the numerator.
2. Write the total number of possible outcomes in the denominator.
3. Simplify.

Note that this does not mean that if your friend wins one time, you are guaranteed to win the next time. In fact, your friend could have a streak of several wins before you win, if at all. On any given day, the number of times you win may be more or less than $\frac{1}{2}$ of the time. However, over a lifetime, if you were to look at how many times you won, it would be close to $\frac{1}{2}$ of the time. Taken to an extreme, if you could play the game forever, you would win exactly $\frac{1}{2}$ of the time.

Answer to Your Turn 3

$\dfrac{5}{1}$

Some of the problems we will explore involve things like dice and cards. A six-sided game die is a cube with dots numbering each of the six sides.

A standard deck of cards contains 52 cards with four suits of 13 cards. There are two red suits, diamonds and hearts, and two black suits, spades and clubs. The 13 cards are as follows: Ace, 2, 3, 4, 5, 6, 7, 8, 9, 10, Jack, Queen, King. The Jack, Queen, and King are referred to as *face cards*.

Example 4   Write each probability in simplest form.

a. What is the probability of rolling a 2 on a six-sided die?

Solution: Write the ratio of the number of 2's to the total number of outcomes on a six-sided die. There is only one 2 on a six-sided die.

$$P = \frac{1}{6} \begin{array}{l}\leftarrow \text{ number of 2's on a six-sided die} \\ \\ \leftarrow \text{ number of possible outcomes}\end{array}$$

b. What is the probability of tossing a 7 on a six-sided die?

Solution: Write the ratio of the number of 7's to the total number of outcomes on a six-sided die. There are no 7's on a normal six-sided die.

$$P = \frac{0}{6} = 0$$

c. What is the probability of tossing a 1, 2, 3, 4, 5, or 6 on a six-sided die?

Solution: There are six favorable outcomes out of six possible outcomes.

$$P = \frac{6}{6} = 1$$

d. What is the probability of drawing a red King out of a standard deck of 52 cards?

Solution: Write the ratio of the number of red Kings to the total number of cards in the deck. There are two red Kings, the King of diamonds and King of hearts, in a standard deck of 52 cards.

$$P = \frac{2}{52} = \frac{1}{26}$$

Do Your Turn 4 ▶

Your Turn 4

Write each probability in simplest form.

a. What is the probability of selecting a 2 from a standard deck of cards?

b. Suppose 28 males and 30 females put their names in a hat for a drawing. If the winner's name is selected from the hat at random, what is the probability that the winner will be male?

Answers to Your Turn 4
a. $\frac{1}{13}$ b. $\frac{14}{29}$

OBJECTIVE 3 Calculate unit ratios.

Sometimes it is more useful to write a ratio as a **unit ratio**.

DEFINITION   Unit ratio: A ratio in which the denominator is 1.

For example, debt-to-income ratios are more informative when expressed as unit ratios. For example, suppose a person pays $200 toward debt out of every $1000 of income.

$$\text{debt-to-income ratio} = \frac{200}{1000} = \frac{1}{5}$$

The simplified ratio indicates that the person has $1 of debt for every $5 of income. Writing the ratio as a unit ratio will tell us how much debt the person pays for every $1 of income. Because a unit ratio has a denominator of 1, we divide the denominator into the numerator.

$$\text{debt-to-income ratio} = \frac{200}{1000} = \frac{1}{5} = \frac{0.2}{1}$$

The unit ratio $\frac{0.2}{1}$, or 0.2, means the person has a debt of $0.2 for every $1 of income.

PROCEDURE *To calculate a unit ratio, divide the denominator into the numerator.*

Your Turn 5

A family has a total debt of $7842 and a gross income of $34,240. Calculate the unit ratio of debt to income and interpret the results.

Example 5 A family has total debt of $5865 and a gross annual income of $45,200. Calculate the unit ratio of debt-to-income and interpret the results.

Understand: We are to calculate a unit ratio of debt to income.

Plan: Write the ratio of debt to income, then divide to calculate the unit ratio.

Execute: debt-to-income ratio $= \dfrac{5865}{45,200} \approx \dfrac{0.13}{1}$ or 0.13

Answer: The family has a debt-to-income ratio of approximately 0.13, which means that it owes about $0.13 for every $1 of income.

Check: Verify the calculation by multiplying 0.13 by 45,200 to get approximately 5865. You will not get exactly 5865 because 0.13 was an approximation.

◀ **Do Your Turn 5**

Your Turn 6

A certain college has a total of 14,280 students and 788 faculty. Write a unit ratio of students to faculty.

Example 6 A certain college has 9752 students and 460 faculty members. Write the student-to-faculty ratio as a unit ratio.

Understand: We are to calculate a unit ratio of students to faculty.

Plan: Write the ratio of students to faculty, then divide to calculate the unit ratio.

Execute: students-to-faculty ratio $= \dfrac{9752}{460} = \dfrac{21.2}{1}$ or 21.2

Answer: The student-to-faculty ratio of 21.2 means that there are about 21 students for every faculty member at the college.

Check: Verify the calculation by multiplying 21.2 by 460 to get approximately 9752.

◀ **Do Your Turn 6**

OBJECTIVE 4 Write rates in simplest form.

We can now build on what we've learned about unit ratios to discuss **rates.**

DEFINITION Rate: A unit ratio comparing two different measurements.

One example of a rate is velocity or speed, which is the ratio of distance to time. An *average* velocity or rate is the ratio of the total distance traveled to the total time of the trip.

Answer to Your Turn 5
≈ 0.23; This ratio means that the family owes about $0.23 for every $1 of income.

Answer to Your Turn 6
18.1

$$r = \frac{d}{t}$$

◀ **Note:** Units for velocity always have a distance unit over a time unit. For example, miles per hour is written mi./hr. and feet per second is written ft./sec.

Example 7 An athlete is clocked running the 40-yard dash in 4.4 seconds. What is the athlete's average rate in yards per second?

Understand: We are to find the rate in yards per second. The unit yards per second indicates writing a ratio of the number of yards to the number of seconds.

Plan: Using r to represent the rate, we will write a ratio of the number of yards to the number of seconds then divide.

Execute: $r = \dfrac{40 \text{ yd.}}{4.4 \text{ sec.}} = 9.\overline{09}$ yd./sec. ≈ 9.1 yd./sec.

▲
Note: The / is read *per* so that the unit yd./sec. is read *yards per second*.

Answer: The athlete ran at an average rate of approximately 9.1 yd./sec.

Check: Verify that running at a rate of 9.1 yards per second would cover 40 yards in 4.4 seconds. We can use the formula $d = rt$.

$d = (9.1)(4.4) = 40.04$ yd. ◄ **Note:** Our answer is not exactly 40 because we used the approximate value 9.1 for the rate. 40.04 is reasonably close considering our rounding of $9.\overline{09}$.

Do Your Turn 7 ▶

Another common rate is unit price, which is a ratio of price to quantity.

$$u = \frac{p}{q}$$

◄ Units written for unit price always have a money unit over a quantity unit. For example, $ per pound is written $/lb. or cents per ounce is written ¢/oz.

Example 8 A 15-ounce box of cereal costs $2.39. What is the unit price in cents per ounce?

Understand: We are asked to find the unit price in cents per ounce. Because $2.39 is in terms of dollars, we must express the amount in cents. Because there are 100 cents in every dollar, we multiply 2.39 by 100. In Chapter 6, we learned that multiplying a decimal number by 100 moves the decimal point to the right 2 places, so $2.39 = 239$¢.

Plan: Using u to represent the unit price, we will write the ratio of cents to ounces, then divide.

Execute: $u = \dfrac{239 \text{¢}}{15 \text{ oz.}} = 15.9\overline{3}$¢/oz. ≈ 15.93¢/oz.

Answer: The unit price is approximately 15.93¢/oz.

Check: Verify that purchasing 15 ounces at a unit price of 15.93¢ per ounce equals a total price of 239¢. Use the formula $p = uq$.

$p = (15.93)(15) = 238.95$ ◄ **Note:** Our answer is not exactly 239 because we used the approximate value 15.93 for the rate. 238.95 is reasonably close to 239 given our rounding of 15.93.

Do Your Turn 8 ▶

Your Turn 7

Solve.

a. A cyclist rides 20 miles in 48 minutes. What is the cyclist's average rate in miles per minute?

b. Sasha drives 45 miles in $\frac{3}{4}$ hour. After a break she drives another 60 miles in $\frac{4}{5}$ hour. What was her average rate for the entire trip in miles per hour?

Your Turn 8

Solve.

a. A 50-ounce bottle of detergent costs $3.58. What is the unit price in cents per ounce?

b. 4.5 pounds of chicken cost $5.22. What is the unit price in dollars per pound?

Answers to Your Turn 7
a. ≈ 0.42 mi./min.
b. 67.7 mi./hr.

Answers to Your Turn 8
a. 7.16¢/oz. b. $1.16/lb.

OBJECTIVE 5 Use unit price to determine the better buy.

As consumers, we are often faced with a choice between two different sizes of the same item. Unit prices can be used to determine which of two different-size quantities is a better buy.

Your Turn 9

Which is the better buy?

a. a 50-ounce bottle of detergent for $3.58

or

a 100-ounce bottle of detergent for $6.19

b. a 10.5-ounce can of soup for 79¢

or

a 16-ounce can of soup for $1.29

Example 9 Which is the better buy?

15-ounce box of raisin bran for $2.45   or   20-ounce box of raisin bran for $3.05

Understand: We must determine which of the two different sizes is the better buy. We are given the price for each box and the quantity in each box.

Plan: Let F represent the unit price of the 15-ounce box and T represent the unit price of the 20-ounce box. Calculate the unit price for each box, then compare unit prices. The box with the smaller unit price is the better buy.

Execute: **15-oz. box**   **20-oz. box**

$$F = \frac{\$2.45}{15 \text{ oz.}} = \$0.16\overline{3}/\text{oz.} \qquad T = \frac{\$3.05}{20 \text{ oz.}} = \$0.1525/\text{oz.}$$

Answer: Because $0.1525/oz. is the smaller unit price, the 20-ounce box is the better buy. In other words, we would pay less for each ounce of raisin bran by buying the 20-ounce box than we would pay if we bought the 15-ounce box of the same raisin bran.

Check: Verify the calculations by reversing the process. We will leave this to the reader.

> **DISCUSSION**   Under what circumstances might it actually be better to purchase the smaller quantity even if it is not the better buy?

◀ **Do Your Turn 9**

Your Turn 10

Which is the better buy?

a. two 16-ounce cans of beans, for $1.59

or

a 28-ounce can of beans for $1.19

b. three 2-pound bags of store-brand fries for $4.00

or

a 2-pound bag of name-brand fries for $1.30

Example 10 Which is the better buy?

two 10.5-ounce containers of   or   a 32-ounce container of
name-brand yogurt for $1.19   store-brand yogurt for $1.95

Understand: We must determine which is the better buy. Note that buying 2 of the 10.5-ounce yogurt means that we are buying $2(10.5) = 21$ ounces of yogurt for $1.19.

Plan: Let N represent the unit price of the name-brand yogurt and S represent the unit price of the store-brand yogurt. Calculate the unit prices, then compare.

Execute: **Name brand**   **Store brand**

$$N = \frac{\$1.19}{21 \text{ oz.}} = \$0.05\overline{6}/\text{oz.} \qquad S = \frac{\$1.95}{32 \text{ oz.}} \approx \$0.061/\text{oz.}$$

Answer: Because $0.05\overline{6}/oz. is the smaller unit price, buying 2 of the 10.5-ounce name-brand yogurt is a better buy than buying 32 ounces of the store brand.

Check: Verify the calculations by reversing the process. We will leave this to the reader.

◀ **Do Your Turn 10**

Answers to Your Turn 9
a. 100-oz. bottle **b.** 10.5-oz. can

Answers to Your Turn 10
a. 28-oz. can
b. 2-lb. bag of name-brand fries

For
Extra
Help

Videotape
DVT

Tutor
Center
Addison-Wesley
Tutor Center

Math XL
Math XL

MyMathLab

Student Solutions
Manual

1. What is a ratio?

2. What is a unit ratio?

3. How do you calculate a unit ratio?

4. What is a rate?

5. What is a unit price?

6. Brand A of a certain product has a unit price of $0.45 per ounce, whereas brand B has a unit price of $0.42 per ounce. Which is the better buy? Why?

For Exercises 7–18, write each ratio in simplest form.

7. There are 62 males and 70 females attending a conference.

   a. What is the ratio of males to total attendance?

   b. What is the ratio of females to total attendance?

   c. What is the ratio of males to females?

   d. What is the ratio of females to males?

8. The following table shows the number of mature trees in a region of forest.

Tree Species	Number of Mature Trees
Pine	488
Maple	264
Oak	114
Other	295

   a. What is the ratio of pine trees to maple trees?

   b. What is the ratio of maple trees to oak trees?

   c. What is the ratio of pine trees to total trees?

   d. What is the ratio of oak trees to total trees?

9. Calcium nitride is a chemical compound that is composed of 3 calcium atoms and 2 nitrogen atoms.

   a. What is the ratio of calcium to nitrogen?

   b. What is the ratio of nitrogen to calcium?

   c. What is the ratio of calcium to total atoms in the compound?

   d. What is the ratio of nitrogen to total atoms in the compound?

LEARNING STRATEGY

When quizzing someone or watching someone work a problem, be careful that you don't offer tips or hints while they speak or work. Let them say or do the entire procedure, even if you note a mistake. Correct only after they've completed the procedure. Remember, nobody will be there to give hints or clues that something is wrong during a test.

OF INTEREST

Chemical compounds are written with subscripts that indicate the number of atoms of each element in the compound. Calcium nitride, for example, is written as Ca_3N_2.

10. Hydrogen phosphate is a chemical compound composed of 3 atoms of hydrogen, 1 atom of phosphorous, and 4 atoms of oxygen.

   a. What is the ratio of hydrogen to oxygen?

   b. What is the ratio of phosphorous to oxygen?

   c. What is the ratio of hydrogen to total atoms in the compound?

   d. What is the ratio of oxygen to total atoms in the compound?

OF INTEREST

Hydrogen phosphate is written as H_3PO_4. Note that there is no subscript written for P. The subscript is an understood 1 just as when there is no exponent written for a particular base in exponential form.

11. A roof is to have a pitch, or slant, of 14 inches vertically to 16 inches horizontally. Write the ratio of vertical distance to horizontal distance in simplest form.

12. The steps on a staircase rise 8 inches and are 10 inches wide. Write the ratio of the rise to the width of each step in simplest form.

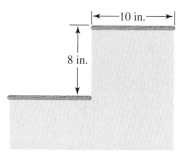

13. The back wheel of a bicycle rotates $3\frac{1}{2}$ times with $2\frac{1}{4}$ rotations of the pedals. Write the ratio of back wheel rotations to pedal rotations in simplest form.

14. At 20 miles per hour in second gear, a car engine is running about 3000 revolutions per minute (rpm). The drive tire is rotating about 260 revolutions per minute. What is the ratio of the engine revolutions per minute to the tire revolutions per minute?

15. In one week, Nina receives 7 bills, 3 letters, 17 advertisements, and 3 credit card offers in her mail.

   a. What is the ratio of bills to total mail?

   b. What is the ratio of advertisements to total mail?

   c. What is the ratio of letters to credit card offers?

   d. What is the ratio of bills and letters to advertisements and credit card offers?

16. Nutrition facts of a cereal bar are shown in the table to the right.

   a. What is the ratio of fiber to total carbohydrates?

   b. What is the ratio of protein to total fat?

   c. What is the ratio of total carbohydrates to the weight of the bar?

   d. What is the ratio of saturated fat to total fat?

Total Bar Weight	37 g
Total Fat	3 g
Saturated Fat	0.5 g
Total Carbohydrates	27 g
Fiber	1 g
Sugars	13 g
Protein	2 g

17. According to the U.S. Bureau of the Census, the populations (in millions) of New England states in 2002 were as shown in the table to the right.

   a. What is the ratio of the Massachusetts population to the New York population?

   b. What is the ratio of the Vermont population to the Rhode Island population?

   c. What is the ratio of the Maine population to the New York population?

   d. If the U.S. population was 280.5 million in 2002, what was the ratio of the Connecticut population to the U.S. population?

State	Population (millions)
Connecticut	3.4
Maine	1.3
Massachusetts	6.2
New Hampshire	1.2
New York	18.6
Rhode Island	1.0
Vermont	0.6

18. The bridge and nut on a standard guitar are set so that the length of each string between them is $25\frac{1}{2}$ inches. The seventh fret is placed on the neck of the guitar exactly 17 inches from the bridge.

   a. What is the ratio of the distance between the bridge and the seventh fret to the total length of the strings?

   b. What is the ratio of the distance between the bridge and the seventh fret to the distance between the seventh fret and the nut?

   c. What is the ratio of the distance between the seventh fret and the nut to the total length of the string?

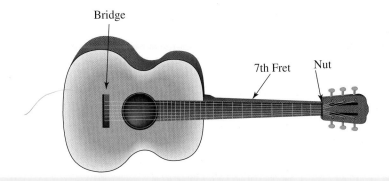

Bridge

7th Fret   Nut

OF INTEREST

Musical tones are created on stringed instruments by changing the string length. Shortening the string length causes the pitch to get higher, which with guitars, violins, and basses is achieved by placing a finger on a string. By placing a finger at the seventh fret on a guitar, the string length is shortened from 25.5 inches to 17 inches, which creates a musical tone that is higher in pitch than the tone created by plucking the string with no finger placement (open). In fact, the musical tone created at the seventh fret is a musical fifth from the open string tone. Musical notes are named after the letters of the alphabet, so if the tone of an open string is named A then the note at the seventh fret is named E, which is the fifth letter from A.

For Exercises 19–28, calculate the probability.

19. What is the probability of selecting a King in a standard deck of cards?

20. What is the probability of selecting a red Ace in a standard deck of cards?

21. What is the probability of selecting an Ace or a King in a standard deck of cards?

22. What is the probability of selecting a Queen or a Jack in a standard deck of cards?

23. What is the probability of tossing a 1 or 6 on a six-sided game die?

24. What is the probability of tossing a 1, 2, or 3 on a six-sided game die?

25. What is the probability of tossing a 7 on a six-sided game die?

26. What is the probability of selecting a 1 in a standard deck of cards?

27. Suppose 12 males and 14 females put their names in a hat for a drawing. What is the probability that the winner will be female?

28. Miwa sends 14 entries in a random drawing contest. If the company receives 894,264 entries, what is the probability that Miwa will win?

29. The following table shows the number of candies of each color in a bag of M&M candies.

Green	42
Red	28
Brown	30
Blue	25

   a. What is the probability that a candy chosen at random will be blue?

   b. What is the probability that a candy chosen at random will be red?

   c. What is the probability that a candy chosen at random will be green or brown?

   d. What is the probability that a candy chosen at random will be a color other than red?

30. The following table shows the home states for people attending a conference.

Georgia	46
South Carolina	42
North Carolina	38
Tennessee	18
Florida	14

   a. What is the probability that a person chosen at random is from North Carolina?

   b. What is the probability that a person chosen at random is from Florida?

   c. What is the probability that a person chosen at random is from Georgia or South Carolina?

   d. What is the probability that a person chosen at random is from a state other than Tennessee?

For Exercises 31–38, calculate each ratio as a unit ratio and interpret the result.

31. A family has a total debt of $4798 and a gross income of $29,850. What is its debt-to-income ratio?

OF INTEREST

Debt-to-income ratios are one of the determining factors in qualifying for a loan. More specifically, lenders consider the ratio of monthly debt to gross monthly income. This ratio is called the *back-end ratio*. We'll say more about this in Section 7.6.

32. A family has a total debt of $9452 and a gross income of $42,325. What is its debt-to-income ratio?

33. A family is seeking a mortgage to purchase a new home. If it is approved for the mortgage, the new payment will be $945. If the gross monthly income is $3650, what is the payment-to-income ratio?

34. A family pays $550 in rent. If its gross monthly income is $2240, what is the rent-to-income ratio?

OF INTEREST

The ratio of mortgage payment to gross monthly income is another determining factor in qualifying for a loan. This ratio is called the *front-end* ratio.

35. At a certain college campus there are 12,480 students and 740 faculty. What is the student-to-faculty ratio?

36. A certain college has 4212 students and 310 faculty. What is the student-to-faculty ratio?

37. At the close of the stock market on a particular day, each share of a computer company's stock was selling for $54\frac{1}{8}$. The annual earnings per share were $4.48. What is the ratio of selling price to annual earnings?

38. At the close of the stock market on a particular day, each share of a textile company's stock was selling for $12\frac{5}{16}$. The annual earnings per share were $3.52. What is the ratio of selling price to annual earnings?

OF INTEREST

The ratio of selling price to annual earnings for a stock is call its *price-to-earnings ratio*, or *P/E ratio* for short. Companies with higher (15+) P/E ratios generally show a trend of increasing profits. Companies with lower (less than 5) P/E ratios are considered high-risk investments.

For Exercises 39–44, calculate the rate.

39. Tara drove 306.9 miles in 4.5 hours. What was her average rate in miles per hour?

40. Li drove 130.2 kilometers in $2\frac{1}{3}$ hours. What was Li's average rate in kilometers per hour?

41. A long-distance phone call lasting 23 minutes costs $2.76. What is the unit price in cents per minute?

42. 16.8 gallons of gas cost $32.76. What is the unit price in dollars per gallon?

43. Gina paid $3.75 for $\frac{1}{2}$ pound of shrimp. What was the unit price in dollars per pound?

44. A 32-ounce bag of frozen french fries costs $1.50. What is the unit price in cents per ounce?

For Exercises 45–50, determine which is the better buy.

45. a 10.5-ounce can of soup for $0.89   or
a 16-ounce can of soup for $1.19

46. a 15-ounce can of fruit cocktail for $1.09   or
a 20-ounce can of fruit cocktail for $1.49

47. a bag containing 16 diapers for $4.89   or
$7.95 for a bag containing 24 diapers

48. a 28-ounce box of rice for $1.59   or
$1.79 for a 32-ounce box of rice

49. two 15.5-ounce boxes of store-brand raisin bran $5.00   or
a 20-ounce box of name-brand raisin bran for $3.45

50. three 15-ounce cans of mixed vegetables for $1.98   or
two 12.5-ounce cans of mixed vegetables for $1.19

Review Exercises

For Exercises 1–3, evaluate.

[6.3]   **1.** $(9.6)(12.5)$

[5.3]   **2.** $5\frac{1}{4} \cdot \frac{6}{7}$

[6.5]   **3.** $4.5\left(2\frac{3}{4}\right)$

[6.4]   **4.** Solve and check. $10.4x = 89.44$

[5.8]   **5.** Solve and check. $\frac{2}{3}y = 5$

7.2 Proportions

OBJECTIVE 1 Determine whether two ratios are proportional.

Now that we have learned about ratios in Section 7.1, we can solve problems involving equal ratios in a **proportion.**

DEFINITION **Proportion:** An equation in the form $\dfrac{a}{b} = \dfrac{c}{d}$, where $b \neq 0$ and $d \neq 0$.

In Chapter 5, we learned that we can write equivalent fractions by multiplying or dividing both the numerator and denominator by the same number. For example, $\dfrac{5}{8}$ and $\dfrac{10}{16}$ are equivalent because $\dfrac{5}{8} = \dfrac{5 \cdot 2}{8 \cdot 2} = \dfrac{10}{16}$, or $\dfrac{10}{16} = \dfrac{10 \div 2}{16 \div 2} = \dfrac{5}{8}$. We also learned a trick for determining whether two fractions are equivalent: Their cross products are equal.

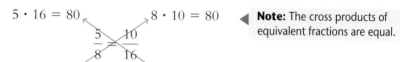

$$5 \cdot 16 = 80 \qquad 8 \cdot 10 = 80$$
$$\dfrac{5}{8} \,\diagup\!\!\!\!\!\diagdown\, \dfrac{10}{16}$$

◀ **Note:** The cross products of equivalent fractions are equal.

We can show that the cross products are equal for any proportion. We begin with the general form of a proportion from the definition: $\dfrac{a}{b} = \dfrac{c}{d}$, where $b \neq 0$ and $d \neq 0$.

We can use the multiplication principle of equality to clear the fractions by multiplying both sides of the proportion by the LCD, which is bd.

$$\dfrac{\overset{1}{\cancel{bd}}}{1} \cdot \dfrac{a}{\cancel{b}} = \dfrac{\overset{1}{\cancel{bd}}}{1} \cdot \dfrac{c}{\cancel{d}}$$

$$ad = bc$$

Note that ad and bc are the cross products.

$$ad \qquad bc$$
$$\dfrac{a}{b} \,\diagup\!\!\!\!\!\diagdown\, \dfrac{c}{d}$$

We can write the following rule for proportions.

RULE *If two ratios are proportional, then their cross products are equal.*

In math language: If $\dfrac{a}{b} = \dfrac{c}{d}$, where $b \neq 0$ and $d \neq 0$, then $ad = bc$.

Example 1 Determine whether the ratios are proportional.

a. $\dfrac{3.5}{8} \overset{?}{=} \dfrac{1.4}{3.2}$

Solution: If the cross products are equal, then the ratios are proportional.

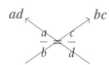

$$(3.2)(3.5) = 11.2 \qquad (8)(1.4) = 11.2$$
$$\dfrac{3.5}{8} \,\overset{?}{\underset{\diagdown}{=}}\, \dfrac{1.4}{3.2}$$

Because the cross products are equal, these ratios are proportional.

b. $\dfrac{\dfrac{3}{4}}{\dfrac{3}{10}} \overset{?}{=} \dfrac{\dfrac{5}{12}}{\dfrac{1}{2}}$

Solution:

$$\dfrac{1}{2} \cdot \dfrac{3}{4} = \dfrac{3}{8} \qquad \dfrac{\overset{1}{\cancel{3}}}{\underset{2}{\cancel{10}}} \cdot \dfrac{\overset{1}{\cancel{5}}}{\underset{4}{\cancel{12}}} = \dfrac{1}{8}$$

$$\dfrac{\dfrac{3}{4}}{\dfrac{3}{10}} \overset{?}{=} \dfrac{\dfrac{5}{12}}{\dfrac{1}{2}}$$

Because the cross products are not equal, these ratios are not proportional.

Do Your Turn 1 ▷

Your Turn 1

Determine whether the ratios are proportional.

a. $\dfrac{14}{16} \overset{?}{=} \dfrac{18}{20}$

b. $\dfrac{3.6}{5} \overset{?}{=} \dfrac{18}{25}$

c. $\dfrac{3\dfrac{1}{4}}{16} \overset{?}{=} \dfrac{6}{29\dfrac{7}{13}}$

OBJECTIVE 2 Solve for a missing number in a proportion.

Cross products can be used to solve for a missing number in a proportion.

PROCEDURE *To solve a proportion using cross products:*
1. Calculate the cross products.
2. Set the cross products equal to one another.
3. Use the multiplication/division principle of equality to isolate the variable.

LEARNING STRATEGY

If you are a visual learner, draw the arrows as we have done in Example 2 to help visualize the pattern of cross products.

Example 2   Solve. $\dfrac{5}{12} = \dfrac{x}{8}$

Solution: Because this is a proportion, the cross products must be equal.

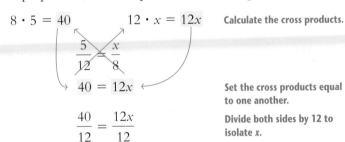

$8 \cdot 5 = 40 \qquad 12 \cdot x = 12x$ **Calculate the cross products.**

$$\dfrac{5}{12} = \dfrac{x}{8}$$

$$40 = 12x \qquad \text{Set the cross products equal to one another.}$$

$$\dfrac{40}{12} = \dfrac{12x}{12} \qquad \text{Divide both sides by 12 to isolate } x.$$

$$\dfrac{10}{3} = x$$

$$3\dfrac{1}{3} = x$$

Note: We could have expressed the answer as $3.\overline{3}$.

DISCUSSION
How can we check the solution?

Your Turn 2

Solve.

a. $\dfrac{5}{9} = \dfrac{n}{27}$

b. $\dfrac{y}{16} = \dfrac{7}{10}$

Do Your Turn 2 ▷

Answers to Your Turn 1
a. no
b. yes
c. yes

Answers to Your Turn 2
a. $n = 15$ b. $y = 11.2$

Solve.

a. $\dfrac{4\frac{2}{3}}{m} = \dfrac{\frac{3}{4}}{-9}$

b. $\dfrac{-6.5}{y} = \dfrac{-13}{14}$

Example 3 Solve. $\dfrac{2\frac{1}{4}}{8} = \dfrac{-9}{b}$

Solution: Because this is a proportion, the cross products must be equal.

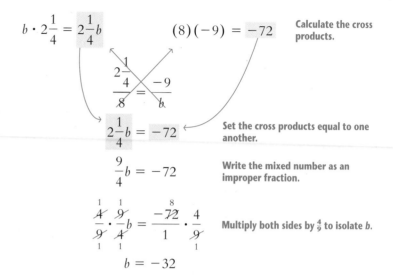

$$b \cdot 2\frac{1}{4} = 2\frac{1}{4}b \qquad\qquad (8)(-9) = -72 \qquad \text{Calculate the cross products.}$$

$$2\frac{1}{4}b = -72 \qquad \text{Set the cross products equal to one another.}$$

$$\frac{9}{4}b = -72 \qquad \text{Write the mixed number as an improper fraction.}$$

$$\overset{1}{\underset{1}{\cancel{\frac{4}{9}}}} \cdot \overset{1}{\underset{1}{\cancel{\frac{9}{4}}}}b = \frac{\overset{8}{\cancel{-72}}}{1} \cdot \frac{4}{\underset{1}{\cancel{9}}} \qquad \text{Multiply both sides by } \frac{4}{9} \text{ to isolate } b.$$

$$b = -32$$

◀ **Do Your Turn 3**

OBJECTIVE 3 Solve proportion problems.

In the typical proportion problem, you will be given a ratio. You will then be asked to upsize or downsize from that ratio. The key to solving proportion problems is identifying the two equivalent ratios and then equating them.

Example 4 Bobbie can read 40 pages in 50 minutes. How long will it take her to read 220 pages?

Understand: We must calculate the time it takes Bobbie to read 220 pages. We are given a rate of 40 pages in 50 minutes.

$$\text{A rate of 40 pages in 50 minutes} = \frac{40 \text{ pages}}{50 \text{ min.}}.$$

Let *x* represent the unknown number of minutes.

If we translate the rate with the unknown minutes in the same way, we have

$$\text{A rate of 220 pages in } x \text{ minutes} = \frac{220 \text{ pages}}{x \text{ min.}}.$$

Plan: Because the rates are equivalent, we can equate them in a proportion. Then solve.

Execute:
$$\frac{40 \text{ pages}}{50 \text{ min.}} = \frac{220 \text{ pages}}{x \text{ min.}}$$

$$x \cdot 40 = 40x \qquad 50 \cdot 220 = 11{,}000$$

$$\frac{40}{50} \times \frac{220}{x}$$

$$40x = 11{,}000$$

$$\frac{40x}{40} = \frac{11{,}000}{40}$$

$$x = 275$$

CONNECTION When setting up the proportion, write the given rate any way you wish. Then write the rate with the unknown in logical correspondence to the given rate. One way to ensure logical correspondence is to match the units straight across.

Notice that the units in the numerators match and the units in the denominators match.

$$\frac{40 \text{ pages}}{50 \text{ min.}} = \frac{220 \text{ pages}}{x \text{ min.}}$$

Answer: It will take Bobbie 275 minutes to read 220 pages.

Check: Do a quick estimate. Bobbie reads 40 pages in 50 minutes. 220 pages is a little over 5 times the number of pages, therefore, it ought to take her a little over 5 times the amount of time. $5 \cdot 50 = 250$ minutes, so a little more than that would make 275 minutes a very reasonable answer.

If you set up the proportion incorrectly, you will get an unreasonable answer. Checking by estimation is a quick way to identify unreasonable answers.

For example, suppose we had set up Example 4 this way:

$$\frac{40 \text{ pages}}{50 \text{ min.}} = \frac{x \text{ min.}}{220 \text{ pages}}$$

WARNING This setup is incorrect because the numerators and denominators do not correspond.

Solving this incorrect setup, we get

$$40 \cdot 220 = 8800 \qquad 50 \cdot x = 50x$$

$$\frac{40}{50} \diagdown \frac{x}{220}$$

$$8800 = 50x$$

$$\frac{8800}{50} = \frac{50x}{50}$$

$$176 = x$$

WARNING 176 minutes is unreasonable because it is a little over 3 times the amount of time ($3 \cdot 50 = 150$ minues), while she is reading more than 5 times the number of pages ($5 \cdot 40 = 200$ pages).

There are many correct ways to set up a proportion for a particular problem. Following are a few other correct ways we could have set up the proportion for Example 4. Note that they all have logical correspondence in the rates.

$$\frac{50 \text{ min.}}{40 \text{ pages}} = \frac{x \text{ min.}}{220 \text{ pages}} \qquad \frac{40 \text{ pages}}{220 \text{ pages}} = \frac{50 \text{ min.}}{x \text{ min.}} \qquad \frac{50 \text{ min.}}{x \text{ min.}} = \frac{40 \text{ pages}}{220 \text{ pages}}$$

Think:	Think of this version as an analogy:	This version is a rearrangement of the previous version:
In 50 min. in x min. she reads so she reads 40 pages, 220 pages.	If 40 pages in 50 min., then 220 pages in x min.	If 50 min. yields 40 pages, then x min. yields 220 pages.

PROCEDURE *To solve proportion problems:*
1. Set up the given ratio any way you wish.
2. Set the given ratio equal to the other ratio with the unknown so that the numerators and denominators correspond logically.
3. Solve using cross products.

Do Your Turn 4 ▶

Your Turn 4

Solve.

a. Erica drove 303.8 miles using 12.4 gallons of gasoline. At this rate, how much gasoline would it take for her to drive 1000 miles?

DISCUSSION How did she determine the 303.8 miles and the 12.4 gallons?

b. Xion can mow a 2500-square-foot lawn in 20 minutes. At this rate, how long will it take him to mow a 12,000-square-foot lawn?

Answers to Your Turn 4
a. 40.8 gal. **b.** 96 min.

Consider triangles *ABC* and *DEF*.

CONNECTION In Section 4.5, we learned that congruent angles have the same measure.

Though the triangles are different in size, their corresponding angle measurements are congruent. In symbols, we write: $\angle B \cong \angle E$, $\angle A \cong \angle D$, and $\angle C \cong \angle F$.

Besides having all angles congruent, notice that the sides of triangle *DEF* are twice as long as the corresponding sides of triangle *ABC*. Or, we could say that the sides of triangle *ABC* are all half as long as the corresponding sides of triangle *DEF*. In other words, the corresponding side lengths are proportional.

$$\text{triangle } DEF \rightarrow \atop \text{triangle } ABC \rightarrow \quad \frac{6}{3} = \frac{10}{5} = \frac{8}{4} = \frac{2}{1}$$

$$\text{triangle } ABC \rightarrow \atop \text{triangle } DEF \rightarrow \quad \frac{3}{6} = \frac{5}{10} = \frac{4}{8} = \frac{1}{2}$$

Because the corresponding angles in the two triangles are congruent and the side lengths are proportional, we say these are **similar figures.**

DEFINITION **Similar figures:** Figures with congruent angles and proportional side lengths.

Example 5 The triangles shown are similar. Find the missing length.

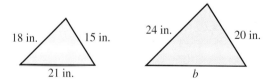

Understand: Because the triangles are similar, the lengths of the corresponding sides are proportional. We can write ratios of the lengths in the smaller triangle to the corresponding lengths in the larger triangle this way:

$$\text{smaller triangle} \rightarrow \atop \text{larger triangle} \rightarrow \quad \frac{18}{24} = \frac{15}{20} = \frac{21}{b}$$

Plan: Write a proportion, then solve. We can use either $\frac{18}{24}$ or $\frac{15}{20}$ as the given ratio in our proportion.

Execute:
$$\frac{18}{24} = \frac{21}{b} \qquad \text{or} \qquad \frac{15}{20} = \frac{21}{b}$$

$$18 \cdot b = 18b \quad \overset{18}{\underset{24}{\diagup}}\overset{21}{\underset{b}{\diagdown}} \quad 24 \cdot 21 = 504 \qquad 15 \cdot b = 15b \quad \overset{15}{\underset{20}{\diagup}}\overset{21}{\underset{b}{\diagdown}} \quad 20 \cdot 21 = 420$$

$$18b = 504 \qquad\qquad 15b = 420$$

$$\frac{18b}{18} = \frac{504}{18} \qquad\qquad \frac{15b}{15} = \frac{420}{15}$$

$$b = 28 \qquad\qquad\qquad b = 28$$

Answer: The unknown length is 28 in.

Check: Verify that 28 is in the same ratio using a different pair of corresponding sides. In other words, if you used $\frac{18}{24}$ as the given ratio to solve for b, then use $\frac{15}{20}$ as a check.

> **CONNECTION** There are many correct ways to write the proportions. Here is another correct way:
>
> $$\begin{array}{l}\text{the upper left side} \rightarrow \\ \text{to} \\ \text{the base of each triangle} \rightarrow\end{array} \frac{18}{21} = \frac{24}{b}$$
>
> Here we relate the corresponding sides across instead of in the same ratio.

Do Your Turn 5 ▶

Sometimes, more than one length is missing in problems involving similar figures. In order to solve these problems, we must be given the length of at least one pair of corresponding sides.

Example 6 The figures shown are similar. Find the missing lengths.

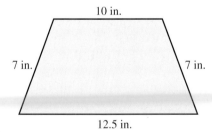

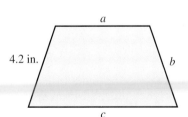

Understand: Since the figures are similar, the ratios of the lengths of the corresponding sides are proportional. The corresponding sides can be shown as follows:

$$\begin{array}{l}\text{larger trapezoid} \rightarrow \\ \text{to} \\ \text{smaller trapezoid} \rightarrow\end{array} \frac{7}{4.2} = \frac{10}{a} = \frac{7}{b} = \frac{12.5}{c}$$

Plan: Write a proportion to find each missing side length.

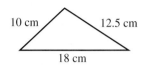
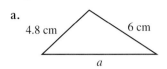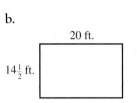
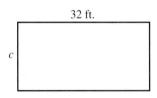

Find the missing lengths in the similar shapes.

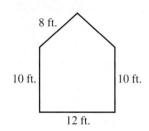

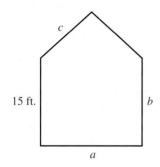

Execute: It is not necessary to write a proportion to find b. Since two sides in the larger trapezoid measure 7 inches, their corresponding sides in the smaller trapezoid must be equal in length, so we can conclude that $b = 4.2$ inches.

To find a: $\dfrac{7}{4.2} = \dfrac{10}{a}$

$a \cdot 7 = 7a \qquad 4.2 \cdot 10 = 42$

$\dfrac{7}{4.2} = \dfrac{10}{a}$

$7a = 42$

$\dfrac{7a}{7} = \dfrac{42}{7}$

$a = 6$

To find c: $\dfrac{7}{4.2} = \dfrac{12.5}{c}$

$c \cdot 7 = 7c \qquad 4.2 \cdot 12.5 = 52.5$

$\dfrac{7}{4.2} = \dfrac{12.5}{c}$

$7c = 52.5$

$\dfrac{7c}{7} = \dfrac{52.5}{7}$

$c = 7.5$

Answer: The missing lengths are: $a = 6$ in., $b = 4.2$ in., and $c = 7.5$ in.

Check: We can verify the reasonableness of the answers by doing quick estimates of the missing side lengths. Comparing the smaller trapezoid to the larger, we see that the 4.2-inch side is a little more than half of the corresponding 7-inch side in the larger trapezoid. Each of the corresponding sides should follow this same rule. The 10-inch side of the larger trapezoid should correspond to a length of a little more than 5 for side a of the smaller trapezoid, and it does at 6 inches. The 12.5-inch side should correspond to a length of a little more than 6.25 for side c on the smaller trapezoid, and it does at 7.5 inches.

◀ **Do Your Turn 6**

Answers to Your Turn 6
$a = 18$ ft., $b = 15$ ft.,
$c = 12$ ft.

7.2 Exercises

For Extra Help

Videotape DVT

Addison-Wesley Tutor Center

Math XL

MyMathLab

Student Solutions Manual

1. What is a proportion?

2. If two ratios are equivalent, then their _____ are equal.

3. How do you solve for an unknown amount in a proportion?

4. What are similar figures?

For Exercises 5–18, determine whether the ratios are proportional.

5. $\dfrac{3}{4} \stackrel{?}{=} \dfrac{9}{12}$

6. $\dfrac{15}{25} \stackrel{?}{=} \dfrac{3}{5}$

7. $\dfrac{15}{4} \stackrel{?}{=} \dfrac{7}{2.4}$

8. $\dfrac{7}{4} \stackrel{?}{=} \dfrac{15}{6.2}$

9. $\dfrac{15}{24} \stackrel{?}{=} \dfrac{2.5}{3.2}$

10. $\dfrac{10}{14} \stackrel{?}{=} \dfrac{2.5}{3.5}$

11. $\dfrac{9.5}{14} \stackrel{?}{=} \dfrac{28.5}{42}$

12. $\dfrac{6.5}{14} \stackrel{?}{=} \dfrac{10.5}{42}$

13. $\dfrac{16.8}{40.2} \stackrel{?}{=} \dfrac{17.5}{42.5}$

14. $\dfrac{15.2}{40.2} \stackrel{?}{=} \dfrac{60.8}{120.4}$

15. $\dfrac{\frac{4}{5}}{\frac{7}{10}} \stackrel{?}{=} \dfrac{12}{10\frac{1}{2}}$

16. $\dfrac{\frac{2}{5}}{\frac{3}{10}} \stackrel{?}{=} \dfrac{15}{10\frac{1}{2}}$

17. $\dfrac{2\frac{1}{3}}{3\frac{1}{2}} \stackrel{?}{=} \dfrac{5\frac{1}{2}}{7\frac{1}{3}}$

18. $\dfrac{3\frac{1}{4}}{5\frac{1}{2}} \stackrel{?}{=} \dfrac{4\frac{2}{3}}{6\frac{1}{2}}$

For Exercises 19–34, solve for the missing number.

19. $\dfrac{x}{8} = \dfrac{20}{32}$

20. $\dfrac{28}{42} = \dfrac{y}{6}$

21. $\dfrac{-3}{8} = \dfrac{n}{20}$

22. $\dfrac{m}{12} = \dfrac{2}{-5}$

23. $\dfrac{2}{n} = \dfrac{7}{21}$

24. $\dfrac{5}{k} = \dfrac{20}{4}$

25. $-\dfrac{4}{28} = \dfrac{5}{m}$

26. $-\dfrac{26}{4} = \dfrac{130}{t}$

27. $\dfrac{18}{h} = \dfrac{21.6}{30}$

28. $\dfrac{20}{32.8} = \dfrac{8}{k}$

29. $\dfrac{-14}{b} = \dfrac{3.5}{6.25}$

30. $\dfrac{4.9}{6.4} = \dfrac{-24.5}{c}$

31. $\dfrac{4\frac{3}{4}}{5\frac{1}{2}} = \dfrac{d}{16\frac{1}{2}}$

32. $\dfrac{\frac{3}{5}}{6\frac{1}{2}} = \dfrac{1\frac{2}{3}}{j}$

33. $\dfrac{-9.5}{22} = \dfrac{-6\frac{1}{3}}{t}$

34. $\dfrac{u}{-\frac{3}{8}} = \dfrac{-4.2}{5}$

For Exercises 35–46, solve.

35. William drove 358.4 miles using 16.4 gallons of gasoline. At this rate, how much gasoline would it take for him to drive 750 miles?

36. A carpet cleaning company claims they will clean 400 square feet for $54.95. At this rate how much should they charge to clean 600 square feet?

37. Gunther estimates that a 12.5-pound turkey will provide 30 servings. At this rate, how many servings would a 15-pound turkey provide?

38. A 20-pound bag of fertilizer covers 5000 square feet. How many pounds of fertilizer should be used for a lawn that is 12,000 square feet? If the fertilizer only comes in 20-pound bags, how many bags must be purchased?

39. Nathan notes that he consumes 16 ounces of yogurt in five days. At this rate, how many ounces will he consume in a year? (1 year = $365\frac{1}{4}$ days)

40. In taking a patient's pulse, a nurse counts 19 heartbeats in 15 seconds. How many heartbeats is this in 1 minute? (1 minute = 60 seconds)

41. A recipe for bran muffins calls for $1\frac{1}{2}$ cups of bran flakes. The recipe yields 8 muffins. How much bran flakes should be used to make a dozen muffins?

42. The same recipe for bran muffins calls for $2\frac{1}{2}$ teaspoons of baking powder. How much baking powder should be used to make a dozen muffins?

43. A $42\frac{1}{2}$-foot wall measures $8\frac{1}{2}$ inches on a scale drawing. How long should a $16\frac{1}{4}$-foot wall measure on the drawing?

44. Corrine is drafting the blueprints for a house. The scale for the drawing is $\frac{1}{4}$ inch = 1 foot. How long should she draw the line representing a wall that is $13\frac{1}{4}$ feet long?

45. A building company estimates that construction on a high-rise building will move at a pace of 18 vertical feet every 30 days. At this rate, how long will it take them to complete a building that is 980 feet tall?

46. A water treatment plant can treat 40,000 gallons of water in 21 hours. How many gallons does the plant treat in a week? (1 week = 168 hours)

For Exercises 47–55, find the missing lengths in the similar shapes.

47.

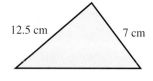
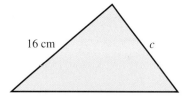

48.

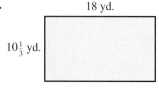
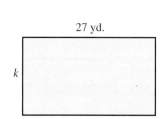

49.

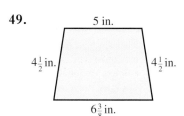 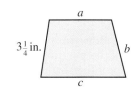

50.

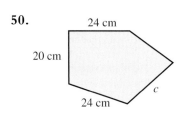 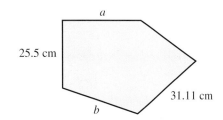

51. Yury wants to estimate the height of his house on a cloudy day. He places a small mirror on the ground 16.5 meters away from the house and walks back (away from the mirror and the house) for 3 meters until he can see the reflection of the top of the house in the mirror. If the distance from the ground to Yury's eyes is 1.8 meters, how tall is the house? (*Note:* The two triangles are similar.)

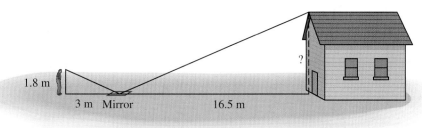

52. To estimate the height of a tree, a forester uses the concept of similar triangles. A particular tree is found to have a shadow measuring 84 feet in length. The forester has her own shadow measured at the same time. Her shadow measures $7\frac{1}{2}$ feet. If she is $5\frac{1}{2}$ feet tall, how tall is the tree?

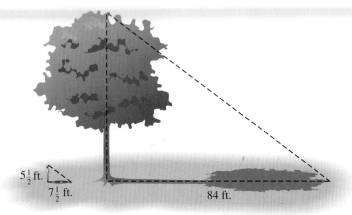

53. To estimate the height of the great pyramid in Egypt, the Greek mathematician Thales performed a procedure similar to that in Exercise 52, except that he supposedly used a staff instead of his body. Suppose the shadow from the staff measured to be 6 feet and, at the same time, the shadow of the pyramid measured to be $524\frac{2}{3}$ feet from the center of the pyramid to the tip of the shadow. If the staff was $5\frac{1}{2}$ feet tall, what was the height of the pyramid?

OF INTEREST

Thales was a Greek mathematician who lived from 640 B.C. to 546 B.C. He was born in Miletus, a city on the coast of what is now Turkey. Thales is considered to be the father of science and mathematics in the Greek culture. He is the first to have proposed the importance of proof. It is said that he asserted that water is the origin of all things and that matter is infinitely divisible.

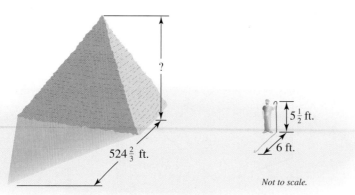

$524\frac{2}{3}$ ft.

$5\frac{1}{2}$ ft.

6 ft.

Not to scale.

54. To estimate the distance across a river an engineer creates similar triangles. The idea is to create two similar right triangles, as shown in the figure. Calculate the width of the river.

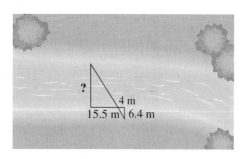

?

4 m

15.5 m   6.4 m

DISCUSSION How would one create the similar right triangles?

55. The method described in Exercise 54 could be used to determine the width of the Grand Canyon at a given point. Consider the measurements shown. Calculate the width of the canyon.

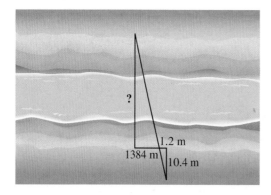

?

1.2 m

1384 m   10.4 m

OF INTEREST

Located in northwest Arizona, the Grand Canyon is a gorge created by the erosive action of the Colorado River. The canyon is about 277 miles (446 kilometers) long and ranges in width from 4 to 18 miles (6.4 to 29 kilometers). The canyon reaches depths of up to 1 mile (1.6 kilometers).

PUZZLE PROBLEM It takes a clock 5 seconds to strike 6 o'clock. How many seconds will it take for the clock to strike 12 o'clock?

Review Exercises

For Exercises 1–5, evaluate.

[5.3]   **1.** $\dfrac{4}{9} \cdot \dfrac{5}{12}$

[6.5]   **2.** $59.2 \cdot \dfrac{1}{2} \cdot \dfrac{1}{2} \cdot \dfrac{1}{4}$

[5.3]   **3.** $9\dfrac{1}{3}(144)$

[6.4]   **4.** $70.5 \div 12$

[5.4]   **5.** $60\dfrac{3}{4} \div 3$

7.3 American Measurement

OBJECTIVE 1 Convert American units of length.

To convert from one unit of measurement to another, we will use a method known as *dimensional analysis* or *unit analysis*. First, consider units of length. Following are some facts about American units of length.

$$1 \text{ foot} = 12 \text{ inches}$$
$$1 \text{ yard} = 3 \text{ feet}$$
$$1 \text{ mile} = 5280 \text{ feet}$$

We can express these relations as ratios.

$$\frac{1 \text{ ft.}}{12 \text{ in.}} \quad \text{or} \quad \frac{12 \text{ in.}}{1 \text{ ft.}} \qquad \frac{1 \text{ yd.}}{3 \text{ ft.}} \quad \text{or} \quad \frac{3 \text{ ft.}}{1 \text{ yd.}} \qquad \frac{1 \text{ mi.}}{5280 \text{ ft.}} \quad \text{or} \quad \frac{5280 \text{ ft.}}{1 \text{ mi.}}$$

Because the numerator and denominator of each of these ratios represents the same amount, these ratios are all equivalent to the number 1 and are called **unit fractions.**

DEFINITION Unit fraction: A fraction with a value equivalent to 1.

Because the value of any given unit fraction is 1, multiplying by unit fractions is a way of finding an equivalent amount. To use dimensional analysis, we multiply the given measurement by unit fractions so that the undesired units divide out, leaving the desired unit.

Suppose we want to convert 4 feet to units of inches. Because 4 feet has the undesired unit feet in a numerator, we must multiply by a unit fraction that has the unit feet appearing in the denominator. We can then divide out the undesired **feet** units, leaving us with the desired unit, **inches**.

$$4 \text{ ft.} = \frac{4 \text{ ft.}}{1} \cdot \frac{12 \text{ in.}}{1 \text{ ft.}} = 48 \text{ in.}$$

PROCEDURE *To convert units using dimensional analysis, multiply the given measurement by unit fractions so that the undesired units divide out, leaving the desired units.*

Note:

If an undesired unit is in a numerator, that unit should appear in the denominator of a unit fraction.

If an undesired unit is in the denominator, that unit should appear in the numerator of a unit fraction.

Example 1 Convert using dimensional analysis.

a. 3.5 miles to feet

Solution: Because 3.5 miles can be expressed in a numerator over 1, we will use a unit fraction with miles in the denominator. Because the desired unit is feet, we will use the fact that there are 5280 feet to 1 mile.

$$3.5 \text{ mi.} = \frac{3.5 \text{ mi.}}{1} \cdot \frac{5280 \text{ ft.}}{1 \text{ mi.}} = 18{,}480 \text{ ft.}$$

b. $8\frac{2}{3}$ yards to feet

Solution: Because $8\frac{2}{3}$ yards can be expressed in a numerator over 1, we will use a unit fraction with yards in the denominator. Because the desired unit is feet, we will use the fact that there are 3 feet to 1 yard.

$$8\frac{2}{3} \text{ yd.} = \frac{26 \text{ yd.}}{3} = \frac{26 \cancel{\text{ yd.}}}{\cancel{3}} \cdot \frac{\overset{1}{\cancel{3}} \text{ ft.}}{1 \cancel{\text{ yd.}}} = 26 \text{ ft.}$$

c. 78 inches to feet

Solution: Because 78 inches can be expressed in a numerator over 1, we will use a unit fraction with inches in the denominator. Because the desired unit is feet, we will use the fact that there are 12 inches to 1 foot.

$$78 \text{ in.} = \frac{\overset{13}{\cancel{78} \text{ in.}}}{1} \cdot \frac{1 \text{ ft.}}{\underset{2}{\cancel{12} \text{ in.}}} = \frac{13}{2} \text{ ft.} = 6\frac{1}{2} \text{ ft. or } 6.5 \text{ ft.}$$

Do Your Turn 1 ▶

Sometimes, we may not know a direct relation between the given unit and the desired unit. In such cases, we must convert to one or more intermediate units to get to the desired unit. This means we will use more than one unit fraction to convert to the desired unit.

Example 2 Convert using dimensional analysis.

a. 10 yards to inches

Solution: If we do not know a direct relation of yards to inches, we can first convert yards to feet, then feet to inches. To convert from yards to feet, we use the fact that there are 3 feet to 1 yard. Then to convert from feet to inches, we use the fact that there are 12 inches to 1 foot.

$$10 \text{ yd.} = \frac{10 \cancel{\text{ yd.}}}{1} \cdot \frac{3 \cancel{\text{ ft.}}}{1 \cancel{\text{ yd.}}} \cdot \frac{12 \text{ in.}}{1 \cancel{\text{ ft.}}} = 360 \text{ in.}$$

Note: Multiplying these two ratios gives the direct fact that there are 36 inches to 1 yard.

b. 4400 yards to miles

Solution: Because we do not have a direct relation of yards to miles, we can convert yards to feet, then convert feet to miles. We use the fact that there are 3 feet to 1 yard and then use 1 mile to 5280 feet.

$$4400 \text{ yd.} = \frac{4400 \cancel{\text{ yd.}}}{1} \cdot \frac{3 \cancel{\text{ ft.}}}{1 \cancel{\text{ yd.}}} \cdot \frac{1 \text{ mi.}}{5280 \cancel{\text{ ft.}}} = \frac{13{,}200}{5280} \text{ mi.} = 2\frac{1}{2} \text{ mi. or } 2.5 \text{ mi.}$$

DISCUSSION What is the direct relation of yards and miles?

If we had known the direct relations in Example 2, we would not have had to use several unit fractions. The more relations you know, the fewer unit fractions you have to use. However, this means more to remember. So there's a trade-off: Remember more relations and have simpler calculations, or remember fewer relations and be prepared to use more unit fractions. Each has its merits, but in the end it's up to the individual to choose which is better.

Do Your Turn 2 ▶

Your Turn 1

Convert using dimensional analysis.

a. 20 feet to inches

b. $4\frac{3}{10}$ miles to feet

c. 62 feet to yards

Your Turn 2

Convert using dimensional analysis.

a. 0.8 yards to inches

b. $\frac{7}{10}$ mile to yards

Answers to Your Turn 1
a. 240 in. **b.** 22,704 ft.
c. $20\frac{2}{3}$ ft. (or 20.6 ft.)

Answers to Your Turn 2
a. 28.8 in. **b.** 1232 yd.

OBJECTIVE 2 Convert American units of area.

Now that we have discussed dimensional analysis, we can apply this method to other types of measurement. Consider units of area. Recall that area is always represented in square units, like square inches (in.^2), square feet (ft.^2), square yards (yd.^2), or square miles (mi.^2).

A square foot is a 1-foot by 1-foot square. If we subdivide that square into inches, we have a 12-inch by 12-inch square.

Calculating the area of a 1-foot by 1-foot square as a 12-inch by 12-inch square, we find that it contains 144 square inches. Note that if we square both sides of the relation for length, we get the relation for area.

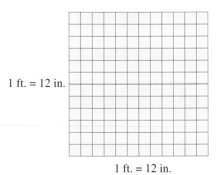

1 ft. = 12 in.

1 ft. = 12 in.

$$1 \text{ ft.} = 12 \text{ in.} \quad \leftarrow \text{ relation for length}$$
$$(1 \text{ ft.})^2 = (12 \text{ in.})^2 \quad \text{Square both sides.}$$
$$1 \text{ ft.}^2 = 144 \text{ in.}^2 \quad \leftarrow \text{ relation for area}$$

Conclusion: We can derive the relations for area by squaring the relations for length.

Because 1 yd. = 3 ft., we can conclude that $1 \text{ yd.}^2 = 9 \text{ ft.}^2$.

Because 1 mi. = 5280 ft., we can conclude that $1 \text{ mi.}^2 = 27{,}878{,}400 \text{ ft.}^2$.

Your Turn 3

Convert using dimensional analysis.

a. 12 square yards to square feet

b. 6000 square inches to square feet

c. 8 square yards to square inches

Example 3   Convert using dimensional analysis.

a. 8 square feet to square inches

Solution: Because the given measurement is in square feet, we will use a unit fraction with square feet in the denominator. Because we are to convert to square inches, we will use the ratio of 144 square inches to 1 square foot.

$$8 \text{ ft.}^2 = \frac{8 \cancel{\text{ft.}^2}}{1} \cdot \frac{144 \text{ in.}^2}{1 \cancel{\text{ft.}^2}} = 1152 \text{ in.}^2$$

b. 150 square feet to square yards

Solution: Because the given measurement is in square feet, we will use a unit fraction that has square feet in the denominator. Because we are to convert to square yards, we use the ratio of 1 square yard to 9 square feet.

$$150 \text{ ft.}^2 = \frac{\overset{50}{\cancel{150}} \cancel{\text{ft.}^2}}{1} \cdot \frac{1 \text{ yd.}^2}{\underset{3}{\cancel{9}} \cancel{\text{ft.}^2}} = \frac{50}{3} \text{ yd.}^2 = 16\frac{2}{3} \text{ yd.}^2 \quad \text{or} \quad 16.\overline{6} \text{ yd.}^2$$

Do Your Turn 3 ▶

Answers to Your Turn 3
a. 108 ft.^{2} b. 41$\frac{2}{3}$ ft.^{2}
or 41.$\overline{6}$ ft.^{2} c. 10,368 in.^{2}

Example 4   A room that is 25.5 feet long by 18 feet wide is to be carpeted. Calculate the area of the room in square yards.

Understand: We must calculate the area of a rectangular room. The formula for the area of a rectangle is $A = lw$.

The given measurements are in terms of feet. If we calculate the area with these measurements, the area unit will be square feet. The desired unit is square yards. There are two approaches to this problem:

1. Calculate the area in terms of square feet, then convert square feet to square yards.

2. Convert feet to yards and calculate the area with the measurements in terms of yards so that the outcome is in square yards.

Plan: Let's use the first approach. We will calculate the area using the given measurements so that the area will be in square feet. We will then convert square feet to square yards.

Execute: $A = lw$

$A = (25.5)(18)$ Replace *l* with 25.5 and *w* with 18.

$A = 459$ Multiply.

Now convert 459 square feet to square yards.

$$459 \text{ ft.}^2 = \frac{459 \text{ ft.}^2}{1} \cdot \frac{1 \text{ yd.}^2}{9 \text{ ft.}^2} = \frac{459}{9} \text{ yd.}^2 = 51 \text{ yd.}^2$$

Answer: The room will require 51 square yards of carpet.

Check: Let's use the second method we described as a check. We will convert the given measurements to yards, then calculate the area.

Length: $25.5 \text{ ft.} = \dfrac{25.5 \text{ ft.}}{1} \cdot \dfrac{1 \text{ yd.}}{3 \text{ ft.}} = \dfrac{25.5}{3} \text{ yd.} = 8.5 \text{ yd.}$

Width: $18 \text{ ft.} = \dfrac{18 \text{ ft.}}{1} \cdot \dfrac{1 \text{ yd.}}{3 \text{ ft.}} = \dfrac{18}{3} \text{ yd.} = 6 \text{ yd.}$

Area: $A = (8.5)(6)$ Replace *l* with 8.5 and *w* with 6.

$A = 51 \text{ yd.}^2$

Do Your Turn 4 ▶

Your Turn 4

A room is 24 feet long by $16\frac{1}{2}$ feet wide. Find its area in square yards.

OBJECTIVE 3 Convert American units of capacity.

Now consider **capacity.**

DEFINITION **Capacity:** A measure of the amount of liquid a container holds.

In the American system, we use units like gallons (gal.), quarts (qt.), pints (pt.), cups (c.), and ounces (oz.) to measure the capacity of a container. Following are some facts about capacity units.

1 cup = 8 ounces
1 pint = 2 cups
1 quart = 2 pints
1 gallon = 4 quarts

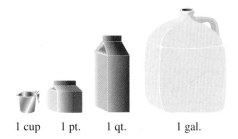

1 cup   1 pt.   1 qt.   1 gal.

Answer to Your Turn 4
44 yd.^{2}

Your Turn 5

Convert using dimensional analysis.

a. 6 quarts to cups

b. 20 ounces to pints

c. 3 gallons to cups

Example 5 Convert using dimensional analysis.

a. 5 pints to ounces

Solution: Because we do not have a direct relation of pints to ounces, we first convert pints to cups and then cups to ounces. Because the given measurement is in pints, the first unit fraction should have pints in the denominator. We use the ratio of 2 cups to 1 pint. The next unit fraction will have to have cups in the denominator, so we use the ratio of 8 ounces to 1 cup.

$$5 \text{ pt.} = \frac{5 \text{ pt.}}{1} \cdot \frac{2 \text{ c.}}{1 \text{ pt.}} \cdot \frac{8 \text{ oz.}}{1 \text{ c.}} = 80 \text{ oz.}$$

b. 240 ounces to gallons

Solution: Because we do not have a direct relation of ounces to gallons, we first convert ounces to cups. Because the given measurement is in ounces, the first ratio will have ounces in the denominator. We use the ratio 8 ounces to 1 cup. We then use 2 cups to 1 pint to convert from cups to pints. Next we use the ratio 2 pints to 1 quart to convert from pints to quarts. Finally, to convert from quarts to gallons we use the ratio 4 quarts to 1 gallon.

$$240 \text{ oz.} = \frac{240 \text{ oz.}}{1} \cdot \frac{1 \text{ c.}}{8 \text{ oz.}} \cdot \frac{1 \text{ pt.}}{2 \text{ c.}} \cdot \frac{1 \text{ qt.}}{2 \text{ pt.}} \cdot \frac{1 \text{ gal.}}{4 \text{ qt.}} = \frac{240}{128} \text{ gal.} = 1\frac{7}{8} \text{ gal. or } 1.875 \text{ gal.}$$

◀ **Do Your Turn 5**

OBJECTIVE 4 Convert American units of weight.

1 ton

Now let's convert units of weight. In the American system, weight can be expressed in terms of ounces (oz.), pounds (lb.), and tons (T). Following is a list of two facts about weight.

1 lb.

1 pound = 16 ounces
1 ton = 2000 pounds

GRAHAM CRACKERS

1 lb. (16 oz.)

Your Turn 6

Convert using dimensional analysis.

a. 12 pounds to ounces

b. 5000 pounds to tons

Example 6 Convert using dimensional analysis.

a. 12 ounces to pounds

Solution: Because the given measurement is in ounces, we multiply by the ratio of 1 pound to 16 ounces.

$$12 \text{ oz.} = \frac{\overset{3}{\cancel{12} \text{ oz.}}}{1} \cdot \frac{1 \text{ lb.}}{\underset{4}{\cancel{16} \text{ oz.}}} = \frac{3}{4} \text{ lb. or } 0.75 \text{ lb.}$$

b. 0.2 tons to ounces

Solution: Because we do not have a direct relation of tons to ounces, we convert tons to pounds, then convert pounds to ounces.

$$0.2 \text{ T} = \frac{0.2 \text{ T}}{1} \cdot \frac{2000 \text{ lb.}}{1 \text{ T}} \cdot \frac{16 \text{ oz.}}{1 \text{ lb.}} = 6400 \text{ oz.}$$

Answers to Your Turn 5
a. 24 c. b. 1.25 pt. c. 48 c.

Answers to Your Turn 6
a. 192 oz. b. 2.5 T

◀ **Do Your Turn 6**

OBJECTIVE 5 Convert units of time.

Now let's consider units of time. Time is measured in units of years (yr.), months (mo.), weeks (wk.), days (d.), hours (hr.), minutes (min.), and seconds (sec.). We will focus our attention on converting between years, days, hours, minutes, and seconds.

$$1 \text{ year} = 365\frac{1}{4} \text{ or } 365.25 \text{ days}$$

$$1 \text{ day} = 24 \text{ hours}$$
$$1 \text{ hour} = 60 \text{ minutes}$$
$$1 \text{ minute} = 60 \text{ seconds}$$

◄ **Note:** Unit abbreviations for time vary. Years can be y. or yr. Days can be d. or da. Hours can be h. or hr. Seconds can be s. or sec. However, the only abbreviation for minutes is min., since mi. is miles and m is meters.

Example 7 Convert using dimensional analysis.

a. $4\frac{1}{2}$ hours to seconds

Solution: Because we do not have a direct relation of hours to seconds, we will convert hours to minutes, then minutes to seconds.

$$4\frac{1}{2} \text{ hr.} = \frac{\overset{9}{\cancel{\text{hr.}}}}{\underset{1}{\cancel{2}}} \cdot \frac{\overset{30}{\cancel{60} \text{ min.}}}{1 \text{ hr.}} \cdot \frac{60 \text{ sec.}}{1 \text{ min.}} = 16{,}200 \text{ sec.}$$

b. 15 minutes to days

Solution: Because we do not have a direct relation of minutes to days, we will convert minutes to hours, then hours to days.

$$15 \text{ min.} = \frac{\overset{1}{\cancel{15} \text{ min.}}}{1} \cdot \frac{1 \text{ hr.}}{\underset{4}{\cancel{60} \text{ min.}}} \cdot \frac{1 \text{ d.}}{24 \text{ hr.}} = \frac{1}{96} \text{ d.} \quad \text{or} \quad 0.01041\overline{6} \text{ d.}$$

Do Your Turn 7 ▷

Your Turn 7

Convert using dimensional analysis.

a. 8.5 minutes to seconds

b. 5 days to minutes

c. 1 year to hours

OBJECTIVE 6 Convert American units of speed.

The final type of measurement that we will consider is speed. Speed is the ratio of distance to time like feet per second (ft./sec.) or miles per hour (mi./hr.). We can convert either the distance unit or the time or both within the speed unit.

Example 8 Convert using dimensional analysis.

a. 90 feet per second to feet per minute

Solution: Note that the distance unit, feet, is the same in the given speed unit and the desired speed unit. This means that we need to convert only the time part of the speed unit from seconds to minutes. Because seconds is in the denominator of the given unit, we must place seconds in the numerator of the unit fraction. This means we must use the ratio of 60 seconds to 1 minute.

$$90 \text{ ft./sec.} = \frac{90 \text{ ft.}}{1 \text{ sec.}} \cdot \frac{60 \text{ sec.}}{1 \text{ min.}} = \frac{5400 \text{ ft.}}{1 \text{ min.}} = 5400 \text{ ft./min.}$$

Answers to Your Turn 7
a. 510 sec. **b.** 7200 min.
c. 8766 hr.

Convert using dimensional analysis.

a. 500 feet per minute to feet per second

b. 28 feet per second to miles per hour

c. 9 yards per second to feet per minute

b. 60 miles per hour to feet per second

Solution: Both the distance and time units must be converted. We need to convert the distance units from miles to feet and the time units from hours to seconds. Because the given speed has miles in the numerator, we use a unit fraction with miles in the denominator. Because the given speed has hours in the denominator, we use a unit fraction with hours in the numerator.

We can carry out the two conversions in two stages:

$$60 \text{ mi./hr.} = \frac{60 \text{ mi.}}{1 \text{ hr.}} \cdot \frac{5280 \text{ ft.}}{1 \text{ mi.}} = \frac{316{,}800 \text{ ft.}}{1 \text{ hr.}} = 316{,}800 \text{ ft./hr.}$$

$$316{,}800 \text{ ft./hr.} = \frac{316{,}800 \text{ ft.}}{1 \text{ hr.}} \cdot \frac{1 \text{ hr.}}{60 \text{ min.}} \cdot \frac{1 \text{ min.}}{60 \text{ sec.}} = \frac{316{,}800 \text{ ft.}}{3600 \text{ sec.}} = 88 \text{ ft./sec.}$$

Or we can perform both conversions in one long line.

$$60 \text{ mi./hr.} = \frac{60 \text{ mi.}}{1 \text{ hr.}} \cdot \frac{5280 \text{ ft.}}{1 \text{ mi.}} \cdot \frac{1 \text{ hr.}}{60 \text{ min.}} \cdot \frac{1 \text{ min.}}{60 \text{ sec.}} = \frac{316{,}800 \text{ ft.}}{3600 \text{ sec.}} = 88 \text{ ft./sec.}$$

◀ **Do Your Turn 8**

Answers to Your Turn 8

a. $8\frac{1}{3}$ ft./sec. or $8.\overline{3}$ ft./sec.

b. $19\frac{1}{11}$ mi./hr. or $19.\overline{09}$ mi./hr.

c. 1620 ft./min.

Videotape DVT

 Addison-Wesley Tutor Center

Math XL

 MyMathLab

Student Solutions Manual

1. What is a unit fraction?

2. How do you convert units using dimensional analysis?

3. Complete the relations about American units for distance.

   1 ft. = _____ in.

   1 yd. = _____ ft.

   1 mi. = _____ ft.

4. Complete the relations for American units of area.

   1 ft.² = _____ in.²

   1 yd.² = _____ ft.²

5. Complete the relations for American units of capacity.

   1 cup = _____ oz.

   1 pt. = _____ cups.

   1 qt. = _____ pt.

   1 gal. = _____ qt.

6. Complete the relations for American units of time.

   1 year = _____ days

   1 day = _____ hours

   1 hour = _____ minutes

   1 minute = _____ seconds

For Exercises 7–22, convert using dimensional analysis.

7. 52 yards to feet

8. 6 feet to inches

9. 90 inches to feet

10. 45 feet to yards

11. 6.5 miles to feet

12. 7392 feet to miles

13. $5\frac{1}{2}$ feet to inches

14. $15\frac{2}{3}$ yards to feet

15. 20.8 miles to yards

16. 0.2 yards to inches

17. 99 inches to yards

18. 3168 yards to miles

19. 30 square yards to square feet

20. 648 square inches to square feet

21. $10\frac{1}{2}$ square yards to square inches

22. 40 square feet to square yards

23. A room that is 13 feet long by 12.5 feet wide is to be carpeted. Calculate the area in square yards.

24. A room that is 16.5 feet long by 14 feet wide is to be carpeted. Calculate the area in square yards.

For Exercises 25–60, convert using dimensional analysis.

25. 9 pints to cups

26. 6 quarts to pints

27. 2.5 cups to ounces

28. 16.4 gallons to quarts

29. 240 ounces to pints

30. 12 cups to pints

31. $8\frac{1}{2}$ pints to quarts

32. 60 ounces to quarts

33. 0.4 gallons to ounces

34. 5.2 gallons to pints

35. 40.8 pints to gallons

36. $6\frac{1}{3}$ cups to pints

37. $12\frac{1}{4}$ pounds to ounces

38. $9\frac{1}{2}$ pounds to ounces

39. 180 ounces to pounds

40. 124 ounces to pounds

41. 1200 pounds to tons

42. 6200 pounds to tons

43. 4.2 tons to pounds

44. 6.4 tons to pounds

45. 150 minutes to hours

46. 200 minutes to hours

47. 90 seconds to minutes

48. 20 seconds to minutes

49. $10\frac{1}{2}$ minutes to seconds

50. 12.2 minutes to seconds

51. $2\frac{1}{4}$ hours to minutes

52. $8\frac{1}{2}$ hours to minutes

53. 0.2 hours to seconds

54. 1.5 hours to seconds

55. 30 days to years

56. 90 days to years

57. 1 day to minutes

58. 7 days to seconds

59. 72 years to minutes

60. 85 years to seconds

For Exercises 61–66, solve using dimensional analysis.

61. The speed of sound in dry air at 0°C is 1088 feet per second. Calculate the speed of sound in miles per minute.

62. Calculate the speed of sound in miles per hour.

63. An investigating officer examining skid marks at the scene of an accident estimates that the speed of the vehicle was 80 feet per second. The driver of the vehicle claims to have been going 40 miles per hour. Are they in agreement?

64. A defendant in a collision case claims to have been traveling at 35 miles per hour around a blind curve and did not have time to swerve to miss rear-ending the plaintiff, who had just pulled into the road. The investigating officer determines that any speed less than 60 feet per second would have allowed time to brake or swerve to avoid the collision. How would you judge the defendant?

65. The escape velocity for Earth is 36,687.6 feet per second. Calculate this speed in miles per hour.

66. The minimum speed required to achieve orbit around Earth is 25,957 feet per second. Calculate this speed in miles per hour.

> **OF INTEREST**
>
> Escape velocity is the speed that an object must achieve to escape the gravitational pull of a planet. Any object traveling slower than the escape velocity will either return to the planet or orbit it.

> **OF INTEREST**
>
> An object traveling less than the minimum speed to achieve orbit will return to the planet. Objects traveling at speeds greater than the minimum speed to achieve orbit but less than the escape velocity will orbit the planet. For Earth, objects traveling at exactly 25,957 feet per second will orbit in a circle. As speed is increased, the path becomes more elliptical.

PUZZLE PROBLEM A light-year (lt-yr) is the distance that light travels in one year. If light travels at a rate of 186,282 miles per second, how far is 1 light-year in miles? *Proxima Centauri* is the next nearest star to our star (the Sun) at a distance of about 4.2 light-years. What is this distance in miles?

Review Exercises

For Exercises 1–6, perform the indicated operation.

[6.3] **1.** 8.4(100)

[6.3] **2.** 0.95(1000)

[6.3] **3.** (0.01)(32.9)

[6.4] **4.** 45.6 ÷ 10

[6.4] **5.** 9 ÷ 1000

[6.4] **6.** 4800 ÷ 1000

7.4 Metric Measurement

OBJECTIVE 1 Convert units of metric length.

OBJECTIVES

1 Convert units of metric length.

2 Convert units of metric capacity.

3 Convert units of metric mass.

4 Convert units of metric area.

The problem with the American system of measurement is the use of strange numbers that are not easy to remember. The metric system was designed to be easy to use and remember because it is a base-10 system. This means that different units within a type of measurement are larger or smaller by a power of 10.

Because our decimal notation is a base-10 system, when we increase or decrease the size of the measurement in the metric system, we will multiply or divide by powers of 10, which, as we have seen, merely moves the decimal point either right or left. Thus, converting within the metric system will involve moving the decimal point either right or left an appropriate number of places.

Each type of measurement has a **base unit**.

DEFINITION Base unit: A basic unit; other units are named relative to it.

To the right is a list of the base units for common metric measurements:

Type of Measurement	Base Unit
Length	meter (m)
Capacity	liter (l)
Mass	gram (g)

To further simplify the system, prefixes indicate the size of the unit within each type of measurement relative to the base unit. Following is a unit chart showing the most common prefixes. In the chart the units are listed in decreasing order. The units to the left of the base unit are larger than the base unit, and the units to right of the base unit are smaller than the base unit.

kilo- (k)	hecto- (h)	deka- (da)	base unit	deci- (d)	centi- (c)	milli- (m)

A great feature of the system is that conversions work the same whether converting length units, capacity units, or mass units. Let's first consider metric length. As we've mentioned, the base unit for length is the meter (m), which is a little larger than a yard. Each of the other units for length is a power of 10 larger or smaller than the base unit. Putting the base unit with the prefixes, we have

kilometer (km)	hectometer (hm)	dekameter (dam)	meter (m)	decimeter (dm)	centimeter (cm)	millimeter (mm)

1 km = 1000 m   1 hm = 100 m   1 dam = 10 m   1 m   1 m = 10 dm   1 m = 100 cm   1 m = 1000 mm

Here are some comparisons to help you visualize the size of these units.

A kilometer is about 0.6 miles.

A hectometer is about 110 yards, which is the length of a football field including one end zone.

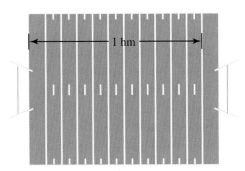

A dekameter is about the length of a school bus.

A decimeter is about the width of your hand across your palm (include your thumb).

A centimeter is about the width of your pinky finger at the tip.

A millimeter is about the thickness of your fingernail.

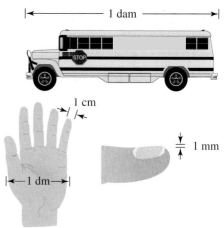

We can use these facts to convert metric length using dimensional analysis. Suppose we want to convert 2.8 meters to centimeters. We use the fact that there are 100 centimeters to 1 meter.

$$2.8 \text{ m} = \frac{2.8 \text{ m}}{1} \cdot \frac{100 \text{ cm}}{1 \text{ m}} = 280 \text{ cm}$$

Note that we multiplied by 100, which causes the decimal point to move 2 places to the right of its position in 2.8. Also notice that from the base unit to centi- is two places to the right on the unit chart.

kilo-	hecto-	deka-	Base unit	deci-	centi-	milli-
(k)	(h)	(da)		(d)	(c)	(m)

Suppose we want to convert 2.8 meters to kilometers. We use the fact that 1 kilometer is equal to 1000 meters. Based on what we just saw, what do you expect to happen?

$$2.8 \text{ m} = \frac{2.8 \text{ m}}{1} \cdot \frac{1 \text{ km}}{1000 \text{ m}} = \frac{2.8}{1000} \text{ km} = 0.0028 \text{ km}$$

Note that we divided by 1000, which caused the decimal point to move to the left 3 places. Also notice that from the base unit to kilo on the prefix chart is 3 places to the left.

kilo-	hecto-	deka-	Base unit	deci-	centi-	milli-
(k)	(h)	(da)		(d)	(c)	(m)

Conclusion: The decimal point moves the same number of places and in the same direction as the number of units and direction taken to get from the given unit to the desired unit on the unit chart.

Example 1 Convert.

a. 0.15 meters to decimeters

Solution: Let's first use dimensional analysis. Because the given measurement is in meters, we will use a ratio with meters in the denominator. We use the ratio 10 decimeters to 1 meter.

$$0.15 \text{ m} = \frac{0.15 \text{ m}}{1} \cdot \frac{10 \text{ dm}}{1 \text{ m}} = 1.5 \text{ dm}$$

Note that we multiplied 0.15 by 10, which causes the decimal point to move 1 place to the right. Using the unit chart, decimeter is the first unit to the right of meter. To convert meters to decimeters, we move the decimal point 1 place to the right of its position in 0.15.

$$0.\underset{\curvearrowright}{1}5 \text{ m} = 1.5 \text{ dm}$$

b. 56.1 dekameters to centimeters

Solution: Using dimensional analysis, because we do not know a direct relation of dekameters to centimeters, we first convert dekameters to meters. We then convert meters to centimeters.

$$56.1 \text{ dam} = \frac{56.1 \text{ dam}}{1} \cdot \frac{10 \text{ m}}{1 \text{ dam}} \cdot \frac{100 \text{ cm}}{1 \text{ m}} = 56{,}100 \text{ cm}$$

Note that we are multiplying 56.1 by 1000, which causes the decimal point to move 3 places to the right from its original position in 56.1. Using the unit chart, centimeter is the third unit to the right of dekameter (including the base unit). To convert from dekameters to centimeters, we move the decimal point 3 places to the right of its position in 56.1.

$$56.\underset{\curvearrowright}{1\,0\,0} \text{ dam} = 56{,}100 \text{ cm}$$

Do Your Turn 1 ▷

Your Turn 1

Convert.

a. 9.74 meters to millimeters

b. 0.075 hectometers to meters

c. 0.61 dekameters to decimeters

Example 2 Convert.

a. 2500 millimeters to meters

Solution: Using dimensional analysis, because the initial unit is millimeters, we multiply by a ratio that has millimeters in the denominator. We use the ratio of 1 meter to 1000 millimeters.

$$2500 \text{ mm} = \frac{2500 \text{ mm}}{1} \cdot \frac{1 \text{ m}}{1000 \text{ mm}} = \frac{2500}{1000} \text{ m} = 2.5 \text{ m}$$

Note that we divided 2500 by 1000, which causes the decimal point to move to the left 3 places from its original position in 2500. Using the unit chart, we see that meter is the third unit to the left of millimeter, so we move the decimal point 3 places to the left of its position in 2500.

$$2\underset{\curvearrowleft}{5\,0\,0} \text{ mm} = 2.5 \text{ m}$$

Note: The decimal point is understood to be to the right of the last digit in a whole number.

Note: 2.5 is the same as 2.500. However, the 0 digits are no longer needed because they are to the right of the last nonzero decimal digit.

Your Turn 2

Convert.

a. 6800 millimeters to meters

b. 456 meters to hectometers

c. 20,800 centimeters to kilometers

b. 970 centimeters to hectometers

Solution: Using dimensional analysis, because we do not know a direct relation of centimeters to hectometers, we first convert centimeters to meters. We then convert meters to centimeters.

$$970 \text{ cm} = \frac{970 \text{ cm}}{1} \cdot \frac{1 \text{ m}}{100 \text{ cm}} \cdot \frac{1 \text{ hm}}{100 \text{ m}} = \frac{970}{10{,}000} \text{ hm} = 0.097 \text{ hm}$$

Note that we divided 970 by 10,000, which causes the decimal point to move 4 places to the left from its original position in 970. On the unit chart, we see that hectometer is four units to the left of centimeter, so we move the decimal point 4 places to the left of its position in 970.

$$\underset{\curvearrowleft}{0\,9\,7\,0} \text{ cm} = 0.097 \text{ hm}$$

Do Your Turn 2 ▷

Answers to Your Turn 1
a. 9740 mm b. 7.5 m c. 61 dm

Answers to Your Turn 2
a. 6.8 m b. 4.56 hm c. 0.208 km

From Examples 1 and 2, we see that the beauty of the metric system is that the movement from one unit to another on the unit chart mirrors the movement of the decimal point in the number. From here on, we'll use the chart to convert within the metric system.

PROCEDURE *To convert metric units of length, capacity, or mass, use the following chart:*

kilo- (k)	hecto- (h)	deka- (da)	base unit	deci- (d)	centi- (c)	milli- (m)

1. Locate the given unit on the chart.
2. Count the number of units and note the direction (right or left) to get to the desired unit.
3. Move the decimal point in the given measurement the same number of places and in the same direction.

LEARNING STRATEGY

Many people remember the order of units by making up a sentence with the initial letters of each unit. Here's an example:

King **H**ector's **D**ata **B**ase **D**ecidedly **C**entralized **M**illions

OBJECTIVE 2 Convert units of metric capacity.

Consider metric capacity. The base unit for metric capacity is the liter (l). A liter is the amount of liquid that would fill a 10-centimeter by 10-centimeter by 10-centimeter cube. In other words, a liter of liquid occupies the same space as 1000 cubic centimeters.

We need to make an important connection. There are 1000 milliliters to 1 liter. If there are 1000 cubic centimeters and 1000 milliliters to 1 liter, then we can conclude that 1 milliliter must occupy 1 cubic centimeter of space. The common abbreviation for cubic centimeter is cc.

Conclusion: $1 \text{ ml} = 1 \text{ cm}^3$ or 1 cc

This is an extremely important connection for people entering health-related fields. Medications are often prescribed in units of milliliters, yet the packaging for the medication is expressed in cubic centimeters or vice versa. The person administering the medication needs to recognize that milliliters and cubic centimeters are used interchangeably. We can alter the chart for capacity slightly by including the cubic centimeters with milliliters.

kiloliter (kl)	hectoliter (hl)	dekaliter (dal)	liter (l)	deciliter (dl)	centiliter (cl)	milliliter or cubic centimeter (ml or cm^3 or cc)
1kl = 1000 l	1 hl = 100 l	1 dal = 10 l	1 l	1 l = 10 dl	1 l = 100 cl	1 l = 1000 ml
						1 l = 1000 cm^3 or cc

A liter is slightly more than a quart.

A milliliter (or cubic centimeter) is a little less than a quarter of a teaspoon.

WARNING Be careful that you understand the difference between centiliter and cubic centimeter.

Example 3 Convert.

a. 680 deciliters to hectoliters

Solution: Because hectoliter is the third unit to the left of deciliter, we move the decimal point 3 places to the left in 680.

$$680 \text{ dl} = 0.68 \text{ hl}$$

b. 0.0184 dekaliters to milliliters

Solution: Because milliliter is four units to the right of dekaliter, we move the decimal point 4 places to the right in 0.0184.

$$0.0184 \text{ dal} = 184 \text{ ml}$$

c. 0.0085 liters to cubic centimeters

Solution: Because cubic centimeters and milliliters are interchangeable, we consider the conversion as moving from liters to milliliters on the chart. Because the milliliter is the third unit to the right from liter, we move the decimal point 3 places to the right in 0.0085.

$$0.0085 \text{ l} = 8.5 \text{ cc}$$

Do Your Turn 3 ▶

Your Turn 3

Convert.

a. 1.2 kiloliters to dekaliters

b. 7800 centiliters to liters

c. 0.17 centiliters to cubic centimeters

OBJECTIVE 3 Convert units of metric mass.

Recall that mass and weight are different measurements. Weight is the force due to gravity and is dependent on the mass of an object and the gravitational pull on that object by a planet or other object. Mass, on the other hand, is a measure of the amount of matter that makes up an object.

As we have mentioned, the base unit for metric mass is the gram (g).

kilogram (kg)	hectogram (hg)	dekagram (dag)	gram (g)	decigram (dg)	centigram (cg)	milligram (mg)
1 kg = 1000 g	1 hg = 100 g	1 dag = 10 g	1 g	1 g = 10 dg	1 g = 100 cg	1 g = 1000 mg

A large paper clip has a mass of about 1 gram. Because this is quite small, most objects are measured in terms of kilograms. 1 liter of water has a mass of 1 kilogram. So, a 2-liter bottle of water has a mass of 2 kilograms. You are probably familiar with the milligram as well. Medicines in tablet form are usually measured in milligram. Common over-the-counter pain relief medicines are usually in doses of around 200 milligrams.

CONNECTION Because 1 liter is the amount of liquid contained in a 10-centimeter by 10-centimeter by 10-centimeter cube, we can say that that same cube filled with water has a mass of 1 kilogram.

Example 4 Convert.

a. 250 milligrams to grams

Solution: Because gram is the third unit to the left of milligram, we move the decimal point 3 places to the left in 250.

$$250 \text{ mg} = 0.25 \text{ g}$$

b. 0.062 hectograms to grams

Solution: Because gram is the second unit to the right of hectogram, we move the decimal point 2 places to the right in 0.062.

$$0.062 \text{ hg} = 6.2 \text{ g}$$

Do Your Turn 4 ▶

Your Turn 4

Convert.

a. 6500 milligrams to dekagrams

b. 0.15 kilograms to grams

c. 2.6 centigrams to decigrams

Answers to Your Turn 3
a. 120 dal b. 78 l c. 1.7 cc

Answers to Your Turn 4
a. 0.65 dag b. 150 g c. 0.26 dg

For very large measurements of mass we use the metric ton (t).

$$1 \text{ t} = 1000 \text{ kg}$$

How can we convert to metric tons? If we can get to kilograms on the unit chart, then we can use the preceding fact to convert kilograms to metric tons.

CONNECTION We have said that a 10-centimeter by 10-centimeter by 10-centimeter cube full of water has a capacity of 1 liter and a mass of 1 kilogram. A metric ton would be the mass of a cube of water 1000 times the volume of the 10-centimeter by 10-centimeter by 10-centimeter cube. This cube would be 100 centimeters by 100 centimeters by 100 centimeters, which is actually 1 meter by 1 meter by 1 meter. In other words, 1 metric ton is the mass of a 1 meter by 1 meter by 1 meter cube full of water. The volume of this cube is 1 cubic meter and the capacity is 1000 liters, or 1 kiloliter.

Suppose we want to convert 4500 kilograms to metric tons. We could use the preceding fact in a dimensional analysis setup.

$$4500 \text{ kg} = \frac{4500 \text{ kg}}{1} \cdot \frac{1 \text{ t}}{1000 \text{ kg}} = \frac{4500}{1000} \text{ t} = 4.5 \text{ t}$$

Note that we divided by 1000 to convert kilograms to metric tons, which caused the decimal point to move left 3 places in 4500. This corresponds to moving three steps to the left on the chart.

Conclusion: We can think of the metric ton as being three units to the left of kilogram.

Example 5 Convert.

a. 64,000 kilograms to metric tons

Solution: We can think of metric ton as being the third unit to the left of kilogram, so we move the decimal point 3 places to the left in 64,000.

$$64,000 \text{ kg} = 64 \text{ t}$$

b. 0.0058 metric tons to kilograms

Solution: If metric ton is the third unit to the left of kilogram, then kilogram is the third unit to the right of metric ton, so we move the decimal point 3 places to the right in 0.0058.

$$0.0058 \text{ t} = 5.8 \text{ kg}$$

c. 240,000 dekagrams to metric tons

Solution: Kilogram is two units to the left of dekagram and metric ton is three units to the left of kilogram. Therefore, metric ton is a total of five units to the left of dekagram, so we move the decimal point 5 places to the left in 240,000.

$$240,000 \text{ dag} = 2.4 \text{ t}$$

◁ **Do Your Turn 5**

OBJECTIVE 4 Convert units of metric area.

To discover the facts about area in the American system, we determined how many square inches fit in 1 square foot. We can perform the same analysis with the metric system. Let's find the number of square decimeters that fit in 1 square meter. Because there are 10 decimeters to 1 meter, a 1-meter by 1-meter square is a 10-decimeter by 10-decimeter square.

10 dm

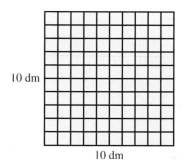

10 dm

If we calculate the area of this square meter in terms of decimeters, we have $10 \cdot 10 = 100$ square decimeters.

$$1 \text{ m}^2 = 100 \text{ dm}^2$$

We concluded in our discussion of area facts for the American system that we could simply square both sides of the length relations. Notice that the same applies with the metric relations.

$$1 \text{ m} = 10 \text{ dm} \quad \leftarrow \text{ relation for length}$$
$$(1 \text{ m})^2 = (10 \text{ dm})^2 \quad \text{Square both sides.}$$
$$1 \text{ m}^2 = 100 \text{ dm}^2 \quad \leftarrow \text{ relation for area}$$

Deriving the other relations in this way, we have

square kilometer (km²)	square hectometer (hm²)	square dekameter (dam²)	square meter (m²)	square decimeter (dm²)	square centimeter (cm²)	square millimeter (mm²)
1 km² = 1,000,000 m²	1 hm² = 10,000 m²	1 dam² = 100 m²	1 m²	1 m² = 100 dm²	1 m² = 10,000 cm²	1 m² = 1,000,000 mm²

We can still use the chart to convert units of metric area, but we have to alter our thinking a bit. Consider the following conversions and look for a pattern.

Suppose we want to convert 2.4 square meters to square decimeters.

$$2.4 \text{ m}^2 = \frac{2.4 \text{ m}^2}{1} \cdot \frac{100 \text{ dm}^2}{1 \text{ m}^2} = 240 \text{ dm}^2$$

Note: When we converted length measurements of meters to decimeters, we multiplied by 10, causing the decimal point to move 1 place to the right. When converting square meters to square decimeters, we multiply by 100, or 10^2, which moves the decimal point 2 places to the right.

Now let's convert 2.4 square meters to square centimeters.

$$2.4 \text{ m}^2 = \frac{2.4 \text{ m}^2}{1} \cdot \frac{10,000 \text{ cm}^2}{1 \text{ m}^2} = 24,000 \text{ cm}^2$$

Note: When we converted length measurements of meters to centimeters, we multiplied by 100, causing the decimal point to move 2 places to the right. When converting square meters to square centimeters, we multiply by 10,000, or 10^4, which moves the decimal point 4 places to the right.

Do you see a pattern? Consider 2.4 square meters to square millimeters.

$$2.4 \text{ m}^2 = \frac{2.4 \text{ m}^2}{1} \cdot \frac{1,000,000 \text{ mm}^2}{1 \text{ m}^2} = 2,400,000 \text{ mm}^2$$

Note: When we converted length measurements of meters to millimeters, we multiplied by 1000, causing the decimal point to move 3 places to the right. When converting square meters to square millimeters, we multiply by 1,000,000, or 10^6, which moves the decimal point 6 places to the right.

PROCEDURE *To convert units of area, move the decimal point twice the number of places as units from the initial area unit to the desired area unit on the chart.*

LEARNING STRATEGY

Use the exponent of 2 in the area units as a reminder to move the decimal point twice the number of units from the initial unit to the desired unit on the chart.

Your Turn 6

Convert.

a. 0.25 square meters to square centimeters

b. 4750 square meters to square decimeters

c. 95,100,000 square centimeters to square hectometers

Example 6 Convert.

a. 0.0081 square meters to square millimeters

Solution: Square millimeter is the third unit to the right of square meter. With area units we move the decimal point twice the number of places, so the decimal point moves to the right 6 places.

$$0.0081 \text{ m}^2 = 8100 \text{ mm}^2$$

b. 45,100 square meters to square hectometers

Solution: Square hectometer is the second unit to the left of square meter. Moving the decimal point twice that number of places means it moves to the left 4 places.

$$45,100 \text{ m}^2 = 4.51 \text{ hm}^2$$

c. 0.000075 square kilometers to square decimeters

Solution: Square decimeter is the fourth unit to the right of square kilometer. Moving the decimal point twice that number of places means it moves to the right 8 places.

$$0.000075 \text{ km}^2 = 7500 \text{ dm}^2$$

◀ **Do Your Turn 6**

CONNECTION We can connect metric prefixes with powers of 10 and word names. There are more unit prefixes in the metric system than we've presented here. These unit prefixes are used in science and the computer field. You may have heard of computer hard drives with storage capacity in gigabytes. The prefix *giga* is a metric prefix. We can use powers of 10 to indicate the size of a unit with a particular prefix relative to the base unit. For example, because a kilometer is 1000 times the size of 1 meter, we can say 1 km = 1000 m = 10^3 m. Or, because there are 1000 millimeters in 1 meter , we can say that 1 millimeter is 1/1000 or $\frac{1}{10^3}$ of 1 meter. Below is a list of the prefixes, their size relative to the base unit, and the common word name.

Prefix	Factor	Word Name
Tera- (T)	$= 10^{12}$	Trillion
Giga- (G)	$= 10^9$	Billion
Mega- (M)	$= 10^6$	Million
Kilo- (k)	$= 10^3$	Thousand
Hecto- (h)	$= 10^2$	Hundred
Deka- (da)	$= 10^1$	Ten
Base unit	$= 10^0 = 1$	
Deci- (d)	$= \frac{1}{10}$	Tenth
Centi- (c)	$= \frac{1}{10^2}$	Hundredth
Milli- (m)	$= \frac{1}{10^3}$	Thousandth
Micro- (μ)	$= \frac{1}{10^6}$	Millionth
Nano- (n)	$= \frac{1}{10^9}$	Billionth
Pico- (p)	$= \frac{1}{10^{12}}$	Trillionth

We can connect these powers of 10 with scientific notation. For example, 6.5 gigabytes can be expressed as

$$6.5 \text{ GB} = 6.5 \text{ billion bytes} = 6.5 \times 10^9 \text{ bytes} = 6,500,000,000 \text{ bytes.}$$

On the smaller side, X-rays have a wavelength of about 3 nanometers.

$$3 \text{ nm} = 3 \text{ billionths of a meter} = 3 \times \frac{1}{10^9} \text{ m} = \frac{1}{1,000,000,000} \text{ m} = 0.000000003 \text{ m}$$

Answers to Your Turn 6
a. 2500 cm^2 b. 475,000 dm^2
c. 0.951 hm^2

7.4 Exercises

For Extra Help
 Videotape DVT
 Addison-Wesley Tutor Center
 Math XL
 MyMathLab
 Student Solutions Manual

1. What is a base unit?

2. List the six basic prefixes in the metric system.

3. What is the base unit for distance in the metric system?

4. What is the base unit for capacity in the metric system?

5. Define mass.

6. What is the base unit for mass in the metric system?

For Exercises 7 to 50, convert.

7. 4.5 meters to centimeters

8. 0.81 meters to millimeters

9. 0.07 kilometers to meters

10. 1.3 hectometers to meters

11. 3800 meters to kilometers

12. 125 centimeters to meters

13. 9500 millimeters to meters

14. 3.48 dekameters to meters

15. 1540 decimeters to dekameters

16. 0.45 centimeters to millimeters

17. 112 meters to hectometers

18. 79 millimeters to decimeters

19. 0.12 liters to milliliters

20. 0.05 liters to milliliters

21. 0.005 kiloliters to liters

22. 145 deciliters to liters

23. 9.5 centiliters to dekaliters

24. 0.248 hectoliters to liters

25. 0.08 liters to cubic centimeters

26. 6 cubic centimeters to liters

27. 12 cubic centimeters to milliliters

28. 0.085 dekaliters to cubic centimeters

29. 0.4 liters to cubic centimeters

30. 145 centiliters to cubic centimeters

31. 12 grams to milligrams

32. 4500 grams to kilograms

33. 0.009 kilograms to grams

34. 900 milligrams to grams

35. 0.05 dekagrams to centigrams

36. 0.045 hectograms to milligrams

37. 3600 kilograms to metric tons

38. 850 kilograms to metric tons

39. 0.106 metric tons to kilograms

40. 5.07 metric tons to kilograms

41. 65,000 hectograms to metric tons

42. 0.008 metric tons to dekagrams

43. 1.8 square meters to square centimeters

44. 580,000 square meters to square kilometers

45. 2200 square centimeters to square meters

46. 19,000 square meters to square hectometers

47. 0.05 square centimeters to square millimeters

48. 2.4 square decimeters to square centimeters

49. 44,300 square decimeters to square hectometers

50. 0.068 square kilometers to square meters

PUZZLE PROBLEM A light-year (lt-yr) is the distance that light travels in 1 year. The speed of light is 3×10^8 meters per second. Calculate the distance of 1 light-year in kilometers. The Magellanic Cloud is the nearest galaxy to our own Milky Way galaxy at a distance of 150,000 light-years. Calculate this distance in kilometers.

Review Exercises

For Exercises 1 and 2, evaluate.

[6.5] **1.** $8.4(12) \div 100$

[2.5] **2.** $6\dfrac{1}{2} \div 3 \cdot 10$

[6.2] **3.** Combine like terms. $6.5x^2 - 8x + 14.4 - 10.2x - 20$

[6.3] **4.** Distribute. $100(4.81y^2 - 2.8y + 7)$

[6.6] **5.** Solve and check. $7x - 3.98 - 2.4x + 8.9$

Converting Between the American and Metric Systems

OBJECTIVE 1 Convert units of length.

When converting between systems of measurement, we use dimensional analysis. The key is to have at least one relation that acts as a bridge from one system to the other. However, the more relations we know, the more likely we'll be able to convert directly to the desired unit.

Let's consider units of length. The table to the right lists common relations for converting metric and American units of length. Note that most conversions are approximations.

American to Metric	Metric to American
1 in. = 2.54 cm	1 m ≈ 3.282 ft.
1 yd. ≈ 0.914 m	1 m ≈ 1.094 yd.
1 mi. ≈ 1.609 km	1 km ≈ 0.621 mi.

Example 1   Convert.

a. 100 yards to meters

Solution: Because the initial unit is yards, we will multiply by a unit fraction that has yards in the denominator. Because the desired unit is meters, we need to use a relation of yards to meters. Note that we can choose to use 0.914 meters ≈ 1 yard or 1 meter ≈ 1.094 yards.

$$100 \text{ yd.} \approx \frac{100 \text{ yd.}}{1} \cdot \frac{0.914 \text{ m}}{1 \text{ yd.}} \approx 91.4 \text{ m} \quad \text{or}$$

$$100 \text{ yd.} \approx \frac{100 \text{ yd.}}{1} \cdot \frac{1 \text{ m}}{1.094 \text{ yd.}} \approx \frac{100}{1.094} \text{ m} \approx 91.4 \text{ m}$$

b. 5 kilometers to miles

Solution: Because the initial unit is kilometers, we will multiply by a unit fraction that has kilometers in the denominator. Because the desired unit is miles, we need to use a relation of kilometers to miles. Note that we can choose to use 1 mile ≈ 1.609 kilometers or 0.621 miles ≈ 1 kilometer.

$$5 \text{ km} \approx \frac{5 \text{ km}}{1} \cdot \frac{1 \text{ mi.}}{1.609 \text{ km}} \approx \frac{5}{1.609} \text{ mi.} \approx 3.1 \text{ mi.} \quad \text{or}$$

$$5 \text{ km} \approx \frac{5 \text{ km}}{1} \cdot \frac{0.621 \text{ mi.}}{1 \text{ km}} \approx 3.1 \text{ mi.}$$

Do Your Turn 1 ▶

Your Turn 1

Convert.

a. 60 miles to kilometers

b. 42 meters to feet

c. $9\frac{1}{2}$ inches to centimeters

Sometimes, we may not have a direct relation from the initial unit to the desired unit. In these situations, we simply select a relation to a unit that is close to the desired unit in the target system, then convert from that unit to the desired unit.

Example 2   Convert.

a. 0.3 miles to meters

Solution: The relations that we have relate miles to kilometers. We will use the fact that 1.609 kilometers ≈ 1 mile. Once we have an answer in kilometers, we will then use the unit chart and move the decimal point 3 places to the right to get to meters.

$$0.3 \text{ mi.} \approx \frac{0.3 \text{ mi.}}{1} \cdot \frac{1.609 \text{ km}}{1 \text{ mi.}} \approx 0.4827 \text{ km}$$

Answers to Your Turn 1
a. ≈ 96.54 km **b.** ≈ 137.84 ft.
c. 24.13 cm

Because meter is the third unit to the right of kilometer, we move the decimal point 3 places to the right in 0.4827.

$$0.4827 \text{ km} = 482.7 \text{ m}$$

b. 64 millimeters to inches

Solution: The only relation we have to inches is 1 inch = 2.54 centimeters. To use this relation, we need the initial unit to be in centimeters. Therefore, we first convert 64 millimeters to centimeters.

Because centimeter is one unit to the left of millimeter on the metric chart, we move the decimal point 1 place to the left in 64.

$$64 \text{ mm} = 6.4 \text{ cm}$$

Now we can use the fact that 1 inch = 2.54 cm.

$$6.4 \text{ cm} = \frac{6.4 \text{ cm}}{1} \cdot \frac{1 \text{ in.}}{2.54 \text{ cm}} = \frac{6.4}{2.54} \text{ in.} \approx 2.52 \text{ in.}$$

◀ **Do Your Turn 2**

OBJECTIVE 2 Convert units of capacity.

Though the units and numbers change, the method for converting units of capacity does not change.

American to Metric	Metric to American
1 qt. ≈ 0.946 l	1 l ≈ 1.057 qt.

Your Turn 3

Convert.

a. 6 quarts to liters

b. 30 gallons to liters

c. 60 milliliters to ounces

d. 20 cubic centimeters to ounces

Example 3 Convert.

a. 2 gallons to liters

Solution: The relations that we have relate quarts to liters. Because the initial unit is gallons, we must first convert gallons to quarts. We can then use the fact that 1 quart ≈ 0.946 liters.

$$2 \text{ gal.} \approx \frac{2 \text{ gal.}}{1} \cdot \frac{4 \text{ qt.}}{1 \text{ gal.}} \cdot \frac{0.946 \text{ l}}{1 \text{ qt.}} \approx 7.568 \text{ l}$$

Note: Remember, there are 4 quarts (qt.) to 1 gallon (gal.).

b. 40 milliliters to ounces

Solution: The relations that we have relate quarts to liters. Because the initial unit is milliliters, we must first convert milliliters to liters. Because liter is three units to the left of milliliter, we move the decimal point 3 places to the left.

$$40 \text{ ml} = 0.04 \text{ l}$$

Note: Because milliliters and cubic centimeters are the same size, we would follow the exact same procedure to convert cubic centimeters to ounces.

We can now use the fact that 1 l ≈ 1.057 qt. However, our target unit is ounces, so we'll need to convert quarts to ounces.

$$0.04 \text{ l} \approx \frac{0.04 \text{ l}}{1} \cdot \frac{1.057 \text{ qt.}}{1 \text{ l}} \cdot \frac{2 \text{ pt.}}{1 \text{ qt.}} \cdot \frac{2 \text{ c.}}{1 \text{ pt.}} \cdot \frac{8 \text{ oz.}}{1 \text{ c.}} \approx 1.35 \text{ oz.}$$

Answers to Your Turn 2
a. ≈ 1255.33 dm b. ≈ 19.7 in.
c. ≈ 2188 yd.

Answers to Your Turn 3
a. ≈ 5.676 l b. ≈ 113.52l
c. ≈ 2.03 oz. d. ≈ 0.676 oz.

◀ **Do Your Turn 3**

OBJECTIVE 3 Convert units of mass/weight.

The primary conversion relations that link American weight to metric mass are shown to the right:

American to Metric	Metric to American
1 lb. ≈ 0.454 kg	1 kg ≈ 2.2 lb.
1 T ≈ 0.907 t	1 t ≈ 1.1 T

Example 4 Convert.

a. 180 pounds to kilograms

Solution: We have a direct relation of pounds to kilograms. We can then use the fact that 1 pound ≈ 0.454 kilograms.

$$180 \text{ lb.} \approx \frac{180 \text{ lb.}}{1} \cdot \frac{0.454 \text{ kg}}{1 \text{ lb.}} \approx 81.72 \text{ kg}$$

b. $6\frac{1}{2}$ tons to metric tons

Solution: We have a direct relation of American tons to metric tons. We can then use the fact that 1 American ton ≈ 0.907 metric tons.

$$6\frac{1}{2} \text{ T} \approx \frac{6\frac{1}{2} \text{ T}}{1} \cdot \frac{0.907 \text{ t}}{1 \text{ T}} \approx 5.9 \text{ t}$$

c. 250 grams to ounces

Solution: The relations that we have relate kilograms to pounds. Because the initial unit is grams, we must first convert grams to kilograms. Because kilogram is three units to the left of gram, we move the decimal point 3 places to the left.

$$250 \text{ g} = 0.25 \text{ kg}$$

We can now use the fact that 1 kilogram ≈ 2.2 pounds. However, our target unit is ounces, so we will need to convert pounds to ounces.

$$0.25 \text{ kg} \approx \frac{0.25 \text{ kg}}{1} \cdot \frac{2.2 \text{ lb.}}{1 \text{ kg}} \cdot \frac{16 \text{ oz.}}{1 \text{ lb.}} \approx 8.8 \text{ oz.}$$

Do Your Turn 4 ▷

Note: Remember, there are 16 ounces (oz.) in 1 pound (lb.).

Your Turn 4

Convert.

a. 50 pounds to kilograms

b. 4.2 metric tons to American tons

c. 40 ounces to grams

OBJECTIVE 4 Convert units of temperature.

Temperature is a measure of the heat of a substance. In the American system, temperature is measured in degrees Fahrenheit (°F). In the metric system, temperature is measured in degrees Celsius (°C). To relate the two systems, we need two reference points. For simplicity, we will use the freezing point and the boiling point of water.

On the Fahrenheit scale, the freezing point of water is 32°F and the boiling point of water is 212°F. Note that the difference between these two points is 212°F − 32°F = 180°F.

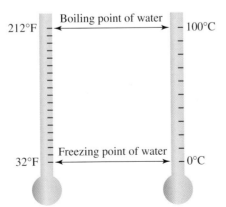

On the Celsius scale, the freezing point of water is 0°C and the boiling point of water is 100°C. Note that the difference between these two points is 100°C.

Answers to Your Turn 4
a. ≈ 22.7 kg **b.** ≈ 4.62 T
c. ≈ 1135 g

Notice that a change of 180°F is the same as a change of 100°C. We can write this relation as a ratio, which we can use to convert between the two temperature scales.

$$\begin{array}{c} \text{degrees Fahrenheit} \rightarrow \\ \text{to} \\ \text{degrees Celsius} \rightarrow \end{array} \quad \frac{180°F}{100°C} = \frac{9°F}{5°C}$$

Let's try using our ratio to convert 100°C to degrees Fahrenheit. We should get 212°F.

$$100°C = \frac{100°\cancel{C}}{1} \cdot \frac{9°F}{5°\cancel{C}} = 180°F \quad \blacktriangleleft \quad \boxed{\text{But 180°F isn't correct. We should get 212°F.}}$$

The reason multiplying by $\frac{9}{5}$ alone does not produce the correct temperature is because the two scales do not have the same reference point for 0. The Fahrenheit scale is shifted by 32°F. Therefore, to convert degrees Celsius to degrees Fahrenheit, after multiplying by $\frac{9}{5}$, we need to *add* 32°F.

$$100°C = \frac{100°\cancel{C}}{1} \cdot \frac{9°F}{5°\cancel{C}} + 32°F$$
$$= 180°F + 32°F$$
$$= 212°F$$

We can write a formula to perform this conversion. To convert from degrees Celsius to degrees Fahrenheit, multiply degrees Celsius by $\frac{9}{5}$, then add 32.

Conclusion: $F = \dfrac{9}{5}C + 32$

Your Turn 5

Convert.

a. 60°C to degrees Fahrenheit

b. 120.5°C to degrees Fahrenheit

c. 20°C to degrees Fahrenheit

d. −5°C to degrees Fahrenheit

Example 5 Convert 45°C to degrees Fahrenheit.

Solution: Because the desired unit is degrees Fahrenheit, we use the formula $F = \frac{9}{5}C + 32$.

$$F = \frac{9}{5}(45) + 32 \qquad \text{Replace } C \text{ with 45.}$$

$$F = \frac{9}{\cancel{5}}\left(\frac{\overset{9}{\cancel{45}}}{1}\right) + 32 \qquad \text{Divide out the common factor, 5.}$$

$$F = 81 + 32 \qquad \text{Multiply.}$$
$$F = 113°F \qquad \text{Add.}$$

◀ **Do Your Turn 5**

Now let's convert degrees Fahrenheit to degrees Celsius. To derive a formula, let's convert 86°F to degrees Celsius using $F = \dfrac{9}{5}C + 32$. We will replace F with 86, and then solve for C. Note the order of operations as we solve.

Answers to Your Turn 5
a. 140°F **b.** 248.9°F **c.** 68°F
d. 23°F

$$86 = \frac{9}{5}C + 32 \qquad \text{Replace } F \text{ with 86.}$$

$$86 = \frac{9}{5}C + 32$$

$$\underline{-32 \qquad\qquad -32} \qquad \text{Subtract 32 from both sides.}$$

$$54 = \frac{9}{5}C + 0$$

$$\frac{5}{\overset{1}{\underset{1}{9}}} \cdot \frac{\overset{6}{54}}{1} = \frac{\overset{1}{5}}{\overset{9}{1}} \cdot \frac{\overset{1}{9}}{\overset{5}{1}}C \qquad \text{Multiply both sides by } \tfrac{5}{9} \text{ to isolate C.}$$

$$30°C = C$$

In the solution, we subtracted 32 from 86 to get 54, and then multiplied that difference by $\frac{5}{9}$. This suggests that to convert degrees Fahrenheit to degrees Celsius, we subtract 32 from the given degrees Fahrenheit, and then multiply that difference by $\frac{5}{9}$.

Conclusion: $C = \frac{5}{9}(F - 32)$

Note: The parentheses are needed so that the subtraction occurs *before* the multiplication.

Example 6 Convert 80°F to degrees Celsius.

Solution: Because the desired unit is degrees Celsius, we use the formula $C = \frac{5}{9}(F - 32)$.

$$C = \frac{5}{9}(80 - 32) \qquad \text{Replace } F \text{ with 80.}$$

$$C = \frac{5}{9}(48) \qquad \text{Subtract.}$$

$$C = \frac{5}{\underset{3}{9}}\left(\frac{\overset{16}{48}}{1}\right) \qquad \text{Divide out the common factor, 3.}$$

$$C = 26\frac{2}{3}°C \text{ or } 26.\overline{6}°C \qquad \text{Multiply.}$$

Do Your Turn 6 ▶

Your Turn 6

Convert.

a. 77°F to degrees Celsius

b. 40.1°F to degrees Celsius

c. −28°F to degrees Celsius

d. 0°F to degrees Celsius

Answers to Your Turn 6
a. 25°C **b.** 4.5°C **c.** −33.$\overline{3}$°C
d. −17.$\overline{7}$°C

7.5 Exercises

For Extra Help

Videotape DVT

Tutor Center
Addison-Wesley
Tutor Center

Math XL
Math XL

MyMathLab

Student Solutions Manual

1. What conversion fact correctly completes the following dimensional analysis?

$$9 \text{ in.} \times \frac{?}{1 \text{ in.}} = 22.86 \text{ cm}$$

2. What conversion fact correctly completes the following dimensional analysis?

$$2.5 \text{ kg} \times \frac{?}{1 \text{ kg}} \approx 5.5 \text{ lb.}$$

3. What is the formula for converting degrees Celsius to degrees Fahrenheit?

4. What is the formula for converting degrees Fahrenheit to degrees Celsius?

For Exercises 5–44, convert.

5. A football field measures 100 yards. Convert the length of the field to meters.

6. A circular table has a diameter of 44 inches. Convert the diameter to centimeters.

7. The Boston Marathon is 26.2 miles. Convert to kilometers.

8. A mechanic uses a $\frac{3}{4}$-inch wrench. Convert to millimeters.

9. A carpenter uses a $\frac{3}{8}$-inch drill bit. Convert to millimeters.

10. A family builds a 4-foot fence. Convert to centimeters.

11. An athlete prepares for the 400-meter dash. Convert to yards.

12. A swimmer prepares for the 100-meter freestyle. Convert to feet.

13. A sign in Germany indicates that it is 40 kilometers to Berlin. Convert to miles.

14. A sign in France indicates that road construction is 0.4 kilometers ahead. Convert to feet.

15. An oceanographer measures a fish to be 45 centimeters. What is this in inches?

16. The width of a pencil is about 7 millimeters. What is this in inches?

17. How many liters are in a 16-ounce bottle of soda?

18. How many milliliters are in an 8-ounce can of soda?

19. A recipe calls for $2\frac{1}{2}$ cups of milk. Convert to milliliters.

20. How many liters are in a half-gallon of milk?

21. How many gallons are in a 3-liter cola?

22. How many ounces are in a 2-liter cola?

23. The instructions on a pain reliever for children indicate to give the child 20 milliliters. How many ounces is this?

24. A child is to receive 40 milliliters of liquid medicine. How many ounces is this?

25. A woman weighs 135 pounds. How many kilograms is this?

26. A man weighs 168 pounds. How many kilograms is this?

27. How many grams is a 12-ounce steak?

28. How many grams is a 2.5-ounce serving of rice?

29. A pill contains 500 milligrams of medicine. How many ounces is this?

30. A pill contains 300 milligrams of aspirin. How many ounces is this?

31. A person from Russia claims that her mass is 55 kilograms. What is her weight in pounds?

32. A person from Japan claims to have a mass of 68 kilograms. What is his weight in pounds?

33. Temperatures in the Sahara desert can reach $54\frac{4}{9}°$C. What is this in degrees Fahrenheit?

34. Temperatures on the moon can reach 127°C at noon. What is this in degrees Fahrenheit?

35. On a cold day in December, the temperature is reported to be 6°C. What is this in degrees Fahrenheit?

36. The lowest temperatures on the moon can reach -173°C just before dawn. What is this in degrees Fahrenheit?

37. Absolute zero is the temperature at which molecular energy is at a minimum. This temperature is -273.15°C. What is this in degrees Fahrenheit?

38. Liquid oxygen boils at a temperature of -182.96°C. What is this in degrees Fahrenheit?

DISCUSSION What is the boiling point of oxygen on the Kelvin scale?

39. Normal body temperature is 98.6°F. What is normal body temperature in degrees Celsius?

40. A recipe calls for the oven to be preheated to 350°F. What is this temperature in degrees Celsius?

41. On a hot summer day, the temperature is reported to be 102°F. What is this in degrees Celsius?

42. During the summer, it is suggested to keep the thermostat at 78°F. What is this in degrees Celsius?

43. During January, the temperature is reported to be -5°F. What is this in degrees Celsius ?

44. At the South Pole, the temperatures can get down to -40°F. What is this in degrees Celsius?

PUZZLE PROBLEM A common unit of antiquity was the cubit, which was the length of the forearm from the tip of the middle finger to the elbow. In the Biblical story of Noah's ark, the ark was to be built 300 cubits long, 50 cubits wide, and 30 cubits high. If the cubit was about 1.5 feet in length, what were the dimensions of the ark in feet? Yards? Meters?

Review Exercises

For Exercises 1 and 2, evaluate.

[6.4] **1.** $896 \div 3200$

[6.5] **2.** $(90 + 110 + 84) \div 2400$

[5.7] **3.** Combine like terms. $-\frac{1}{3}x + \frac{2}{5}y - x - \frac{3}{4}y$

For Exercises 4 and 5, solve and check.

[5.8] **4.** $2(t + 1) = 5t$

[7.2] **5.** $\frac{m}{20.4} = \frac{8}{15}$

7.6 Applications and Problem Solving

OBJECTIVE **1** Use debt-to-income ratios to decide loan qualification.

In Section 7.1, we considered ratios of debt to income. Lenders consider debt-to-income ratios to determine whether an applicant qualifies for a loan. The two ratios commonly used are **front-end** and **back-end ratios.**

DEFINITIONS

Front-end ratio: The ratio of the total monthly house payment to gross monthly income.

Back-end ratio: The ratio of the total monthly debt payments to gross monthly income.

The total monthly house payment includes the principal and interest for the mortgage, property taxes, and homeowners insurance premiums. It is usually referred to as the PITI payment.

Suppose a family with a gross monthly income of $4000 is trying to qualify for a loan. If it were to get the loan, the monthly PITI payment would be approximately $1000.

$$\text{front-end ratio} = \frac{1000}{4000} = 0.25$$

Because the general guideline for lenders on a conventional loan is that the front-end ratio should not be greater than 0.28, this family would meet the qualification for the front-end ratio.

Now, let's consider the back-end ratio. The total monthly debt includes the monthly PITI payment, minimum payments for all credit cards, all auto loan payments, and student loan payments. Sometimes student loans and car loans are not considered if they are to be paid off within a certain time frame. Suppose the family has the following monthly debts:

$$\text{monthly PITI payment} = \$1000$$
$$\text{credit card 1 minimum payment} = \$50$$
$$\text{credit card 2 minimum payment} = \$40$$
$$\text{car loan} = \$253$$
$$\underline{\text{student loan} = \$104}$$
total monthly debt = $1447

Now we can calculate the back-end ratio.

$$\text{back-end ratio} = \frac{1447}{4000} = 0.36175.$$

The general guideline for lenders on a conventional loan is that the back-end ratio should not be greater than 0.36. This family's back-end ratio is just slightly greater than the maximum allowed value for the back-end ratio.

In summary, the front-end ratio for this family tells us that it has enough income to make the monthly PITI payment, but the back-end ratio indicates that it has a bit too much total debt.

These two ratios are only part of what lenders consider in deciding whether a person qualifies for a loan or mortgage. Two other factors considered are credit score and the type of loan.

Credit scores are calculated by national credit bureaus. Generally, lenders will request credit reports from several different bureaus. For example, it is common to request three reports and use the middle of the three scores in the decision process. Credit scores range from about 450 up to 900. A score of 650 is generally considered acceptable, greater than 650 is good, and less than 650 is considered risky.

OF INTEREST

Some factors that impact a credit score in order of importance are

1. Payment history or unresolved collections (Nonpayment and late payment have a negative impact.)

2. Excessive lines of credit (More than three different lines of credit has a negative impact.)

3. Balances (A balance at or near the maximum has a negative impact.)

The type of loan affects the factors considered as well. For example, with VA (Veteran Affairs) loans, the front-end ratio and credit score are not considered. The back-end ratio can be up to 0.41. Of course, the primary qualification with a VA loan is that one must be a veteran. With an FHA (Federal Housing Authority) loan, the credit score is not considered. The front-end ratio can be up to 0.29 and back-end ratio up to 0.41.

Table 7.1 Factors for loan qualification

	Conventional	VA	FHA
1. Front-end ratio should not exceed	0.28	N/A	0.29
2. Back-end ratio should not exceed	0.36	0.41	0.41
3. Credit score should be	650 or higher	N/A	N/A

Example 1   The Jones family is trying to qualify for a conventional loan. Its gross monthly income is $4200. If it gets the loan, the monthly PITI payment will be approximately $985. The total monthly debt excluding the PITI payment is $535. The credit score is 720. Does the family qualify?

Understand: We are to decide whether the Jones family qualifies for the loan. Because the family is applying for a conventional loan, we must consider the front-end ratio, back-end ratio, and credit score.

Plan: 1. Calculate the front-end ratio. The front-end ratio should not exceed 0.28.

$$\text{front-end ratio} = \frac{\text{monthly PITI payment}}{\text{gross monthly income}}$$

2. Calculate the back-end ratio. The back-end ratio should not exceed 0.36.

$$\text{back-end ratio} = \frac{\text{total monthly debt payments}}{\text{gross monthly income}}$$

3. Consider the credit score.

Execute: front-end ratio $= \dfrac{985}{4200} \approx 0.235$

back-end ratio $= \dfrac{1520}{4200} \approx 0.362$

credit score $= 720$

Note: The total monthly debt payment is found by adding the monthly PITI payment to the current monthly debt.

$985 + 535 = 1520$

Answer: The back-end ratio is just slightly greater than 0.36. However, the Jones family has a good credit score at 720. The good credit score should be taken into consideration to offset the 0.362 back-end ratio. The family would most likely qualify for the loan.

Check: Verify the ratio calculations by reversing the process. This will be left to the reader.

Do Your Turn 1 ▷

Your Turn 1

Use Table 7.1 to solve.

a. The Cicone family is trying to qualify for a conventional loan. The gross monthly income is $2870. If it gets the loan, the monthly PITI payment will be approximately $897. The total monthly debt excluding the monthly PITI payment is $223. The family has a credit score of 700. Does the family qualify?

b. The Mathias family is trying to qualify for an FHA loan. Its gross monthly income is $2580. If the family gets the loan, the monthly PITI payment will be approximately $710. The family has the following monthly debt payments:

credit card 1 = $25

car loan 1 = $265

student loan = $78

Their credit score is 750. Does the Cicone family qualify?

Answers to Your Turn 1
a. front-end = 0.313; back-end = 0.39; The family does not qualify.
b. front-end = 0.275; back-end = 0.418; Though the back-end ratio is just slightly greater than the required 0.41, the other factors warrant qualification.

One useful piece of information in looking to buy or build a home is the maximum monthly PITI payment for which you can qualify.

Your Turn 2

Suppose a person has a gross monthly income of $1680. What is the maximum monthly PITI payment that would meet the front-end ratio qualification for a FHA loan?

Example 2 Suppose a family has a gross monthly income of $3000. What is the maximum monthly PITI payment that would meet the front-end ratio qualification for a conventional loan?

Understand: We are to calculate the maximum monthly PITI payment that would meet the front-end ratio qualification for a conventional loan. The general guideline is that the front-end ratio should not exceed 0.28. The front-end ratio is calculated as follows:

$$\text{front-end ratio} = \frac{\text{monthly PITI payment}}{\text{gross monthly income}}$$

Plan: Let p represent the monthly PITI payment. Replace each part of the front-end ratio formula with appropriate values and solve for p.

Execute.

$$0.28 \quad \frac{p}{3000}$$

$$\frac{3000}{1} \cdot \frac{0.28}{1} = \frac{p}{3000} \cdot \frac{3000}{1} \qquad \text{Multiply both sides by 3000 to isolate } p.$$

$$840 = p$$

Answer: The maximum PITI payment that would meet the front-end ratio qualification for a conventional loan with a gross monthly income of $3000 is $840.

Check: Verify that the front-end ratio with a monthly PITI payment of $840 and gross monthly income of $3000 is in fact 0.28.

$$\text{front-end ratio} = \frac{840}{3000} = 0.28$$

◁ **Do Your Turn 2**

In addition to finding the maximum monthly PITI payment to qualify with the front-end ratio, it is helpful to also calculate the maximum debt one can have yet still satisfy the back-end ratio qualification.

Example 3 Suppose a family has a gross monthly income of $3200. What is the maximum debt that would meet the back-end ratio qualification for an FHA loan?

Understand: We are to calculate the maximum debt that would meet the back-end ratio qualification for an FHA loan. The general guideline is that the back-end ratio should not exceed 0.41. The back-end ratio is calculated as follows:

$$\text{back-end ratio} = \frac{\text{total monthly debt payments}}{\text{gross monthly income}}$$

Answer to Your Turn 2
$487.20

Plan: Let d represent the total monthly debt. Replace each part of the back-end ratio formula with appropriate values and solve for d.

Execute:
$$0.41 = \frac{d}{3200}$$

$$\frac{3200}{1} \cdot \frac{0.41}{1} = \frac{d}{3200} \cdot \frac{\overset{}{3200}}{1} 6$$

$$1312 = d$$

Answer: The maximum monthly debt that would meet the back-end ratio qualification for an FHA loan with a gross monthly income of $3200 is $1312.

Check: Verify that the back-end ratio with a total monthly debt of $1312 and gross monthly income of $3200 is in fact 0.41.

$$\text{back-end ratio} = \frac{1312}{3200} = 0.41$$

Do Your Turn 3 ▶

Your Turn 3

Suppose a family has a gross monthly income of $2350. What is the maximum debt that would meet the back-end ratio qualification for a VA loan?

OBJECTIVE 4 Solve problems involving two objects moving in the same direction.

In Section 5.8, we considered distance-rate-time problems in which objects were moving in opposite directions. What if the objects are traveling in the same direction? The usual objective is to calculate the time for one object to catch up to another object ahead. We will use the same table.

Categories	Rate	Time	Distance

Example 4 John and Karen are traveling south in separate cars on the same interstate. Karen is traveling at 65 miles per hour and John at 70 miles per hour. Karen passes Exit 38 at 1:15 P.M. John passes the same exit at 1:30 P.M. At what time will John catch up to Karen?

Understand: To find the time that John catches up to Karen, we must calculate the amount of time it will take him to catch up to her. We can then add that time to 1:30 P.M.

John passed Exit 38 15 minutes after Karen did. In order to use $d = rt$, the time units must match the time units within the rate. Because 65 and 70 are in miles per hour, the time units must be in hours, so we must convert 15 minutes to hours.

$$15 \text{ min.} = \frac{\overset{1}{\cancel{15}} \text{ min.}}{1} \cdot \frac{1 \text{ hr.}}{\underset{4}{\cancel{60}} \text{ min.}} = \frac{1}{4} \text{ hr.} \quad \text{or} \quad 0.25 \text{ hr.}$$

Let's complete the table:

We will let t represent the amount of time it takes John to catch up to Karen. But, Karen's travel time is a bit tricky. Suppose John catches up at 2:15 P.M. Notice that this is 1 hour after Karen passed Exit 38. For John, it is only 45 minutes after he passed Exit 38.

Conclusion: At the time John catches up, Karen's travel time from Exit 38 is 15 minutes (0.25 hour) more than John's travel time from the same exit. This will be the case no matter what time he catches up, so Karen's travel time is $t + 0.25$.

Answer
$963.50

Carlos and Rita are running in a marathon. Carlos passes a checkpoint at 10:30 A.M. Rita passes the same checkpoint at 11:00 A.M. Carlos is pacing himself at 4 miles per hour while Rita is pushing at 5 miles per hour. How much time will it take Rita to catch up with Carlos?

Categories	Rate	Time	Distance
Karen	65	$t + 0.25$	$65(t + 0.25)$
John	70	t	$70t$

Note: Because $d = rt$, we multiply the rate and time values to equal the distance values.

What can we conclude about the distances they each will have traveled when John catches up? Notice that Exit 38 is the common reference point from which we describe the time for John to catch up. When John catches up, he and Karen will be the same distance from Exit 38, so their distances are equal.

Plan: Because they will have gone the same distance past Exit 38 when John catches up, we set the expressions of their individual distances equal and solve for t.

Execute: John's distance = Karen's distance (at the time he catches up)

Note: The coefficients of the like terms are the individual rates, so $70 - 65 = 5$ means that John's rate relative to Karen's is 5 miles per hour. This means he must make up the 16.25 miles that separates them at a rate of 5 miles per hour.

$$70t = 65(t + 0.25)$$
$$70t = 65t + 16.25$$
$$70t = 65t + 16.25$$
$$\underline{-65t \quad -65t}$$
$$5t = 0 + 16.25$$
$$\frac{5t}{5} = \frac{16.25}{5}$$
$$t = 3.25$$

Note: Since 65 is the rate and 0.25 is the additional amount of time, their product, 16.25 miles, is the distance that Karen is ahead of John. John must make up this distance to catch her.

Note: Dividing 16.25 miles by 5 miles per hour calculates the time it takes John to make up the distance between Karen and him.

Answer: It will take John 3.25 hours, which is 3 hours and 15 minutes, after passing Exit 38 to catch up to Karen. Since he passed Exit 38 at 1:30 P.M., he catches up to Karen at 4:45 P.M.

Check: Use $d = rt$ to verify that John and Karen are equal distances from Exit 38 after traveling their respective amounts of time. If John travels for 3.25 hours after Exit 38, then Karen travels for $3.25 + 0.25 = 3.5$ hours after Exit 38.

John's distance: $d = (70)(3.25)$
$d = 227.5$ mi.

Karen's distance: $d = (65)(3.5)$
$d = 227.5$ mi.

◁ **Do Your Turn 4**

Answer
2 hr.

OBJECTIVE 5 Calculate medical dosages.

We can use the method of dimensional analysis to determine appropriate medical dosages. Recall that the main idea is to note the initial unit and the desired unit. Then multiply by appropriate unit fractions so that the undesired units divide out, leaving the desired units.

Example 5 A patient with a mass of 42 kilograms is to receive an antibiotic. The order is to administer 20 milligrams per kilogram. How much of the antibiotic should be given?

Understand: The order of 20 milligrams per kilogram means that the patient should receive 20 milligrams of the antibiotic for every 1 kilogram of the patient's mass.

Plan: Multiply the patient's mass by the dosage per unit mass.

Execute: total dose = patient's mass · dosage per unit of mass

$$\text{total dose} = \frac{42 \text{ kg}}{1} \cdot \frac{20 \text{ mg}}{1 \text{ kg}} = 840 \text{ mg}$$

Answer: The patient should receive 840 milligrams of the antibiotic.

Check: Verify that a 42-kilogram patient receiving 840 milligrams of antibiotic receives 20 milligrams of antibiotic per kilogram of mass.

$$\frac{840 \text{ mg}}{42 \text{ kg}} = 20 \text{ mg/kg}$$

Do Your Turn 5 ▷

Your Turn 5

A patient with a mass of 68 kilograms is to receive isoniazid, an antitubercular drug. The order is to administer 10 milligrams per kilogram. How much isoniazid should be given?

Many medications are administered in intravenous (IV) solutions. A regulator is used to regulate the number of drops of the solution the patient receives over a period of time (usually 1 minute). Medical orders are usually not written in terms of the number of drops per minute so, to properly set the regulator, they must be converted to number of drops per minute.

Example 6 A patient is to receive 500 milliliters of 5% D/W solution IV over 4 hours. The label on the IV bag indicates that 10 drops equals to 1 milliliter of the solution. How many drops should the patient receive each minute?

Understand: The order is for 500 milliliters over 4 hours. With this particular IV, 10 drops of the solution is equal to 1 milliliter. We can write these rates as:

$$\frac{500 \text{ ml}}{4 \text{ hr.}} \quad \text{and} \quad \frac{10 \text{ drops}}{1 \text{ ml}}$$

We want to end up with a description of the number of drops per minute. This means that milliliters and hours are undesired units. We need to convert hours to minutes and divide out milliliters completely.

Plan: Multiply the rates to eliminate milliliters. Use the fact that 1 hour is 60 minutes to convert hours to minutes.

Execute: $\dfrac{500 \text{ ml}}{4 \text{ hr.}} \cdot \dfrac{10 \text{ drops}}{1 \text{ ml}} \cdot \dfrac{1 \text{ hr.}}{60 \text{ min}} = \dfrac{5000 \text{ drops}}{240 \text{ min.}} = 20.8\overline{3} \text{ drops/min.}$

Answer: The patient should receive about 20 to 21 drops per minute.

Answer to Your Turn 5
680 mg

A patient is to receive 250 cubic centimeters of 5% D/W solution IV over 4 hours. The label on the box of the IV indicates that 15 drops dissipates 1 cubic centimeter of the solution. How many drops should the patient receive each minute?

Check: Verify that a rate of 21 drops per minute with every 10 drops equal to 1 milliliter is a total of 500 milliliters after 4 hours.

$$\frac{4 \text{ hr.}}{1} \cdot \frac{60 \text{ min.}}{1 \text{ hr.}} \cdot \frac{21 \text{ drops}}{1 \text{ min.}} = 5040 \text{ drops}$$

If the patient recieves 5040 drops in 4 hours and every 10 drops is 1 milliliter, we can calculate the total number of milliliters that the patient receives:

$$\frac{5040 \text{ drops}}{1} \cdot \frac{1 \text{ ml}}{10 \text{ drops}} = \frac{5040}{10} \text{ ml} = 504 \text{ ml}$$

Because we rounded the number of drops to whole amounts, this result does not match 500 milliliters exactly. However, it is reasonable.

◄ **Do Your Turn 6**

Answer to Your Turn 6
15 to 16 drops per minute

7.6 **Exercises**

For Extra Help					
Videotape DVT		Addison-Wesley Tutor Center	Math XL	MyMathLab	Student Solutions Manual

1. What is a front-end ratio?

2. What is a back-end ratio?

3. What is the PITI payment?

4. A car passes under a bridge on a highway. A half-hour later, a second car passes under the same bridge traveling in the same direction at a speed greater than the first car. Suppose the second car catches up to the first car in t hours. Write an expression for the first car's travel time (in hours) since passing under the bridge.

For Exercises 5–20, use the table to solve.

Factors for loan qualification

	Conventional	VA	FHA
1. Front-end ratio should not exceed	0.28	N/A	0.29
2. Back-end ratio should not exceed	0.36	0.41	0.41
3. Credit score should be	650 or higher	N/A	N/A

5. The Wu family is trying to qualify for a conventional loan. Its gross monthly income is $3850. If the family gets the loan, the monthly PITI payment will be approximately $845. Its monthly debt excluding the monthly PITI payment is $435. The family has a credit score of 700. Does the Wu family qualify?

6. The Deas family is trying to qualify for a conventional loan. Its gross monthly income is $3240. If the family gets the loan, the monthly PITI payment will be approximately $925. The total monthly debt excluding the monthly PITI payment is $495. The credit score is 680. Does the Deas family qualify?

7. The Rivers family is trying to qualify for a VA loan. The gross monthly income is $2845. If the family gets the loan, the monthly PITI payment will be approximately $756. The credit score is 660. The family has the following monthly debt payments:

$$\text{credit card } 1 = \$30$$
$$\text{credit card } 2 = \$45$$
$$\text{car loan } 1 = \$279$$
$$\text{student loan} = \$93$$

Does the Rivers family qualify?

8. The Santana family is trying to qualify for a VA loan. The gross monthly income is $4200. If the family gets the loan, the monthly PITI payment will be approximately $987. The credit score is 650. The family has the following monthly debt payments:

$$\text{credit card 1} = \$35$$
$$\text{credit card 2} = \$85$$
$$\text{car loan 1} = \$305$$
$$\text{car loan 2} = \$289$$

Does the Santana family qualify?

9. The Bishop family is trying to qualify for an FHA loan. The gross monthly income is $2530. If the family gets the loan, the monthly PITI payment will be approximately $785. The family has the following monthly debt payments:

$$\text{credit card 1} = \$40$$
$$\text{car loan} = \$249$$
$$\text{student loan} = \$65$$

The credit score is 700. Does the Bishop family qualify?

10. The Nicks family is trying to qualify for an FHA loan. The gross monthly income is $2760. If the family gets the loan, the monthly PITI payment will be approximately $805. The family has the following monthly debt payments:

$$\text{credit card 1} = \$25$$
$$\text{credit card 2} = \$35$$
$$\text{car loan 1} = \$279$$

The credit score is 650. Does the Nicks family qualify?

11. Suppose a family has a gross monthly income of $4800. What is the maximum monthly PITI payment that would meet the front-end ratio qualification for a conventional loan?

12. Suppose a family has a gross monthly income of $2980. What is the maximum monthly PITI payment that would meet the front-end ratio qualification for a conventional loan?

13. Suppose a person has a gross monthly income of $1840. What is the maximum monthly PITI payment that would meet the front-end ratio qualification for a FHA loan?

14. Suppose a person has a gross monthly income of $1580. What is the maximum monthly PITI payment that would meet the front-end ratio qualification for a FHA loan?

15. Suppose a family has a gross monthly income of $3480. What is the maximum debt that would meet the back-end ratio qualification for a conventional loan?

16. Suppose a family has a gross monthly income of $2240. What is the maximum debt that would meet the back-end ratio qualification for a conventional loan?

17. Suppose a family has a gross monthly income of $1675. What is the maximum debt that would meet the back-end ratio qualification for a VA loan?

18. Suppose a family has a gross monthly income of $2110. What is the maximum debt that would meet the back-end ratio qualification for a VA loan?

19. Suppose a family has a gross monthly income of $2375. What is the maximum debt that would meet the back-end ratio qualification for a FHA loan?

20. Suppose a family has a gross monthly income of $2680. What is the maximum debt that would meet the back-end ratio qualification for a FHA loan?

21. Terese and Brian are traveling south in separate cars on the same interstate. Terese is traveling at 60 miles per hour and Brian at 70 miles per hour. Terese passes Exit 62 at 2:30 P.M. Brian passes the same exit at 3:30 P.M. At what time will Brian catch up to Terese?

22. Sherryl and Darrin are traveling west in separate cars on the same interstate. Darrin is traveling at 65 miles per hour and Sherryl at 70 miles per hour. Darrin passes a rest area at 10:05 A.M. Sherryl passes the same rest area at 10:25 P.M. At what time will Sherryl catch up to Darrin?

23. Liz and Tyler are riding bicycles in the same direction on the same trail. Liz passes a marker at 12:18 P.M. Tyler passes the same marker at 12:33 P.M. If Liz is averaging 8 miles per hour and Tyler is averaging 10 miles per hour, at what time will Tyler catch up to Liz?

24. Nikki and Berry are running in the same direction on the same trail. Berry passes a marker at 8:10 A.M. Nikki passes the same marker at 8:20 A.M. If Berry is averaging 4 miles per hour and Nikki is averaging 6 miles per hour, at what time will Nikki catch up to Berry?

25. A patient with a mass of 55 kilograms is to receive an antibiotic. The order is to administer 9.5 milligrams per kilogram. How much of the antibiotic should be given?

26. A patient with a mass of 81 kilograms is to receive 1.4 milligrams per kilogram of metroprolol tartrate (an antihypertensive drug). How much of the drug should the patient receive?

27. A patient weighing 132 pounds is to receive 0.04 milligram per kilogram of clonazepam (an anticonvulsant drug). If each tablet contains 0.5 milligram, how many tablets should the patient receive?

28. A patient weighing 165 pounds is to receive 7.2 milligrams per kilogram of azithromycin. If each capsule contains 300 milligrams, how many capsules should the patient receive?

29. A patient is to receive 250 milliliters of 5% D/W solution IV over 6 hours. The label on the box of the IV indicates that 15 drops dissipate 1 milliliter of the solution. How many drops should the patient receive each minute?

30. A patient is to receive 500 milliliters of 5% D/W solution IV over 2 hours. The label on the box of the IV indicates that 15 drops dissipate 1 milliliter of the solution. How many drops should the patient receive each minute?

31. 500 milliliters of 5% D/W solution IV contains 20,000 units of heparin. If the patient receives 10 milliliters per hour, how many units is the patient receiving each hour?

32. 500 milliliters of 5% D/W solution IV contains 20 units of pitocin. If the patient receives 0.002 unit per minute, how many milliliters is she receiving in an hour?

Review Exercises

[7.1]   **1.** Simplify the ratio 0.14 to 0.5.

[7.1]   **2.** What is the probability of randomly selecting a King or Jack from a shuffled standard deck of cards?

[7.1]   **3.** A 16-ounce bottle of apple juice costs $2.29. What is the unit price in dollars per ounce?

[7.3]   **4.** Convert 75 feet to yards.

[7.4]   **5.** Convert 8.5 grams to milligrams.

Defined Terms

Review the following terms and for those you do not know, study its definition on the page number next to it.

Section 7.1
Ratio *(p. 488)*
Theoretical probability of
   equally likely outcomes
   (p. 491)

Unit ratio *(p. 491)*
Rate *(p. 492)*

Section 7.2
Proportion *(p. 500)*
Similar figures *(p. 504)*

Section 7.3
Unit fraction *(p. 512)*
Capacity *(p. 515)*

Section 7.4
Base unit *(p. 521)*

Section 7.6
Front-end ratio *(p. 538)*
Back-end ratio *(p. 538)*

Procedures, Rules, and Key Examples

Procedures/Rules

Section 7.1 Ratios, Probability, and Rates

For two quantities a and b, the ratio of a to b is $\dfrac{a}{b}$.

To simplify a ratio of two fractions or mixed numbers, divide the fractions or mixed numbers.

To clear decimals in a ratio, multiply the numerator and denominator by an appropriate power of 10 as determined by the decimal number with the most decimal places.

Key Example(s)

Example 1:

A roof has a vertical rise of 8 inches for every 13 inches of horizontal length. What is the ratio of vertical rise to horizontal length?

Answer: $\dfrac{8}{13}$

Example 2:

A recipe calls for $\frac{1}{2}$ cup of flour to $2\frac{1}{4}$ cups of milk. Write the ratio of flour to milk in simplest form.

$$\frac{\frac{1}{2}}{2\frac{1}{4}} = \frac{1}{2} \div 2\frac{1}{4}$$

$$= \frac{1}{2} \div \frac{9}{4}$$

$$= \frac{1}{\overset{}{\underset{1}{2}}} \cdot \frac{\overset{2}{\cancel{4}}}{9}$$

$$= \frac{2}{9}$$

Example 3:

A small plastic tube has a diameter of 0.08 centimeters and a length of 12.5 centimeters. What is the ratio of length to diameter?

$$\frac{12.5}{0.08} = \frac{12.5(100)}{0.08(100)} = \frac{1250}{8} = \frac{625}{4}$$

Procedures/Rules	Key Example(s)

To write a probability:

1. Write the number of favorable outcomes in the numerator.
2. Write the total number of possible outcomes in the denominator.
3. Simplify.

To calculate a unit ratio, divide the denominator into the numerator.

Example 4:

Write the probability of drawing a red 10 from a standard deck of cards in simplest form.

There are 2 red 10's in a standard deck of 52 cards.

$$P = \frac{2}{52} = \frac{1}{26}$$

Example 5:

The stock for a chemical company sold for $16\frac{1}{4}$ at the close of the market one day. If the company's annual profit on each share of stock was $1.85 the previous year, what is the price-to-earnings ratio?

$$\frac{16\frac{1}{4}}{1.85} = \frac{16.25}{1.85} \approx 8.78$$

Section 7.2 Proportions

If two ratios are proportional, then their cross products are equal.

In math language: If $\frac{a}{b} = \frac{c}{d}$, where $b \neq 0$ and $d \neq 0$, then $ad = bc$.

Example 1:

Determine whether the ratios are proportional.

$$9.6 \cdot 8 = 76.8 \qquad 6.4 \cdot 12 = 76.8$$

$$\frac{8}{6.4} \overset{?}{=} \frac{12}{9.6}$$

The cross products are equal, therefore the ratios are proportional.

To solve a proportion using cross products:

1. Calculate the cross products.
2. Set the cross products equal to one another.
3. Use the multiplication/division principle of equality to isolate the variable.

Example 2:

Solve. $\dfrac{2.8}{-9} = \dfrac{m}{15}$

$$15(2.8) = 42 \qquad\qquad -9 \cdot m = -9m$$

$$\frac{2.8}{-9} \overset{?}{=} \frac{m}{15}$$

$$42 = -9m$$

$$\frac{42}{-9} = \frac{-9m}{-9}$$

$$-4.\overline{6} = m$$

To solve proportion problems:

1. Set up the given ratio any way you wish.
2. Set the given ratio equal to the other ratio with the unknown so that the numerators and denominators correspond logically.
3. Solve using cross products.

Example 3:

A garden plan shows 30 plants covering 96 square feet. How many plants would be needed to cover a garden area that is 140 square feet?

$$\frac{30 \text{ plants}}{96 \text{ ft.}^2} = \frac{n \text{ plants}}{140 \text{ ft.}^2}$$

$$4200 = 96n$$

$$\frac{4200}{96} = \frac{96n}{96}$$

$$43.75 = n$$

Answer: Only whole plants can be purchased, so the larger garden would require about 44 plants.

Section 7.3 American Measurement

To convert units using dimensional analysis, multiply the given measurement by unit fractions so that the undesired units divide out, leaving the desired units.

Notes: If an undesired unit is in a numerator, that unit should appear in the denominator of a unit fraction.

If an undesired unit is in the denominator, that unit should appear in the numerator of a unit fraction.

Relations for conversions involving American units:

Length:
1 ft. = 12 in.
1 yd. = 3 ft.
1 mi. = 5280 ft.

Area:
1 ft.^{2} = 144 in.^{2}
1 yd.^{2} = 9 ft.^{2}

Capacity:
1 c. = 8 oz.
1 pt. = 2 c.
1 qt. = 2 pt.
1 gal. = 4 qt.

Weight:
1 lb. = 16 oz.
1 T = 2000 lb.

Time:
1 yr. = $365\frac{1}{4}$ d.
1 d. = 24 hr.
1 hr. = 60 min.
1 min. = 60 sec.

Example 1: Convert.

a. 4.5 feet to inches
$$4.5 \text{ ft.} = \frac{4.5 \text{ ft.}}{1} \cdot \frac{12 \text{ in.}}{1 \text{ ft.}} = 54 \text{ in.}$$

b. 20 cups to quarts
$$20 \text{ c.} = \frac{20 \text{ c.}}{1} \cdot \frac{1 \text{ pt.}}{2 \text{ c.}} \cdot \frac{1 \text{ qt.}}{2 \text{ pt.}} = \frac{20}{4} \text{qt.} = 5 \text{ qt.}$$

Section 7.4 Metric Measurement

To convert metric units of length, capacity, or mass, use the following chart:

kilo-	hecto-	deka-	base unit	deci-	centi-	milli-
(k)	(h)	(da)		(d)	(c)	(m)

Memory tip:
King **H**ector's **D**ata **B**ase **D**ecidedly **C**entralized **M**illions

1. Locate the given unit on the chart.
2. Count the number of units and note the direction (right or left) to get to the desired unit.
3. Move the decimal point in the given measurement the same number of places and in the same direction.

Notes:
In capacity, the cubic centimeter (cc) represents the same amount as the milliliter (ml).

The metric ton is 1000 kilograms, or 3 steps on the chart to the left of kilograms.

To convert units of area:

Move the decimal point twice the number of places as units from the initial unit to the desired unit on the chart.

Example 1: Convert.

a. 2.8 meters to centimeters

(Centimeter is 2 places to the right of meter on the chart, so the decimal point moves right 2 places.)
$$2.8 \text{ m} = 280 \text{ cm}$$

b. 97,500 decigrams to kilograms

(Kilogram is 4 places to the left of decigram on the chart, so the decimal point moves 4 places to the left.)
$$97,500 \text{ dg} = 9.75 \text{ kg}$$

c. 0.005 square centimeters to square millimeters
$$0.005 \text{ cm}^2 = 0.5 \text{ mm}^2$$

Section 7.5 Converting Between the American and Metric Systems

Facts for converting between the American and metric systems:

Length:

1 in. = 2.54 cm   1 m ≈ 3.282 ft.

1 yd. ≈ 0.914 m   1 m ≈ 1.094 yd.

1 mi. ≈ 1.609 km   1 km ≈ 0.621 mi.

Capacity:

1 qt. ≈ 0.946 l

1 l ≈ 1.057 qt.

Weight/Mass:

1 lb. ≈ 0.454 kg   1 kg ≈ 2.2 lb.

1 T ≈ 0.907 t   1 t ≈ 1.1 T

Example 1: Convert.

a. 50 yards to meters

$$50 \text{ yd.} \approx \frac{50 \cancel{\text{ yd.}}}{1} \cdot \frac{0.914 \text{ m}}{1 \cancel{\text{ yd.}}} \approx 45.7 \text{ m.}$$

b. 3 liters to pints

$$3 \text{ l.} \approx \frac{3 \cancel{l}}{1} \cdot \frac{1.057 \cancel{\text{ qt.}}}{1 \cancel{l}} \cdot \frac{2 \text{ pt.}}{1 \cancel{\text{ qt.}}} \approx 6.342 \text{ qt.}$$

c. 192 ounces to kilograms

$$192 \text{ oz.} \approx \frac{192 \cancel{\text{ oz.}}}{1} \cdot \frac{1 \cancel{\text{ lb.}}}{16 \cancel{\text{ oz.}}} \cdot \frac{0.454 \text{ kg.}}{1 \cancel{\text{ lb.}}} \approx 5.448 \text{ kg.}$$

Formulas

Convert °C to °F: $F = \dfrac{9}{5}C + 32$

Convert °F to °C: $C = \dfrac{5}{9}(F - 32)$

$$\text{front-end ratio} = \frac{\text{monthly PITI payment}}{\text{gross monthly income}}$$

$$\text{back-end ratio} = \frac{\text{total monthly debt payments}}{\text{gross monthly income}}$$

For Exercises 1–6, answer true or false.

[7.1]   **1.** Every ratio is a rational number.

[7.1]   **2.** The ratio 0.9 to 0.5 can be expressed as $\dfrac{9}{5}$.

[7.2]   **3.** The cross products of a proportion are equal.

[7.1]   **4.** A unit price is the ratio of quantity to price.

[7.1]   **5.** The probability of rolling a 7 on a die with six sides is 0.

[7.4]   **6.** 1 cubic centimeter is the same amount as 1 centiliter.

[7.1]   **7.** To calculate probability, write a ratio with the number of _____ outcomes in the numerator and the total number of _____ outcomes in the denominator.

[7.1]   **8.** The denominator of a unit ratio is always equal to ___.

[7.1]   **9.** To calculate a unit ratio, divide the _____ into the _____.

[7.3]   **10.** To convert a given measurement using dimensional analysis, multiply the given measurement by a unit fraction that has the given unit in the _____.

For Exercises 11 and 12, write each ratio in simplest form.

[7.1]   **11.** A company has 32 male employees and 22 female employees. Write the ratio of female to male employees.

[7.1]   **12.** The shortest side on a right triangle measures 2.5 inches. The hypotenuse measures 10.8 inches. Write the ratio of the shortest side to the hypotenuse.

[7.1] *For Exercises 13 and 14, calculate the probability in simplest form.*

13. What is the probability of randomly selecting a Queen or King from a shuffled standard deck of cards?

14. What is the probability of rolling a 1 or a 2 on a die that has six sides?

[7.1] *For Exercises 15 and 16, calculate the unit ratio.*

15. A college has 4500 students and 275 faculty. Write a unit ratio of students to faculty.

16. The price for each share of stock for a communications company closed at $12\frac{5}{8}$. The previous year's annual earning for the stock was $1.78. Calculate the price-to-earnings ratio.

[7.1] *For Exercises 17 and 18, calculate the rate.*

17. Sekema drove 210.6 miles in 3 hours. What was her average rate in miles per hour?

18. A phone call lasting 28 minutes costs $2.52. What is the rate in cents per minute?

[7.1] *For Exercises 19 and 20, calculate the unit price.*

19. A 40-ounce bag of sugar costs $1.79. What is the unit price in cents per ounce?

20. A 64-ounce bottle of juice costs $2.49. What is the unit price in cents per ounce?

[7.1] *For Exercises 21 and 22, determine which is the better buy.*

21. $0.79 for a 32-ounce cup of cola
or
$0.69 for a 20-ounce cup of cola

22. two 12-ounce cans of name brand frozen juice for $0.99
or
four 8-ounce cans of store brand frozen juice for $1.45

[7.2] *For Exercises 23–26, solve.*

23. $\dfrac{1.8}{-5} = \dfrac{k}{12.5}$

24. $\dfrac{2}{m} = \dfrac{3\frac{1}{4}}{15}$

25. The instructions on the back of a can of wood sealant indicate that 3.78 liters will cover about 25 square meters of mildly porous wood. If the can contains 4.52 liters of sealant, how many square meters of mildly porous wood will it cover?

26. On a map, $\frac{1}{4}$ inch represents 20 miles. Suppose two cities are $3\frac{1}{2}$ inches apart on the map. How far apart are the two cities in miles?

[7.3] *For Exercises 27–32, use dimensional analysis to convert.*

27. 12 yards to inches

28. 40 ounces to pounds

29. 6 quarts to ounces

30. 2.25 hours to minutes

31. A room that is 14 feet wide by 15.5 feet long is to be carpeted. Calculate the area of the room in square yards.

32. A meteor streaks through Earth's upper atmosphere at a speed of 116,160 feet per second. Calculate the speed in miles per hour.

[7.4, 7.5] *For Exercises 33–42, convert.*

33. 5 meters to decimeters

34. 0.26 kilograms to grams

35. 950,000 grams to metric tons

36. 0.075 liters to cubic centimeters

37. 20 feet to meters

38. 12 miles to kilometers

39. 145 pounds to kilograms

40. 3 liters to gallons

41. 90°F to degrees Celsius

42. −4°C to degrees Fahrenheit

[7.6] **43.** A patient is to receive 450 milliliters of NS (normal saline) IV in 6 hours. If the IV dissipates 1 milliliter every 10 drops, what is the drip rate in drops per minute?

[7.6] **44.** A patient is to receive an antibiotic. The patient has a mass of 32 kilograms. The order for the antibiotic is 2.9 milligrams per kilogram. How much should the patient receive?

[7.6] **45.** A family applying for a loan has a gross monthly income of $3450. If the monthly PITI payment will be $978, what is the ratio of monthly PITI payment to gross monthly income (front-end ratio)?

[7.6] **46.** A family has the following monthly debts:

$$\text{monthly PITI} = \$565$$
$$\text{credit card} = \$45$$
$$\text{car loan} = \$225$$
$$\text{student loan} = \$125$$

If the gross monthly income is $2420, what is the ratio of monthly debt to gross monthly income (back-end ratio)?

[7.6] **47.** A family with a gross monthly income of $1970 is applying for a FHA loan. To qualify, the front-end ratio must not exceed 0.29. What is the maximum monthly PITI payment that the family can have and qualify for the loan?

[7.6] **48.** A family with a gross monthly income of $4200 is applying for a VA loan. To qualify, the back-end ratio must not exceed 0.41. What is the maximum monthly debt that the family can have and qualify for the loan?

[7.6] **49.** Dale is traveling 60 miles per hour and passes Exit 50 on a highway. Tanya passes the same exit $\frac{1}{2}$ hour later at 70 miles per hour. If they maintain the same rates of speed, how long will it take Tanya to catch up to Dale?

[7.6] **50.** Vijay is walking at 3 miles per hour in a walk-run charity event and passes a drink station at 10:42 A.M. Micah jogs past the same drink station at 11:02 A.M. at a speed of 5 miles per hour. What time will Micah catch up with Vijay?

[7.1]   **1.** A computer screen is 9.5 inches wide by 13 inches high. Write the ratio of the height to the width in simplest form.

1. _____

[7.1]   **2.** Dianne has 22 entries in a random-drawing contest that has 4578 total entries. What is the probability that Dianne will win? Write the probability in simplest form.

2. _____

[7.1]   **3.** What is the probability of rolling a 9 or a 10 on a die that has 10 sides? Write the probability in simplest form.

3. _____

[7.1]   **4.** A college has 3850 students and 125 faculty. Write a unit ratio of students to faculty.

4. _____

[7.1]   **5.** Laurence drove 158.6 miles in 2.25 hours. What was his average rate in miles per hour?

5. _____

[7.1]   **6.** A long-distance phone call lasting 36 minutes costs $5.04. What is the rate in cents per minute?

6. _____

[7.1]   **7.** A 15-ounce can of mixed vegetables costs $0.59. What is the unit price in cents per ounce?

7. _____

[7.1]   **8.** Which is the better buy: $0.89 for a 20-ounce can of pineapple or $0.69 for a 15-ounce can of pineapple?

8. _____

[7.2]   **9.** Solve and check. $\dfrac{-6.5}{12} = \dfrac{n}{10.8}$

9. _____

[7.2]   **10.** A company charges $49 to install 20 square feet of hardwood floor. How much would it charge to install the same floor in a 210-square-foot room?

10. _____

[7.2]   **11.** On a map, $\frac{1}{4}$ inch represents 50 miles. Suppose two cities are $4\frac{1}{2}$ inches apart on the map. How far apart are the two cities in miles?

11. _____

[7.3]   *For Exercises 12–16, use dimensional analysis to convert.*

12. 14 feet to inches

12. _____

13. 20 pounds to ounces

13. _____

14. 36 pints to gallons

14. _____

15. 150 minutes to hours

15. _____

[7.3]   **16.** A room that is 20 feet long by 12.5 feet wide is to be carpeted. Calculate the area of the room in square yards.

16. _____

17. a. _____

   b. _____

18. a. _____

   b. _____

19. a. _____

   b. _____

20. a. _____

   b. _____

21. _____

22. _____

23. _____

24. _____

25. _____

[7.4] **17.** Convert.

   a. 0.058 meters to centimeters

   b. 420 meters to kilometers

[7.4] **18.** Convert.

   a. 24 grams to kilograms

   b. 0.091 metric tons to grams

[7.4] **19.** Convert.

   a. 280 square centimeters to square meters

   b. 0.8 square meters to square decimeters

[7.4] **20.** Convert.

   a. 6.5 deciliters to liters

   b. 0.8 liters to cubic centimeters

[7.5] **21.** Convert 75 feet to meters.

[7.5] **22.** Convert 30°F to degrees Celsius.

[7.6] **23.** A family has the following monthly debts:

$$\begin{aligned} \text{monthly PITI} &= \$714 \\ \text{credit card 1} &= \$128 \\ \text{credit card 2} &= \$78 \\ \text{car payment 1} &= \$295 \\ \text{car payment 2} &= \$324 \end{aligned}$$

If the gross monthly income is $3420, what is the ratio of monthly debt to gross monthly income (back-end ratio)?

[7.6] **24.** A patient is to receive 750 milliliters of NS IV in 8 hours. If the IV dissipates 1 milliliter every 10 drops, what is the drip rate in drops per minute?

[7.6] **25.** Catrina is traveling 65 miles per hour and passes mile marker 80 on a highway. Darryl passes the same mile marker 1 hour later at 70 miles per hour. If they maintain the same rates, how long will it take Darryl to catch up to Catrina?

For Exercises 1–6, answer true or false.

[6.3] **1.** $-2.4 \times 10^6 > -2.4 \times 10^5$

[3.7, 5.3] **2.** $\left(5\frac{3}{4}\right)^0 = 0$

[3.6] **3.** 127 is a prime number.

[6.4] **4.** $\sqrt{5}$ is a rational number.

[5.4] **5.** $\frac{4}{9}x = 12$ is a linear equation.

[1.3, 3.5] **6.** $5^2 \cdot 5^6 = 5^{12}$

For Exercises 7–9, fill in the blank.

[5.4] **7.** To calculate the quotient of two fractions, we write an equivalent multiplication statement using the _____ of the divisor.

[5.8] **8.** To clear fractions from an equation, we can use the multiplication principle of equality and multiply both sides by the _____ of the denominators.

[7.2] **9.** The _____ products of equal ratios are equal.

[7.2] **10.** Explain in your own words how to solve a proportion.

[6.1] **11.** Write the word name for 49,802.76.

[6.3] **12.** Write 914,000,000 in scientific notation.

[5.1] **13.** Graph $-3\frac{2}{3}$ on a number line.

[6.1] **14.** Round 41.325 to the nearest hundredth.

[3.2] **15.** What is the coefficient of $-x^3$?

[3.2] **16.** What is the degree of $4x^2 - 12x^5 + 10x - 8$?

For Exercises 17–24, simplify.

[2.5] **17.** $[10 - 3(7)] - [16 \div 2(7 + 3)]$

[5.4] **18.** $5\frac{2}{3} \div \left(-2\frac{1}{6}\right)$

[5.6] **19.** $10\frac{1}{2} - 4\frac{3}{5}$

[6.5] **20.** $|12.5 - 18.65| \div 0.2$

[6.5] **21.** $\left(8\frac{2}{3}\right)(-3.6)$

[5.3] **22.** $\left(\frac{3}{4}\right)^2$

[5.4] **23.** $\sqrt{\dfrac{45}{5}}$

[6.3] **24.** $(-0.3)^4$

[6.5] **25.** Evaluate $\frac{1}{2}mv^2$ when $m = 38$, and $v = 0.4$.

[6.2] **26.** Subtract. $(12.4y^3 - 8.2y^2 + 9) - (4.1y^3 + y - 1.2)$

[5.7] **27.** Multiply. $\left(\dfrac{2}{3}a - 1\right)\left(\dfrac{3}{4}a - 7\right)$

[3.6] **28.** Find the prime factorization of 2400.

[3.7] **29.** Find the GCF of $40xy^2$ and $60x^5$.

[3.7] **30.** Divide. $\dfrac{18x^6}{3x^2}$

[3.7] **31.** Factor. $32m^4 + 24m^2 - 16m$

[5.3] **32.** Simplify. $\dfrac{9x^2}{20y} \cdot \dfrac{5y}{12x^6}$

[5.5] **33.** Find the LCM of $8k^2$ and $10k$.

[5.6] **34.** Simplify. $\dfrac{2}{x} + \dfrac{5}{9}$

For Exercises 35–37, solve and check.

[6.6] **35.** $2.5n - 16 = 30$

[5.8] **36.** $\dfrac{3}{4}x - 5 = \dfrac{1}{5}x + 3$

[7.0] **37.** $\dfrac{10.2}{b} = \dfrac{-12.5}{20}$

[7.3] **38.** Convert 12.5 miles to feet.

[7.4] **39.** Convert 0.48 kilograms to grams.

[7.5] **40.** Convert 60°C to degrees Fahrenheit.

For Exercises 41–50, solve.

[6.5] **41.** Find the volume of the shape.

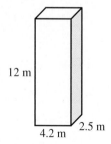

[6.5] **42.** Calculate the area of the shape.

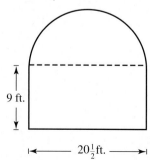

[3.8] **43.** In one month a company had a profit of $45,698, with $96,408 in revenue. Find its costs.

[6.5] **44.** A triangle has an area of 38.75 square meters. If the base is 12.5 meters, what is the height?

[3.8] **45.** Write an expression in simplest form for the area.

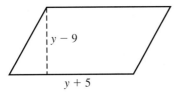

[4.5] **46.** The width of a rectangle is 3 less than the length. If the perimeter is 50 meters, find the length and width.

[7.1] **47.** What is the probability of selecting a face card (Jack, Queen, or King) from a standard deck of cards?

[7.1] **48.** Which is the better buy: 16 ounces of beans for $0.99 or 28 ounces of beans for $1.79?

[7.2] **49.** A family consumes 6 liters of cola every 7 days. At this rate, how many liters of cola will the family consume in 1 year?

[7.2] **50.** A kite is held out on a line that is nearly straight. When 7 meters of line have been let out, the kite is 6 meters off the ground. How high is the kite if 100 meters of line have been let out and the angle of the line has not changed? (*Hint*: Use similar triangles.)

Percents

> "*Nothing valuable can be lost by taking time.*"
>
> —ABRAHAM LINCOLN, U.S. PRESIDENT

> "*Since time is the one immaterial object which we cannot influence— neither speed up nor slow down, add to nor diminish—it is an imponderably valuable gift.*"
>
> —MAYA ANGELOU, POET

> "*Time stays long enough for those who use it.*"
>
> —LEONARDO DA VINCI, SCIENTIST, INVENTOR, AND ARTIST

Time Management

Balancing all the activities that we take on in life can be challenging. Many students overload themselves with too many activities and too many courses. If you have not already done so, do the exercise entitled "How Do I Do It All," which is in the *To The Student* section at the beginning of the text. The exercise will help you determine if you are overloaded and, if so, by how much so that you can adjust your schedule for a more reasonable time commitment.

In managing your time, make every minute count. Many people find it difficult to study in their home environment because there are too many distractions. The library or a vacant room on campus can be good, distraction-free environments. It may help to make a rule that you cannot leave until you have completed your homework and are prepared for the next class meeting.

Further, by completing all your school work on campus, if you need help, you can easily meet with your instructor or go to tutorial services. You'll be amazed at how much more effective your study time is when you put yourself in a distraction-free environment and make every minute count.

8.1 Introduction to Percent

OBJECTIVES

1 Write a percent as a fraction.

2 Write a percent as a decimal number.

3 Write a fraction as a percent.

4 Write a decimal number as a percent.

OBJECTIVE 1 Write a percent as a fraction.

Think about the word *percent*. It is a compound word made from the prefix *per*, which means "for each" or "divide" and the suffix *cent*, coming from the Latin word *centum*, which means 100. Therefore, **percent** means for each 100, or divide by 100.

DEFINITION Percent: Ratio representing some part out of 100.

The symbol for percent is %. For example, 20 percent, which means 20 out of 100, is written as 20%. Note that the definition can be used to write percents as fractions or decimal numbers. Let's first consider writing percents as fractions.

$$20\% = 20 \text{ out of } 100$$
$$= \frac{20}{100}$$
$$= \frac{1}{5}$$

Note: Using the definition of percent, we can replace the percent sign with out of 100 or per 100, which translates to a fraction with a denominator of 100.

PROCEDURE *To write a percent as a fraction:*
1. Write the percent numeral over 100.
2. Simplify.

Example 1 Write each percent as a fraction in simplest form.

a. 12%

Solution: $12\% = \frac{12}{100} = \frac{3}{25}$ **Write 12 over 100, then simplify.**

Note: Recall from Chapter 5 that to simplify a fraction, we divide the numerator and denominator by their GCF. In this case, the GCF for 12 and 100 is 4.

b. 20.5%

Solution: $20.5\% = \frac{20.5}{100}$ **Write 20.5 over 100.**

$$= \frac{20.5(10)}{100(10)}$$ **Clear the decimal by multiplying the numerator and denominator by 10.**

$$= \frac{205}{1000}$$ **Multiply**

$$= \frac{41}{200}$$ **Simplify to lowest terms.**

Note: In Section 7.1, we learned to clear decimals in a ratio by multiplying the numerator and denominator by an appropriate power of 10 as determined by the decimal number with the greatest number of decimal places. In this case, 20.5 has the most decimal places. To clear 1 place, we multiply by 10.

c. $30\frac{1}{4}\%$

Solution: $30\frac{1}{4}\% = \dfrac{30\frac{1}{4}}{100}$ Write $30\frac{1}{4}$ over 100.

$= 30\frac{1}{4} \div 100$ Write the complex fraction in division form.

$= \dfrac{121}{4} \cdot \dfrac{1}{100}$ Write an equivalent multiplication with improper fractions.

$= \dfrac{121}{400}$ Multiply.

Do Your Turn 1 ▷

Your Turn 1

Write each percent as a fraction in simplest form.

a. 48%

b. 6.25%

c. $10\frac{1}{3}\%$

OBJECTIVE 2 Write a percent as a decimal number.

In Section 7.1, we learned to write unit ratios by dividing the denominator into the numerator. We can use this process to write a percent as a decimal number. Because the denominator will always be 100, we will always divide the percent's numeral by 100.

$$20\% = 20 \text{ out of } 100 = \frac{20}{100} = 0.2$$

◁ **Note:** In Chapter 6, we learned that dividing by 100 causes a decimal point to move 2 places to the left from its original position in the dividend.

If the percent contains a mixed number or fraction, as in $84\frac{1}{2}\%$, we can write the fraction or mixed number as a decimal number and then divide by 100.

$$84\frac{1}{2}\% = 84.5\% = \frac{84.5}{100} = 0.845$$

PROCEDURE *To write a percent as a decimal number:*
1. Write the fraction or mixed number percent in decimal form.
2. Divide by 100.

Example 2 Write each percent as a decimal number.

a. 92%

Solution: $92\% = \dfrac{92}{100} = 0.92$ Divide 92 by 100, which causes the decimal to move 2 places to the left.

b. 125%

Solution: $125\% = \dfrac{125}{100} = 1.25$ Divide 125 by 100.

c. 9.2%

Solution: $9.2\% = \dfrac{9.2}{100} = 0.092$ Divide 9.2 by 100.

Answers to Your Turn 1
a. $\dfrac{12}{25}$ b. $\dfrac{1}{16}$ c. $\dfrac{31}{300}$

Write each percent as a decimal number.

a. 14%

b. 250%

c. 1.7%

d. $83\frac{1}{4}\%$

d. $21\frac{2}{3}\%$

Solution: Because $21\frac{2}{3}$ is a mixed number, we first write it as a decimal number.

Note: Remember, to write a fraction as a decimal number, we divide the denominator into the numerator.

$$21\frac{2}{3}\% = 21.\overline{6}\% \qquad \text{Write } 21\frac{2}{3} \text{ as a decimal number.}$$

$$= \frac{21.\overline{6}}{100} \qquad \text{Divide by 100.}$$

$$= 0.21\overline{6}$$

◀ **Do Your Turn 2**

OBJECTIVE 3 Write a fraction as a percent.

We have seen that $20\% = \dfrac{20}{100} = \dfrac{1}{5} = 0.2$. Note that to write a percent as a fraction or as a decimal number, we divide by 100. To reverse this process, we should multiply by 100 and attach a percent sign.

PROCEDURE *To write a fraction as a percent:*
1. Multiply by 100%.
2. Simplify.

CONNECTION Notice that $100\% = \dfrac{100}{100} = 1$. Therefore, when we multiply a number by 100% we are multiplying by 1, which is why the resulting percent is equal to the given number.

Example 3 Write each fraction as a percent.

a. $\dfrac{1}{2}$

Solution: $\dfrac{1}{2} = \dfrac{1}{\underset{1}{\cancel{2}}} \cdot \dfrac{\overset{50}{\cancel{100}}}{1}\%$ Multiply by 100%.

$= 50\%$

b. $\dfrac{5}{8}$

Solution: $\dfrac{5}{8} = \dfrac{5}{\underset{2}{\cancel{8}}} \cdot \dfrac{\overset{25}{\cancel{100}}}{1}\%$ Multiply by 100%.

$= \dfrac{125}{2}\%$ Simplify.

$= 62\dfrac{1}{2}\%$ or 62.5%

c. $\dfrac{2}{3}$

Solution: $\dfrac{2}{3} = \dfrac{2}{3} \cdot \dfrac{100}{1}\%$ Multiply by 100%.

$= \dfrac{200}{3}\%$ Simplify.

$= 66\dfrac{2}{3}\%$ or $66.\overline{6}\%$

Do Your Turn 3 ▶

OBJECTIVE 4 Write a decimal number as a percent.

To write a decimal number as a percent, we follow the same procedure as for fractions: multiply by 100%. Recall that multiplying a decimal number by 100 causes the decimal point to move 2 places to the right.

PROCEDURE *To write a decimal number as a percent, multiply by 100%.*

Note: Multiplying a decimal number by 100 will cause the decimal point to move 2 places to the right.

Example 4   Write each decimal number as a percent.

a. 0.7

Solution: $0.7 = (0.7)(100\%)$ Multiply by 100%.
$= 70\%$

b. 0.018

Solution: $0.018 = (0.018)(100\%)$ Multiply by 100%.
$= 1.8\%$

c. 2

Solution: $2 = (2)(100\%)$ Multiply by 100%.
$= 200\%$

d. $0.\overline{25}$

Solution: $0.\overline{25} = (0.\overline{25})(100\%)$ Multiply by 100%.
$= 25.\overline{25}\%$

Note: Remember, the repeat bar means that the decimal digits beneath it repeat without end, so $0.\overline{25} = 0.252525 \ldots$.

Do Your Turn 4 ▶

8.1 Exercises

For
Extra
Help

Videotape
DVT

Addison-Wesley
Tutor Center

Math XL

MyMathLab

Student Solutions
Manual

1. What is a percent?

2. Explain how to write a percent as a fraction.

3. Explain how to write a percent as a decimal number.

4. Explain how to write a fraction or decimal number as a percent.

For Exercises 5–16, write each percent as a fraction in simplest form.

5. 20%

6. 30%

7. 15%

8. 85%

9. 14.8%

10. 18.6%

11. 3.75%

12. 6.25%

13. $45\frac{1}{2}\%$

14. $65\frac{1}{4}\%$

15. $33\frac{1}{3}\%$

16. $66\frac{2}{3}\%$

For Exercises 17–28, write each percent as a decimal number.

17. 75%

18. 58%

19. 125%

20. 210%

21. 12.9%

22. 11.6%

23. 1.65%

24. 100.5%

25. $53\frac{2}{5}\%$

26. $70\frac{3}{4}\%$

27. $16\frac{1}{6}\%$

28. $6\frac{2}{3}\%$

For Exercises 29–40, write each fraction as a percent.

29. $\frac{1}{2}$

30. $\frac{3}{4}$

31. $\frac{3}{5}$

32. $\frac{1}{5}$

33. $\frac{3}{8}$

34. $\frac{5}{8}$

35. $\frac{1}{6}$

36. $\frac{5}{6}$

37. $\frac{2}{3}$

38. $\frac{1}{3}$

39. $\frac{4}{9}$

40. $\frac{5}{11}$

For Exercises 41–56, write each decimal number as a percent.

41. 0.96

42. 0.42

43. 0.8

44. 0.7

45. 0.09

46. 0.01

47. 1.2

48. 3.58

49. 0.028

50. 0.065

51. 4.051

52. 0.007

53. $0.\overline{6}$

54. $0.\overline{3}$

55. $1.\overline{63}$

56. $0.\overline{47}$

Review Exercises

[6.3]   *For Exercises 1–3, evaluate.*

    1. 0.6(24)

    2. 0.45(3600)

    3. 0.05(76.8)

[6.4]   *For Exercises 4 and 5, solve and check.*

    4. $0.15x = 200$

    5. $0.75y = 28.5$

[5.8]   **6.** Three-fourths of a number is twenty-four. Find the number.

8.2 Translating Percent Sentences Word for Word

OBJECTIVE 1 Solve percent sentences by direct translation to an equation.

OBJECTIVE

1 Solve percent sentences by direct translation to an equation.

To solve problems involving percents, it is often helpful to reduce the problem to the following simple percent sentence:

A **percent** of a **whole** amount is a **part** of the whole.

Note that there are three pieces in the simple sentence: the *percent*, the *whole*, and the *part*. Any one of those three pieces could be unknown in a problem.

	A **percent**	of	a **whole**	is a **part**.
Unknown part:	20%	of	80	is what number?
Unknown whole:	32%	of	what number	is 12?
Unknown percent:	What percent	of	240	is 60?

Keep in mind that the English language allows for variations of wording in each case. Our goal is to translate a variety of these sentences to equations that we can solve. We will consider two methods of translation:

1. Word-for-word translation to an equation.

2. Translation to a proportion.

We will consider word-for-word translation in this section and translation to a proportion in Section 8.3. To translate simple percent sentences word for word, we use key words, paying special attention to the words *of* and *is*.

PROCEDURE *To translate a simple percent sentence word for word:*
1. Select a variable for the unknown.
2. Translate *is* to an equal sign.
3. If *of* is preceded by the percent, translate it to multiplication.
   If *of* is preceded by a whole number, translate it to division.

CONNECTION Percents are fractions. We learned in Chapter 5 that if the word *of* is preceded by a fraction, we multiply. For example, we have learned $50\% = \dfrac{1}{2}$. So, the phrase 50% of 80 is the same as $\dfrac{1}{2}$ of 80. In both phrases, the word *of* indicates to multiply. In Chapter 7, we learned that *of* can be short for *out of*, as in 25 of the 30 students got an A, which is the same as 25 out of the 30 students got an A. When *of* is short for *out of*, it translates to division.

After translating a sentence to an equation, if it contains a percent, we must write the percent as a fraction or a decimal number.

First, let's consider the case where the part is unknown.

Example 1 Translate to an equation, then solve.

a. 20% of 80 is what number?

Solution: The words *what number* indicate an unknown. The unknown is the part of the whole, so we will use c to represent it. Because *of* is preceded by the percent, it indicates multiplication. The word *is* translates to an equal sign.

Note: For consistency, we will always use the variable c, for *change*, to represent the part of a whole.

$$\underset{\downarrow}{20\% \text{ of } 80} \quad \underset{\downarrow}{\text{is}} \quad \underset{\checkmark}{\text{what number?}}$$

$$20\% \cdot 80 = c \qquad \text{Translate to an equation.}$$

$$(0.2)(80) = c \qquad \text{Write the percent as a decimal.}$$

$$16 = c \qquad \text{Multiply.}$$

b. What number is 105% of 60.4?

Solution: What number is 105% of 60.4?

DISCUSSION Why is the result greater than 60.4?

$c = 105\% \cdot 60.4$ **Translate to an equation.**

$c = (1.05)(60.4)$ **Write the percent as a decimal number.**

$c = 63.42$ **Multiply.**

c. What number is 5.4% of $34\frac{1}{3}$?

Solution: What number is 5.4% of $34\frac{1}{3}$?

$c = 5.4\% \cdot 34\frac{1}{3}$ **Translate to an equation.**

$c = (0.054)\left(\dfrac{103}{3}\right)$ **Write the percent as a decimal number and the mixed number as an improper fraction.**

$c = \left(\dfrac{0.054}{1}\right)\left(\dfrac{103}{3}\right)$ **Write the decimal number over 1.**

$c = \dfrac{5.562}{3}$ **Multiply.**

$c = 1.854$ **Divide.**

Do Your Turn 1 ▷

Now let's translate sentences in which the unknown is the whole amount. The whole amount always follows the word *of*.

Example 2 Translate to an equation, then solve.

a. 32% of what number is 12?

Solution: The words *what number* indicate an unknown. The unknown is a whole amount, so we will use *w* to represent it. Because *of* is preceded by the percent, it indicates multiplication. The word *is* translates to an equal sign.

32% of what number is 12?

Note: For consistency, we will always use the variable *w*, for *whole*, to represent unknown whole amounts.

▷ $32\% \cdot w = 12$ **Translate to an equation.**

$0.32w = 12$ **Write the percent as a decimal number.**

$\dfrac{0.32w}{0.32} = \dfrac{12}{0.32}$ **Divide both sides by 0.32 to isolate *w*.**

$w = 37.5$

b. $16\frac{1}{2}$ is 2% of what number?

Solution: $16\frac{1}{2}$ is 2% of what number?

$16\frac{1}{2} = 2\% \cdot w$ **Translate to an equation.**

$16.5 = 0.02w$ **Write the percent as a decimal number.**

$\dfrac{16.5}{0.02} = \dfrac{0.02w}{0.02}$ **Divide both sides by 0.02 to isolate *w*.**

$825 = w$

Do Your Turn 2 ▷

Last, we will translate sentences in which the percent is the unknown amount.

Example 3   Translate to an equation, then solve.

a. What percent of 240 is 60?

Solution: *What percent* indicates that the unknown is the percent, so we will use p to represent it. Because *of* is preceded by the percent, it indicates multiplication. The word *is* translates to an equal sign.

What percent of 240 is 60?

$$p \cdot 240 = 60 \qquad \text{Translate to an equation.}$$

$$240p = 60$$

$$\frac{240p}{240} = \frac{60}{240} \qquad \text{Divide both sides by 240 to isolate } p.$$

$$p = 0.25$$

> **Note:** For consistency, we will always use the variable p to represent unknown *percents*.

Note that the answer is a decimal number. To write the result as a percent, we multiply by 100%.

$$p = (0.25)(100\%) = 25\%$$

b. 82.4 is what percent of 164.8?

Solution: 82.4 is what percent of 164.8?

$$82.4 = \qquad p \quad \cdot \quad 164.8 \qquad \text{Translate to an equation.}$$

$$82.4 = 164.8p$$

$$\frac{82.4}{164.8} = \frac{164.8p}{164.8} \qquad \text{Divide both sides by 164.8 to isolate } p.$$

$$0.5 = p$$

$$(0.5)(100\%) = p \qquad \text{Multiply by 100\% to write 0.5 as a percent.}$$

$$50\% = p$$

c. What percent is 15 out of 40?

Solution: *What percent* indicates that the unknown is the percent, so we will use p to represent it. The key phrase *out of* indicates a ratio. The word *is* translates to an equal sign. After translating, there are two ways of solving the equation.

What percent is 15 out of 40?

$$p = \frac{15}{40}$$

Divide 15 by 40. $p = 0.375$ or $p = \dfrac{3}{8}$ Simplify $\frac{15}{40}$.

Multiply by 100% to write as a percent. $p = (0.375)(100\%)$ $p = \dfrac{3}{\underset{2}{8}} \cdot \dfrac{\overset{25}{100}}{1}\%$ Multiply by 100% to write as a percent.

$$p = 37.5\% \qquad\qquad p = \frac{75}{2}\%$$

> **Note:** Since $37.5\% = 37\frac{1}{2}\%$, either form is acceptable as the answer.

$$p = 37\frac{1}{2}\%$$

d. What percent is 12 of 18?

Solution: *What percent* indicates that the unknown is the percent, so we will use p to represent it. The word *of* is preceded by a whole number, which means it is an abbreviation of *out of.* It is the same as if we had been asked: What percent is 12 out of 18? Therefore, the *of* in this case indicates a ratio.

What percent is 12 of 18?

$$p = \frac{12}{18}$$

Divide 12 by 18. $p = 0.\overline{6}$ or $p = \frac{2}{3}$ **Simplify $\frac{12}{18}$.**

Multiply by 100% to write as a percent. $p = (0.\overline{6})(100\%)$ $p = \frac{2}{3} \cdot \frac{100}{1}\%$ **Multiply by 100% to write as a percent.**

$p = 66.\overline{6}\%$ $p = \frac{200}{3}\%$

$p = 66\frac{2}{3}\%$

Do Your Turn 3 ▶

Your Turn 3

Translate to an equation, then solve.

a. What percent of 400 is 50?

b. What percent is 19 out of 40?

c. 32 of 72 is what percent?

d. 15.37 is what percent of 14.5?

Answers to Your Turn 3
a. $p \cdot 400 = 50$; 12.5%

b. $p = \frac{19}{40}$; 47.5%

c. $\frac{32}{72} = p$; 44.$\overline{4}$%

d. $15.37 = p \cdot 14.5$; 106%

8.2 Exercises

For Extra Help

 Videotape DVT

 Addison-Wesley Tutor Center

 Math XL

 MyMathLab

Student Solutions Manual

1. A percent of a _____ is the

_____ .

2. What does the word *of* indicate when preceded by a percent sign as in 12% of 30 is what number?

3. What does the word *of* indicate if preceded by a whole number as in what percent is 15 of 40?

4. What does the word *is* indicate?

For Exercises 5–44, translate to an equation, then solve.

5. 40% of 350 is what number?

6. 60% of 400 is what number?

7. What number is 15% of 78?

8. What number is 5% of 44?

9. 150% of 60 is what number?

10. What number is 105% of 84?

11. What number is 16% of $30\frac{1}{2}$?

12. 95% of $40\frac{3}{5}$ is what number?

13. 110% of 46.5 is what number?

14. What number is 120% of 36.82?

15. What number is $5\frac{1}{2}$% of 280?

16. $10\frac{1}{4}$% of 68 is what number?

17. $66\frac{2}{3}$% of 56.8 is what number?

18. $16\frac{1}{6}$% of 365 is what number?

19. What number is 8.5% of 54.82?

20. What number is 3.5% of 24,500?

21. 12.8% of 36,000 is what number?

22. 19.8% of 84.5 is what number?

23. 35% of what number is 77?

24. 70% of what number is 45.5?

25. 4800 is 40% of what number?

26. 1950 is 65% of what number?

27. 605 is 2.5% of what number?

28. 168.3 is 19.8% of what number?

29. 105% of what number is 48.3?

30. 120% of what number is 101.52?

31. 47.25 is $10\frac{1}{2}$% of what number?

32. 2000 is $6\frac{1}{4}$% of what number?

33. What percent of 68 is 17?

34. What percent of 120 is 48?

35. 2.142 is what percent of 35.7?

36. 25.12 is what percent of 125.6?

37. What percent of $24\frac{1}{2}$ is $7\frac{7}{20}$?

38. $5\frac{5}{8}$ is what percent of $9\frac{3}{8}$?

39. What percent is 21 out of 60?

40. What percent is 114 out of 200?

41. 25 out of 35 is what percent?

42. 38 out of 42 is what percent?

43. 43 of 65 is what percent?

44. What percent is 29 of 30?

Review Exercises

For Exercises 1–3, evaluate.

[6.3]   **1.** 12.8(100)

[5.3]   **2.** $5\frac{3}{4} \cdot 100$

[5.3]   **3.** $8\frac{1}{4} \cdot 10\frac{1}{2}$

[7.2]   *For Exercises 4 and 5, solve and check.*

4. $\dfrac{x}{14} = \dfrac{3}{4}$

5. $\dfrac{n}{100} = \dfrac{5}{8}$

[5.3]   **6.** An assembly line manager finds that 3 out of 20 of the company's product finish with a minor cosmetic defect. If the factory produces 2400 units each week, how many units have the defect?

8.3 Translating Percent Sentences to Proportions

OBJECTIVE 1 Solve percent sentences by proportion.

OBJECTIVE

1 Solve percent sentences by proportion.

In Section 8.2, we translated simple percent sentences word for word. Let's recall the structure of those sentences.

A **percent** of a **whole** amount is the **part** of the whole.

In this section, we will translate simple percent sentences to a proportion.

Recall that a proportion is two ratios set equal. In a simple percent sentence, one of the ratios is the percent, because by definition, it is a ratio of a quantity to 100. The other ratio is the ratio of the part to the whole. For example, the sentence 20% of 80 is 16 translates to the following proportion.

20% expressed as a ratio in fraction form
$$\frac{20}{100} = \frac{16}{80} \begin{array}{l} \leftarrow \text{part} \\ \text{to} \\ \leftarrow \text{whole} \end{array}$$

Note: In terms of equivalent ratios, we can say that 16 is the part of 80 that is in the same ratio as 20 to 100. The ratios 20 to 100 and 16 to 80 are equivalent because both ratios simplify to $\frac{1}{5}$.

In general, the percent ratio is equivalent to the ratio of the part to the whole amount, which suggests the following procedure.

PROCEDURE *To translate a simple percent sentence to a proportion, write the proportion in the following form:*

$$\text{percent} = \frac{\text{part}}{\text{whole}}$$

where the percent is expressed as a fraction with a denominator of 100.

Note: The whole amount always follows the word *of*.

Once the sentence is translated, we solve the proportion. Recall that we cross multiply to solve proportions.

Example 1 Translate to a proportion, then solve.

a. 60% of 40 is what number?

Solution: To translate the sentence to a proportion, we use $\text{percent} = \frac{\text{part}}{\text{whole}}$. The number 40 follows the word *of*, so it is the whole amount. The part is unknown, so we will represent it with c.

60% expressed as a ratio
$$\frac{60}{100} = \frac{c}{40} \begin{array}{l} \leftarrow \text{part} \\ \leftarrow \text{whole} \end{array}$$

$$40 \cdot 60 = 2400 \qquad 100 \cdot c = 100c$$

$$\frac{60}{100} \underset{\times}{=} \frac{c}{40} \qquad \text{Cross multiply.}$$

$$2400 = 100c \qquad \text{Equate the cross products.}$$

$$\frac{2400}{100} = \frac{100c}{100} \qquad \text{Divide both sides by 100 to isolate } c.$$

$$24 = c$$

b. What number is 120% of 48.5?

Solution: Translate using $percent = \dfrac{part}{whole}$.

The number 48.5 follows *of*, so it is the whole amount. The part is unknown.

$$120\% \text{ expressed as a} \atop \text{ratio} \left\{ \frac{120}{100} = \frac{c}{48.5} \begin{array}{l} \leftarrow \textbf{part} \\ \leftarrow \textbf{whole} \end{array} \right.$$

$$(48.5)(120) = 5820 \qquad\qquad 100 \cdot c = 100c$$

$$\frac{120}{100} \diagup\!\!\!\!\diagdown \frac{c}{48.5} \qquad \textbf{Cross multiply.}$$

$$5820 = 100c \qquad \textbf{Equate the cross products.}$$

$$\frac{5820}{100} = \frac{100c}{100} \qquad \textbf{Divide both sides by 100 to isolate } c.$$

DISCUSSION Why is the result greater than 48.5?

$$58.2 = c$$

LEARNING STRATEGY

If you are an auditory learner, imagine or say to yourself "of goes underneath" to help you remember that the number that follows *of* is the denominator.

c. What number is 15.2% of $20\frac{1}{3}$?

Solution: Translate using $percent = \dfrac{part}{whole}$.

The number $20\frac{1}{3}$ follows *of*, so it is the whole amount. The part is unknown.

$$15.2\% \text{ expressed} \atop \text{as a ratio} \left\{ \frac{15.2}{100} = \frac{c}{20\frac{1}{3}} \begin{array}{l} \leftarrow \textbf{part} \\ \leftarrow \textbf{whole} \end{array} \right.$$

$$\frac{15.2}{1} \cdot \frac{61}{3} = \frac{927.2}{3} = 309.0\overline{6} \qquad\qquad 100 \cdot c = 100c$$

$$\frac{15.2}{100} \diagup\!\!\!\!\diagdown \frac{c}{\frac{61}{3}} \qquad \begin{array}{l}\textbf{Write the mixed number as} \\ \textbf{a fraction and cross multiply.}\end{array}$$

$$309.0\overline{6} = 100c \qquad \textbf{Equate the cross products.}$$

$$\frac{309.0\overline{6}}{100} = \frac{100c}{100} \qquad \textbf{Divide both sides by 100 to isolate } c.$$

$$3.090\overline{6} = c$$

DISCUSSION How would you compare the proportion method with the direct translation method when the part is the unknown?

Do Your Turn 1 ▶

Your Turn 1

Translate to a proportion, then solve.

a. 75% of 620 is what number?

b. What number is 8% of $46\frac{1}{2}$?

c. What number is 15.2% of 90?

Now let's translate sentences in which the unknown is the whole amount. Recall that the whole amount always follows the word *of*. In the proportion, the whole amount is placed in the denominator because it is in logical correspondence with the 100 that is the denominator from the percent. We will continue to use the variable w to represent the unknown whole amount.

Answers to Your Turn 1

a. $\dfrac{75}{100} = \dfrac{c}{620}$; 465

b. $\dfrac{8}{100} = \dfrac{c}{46\frac{1}{2}}$; 3.72

c. $\dfrac{15.2}{100} = \dfrac{c}{90}$; 13.68

Translate to a proportion, then solve.

a. 95% of what number is 64.6?

b. $10\frac{1}{5}$ is 20% of what number?

Example 2 Translate to a proportion, then solve.

a. 85% of what number is 352.75?

Solution: Translate using percent $= \dfrac{\text{part}}{\text{whole}}$. The words *what number* follow *of*, so the whole amount is unknown. The number 352.75 is the part.

85% expressed as a ratio
$$\begin{cases} \dfrac{85}{100} = \dfrac{352.75}{w} \quad \begin{array}{l}\leftarrow \textbf{part}\\ \leftarrow \textbf{whole}\end{array}\end{cases}$$

$$w \cdot 85 = 85w \qquad 100 \cdot 352.75 = 35{,}275$$

$$\dfrac{85}{100} \diagup\!\!\!\!\!\diagdown \dfrac{352.75}{w} \qquad \textbf{Cross multiply.}$$

$$85w = 35{,}275 \qquad \textbf{Equate the cross products.}$$

$$\dfrac{85w}{85} = \dfrac{35{,}275}{85} \qquad \textbf{Divide both sides by 85 to isolate } w.$$

$$w = 415$$

b. $18\frac{3}{8}$ is $8\frac{3}{4}$% of what number?

Solution: Translate using percent $= \dfrac{\text{part}}{\text{whole}}$. The words *what number* follow *of*, so the whole amount is unknown. The number $18\frac{3}{8}$ is the part.

$8\frac{3}{4}$% expressed as a ratio
$$\begin{cases} \dfrac{8\frac{3}{4}}{100} = \dfrac{18\frac{3}{8}}{w} \quad \begin{array}{l}\leftarrow \textbf{part}\\ \leftarrow \textbf{whole}\end{array}\end{cases}$$

$$w \cdot \dfrac{35}{4} = \dfrac{35}{4}w \qquad \dfrac{\overset{25}{\cancel{100}}}{1} \cdot \dfrac{147}{\underset{2}{\cancel{8}}} = \dfrac{3675}{2}$$

$$\dfrac{\frac{35}{4}}{\cancel{100}} \diagup\!\!\!\!\!\diagdown \dfrac{\frac{147}{8}}{w} \qquad \begin{array}{l}\textbf{Write the mixed numbers as fractions}\\ \textbf{and cross multiply.}\end{array}$$

$$\dfrac{35}{4}w = \dfrac{3675}{2} \qquad \textbf{Equate the cross products.}$$

$$\dfrac{\overset{1}{\cancel{4}}}{\underset{1}{\cancel{35}}} \cdot \dfrac{\overset{1}{\cancel{35}}}{\underset{1}{\cancel{4}}}w = \dfrac{\overset{105}{\cancel{3675}}}{\underset{1}{\cancel{2}}} \cdot \dfrac{\overset{2}{\cancel{4}}}{\underset{1}{35}} \qquad \textbf{Multiply both sides by } \tfrac{4}{35} \textbf{ to isolate } w.$$

$$w = 210$$

Note: Remember, if the coefficient is a fraction, we can isolate the variable by multiplying both sides of the equation by the reciprocal of the fraction.

◀ **Do Your Turn 2**

Answers to Your Turn 2

a. $\dfrac{95}{100} = \dfrac{64.6}{w}$; 68

b. $\dfrac{20}{100} = \dfrac{10\frac{1}{5}}{w}$; 51

Last, we will translate sentences in which the percent is the unknown amount. We will use the variable P to represent an unknown percent.

Example 3 Translate to a proportion, then solve.

a. What percent of 240 is 60?

Solution: *What percent* indicates that the percent is unknown, so we will use P to represent it. Because 240 follows the word *of*, it is the whole amount. Therefore, 60 must be the part.

the unknown percent expressed as a ratio $\left\{ \dfrac{P}{100} = \dfrac{60}{240} \right.$ ← part
← whole

Note: Although the percent is unknown, we know that when it is written as a ratio, the denominator is 100.

$240 \cdot P = 240P \qquad 100 \cdot 60 = 6000$

$$\dfrac{P}{100} \diagdown \dfrac{60}{240}$$ **Cross multiply.**

$240P = 6000$ **Equate the cross products.**

$\dfrac{240P}{240} = \dfrac{6000}{240}$ **Divide both sides by 240 to isolate P.**

$P = 25$

Note: In calculating the cross products, we multiplied by 100, so this result is a percent.

Answer: 25%

Note: This example is the same as Example 3a in Section 8.2 where we illustrated a point about the outcome. When we solved this problem in Section 8.2 using word-for-word translation, the result was 0.25, a decimal number, and we had to multiply by 100% to write this result as a percent. With the proportion method, the result is the correct percent because when we cross multiply, we multiply by 100. Multiplying by 100 is already part of the procedure. As a reminder of this, we will use uppercase P for an unknown percent in a proportion and lowercase p in equations from word-for-word translations.

b. 17.64 is what percent of 84?

Solution: *What percent* indicates that the unknown is the percent. Because 84 follows the word *of*, it is the whole amount. Therefore, 17.64 must be the part.

the unknown percent expressed as a ratio $\left\{ \dfrac{P}{100} = \dfrac{17.64}{84} \right.$ ← part
← whole

$84 \cdot P = 84P \qquad 100 \cdot 17.64 = 1764$

$$\dfrac{P}{100} \diagdown \dfrac{17.64}{84}$$ **Cross multiply.**

$84P = 1764$ **Equate the cross products.**

$\dfrac{84P}{84} = \dfrac{1764}{84}$ **Divide both sides by 84 to isolate P.**

$P = 21$

Answer: 21%

c. What percent is 15 out of 40?

Solution: *What percent* indicates that the unknown is the percent. Because 40 follows the word *of*, it is the whole amount. Therefore, 15 must be the part.

$$\text{the unknown percent expressed as a ratio}\begin{cases} & \end{cases} \frac{P}{100} = \frac{15}{40} \begin{matrix} \leftarrow \textbf{part} \\ \leftarrow \textbf{whole} \end{matrix}$$

$$40 \cdot P = 40P \qquad 100 \cdot 15 = 1500$$

$$\frac{P}{100} \diagdown \frac{15}{40} \qquad \textbf{Cross multiply.}$$

$$40P = 1500 \qquad \textbf{Equate the cross products.}$$

$$\frac{40P}{40} = \frac{1500}{40} \qquad \textbf{Divide both sides by 40 to isolate } P.$$

$$P = 37.5$$

Answer: 37.5%

d. What percent is 34 of 102?

Solution: *What percent* indicates that the unknown is the percent. Because 102 follows the word *of*, it is the whole amount. Therefore, 34 is the part.

$$\text{the unknown percent expressed as a ratio}\begin{cases} & \end{cases} \frac{P}{100} = \frac{34}{102} \begin{matrix} \leftarrow \textbf{part} \\ \leftarrow \textbf{whole} \end{matrix}$$

$$102 \cdot P = 102P \qquad 100 \cdot 34 = 3400$$

$$\frac{P}{100} \diagdown \frac{34}{102} \qquad \textbf{Cross multiply.}$$

$$102P = 3400 \qquad \textbf{Equate the cross products.}$$

$$\frac{102P}{102} = \frac{3400}{102} \qquad \textbf{Divide both sides by 102 to isolate } P.$$

$$P = 33.\overline{3}$$

Answer: 33.$\overline{3}$%

CONNECTION The word *of* is preceded by a whole number, which means it is an abbreviation for *out of*. It is the same as if we had been asked: What percent is 34 *out of* 102? 34 of 102 translates to the ratio $\frac{34}{102}$.

◄ **Do Your Turn 3**

Note: This example is the same as Example 3c in Section 8.2. Comparing the approach here to the method in Section 8.2, many people think the proportion method is superior for this type of percent sentence because *out of* translates to a ratio.

Your Turn 3

Translate to a proportion, then solve.

a. What percent of 360 is 45?

b. 58.75 is what percent of 94?

c. What percent is 27 out of 162?

d. 58 of 72 is what percent?

Answers to Your Turn 3

a. $\dfrac{P}{100} = \dfrac{45}{360}$; 12.5%

b. $\dfrac{P}{100} = \dfrac{58.75}{94}$; 62.5%

c. $\dfrac{P}{100} = \dfrac{27}{162}$; $16\dfrac{2}{3}$%

   or 16.$\overline{6}$%

d. $\dfrac{P}{100} = \dfrac{58}{72}$; $80\dfrac{5}{9}$%

   or 80.$\overline{5}$%

Videotape DVT   Addison-Wesley Tutor Center   Math XL   MyMathLab   Student Solutions Manual

1. How do you translate a simple percent sentence to a proportion?

2. When using word-for-word translation as we did in Section 8.2 to find an unknown percent, the calculation will not be a percent, whereas if we use a proportion, the result is a percent. Why?

For Exercises 3–44, translate to a proportion, then solve.

3. 30% of 560 is what number?

4. 80% of 420 is what number?

5. What number is 12% of 85?

6. What number is 7% of 62?

7. 125% of 90 is what number?

8. What number is 106% of 77?

9. What number is 15% of $9\frac{1}{2}$?

10. 98% of $60\frac{1}{4}$ is what number?

11. 102% of 88.5 is what number?

12. What number is 150% of 45.25?

13. What number is $12\frac{1}{2}$% of 440?

14. $5\frac{1}{4}$% of 164 is what number?

15. $33\frac{1}{3}$% of 87.6 is what number?

16. $10\frac{1}{2}$% of 24.8 is what number?

17. What number is 6.5% of 22,800?

18. What number is 9.5% of 288?

19. 16.9% of 2450 is what number?

20. 14.8% of 160 is what number?

21. 65% of what number is 52?

22. 90% of what number is 133.2?

23. 4480 is 80% of what number?

24. 2400 is 75% of what number?

25. 1143 is 4.5% of what number?

26. 645 is 12.9% of what number?

27. 120% of what number is 65.52?

28. 110% of what number is 251.35?

29. 13.53 is $5\frac{1}{2}$% of what number?

30. 461.25 is $10\frac{1}{4}$% of what number?

31. What percent of 80 is 24?

32. What percent of 180 is 81?

33. 2.226 is what percent of 74.2?

34. 7.328 is what percent of 91.8?

35. What percent of $30\frac{1}{3}$ is $18\frac{1}{5}$?

36. $5\frac{1}{20}$ is what percent of $12\frac{5}{8}$?

37. What percent is 76 out of 80?

38. What percent is 186 out of 300?

39. 32 out of 45 is what percent?

40. 47 out of 60 is what percent?

41. 17 of 19 is what percent?

42. What percent is 169 of 180?

43. What percent of 75 is 80?

44. What percent of 90 is 110?

Review Exercises

For Exercises 1–3, evaluate.

[6.3] **1.** 0.48(560)

[6.3] **2.** 0.05(24.8)

[5.3] **3.** $\dfrac{9}{16} \cdot \dfrac{20}{27}$

[5.3] **4.** $\frac{5}{8}$ of the respondents in a survey disagreed with a given statement. If 400 people were surveyed, how many disagreed with the given statement?

[5.3] **5.** $\frac{3}{4}$ of the respondents in a survey agreed with a given statement. Of these, $\frac{4}{5}$ were female. What fraction of all respondents were females that agreed with the particular statement?

Solving Percent Problems (Portions)

OBJECTIVE 1 Solve for the part in percent problems.

Now that we have seen how to translate simple percent sentences, we can solve percent problems that are more complex. Our strategy with these more complex problems is to use simple sentences as a guide to translate those problems to an equation or a proportion that we can solve.

The problems in this section will be phrased in terms of whole amounts and portions (parts) of those whole amounts.

OBJECTIVES

1 Solve for the part in percent problems.

2 Solve for the whole in percent problems.

3 Calculate the percent.

PROCEDURE *To solve problems involving percents:*
1. Identify the percent, whole amount, and part.
2. Write these pieces in a simple percent sentence (if needed).
3. Translate to an equation or a proportion.
4. Solve.

First, we consider problems in which the part is unknown.

DISCUSSION When is translating to a proportion advantageous compared to word-for-word translation and vice-versa?

Example 1 30% of a 400-milliliter solution is acetone. How many milliliters of the solution is acetone?

Understand: We must find the part of the solution that is acetone given the percent and the capacity of the whole solution.

Plan: Write a simple percent sentence, translate to an equation or a proportion, and then solve.

Execute: Percent of the whole amount is the part.

$$30\% \quad \text{of} \quad 400 \quad \text{is what number?}$$
$$30\% \cdot 400 = c \qquad \text{Translate to an equation.}$$
$$0.3(400) = c \qquad \text{Write the percent as a decimal.}$$
$$120 = c \qquad \text{Multiply.}$$

Answer: The solution contains 120 ml of acetone.

Check: We will repeat the solution using the proportion method.

30% expressed as a ratio
$$\frac{30}{100} = \frac{c}{400} \qquad \begin{array}{l} \leftarrow \text{acetone (part)} \\ \leftarrow \text{whole solution} \end{array}$$
$$12,000 = 100c$$
$$\frac{12,000}{100} = \frac{100c}{100}$$
$$120 = c$$

Your Turn 1

Sabrina scored 80% on a test with 60 total questions. How many questions did she answer correctly?

Note: We chose to translate word for word to an equation because it seemed easier since we were given the percent and whole amount. The proportion method would have required more steps.

Do Your Turn 1 ▷

Salespeople often earn a **commission** based on sales. The commission can be a percent of the total sales amount or a percent of the profit.

_____

DEFINITION Commission: A portion of sales earnings that a salesperson receives.

Answer to Your Turn 1
48

Solve.

a. Al is a salesperson at a computer store and earns 10% of total sales in commission. If in one month he sold $15,294 worth of merchandise, how much will he receive as commission?

b. Brin sells scanning equipment. She earns 20% of the net profit in commission. If the net profit was $16,148, what was her commission?

Example 2 Carrie earns 15% of total sales in commission. If she sells $2485 in merchandise over a two-week period, what is her commission?

Understand: We must find Carrie's commission given her commission rate and total sales.

Plan: Write a simple percent sentence, translate to an equation or a proportion, and then solve.

Execute: Her commission is 15% of her total sales.

What number is 15% of 2485?

$c = 15\% \cdot 2485$	Translate to an equation.
$c = 0.15(2485)$	Write the percent as a decimal number.
$c = 372.75$	Multiply.

Answer: Carrie's commission will be $372.75.

Check: We will repeat the solution using the proportion method.

15% expressed as a ratio
$$\frac{15}{100} = \frac{c}{2485} \quad \begin{array}{l} \leftarrow \text{commission (part)} \\ \leftarrow \text{total sales (whole)} \end{array}$$

$$37,275 = 100c$$

$$\frac{37,275}{100} = \frac{100c}{100}$$

$$372.75 = c$$

DISCUSSION If Carrie receives 15% of total sales in commission, what percent does the company receive? How much of the $2485 does the company receive?

◀ Do Your Turn 2

OBJECTIVE 2 Solve for the whole in percent problems.

Now let's consider problems in which the whole amount is unknown.

Example 3 Tedra sells real estate and receives a commission of $5445.60. If her commission rate is 6%, what was the total sale?

Understand: We must find Tedra's total sale given her commission rate and commission.

Plan: Write a simple percent sentence, translate to an equation or a proportion, and then solve.

Execute: 5445.60 is 6% of the total sale.

5445.60 is 6% of what number?

$5445.60 = 6\% \cdot w$	Translate.
$5445.60 = 0.06w$	Write the percent as a decimal number.
$\dfrac{5445.60}{0.06} = \dfrac{0.06w}{0.06}$	Divide both sides by 0.06 to isolate *w*.
$90,760 = w$	

Answer: Tedra's sale was $90,760.

Check: We will repeat the solution using the proposition method.

6% expressed as a ratio

$$\dfrac{6}{100} = \dfrac{5445.60}{w} \quad \begin{array}{l}\leftarrow \text{commission (part)} \\ \leftarrow \text{total sales (whole)}\end{array}$$

$$6w = 544{,}560$$

$$\dfrac{6w}{6} = \dfrac{544{,}560}{6}$$

$$w = 90{,}760$$

Do Your Turn 3 ▷

Your Turn 3

Todd earns 20% commission on the sale of medical equipment. If Todd receives $3857.79 as a commission, what was his total sale?

Example 4   A reporter states that 456 people in a survey indicated that they felt their schools were safe. The reporter went on to say that this was 48% of the respondents. How many respondents were involved in the survey?

Understand: We must find the total respondents in a survey given that 456 people represent 48% of the total respondents.

Plan: Write a simple percent sentence, translate to an equation or a proportion, and then solve.

Execute: 456 is 48% of the total respondents.

456 is 48% of what number?

$456 = 48\% \cdot w$	Translate.
$456 = 0.48w$	Write the percent as a decimal number.
$\dfrac{456}{0.48} = \dfrac{0.48w}{0.48}$	Divide both sides by 0.48 to isolate w.
$950 = w$	

Answer: There were 950 people who responded in the survey.

Check: We will repeat the solution using the proportion method.

48% expressed as a ratio

$$\dfrac{48}{100} = \dfrac{456}{w} \quad \begin{array}{l}\leftarrow \text{part that felt safe} \\ \leftarrow \text{total responded (whole)}\end{array}$$

$$48w = 45{,}600$$

$$\dfrac{48w}{48} = \dfrac{45{,}600}{48}$$

$$w = 950$$

DISCUSSION   If 48% of the respondents felt schools are safe, what percent did not feel schools are safe? How many of the respondents did not feel schools are safe?

Do Your Turn 4 ▷

Your Turn 4

A power amplifier is operating at 80% of its full power capability. If its current output is 400 W (watts), what is its full power capability?

OBJECTIVE 3 Calculate the percent.

Lastly, we will solve problems in which the percent is unknown. Recall that if we use the proportion method to find the percent, the answer will be in percent form.

Example 5   Latasha answered 52 questions correctly on a test with a total of 60 questions. What percent of the questions did she answer correctly?

Understand: We must find the percent of the questions that Latasha answered correctly given that she answered 52 correctly out of 60 total questions.

Plan: Write a simple percent sentence, translate to a proportion, and then solve.

Answer to Your Turn 3
$19,288.95

Answer to Your Turn 4
500 W

Out of the 286 students
taking foreign language at a
certain school, 234 passed.
What percent passed?

Execute: What percent is 52 out of 60?

the unknown percent
expressed as a ratio $\left\{ \dfrac{P}{100} = \dfrac{52}{60} \right.$ ← number correct (part)

← total questions (whole)   **Write a proportion.**

$$60P = 5200 \qquad \text{Equate the cross products.}$$

$$\frac{60P}{60} = \frac{5200}{60} \qquad \text{Divide both sides by 60 to isolate } P.$$

$$P = 86.\overline{6}$$

Answer: Latasha answered $86.\overline{6}\%$ of the questions correctly. Her score on the test would most likely be rounded to 87%.

Check: We will repeat the solution using word-for-word translation.

Note: Remember, when we use word-for-word translation to find an unknown percent, the result will not be a percent. To express the result as a percent we multiply by 100.

What percent is 52 out of 60?

$$p = \frac{52}{60}$$

$$p = 0.8\overline{6} \qquad \text{Divide 52 by 60.}$$

$$p = 0.8\overline{6}(100\%) \qquad \text{Multiply by 100\% to write the decimal as a percent.}$$

$$p = 86.\overline{6}\%$$

◀ **Do Your Turn 5**

After 4 months of con-
struction, 26 floors of the
55 floors in a skyscraper are
complete. What percent
remains to be completed?

Example 6 A manufacturing plant produces 485 units of its product in 1 week. If 3 units are found to be defective, what percent of the units produced are *without* defect?

Understand: We must find the percent of the units produced that are without defect given that 3 units are defective out of 485 units produced.

Plan: We must first calculate the number of units that are without defect by subtracting the number of defective units from the total number of units. We can then write a simple percent sentence, translate to a proportion, and solve.

Execute: Number of units produced without defect = $485 - 3 = 482$.

What percent is 482 out of 485?

the unknown percent
expressed as a ratio $\left\{ \dfrac{P}{100} = \dfrac{482}{485} \right.$ ← units without defect (part)

← total units produced (whole)   **Write a proportion.**

$$485P = 48{,}200 \qquad \text{Equate the cross products.}$$

$$\frac{485P}{485} = \frac{48{,}200}{485} \qquad \text{Divide both sides by 485 to isolate } P.$$

$$P \approx 99.4$$

Answer: Approximately 99.4% of the products manufactured in a week are without defect.

Check: We will repeat the solution using word-for-word translation.

What percent is 482 out of 485?

$$P = \frac{482}{485}$$

$$P \approx 0.994 \qquad \text{Divide.}$$

$$P \approx 0.994(100\%) \qquad \text{Write the decimal as a percent.}$$

$$P \approx 99.4\%$$

◀ **Do Your Turn 6**

8.4 Exercises

For Extra Help

 Videotape DVT

 Addison-Wesley Tutor Center

 Math XL

 MyMathLab

 Student Solutions Manual

1. What is the general approach to solving problems involving percents?

2. In the following problem, identify the percent, the whole amount, and the part.

   20% of the students in a class received an A on their first test. If there are 24 students in the class, how many received an A?

   Percent: _____

   Whole amount: _____

   Part: _____

3. In the following problem, identify the percent, the whole amount, and the part.

   In a survey, 4000 people were asked if they agree or disagree with a legislative decision. 2400 agreed with the decision. What percent agreed?

   Percent: _____

   Whole amount: _____

   Part: _____

4. In the following problem, identify the percent, the whole amount, and the part.

   Gary earns a commission rate of 12%. If he earned $4800 in commission, what were his total sales?

   Percent: _____

   Whole amount: _____

   Part: _____

For Exercises 5–36, write an equation or a proportion. Then solve.

5. 60% of an 800-milliliter mixture is HCl (hydrochloric acid). How many milliliters of the solution are HCl?

6. The label on a bottle of rubbing alcohol indicates that it is 70% isopropyl alcohol. If the bottle contains 473 milliliters, how many milliliters of isopropyl alcohol does it contain?

OF INTEREST

Here's a case where 1 + 1 does *not* equal 2. Measure 1 cup of water and mix with 1 cup of rubbing alcohol. This will not equal exactly 2 cups of liquid. The reason is that the water and alcohol molecules are different sizes and shapes, allowing the molecules to *squeeze* together so that the total volume is less than 2 cups. The purer the alcohol, the more evident this effect is.

7. Sabrina scored 90% on a test with 30 total questions.
   a. How many questions did she answer correctly?

   b. What percent did she answer incorrectly?

   c. How many questions did she answer incorrectly?

8. Willis scored 85% on a test with 40 questions.
   a. How many questions did he answer correctly?

   b. What percent did he answer incorrectly?

   c. How many questions did he answer incorrectly?

9. According to a study by Columbia University's National Center on Addiction and Substance Abuse, drug and alcohol abuse played a part in the crimes committed by 80% of all people incarcerated. If there are about 2.0 million people in America's prisons, for how many of these did drug and alcohol abuse play a part in their crime? (*Source:* Bureau of Justice Statistics (2002))

10. According to the Bureau of Justice Statistics, in 2002 93% of America's inmates were men. If there were about 1.7 million inmates in America, then how many were men? (*Source:* Bureau of Justice Statistics (2002))

11. Wright earns 10% of total sales in commission. If he sells $3106 in merchandise over a 2-week period, what is his commission?

12. Kera is a salesperson at a car dealership and earns 25% of the profit in commission. If in 1 month the dealership made a profit of $9680 from her sales, how much will she receive as commission?

13. Liz earns 8% of total sales in commission and $6 per hour for working. If she works a total of 60 hours in 2 weeks and her total sales are $4249, what will her gross pay be for the 2-week period?

14. Vastine earns 10% of total sales in commission and $5 per hour for working. If he works a total of 80 hours over 2 weeks and has total sales of $5827, what is his gross pay?

15. Cush sells cars and earns 25% of the profit in commission. In 1 month the dealership grossed $78,950. If the total cost to the company was $62,100, what was Cush's commission?

16. A writer earns 15% of the publishing company's net profit on the sale of her book in royalties (commission). If the company sells 13,000 copies of the book and makes a profit of $6 on each book, what is the writer's royalty?

17. Rosita sells real estate and receives a commission of $750. If her commission rate is 6% of the seller's net, what is the seller's net?

18. Andre earns 20% of the net profit in commission on the sale of cash registers. If he receives $2675.92 as a commission, what is the net profit on his sales?

19. A basketball team in one game made 42% of the total number of attempted shots. If they made a total of 39 baskets, how many shot attempts did they have in the game?

20. A football quarterback in one game completes 34% of his pass attempts. If he completed 14 passes, how many pass attempts did he have in the game?

21. At a certain company, 43 employees were hired in the last 5 years. If this represents 38% of the company's current employees, how many current employees does the company have?

22. $5,710,900 of the budget at a school comes from the state. If this represents 65% of the school's total budget, what is the school's total budget?

23. Hunter answered 58 questions correctly out of a total of 65 questions. What percent of the total questions did Hunter answer correctly?

24. Corrina answered 41 questions correctly out of a total of 45 questions. What percent of the total questions did she answer correctly?

25. A baseball player has 48 hits in a season where he has been at bat 110 times. What percent of his times at bat did he get a hit?

OF INTEREST

In baseball, a player's batting average is the ratio of the number of hits to the number of times at bat. It is expressed as a decimal number to the nearest thousandth. The batting average can be expressed as a percent by multiplying by 100. For example, a player with a batting average of 0.345 gets a hit 34.5% (a little more than $\frac{1}{3}$) of the times he goes to bat.

26. In one season a football kicker scores 33 extra points in 35 attempts. What percent of his attempts did he score?

27. Of the world's 6.3 billion people, about 3 billion live on less than $2 per day. What percent of the world's population live on less than $2 per day? (*Source:* U.S. Bureau of the census, 2004)

28. In 2000, Americans spent $65 billion on illegal drugs. $35 billion of that total went toward cocaine. What percent of the total money spent on illegal drugs was spent on cocaine? (*Source:* Office of National Drug Control Policy, 2001)

OF INTEREST

That same $65 billion could have put about 16 million American students through 4 years of college. (*Source:* The College Board, 2002)

29. Suppose a person spends 8 hours sleeping, 7 hours working, 1 hour commuting, and 1 hour eating in a 24-hour period. What percent of the day is left?

30. If Mozart composed his first music at the age of 4 and continued composing and performing until he died at the age of 35, what percent of his life was spent involved in music?

OF INTEREST

Wolfgang Amadeus Mozart (1756–1791) is considered to be one of the greatest classical composers. By the age of 6, he was an accomplished performer on clavier, organ, and violin and had already composed several pieces of high quality. It is said that he completed his compositions in his head and then merely copied the completed work to paper with few or no corrections. He composed nearly 600 pieces in his lifetime and performed throughout Europe for royalty as well as common folk. It is believed that he died of typhoid fever, but some believe that he may have been poisoned. Because of his poverty, he was buried in a mass unmarked grave.

31. Below are the results of a survey. Respondents could respond in three ways: agree, disagree, or no opinion.

> 248 agreed
> 562 disagreed
> 89 said no opinion

What percent of all respondents either agreed or disagreed?

32. Below are the results of a survey. Respondents could respond in three ways: agree, disagree, or no opinion.

> 612 agreed
> 145 disagreed
> 94 said no opinion

What percent of all respondents either agreed or disagreed?

33. The Sun is a sphere with a radius of about 695,990 kilometers. The core of the Sun is a sphere with a radius of about 170,000 kilometers. What percent of the Sun's total volume is made up by the core?

34. In a 1860-square-foot house, the living room is 16 feet wide by 20 feet long. What percent of the total area of the house is the living room area?

35. Following is a list of the Geiger family's net monthly income and expenses.

Income	Expenses
Mr. Geiger = $3025.85	Mortgage = $948.75
Mrs. Geiger = $1050.72	Childcare = $800
	Car payment 1 = $294.88
	Car payment 2 = $275.96
	Credit card payments = $150
	Utilities = $285
	Groceries = $475
	Entertainment = $260

DISCUSSION   What financial advice would you give the Geiger family?

a. What percent of their income goes to each expense?

b. What is the total percentage of the family income paid toward expenses?

c. What percent of their income is left after expenses?

36. Following is a list of the Sharp family's net monthly income and expenses.

Income	Expenses
Mr. Sharp = $2625.85	Mortgage = $752.45
	Car payment = $385.95
	Credit card payments = $85
	Utilities = $210
	Groceries = $450
	Entertainment = $360

DISCUSSION What financial advice would you give the Sharp family?

a. What percent of their income goes to each expense?

b. What is the total percentage of the family income paid toward expenses?

c. What percent of their income is left after expenses?

PUZZLE PROBLEM There are approximately 2×10^{11} stars in our galaxy. Suppose 10% of those stars have planets orbiting them. Suppose 1% of those planetary systems have a planet with a climate that could support life. Suppose that 1% of those planets that could support life actually have living organisms. Suppose that 1% of those planets with life have intelligent life. Suppose that 1% of the intelligent life have developed civilizations with communication devices. Suppose that 1% of those civilizations with communications devices are within a 500 light-year radius from Earth. Based on these suppositions, how many stars might we expect to have planets with intelligent life that have communication devices within a 500 light-year radius?

DISCUSSION In the Puzzle Problem, what flaws may exist in the suppositions? If the suppositions are reasonable, then what are the implications about contact or communication with other intelligent life forms?

Review Exercises

[6.2]   **1.** Combine like terms. $5.1x + 9.8 + x - 12.4$

For Exercises 2–4, solve and check.

[6.6]   **2.** $14.7 = x + 0.05x$

[6.6]   **3.** $24.48 = y - 0.2y$

[7.2]   **4.** $\dfrac{m}{2.4} = \dfrac{5.6}{8.4}$

[1.2]   **5.** Amelia receives a raise so that her new salary is $28,350. If her former salary was $25,800, what was the amount of the raise?

8.5 Solving Problems Involving Percent of Increase or Decrease

OBJECTIVES

1 Solve problems involving a percent of increase.

2 Solve problems involving a percent of decrease.

3 Calculate the percent of increase or decrease.

OBJECTIVE 1 Solve problems involving a percent of increase.

In Section 8.4, we solved percent problems that were phrased in terms of parts and wholes. In this section, we will explore problems where an initial amount (whole) is increased or decreased by an amount (part) as determined by the percent. To help with translating the information in these problems, we will rephrase our simple percent sentence. The whole amount will be the initial amount. The part will be the amount of increase or decrease.

The **percent** of an **initial amount** is the **amount of increase or decrease**
whole   part

Situations like sales tax and salary raises increase an initial amount by an amount that is a percent of the initial amount. Discounts and deductions from one's paycheck decrease an initial amount by an amount that is a percent of the initial amount.

It does not matter whether the problem is an increase situation or a decrease situation, the percent sentence is the same. However, after calculating the amount of the increase or decrease, we either add or subtract accordingly.

Let's first explore examples of percent of increase. Cities and states collect money by requiring that a sales tax be added to the price of items purchased in the city or state. The tax amount is a percent of the price of an item.

Example 1 A software package costs $35. If the sales tax rate is 5%, calculate the sales tax and the total amount of the purchase.

Understand: We must calculate the sales tax and the total amount of the purchase given the tax rate and initial amount.

Plan: Write a simple percent sentence, translate to an equation, and solve for the sales tax. Then to calculate the final amount, add the sales tax to the price.

Execute: The sales tax is 5% of the price.

What number is 5% of 35?

$$c = 5\% \cdot 35 \qquad \text{Translate.}$$
$$c = 0.05(35) \qquad \text{Write the percent as a decimal number.}$$
$$c = 1.75 \qquad \text{Multiply.}$$

Answer: The sales tax is $1.75. Because sales tax increases the initial price, we add the tax to the price to get the total amount:

$$\text{total} = \$35 + \$1.75 = \$36.75$$

Check: We can verify the final addition step by subtracting the initial amount from the total amount.

$$\$36.75 - \$35 = \$1.75$$

Now we can set up a proportion to verify that the tax rate is in fact 5%.

the unknown percent
expressed as a ratio
$$\frac{P}{100} = \frac{1.75}{35} \quad \begin{array}{l} \leftarrow \text{tax (part)} \\ \leftarrow \text{initial amount (whole)} \end{array}$$
$$35P = 175$$
$$\frac{35P}{35} = \frac{175}{35}$$
$$P = 5$$

CONNECTION To calculate the amount of increase or decrease mentally when the percent is a multiple of 5%, we can calculate the 10% amount, then adjust by an appropriate factor. Because 10% is 0.1, when we multiply a decimal number by 10%, the decimal point moves 1 place to the left.

10% of 35 is 3.5 (Move the decimal point left 1 place in 35.)

Because 5% is half of 10%, the 5% amount is half of the 10% amount. Because 10% of 35 is 3.5, we can say that 5% of 35 will be half of 3.5, which is 1.75.

We could also calculate percents that are multiples of 10%. For example, 20% is twice the 10% amount. Because 10% of 35 is 3.5, we can say that 20% is twice 3.5, which is 7.

The process we used to check Example 1 illustrates how we will solve problems in which the percent of increase or decrease is unknown. Look at Example 10 in this section.

There is another way we can calculate the total amount. Consider the fact that we pay the entire price of the item, which is 100% of the price, plus tax. In our example, if we paid 100% of $35 plus 5% of $35 in tax, then our total is 105% of the $35.

$$\text{total} = 105\% \cdot 35$$
$$= 1.05 \cdot 35$$
$$= 36.75$$

Look at the multiplication process to see why this works.

$$
\begin{array}{r}
35 \\
\times\ 1.05 \\
\hline
175 \\
000 \\
+\ 35 \\
\hline
36.75
\end{array}
$$

Multiplying 35 by the 5 digit in 1.05 places the tax in the sum.

Multiplying 35 by the digit 1 in the ones place of 1.05 places the initial price in the sum.

The tax and initial price are added within the multiplication procedure to end up with the total amount.

Conclusion: In general to calculate the final amount given the initial amount and the percent of increase, we can multiply the initial amount by the sum of 100% and the percent of increase.

Do Your Turn 1 ▷

Another example of percent of increase is when raises are given based on a percent of the salary. Since we keep 100% of the former salary and add a percent of the former salary as a raise, we can apply the alternative method we learned at the end of Example 1. That is, we can calculate the new salary by multiplying the old salary by the sum of 100% and the percent of increase.

Example 2 Carlita's current annual salary is $23,500. If she gets a 4% raise, what is her new salary?

Understand: We must calculate Carlita's new salary given her current salary and the percent of the raise. Because the 4% is added to 100% of her current salary, her new salary will be 104% of her current salary.

Plan: Write a simple percent sentence, translate to an equation, and solve for the new salary. Because we are calculating a final amount, we will use the variable a.

Execute: The new salary is 104% of the current salary.

What number is 104% of 23,500?
$$a = 104\% \cdot 23{,}500$$
$$a = 1.04(23{,}500) \qquad \text{Write the percent as a decimal number.}$$
$$a = 24{,}440 \qquad \text{Multiply.}$$

Answer: Carlita's new salary will be $24,440. Note that if we need to know the amount of the raise, we can subtract the initial salary from the new salary.

$$\text{raise amount} = \$24{,}440 - \$23{,}500 = \$940$$

Your Turn 1

Solve.

a. A mountain bike is priced at $365.95. Calculate the sales tax if the tax rate is 7%.

b. A suit is priced at $128.95. Calculate the sales tax and total amount if the tax rate is 6%.

◀ **Note:** Although we can continue using c and w for part and whole, respectively, variables are often chosen to represent the context of the problem. For instance, in Example 2, we can use c because we are looking for the part, or a to represent *amount*.

Answers to Your Turn 1
a. $25.62 b. $7.74; $136.69

Benjamin has a current salary of $22,600. If he receives a 3% raise, what will be his new salary?

Check: We will use a proportion to verify that the raise of $940 is indeed 4% of her current salary.

the unknown percent expressed as a ratio $\left\{ \dfrac{P}{100} = \dfrac{940}{23,500} \right.$ ← raise amount (part)
← initial salary (whole)

$$23,500P = 94,000$$

$$\frac{23,500P}{23,500} = \frac{94,000}{23,500}$$

$$P = 4$$

CONNECTION Note how we are calculating the percent of the increase in the check. The percent of increase is equal to the ratio of the raise amount to the initial salary.

◁ **Do Your Turn 2**

What if the initial amount is unknown? We must look carefully at the information given. First, let's consider an example in which we are given the percent of increase and the amount of the increase.

Example 3 Carmine recieves a raise of $1950, which is a 6% increase to his salary. What was his former salary?

Understand: We must determine Carmine's former salary given the percent of increase and the amount of the raise.

Plan: Write a simple percent sentence, translate to an equation, and solve for the new salary. We will use the variable f to represent the former salary.

Julia indicates that she received a 5% raise in the amount of $2260. What was her former salary?

Execute: 6% of the former salary is the raise amount.

6% of what salary is 1950?

$$6\% \cdot \quad f \quad = 1950 \qquad \text{Translate.}$$

$$0.06f = 1950 \qquad \text{Write the percent as a decimal number.}$$

$$\frac{0.06f}{0.06} = \frac{1950}{0.06} \qquad \text{Divide both sides by 0.06 to isolate } f.$$

$$f = 32,500$$

Answer: Carmine's former salary was $32,500.

Check: Verify that 6% of $32,500 equals a raise of $1950.

6% of 32,500 is what raise?

$$6\% \cdot 32,500 = r$$

$$0.06(32,500) = r$$

$$1950 = r$$

◁ **Do Your Turn 3**

Answer to Your Turn 2
$23,278

Answer to Your Turn 3
$45,200

What if we are given the percent of increase and the final amount, after the increase?

Example 4 After an 8% raise, Angela's new salary is $45,360. What was her former salary?

Understand: We must find Angela's former salary given her new salary after an 8% raise.

Plan: Write an equation, then solve.

Execute: Angela's new salary is her former salary plus the raise amount, which is 8% of her former salary.

new salary = former salary + raise amount

new salary = former salary + 8% · former salary

Note: The expression $f + 0.08f$ is the same as $1f + 0.08f$, so when we combine those two like terms, we add the coefficients 1 and 0.08 to get $1.08f$.

$45,360 = f + 0.08f$ **Translate to an equation with f representing her former salary.**

$45,360 = 1.08f$ **Combine like terms.**

$$\frac{45,360}{1.08} = \frac{1.08f}{1.08}$$ **Divide both sides by 1.08 to isolate f.**

$42,000 = f$

Answer: Angela's former salary was $42,000. Notice that this checks with Example 3.

Do Your Turn 4 ▷

OBJECTIVE 2 Solve problems involving a percent of decrease.

Now let's explore problems with percent of decrease, which involve *subtracting* an amount of decrease from an initial amount to get a final amount.

Discount prices and pay deductions are two common situations in which percent of decrease is used. A discount occurs when a percent of the price of an item is deducted from the item's price.

Example 5 A 20% discount is to be applied to a sweater with an initial price of $49.95. What are the discount amount and final price?

Understand: We must calculate the discount amount and final price given the discount rate and the initial amount.

Plan: Write a simple percent sentence, translate to an equation, and solve for the discount amount. Then calculate the final price by subtracting the discount amount from the initial price.

Execute: The discount amount is 20% of the initial price.

What amount is 20% of 49.95?

$d = 20\% \cdot 49.95$ **Translate.**

$d = 0.2(49.95)$ **Write the percent as a decimal number.**

$d = 9.99$ **Multiply.**

Answer: The discount amount is $9.99. Because this discount is an amount of decrease, we subtract it from the initial price to get the final price.

final price = $49.95 − $9.99 = $39.96

Your Turn 4

Solve.

a. Robert received a 7% raise. If his new salary is $37,450, what was his former salary?

b. Josh recalls paying a total of $16.75 for a music CD. He knows the sales tax rate in his state is 5%. What was the initial price of the CD?

CONNECTION The equations in Example 4 agree with our alternative method for finding the final amount given the initial amount and the percent of increase. The equations $45,360 = f + 0.08f$ indicates that 45,360 is the sum of 100% of the former salary and 8% of the former salary. After combining like terms, we have the equation $45,360 = 1.08f$, which indicates that 45,360 is 108% of the former salary.

Answers to Your Turn 4
a. $35,000 b. $15.95

A coat with an initial price of $88.95 is marked 40% off. What are the discount amount and final price?

Check: Verify the discount amount by subtracting the final price from the initial price.

$$\$49.95 - \$39.96 = \$9.99$$

We can use a proportion to verify that the discount rate is 20%.

the unknown percent expressed as a ratio $\left\{ \dfrac{P}{100} = \dfrac{9.99}{49.99} \right.$ ← discount amount (part) ← initial price (whole)

$$49.95P = 999$$

$$\frac{49.95P}{49.95} = \frac{999}{49.95}$$

$$P = 20$$

CONNECTION Notice that the pattern we've established in the checks continues. The percent of decrease is equal to the ratio of the amount of decrease (discount amount) to the initial amount (initial price).

We can find the final amount after a percent of decrease is applied to an initial amount using an alternative method similar to the alternative method that we learned for percent of increase problems. Consider the fact that with discounts, we are deducting the discount amount from the initial amount. In our example, we deducted 20% of the price from 100% of the price so that the final amount is 80% of the price.

The final price is 80% of the initial price.
$$\text{final price} = 80\% \cdot 49.95$$
$$\text{final price} = 0.8(49.95)$$
$$\text{final price} = 39.96$$

Conclusion: In general, to calculate the final amount given the initial amount and the percent of decrease, we can multiply the initial amount by a percent found by subtracting the percent of decrease from 100%.

◀ **Do Your Turn 5**

Income deductions also use percent of decrease. FICA, Medicare, and retirement are deductions that are based on a percent of gross wages. A person's *net pay,* or take-home pay, is the amount that is left after all deductions.

Example 6 Approximately 28% of Deanne's gross monthly income is deducted for Social Security, federal withholdings, state withholdings, and insurance. If Deanne's gross monthly salary is $2111.67, what is her net pay?

Understand: If 28% of her gross monthly salary is deducted from her gross salary, then her net pay will be 72% of her gross salary $(100\% - 28\% = 72\%)$.

Plan: Write a simple percent sentence, translate to an equation or proportion, and then solve. We will use the variable n to represent net pay.

Execute: Her net pay is 72% of her gross monthly salary

What amount is 72% of 2111.67?

DISCUSSION What were the total deductions?

$n = 72\% \cdot 2111.67$ Translate.

$n = 0.72(2111.67)$ Write the percent as a decimal number.

$n \approx 1520.40$ Multiply.

Answer: Her net pay is $1520.40.

Answer to Your Turn 5
$35.58; $53.37

Check: We will verify the deduction percent. The deduction percent is equal to the ratio of the deduction amount to the gross pay. We first need the deduction amount by subtracting net pay from gross pay.

$$\$2111.67 - \$1520.40 = \$591.27$$

We can use a proportion to verify that the deduction is 28%.

the unknown percents expressed as a ratio $\left\{ \dfrac{P}{100} = \dfrac{591.27}{2111.67} \right.$ ← deduction amount ← gross pay

$$2111.67P = 59,127$$

$$\frac{2111.67P}{2111.67} = \frac{59,127}{2111.67}$$

$$P \approx 28$$

Do Your Turn 6 ▶

Suppose the initial amount is unknown in a percent of decrease problem. If we are given the percent and the amount of decrease, the problem is the same as Example 3, when we were given the percent and the amount of increase.

Example 7 If 32% of Darlene's gross monthly salary results in total deductions of $918.4, what is her gross monthly salary?

Understand: We must find Darlene's gross monthly salary given the percent of decrease and the total deductions.

Plan: Write a simple percent sentence, translate to an equation, and solve for the salary. We will use the variable g to represent her gross monthly salary.

Execute: 32% of her gross monthly salary is total deductions.

32% of what salary is 918.4?

$$32\% \cdot \quad g \quad = 918.4 \qquad \text{Translate.}$$

$$0.32g = 918.4 \qquad \text{Write the percent as a decimal number.}$$

$$\frac{0.32g}{0.32} = \frac{918.4}{0.32} \qquad \text{Divide both sides by 0.27 to isolate } g.$$

$$g = 2870$$

Answer: Darlene's gross monthly salary is $2870.

Check: Verify that 32% of $2870 equals the total deductions of $918.4.

32% of 2870 is what amount of decrease?

$$32\% \cdot 2870 = d$$

$$0.32(2870) = d$$

$$918.4 = d$$

Do Your Turn 7 ▶

What if we are given net salary and the percent of decrease? This situation is similar to Example 4.

[Your Turn 6]

Solve.

a. 1.45% of gross monthly salary goes to Medicare. If a person's gross pay is $1187.50, what is that individual's Medicare contribution?

b. Sandra's total deductions come to about 31% of her gross pay. If her gross pay is $1337.50, what is her net pay?

[Your Turn 7]

Clayton purchased a suitcase on sale for 20% off. If the discount amount was $37.09, what was the initial price?

Answers to Your Turn 6
a. $17.22 b. $922.88

Answer to Your Turn 7
$185.45

Clarece purchased a robe discounted by 25%. If the price after the discount was $59.96 what was the initial price?

CONNECTION The equations in Example 8 agree with our alternative method for finding the final amount given an initial amount and a percent of decrease. The equation $2516 = g - 0.26g$ indicates that 2516 is the amount left after subtractions 26% of the gross pay from 100% of the gross pay. After combining like terms, we have the equation $2516 = 0.74g$, which indicates that 2516 is 74% of Rodney's gross pay.

Example 8   After deductions, Rodney's net pay is $2516. If the total of the deductions was 26% of his gross monthly salary, what was his gross monthly salary?

Understand: We must find Rodney's gross monthly salary given the percent of decrease and the net pay.

Plan: Write an equation, then solve.

Execute: Rodney's net pay is gross pay minus the total deductions, which are 26% of his gross pay.

$$\text{net pay} = \text{gross monthly salary} - \text{total deductions}$$

$$\text{net pay} = \text{gross monthly salary} - \overbrace{26\% \cdot \text{gross pay}}$$

Note: The expression $g - 0.26g$ is the same as $1g - 0.26g$, so when we combine those two like terms, we subtract the coefficients 1 and 0.26 to get $0.74g$.

$2516 = g - 0.26g$ → Translate to an equation with g representing his gross pay.

$2516 = g - 0.26g$

$2516 = 0.74g$ Combine like terms.

$\dfrac{2516}{0.74} = \dfrac{0.74g}{0.74}$ Divide both sides by 0.74 to isolate g.

$3400 = g$

Answer: Rodney's gross monthly salary is $3400.

Check: Verify that deducting 26% of $3400 from $3400 results in a net pay of $2516. Or, using the alternative method, verify that 74% of $3400 is $2516.

$$n = 74\% \cdot 3400$$
$$n = 0.74 \cdot 3400$$
$$n = 2516$$

◄ **Do Your Turn 8**

OBJECTIVE 3 Calculate the percent of increase or decrease.

What if the percent of increase or decrease is the unknown? We shall see that the given information will affect how we approach the problem. Recall that the basic percent sentence is:

The **percent** of an **initial amount** is the **amount of increase or decrease.**

If we are given the initial amount and the amount of increase or decrease, we can write an equation or proportion and solve for the percent.

Example 9   Tiera is visiting a different state and purchases a shirt that is priced $24.95. She notes that the tax on her receipt is $2.25. What is the sales tax rate?

Understand: We must calculate the sales tax rate, which is a percent of increase. We are given the initial amount (initial price) and the amount of increase (sales tax).

Plan: Write a simple percent sentence, translate to an equation or proportion, and then solve.

Answer to Your Turn 8
$79.95

Execute: The percent of an initial amount is the amount of increase or decrease.

What percent of 24.95 is 2.25?

the unknown percent expressed as a ratio $\left\{ \dfrac{P}{100} = \dfrac{2.25}{24.95} \right.$ ← amount of increase (part)
← initial amount (whole)

$24.95P = 225$ **Equate the cross products.**

$\dfrac{24.95P}{24.95} = \dfrac{225}{24.95}$ **Divide both sides by 24.95.**

$P \approx 9$

Note: It is most likely that the actual tax rate is exactly 9%, and when the tax amount was calculated, it was rounded to the nearest cent.

Answer: The tax rate is 9%.

Check: Verify that 9% of $24.95 gives a sales tax of $2.25.

$$\text{sales tax} = 9\% \cdot 24.95$$
$$= 0.09(24.95)$$
$$\approx 2.25 \qquad \text{It checks.}$$

Do Your Turn 9 ▶

Now suppose we are given the initial amount and final amount after the increase or decrease has been applied. Because the percent sentence includes the initial amount and the amount of increase or decrease, we must first calculate the increase or decrease. Look back over the checks for Examples 1, 2, 5, and 6. You should notice that in each case, we first had to calculate the amount of the increase or decrease in order to find the percent. We did this by calculating the difference between the final amount and initial amount. Also notice that the percent of increase or decrease is equal to the ratio of the amount of increase or decrease to the initial amount.

$$\dfrac{P}{100} = \dfrac{\textit{amount of increase or decrease}}{\textit{initial amount}}$$

Example 10 Jerod was making $12.50 per hour. He is given a raise so that now he is making $14.00 per hour. What was the percent of the increase?

Understand: We must calculate the percent of the increase given Jerod's initial hourly wage and his new hourly wage.

Plan: Because the percent of the increase is equal to the ratio of the amount of increase to the initial amount, we must first calculate the amount of increase by subtracting the initial hourly wage from the new hourly wage. Then we can write a proportion and solve.

Execute: amount of increase = new hourly wage − initial hourly wage
$$= \$14.00 - \$12.50$$
$$= \$1.50$$

Solve.

a. Marianne purchases a small rug on sale. The initial price was $185.79. The price after the discount was $157.92. What was the percent of decrease (discount rate)?

b. Quan Li manages a manufacturing plant and has just received the end-of-the-month report on production. The report indicates that the plant produced 2336 units. At the end of the previous month, the plant had produced 2540 units. What is the percent of decrease in production?

Now we can write a proportion and solve for the percent. In terms of the simple percent sentence, we are asking:

What percent of 12.50 is 1.50?

the unknown percent expressed as a ratio $\left\{ \dfrac{P}{100} = \dfrac{1.50}{12.50} \right.$ ← amount of increase (part) ← initial amount (whole)

$$12.50P = 150$$
$$\frac{12.50P}{12.50} = \frac{150}{12.50}$$
$$P = 12$$

Answer: Jerod received a 12% raise.

Check: Verify that a 12% raise applied to an initial hourly wage of $12.50 equals a new hourly wage of $14.00.

The raise amount is 12% of 12.50.

$$R = 0.12(12.50)$$
$$R = 1.50$$

the new hourly wage − initial hourly wage + the raise amount
$$N = \$12.50 + \$1.50$$
$$N = \$14.00$$

CONNECTION We could have used the shortcut for increase problems. Adding the percent of the increase to 100%, we can say that the new hourly wage is 112% of the intial hourly wage.

new hourly wage = 112% · 12.50
$$N = 1.12(12.50)$$
$$N = 14.00$$

◀ Do Your Turn 10

Answers to Your Turn 10
a. 15% b. 8%

8.5 Exercises

For Extra Help

Videotape DVT

Addison-Wesley Tutor Center

Math XL

MyMathLab

Student Solutions Manual

1. When solving percent problems involving percent of increase or decrease, the three elements in our basic percent sentence are renamed. Complete the sentence with the new names for percent of increase or decrease.

   The _____ of an

   _____ is the

   _____ .

2. In the following problem, identify the percent, the initial amount, and the increase or decrease amount.

   Sharon's current salary is $34,000. If she receives a 3% raise, how much will the raise be?

   What will be her new salary?

   Percent: _____

   Initial amount: _____

   Increase or decrease amount: _____

3. In the following problem, identify the percent, the initial amount, and the increase or decrease amount.

   A 25% discount is applied to the price of an appliance resulting in a discount of $80. What was the original price of the appliance?

   Percent: _____

   Initial amount: _____

   Increase or decrease amount: _____

4. In the following problem, identify the percent, the whole amount, and the part.

   An electric bill increases from $40 one month to $50 the next month. What is the percent of increase?

   Percent: _____

   Whole amount: _____

   Part: _____

For Exercises 5–48, solve.

5. A desk is priced at $185.95. The sales tax rate is 5%. Calculate the sales tax and total amount of the purchase.

6. A guitar is priced at $585.95. Calculate the sales tax and the total amount of purchase if the tax rate is 6%.

7. A car is priced at $26,450. If the sales tax rate is 6%, calculate the sales tax and total amount of the purchase.

8. A computer is priced at $1998.95. Calculate the sales tax and total amount if the tax rate is 7%.

9. Michelle's current annual salary is $29,600. If she gets a 3.5% raise, what is her new salary?

10. Lou's current annual salary is $23,800. If he receives a 6.5% raise, what is his new salary?

11. The bill in a restaurant comes to a total of $48.75. The patrons decide to tip the server 20%. What tip amount should be left?

12. The bill in a restaurant comes to a total of $32.85. The patrons decide to tip the server 15%. What tip amount should be left?

OF INTEREST

Many states have a law that sets the maximum sales tax that can be charged on a purchase. It is sometimes beneficial to shop around in different states for large purchases.

When tipping, most people round to whole-dollar amounts. A convenient scale for 15% is to recognize that every $10 of the bill corresponds to a tip of $1.50. The rest of the scale would follow:

Total Bill	Tip
$5	$0.75
$10	$1.50
$20	$3.00

In our example of a total bill of $48.75, we would round to $50. $50 is two $20's and one $10. Each $20 is a $3 tip and the $10 is a $1.50 tip, for a total tip of $7.50.

$$\text{Total Bill} = \$20 + \$20 + \$10 = \$50$$
$$\downarrow \quad \downarrow \quad \downarrow \quad \downarrow$$
$$\text{Tip} = \$3 + \$3 + \$1.50 = \$7.50$$

13. Jack indicates that he received a 4% raise that amounts to $1008. What was his former annual salary?

14. Lucia indicates that she received a 5.5% raise in the amount of $1793. What was her former salary?

15. Huang indicates that he received a 4% raise. His new salary is $30,420. What was his former salary?

16. Denzel indicates that he received a 9% raise. His new salary is $39,828.60. What was his former salary?

17. Rachel paid a total of $45.10 for a pair of shoes. She knows the sales tax rate in her state is 5%. What was the initial price of the shoes?

18. Paul paid $635.95 for a new stereo system. If the sales tax rate in his state is 6%, what was the initial price?

19. A 30% discount is to be applied to a dress with an initial price of $65.99. What are the discount amount and final price?

20. A sweater with an initial price of $42.95 is marked 25% off. What are the discount amount and final price?

21. A sofa with an initial price of $699.99 is marked 35% off. What are the discount amount and final price?

22. A necklace with an initial price of $349.90 is marked 45% off. What are the discount amount and final price?

23. Approximately 27% of Sheila's gross monthly income is deducted for Social Security, federal withholdings, state withholdings, insurance, etc. If her gross monthly salary is $2708.33, what is her net pay?

24. Approximately 29% of Natron's gross monthly income is deducted for Social Security, federal withholdings, state withholdings, insurance, etc. If his gross monthly salary is $2487.50, what is his net pay?

25. 1.45% of gross monthly salary goes to Medicare. If a person's gross pay is $1546.67, what is the Medicare contribution?

26. 6.2% of gross pay goes to FICA. If a person's gross pay is $540.83, what is the FICA deduction?

27. If 26% of Gordon's gross monthly salary results in total deductions of $674.96, what is his gross monthly salary?

28. Marcelle's retirement deduction from her monthly pay is $187.50. If the retirement deduction is 6% of her gross monthly salary, what is her gross monthly salary?

29. Patricia purchased a table on sale for 30% off. If the discount amount was $44.09, what was the initial price?

30. A lamp is on sale for 15% off. If the discount amount is $13.49, what is the initial price?

31. After deductions, Luther's net pay is $1850.40. If the total of the deductions was 28% of his gross monthly salary, what was his gross monthly salary?

32. After deductions, Wendy's net pay is $1016.60. If the total of the deductions was 32% of her gross 2-week salary, what was her gross 2-week salary?

33. Hannah purchased a curio stand discounted 35%. If the price after the discount was $246.68 what was the initial price?

34. Vladimir purchased a set of golf clubs discounted 15%. If the price after the discount was $577.96 what was the initial price?

35. Melliah purchases a tennis racquet that is priced $54.95. She notes that the tax on her receipt is $3.85. What is the sales tax rate?

36. The property tax on a vehicle is $286.45. If the vehicle is assessed at a value of $8500, what is the tax rate?

37. Tina received a raise of $849.10. If her former salary was $24,260, what was the percent of the increase?

38. Jana notes that her July electricity bill is $8.96 higher than last year's amount. If last year's bill was $148.75, what is the percent of increase?

39. A salesperson says he can take $1512 off the price of a car. If the initial price is $18,900, what is the percent of decrease?

40. In 2003, a company's cost of production for 1 unit of a product was $328.50. After taking some cost-cutting measures, the company was able to reduce the cost of production by $14.50. What is the percent of decrease in cost?

41. Lena was making $8.50 per hour. She is given a raise so that now she is making $10.00 per hour. What was the percent of increase?

42. Gloria notes that her water bill was $18.50 in March. Her April water bill is for $32.40. What is the percent of increase?

43. A social worker worked a total of 1259 cases in 2003. The next year he worked 1540 cases. What was the percent of increase in case load?

44. 1996 was the first year of operation for a business. At the end of that first year, the business recorded a profit of $15,280. In 2004, that same business recorded a profit of $80,450. What was the percent of increase in profit?

45. Van purchases a DVD player on sale. The initial price was $175.90. The price after the discount was $149.52. What was the percent of decrease (discount rate)?

46. Risa's gross 2-week salary is $1756.25. If her net pay is $1334.75, what is the percent of decrease?

47. By insulating their water heater, adding a storm door, and installing new double-pane windows, the Jones family decreased their electric and gas bills from a total of $1536.98 in 2003 to $1020.57 in 2004. What was the percent of decrease in the Jones family's electric and gas charges?

48. Gloria is an office manager. Through her suggestions for improving efficiency and cutting excessive spending, the cost of operation dropped from $65,886.35 in 2003 to $60,245.50 in 2004. What was the percent of decrease in operation cost?

Review Exercises

For Exercises 1 and 2, solve and check.

[6.4] **1.** $0.12x = 450$

[4.3] **2.** $20 = 9(y + 2) - 3y$

For Exercises 3 and 4, simplify.

[6.5] **3.** $(1 + 0.08)^2$

[5.7] **4.** $\left(\dfrac{1}{5}\right)^3$

[4.5] **5.** Will has some $5 bills and $10 bills in his wallet. If he has 16 bills worth a total of $105, how many of each bill is in his wallet? Use a table.

Categories	Value	Number	Amount

Solving Problems Involving Interest

OBJECTIVE **1** Solve problems involving simple interest.

Percent calculations are involved in borrowing and investing money. The amount of money borrowed or invested is called **principal.** Lenders charge a fee that is a certain percent of the principal to borrow that principal. This fee is called **interest.** Similarly, investors are paid interest for investing the principal. The percent used to calculate interest is called an **interest rate.**

DEFINITIONS **Principal:** An initial amount of money.

Interest: An amount of money that is a percent of the principal.

Interest rate: A percent used to calculate interest.

There are two ways interest can be calculated and, therefore, two types of interest. The first is **simple interest** and is calculated only on the original principal and is the same every year.

DEFINITION **Simple interest:** Interest calculated using only the original principal and the amount of time the principal earns interest.

For example, if we invest a principal of $100 in a savings account earning 5% simple interest, then after 1 year, we would be paid $(100)(0.05) = \$5$ in interest. In other words, after 1 year, the bank deposits $5 into our account so that the new balance is $105. After 2 years, the interest will be $(\$5)2 = \$10$, and so on. After half a year, we would receive half the amount of annual interest. That is, instead of $5, we would receive $(100)(0.05)\left(\frac{1}{2}\right) = \$2.50$.

Our example suggests the following formula:

$$\text{simple interest} = \text{principal} \cdot \text{simple interest rate} \cdot \text{time}$$

Using I for interest, P for principal, r for simple interest rate, and t for time, we have the following formula:

$$I = Prt$$

DISCUSSION The same letter can represent different things in different contexts. In interest calculations, P is used to represent the principal. For what else have we used P?

Example 1 Graham invests $800 in a savings account earning 3% simple interest. How much interest will be earned after 1 year? What will the final balance be after 1 year?

Understand: We must calculate the interest and final balance after 1 year given the principal, rate, and time.

Plan: Use the simple interest formula $I = Prt$ to calculate interest. Then add the interest to the principal to calculate the final balance.

Execute: $I = Prt$

$I = (800)(3\%)(1)$ **Replace P with 800, r with 3%, and t with 1.**

$I = (800)(0.03)(1)$ **Write 3% as a decimal number.**

$I = 24$ **Multiply.**

Answer: Graham will earn $24 in interest after 1 year. To find the final balance, we add the interest to the principal.

$$\text{final balance} = \$800 + \$24 = \$824$$

Check: Verify that $24 is in fact 3% of $800. We will leave this to the reader.

Do Your Turn 1 ▷

Suppose the time is expressed in units other than years, as in 9 months. Because the interest rate is annual, the time must be expressed in terms of years. We can use dimensional analysis from Chapter 7 to convert the units of time. To convert 9 months to years, we use the fact that 1 year is 12 months, and write:

$$9 \text{ mo.} = \frac{9 \text{ mo.}}{1} \cdot \frac{1 \text{ yr.}}{12 \text{ mo.}} = \frac{9}{12} \text{ yr.} = \frac{3}{4} \text{ yr.}$$

Suppose the time is given in days, as in 60 days. To simplify the calculation, financial institutions will use 365 days to a year instead of $365\frac{1}{4}$.

$$60 \text{ days} = \frac{60 \text{ days}}{1} \cdot \frac{1 \text{ yr.}}{365 \text{ days}} = \frac{60}{365} \text{ yr.}$$

▲

Note: Because we will use the fractions in the following steps, it isn't necessary to simplify them. The simplification will occur in the calculation.

Let's now consider some problems where the time is given in units other than years.

Example 2 Calvin borrows $2000 earning 15% simple interest. If he negotiates to repay the loan after 90 days, how much will he have to pay back?

Understand: We must calculate the final balance after 90 days. Because the loan is for 90 days, and the time must be in terms of years, we must convert 90 days to years.

$$90 \text{ days} = \frac{90 \text{ days}}{1} \cdot \frac{1 \text{ yr.}}{365 \text{ days}} = \frac{90}{365} \text{ yr.}$$

Plan: Use the simple interest formula $I = Prt$ to calculate interest. Then add the interest to the principal to calculate the final balance.

Execute: $I = Prt$

$$I = (2000)(0.15)\left(\frac{90}{365}\right) \quad \text{Substitute for } P, r, \text{ and } t. \text{ Remember } 15\% = 0.15.$$

$$I = \frac{2000}{1} \cdot \frac{0.15}{1} \cdot \frac{90}{365} \quad \text{Write 2000 and 0.15 over 1.}$$

$$I = \frac{27,000}{365} \quad \text{Multiply.}$$

$$I \approx 73.97 \quad \text{Divide and round to the nearest hundredth.}$$

Answer: In 90 days, Calvin will have to pay back the $2000 principal plus $73.97 in interest for a total of $2073.97.

Check: Verify that $73.97 is in fact 15% of $2000 for 90 days. We will leave this to the reader.

Do Your Turn 2 ▷

Your Turn 2

Calculate the interest and the final balance.

a. $4500 earning 12% simple interest for 1 quarter ($\frac{1}{4}$ of a year)

b. $900 earning 5.5% simple interest for 30 days

Note: We round to the nearest hundredth because interest is money. ◀

Answers to Your Turn 1
a. $51.20; $691.20
b. $137.50; $2637.50

Answers to Your Turn 2
a. $135; $4635
b. $4.07; $904.07

What if the principal is the missing amount? Like the problems of percent of increase, our approach will depend on the given information. If we are given the percent and the interest, we can use $I = Prt$ and solve for P.

Your Turn 3

Solve.

a. Ingrid notes that her savings account earned $42.32 in 1 year. If the account earns 6.2% simple interest, what was her principal?

b. Lawrence notes that his account earned $59.50 in half a year. If the account earns 14% simple interest, what was his principal?

Example 3 Sandy notes that $6.80 was deposited in her savings account after 3 months. The account earns 5.5% simple interest. What was the principal?

Understand: We must calculate the principal. Because the time is 3 months and the time must be in terms of years, we must convert 3 months to years.

$$3 \text{ mo.} = \frac{3 \text{ mo}}{1} \cdot \frac{1 \text{ year}}{12 \text{ mo}} = \frac{3}{12} \text{ yr.}$$

Plan: Use the simple interest formula $I = Prt$ and solve for P.

Execute: $I = Prt$

$$6.80 = P(0.055)\left(\frac{3}{12}\right) \qquad \text{Substitute for } I, r, \text{ and } t.$$

$$6.00 = P\left(\frac{0.055}{1} \cdot \frac{3}{12}\right) \qquad \text{Write 0.055 over 1.}$$

$$6.80 = P\left(\frac{0.165}{12}\right) \qquad \text{Multiply.}$$

$$6.80 = 0.01375P \qquad \text{Divide 0.165 by 12.}$$

$$\frac{6.80}{0.01375} = \frac{0.01375P}{0.01375} \qquad \substack{\text{Divide both sides by} \\ \text{0.01375.}}$$

$$494.55 \approx P$$

Note: We round to the nearest hundredth because principal is money.

Answer: The principal was $494.55.

Check: Verify that a principal of $494.55 earning 5.5% simple interest for 3 months earns $6.80 in interest. We will leave this to the reader.

◁**Do Your Turn 3**

Now suppose the final balance is given instead of the interest. We must remember that the final balance is a combination of principal and interest. We can say:

$$\text{final balance} = \text{principal} + \text{interest}$$

Because interest is calculated by the relationship Prt, we can write this equation as:

$$B = P + Prt$$

We can rewrite this formula by factoring out P.

$$B = P(1 + rt)$$

Example 4 Pedro notes that his new balance at the end of a year is $1441.60. If the account earns 6% simple interest, what was the principal?

Understand: We must calculate the principal given the simple interest rate and the final balance.

Plan: Use the final balance formula $B = P(1 + rt)$.

Answers to Your Turn 3
a. $682.58 b. $850

Execute:

$$B = P(1 + rt)$$
$$1141.60 = P[1 + 0.06(1)]$$ Replace *B* with 1441.60, *r* with 0.06, and *t* with 1.
$$1441.60 = 1.06P$$ In the bracket, multiply and then add.
$$\frac{1441.60}{1.06} = \frac{1.06P}{1.06}$$ Divide both sides by 1.06 to isolate *P*.
$$1360 = P$$

CONNECTION This is the same type of problem as Example 4 in Section 8.5. Consider the meaning of the equation $1441.60 = 1.06P$. Because 1.06 is 106%, this equation is saying that 1441.60 is 106% of the principal. This should make sense because 1441.60 is 100% (all) of the principal combined with the interest, where the interest is 6% of the principal.

Answer: Pedro's principal was $1360.

Check: Verify that a principal of $1360 earning 6% simple interest will come to a final balance of $1441.60 at the end of 1 year. We will leave this to the reader.

Do Your Turn 4 ▶

OBJECTIVE 2 Solve problems involving compound interest.

Now that we've explored simple interest, we can consider another way that interest is calculated: by compounding. Recall that simple interest is calculated based solely on the original principal. **Compound interest** is calculated using the principal *and* prior earned interest.

DEFINITION   Compound interest: Interest that is calculated based on principal and prior earned interest.

In compound interest calculations, the annual interest rate is called **annual percentage rate (APR).**

DEFINITION   Annual percentage rate (APR): An interest rate that is used to calculate compound interest.

In order to develop a formula for compound interest, let's work through Example 5 the "*long*" way and look for a pattern in the process.

Example 5 Chan invests $500 in a savings account with an APR of 8%. If the interest is compounded quarterly, how much will he have after 1 year?

Understand: We must calculate Chan's final balance after 1 year if the interest is compounded quarterly. This means every 3 months the bank will make a deposit into Chan's account. Therefore, he will start each quarter with a little more money than the prior quarter and therefore earn slightly more interest.

Plan: Calculate the final balance after each quarter in the year. This means we will perform four calculations. Recall that final balance can be calculated using the formula $B = P(1 + rt)$.

Your Turn 4

Solve.

a. The total amount that Nala must repay on a 1-year loan with 8% simple interest is $12,960. What was the principal?

b. Ron notes that his new balance at the end of 6 months is $1221. If the account earns 4% simple interest, what was the principal?

Answers to Your Turn 4
a. $12,000 b. $1197.06

Execute: Since the money is compounded quarterly, we first need to convert the 8% annual percentage rate to a quarterly rate. Since there are 4 quarters in a year, we find the quarterly rate by dividing the APR by 4:

$$\text{quarterly rate} = \frac{0.08}{4} = 0.02 \quad (\text{or } 2\%)$$

CONNECTION We could have also used dimensional analysis that we learned in Chapter 7.

$$8\% \text{ APR} = \frac{8\%}{1 \text{ year}} \cdot \frac{1 \text{ year}}{4 \text{ quarters}} = \frac{8\%}{4 \text{ quarters}} = \frac{2\%}{1 \text{ quarters}}$$
$$= 2\% \text{ per quarter}$$

So, at the end of each quarter, 2% of the principal will be deposited into the account, which means the balance is 102% of the principal.

First quarter: $B = 500(1.02) = 510$

After the first quarter, the balance is $510. For the second quarter, the bank will calculate the balance based on a principal of $510. Then, that balance will be multiplied by 1.02 for the third quarter, and so on.

Second quarter: $B = 510(1.02) = 520.2$

Third quarter: $B = 520.2(1.02) = 530.604$

Fourth quarter (1 year): $B = 530.604(1.02) \approx 541.22$

Note: Each calculation ◀ is 1.02 times the previous balance.

Answer: After 1 year, the balance will be $541.22.

Notice that finding the balance for each quarter is tedious. It is also unnecessary. Since each new balance is found by multiplying the previous balance by 1.02, we are repeatedly multiplying by 1.02, which suggests that we can use an exponent.

$$500 \cdot \overbrace{1.02 \cdot 1.02 \cdot 1.02 \cdot 1.02} = 500(1.02)^4$$
$$= 541.21608$$
$$\approx 541.22$$

Note: Four factors of 1.02 ◀ is the same as 1.02 raised to the fourth power.

Now, suppose we had to find the balance after 2 years. In 2 years, there are 8 quarters, which we find by multiplying 2 years by 4 quarters per year. So, now we would multiply the $500 principal by 8 factors of 1.02, which means we can raise it to the 8th power. So the expression for the balance after two years is $500(1.02)^8$.

Chan's balance after two years: $500(1.02)^8 \approx 585.83$

In three years, we have $3 \cdot 4 = 12$ factors of 1.02, so the exponent would be 12, and so on.

Think about the order of operations involved here. We first converted the APR to a rate that matches the compound period by dividing the APR by the number of compounds, which we can represent with $\frac{r}{n}$. We then added 1 to that adjusted rate, so we now have $1 + \frac{r}{n}$. We then raised that sum to a power which is the product of the time and the number of times the principal is compounded per year, so we have $\left(1 + \frac{r}{n}\right)^{nt}$. Last, we multiplied that result by the principal, so we have the following formula for calculating the final balance if a principal amount P is invested for t years at an APR of r compounded n times per year.

$$B = P\left(1 + \frac{r}{n}\right)^{nt}$$

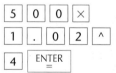

Calculator Tips

These exponent calculations are best done with a calculator. On a scientific calculator, the calculation to the right would proceed as follows:

| 5 | 0 | 0 | × |

| 1 | . | 0 | 2 | ^ |

| 4 | ENTER = |

Next we list some common compound time lengths and their corresponding values for n.

Annually: compounded once per year, so $n = 1$
Semiannually: compounded twice per year, so $n = 2$
Quarterly: compounded 4 times per year, so $n = 4$
Daily: compounded 365 times per year, so $n = 365$

Example 6   Bjorn invests $15,000 at 12% APR compounded semiannually. What will be his final balance after 4 years?

Understand: We must calculate the final balance given the principal, the APR, and that it is compounded semiannually for 4 years. Compounding semiannually means the principal is compounded twice each year, so $n = 2$.

Plan: Use the compound interest formula $B = P\left(1 + \dfrac{r}{n}\right)^{nt}$

$$P = 15,000$$
$$r = 12\% = 0.12$$
$$t = 4$$
$$n = 2$$

Execute: $B = P\left(1 + \dfrac{r}{n}\right)^{nt}$

$B = 15,000\left[1 + \dfrac{0.12}{2}\right]^{2 \cdot 4}$ Replace *P* with 15,000, *r* with 0.12, *n* with 2, and *t* with 4.

$B = 15,000[1 + 0.06]^8$ Divide 0.12 by 2 and multiply 2 · 4.

$B = 15,000[1.06]^8$ Add 1 and 0.06.

$B \approx 23,907.72$ Evaluate the exponential expression, then multiply.

DISCUSSION   How much would Bjorn have in 4 years if he had invested in a simple interest plan earning 12% per year?

Answer: Bjorn will have $23,907.72 at the end of 4 years.

Do Your Turns 5 and 6 ▶

Example 7   Paula has a balance of $685.72 on a credit card that has a 14.9% APR. The interest with this particular card is compounded daily. If she does not charge any money to this account until after she receives her next bill (30 days), what will be her balance on the next bill?

Understand: We must calculate the final balance given the principal, the APR and that the principal is compounded daily for 30 days. Compounding daily means the principal is compounded 365 times per year, so $n = 365$.

Plan: Use the compound interest formula $B = P\left(1 + \dfrac{r}{n}\right)^{nt}$.

$$P = \$685.72$$
$$r = 14.9\% = 0.149$$
$$t = \dfrac{30}{365}$$
$$n = 365$$

◀ **Note:** Since *t* must be in years, we had to convert 30 days to years.

$30 \text{ days} \times \dfrac{1 \text{ year}}{365 \text{ days}} = \dfrac{30}{365} \text{ years}$

Your Turn 5

Deanne invests $12,000 at 14% APR.

a. Suppose the interest is compounded annually. What would be her final balance after 2 years?

b. Suppose the interest is compounded semiannually. What will be her final balance after 2 years?

Your Turn 6

Kyle invests $8000 at 8% APR.

a. Suppose the interest is compounded semi-annually. What would be his final balance after 3 years?

b. Suppose the interest is compounded quarterly. What will be his final balance after 3 years?

Answers to Your Turn 5
a. $15,595.20 b. $15,729.55

Answers to Your Turn 6
a. $10,122.55 b. $10,145.93

Calculator Tips

After replacing the variables with numbers, we can enter the entire numeric expression using a scientific calculator. The key sequence is as follows:

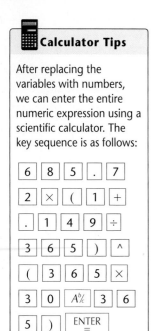

Execute: $B = P\left(1 + \dfrac{r}{n}\right)^{nt}$

$B = 685.72\left[1 + \dfrac{0.149}{365}\right]^{365 \cdot \frac{30}{365}}$ Replace *P* with 685.72, *r* with 0.149, *n* with 365, and *t* with $\frac{30}{365}$.

$B = 685.72[1 + 0.000408219]^{30}$ Divide 0.149 by 365 and multiply $365 \cdot \frac{30}{365}$.

$B = 685.72[1.000408219]^{30}$ Add.

$B \approx 694.17$ Evaluate the exponential expression, then multiply.

DISCUSSION What is the interest or finance charge that Paula owes?

Answer: Paula's balance will be $694.17 on her next bill.

OF INTEREST

Read the fine print of a credit card statement to find out how the interest is calculated for that card. Sometimes, there are several methods listed for the same type of card. The method used will be indicated somewhere on the statement. In addition to daily compounding as we saw in Example 7, here are a few other common methods:

Method 1: Apply the monthly percentage rate to the average daily balance. Applying this method to Example 7:

monthly percentage rate = 14.9% ÷ 12 = 1.242% = 0.01242

finance charge (interest) = 685.72(0.01242) = $8.52

final balance = 685.72 + 8.52 = $694.24

Method 2: Apply the daily percentage rate to the average daily balance. Then multiply the result times the number of days in the billing cycle (simple interest). Applying this method to Example 7:

daily percentage rate = 14.9% ÷ 365 = 0.04082% = 0.0004082

finance charge = 685.72 (0.0004082)^{30} = $8.40

final balance = 685.72 + 8.40 = $694.12   There were 30 days in the billing cycle.

◁ **Do Your Turn 7**

Your Turn 7

Marla has a balance of $426.85 on a credit card that has a 12.9% APR. The interest with this particular card is compounded daily. If she does not charge any money to this account until after she receives her next bill (30 days), what will be her balance on the next bill?

OBJECTIVE 3 Solve problems involving amortization.

In Example 7, notice that the interest charged during the month was $8.45 (694.17 − 685.72 = 8.45). If Paula just paid $8.45, her balance goes back to the initial balance, $685.72, and the next month's balance will be back to $694.17. Thus, if all she does is pay the finance charge (interest), then she will never pay off the debt.

In order to pay off a debt, we must pay more than the interest. Any amount that we pay in addition to the interest will decrease the principal. This will then decrease the amount of interest for the next payment.

For example, suppose we borrow $1000 at 12% APR and make a $200 payment each month. We will need to calculate the balance at the end of each month. Because the interest rate is 12% APR, the monthly rate will be 1%.

Answer to Your Turn 7
$431.40

First month balance: $B = 1000\left[1 + 0.12\left(\dfrac{1}{12}\right)\right] = 1000(1.01)$

$$= \$1010$$

Applying the $200 payment: $1010 - \$200 = \$810$

▲

Note: $10 of the $200 payment is going toward interest while the other $190 is going toward the principal. In our next payment, we will owe less interest because the principal is now less.

Second month balance: $B = 810(1.01) = \$818.10$

Applying the $200 payment: $818.10 - \$200 = \$618.10$

▲

Note: Now only $8.10 is interest, so the rest of the $200 payment, which amounts to $191.90, goes toward principal. In the next payment, we will pay even less interest and more of the $200 can go toward principal.

This process would continue until we pay off the loan. The payments we make to pay off a loan are called *installments*. When we pay off a loan in installments, we **amortize** the loan.

DEFINITION **Amortize:** To pay off a loan or debt in installments.

In amortizing a loan we are interested in the payment that we must make. The factors that affect this payment are the principal, percentage rate, the number of payments per year, and the agreed-upon time to pay off the loan.

Because the calculations can be rather involved, it is common to use amortization tables to determine the payment. Most bookstores carry small amortization books that list tables by percentage rate. Table 8.1 (next page) shows the monthly payment for various principal amounts borrowed at a rate of 9% and paid off in 3, 4, 5, 15, or 30 years.

Notice that under the 30-year column and the row for 90,000 the payment is $724.16. Note that the table does not list every possible principal amount. However, we can use combinations of the listed principal values to determine the payment for a principal that is not listed.

Example 8 Use Table 8.1 to find the monthly payment for $72,300 at 9% APR for 15 years.

Understand: We must determine the monthly payment using the amortization table. Because 72,300 is not listed, we must use combinations of principals that are listed. We can use any of the principals whose sum is 72,300.

Plan: Locate principal amounts whose sum is 72,300, find each of their monthly payments in the 15-year column, and then add all the monthly payments to get the total monthly payment for 72,300.

Table 8.1 9% APR Monthly Payment Schedule

Amount	3 Years	4 Years	5 Years	15 Years	30 Years
100	3.18	2.49	2.08	1.01	0.80
200	6.36	4.98	4.15	2.03	1.61
500	15.90	12.44	10.38	5.07	4.02
1,000	31.80	24.89	20.76	10.14	8.05
2,000	63.60	49.77	41.52	20.29	16.09
5,000	159.00	124.43	103.79	50.71	40.23
6,000	190.80	149.31	124.55	60.86	48.28
7,000	222.60	174.20	145.31	71.00	56.32
8,000	254.40	199.08	166.07	81.14	64.37
9,000	286.20	223.97	186.83	91.28	72.42
10,000	318.00	248.85	207.58	101.43	80.46
15,000	477.00	373.28	311.38	152.14	120.69
20,000	635.99	497.70	415.17	202.85	160.92
25,000	794.99	622.13	518.96	253.57	201.16
30,000	953.99	746.55	622.75	304.28	241.39
35,000	1112.99	870.98	726.54	354.99	281.62
36,000	1144.79	895.86	747.30	365.14	289.66
37,000	1176.59	920.75	768.06	375.28	297.71
38,000	1208.39	945.63	788.82	385.42	305.76
39,000	1240.19	970.52	809.58	395.56	313.80
40,000	1271.99	995.40	830.33	405.71	321.85
41,000	1303.79	1020.29	851.09	415.85	329.90
42,000	1335.59	1045.17	871.85	425.99	337.94
43,000	1367.69	1070.06	892.61	436.13	345.99
44,000	1399.19	1094.94	913.37	446.28	354.03
45,000	1430.99	1119.83	934.13	456.42	362.08
46,000	1462.79	1144.71	954.88	466.56	370.13
47,000	1494.59	1169.60	975.64	476.71	378.17
48,000	1526.39	1194.48	996.40	486.85	386.22
49,000	1558.19	1219.37	1017.16	496.99	394.27
50,000	1589.99	1244.25	1037.92	507.13	402.31
51,000	1621.79	1269.14	1058.68	517.28	410.36
52,000	1653.59	1294.02	1079.43	527.42	418.40
53,000	1685.39	1318.91	1100.19	537.56	426.45
54,000	1717.19	1343.79	1120.95	547.70	434.50
55,000	1748.99	1368.68	1141.71	557.85	442.54
56,000	1780.79	1393.56	1162.47	567.99	450.59
57,000	1812.58	1418.45	1183.23	578.13	458.63
58,000	1844.38	1443.33	1203.98	588.27	466.68
59,000	1876.18	1468.22	1224.74	598.42	474.73
60,000	1907.98	1493.10	1245.50	608.56	482.77
65,000	2066.98	1617.53	1349.29	669.27	523.00
70,000	2225.98	1741.95	1453.08	709.99	563.24
75,000	2384.98	1866.38	1556.88	760.70	603.47
80,000	2543.98	1990.80	1660.67	811.41	643.70
85,000	2702.98	2115.23	1764.46	862.13	683.93
90,000	2861.98	2239.65	1868.25	912.84	724.16
95,000	3020.97	2364.08	1972.04	963.55	764.39
100,000	3179.97	2488.50	2075.84	1014.27	804.62
105,000	3338.97	2612.93	2179.63	1064.98	844.85
110,000	3497.97	2737.35	2283.42	1115.69	885.08
120,000	3815.97	2986.21	2491.00	1217.12	965.55
130,000	4133.97	3235.06	2698.59	1318.55	1046.01
140,000	4451.96	3483.91	2906.17	1419.97	1126.47
150,000	4769.96	3732.76	3113.75	1521.40	1206.93
175,000	5564.95	4354.88	3632.71	1774.97	1408.09
200,000	6359.95	4977.01	4151.67	2028.53	1609.25
225,000	7154.94	5599.13	4670.63	2282.10	1810.40
250,000	7949.93	6221.26	5189.59	2535.67	2011.56

Execute: We can make 72,300 by adding the following listed principals:

$$70,000 + 2000 + 200 + 100 = 72,300$$

Now, using the 15-year column, we find the monthly payment for each of our listed principals, and then add those monthly payments.

$$70,000 + 2000 + 200 + 100 = 72,300$$
$$\downarrow \qquad \downarrow \qquad \downarrow \qquad \downarrow \qquad \downarrow$$
$$709.99 + 20.29 + 2.03 + 1.01 = 733.32$$

Answer: The monthly payment will be $733.32.

Do Your Turn 8 ▶

Your Turn 8

Use Table 8.1 to find the monthly payment.

a. $15,000 for 5 years

b. $125,000 for 30 years

c. $97,500 for 15 years

Answers to Your Turn 8
a. $311.38 b. $1005.78
c. $988.91

8.6 Exercises

For Extra Help

  Videotape DVT

 Addison-Wesley Tutor Center

 Math XL

 MyMathLab

 Student Solutions Manual

1. What is principal?

2. What is interest?

3. What is the formula for calculating simple interest and what does each variable in it represent?

4. What is an APR?

5. What is the formula for calculating the final balance when interest is compounded and what does each variable represent?

6. What does amortize mean?

For Exercises 7–16, calculate the simple interest and final balance.

7. $4000 at 3% for 1 year

8. $500 at 4% for 1 year

9. $350 at 5.5% for 2 years

10. $1200 at 6.5% for 3 years

11. $12,250 at 12.9% for 5 years

12. $18,000 at 14.5% for 4 years

13. $2400 at 8% for six months $\left(\frac{1}{2} \text{ year}\right)$

14. $840 at 12% for 3 months $\left(\frac{1}{4} \text{ year}\right)$

15. $2000 at 6.9% for 60 days

16. $600 at 4.9% for 30 days

For Exercises 17–20, solve for the principal.

17. Nuno notes that his savings account earned $331.12 in 1 year. If the account earns 6.9% simple interest, what was his principal?

18. Sharice notes that her savings account earned $21.28 in 1 year. If the account earns 5.5% simple interest, what was her principal?

19. Kelly notes that her savings account earned $15.57 in 3 months. If the simple interest rate for her account is 3.2%, what was her principal?

20. Roger notes that his savings account earned $24.56 in 9 months. If the simple interest rate for his account is 4.8%, what was his principal?

For Exercises 21–26, calculate the final balance.

21. $5000 at 8% compounded annually for 2 years

22. $6400 at 12% compounded annually for 2 years

23. $840 at 6% compounded annually for 2 years

24. $1450 at 10% compounded annually for 2 years

25. $14,000 at 8% compounded semiannually for 1 year

26. $20,000 at 10% compounded semiannually for 1 year

For Exercises 27–40, use a calculator to solve.

27. $400 at 9% compounded annually for 3 years

28. $650 at 5% compounded annually for 3 years

29. $1600 at 6% compounded semiannually for 3 years

30. $960 at 14% compounded semiannually for 3 years

31. $290 at 9% compounded semiannually for 4 years

32. $500 at 11% compounded semiannually for 5 years

33. $1300 at 12% compounded quarterly for 2 years

34. $900 at 8% compounded quarterly for 3 years

35. $450 at 6% compounded quarterly for 4 years

36. $2400 at 10% compounded quarterly for 4 years

37. Kevin has a balance of $860.20 on a credit card that has a 16.9% APR. The interest is compounded daily. If he does not charge any money to this account until after he receives his next bill (30 days), what will be his balance on the next bill?

38. Phyllis has a balance of $285.32 on a credit card that has a 7.9% APR. The interest is compounded daily. If she does not charge any money to this account until after she receives her next bill (30 days), what will be her balance on the next bill?

39. Lindsey has a balance of $694.75 on a credit card that has a 19.8% APR. The interest is compounded daily. If she does not charge any money to this account until after she receives her next bill (30 days), what will be her balance on the next bill?

40. Scott has a balance of $1248.56 on a credit card that has a 21.9% APR. The interest is compounded daily. If she does not charge any money to this account until after she receives her next bill (30 days), what will be her balance on the next bill?

For Exercises 41–50, use Table 8.1 to find the monthly payment.

41. $15,000 for 5 years

42. $9000 for 3 years

43. $17,000 for 4 years

44. $21,000 for 5 years

45. $92,500 for 30 years

46. $78,000 for 15 years

47. $84,100 for 15 years

48. $103,500 for 30 years

49. $115,200 for 30 years

50. $147,200 for 15 years

Review Exercises

[5.1]   **1.** Graph $-6\dfrac{3}{4}$ on a number line.

[6.4]   **2.** Write $\dfrac{5}{6}$ as a decimal number.

[6.4]   **3.** Simplify. $\dfrac{65 + 90 + 72 + 84}{4}$

[3.1]   **4.** Evaluate $2x + 3y$ when $x = -3$ and $y = 4$.

[5.8]   **5.** Solve and check. $\dfrac{3}{8}x = -\dfrac{21}{30}$

Defined Terms

Review the following terms and for those you do not know, study its definition on the page number next to it.

Section 8.1
Percent *(p. 562)*

Section 8.4
Commission *(p. 581)*

Section 8.6
Principal *(p. 602)*
Interest *(p. 602)*
Interest rate *(p. 602)*

Simple interest *(p. 602)*
Compound interest
  (p. 605)
Annual percentage rate
  (APR) *(p. 605)*

Amortize *(p. 609)*

Procedures, Rules, and Key Examples

Procedures/Rules	*Key Example(s)*

Section 8.1 Introduction to Percent

To write a percent as a fraction:
1. Write the percent numeral over 100.
2. Simplify.

Example 1: Write each percent as a fraction.

a. $45\% = \dfrac{45}{100} = \dfrac{9}{20}$

b. $16.2\% = \dfrac{16.2}{100} = \dfrac{16.2(10)}{100(10)} = \dfrac{162}{1000} = \dfrac{81}{500}$

c. $9\dfrac{1}{2}\% = \dfrac{9\dfrac{1}{2}}{100} = \dfrac{19}{2} \div 100 = \dfrac{19}{2} \cdot \dfrac{1}{100} = \dfrac{19}{200}$

To write a percent as a decimal number:
1. Write the fraction or mixed number percent in decimal form.
2. Divide by 100.

Example 2: Write each percent as a decimal number.

a. $62\% = 62 \div 100 = 0.62$

b. $4.5\% = 4.5 \div 100 = 0.045$

c. $33\dfrac{1}{3}\% = 33.\overline{3}\% = 33.\overline{3} \div 100 = 0.\overline{3}$

To write a fraction as a percent:
1. Multiply by 100%.
2. Simplify.

Example 3: Write each fraction as a percent.

a. $\dfrac{3}{5} = \dfrac{3}{\cancel{5}} \cdot \dfrac{\cancel{100}^{20}}{1}\% = 60\%$

b. $\dfrac{2}{3} = \dfrac{2}{3} \cdot \dfrac{100}{1}\% = \dfrac{200}{3}\% = 66\dfrac{2}{3}\%$

To write a decimal number as a percent, multiply by 100%.

Note: Multiplying a decimal number by 100% causes the decimal point to move 2 places to the right.

Example 4: Write each decimal number as a percent.

a. $0.49 = 0.49 \cdot (100\%) = 49\%$

b. $0.067 = 0.067 \cdot (100\%) = 6.7\%$

c. $1.03 = 1.03 \cdot (100\%) = 103\%$

Section 8.2 *Translating Percent Sentences Word for Word*

The simple percent sentence is: A **percent** of a **whole** amount is the **part** of the whole.

To translate a percent sentence word for word:

1. Select a variable for the unknown.
2. Translate the word *is* to an equal sign.
3. If *of* is preceded by the percent, translate it to multiplication. If *of* is preceded by a whole number, translate it to division.

Notes: The first step in solving any of the translated equations is to write the percent as a decimal or fraction.

When using this method to calculate a missing percent, the outcome will be a decimal number or fraction. You must write the decimal number or fraction as a percent.

Example 1: Translate to an equation.

a. 40% of 70 is what number?

Translation: $40\% \cdot 70 = c$

b. 15% of what number is 65?

Translation: $15\% \cdot w = 65$

c. What percent of 200 is 50?

Translation: $p \cdot 200 = 50$

d. What percent is 20 out of 30?

Translation: $p = 20 \div 30$

e. What percent is 15 of 20?

Translation: $p = 15 \div 20$

Section 8.3 *Translating Percent Sentences to Proportions*

To translate a simple percent sentence to a proportion, write the proportion in the following form:

$$\text{percent} = \frac{\text{part}}{\text{whole}},$$

where the percent is expressed as a fraction with a denominator of 100.

Notes: The whole amount always follows the word *of*.

When using this method to calculate a missing percent, the result will be a percent. You must simply write the percent sign beside the result.

Example 1: Translate to a proportion.

a. 40% of 70 is what number?

Translation: $\dfrac{40}{100} = \dfrac{c}{70}$

b. 15% of what number is 65?

Translation: $\dfrac{15}{100} = \dfrac{65}{w}$

c. What percent of 200 is 50?

Translation: $\dfrac{P}{100} = \dfrac{50}{200}$

d. What percent is 20 out of 30?

Translation: $\dfrac{P}{100} = \dfrac{20}{30}$

e. What percent is 15 of 20?

Translation: $\dfrac{P}{100} = \dfrac{15}{20}$

Sections 8.4 and 8.5 *Solving Percent Problems*

To solve problems involving percents:

1. Identify the percent, whole amount, and part.
2. Write these pieces in a simple percent sentence (if needed).
3. Translate to an equation or a proportion.
4. Solve.

Example 1: Angela answered 80% of the 40 questions on an exam correctly. How many questions did she answer correctly?

Simple sentence: 80% of 40 is what number?

$$0.8 \cdot 40 = n$$
$$32 = n$$

Answer: Angela answered 32 questions correctly.

Procedures/Rules	Key Example(s)

Example 2: A book costs $6.95. If the tax rate is 5%, what is the final price?

Simple sentence: 5% of $6.95 is tax.

$$0.05 \cdot 6.95 = t$$
$$0.35 \approx t$$

Answer: The final price is $6.95 + 0.35 = \$7.30$.

Example 3: Marcus notices his electric bill increased from $120 in January of 2003 to $140 in January of 2004. What was the percent of the increase?

Amount of increase: $140 - 120 = \$20$

Simple sentence: What percent of 120 is 20?

$$\frac{P}{100} = \frac{20}{120}$$
$$120P = 2000$$
$$\frac{120P}{120} = \frac{2000}{120}$$
$$P = 16.\overline{6}$$

Answer: His electric bill increased $16.\overline{6}\%$.

Section 8.6 Solving Problems Involving Interest

To solve problems involving interest, use the appropriate formula:

$$I = Prt, \ B = P(1 + rt), \ \text{or} \ B = P\left(1 + \frac{r}{n}\right)^{nt}.$$

Example 1: Karen invests $600 at a simple interest rate of 4%. Find the amount of interest she earns and her balance after nine months.

$$I = Prt$$
$$I = (600)(0.04)\left(\frac{3}{4}\right)$$
$$I = 18$$

Answer: After nine months, the interest is $18, which makes the balance $618.

Example 2: Darryl invests $5000 at 8% APR compounded quarterly. Find his balance after two years.

$$B = P\left(1 + \frac{r}{n}\right)^{nt}$$
$$B = 5000\left(1 + \frac{0.08}{4}\right)^{4 \cdot 2}$$
$$B = 5000(1.02)^{8}$$
$$B \approx 5858.30$$

Answer: Darryl's balance will be $5858.30.

To determine the monthly payment for a given principal, use Table 8.1.

Example 3: Use Table 8.1 to determine the monthly payment if a principal of $92,500 is amortized at 9% APR for 30 years.

$$90,000 + 2000 + 500 = 92,500$$
$$\downarrow \qquad \downarrow \qquad \downarrow$$
$$724.16 + 16.09 + 4.02 + 744.27$$

Answer: The payment will be $744.27.

Formulas

Simple interest

$$I = Prt$$
$$B = P(1 + rt)$$

where P represents the principal
r represents the simple interest rate
t represents time length, in years

Compound interest

$$B = P\left(1 + \frac{r}{n}\right)^{nt}$$

where P represents the principal
r represents the APR
t represents the time in years
n represents the number of compounds per year
B represents the final balance

For Exercises 1–6, answer true of false.

[8.1]   **1.** A percent cannot represent an amount more than 1.

[8.2]   **2.** 20% of 60 is greater than 30.

[8.2]   **3.** 60% of 300 is greater than 150.

[8.1]   **4.** 28 is the same as 28%.

[8.1]   **5.** 0.6 is the same as 6%.

[8.5]   **6.** The more often a given principal is compounded at the same rate, the larger the final balance will be.

[8.1]   **7.** In general, to write a percent as a decimal number, we _____ by 100.

[8.1]   **8.** In general, to write a decimal number or fraction as a percent, we _____ by 100%.

[8.2]   **9.** Explain in your own words how to translate the basic percent sentence word-for-word.

[8.3]   **10.** Explain in your own words how to translate the simple percent sentence to a proportion.

[8.1]   **11.** Write as a fraction in simplest form.

    a. 40%

    b. 26%

    c. 6.5%

    d. $24\frac{1}{2}\%$

[8.1]   **12.** Write as a decimal number.

    a. 16%

    b. 150%

    c. 3.2%

    d. $40\frac{1}{3}\%$

[8.1] *For Exercises 13–16, write each number as a percent.*

    13. 0.54   **14.** 1.3   **15.** $\dfrac{3}{8}$ **16.** $\dfrac{4}{9}$

[8.2] *For Exercises 17–20, translate word for word, then solve.*

    17. What number is 15% of 90?

    18. 12.8% of what number is 5.12?

    19. 12.5 is what percent of 20?

    20. What percent is 40 of 150?

For Exercises 21–24, translate to a proportion, then solve.

21. What number is 40.5% of 800?

22. $10\frac{1}{2}$ is 15% of what number?

23. 8.1 is what percent of 45?

24. What percent is 16 of 30?

For Exercises 25–32, write a word-for-word equation or a proportion, then solve.

[8.4] **25.** 35% of a mixture is soluble fat. If the total volume of the mixture is 600 milliliters, how many milliliters is soluble fat?

[8.4] **26.** In a survey of 2000 adults, 540 believed the Sun revolves around Earth. What percent of those surveyed believed the Sun revolves around Earth?

[8.5] **27.** A stereo receiver is priced at $285.75. The sales tax rate is 6%. Calculate the sales tax and total amount of the purchase.

[8.5] **28.** An outfit is on sale for 25% off. If the initial price is $56.95, what is the price after the discount?

[8.4] **29.** After traveling 180 miles, Boris has completed only 35% of a trip. What is the total distance of the trip?

[8.5] **30.** Kat indicates that she received a 3.5% raise that amounted to $1102.50. What was her former salary?

[8.5] **31.** The Johnsons' power bill increased $27.68 from last month. If their bill was $86.50 last month, what was the percent of increase?

[8.5] **32.** Dave's hourly wage is raised from $12 per hour to $15.50 per hour. What is the percent of increase?

[8.6] *For Exercises 33–38, solve.*

33. $480 is invested at 6% simple interest rate for 1 year. Calculate the interest.

34. $8000 is deposited into savings earns 5.2% simple interest. What will be the balance after 3 months?

35. After 1 year at 4.5% simple interest, the balance in an account is $627. If no deposits or withdrawals were made during the year, what was the principal?

36. $5000 is invested at 8% APR compounded annually. Calculate the balance after 3 years.

37. $1800 is invested at 12.4% APR compounded quarterly. Calculate the balance after a year.

38. A family borrows $90,500 at 9% APR to be amortized over 15 years. Use Table 8.1 to find their monthly payment.

[8.1] **1.** Write 24% as a fraction in simplest form.

1. _____

[8.1] **2.** Write 4.2% as a decimal number.

2. _____

[8.1] **3.** Write $12\frac{1}{2}$% as a decimal number.

3. _____

[8.1] **4.** Write $40\frac{3}{4}$% as a fraction in simplest form.

4. _____

[8.1] *For Exercises 5–8, write each number as a percent.*

5. 0.26

5. _____

6. 1.2

6. _____

7. $\frac{2}{5}$

7. _____

8. $\frac{5}{9}$

8. _____

[8.2] *For Exercises 9 and 10, translate word for word, then solve.*

9. What number is 15% of 76?

9. _____

10. 6.5% of what number is 8.32?

10. _____

[8.3] *For Exercises 11 and 12, translate to a proportion, then solve.*

11. 14 is what percent of 60?

11. _____

12. What percent is 12 of 32?

12. _____

For Exercises 13–19, write a word-for-word equation or a proportion, then solve.

[8.4] **13.** 80% of a class got an A on a particular test. If there were 30 students in the class, how many got an A?

13. _____

[8.4] **14.** Carolyn earns 25% of total sales in commission. If she sold a total of $1218 in merchandise in 1 week, what is her commission?

14. _____

15. _____

[8.4] **15.** The Morgan family has a net monthly income of $2786.92. If they have a $345.75 car payment, what percent of their net monthly income goes toward paying for the car?

16. _____

[8.5] **16.** A microwave oven is priced at $295.75. The sales tax rate is 5%. Calculate the sales tax and total amount of the purchase.

17. _____

[8.5] **17.** A pair of boots is on sale for 30% off. If the initial price of the boots is $84.95, what is the price after the discount?

18. _____

[8.5] **18.** Barbara indicates that she received a 2.5% raise that amounted to $586.25. What was her former salary?

19. _____

[8.5] **19.** Andre's hourly wage is raised from $8.75 per hour to $10.50 per hour. What is the percent of increase?

For Exercises 20–25, solve.

20. _____

[8.6] **20.** $5000 is invested at 6% simple interest rate for 1 year. Calculate the interest.

21. _____

[8.6] **21.** $800 is deposited into savings earning 4% simple interest. What will be the balance after 6 months?

22. _____

[8.6] **22.** After 1 year at 6% simple interest rate, the balance in an account is $2544. If no deposits or withdrawals were made during the year, what was the principal?

23. _____

[8.6] **23.** $2000 is invested at 9% APR compounded annually. Calculate the balance after 3 years.

24. _____

[8.6] **24.** $1200 is invested at 4% APR compounded quarterly. Calculate the balance after a year.

25. _____

[8.6] **25.** The Ramsey family borrows $112,000 at 9% APR to be amortized over 30 years. Use Table 8.1 to find their monthly payment.

For Exercises 1–6, answer true or false.

[2.4] **1.** There are two solutions for $(?)^2 = 81$: $+9$ and -9.

[3.6] **2.** 91 is composite.

[6.4] **3.** π is an irrational number.

[3.2] **4.** $4x^2y^3$ is a monomial.

[5.5] **5.** 12 is the LCM of 36 and 24.

[8.1] **6.** $9\% > 0.1$

[3.5] **7.** Explain in your own words how to multiply two polynomials.

[5.6] **8.** Explain in your own words how to add or subtract fractions.

[6.6] **9.** Explain in your own words how to clear decimal numbers from an equation.

[7.3] **10.** Explain in your own words how to use dimensional analysis to convert units.

[6.3] **11.** Write 27,500,000,000 in scientific notation.

[6.1] **12.** Graph -5.6 on a number line.

[1.3] **13.** Estimate $7805 \cdot 246$ by rounding so that there is only one nonzero digit in each number.

[3.2] **14.** What is the degree of $-5y$?

For Exercises 15–21, simplify.

[2.5] **15.** $[12 + 4(6 - 8)] - (-2)^3 + \sqrt{100 - 36}$

[5.6] **16.** $8\frac{3}{4} - \left(-5\frac{2}{3}\right)$

[5.7] **17.** $10\frac{5}{8} \div \left(-\frac{3}{16}\right) - \frac{1}{2}$

[6.5] **18.** $4.86 \div 0.8 + 58.9$

[6.5] **19.** $\left(4\frac{3}{8}\right)(-5.6)$

[6.5] **20.** $\frac{3}{4} - (0.6)^2$

[6.5] **21.** $\sqrt{(9)(0.25)}$

[6.4] **22.** Approximate $\sqrt{72}$ to the nearest hundredth.

[8.1] **23.** Write $12\frac{1}{2}\%$ as a decimal number.

[8.1] **24.** Write 0.71 as a percent.

$\begin{bmatrix} 5.7 \\ 6.2 \end{bmatrix}$ **25.** Subtract. $\left(x^4 - \dfrac{1}{3}x^2 + 9.6\right) - \left(4x^3 + \dfrac{3}{5}x^2 - 14.6\right)$

[6.3] **26.** Simplify. $(7.8x^6)(5.9x^8)$ [6.3] **27.** Multiply. $(6.2x - 1)(4x + 5)$

[3.6] **28.** Find the prime factorization of 540.   [5.5] **29.** Find the LCM of $24y^2$ and $30x$.

[3.7] **30.** Divide. $\dfrac{18x^6}{3x^2}$ [3.7] **31.** Factor. $32m^4 + 24m^2 - 16m$

[5.4] **32.** Simplify. $\dfrac{10m}{9n^2} \div \dfrac{5}{12n}$ [5.6] **33.** Simplify. $\dfrac{3}{5} - \dfrac{y}{4}$

For Exercises 34–36, solve and check.

[5.8] **34.** $\dfrac{5}{8}y - 3 = \dfrac{3}{4}$ [6.6] **35.** $3.5x - 12.1 = 6.8x + 3.08$

[7.2] **36.** $\dfrac{4\dfrac{1}{3}}{9} = \dfrac{1\dfrac{2}{3}}{n}$

[7.3] **37.** Convert 9.5 pounds to ounces.   [7.4] **38.** Convert 80 milliliters to liters.

[7.5] **39.** Convert 40°F to degrees Celsius.

For Exercises 40–50, solve.

$\begin{bmatrix} 6.4 \\ 5.7 \end{bmatrix}$ **40.** A parallelogram has an area of 201.3 square feet. If the base is 16.5 feet, what is the height? Convert the height to meters.

[6.5]   **41.** Write an expression in simplest form for the area. Calculate the area if h is 9 inches.

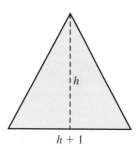

$h + 1$

[4.5] 42. The sum of two integers is 123. One integer is three more than twice the other. What are the integers?

[6.5] 43. Find the area of the shaded region.

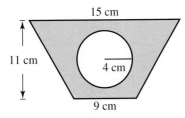

15 cm

11 cm

4 cm

9 cm

[6.5] 44. Find the volume of a soup can that is 4 inches tall and has a 1.2-inch radius.

[6.5] 45. Calculate the volume of a ball with a 16-inch diameter.

[7.1] 46. On a popular talk show, a studio-audience participant is selected at random by seat number. Suppose there are 480 seats. If 12 people from the same family are in the studio audience, what is the probability that someone from that family is selected? Write the probability in simplest form.

[7.2] 47. The following triangles are similar. Find the missing side lengths.

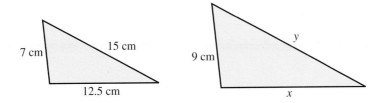

7 cm   15 cm

12.5 cm

9 cm y

x

[5.8]
[6.6] 48. Liam is jogging at 6 miles per hour. Starnell is walking at 4 miles per hour toward Liam on the same trail. If they are 0.25 mile apart, how long until they meet? Express the time in minutes.

[8.4] 49. Of the 450 people in attendance at a retirement seminar, 384 are over 50. What percent of the people in attendance are over 50?

[8.6] 50. $3200 is invested at 6% APR compounded semiannually. What will be the balance after 2 years?

CHAPTER 9

Statistics and Graphs

9.1 Mean, Median, and Mode
9.2 Tables, Charts, and Graphs
9.3 The Rectangular Coordinate System
9.4 Graphing Linear Equations
9.5 Applications with Graphing

"The will to win is important but the will to prepare is vital."

—JOE PATERNO,
COLLEGE FOOTBALL COACH

"You hit home runs not by chance but by preparation."

—ROGER MARIS,
NEW YORK YANKEES OUTFIELDER

Preparing for a Final Exam

Preparing for a cumulative final exam can seem overwhelming. However, if you break up your preparation over at least several days, you'll find the task much less daunting.

First, go back through your notes, from beginning to end, to see the progression of topics. You'll find that most topics keep reappearing throughout the course. Also, review your study sheets and go through the tests you have taken. Often, instructors take final exam questions directly from the tests.

In the second stage of preparation, do each cumulative review in the text. Try to review these over a few days, working through two cumulative reviews each day. Treat the last cumulative review as a practice final exam.

Try to complete all of the above preparation two days before the final exam. During the day before the final, go through the whole process one more time. Make a quick pass through your notes and study sheets, then work the final cumulative review again. Finally, be sure to get plenty of rest the night before, eat healthy, and drink plenty of fluids.

9.1 Mean, Median, and Mode

OBJECTIVES

OBJECTIVES

1 Find the mean.
2 Find weighted means.
3 Find the median.
4 Find the mode.

OBJECTIVE 1 Find the mean.

Numbers can be used to describe characteristics of data. A number used in this way is called a **statistic.**

DEFINITION **Statistic:** A number used to describe some characteristic of a set of data.

Oftentimes we use statistics, such as averages, which describe the *middle* or *central tendency* in a set of data. One such statistic is the **arithmetic mean.** Many people use the words *average* and *mean* interchangeably. Actually, the mean is only one type of average, called the **arithmetic average.** Other averages are the median and mode, which we will discuss later in this chapter.

DEFINITION **Mean** or **arithmetic average:** The sum of all given numbers divided by the number of numbers.

PROCEDURE *To find the mean, or arithmetic average, of a given set of numbers:*
1. Calculate the sum of all the given numbers.
2. Divide the sum by the number of numbers.

The common variable used to indicate a mean is $\bar{x}$.

Your Turn 1

a. Latonya has the following test scores. What is the mean of her test scores?

82, 74, 95, 86, 90

b. The ages of the employees in a particular department of a company follow. What is the mean age of the employees in the department?

23, 25, 32, 40, 36, 35

Example 1   Clerical specialist salaries within a company are listed below. What is the mean salary of a clerical specialist within the company?

$24,000	$21,900
$19,500	$19,500
$20,400	$25,200

Understand: We must calculate the arithmetic average, or mean, salary. We are given a set of salaries.

Plan: Divide the sum of the salaries by the number of salaries in the list.

Execute: $\bar{x} = \dfrac{24{,}000 + 19{,}500 + 20{,}400 + 21{,}900 + 19{,}500 + 25{,}200}{6}$

$\bar{x} = \dfrac{130{,}500}{6}$

$\bar{x} = 21{,}750$

Answer: The mean salary of the clerical specialists at this company is $21,750.

Check: We can verify the calculations by inverse operations. We will leave this to the reader.

◁ **Do Your Turn 1**

Answers to Your Turn 1
a. 85.4 b. 31.8$\overline{3}$

Example 2 Juan is taking a course in which the grade is determined by the average (mean) of four tests. He has the following test scores in the course: 68, 85, and 90. To get a B, he must have a test average (mean) of 80 on the four tests. What is the minimum score he needs on the last test to get a B?

Note: In grade calculations, the word *average* is often incorrectly used to represent the *arithmetic* average or mean.

Understand: We must calculate the minimum test score on the last of four tests that will produce an 80 mean for the four tests. We are given the first three test scores.

Plan: Write an equation and solve.

Execute: To calculate a mean, we divide the sum of the scores by the number of scores.

$$\frac{\text{sum of the scores}}{\text{number of the scores}} = \text{mean}$$

Let x represent the unknown fourth test score.

$$\frac{68 + 85 + 90 + x}{4} = 80$$

$$\frac{243 + x}{4} = 80 \qquad \text{Add in the numerator.}$$

$$\cancel{4} \cdot \frac{243 + x}{\cancel{4}} = 80 \cdot 4 \qquad \begin{array}{l}\text{Clear the 4 by multiplying}\\ \text{both sides by 4.}\end{array}$$

$$\begin{array}{r} 243 + x = 320 \\ -243 \qquad -243 \\ \hline 0 + x = 77 \end{array} \qquad \text{Subtract 243 from both sides.}$$

$$x = 77$$

Answer: If Juan scores 77 on the last test, the mean of the four tests will be exactly 80 and he will receive a B.

Check: Verify that a 77 on the last test will produce a mean of 80.

$$\frac{68 + 85 + 90 + 77}{4} = \frac{320}{4} = 80$$

DISCUSSION What if Juan knows the instructor would round a 79.5 up to 80? What would be the lowest score he could make on the last test and still receive a B? If the test is a 50-question test with each question worth 2 points and no partial credit given, how many questions can Juan miss and receive a B?

Do Your Turn 2 ▶

Your Turn 2

Trina's grade in a course is determined by the average (mean) of five tests. Her scores on the first four tests are: 80, 92, 88, and 94. To get an A, she must have a final average (mean) of 90 or higher. What is the minimum score she needs on the fifth test to receive an A?

OBJECTIVE 2 Find weighted means.

Sometimes when we calculate a mean, some scores in the data set have more weight than others. An example of a weighted mean is **grade point average (GPA),** where each letter grade has a **grade point** value.

DEFINITIONS **Grade point average (GPA):** The sum of the total grade points earned divided by the total number of credit hours.

Grade point: A numerical value assigned to a letter grade.

Answer to Your Turn 2
96

The following list shows commonly used grade point equivalents for each letter grade.

$$
\left.
\begin{array}{l}
A = 4 \\
B+ = 3.5 \\
B = 3 \\
C+ = 2.5 \\
C = 2 \\
D+ = 1.5 \\
D = 1 \\
F = 0
\end{array}
\right\}
$$

◄ Number of grade points per credit hour of the course.

Note: Some schools also use − grades like A−, B−, C− and D−. Other schools do not use + or − grades at all.

We say GPA is a *weighted* mean because we determine the total number of grade points for a course by multiplying the letter grade's equivalent grade point value by the course's number of credit hours. Suppose you receive an A in a 3 credit hour course. Because an A is worth 4 grade points for each credit of the course, we multiply $4 \cdot 3$ to equal 12 grade points for that course.

To calculate the GPA, we must calculate the total grade points for all courses, then divide by the total number of credit hours.

$$
GPA = \frac{\text{total grade points}}{\text{total credits}}
$$

CONNECTION We can use dimensional analysis:

$$
\frac{4 \text{ grade points}}{\text{credit hour}} \cdot \frac{3 \text{ credit hours}}{1} = 12 \text{ grade points}
$$

Example 3 A student's grade report is shown below. Calculate the student's GPA.

Course	Credits	Grade
MATH 100	5.0	B+
ENG 101	3.0	C+
PSY 101	3.0	C
BIO 101	4.0	A

Understand: We must calculate the student's GPA given the credit hours for each course and the letter grade received. GPA is the ratio of total grade points to total credit hours.

Plan: First, we calculate the total grade points, by multiplying the credits for each course by the value of the letter grade received, then add all those results. The grade points for the grades involved are

$$
B+ = 3.5 \text{ grade points per credit}
$$
$$
C+ = 2.5 \text{ grade points per credit}
$$
$$
C = 2.0 \text{ grade points per credit}
$$
$$
A = 4.0 \text{ grade points per credit}
$$

Execute: After finding the total grade points, we calculate the GPA by dividing the total grade points by the total number of credit hours.

Course	Credits	Grade	Grade Points
MATH 100	5	B+	$(5)(3.5) = 17.5$
ENG 101	3	C+	$(3)(2.5) = 7.5$
PSY 101	3	C	$(3)(2.0) = 6.0$
BIO 101	4	A	$(4)(4.0) = 16.0$
Totals	15		47.0

$$\text{GPA} = \frac{\text{total grade points}}{\text{total credits}} = \frac{47}{15} = 3.1\overline{3}$$

Answer: Rounded to the nearest thousandth, this student's GPA is 3.133.

Check: Verify each calculation using an inverse operation.

Do Your Turn 3 ▶

OBJECTIVE 3 Find the median.

Another average or statistic that describes the *middle* or *central tendency* of a set of numbers is the **median**.

DEFINITION Median: The middle score in an ordered set of scores.

Consider the set of numbers: 4, 5, 9, 3, 2, 10, 9

The definition of *median* indicates that the set must be ordered. This means that we must put the numbers in order from smallest to largest. We must also write each repetition of a score.

2, 3, 4, 5, 9, 9, 10
↑
median

Note: Since 5 is the middle number in this ordered set, ◀ it is the median.

Notice that the median is easy to find in a set that contains an odd number of scores. Suppose the set contains an even number of scores as in 2, 3, 4, 5, 9, 10.

2, 3, 4, 5, 9, 10
↑
median

Note: The middle of this set of ordered scores is between 4 and 5. In fact, the number exactly ◀ halfway between 4 and 5 is the mean of 4 and 5.

$$\text{median} = \text{mean of 4 and 5} = \frac{4 + 5}{2} = 4.5$$

PROCEDURE *To find the median of a set of scores:*
1. Arrange the scores in order from least to greatest.
2. Locate the middle score of the ordered set of scores.

Note: If there are an even number of scores in the set, the median will be the mean of the two middle scores.

Your Turn 3

A student's grade report is shown below. Calculate the student's GPA. Round to the nearest thousandth.

Course	Credits	Grade
MATH 100	5.0	C
ENG 101	3.0	A
PSY 101	3.0	B+
BIO 101	4.0	C+

DISCUSSION   Note that the courses are the same as in Example 3. The letter grades are also the same. However, the C grades are now in BIO and MATH, which have more credit hours than ENG and PSY. How does this affect the GPA?

Answer to Your Turn 3
2.833

Solve.

a. A class of 9 students has the following final scores. What is the median score?

80, 82, 94, 78, 95, 76, 92, 95, 64

b. Following is a list of the Jones family's monthly long-distance charges for the past year. What is the median monthly long-distance charge for the year?

Jan. = $42.70

Feb. = $32.90

Mar. = $47.50

Apr. = $64.85

May = $71.54

Jun. = $41.23

Jul. = $55.21

Aug. = $64.72

Sep. = $53.91

Oct. = $45.90

Nov. = $85.54

Dec. = $98.91

Example 4 Find the median salary for clerical specialists at the company described in Example 1. The salaries were

$24,000	$21,900
$19,500	$19,500
$20,400	$25,200

Understand: We must find the median of the given salaries.

Plan: Arrange the salaries in order from least to greatest, then locate the middle salary in the ordered list. Because there is an even number of salaries, the median salary will be the mean of the middle two salaries.

Execute: 19,500   19,500   20,400   21,900   24,000   25,200

Note: The median is halfway between 20,400 and 21,900, so we must find the mean of 20,400 and 21,900.

$$\text{median} = \frac{20,400 + 21,900}{2} = \frac{42,300}{2} = 21,150$$

Answer: The median income for the clerical specialists at the company is $21,150.

Check: Verify the calculations using inverse operations.

In Example 4, notice that half of the people have salaries greater than the median salary and half have salaries less than the median salary. Median does not indicate how much greater or less the other salaries might be. Also, it is possible that no one earns exactly the mean or median salary.

◀ **Do Your Turn 4**

DISCUSSION When would the median be better at describing data than the mean?

OBJECTIVE 4 Find the mode.

Another statistic we may consider with data sets is called the **mode.**

DEFINITION Mode: The score that occurs most often in a set of scores.

Consider the set of scores: 2, 5, 9, 12, 9, 15, 9, 12

Since 9 is the score that occurs most often, it is the mode of this set. If no score is repeated, then there is no mode. If there is a tie between scores, then we list each as a mode.

PROCEDURE *To find the mode of a set of scores, count the number of repetitions of each score. The score with the most repetitions is the mode.*

Note: Ordering the data helps spot repeated scores and the mode(s). If no score is repeated, then there is no mode. If there is a tie, then list each score as a mode.

Answers to Your Turn 4
a. 82 b. $54.56

Example 5 Find the mode.

a. Twelve players in a golf tournament post the following scores:

$$69, 66, 74, 72, 69, 70, 72, 71, 75, 65, 72, 71$$

Understand: We must find the mode in the set of scores. The mode is the score that occurs most frequently.

Plan: Order the data and count the number of repetitions of each score. The score with the most repetitions is the mode.

Execute: 65, 66, 69, 69, 70, 71, 71, 72, 72, 72, 74, 75

                   twice   twice   three times

72 is the score with the most repetitions, so it is the mode.

Answer: The mode is 72.

b. Final grades for a class of 15 students:

$$85, 86, 72, 65, 80, 91, 62, 76, 80, 85, 76, 78, 80, 96, 85$$

Understand: We must find the mode in the set of scores. The mode is the score that occurs most frequently.

Plan: Order the data and count the number of repetitions of each score. The score with the most repetitions is the mode.

Execute: 62, 65, 72, 76, 76, 78, 80, 80, 80, 85, 85, 85, 86, 91, 96

             twice   three times three times

85 and 80 both occur three times. This means they are *both* modes.

Answer: 85 and 80 are modes.

c. Birth weights for a group of 6 siblings:

  7.4 pounds   7.5 pounds   8.2 pounds   7.6 pounds   7.8 pounds   8.1 pounds

Understand: We must find the mode in the set of weights. The mode is the weight that occurs most frequently.

Plan: Order the data and count the number of repetitions of each weight. The weight with the most repetitions is the mode.

Execute: 7.4 pounds, 7.5 pounds, 7.6 pounds, 7.8 pounds, 8.1 pounds, 8.2 pounds. Because no weight is repeated, there is no mode.

Answer: There is no mode for this set of weights.

Do Your Turn 5 ▶

Your Turn 5

Solve.

a. Following is a list of ages for people at a workshop. Find the mode(s).

  25 31 30 38 28

  25 31 36 25 35

b. Following is a list of canned vegetable prices. Find the mode(s).

  $0.79 $0.89 $0.49

  $0.79 $1.09 $0.89

  $0.99

Answers to Your Turn 5
a. 25 b. $0.79, $0.89

9.1 **Exercises**

For Extra Help

Videotape DVT

Tutor Center
Addison-Wesley Tutor Center

Math XL

MyMathLab

Student Solutions Manual

1. What is the mean of a set of data?

2. What is the median of a set of data?

3. Explain how to find the median of a set containing an even number of scores.

4. What is the mode of a set of data?

For Exercises 5–14, find the mean, median, and mode(s).

5. Following is a list of nurse salaries on the surgical floor of a hospital. Find the mean, median, and mode of the salaries.

$25,500	$28,700
$28,700	$27,500
$26,450	$24,200

6. Following is a list of the hourly wages for employees at a packaging plant. Find the mean, median, and mode of the wages.

$11.00	$12.00	$11.00
$15.00	$18.25	$12.50
$13.50	$16.25	$18.00

7. Find the mean, median, and mode for the test scores of students in a history class.

80	92	64	78	88
80	82	74	72	60
55	96	100	71	82
75	82	90	86	58

8. A basketball team has the following final scores. Find the mean, median, and mode of the scores.

76	82	80	78	85	75
78	80	72	70	84	88

9. A marine biologist studying leatherback sea turtles measures and records their lengths. Following is a list of lengths for the last 20 turtles. Find the mean, median, and mode of the lengths.

2.2 m	1.8 m	1.9 m	2.3 m	2.1 m
1.6 m	1.5 m	1.8 m	1.2 m	2.0 m
2.1 m	1.4 m	1.2 m	2.1 m	1.7 m
2.2 m	2.0 m	1.6 m	1.8 m	1.9 m

10. Following is a list of heights of players on a basketball team. Find the mean, median, and mode of the heights.

6.5 ft.	6.75 ft.	6.25 ft.	6.5 ft.	6.8 ft.	6.75 ft.
6.25 ft.	7 ft.	6.75 ft.	6.25 ft.	6 ft.	6.2 ft.

11. Following is a list of rainfall amounts for one month. Each amount is the total of rainfall in one 24-hour period. Calculate the mean, median, and mode.

0.4 in.	0.8 in.	1.2 in.	3.4 in.
0.6 in.	1.5 in.	0.6 in.	1.0 in.
1.4 in.	2.3 in.	1.3 in.	0.5 in.

12. The table lists daily high and low temperatures for the last two weeks. Find the mean, median, and mode of the high temperatures. Then find the mean, median, and mode of the low temperatures.

	Sun.	Mon.	Tue.	Wed.	Thu.	Fri.	Sat.
High:	91	92	95	93	90	89	88
Low:	72	74	78	75	72	70	71
High:	90	94	96	95	92	90	91
Low:	72	74	78	76	73	70	72

13. The table lists each month's electric and gas charges for a family over a two-year period. Find the mean, median, and mode of the electric and gas charges each year.

Month	Year 1	Year 2
January	$158.92	$165.98
February	$147.88	$162.85
March	$125.90	$130.45
April	$108.40	$112.55
May	$87.65	$90.45
June	$114.58	$125.91
July	$145.84	$137.70
August	$142.78	$140.19
September	$90.25	$96.15
October	$104.12	$115.75
November	$136.62	$145.21
December	$158.18	$160.25

DISCUSSION What can you conclude in comparing the mean, median, and mode for the two years?

14. The table lists water consumption, in cubic feet, by a family over a two-year period. Find the mean, median, and mode for each year.

Month	Year 1	Year 2
January	500	550
February	525	540
March	600	600
April	650	650
May	840	900
June	1100	1430
July	1000	1200
August	1400	1250
September	840	750
October	620	700
November	600	550
December	550	500

DISCUSSION What can you conclude in comparing the mean, median, and mode for the two years?

15. Daniel's grade in a course is determined by the average (mean) of five tests. His scores on the first four tests are: 86, 96, 90, and 88. To get an A he must have a final average of 90 or higher. What is the minimum score he needs on the fifth test to receive an A?

16. Heather's grade in a course is determined by the average (mean) of six tests. Her scores on the first five tests are: 83, 78, 82, 70, and 76. To get a B she must have a final average of 80 or higher. What is the minimum score she needs on the sixth test to receive a B?

17. If Lonnie has a test average (mean) of 94 or higher, she will be exempt from the final exam. She has taken three out of the four tests for the course. Her scores on the three tests are: 88, 92, and 90. What is the lowest she can score on the last test to be exempt from the final exam?

18. If Fernando has a test average (mean) of 90 or higher, he will be exempt from the final exam. He has taken four out of the five tests for the course. His scores on those four tests are: 85, 88, 94, and 86. What is the lowest he can score on the fifth test to be exempt from the final exam?

19. A student's grade report is shown below. Calculate the student's GPA.

Course	Credits	Grade
MATH 101	3.0	B
ENG 101	3.0	C
HIS 102	3.0	B+
CHM 101	4.0	A

20. A student's grade report is shown below. Calculate the student's GPA.

Course	Credits	Grade
MATH 100	5.0	B
ENG 100	3.0	C+
PSY 101	3.0	F
BIO 101	4.0	C

DISCUSSION   What if this student had received an F in MATH 100 and a B in PSY 101?

21. A student's grade report is shown below. Calculate the student's GPA.

Course	Credits	Grade
MATH 102	3.0	B
ENG 100	3.0	B+
PSY 101	3.0	D+
BIO 101	4.0	C
COL 101	2.0	A

DISCUSSION   What if the student had made an A in BIO and a C in COL 101?

22. A student's grade report is shown below. Calculate the student's GPA.

Course	Credits	Grade
MATH 101	3.0	A
ENG 101	3.0	C
SPA 101	3.0	A
PHY 101	4.0	D+
COL 101	2.0	B

Use the grading scale below for Exercises 23 and 24.

A = 90 − 100   B = 80 − 89   C = 70 − 79   D = 60 − 69   F below 60

23. Tamara has the following scores:

Tests: 88, 91, 90, 94
Homework: 70, 90, 80, 70, 80, 90, 100, 90
Final exam: 84

The instructor has decided that the test average (mean) is 40% of the final score, the homework average (mean) is 25% of the final score, and the final exam is the remaining 35% of the final score. What is Tamara's final grade?

24. Justin has the following scores:

Tests: 80, 74, 78, 65, 72
Homework: 62, 74, 85, 91, 78, 54, 72, 81
Project: 90
Final exam: 70

The instructor has decided that the test average (mean) is 35% of the final score, the homework average (mean) is 20% of the final score, the project is 20% of the final score, and the final exam is 25% of the final score. What is his final grade?

Review Exercises

[8.2]   **1.** What percent of 400 is 300?

[8.1]   **2.** Write 12.5% as a decimal number.

[5.6]   **3.** Add. $4\frac{3}{5} + 6\frac{3}{4}$

[3.3]   **4.** Simplify. $-5x + 8y - 9x + 12 - 14y$

[5.8]   **5.** Solve and check. $\frac{4}{5}b = -12$

9.2 Tables, Charts, and Graphs

OBJECTIVES

1. Solve problems using data from tables.
2. Solve problems using data from charts.
3. Solve problems using data from bar graphs.
4. Solve problems using data from line graphs.

OBJECTIVE 1 Solve problems using data from tables.

We use tables, charts, and graphs to represent information in different ways. Tables are generally used to organize lists of data into common groupings. The following table lists the number of crashes, injuries, and fatalities for the years 1988–2002. They are estimates based on samples.

Table 9.1 Motor Vehicle Traffic Data, 1988–2002

Year	Fatalities	Injuries	Crashes
1988	42,130	2,233,000	6,887,000
1989	40,741	2,153,000	6,653,000
1990	39,836	2,122,000	6,471,000
1991	36,937	2,008,000	6,117,000
1992	34,942	1,991,000	6,000,000
1993	35,780	2,022,000	6,106,000
1994	36,254	2,123,000	6,496,000
1995	37,241	2,217,000	6,699,000
1996	37,494	2,238,000	6,770,000
1997	37,324	2,149,000	6,624,000
1998	37,107	2,029,000	6,335,000
1999	37,140	2,054,000	6,279,000
2000	37,526	2,070,000	6,394,000
2001	37,862	2,003,000	6,323,000
2002	38,309	1,929,000	6,316,000

Source: U.S. Department of Transportation, National Highway Traffic Safety Administration, *Traffic Safety Facts 2002*

Example 1 Use Table 9.1 to answer the following questions.

a. How many crashes occurred in 2001?

Answer: 6,323,000

b. How many injuries occurred in 2001?

Answer: 2,003,000

c. How many fatalities in 2001?

Answer: 37,862

d. In what year was the number of fatalities greatest?

Answer: 1988

e. What is the unit ratio of fatalities to injuries for 2000? Interpret the result.

Solution: To calculate a unit ratio, we divide the numerator (the number of fatalities) by the denominator (the number of injuries).

$$\frac{37,526}{2,070,000} \approx 0.018$$

Answer: There were 0.018 fatalities for every 1 injury in 2000. Or, because 0.018 is equivalent to the fraction $\frac{18}{1000}$, it means there were 18 fatalities out of every 1000 injuries.

f. What percent of the crashes in 2002 resulted in injuries?

Solution: We can write a basic percent sentence using the data from the table. There were 1,929,000 injuries out of a total of 6,316,000 crashes.

What percent of 6,316,000 is 1,929,000?

Because we are to find a percent we will use the proportion method.

$$\text{the unknown percent expressed as a ratio} \left\{ \frac{P}{100} = \frac{1,929,000}{6,316,000} \right. \begin{array}{l} \leftarrow \text{ injuries (part)} \\ \leftarrow \text{ crashes (whole)} \end{array}$$

$$6,316,000P = 192,900,000$$

$$\frac{6,316,000P}{6,316,000} = \frac{192,900,000}{6,316,000}$$

$$P \approx 30.5$$

Answer: In 2002, about 30.5% of crashes resulted in injuries.

g. Calculate the percent of the decrease in the number of crashes from 1988 to 2002.

Solution: Because the percent of the decrease is equal to the ratio of the amount of decrease to the initial amount, we must first calculate the amount of decrease. The amount of decrease can be calculated by subtracting the number of crashes in 2002 from the number in 1988. Then we can write the proportion and solve.

$$\begin{aligned} \text{amount of decrease} &= \text{number of crashes in 1988} - \text{number of crashes in 2002} \\ &= 6,887,000 - 6,316,000 \\ &= 571,000 \end{aligned}$$

Now we can put together the proportion to solve for the percent. As a basic percent sentence, we are asking: What percent of 6,887,000 is 571,000?

$$\text{the unknown percent expressed as a ratio} \left\{ \frac{P}{100} = \frac{571,000}{6,887,000} \right. \begin{array}{l} \leftarrow \text{ amount of decrease} \\ \leftarrow \text{ initial amount} \end{array}$$

$$6,887,000P = 57,100,000$$

$$\frac{6,887,000P}{6,887,000} = \frac{57,100,000}{6,887,000}$$

$$P \approx 8.3$$

Answer: The number of crashes decreased by about 8.3% from 1988 to 2002.

h. Determine the probability that a crash results in a fatality.

Solution: Probability is the ratio of the number of *favorable* outcomes to the total number of possible outcomes. In this case, *favorable* outcomes are the number of fatalities and the total number of possible outcomes is the total number of crashes. First, we calculate those two totals.

Fatalities = 42,130 + 40,741 + 39,836 + 36,937 + 34,942 + 35,780 + 36,254 + 37,241 + 37,494 + 37,324 + 37,107 + 37,140 + 37,526 + 37,862 + 38,309
= 566,623

Crashes = 6,887,000 + 6,653,000 + 6,471,000 + 6,117,000 + 6,000,000 + 6,106,000 + 6,496,000 + 6,699,000 + 6,770,000 + 6,624,000 + 6,335,000 + 6,279,000 + 6,394,000 + 6,323,000 + 6,316,000
= 96,470,000

$$P = \frac{\text{fatalities}}{\text{crashes}} = \frac{566,623}{96,470,000} \approx 0.0059$$

Answer: Approximately 59 out of every 10,000 crashes result in fatality. Or, we could say about 5.9 out of every 1000 crashes result in a fatality.

◀ **Do Your Turn 1**

OBJECTIVE 2 Solve problems using data from charts.

Often, data is organized in charts. One common chart is the pie chart, which is often used to show portions. The pie chart in Figure 9.1 below shows how William's monthly net income was spent.

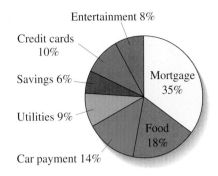

Figure 9.1

Example 2 Use the pie chart in Figure 9.1 to answer the following.

a. If William's net income was $2458.75, how much went toward food?

Solution: William spent 18% of his net income on food. Because $2458.75 is his net income, we can say:

18% of $2458.75 is the amount spent on food.

Using f to represent the amount spent on food, we can translate the preceding sentence to an equation, then solve.

$$0.18(2458.75) = f$$
$$442.58 = f$$

Answer: $442.58 went toward food.

b. If William's net income was $2458.75, how much went toward mortgage and utilities combined?

Solution: 35% of his net income went toward the mortgage payment and 9% went toward utilities for a total of $9 + 35 = 44\%$ going toward mortgage and utilities combined.

Because $2458.75 is William's total net income, we can say:

44% of $2458.75 is the amount spent on mortgage and utilities combined.

Using x to represent the amount spent on mortgage and utilities combined, we can translate the preceding sentence to an equation, then solve.

$$0.44(2458.75) = x$$
$$1081.85 = x$$

Answer: $1081.85 went toward mortgage and utilities combined.

Do Your Turn 2 ▶

Your Turn 2

Use the pie chart in Figure 9.1 to answer the following questions.

a. If William's net income was $2458.75, how much went toward entertainment?

b. What percent of William's income went toward credit cards and car payment combined?

c. What fraction of William's income went toward savings? Express the fraction in lowest terms.

Answers to Your Turn 2
a. $196.70
b. 24%
c. $\dfrac{3}{50}$

The results from a survey are listed below. Construct a pie chart that shows each response as a percentage of the total responses.

$$\text{agree} = 724$$

$$\text{disagree} = 258$$

$$\text{no opinion} = 98$$

Example 3 A company budgets $50,000 each quarter for equipment/supplies, maintenance, lease, and utilities. The following list contains the company's expenditures for one quarter. Construct a pie chart showing the percent of the budget that was spent in each category. Consider any remaining funds surplus.

Equipment and supplies:	$24,500
Maintenance:	$3400
Lease:	$11,845
Utilities:	$6250

Understand: We must construct a pie chart showing the percentage of total budget that went toward each item.

Plan: Calculate the percentage for each item. We can then develop the pie chart.

Execute:

Equipment and supplies	Maintenance	Lease	Utilities
$\dfrac{E}{100} = \dfrac{24,500}{50,000}$	$\dfrac{M}{100} = \dfrac{3400}{50,000}$	$\dfrac{L}{100} = \dfrac{11,845}{50,000}$	$\dfrac{U}{100} = \dfrac{6250}{50,000}$

$$50,000E = 2,450,000 \quad 50,000M = 340,000 \quad 50,000L = 1,184,500 \quad 50,000U = 625,000$$

$$\frac{50,000E}{50,000} = \frac{2,450,000}{50,000} \quad \frac{50,000M}{50,000} = \frac{340,000}{50,000} \quad \frac{50,000L}{50,000} = \frac{1,184,500}{50,000} \quad \frac{50,000U}{50,000} = \frac{625,000}{50,000}$$

$$E = 49\% \qquad M = 6.8\% \qquad L \approx 23.7\% \qquad U = 12.5\%$$

What percent of the budget remains as surplus? We calculate the percent remaining by subtracting the total percent of equipment/supplies, maintenance, lease, and utilities from 100%.

$$\begin{aligned} \text{surplus} &= 100\% - (49\% + 6.8\% + 23.7\% + 12.5\%) \\ &= 100\% - 92\% \\ &= 8\% \end{aligned}$$

Answer: We can now construct the pie chart.

In constructing the chart, we think about what fraction each percent represents. 49% is nearly 50%, which would be half of the chart. 23.7% is nearly 25%, which would be one-quarter of the chart. 12.5% is equivalent to one-eighth. 6.8% and 8% are both less than one-tenth.

Check: Verify that each percentage of the $50,000 total budget equals the correct amount that was spent on that item. We will leave this to the reader.

◁ **Do Your Turn 3**

Answer to Your Turn 3

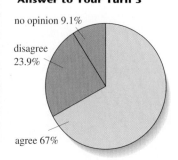

OBJECTIVE 3 Solve problems using data from bar graphs.

Another way to organize data is in bar graphs that may have horizontal or vertical bars. Bar graphs are useful in comparing amounts or percent of data for different categories. Figure 9.2 shows an example of a vertical bar graph.

Figure 9.2 **Unemployment rates of persons 25 years old and over, by highest degree attained: January 2004**

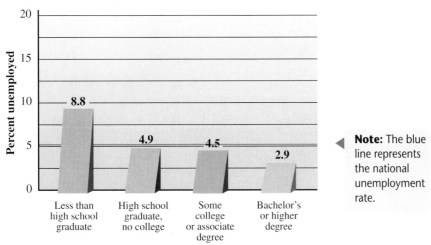

Source: U.S. Department of Labor, Bureau of Labor Statistics, household data, 2004

Notice that two reference lines form the frame of the graph. A horizontal or vertical line used for reference in a graph is called an **axis.**

DEFINITION **Axis:** A line used for reference in a graph.

In Figure 9.2, the horizontal axis indicates the educational levels of unemployed persons. The vertical axis displays the percent of people who are unemployed.

Example 4 Use Figure 9.2 to answer the following questions.

a. What was the unemployment rate for people who did not complete high school?

Answer: On the graph, the left-most bar indicates that 8.8% of people 25 years or older who had not completed high school were unemployed.

b. Which group of people had the lowest unemployment rate?

Answer: Those with a bachelor's or higher degree.

c. Suppose about 90,000,000 people in the labor force are high school graduates with no college education and 25 years of age or older. How many of these people were unemployed in January of 2004?

Solution: From Figure 9.2, 4.9% of high school graduates with no college education are unemployed. If we let N represent the number of high school graduates with no college education who were unemployed, then we can write:

$$N \text{ is } 4.9\% \text{ of } 90,000,000.$$
$$N = 0.049(90,000,000)$$
$$N = 4,410,000$$

Answer: 4,410,000 high school graduates with no college education and 25 years of age or older were unemployed.

d. Suppose about 12,000,000 in the labor force did not complete high school and are 25 years of age or older. How many people who did not complete high school were unemployed in January of 2004?

Your Turn 4

Use the bar graph in Figure 9.2 to answer the following.

a. What percent of people 25 years or older with some college education or an associate degree were unemployed in January of 2004?

b. Which group had the greatest unemployment rate?

c. Suppose 40,000,000 people in the labor force were 25 years of age or older and had a bachelor's degree or higher. How many of those people were unemployed in 2004?

Your Turn 5

A general science test was given to 15-year-old students in each industrialized country. Average scores for some of the countries are listed below. Construct a vertical bar graph with each country listed along the horizontal axis and the scores listed along the vertical axis.

Sweden: 512
Canada: 529
Australia: 528
Germany: 487
France: 500
United States: 499
Hungary: 496
Russian Federation: 460

Source: Organization for Economic Cooperation and Development, *Education at a Glance 2003*

Answers to Your Turn 4
a. 4.5%
b. less than high school graduate
c. 1,160,000

Answer to Your Turn 5 ▶

Solution: 8.8% of people who did not complete high school were unemployed. If we let *H* represent the number of people 25 years of age or older who did not complete high school and were unemployed in January of 2004, then we can write:

H is 8.8% of 12,000,000
$H = 0.088(12,000,000)$
$H = 1,056,000$

DISCUSSION Compare the answers to questions c and d. 8.8% is a larger percentage than 4.9%, so why were there more high school graduates with no college education unemployed than people who had not completed high school?

Answer: 1,056,000 people 25 years of age or older who did not complete high school were unemployed.

◀ **Do Your Turn 4**

Example 5 Following is a list of unemployment rates for January 2004. Construct a vertical bar graph with each category along the horizontal axis and the percentage along the vertical axis.

All workers: 5.6%
Adult men: 5.1%
Adult women: 5.0%
Teenagers: 16.7%

Solution: Draw two axes perpendicular to one another with categories along the horizontal axis and percentages along the vertical axis. Approximate the height of each bar using the corresponding percentage along the vertical axis.

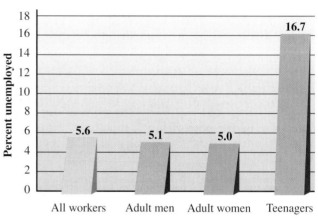

Source: Bureau of Labor Statistics, Employment situation summary 2004

◀ **Do Your Turn 5**

OBJECTIVE **4** **Solve problems using data from line graphs.**

When we wish to show how data have changed over time, we generally use a line graph. The line graph in Figure 9.3 shows the change in unemployed persons from 1989 to 2004. Notice that the number of people is in millions.

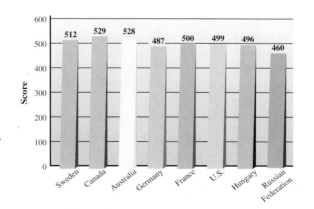

642 Chapter 9 Statistics and Graphs

Figure 9.3 Unemployed persons, 1989–2004 (seasonally adjusted)

Note: The shaded areas represent recessions. The break in the series in January 1994 is due to the redesign of the survey.

Source: Bureau of Labor Statistics, Current Population Survey, 2004

Example 6 Use Figure 9.3 to answer the following.

a. How many people were unemployed at the beginning of 1996?

Answer: about 7,400,000 people

Explanation: If we follow a vertical line straight up from 1996 to the graph, we would reach point A on the graph. If we follow a horizontal line from point A over to the vertical axis, we would reach point B. There we can read or approximate the number of people who are unemployed.

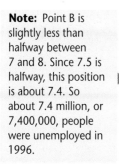

Note: Point B is slightly less than halfway between 7 and 8. Since 7.5 is halfway, this position is about 7.4. So about 7.4 million, or 7,400,000, people were unemployed in 1996.

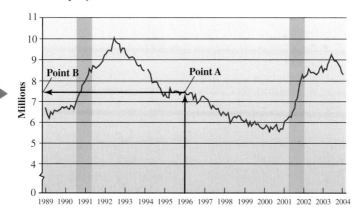

Note: The two blue shaded areas represent recessions. The break in the series in January 1994 is due to the redesign of the survey.

Source: Bureau of Labor Statistics, Current Population Survey, 2004

b. In what year were the greatest number of people unemployed?

Answer: sometime during 1992

Explanation: Each mark represents the beginning of a year. Because the highest point, or peak, for the graph occurred between the marks for 1992 and 1993, the greatest unemployment occurred sometime during 1992.

c. If the labor force was approximately 138,000,000 at the beginning of 1996, what was the unemployment rate at the beginning of 1996?

Solution: The unemployment rate is a percent of the labor force that is unemployed. To calculate this percentage, we need the total labor force and the number of people unemployed.

The number of people who were unemployed at the beginning of 1996 is the answer from question a, 7,400,000. We can write a simple percent sentence, then solve. Because we are to find percent, we will use the proportion method.

Your Turn 6

Use Figure 9.3 to answer the following questions.

a. The lowest number of unemployed people occurred during which year?

b. How many people were unemployed during June 1997? (*Hint:* June is the 6th month, which is halfway to 1998.)

c. If the number of people in the labor force at the beginning of 2000 was approximately 139,000,000, what was the unemployment rate at the beginning of 2000?

Your Turn 7

Following is a list of the Wu family's electric charges for 2003. Construct a line graph with each month along the horizontal axis and the charges along the vertical axis.

January:	$62.50
February:	$68.25
March:	$44.82
April:	$56.78
May:	$82.95
June:	$110.80
July:	$125.75
August:	$138.58
September:	$112.45
October:	$72.35
November:	$54.23
December:	$56.85

What percent of 138,000,000 is 7,400,000?

$$\text{the unknown percent expressed as a ratio} \begin{cases} \dfrac{P}{100} = \dfrac{7,400,000}{138,000,000} & \leftarrow \text{number unemployed} \\ & \leftarrow \text{total labor force} \end{cases}$$

$$138,000,000P = 740,000,000$$

$$\frac{138,000,000P}{138,000,000} = \frac{740,000,000}{138,000,000}$$

$$P \approx 5.4$$

Answer: The unemployment rate at the beginning of 1996 was about 5.4%.

◄ **Do Your Turn 6**

Example 7 Following is a list of revenues for each month during 2004 for a small business. Construct a line graph with each month listed on the horizontal axis and the revenue along the vertical axis.

January:	$20,356	April:	$17,450	July:	$24,678	October:	$19,400
February:	$19,842	May:	$18,540	August:	$23,345	November:	$24,728
March:	$17,785	June:	$22,258	September:	$21,628	December:	$28,258

Solution: Draw axes perpendicular to one another with the months along the horizontal axis and the revenue along the vertical axis. Because the revenues range from $17,450 to $28,258, we will begin the numbers along the vertical axis at $17,000 and make each mark $1000, that is, we'll break the vertical axis and skip $1000 to $16,000 to begin at $17,000.

Imagine a vertical line extending upward for each month and a horizontal line extending out from each revenue value. Where these two imaginary lines intersect, we draw a point. We have drawn these lines for January. Notice that we had to approximate where $20,356 is on the revenue line between 20,000 and 21,000. After drawing each point, we connect the points with straight lines to form the line graph shown in Figure 9.4.

Figure 9.4   2004 Revenue summary

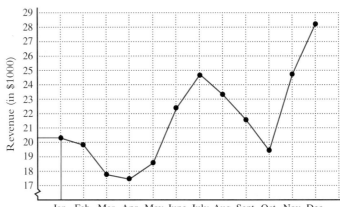

◄ **Do Your Turn 7**

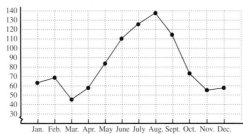

Answers to Your Turn 6

a. 2000

b. about 6,700,000

c. ≈ 4.1%

Answer to Your Turn 7 ►

9.2 Exercises

For Extra Help	Videotape DVT	Addison-Wesley Tutor Center	Math XL	MyMathLab	Student Solutions Manual

1. What is a pie chart (or circle graph) most often used to show?

2. What are bar graphs most commonly used to show?

3. What is an axis?

4. What is a line graph most commonly used to show?

For Exercises 5–14, use Table 9.2.

Table 9.2

American Housing Survey	Occupied Households	Median Household Income	Percent Owner-Occupied
United States	106,406,950	$42,409	68.0
Anaheim–Santa Ana, CA	989,400	$59,403	60.4
Buffalo, NY	461,300	$38,683	68.1
Charlotte, NC	593,700	$46,553	71.5
Columbus, OH	613,200	$47,241	65.5
Dallas, TX	1,235,300	$52,161	63.5
Fort Worth–Arlington, TX	585,900	$48,356	67.0
Kansas City, MO	697,400	$46,116	69.8
Miami–Ft. Lauderdale, FL	1,434,200	$36,907	64.6
Milwaukee, WI	584,600	$47,421	63.5
Phoenix, AZ	1,165,700	$45,153	69.5
Portland, OR	747,800	$48,288	66.5
Riverside–San Bernardino–Ontario, CA	1,083,900	$43,850	70.8
San Diego, CA	999,100	$49,868	58.7

Source: 2002 American Housing Survey

5. How many occupied households were in the United States?

6. How many occupied households were in Milwaukee, WI?

7. What was the median household income in Dallas, TX?

8. What was the median household income in Columbus, OH?

9. Which of the listed cities had the highest median household income?

10. Which of the listed cities had the lowest median household income?

11. Which of the cities listed had the highest percentage of its households owner-occupied?

12. Which of the cities listed had the lowest percentage of its households owner-occupied?

13. How many households were owner-occupied in Buffalo, NY?

14. How many households were owner-occupied in San Diego, CA?

DISCUSSION   Compare your answers to 13 and 14. What are some possible reasons why the number of owner-occupied homes in San Diego, CA, is greater than the number in Buffalo, NY?

For Exercises 15–20, use the pie chart in Figure 9.5.

Figure 9.5 Sources of current-fund revenue for public degree-granting institutions: 1999–2000

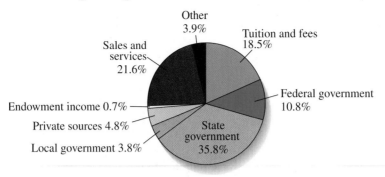

Total revenues = $157.3 billion

Source: U.S. Department of Education, National Center for Education Statistics, *Digest of Educational Statistics*, 2002

15. Which source contributed the lowest percentage of total revenue?

16. Which source contributed the greatest percentage of total revenue?

17. What percent of the total revenue was from local governments?

18. What percent of the total revenue was from the federal government?

19. How much of the $157.3 billion total revenues was from tuition and fees?

20. How much of the $157.3 billion total revenues was from sales and services?

 21. The table shows a list of deductions from a person's monthly pay. Construct a pie chart showing each deduction and the net pay as a percent of the gross salary.

Income	Deductions	
Gross monthly salary: $3220.50	FICA:	$199.67
	Medicare:	$ 46.70
	Fed W/H:	$225.44
	State W/H:	$161.03
	Health plan:	$149.76
	Retirement:	$193.23
	Life insurance:	$ 48.31

22. Respondents in a poll could strongly agree, agree, be neutral, disagree, or strongly disagree. Following is the number of people that responded each way. Construct a pie chart showing each response as a percent of total respondents.

Strongly agree:	68
Agree:	198
Neutral:	25
Disagree:	94
Strongly disagree:	35

For Exercises 23–30, use the bar graph in Figure 9.6.

Figure 9.6   Median annual salary of bachelor's degree recipients employed full-time 1 year after graduation, by field of study: 1980, 1990, and 2000 (in constant 2001 dollars)

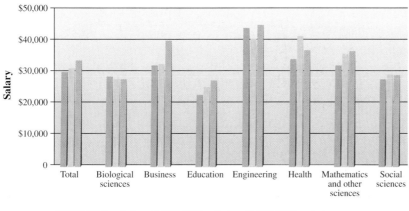

■ 1979–1980 graduates   ■ 1989–1990 graduates   ■ 1999–2000 graduates

Source: U.S. Department of Education, National Center for Education Statistics, Recent College Graduates surveys and Baccalaureate and Beyond Longitudinal Study, First Follow-up survey, 2001

23. For the 1979–1980 graduates, which field had the highest median salary?

24. For the 1989–1990 graduates, which field had the highest median salary?

25. For graduates from 1979 through 2000, which field had the lowest median salary?

26. For 1999–2000 graduates, which field had the lowest median salary?

27. What was the median salary of 1999–2000 graduates in business?

28. What was the median salary of 1999–2000 graduates in mathematics?

29. Estimate the average median salary of all three groups of graduates in social sciences.

30. Estimate the median salary of all three groups of graduates in education.

31. In Exercise 13 from Section 9.1, we listed a family's monthly electric and gas charges for two years. Construct a vertical bar graph with the months along the horizontal axis and the dollar amounts along the vertical axis. Show year 1 and year 2 as separate bars, side-by-side for each month.

Month	Year 1	Year 2
January	$158.92	$165.98
February	$147.88	$162.85
March	$125.90	$130.45
April	$108.40	$112.55
May	$ 87.65	$ 90.45
June	$114.58	$125.91
July	$145.84	$137.70
August	$142.78	$140.19
September	$ 90.25	$ 96.15
October	$104.12	$115.75
November	$136.62	$145.21
December	$158.18	$160.25

32. In Exercise 14 from Section 9.1, we listed a family's monthly water consumption (in cubic feet) for two years. Construct a vertical bar graph with the months along the horizontal axis and the cubic feet consumed along the vertical axis. Show year 1 and year 2 as separate bars, side-by-side for each month.

Month	Year 1	Year 2
January	500	550
February	525	540
March	600	600
April	650	650
May	840	900
June	1100	1430
July	1000	1200
August	1400	1250
September	840	750
October	620	700
November	600	550
December	550	500

For Exercises 33–38, use the bar graph in Figure 9.7.

Figure 9.7   Trends in bachelor's degrees conferred in selected fields of study: 1990–91, 1995–96, and 2000–2001

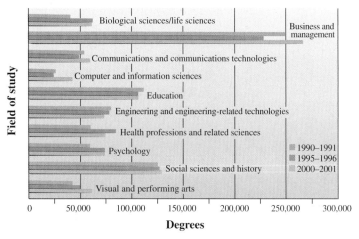

Source: U.S. Department of Education, National Center for Education Statistics, Higher Education General Information Survey (HEGIS), Degrees and Other Formal Awards Conferred survey, and Integrated Postsecondary Education Data System (IPEDS), Completions surveys

33. Which field had the largest number of degree recipients in all three groups of graduates?

34. Which field had the smallest number of degree recipients in 2000–2001?

35. How many degrees were conferred in visual and performing arts in 2000–2001?

36. How many degrees were conferred in education in 1995–1996?

37. How many more degrees were awarded in health professions and related fields in 1995–1996 than in 1990–1991?

38. How many more degrees were awarded in computer and information sciences in 2000–2001 than in 1995–1996?

For Exercises 39–44, use the line graph in Figure 9.8.

Figure 9.8   Issues most frequently cited by the public as major problems facing the local public schools: 1980–2003

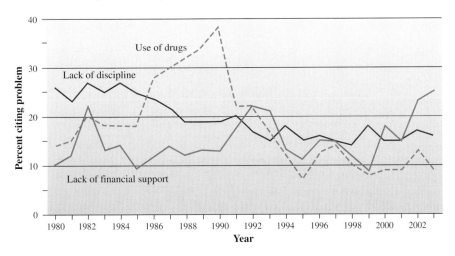

Source: Annual Gallup Poll of the Public's Attitudes toward the Public Schools, *Phi Delta Kappan*, various years

39. What percent of those polled felt that use of drugs was a major problem in 1990?

40. What percent of those polled said that lack of financial support was a major problem in 1989?

41. In what year did use of drugs overtake lack of discipline as the more popular response?

42. Over what years was lack of discipline the most popular response?

43. List the percent for each response in 2002.

44. Based on the directions of the lines after 2003, what would you forecast the responses to be in 2004?

45. Following is a list of a student's GPA each semester until she graduated. Construct a line graph with each semester along the horizontal axis and GPA along the vertical axis.

Fall 2000:	3.25	Fall 2002:	2.50
Spring 2001:	3.50	Spring 2003:	2.25
Fall 2001:	3.60	Fall 2003:	3.00
Spring 2002:	2.75	Spring 2004:	3.50

46. Following is a list of revenue and cost totals at the end of each year over six years for a small business. Construct a line graph with each year along the horizontal axis and the dollar amounts along the vertical axis. Draw one line for revenue and a different line for cost.

Year	Revenue	Cost
1998	$ 45,450	$ 55,000
1999	$ 65,000	$ 62,000
2000	$ 88,000	$ 64,000
2001	$112,000	$ 75,000
2002	$248,000	$ 86,000
2003	$282,000	$124,000

Review Exercises

[2.1] **1.** Graph -5 on a number line.

[5.1] **2.** Graph $6\frac{1}{2}$ on a number line.

For Exercises 3 and 4, simplify.

[2.5] **3.** $|25 - 32|$

[2.5] **4.** $(6 - 10)^2$

[6.6] **5.** Find the missing side length in the right triangle shown.

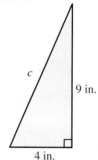

c

9 in.

4 in.

9.3 The Rectangular Coordinate System

OBJECTIVE 1 Determine the coordinates of a given point.

OBJECTIVES

1 Determine the coordinates of a given point.

2 Plot points in the coordinate plane.

3 Determine the quadrant for a given coordinate.

4 Find the midpoint of two points in the coordinate plane.

In 1619, René Descartes, the French philosopher and mathematician, recognized that positions of points in a plane could be described using two axes that intersect at a right angle.

CONNECTION We learned about axes in Section 9.2 when we constructed the horizontal and vertical reference lines on bar and line graphs.

Two perpendicular axes form the *rectangular coordinate system*, or *Cartesian coordinate system*, named in honor of René Descartes. We call the horizontal axis the *x*-axis and the vertical axis the *y*-axis.

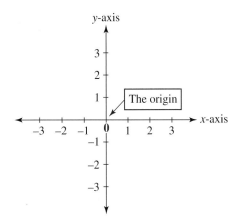

The point where the axes intersect is 0 for both the *x*-axis and *y*-axis. This position is called the *origin*. The positive numbers are to the right and up from the origin, whereas negative numbers are to the left and down from the origin.

Any point in the plane can be described using two numbers, one number from each axis. To avoid confusion, these two numbers are written in a specific order. The number representing a point's horizontal distance from the origin is given first, and the number representing a point's vertical distance from the origin is given second. Because the order in which we say or write these two numbers matters, we say that they form an *ordered pair*. Each number in an ordered pair is called a *coordinate* of the ordered pair.

The notation for an ordered pair is: (horizontal coordinate, vertical coordinate)

Consider the point labeled *A* in the coordinate plane shown. The point is drawn at the intersection of the 3rd line to the right of the origin and the 4th line up from the origin. The ordered pair that describes point *A* is (3, 4).

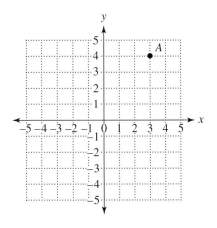

LEARNING STRATEGY

Think of the intersecting lines in the grid as avenues and streets in a city such as New York, where avenues run north-south and streets run east-west. To describe an intersection, a person would say the avenue first, then the street. Point *A* on the figure is at the intersection of 3rd avenue and 4th street. Or, as a New Yorker would say, "3rd and 4th."

PROCEDURE *To determine the coordinates of a given point in the rectangular system:*

1. Follow a vertical line from the point to the *x*-axis (horizontal axis). The number at this position on the *x*-axis is the first coordinate.
2. Follow a horizontal line from the point to the *y*-axis (vertical axis). The number at this position on the *y*-axis is the second coordinate.

OF INTEREST

Born in 1596 near Tours, France, René Descartes became skeptical of the teachings of his day and began formulating his own methods for reasoning. One of his most famous philosophical conclusions is "I think, therefore I am." In 1637, after some urging by his friends, he allowed one work known as the *Method* to be printed. It was in this book that the rectangular coordinate system and analytical geometry was given to the world.

Your Turn 1

Write the coordinates for each point shown.

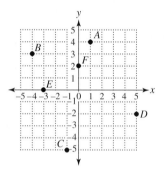

Example 1   Write the coordinates for each point shown.

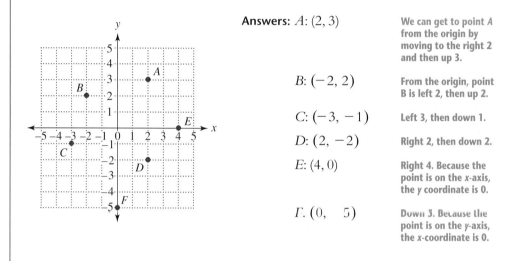

Answers: *A*: (2, 3)

> We can get to point *A* from the origin by moving to the right 2 and then up 3.

B: (−2, 2)

> From the origin, point B is left 2, then up 2.

C: (−3, −1)

> Left 3, then down 1.

D: (2, −2)

> Right 2, then down 2.

E: (4, 0)

> Right 4. Because the point is on the *x*-axis, the *y* coordinate is 0.

F: (0, −5)

> Down 5. Because the point is on the *y*-axis, the *x*-coordinate is 0.

◁ **Do Your Turn 1**

OBJECTIVE 2 Plot points in the coordinate plane.

Now, let's graph or plot points in the rectangular coordinate system given the coordinates. Recall that the first coordinate in an ordered pair describes a point's horizontal distance from the origin and the second coordinate describes that point's vertical distance from the origin. For example, to locate (3, 4), we move 3 units to the right, then 4 units up.

$$(3, 4)$$

right 3   up 4

Answers to Your Turn 1
A: (1, 4) *B*: (−4, 3)
C: (−1, −5) *D*: (5, −2)
E: (−3, 0) *F*: (0, 2)

PROCEDURE *To graph or plot a point given a coordinate pair:*

1. Beginning at the origin (0, 0), move right or left along the *x*-axis the amount indicated by the first coordinate.
2. From that position on the *x*-axis, move up or down the amount indicated by the second coordinate.
3. Draw a dot to represent the point described by the coordinates.

Example 2 Plot the point described by the coordinates.

a. $(4, -2)$ b. $(-3, -4)$ c. $(0, 5)$

Solution:

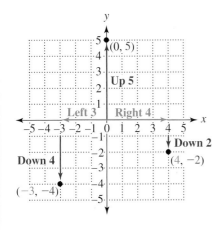

Do Your Turn 2 ▷

LEARNING STRATEGY

If you are a tactile learner, when plotting an ordered pair, try using your pencil or finger to trace the path to the point. More specifically, move your pencil point or fingertip along the x-axis the amount indicated by the x-coordinate (blue arrows in Example 2). Then, from that position on the x-axis, move up or down the amount indicated by the y-coordinate (red arrows in Example 2). The resulting position is the point described by the ordered pair.

Your Turn 2

Plot the point described by the coordinates.

a. $(-4, 2)$

b. $(3, 1)$

c. $(0, -4)$

d. $(-2, 0)$

e. $(-1, -5)$

f. $(3, -2)$

OBJECTIVE 3 Determine the quadrant for a given coordinate.

The two axes divide the coordinate plane into four regions called **quadrants.**

DEFINITION **Quadrant:** One of four regions created by the intersection of the axes in the coordinate plane.

The quadrants are numbered using roman numerals, as shown next:

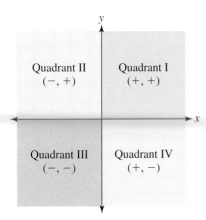

We can use the signs of the coordinates to determine the quadrant in which a point lies. For points in quadrant I, both coordinates are positive. Every point in quadrant II has a negative first coordinate (horizontal) and a positive second coordinate (vertical). Points in quadrant III have both coordinates negative. Every point in quadrant IV has a positive first coordinate and a negative second coordinate.

PROCEDURE *To determine the quadrant for a given coordinate, consider the signs of the numbers in the coordinate.*

    $(+, +)$ means the point is in quadrant I

    $(-, +)$ means the point is in quadrant II

    $(-, -)$ means the point is in quadrant III

    $(+, -)$ means the point is in quadrant IV

Answers to Your Turn 2

Example 3 State the quadrant in which each point is located.

a. (−45, 19)

Answer: Quadrant II (upper left), because the first coordinate is negative and the second coordinate positive.

b. (102, −68)

Answer: Quadrant IV (lower right), because the first coordinate is positive and the second coordinate negative.

c. (0, 91)

Answer: Since the *x*-coordinate is 0, this point is on the *y*-axis and is not in a quadrant.

◀ **Do Your Turn 3**

OBJECTIVE 4 Find the midpoint of two points in the coordinate plane.

Suppose we draw a straight line connecting two points in the coordinate plane, such as (2, 3) and (6, 9) as shown in the graph. The point at (4, 6) is the **midpoint** between those two points.

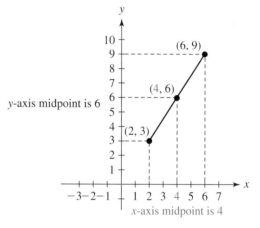

DEFINITION **Midpoint of two points:** The point that lies halfway between the two points on a straight line.

Notice that the *x*-coordinate of the midpoint, 4, is the mean of the *x*-coordinates of the two points it is between.

$$\frac{2 + 6}{2} = \frac{8}{2} = 4$$

Similarly, the *y*-coordinate of the midpoint, 6, is the mean of the *y*-coordinates of the two points.

$$\frac{3 + 9}{2} = \frac{12}{2} = 6$$

We can check that (4, 6) is in fact the midpoint by constructing two right triangles as shown. Since the two triangles are identical, the distance from (2, 3) to (4, 6) is equal to the distance from (4, 6) to (6, 9). Therefore, (4, 6) is exactly halfway between (2, 3) and (4, 6).

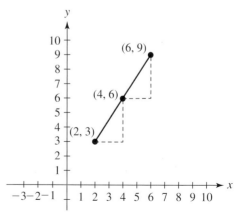

Answers to Your Turn 3
a. I b. III c. on the *x*-axis d. II

Our example suggests the following rule.

RULE *Midpoint*

Given two points with coordinates (x_1, y_1) and (x_2, y_2), the coordinates of their midpoint are $\left(\dfrac{x_1 + x_2}{2}, \dfrac{y_1 + y_2}{2} \right)$.

Example 4 Find the midpoint of the given points.

a. $(2, 6)$ and $(-4, -14)$

Solution: Use the midpoint formula. We will consider $(2, 6)$ to be (x_1, y_1) and $(-4, -14)$ to be (x_2, y_2).

$$\text{midpoint} = \left(\frac{2 + (-4)}{2}, \frac{6 + (-14)}{2} \right)$$

Using $\left(\dfrac{x_1 + x_2}{2}, \dfrac{y_1 + y_2}{2} \right)$, replace x_1 with 2, x_2 with −4, y_1 with 6, and y_2 with −14.

$$= \left(\frac{-2}{2}, \frac{-8}{2} \right)$$

Add in the numerators.

$$= (-1, -4)$$

Simplify.

Note: Because of the commutative property of addition, it does not matter which point we consider to be (x_1, y_1) and which we consider to be (x_2, y_2). To illustrate, let's work through Example 4 again with $(-4, -14)$ as (x_1, y_1) and $(2, 6)$ as (x_2, y_2).

$$\text{midpoint} = \left(\frac{-4 + 2}{2}, \frac{-14 + 6}{2} \right)$$

$$= \left(\frac{-2}{2}, \frac{-8}{2} \right) = (-1, -4)$$

b. $(-4, 8)$ and $(9, -1)$

Solution: Use the midpoint formula. We will consider $(-4, 8)$ to be (x_1, y_1) and $(9, -1)$ to be (x_2, y_2).

$$\text{midpoint} = \left(\frac{-4 + 9}{2}, \frac{8 + (-1)}{2} \right)$$

Using $\left(\dfrac{x_1 + x_2}{2}, \dfrac{y_1 + y_2}{2} \right)$, replace x_1 with −4, x_2 with 9, y_1 with 8, and y_2 with −1.

$$= \left(\frac{5}{2}, \frac{7}{2} \right)$$

Add in the numerators.

Note: Though it is not preferred, we could also express the coordinates as mixed numbers $\left(2\frac{1}{2}, 3\frac{1}{2} \right)$ or as decimals (2.5, 3.5).

Your Turn 4

Find the midpoint of the given points.

a. $(9, 1)$ and $(5, -3)$

b. $(1, -6)$ and $(8, 4)$

c. $(-2, 7)$ and $(-5, -2)$

Answers to Your Turn 4

a. $(7, -1)$ **b.** $\left(\dfrac{9}{2}, -1 \right)$

c. $\left(-\dfrac{7}{2}, \dfrac{5}{2} \right)$

Do Your Turn 4 ▶

For Extra Help

Videotape DVT

Tutor Center Addison-Wesley Tutor Center

Math XL Math XL

MyMathLab

Student Solutions Manual

1. Draw and label the two axes that form the rectangular coordinate system. Label the origin.

2. When writing an ordered pair, which is written first, the horizontal-axis coordinate or the vertical-axis coordinate?

3. Describe how to use the coordinates in an ordered pair to plot a point in the rectangular coordinate system.

4. Which quadrant is the upper-left quadrant?

5. What are the signs of the coordinates of a point in the third quadrant?

6. Given two points with coordinates (x_1, y_1) and (x_2, y_2) the coordinates of their

   midpoint are _____.

For Exercises 7–10, write the coordinates for each point.

7.

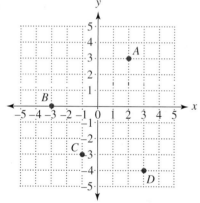

8.

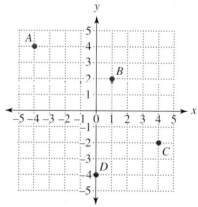

9.

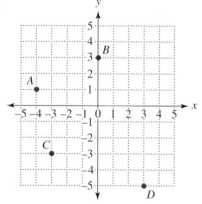

10.

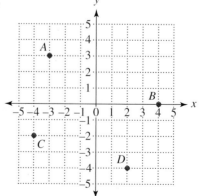

For Exercises 11–14, plot and label the points indicated by the coordinate pairs.

11. $(4, 5), (3, -2), (-4, -1), (0, 2)$

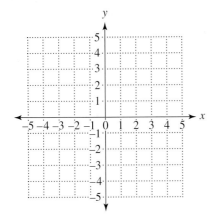

12. $(-1, 0), (2, 1), (-3, -3), (5, -2)$

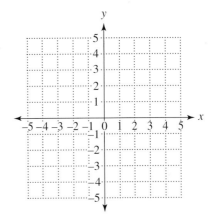

13. $(-4, 0), (1, -4), (-3, -5), (4, 4)$

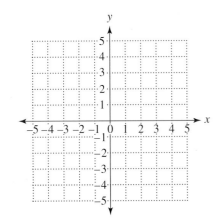

14. $(-2, -4), (0, -3), (5, -3), (-4, 2)$

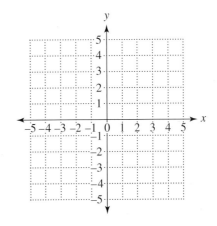

For Exercises 15–22, state the quadrant in which the point is located.

15. $(-99, 201)$ **16.** $(-61, -78)$ **17.** $(105, 50)$ **18.** $(-84, 56)$

19. $(42, -60)$ **20.** $(78, 65)$ **21.** $(-92, -106)$ **22.** $(28, -59)$

For Exercises 23–32, find the midpoint of the given points.

23. $(0, 6)$ and $(8, 0)$

24. $(1, 7)$ and $(9, -8)$

25. $(3, -1)$ and $(7, 2)$

26. $(1, -2)$ and $(10, -14)$

27. $(2, -6)$ and $(-3, 8)$

28. $(-4, 9)$ and $(-1, -1)$

29. $(0, -8)$ and $(-7, 11)$

30. $(-12, -8)$ and $(0, -3)$

31. $(-9, -3)$ and $(-5, -2)$

32. $(-11, -4)$ and $(-17, -7)$

33. $(1.5, -9)$ and $(-4.5, 2.2)$

34. $(-3.6, 12.4)$ and $(1.2, 4.8)$

35. $\left(3\frac{1}{4}, 8\right)$ and $\left(5, 2\frac{1}{2}\right)$

36. $\left(-14, 6\frac{2}{3}\right)$ and $\left(7, -2\frac{5}{6}\right)$

Review Exercises

[5.1]   **1.** Graph $-3\frac{1}{2}$ on a number line.

For Exercises 2 and 3, evaluate.

[3.1]   **2.** $2x - 3y;\ x = 4,\ y = 2$

[3.1]   **3.** $-4x^2 - 5y;\ x = -3,\ y = 0$

For Exercises 4 and 5, solve and check.

[4.3]   **4.** $2x - 5 = 10$

[5.8]   **5.** $-\frac{5}{6}y = \frac{3}{4}$

9.4 Graphing Linear Equations

OBJECTIVE 1 Determine whether a given pair of coordinates is a solution to a given equation with two unknowns.

OBJECTIVES

1 Determine whether a given pair of coordinates is a solution to a given equation with two unknowns.

2 Find three solutions for an equation in two unknowns.

3 Graph linear equations, in x and y.

4 Graph horizontal and vertical lines.

5 Given an equation, find the coordinates of the x- and y-intercepts.

In the past, we've considered equations with one variable. Now let's consider equations that have two variables, such as $x + y = 5$ and $y = 3x - 4$.

Recall that a solution for an equation with one variable is a number that can replace the variable and make the equation a true statement. For an equation with two variables, a solution will be a pair of numbers, one number for each variable, that can replace the corresponding variables and make the equation true. Since these solutions are ordered pairs, we can write them using coordinates. The ordered pair $(1, 4)$ is a solution for $x + y = 5$ because replacing x with 1 and y with 4 makes the equation true.

Note: Recall from Section 9.3 that $(1, 4)$ means $x = 1$ and $y = 4$.

$$x + y = 5$$
$$\downarrow \quad \downarrow$$
$$1 + 4 = 5$$

This equation is true, so the ordered pair is a solution.

PROCEDURE *To determine whether a given ordered pair is a solution for an equation with two variables:*

1. Replace the variables in the equation with the corresponding coordinates.
2. Verify that the equation is true.

Example 1 Determine whether the ordered pair is a solution for the equation.

a. $(2, -5)$; $3x + y = 1$

Solution: $3(2) + (-5) \stackrel{?}{=} 1$ Replace x with **2** and y with **−5**, and see if the equation is true.

$\qquad 6 + (-5) \stackrel{?}{=} 1$

$\qquad\qquad\qquad 1 = 1$ True.

Answer: Because the equation is true, $(2, -5)$ is a solution for $3x + y = 1$.

b. $(-4, -7)$; $y = 2x - 1$

Solution: $-7 \stackrel{?}{=} 2(-4) - 1$ Replace x with **−4** and y with **−7**, and see if the equation is true.

$\qquad -7 \stackrel{?}{=} -8 - 1$

$\qquad -7 = -9$ Not true.

Answer: Because the equation is not true, $(-4, -7)$ is not a solution for $y = 2x - 1$.

Do Your Turn 1 ▷

Fractions and decimal numbers can also be solutions.

Your Turn 1

Determine whether the coordinate pair is a solution for the equation.

a. $(5, 0)$; $4x + y = 20$

b. $(-1, 6)$; $2y - 3x = -9$

c. $(-3, -14)$; $y = x + 11$

d. $(0, -1)$; $y = 3x - 1$

Answers to Your Turn 1
a. yes   b. no   c. no   d. yes

Your Turn 2

Determine whether the ordered pair is a solution for the equation.

a. $(3, 1.5)$; $y = \dfrac{1}{2}x$

b. $\left(5, 2\dfrac{1}{3}\right)$; $y = \dfrac{x}{3} + 1$

Example 2 Determine whether $\left(-\dfrac{1}{3}, 3.5\right)$ is a solution for $y = \dfrac{3}{2}x + 4$.

Solution: $3.5 \overset{?}{=} \dfrac{\overset{1}{\cancel{3}}}{2}\left(-\dfrac{1}{\underset{1}{\cancel{3}}}\right) + 4$ **Replace x with $-\dfrac{1}{3}$, y with 3.5, and divide out common factors.**

$3.5 \overset{?}{=} -\dfrac{1}{2} + 4$ **Multiply.**

$3.5 \overset{?}{=} -0.5 + 4$ **Write the fraction as an equivalent decimal number.**

$3.5 = 3.5$ **Add. The equation is true.**

Answer: Because the equation is true, $\left(-\dfrac{1}{3}, 3.5\right)$ is a solution for $y = \dfrac{3}{2}x + 4$.

◀ **Do Your Turn 2**

OBJECTIVE 2 Find three solutions for an equation in two unknowns.

How can we find solutions to equations in two variables? Consider the equation $x + y = 5$. We've already seen that $(1, 4)$ is a solution. But there are other solutions. We list some of those solutions in the following table.

x	y	Ordered Pair
0	5	(0, 5)
1	4	(1, 4)
2	3	(2, 3)
3	2	(3, 2)
4	1	(4, 1)
5	0	(5, 0)
6	−1	(6, −1)

◀ **Note:** Every solution for $x + y = 5$ is an ordered pair of numbers whose sum is 5.

In fact, $x + y = 5$ has an infinite number of solutions. For every x-value, there is a corresponding y-value that will add to the x-value to equal 5, and vice versa. This gives a clue as to how to find solutions. We can simply choose a value for either x or y and solve for the corresponding value of the other variable. (In this case, our equation was easy enough that we could solve for y mentally.)

PROCEDURE *To find a solution to an equation in two variables:*

1. Choose a value (any value) for one of the variables.
2. Replace the corresponding variable with your chosen value.
3. Solve the equation for the value of the other variable.

Answers to Your Turn 2
a. yes b. no

Example 3 Find three solutions for the equation $2x + y = 5$.

Solution: To find a solution, we replace one of the variables with a chosen value, and then solve for the value of the other variable.

For the first solution, we will choose x to be 0.

$$2x + y = 5$$
$$2(0) + y = 5$$
$$0 + y = 5$$
$$y = 5$$

Solution: $(0, 5)$

Note: Choosing x (or y) to be 0 usually makes the equation very easy to solve.

For the second solution, we will choose x to be 1.

$$2x + y = 5$$
$$2(1) + y = 5$$
$$2 + y = 5$$
$$2 + y = 5$$
$$\underline{-2 \qquad -2}$$
$$0 + y = 3$$
$$y = 3$$

Solution: $(1, 3)$

For the third solution, we will choose x to be 2.

$$2x + y = 5$$
$$2(2) + y = 5$$
$$4 + y = 5$$
$$4 + y = 5$$
$$\underline{-4 \qquad -4}$$
$$0 + y = 1$$
$$y = 1$$

Solution: $(2, 1)$

Keep in mind that there are an infinite number of solutions for a given equation in two variables, so you may get different solutions from someone else solving the same equation.

Do Your Turn 3 ▷

Your Turn 3

Find three solutions for each equation. (Answers may vary.)

a. $x + y = 6$

b. $3x + y = 9$

c. $2x - 3y = 12$

Example 4 Find three solutions for the equation $y = \dfrac{1}{3}x - 5$.

Solution: Notice that in this equation, y is isolated. If we select values for x, we will not have to isolate y as we did in Example 3. We will simply calculate the y value. Also notice that the coefficient for x is a fraction. Since we can choose any value for x, let's choose values like 3 and 6 that will divide out nicely with the denominator of 3.

For the first solution, we will choose x to be 0.

$$y = \frac{1}{3}x - 5$$

$$y = \frac{1}{3}(0) - 5$$

$$y = -5$$

Solution: $(0, -5)$

For the second solution, we will choose x to be 3.

$$y = \frac{1}{3}x - 5$$

$$y = \frac{1}{3}(3) - 5$$

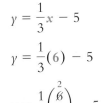

$$y = 1 - 5$$
$$y = -4$$

Solution: $(3, -4)$

For the third solution, we will choose x to be 6.

$$y = \frac{1}{3}x - 5$$

$$y = \frac{1}{3}(6) - 5$$

$$y = \frac{1}{\cancel{3}}\left(\frac{\overset{2}{\cancel{6}}}{1}\right) - 5$$

$$y = 2 - 5$$
$$y = -3$$

Solution: $(6, -3)$

Do Your Turn 4 ▷

Your Turn 4

Find three solutions for each equation. (Answers may vary.)

a. $y = x - 3$

b. $y = -4x$

c. $y = \dfrac{2}{3}x$

d. $y = \dfrac{1}{2}x + 5$

Answers to Your Turn 3
a. $(0, 6), (1, 5), (2, 4)$
b. $(0, 9), (1, 6), (2, 3)$
c. $(0, -4), (6, 0) \left(1, -3\dfrac{1}{3}\right)$

Answers to Your Turn 4
a. $(0, -3), (1, -2), (2, -1)$
b. $(-1, 4), (0, 0), (1, -4)$
c. $(0, 0), (3, 2), (6, 4)$
d. $(0, 5), (2, 6), (4, 7)$

Your Turn 5

Graph each equation. (Note that you found three solutions for each of these equations in Your Turn 3.)

a. $x + y = 6$

b. $3x + y = 9$

c. $2x - 3y = 12$

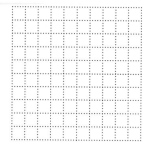

Answers to Your Turn 5

a.

b.

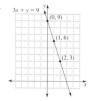

c.

OBJECTIVE 3 Graph linear equations in *x* and *y*.

We've learned that equations in two variables have an infinite number of solutions. Consequently, it is impossible to find all the solutions for an equation in two variables. However, we can represent all the solutions using a graph.

Watch what happens when we plot solutions for $x + y = 5$. Recall some of the solutions: $(0, 5)$, $(5, 0)$, $(1, 4)$, $(4, 1)$, $(2, 3)$, and $(3, 2)$.

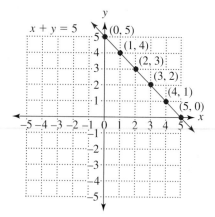

Notice that when we plot the ordered pairs, the points lie on a straight line. In fact, all solutions for $x + y = 5$ are on that same line, which is why we say it is a *linear* equation. By connecting the points, we are graphically representing every possible solution for $x + y = 5$. Placing arrows on either end of the line indicates that the solutions continue beyond our "window" in both directions.

The graph of the solutions of every linear equation will be a straight line. Since two points determine a line in a plane, we need a minimum of two ordered pairs. This means we must find at least two solutions in order to graph a line. However, it is wise to find three solutions, using the third solution as a check. If we plot all three points and they cannot be connected with a straight line, then we know something is wrong.

PROCEDURE *To graph a linear equation:*
1. Find at least two solutions to the equation.
2. Plot the solutions as points in the rectangular coordinate system.
3. Draw a straight line through these points.

Example 5 Graph. $2x + y = 5$

Solution: We found three solutions to this equation in Example 3. Recall those solutions: $(0, 5)$, $(1, 3)$, and $(2, 1)$.

Now we plot each solution as a point in the rectangular coordinate system, then draw a straight line through these points.

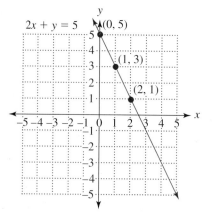

◁ **Do Your Turn 5**

Example 6 Graph. $y = \frac{1}{3}x - 5$

Solution: We found three solutions to this equation in Example 4. Recall those solutions: $(0, -5)$, $(3, -4)$, and $(6, -3)$.

Now we plot each solution as a point in the rectangular coordinate system, then draw a straight line through these points.

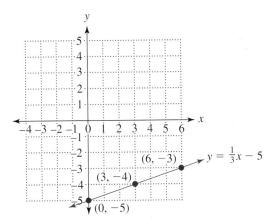

Do Your Turn 6 ▶

OBJECTIVE 4 Graph horizontal and vertical lines.

In Objective 3, we graphed linear equations in two variables. Now let's graph linear equations in one variable, like $y = 3$ or $x = 2$, in which the variable is equal to a constant.

Example 7 Graph.

a. $y = 3$

Solution: The equation $y = 3$ indicates that y is equal to a constant, 3. In other words, y is always 3 no matter what we choose for x. If we choose x to be 0, then y equals 3. If we choose x to be 2, then y is 3. If we choose x to be 4, then y is still 3.

We now have three solutions: $(0, 3)$, $(2, 3)$, and $(4, 3)$.

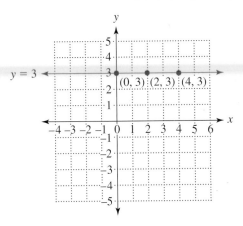

Note: The graph of $y = 3$ is a horizontal line parallel to the x-axis that passes through the y-axis at $(0, 3)$.

b. $x = 2$

Solution: The equation $x = 2$ indicates that x is equal to a constant, 2. In other words, x is always 2 no matter what we choose for y. If we choose y to be 0, then x equals 2. If we choose y to be 1, then x is 2. If we choose y to be 4, then x is still 2.

Your Turn 6

Graph each equation. (Note that you found three solutions for each of these equations in Your Turn 4.)

a. $y = x - 3$ **b.** $y = -4x$

c. $y = \frac{2}{3}x$ **d.** $y = \frac{1}{2}x + 5$

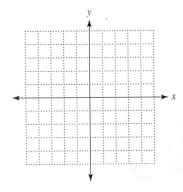

Answers to Your Turn 6

a.

b.

c.

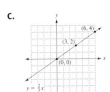

d.

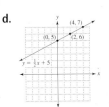

Graph.

a. $y = -4$

b. $x = -3$

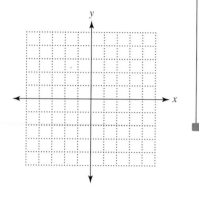

We now have three solutions: $(2, 0)$, $(2, 1)$, and $(2, 4)$.

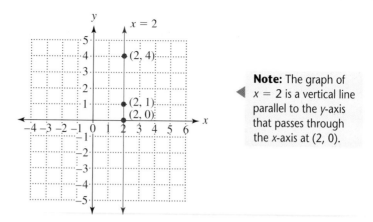

Note: The graph of $x = 2$ is a vertical line parallel to the y-axis that passes through the x-axis at $(2, 0)$.

RULES *Horizontal and Vertical Lines*

The graph of $y = c$, where c is a constant, is a horizontal line parallel to the x-axis and intersects the y-axis at the point with coordinates $(0, c)$.

The graph of $x = c$, where c is a constant, is a vertical line parallel to the y-axis and intersects the x-axis at the point with coordinates $(c, 0)$.

◁ **Do Your Turn 7**

OBJECTIVE 5 Given an equation, find the coordinates of the *x*- and *y*-intercepts.

Consider the following graph. Note that the line intersects the x-axis at the point with coordinates $(4, 0)$. It intersects the y-axis at the point with coordinates $(0, -2)$. These points are called *intercepts*.

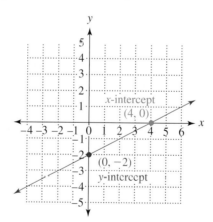

The point where a graph intersects the x-axis is called the **x-intercept.** The point where a graph intersects the y-axis is called the **y-intercept.**

DEFINITIONS *x*-intercept: A point where a graph intersects the *x*-axis.

y-intercept: A point where a graph intersects the *y*-axis.

b.

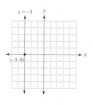

Note that the y-coordinate of any x-intercept will always be 0. Similarly, the x-coordinate of any y-intercept will always be 0. These facts about intercepts suggest the following procedures.

To find an x-intercept:

1. Replace y with 0 in the given equation.
2. Solve for x.

To find a y-intercept:

1. Replace x with 0 in the given equation.
2. Solve for y.

Example 8 Find the coordinates for the x- and y-intercepts.

a. $3x - 4y = 12$

Solution:

x-intercept

For the x-intercept, replace y with 0 and solve for x.
▼

$$3x - 4y = 12$$
$$3x - 4(0) = 12$$
$$3x - 0 = 12$$
$$\frac{3x}{3} = \frac{12}{3}$$ Divide both sides by 3.
$$x = 4$$

y-intercept

For the y-intercept, replace x with 0 and solve for y.
▼

$$3x - 4y = 12$$
$$3(0) - 4y = 12$$
$$0 - 4y = 12$$
$$\frac{-4y}{-4} = \frac{12}{-4}$$ Divide both sides by −4.
$$y = -3$$

Answer: The x-intercept is $(4, 0)$ and the y-intercept is $(0, -3)$.

b. $y = \dfrac{3}{4}x$

Solution:

x-intercept

$$0 = \frac{3}{4}x$$ **Replace y with 0.**

$$\frac{4}{3} \cdot 0 = \frac{4}{3} \cdot \frac{3}{4}x$$ **Multiply both sides by $\frac{4}{3}$.**

$$0 = x$$

y-intercept

$$y = \frac{3}{4}(0)$$ **Replace x with 0.**

$$y = 0$$ **Multiply.**

Answer: The x- and y-intercepts are the same point, $(0, 0)$.

c. $x = -5$

Solution: The graph of $x = -5$ is a vertical line parallel to the y-axis that passes through the x-axis at the point $(-5, 0)$. Notice that this point is the x-intercept. Because the line is parallel to the y-axis, it will never intersect the y-axis. Therefore, there is no y-intercept.

d. $y = 7$

Solution: The graph of $y = 7$ is a horizontal line parallel to the x-axis that passes through the y-axis at the point $(0, 7)$. Notice that this point is the y-intercept. Because the line is parallel to the x-axis it will never intersect the x-axis. Therefore, there is no x-intercept.

Do Your Turn 8 ▷

Your Turn 8

Find the coordinates for the x- and y-intercepts.

a. $x - 4y = 8$

b. $y = -x + 3$

c. $y = -6x$

d. $y = 9$

Answers to Your Turn 8
a. $(8, 0)$, $(0, -2)$
b. $(3, 0)$, $(0, 3)$
c. $(0, 0)$ for both
d. no x-intercept, $(0, 9)$

For
Extra
Help

Videotape
DVT

Addison-Wesley
Tutor Center

Math XL

MyMathLab

Student Solutions
Manual

1. Give an example of a linear equation with two variables.

2. How do you determine if an ordered pair is a solution to an equation with two variables?

3. How do you find a solution to an equation in two variables?

4. How many solutions exist for linear equations in two variables?

5. What will the graph of every linear equation look like?

6. What is the minimum number of solutions that you must find in order to graph a linear equation?

7. What is a y-intercept?

8. What is an x-intercept?

9. What is the x-coordinate of a y-intercept?

10. What is the y-coordinate of an x-intercept?

For Exercises 11–22, determine whether the given pair of coordinates is a solution for the given equation.

11. $(2, 3); x + 2y = 8$

12. $(3, 1); 2x - y = 5$

13. $(-5, 2); y - 4x = 3$

14. $(4, -6); y = 3x - 1$

15. $(9, 0); y = -2x + 18$

16. $(0, -1); y = 2x + 1$

17. $(6, -2); y = -\dfrac{2}{3}x$

18. $(0, 0); y = \dfrac{2}{5}x$

19. $\left(-1\dfrac{2}{5}, 0\right); y - 3x = 5$

20. $\left(\dfrac{2}{3}, -4\dfrac{5}{6}\right); y = \dfrac{1}{4}x - 5$

21. $(2.2, -11.2); y + 6x = -2$

22. $(-1.5, -1.3); y = 0.2x - 1$

For Exercises 23–50, find three solutions for the given equation, then graph.

23. $x - y = 8$

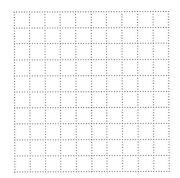

24. $x + y = -5$

25. $2x + y = 6$

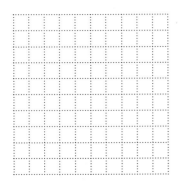

26. $3x - y = 9$

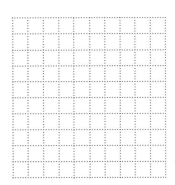

27. $y = x$

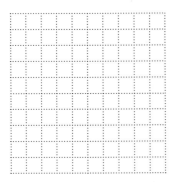

28. $y = -x$

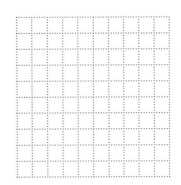

29. $y = 2x$

30. $y = -2x$

31. $y = -5x$

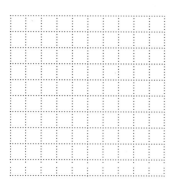

32. $y = 3x$

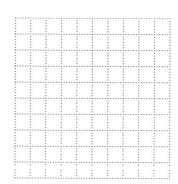

33. $y = x - 3$

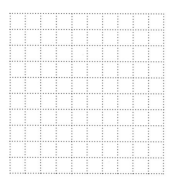

34. $y = -x + 5$

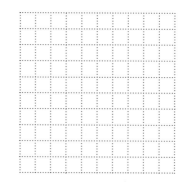

35. $y = -2x + 4$

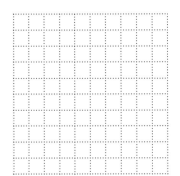

36. $y = 2x - 5$

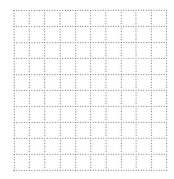

37. $y = 3x + 2$

38. $y = -5x - 1$

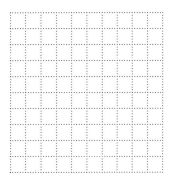

39. $y = \dfrac{1}{2}x$

40. $y = -\dfrac{3}{4}x$

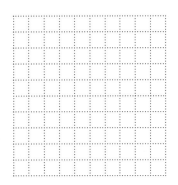

41. $y = -\dfrac{2}{3}x + 4$

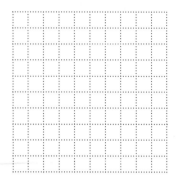

42. $y = \dfrac{4}{5}x - 1$

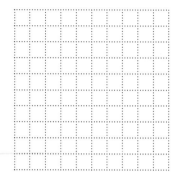

43. $x - \dfrac{1}{4}y = 2$

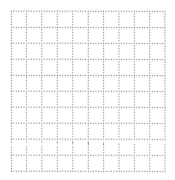

44. $\dfrac{1}{3}x + y = -1$

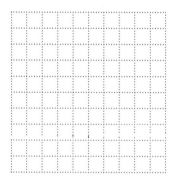

45. $y = 0.4x - 2.5$

46. $1.2x + 0.5y = 6$

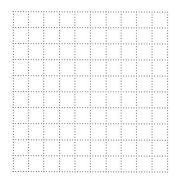

47. $y = -5$

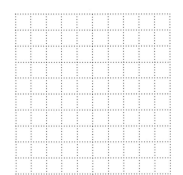

48. $x = -6$

49. $x = 7$

50. $y = 4$

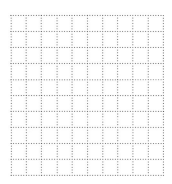

For Exercises 51–66, find the coordinates for the x- and y-intercepts.

51. $x + 2y = 12$

52. $x - 3y = 6$

53. $5x - 4y = 20$

54. $4x + y = 16$

55. $\frac{3}{4}x - y = -6$

56. $-x + \frac{3}{5}y = 9$

57. $6.5x + 2y = 1.3$

58. $4x - 5.6y = 2.8$

59. $y = 6x$

60. $y = -2x$

61. $y = -4x + 1$

62. $y = 2x - 3$

63. $y = \dfrac{1}{5}x - 5$

64. $y = \dfrac{2}{3}x + 6$

65. $x = -9$

66. $y = 3$

67. $y = 4$

68. $x = 2$

69. $y = -3$

70. $x = -6$

PUZZLE PROBLEM Without lifting your pencil from the paper, connect the dots below using only four straight lines.

Review Exercises

[3.1] **1.** Evaluate $-32t + 70$, when $t = 4.2$.

[6.2] **2.** Find the perimeter of the shape.

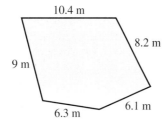

10.4 m

8.2 m

9 m

6.3 m

6.1 m

$\begin{bmatrix} 1.6 \\ 5.3 \end{bmatrix}$ **3.** Find the area of the parallelogram with a base of $6\dfrac{1}{2}$ feet and a height of 7 feet.

$\begin{bmatrix} 5.3 \\ 6.3 \end{bmatrix}$ **4.** Find the area of the triangle with a base of 4.2 inches and a height of 2.7 inches.

$\begin{bmatrix} 5.7 \\ 6.3 \end{bmatrix}$ **5.** Find the area of the trapezoid.

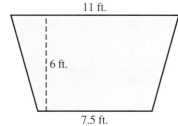

11 ft.

6 ft.

7.5 ft.

9.5 Applications with Graphing

OBJECTIVE 1 Solve problems involving linear equations in two variables.

Linear equations in two variables often describe real situations. Graphing linear equations gives us insights into the situation being described.

OBJECTIVES

1 Solve problems involving linear equations in two variables.

2 Calculate the centroid of a figure given the coordinates of its vertices.

3 Calculate the area of a figure given the coordinates of its vertices.

Example 1 The linear equation $v = -32.2t + 140$ describes the velocity of a ball thrown straight up. The variable t represents the time of the ball's flight in seconds and v represents the velocity of the ball in feet per second (ft./sec.).

a. Calculate the initial velocity of the ball.

Understand: Initial velocity means the velocity at the time the ball is released, which is 0 seconds.

Plan: Let $t = 0$ and solve for v.

Execute: $v = -32.2t + 140$
$\qquad v = -32.2(0) + 140$ Replace t with 0.
$\qquad v = 0 + 140$
$\qquad v = 140$

Answer: Initially the ball is traveling at 140 ft./sec.

Check: Reverse the process. We will leave this to the reader.

DISCUSSION   What is this velocity in miles per hour?

CONNECTION Values for time are usually placed on the horizontal axis and values for velocity on the vertical axis, so the t and v values can be written in an ordered pair of the form (t, v). When $t = 0$, we found that $v = 140$, so that ordered pair is $(0, 140)$. Notice that $(0, 140)$ is the intercept on the vertical axis, which is the v-axis in this case, so $(0, 140)$ is the v-intercept.

b. Calculate the velocity of the ball 2 seconds after being released.

Understand: We are given a time of 2 seconds.

Plan: Let $t = 2$ and solve for v.

Execute: $v = -32.2(2) + 140$ Replace t with 2.
$\qquad v = -64.4 + 140$
$\qquad v = 75.6$

CONNECTION When $t = 2$, $v = 75.6$, so this ordered pair is $(2, 75.6)$.

DISCUSSION   Why is the ball slowing down?

Answer: 2 seconds after being released, the ball is traveling only 75.6 ft./sec. Notice that it has slowed down considerably.

Check: Reverse the process. Again, we will leave this to the reader.

c. How many seconds after the ball is released does it come to a stop before descending?

Understand: We must calculate the time at which the ball stops in midair. After stopping, the ball will then reverse direction and begin falling back toward the ground. When an object is at a stop, it is obviously not moving and therefore has no velocity. We must solve for t when the velocity, v, is 0.

Plan: Let $v = 0$ and solve for t.

Your Turn 1

The equation $p = 0.15r - 30{,}000$ describes the profit for a company, where r represents revenue.

a. Calculate the profit if the revenue is \$320,000

b. Calculate the revenue required to break even (the point at which profit is \$0).

c. Graph the equation.

Execute: $v = -32.2t + 140$

$$0 = -32.2t + 140$$

Replace v with 0.

$$\underline{-140 \qquad\qquad -140}$$

$$-140 = -32.2t + 0$$

$$\frac{-140}{-32.2} = \frac{-32.2t}{-32.2}$$

Divide both sides by −32.2.

$$4.4 \approx t$$

CONNECTION In this case, $v = 0$ and we solve for t. Because t will be shown on the horizontal axis, we have just found the t-intercept. The coordinates of the t-intercept would be written this way:

$$(4.4, 0)$$

Answer: The ball will stop in midair approximately 4.4 sec. after being released.

Check: We could replace t with 4.4 in the original equation and verify that v would equal 0.

d. Graph the equation with t shown on the horizontal axis and v on the vertical axis.

Understand: We must graph the equation. To graph we need at least two solutions. In the answers to questions a, b, and c, we actually have three solutions.

$$(0, 140) \qquad\qquad (2, 75.6) \qquad\qquad (4.4, 0)$$

Plan: Plot each solution in the rectangular coordinate system, then connect the points to form a straight line.

Execute:

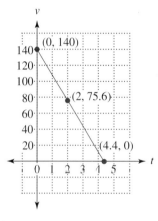

DISCUSSION Follow the line beyond the t-intercept. The line passes beyond the t-axis into quadrant IV. What does this say about the velocity? What does this mean?

◀ **Do Your Turn 1**

OBJECTIVE 2 Calculate the centroid of a figure given the coordinates of its vertices.

We can use coordinates to describe points that define the *corners* of a figure. A corner in a figure is called a *vertex*.

Answers to Your Turn 1
a. \$18,000 **b.** \$200,000
c.

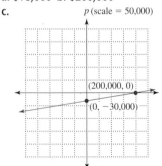

DEFINITION Vertex: A point where two lines join to form an angle.

Points A, B, C, and D are the vertices of the following figure.

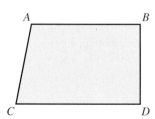

If a figure such as a triangle or parallelogram is drawn in the rectangular coordinate system, we can use the coordinates of its vertices to determine the coordinates of a point in the figure called its **centroid.**

DEFINITION **Centroid:** A point that is the center of gravity of a figure.

To understand center of gravity, imagine cutting the figure shown on the previous page out of cardboard and balancing it on the point of a pencil. Once the figure is balanced, the tip of the pencil would be touching the figure's center of gravity, which is its centroid.

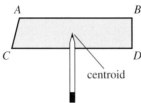

centroid

To see how we use the coordinates of the vertices of a figure in the coordinate plane to find its centroid, let's first consider a line segment, which is like a stick. If we balance a stick on a pencil, the tip would touch a point halfway up the stick, which corresponds to the midpoint of a line segment. Recall from Section 9.3 that we find the coordinates of the midpoint of a line segment connecting two points in the rectangular coordinate system by finding the mean of the x-coordinates and the mean of the y-coordinates of the two points. Similarly, we find the coordinates of the centroid of a figure using the mean of the x-coordinates and the mean of the y-coordinates of the vertices.

RULE

Given the coordinates of the vertices of a figure in the rectangular coordinate system, the coordinates of its centroid are
(mean of the x-coordinates of all vertices, mean of the y-coordinates of all vertices).

Example 2 Find the centroid of the triangle with vertices $(3, 2)$, $(-4, 4)$, and $(-5, 3)$.

Understand: We are to calculate the centroid of a triangle given the coordinates of its vertices.

Plan: Find the mean of the x-coordinates and the mean of the y-coordinates of the three vertices. The mean of the x-coordinates is the sum of the three x-coordinates divided by 3. Similarly, the mean of the y-coordinates is the sum of the three y-coordinates divided by 3.

Execute:

$$\text{centroid} = \left(\frac{x_1 + x_2 + x_3}{3}, \frac{y_1 + y_2 + y_3}{3} \right)$$

$$\underset{\underset{x_1\ y_1}{\uparrow\ \uparrow}}{(3, 2)} \qquad \underset{\underset{x_2\ y_2}{\uparrow\ \uparrow}}{(-4, 4)} \qquad \underset{\underset{x_3\ y_3}{\uparrow\ \uparrow}}{(-5, 3)}$$

$$= \left(\frac{3 + (-4) + (-5)}{3}, \frac{2 + 4 + 3}{3} \right)$$

$$= \left(\frac{-6}{3}, \frac{9}{3} \right)$$

$$= (-2, 3)$$

Answer: The centroid of the triangle is $(-2, 3)$.

Find the centroid of the triangle with vertices $(3, 4)$, $(-4, -1)$, and $(-2, 6)$.

To visualize this, let's plot the vertices and the centroid. You can make this graph on a piece of cardboard and check that the triangle balances on its centroid.

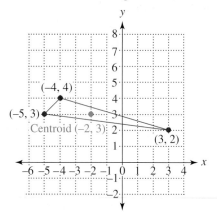

◄ **Do Your Turn 2**

Centroids of figures with four vertices can be found in much the same way.

Find the centroid of the figure with vertices $(2, 3)$, $(4, -5)$, $(-4, 1)$ and $(8, 7)$.

Example 3 Find the centroid of the figure with vertices $(3, 2), (-4, 2)$, $(3, -3)$, and $(-5, -3)$.

Understand: We are to calculate the centroid of a figure given coordinates of its vertices.

Plan: Find the mean of the x-coordinates and the mean of the y-coordinates of the four vertices.

Execute:

$$\text{centroid} = \left(\frac{x_1 + x_2 + x_3 + x_4}{4}, \frac{y_1 + y_2 + y_3 + y_4}{4} \right)$$

$$\underset{\underset{x_1\ y_1}{\uparrow\ \uparrow}}{(3, 2)} \qquad \underset{\underset{x_2\ y_2}{\uparrow\ \uparrow}}{(-4, 2)} \qquad \underset{\underset{x_3\ y_3}{\uparrow\ \uparrow}}{(3, -3)} \qquad \underset{\underset{x_4\ y_4}{\uparrow\ \uparrow}}{(-5, -3)}$$

$$= \left(\frac{3 + (-4) + 3 + (-5)}{4}, \frac{2 + 2 + (-3) + (-3)}{4} \right)$$

$$= \left(\frac{-3}{4}, \frac{-2}{4} \right)$$

$$= \left(-\frac{3}{4}, -\frac{1}{2} \right)$$

Answer: The centroid is $\left(-\dfrac{3}{4}, -\dfrac{1}{2} \right)$.

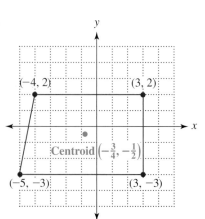

Answer to Your Turn 2
$(-1, 3)$

Answer to Your Turn 3
$\left(\frac{5}{2}, \frac{3}{2} \right)$

◄ **Do Your Turn 3**

OBJECTIVE 3 Calculate the area of a figure given the coordinates of its vertices.

Given the coordinates of the vertices of a figure, we can also find its area.

Example 4 Calculate the area of the figure with vertices at $(4, 3)$, $(-2, 3)$, $(-4, -2)$, and $(2, -2)$.

Understand: This is a four-sided figure. We are to calculate area.

Plan: Plot the points to get a sense of the shape. Then calculate the lengths of the sides that are needed to calculate the area. Last, calculate the area.

Execute:

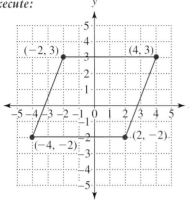

Notice that the figure is a parallelogram. The formula for the area of a parallelogram is $A = bh$, so we need the base and height. Because the base is parallel to the x-axis, we can subtract the x-coordinates.

Base: $|2 - (-4)| = |2 + 4| = |6| = 6$

Recall that height is a vertical measurement from the base to the top of a shape. The line along which height is measured is parallel to the y-axis. Therefore, we can subtract the y-coordinates.

Height: $|3 - (-2)| = |3 + 2| = |5| = 5$

Now we can calculate the area:

$$A = 6 \cdot 5$$
$$A = 30$$

Answer: The area is 30 square units.

Do Your Turn 4 ▷

Your Turn 4

Calculate the area of the figure with the given vertices.

a. $(5, 2)$, $(3, -1)$, $(-1, 2)$, and $(-3, -1)$

b. $(-1, 4)$, $(-3, 4)$ $(-5, -5)$, and $(2, -5)$

Answers to Your Turn 4
a. 18 square units
b. 40.5 square units

9.5 **Exercises**

For
Extra
Help

Videotape
DVT

Addison-Wesley
Tutor Center

Math XL

MyMathLab

Student Solutions
Manual

1. The equation $v = -32.2t + 200$ describes the velocity of an object after being thrown straight up, where t represents the time of the ball's flight in seconds and v represents the velocity in feet per second. How do we determine the amount of time it takes for the object to come to a stop before descending?

2. What is a vertex?

3. A figure has vertices at $(2, 4)$, $(-3, 4)$, $(-3, -1)$, and $(2, -1)$. What is the figure?

4. A triangle has coordinates of (x_1, y_1), (x_2, y_2), and (x_3, y_3). What is the formula for calculating the coordinates of its centroid?

For Exercises 5–14, use the given linear equation to answer the questions.

5. The linear equation $v = -32.2t + 600$ describes the velocity in feet per second of a rocket t seconds after being launched.

  a. Calculate the initial velocity of the rocket.

  b. Calculate the velocity after 3 seconds.

  c. How many seconds after launch will the rocket stop before returning to Earth?

  d. Graph the equation with v as the vertical axis and t as the horizontal axis.

OF INTEREST

At a velocity of 7.9 kilometers per second, an object would just overcome the effects of gravity so that it would enter the lowest possible circular orbit around Earth. An object traveling between 7.9 kilometers per second and 11.18 kilometers per second would travel in an elliptical orbit around Earth. An object traveling 11.18 kilometers per second, or faster, would escape Earth's gravity altogether. What are these velocities in feet per second? Miles per hour?

6. The linear equation $v = -32.2t + 86$ describes the velocity in feet per second of a ball t seconds after being thrown straight up.

  a. Calculate the initial velocity of the ball.

  b. Calculate the velocity after 1.5 seconds.

  c. How many seconds after launch will the ball stop before returning to Earth?

  d. Graph the equation with v on the vertical axis and t on the horizontal axis.

7. The equation $p = 0.24r - 45{,}000$ describes the profit for a company, where r represents revenue in dollars.

   a. Calculate the profit if the revenue is $250,000.

   b. Calculate the revenue required to break even (the point at which profit is $0).

   c. Graph the equation with p on the vertical axis and r on the horizontal axis.

8. The equation $p = 0.18r - 36{,}000$ describes the profit for a company, where r represents revenue in dollars.

   a. Calculate the profit if the revenue is $450,000.

   b. Calculate the revenue required to break even (the point at which profit is $0).

   c. Graph the equation with p on the vertical axis and r on the horizontal axis.

9. The equation $b = 13.5t + 300$ describes the final balance of an account t years after the initial investment is made.

   a. Calculate the initial balance (principal). (*Hint:* $t = 0$.)

   b. Calculate the balance after 5 years.

   c. Calculate the balance after 20 years.

   d. Graph the equation with b on the vertical axis and t on the horizontal axis.

10. The equation $b = 17.5t + 500$ describes the final balance of an account t years after the initial investment is made.

    a. Calculate the initial balance (principal). (*Hint:* $t = 0$.)

    b. Calculate the balance after 5 years.

    c. Calculate the balance after 20 years.

    d. Graph the equation with b on the vertical axis and t on the horizontal axis.

11. The equation $F = \dfrac{9}{5}C + 32$ is used to convert a temperature in °C to temperature in °F.

   a. What is the F-intercept?

   b. What is the C-intercept?

   c. What is F when C is 40?

   d. Graph the equation with F on the vertical axis and C on the horizontal axis.

12. The equation $C = \dfrac{5}{9}(F - 32)$ is used to convert a temperature in °F to temperature in °C.

   a. What is the F-intercept?

   b. What is the C-intercept?

   c. What is C when F is 68?

   d. Graph the equation with C on the vertical axis and F on the khorizontal axis.

13. An HVAC (heating, ventilation, and air-conditioning) service technician charges $9 per hour for labor plus a $58 flat fee for the visit.

   a. Write a linear equation that describes the total cost.

   b. What would be the total cost for $2\dfrac{1}{2}$ hours of labor?

   c. How many hours of labor can the service technician work for a customer who only has $100 available?

   d. Graph the equation.

14. The weekly cost to produce a toy is $0.75 per unit plus a flat $4500 for lease, equipment, supplies, and other expenses. Let C represent the total cost and u represent the number of units produced.

   a. Write a linear equation that describes the total cost.

   b. What would be the total cost to produce 600 units?

   c. What would be the total cost to produce 800 units?

   d. Graph the equation.

For Exercises 15–34, calculate the centroid and area of the figure with the given vertices.

15. $(0, 0)$, $(0, 5)$, $(4, 5)$, $(4, 0)$

16. $(-2, 0)$, $(-2, 7)$, $(0, 7)$, $(0, 0)$

17. $(4, 1)$, $(-3, 1)$, $(-3, -5)$, $(4, -5)$

18. $(2, 5)$, $(-5, 5)$, $(-5, -3)$, $(2, -3)$

19. $(-1, 2)$, $(-6, 2)$, $(-6, -4)$, $(-1, -4)$

20. $(0, -3)$, $(6, -3)$, $(0, -7)$, $(6, -7)$

21. $(0, 0)$, $(2, 6)$, $(7, 6)$, $(5, 0)$

22. $(-8, 0)$, $(-5, 4)$, $(3, 4)$, $(0, 0)$

23. $(2, 5)$, $(3, -1)$, $(7, 5)$, $(8, -1)$

24. $(-5, 3)$, $(-2, 0)$, $(6, 3)$, $(9, 0)$

25. $(-3, 2)$, $(4, 2)$, $(6, -7)$, $(-1, -7)$

26. $(-6, 5)$, $(-4, -3)$, $(3, 5)$, $(5, -3)$

27. $(0, 0)$, $(0, 7)$, $(5, 7)$, $(8, 0)$

28. $(0, 0)$, $(1, 6)$, $(5, 6)$, $(5, 0)$

29. $(-3, 4)$, $(-5, 1)$, $(5, 4)$, $(5, 1)$

30. $(-2, 0)$, $(-2, -4)$, $(4, 0)$, $(7, -4)$

31. $(-3, -2)$, $(-2, 4)$, $(6, 4)$, $(7, -2)$

32. $(-1, 5)$, $(0, 0)$, $(4, 0)$, $(5, 5)$

33. $(0, 3)$, $(-3, -1)$, $(4, 3)$, $(5, -1)$

34. $(-2, -1)$, $(-4, 6)$, $(3, -1)$, $(4, 6)$

PUZZLE PROBLEM   Find the centroid and area of the figure with vertices:

$$(-2, 1), (-2, -5), (6, -5), (6, -2), (0, -2), (0, 1)$$

Review Exercises

[9.1]   **1.** Calculate the mean, median, and mode of the scores.

$$88, 72, 91, 84, 88, 87, 90, 56, 71, 70, 95, 82, 70, 65$$

[8.4]   **2.** 20% of the people in a class received an A. If there were 45 people in the class, how many received an A?

[8.6]   **3.** Calculate the final balance in an account if $400 earning 4.8% APR is compounded quarterly over two years.

[5.3]   **4.** A company allocates $\frac{3}{4}$ of its costs for employee wages; $\frac{2}{3}$ of the employee wages are for nonsalary wages (hourly wages). What portion of the company's total cost goes toward nonsalary wages?

[4.5]   **5.** The sum of two integers is 106. One number is 6 more than three times the other. What are the integers?

Defined Terms

Review the following terms and for those you do not know, study its definition on the page number next to it.

Section 9.1
Statistic *(p. 628)*
Means, or arithmetic
  average *(p. 628)*
Grade point average
  (GPA) *(p. 629)*

Grade point *(p. 629)*
Median *(p. 631)*
Mode *(p. 632)*

Section 9.2
Axis *(p. 641)*

Section 9.3
Quadrant *(p. 653)*
Midpoint of two points
  (p. 654)

Section 9.4
x-intercept *(p. 664)*
y-intercept *(p. 664)*

Section 9.5
Vertex *(674)*
Centroid *(p. 675)*

Procedures, Rules, and Key Examples

Procedures/Rules	Key Example(s)

Section 9.1 Mean, Median, and Mode

To find the mean, or arithmetic average, of a given set of numbers:

1. Calculate the sum of all the given numbers.
2. Divide the sum by the number of numbers.

Example 1: Find the mean, or arithmetic average, of the set of golf scores:

72, 74, 76, 74, 70, 68, 70, 72

$$\bar{x} = \frac{72 + 74 + 76 + 74 + 70 + 68 + 70 + 72}{8}$$

$$\bar{x} = \frac{576}{8}$$

$$\bar{x} = 72$$

To find the median of a set of scores:

1. Arrange the scores in order from least to greatest.
2. Locate the middle score of the ordered set of scores.

Note: If there are an even number of scores in the set, the median will be the mean of the two middle scores.

Example 2: Find the median of the test scores:

75, 90, 80, 92, 95, 72, 60, 84, 88

Arrange in order:

60, 72, 75, 80, 84, 88, 90, 92, 95

median = 84

Suppose there were one more score:

60, 72, 75, 80, 84, 88, 90, 92, 95, 98

$$\text{median} = \frac{84 + 88}{2} = \frac{172}{2} = 86$$

To find the mode of a set of scores:

Count the number of repetitions of each score. The score with the most repetitions is the mode.

Note: If no score is repeated, then there is no mode. If there is a tie, then list each score as a mode.

Example 3: Find the mode of each set.

a. GPAs: 3.5, 3.875, 2.75, 3.0, 3.5, 2.25, 3.5

mode = 3.5

b. Prices:

$24.95, $15.95, $29.99, $21.75

no mode

c. Newborn baby lengths:

21 in., 19.5 in., 18.5 in., 21 in., 19.5 in., 17 in.

modes: 21 in. and 19.5 in.

Section 9.3 The Rectangular Coordinate System

To determine the coordinates of a given point in the rectangular system:

1. Follow a vertical line from the given point to the x-axis (horizontal axis). The number at this position on the x-axis as the first coordinate.

2. Follow a horizontal line from the given point to the y-axis (vertical axis). The number at this position on the y-axis is the second coordinate.

Example 1: Determine the coordinates of the points shown.

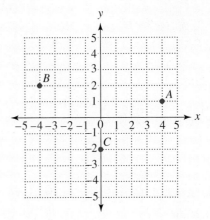

Answers:

A: $(4, 1)$ B: $(-4, 2)$ C: $(0, -2)$

To graph or plot a point given a coordinate pair:

1. Beginning at the origin, $(0, 0)$, move to the right or left along the x-axis the amount indicated by the first coordinate.

2. From that position on the x-axis, move up or down the amount indicated by the second coordinate.

3. Draw a dot to represent the point described by the coordinates.

Example 2: Plot the point described by the coordinates

a. $(-3, -4)$

b. $(2, -3)$

c. $(0, 3)$

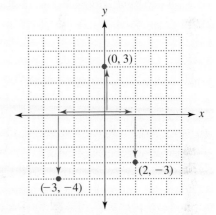

To determine the quadrant for a given coordinate:

Consider the signs of the numbers in the coordinate.

$(+, +)$ means the point is in quadrant I

$(-, +)$ means the point is in quadrant II

$(-, -)$ means the point is in quadrant III

$(+, -)$ means the point is in quadrant IV

Example 3: Determine the quadrant in which each point is located.

a. $(42, 95)$: quadrant I

b. $(-65, 90)$: quadrant II

c. $(-91, -56)$: quadrant III

d. $(75, -102)$: quadrant IV

e. $(-14, 0)$: not in a quadrant; this point is on the x-axis.

To calculate the midpoint of two points with coordinates (x_1, y_1) and (x_2, y_2), use the formula:

$$\text{midpoint} = \left(\frac{x_1 + x_2}{2}, \frac{y_1 + y_2}{2} \right)$$

Example 4: Find the midpoint of $(4, -9)$ and $(-12, -2)$.

$$\text{midpoint} = \left(\frac{4 + (-12)}{2}, \frac{-9 + (-2)}{2} \right)$$

$$= \left(\frac{-8}{2}, -\frac{11}{2} \right)$$

$$= \left(-4, -\frac{11}{2} \right)$$

Section 9.4 Graphing Linear Equations

To determine whether a given ordered pair is a solution for an equation in two variables:

1. Replace the variables in the equation with the corresponding coordinates.
2. Verify that the equation is true.

Example 1: Determine whether $(-6, 2)$ is a solution for $x + 5y = 4$.

$$-6 + 5(2) \stackrel{?}{=} 4$$
$$-6 + 10 \stackrel{?}{=} 4$$
$$4 = 4$$

Since the equation is true, $(-6, 2)$ is a solution.

Example 2: Determine whether $(2.2, 4.8)$ is a solution for $y = 4x - 3$.

$$4.8 \stackrel{?}{=} 4(2.2) - 3$$
$$4.8 \stackrel{?}{=} 8.8 - 3$$
$$4.8 \neq 5.8$$

Since the equation is false, $(2.2, 4.8)$ is not a solution.

To find a solution to an equation in two variables:

1. Choose a value (any value) for one of the variables.
2. Replace the corresponding variable with your chosen value.
3. Solve the equation for the value of the other variable.

Example 3: Find a solution for $2x - y = -3$.

We will choose $x = 1$:

$$2(1) - y = -3$$
$$2 - y = -3$$
$$\underline{2 - y = -3}$$
$$\underline{-2 \qquad -2}$$
$$0 - y = -5$$
$$\frac{-y}{-1} = \frac{-5}{-1}$$
$$y = 5 \qquad \text{Solution: } (1, 5)$$

To graph a linear equation:

1. Find at least two solutions to the equation.
2. Plot the solutions as points in the rectangular coordinate system.
3. Draw a straight line through these points.

Example 4: Graph. $2x - y = -3$

We found one solution above $(1, 5)$.

Two more solutions that can be found the same way as above are $(0, 3)$ and $(-2, -1)$.

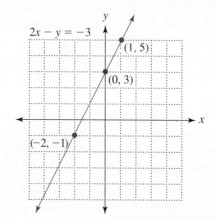

The graph of $y = c$, where c is a constant, is a horizontal line parallel to the x-axis and intersects the y-axis at a point with coordinates $(0, c)$.

Example 5: Graph. $y = 3$

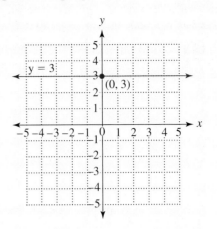

The graph of $x = c$, where c is a constant, is a vertical line parallel to the y-axis and intersects the x-axis at a point with coordinates $(c, 0)$.

Example 6: Graph. $x = -2$

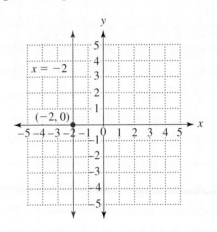

To find the x-intercept:
1. Replace y with 0 in the given equation.
2. Solve for x.

To find the y-intercept:
1. Replace x with 0 in the given equation.
2. Solve for y.

Example 7: Find the coordinates of the x- and y-intercepts for the equation $3x + 5y = 12$.

x-intercept:

$$3x + 5(0) = 12$$
$$3x = 12$$
$$\frac{3x}{3} = \frac{12}{3}$$
$$x = 4$$

x-intercept: $(4, 0)$

y-intercept:

$$3(0) + 5y = 12$$
$$5y = 12$$
$$\frac{5y}{5} = \frac{12}{5}$$
$$y = \frac{12}{5} \text{ or } 2\frac{2}{5}$$

y-intercept: $\left(0, 2\frac{2}{5}\right)$

Section 9.5 Applications with Graphing

Given the coordinates of the vertices of a figure in the rectangular coordinate system, the coordinates of its centroid are
(mean of the x-coordinate of all vertices, mean of the y-coordinates of all vertices)

Example 1: Find the centroid of the triangle with vertices $(4, -3)$, $(-2, -4)$, and $(-5, 1)$.

$$\text{centroid} = \left(\frac{4 + (-2) + (-5)}{3}, \frac{-3 + (-4) + 1}{3}\right)$$
$$= \left(\frac{-3}{3}, \frac{-6}{3}\right)$$
$$= (-1, -2)$$

For Exercises 1–6, answer true or false.

[9.1] **1.** If the median income for a company is $24,000, most people in the company make around $24,000.

[9.1] **2.** If the mode for a set of test scores is 85, 85 was the score received more often than any other score in the set of scores.

[9.1] **3.** There can only be one mode for a set of scores.

[9.3] **4.** $(-45, 60)$ is in quadrant IV.

[9.4] **5.** The equation $x = 2$ is a vertical line with an x-intercept at $(2, 0)$.

[9.4] **6.** The equation $y = -4$ has no x-intercept.

[9.1] **7.** Explain in your own words how to find the mean, or arithmetic average, of a set of numbers.

[9.1] **8.** Explain in your own words how to find the median of a set of numbers.

[9.4] **9.** Explain in your own words how to graph a linear equation.

[9.4] **10.** Explain in your own words how to find the x- and y-intercepts of a linear equation.

[9.1] **11.** Find the mean, median, and mode of the following test scores.

65	94	72	94	88	86
84	87	92	58	76	95
92	94	77	100	62	60
81	80	84	98	61	71

[9.1] **12.** A new housing development is to have 11 different base floor plans to select from. Find the mean, median, and mode of the base prices for the houses in the development.

Savannah	$115,990	Jorden	$127,100
Charlestown	$123,190	Devon	$131,200
Richmond	$123,500	Lexington	$126,800
Dawson	$124,500	Kensington	$134,100
Maguire	$130,000	Karrington	$131,690
Cambridge	$138,250		

[9.1] **13.** Donielle has the following test scores: 86, 91, 92, and 88. Her instructor will allow her to be exempt from the final exam if she has a test average (mean) of 90 or better. What must she score on the fifth test to have a test average (mean) of 90?

[9.1] **14.** Calculate the grade point average.

Course	Credits	Grade
MATH 100	5.0	C
ENG 100	3.0	B+
HIS 101	3.0	A
CHM 101	4.0	A

[9.2] *For Exercises 15–20, use Table 9.3.*

Table 9.3   Percentage of the population ages 25 to 64 that has completed at least secondary and higher education, by age group, gender, and country: 2001

| | 25–64 years old | | 25–34 years old | | | | | |
| | Total | | Total | | Male | | Female | |
Country	Secondary Ed.	Higher Ed.	Secondary Ed.	Higher Ed.	Secondary Ed.	Higher Ed.	Secondary Ed.	Higher Ed.
Canada	82	41	89	51	88	45	91	56
France	64	23	78	34	78	32	78	37
Germany	83	23	85	22	87	23	84	20
Italy	43	16	57	12	55	10	60	13
Japan	83	34	94	48	92	46	95	49
United Kingdom*	63	26	68	29	70	30	65	29
United States	88	37	88	39	87	36	89	42
Country mean	64	23	74	28	73	26	74	29

Source: Organization for Economic Cooperation and Development, Education at a Glance 2003

15. What percent of French people 25–64 years of age completed some form of higher education?

16. What percentage of females ages 25–34 completed secondary education in the United Kingdom?

17. Which country had the highest percentage of its total population 25–64 years of age complete some form of higher education?

18. Which country had the lowest rate of completion of secondary education for people from 25 to 64 years of age?

19. Which country had the highest percentage of its 25–34 age group complete secondary education?

20. Which country had the greatest difference in completion rates for higher education between men and women ages 25–34?

[9.2] *For Exercises 21–26, use the pie chart in Figure 9.9.*

Figure 9.9   Highest level of education attained by persons age 25 and over: 2001

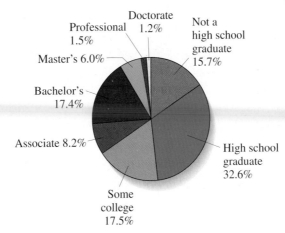

Total persons age 25 and over = 177.0 million

Source: U.S. Department of Commerce, Bureau of the Census, Current Population Survey, unpublished data. 2002

21. What percent of people 25 years and older completed some college?

22. What percent of people 25 years and older did not complete high school?

23. How many people were 25 years or older in 2001?

24. How many people 25 years and older had completed a bachelor's degree as their highest level of education?

25. How many people 25 years and older had completed a master's degree as their highest level of education?

26. What is the combined number of doctorate degrees and professionals who are 25 years or older?

[9.2] **27.** The following is an itemized list of a family's house payment. Use the data to construct a pie chart showing the percent of the total payment that goes toward each item.

Principal: $107.04

Interest: $672.58

Taxes: $129.74

Insurance: $50.08

[9.2] *For Exercises 28–32, use the bar graph in Figure 9.10.*

Figure 9.10   Median annual income of persons 15 years old and over, by highest degree attained and gender: 2003

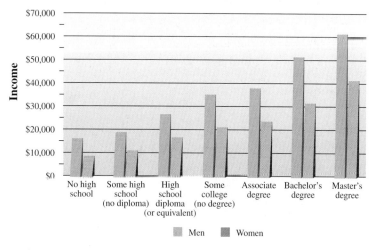

Source: U.S. Census Bureau, 2003

28. What is the median income for men with some high school, but no diploma?

29. What is the median income for women with a master's degree?

30. Which group of people had the lowest median income?

31. Which group of people had the highest median income?

32. Which group of people had a median income of about $30,000?

[9.2] **33.** Refer to the data in Table 9.3 for the percent of people in each country who are 25–64 years of age and who completed secondary school. Use these data to construct a bar graph with each country listed along the horizontal axis and the percentage that completed secondary education along the vertical axis.

[9.2] *For Exercises 34–38, use the line graph in Figure 9.11.*

Figure 9.11   Enrollment in institutions of higher education, by age: Fall 1988 to Fall 2013

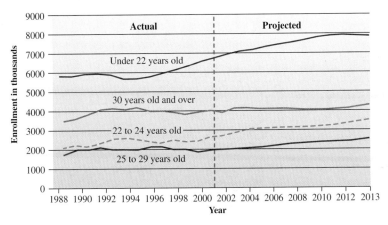

Source: U.S. Department of Education

34. How many people under 22 enrolled in higher education in 1995?

35. How many people 25 to 29 years of age enrolled in higher education in 2000?

36. What is the projected number of people over 30 who will be enrolled in higher education in the year 2010?

37. Which age group had the greatest growth in enrollment from 1990 to 2000?

38. What is the difference between the number of students under 22 who were enrolled in higher education in 1990 and the number of students projected to be enrolled in 2013?

[9.2] **39.** Shown are the mean daytime temperatures for each month in a certain city. Use the data to construct a line graph with each month along the horizontal axis and the temperatures along the vertical axis.

January:	35°F	May:	78°F	September:	80°F
February:	48°F	June:	83°F	October:	72°F
March:	65°F	July:	85°F	November:	54°F
April:	70°F	August:	88°F	December:	46°F

[9.3] **40.** State the coordinates for each point shown on the graph.

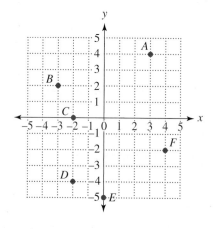

[9.3] **41.** Plot the points in the rectangular coordinate system.

$A\,(5, 2)$ $\quad B\,(2, 0)$ $\quad C\,(-4, -3)$ $\quad D\,(3, -5)$ $\quad E\,(0, -4)$

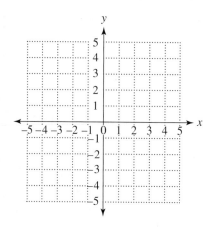

[9.3] 42. State the quadrant in which each point is located.

    a. $(-105, 68)$ **b.** $(-47, -158)$

    c. $(95, 72)$ **d.** $(58, 0)$

[9.3] 43. Find the midpoint of $(2, 9)$ and $(6, 1)$.

[9.3] 44. Find the midpoint of $(-4, 2)$ and $(6, -3)$.

[9.4] 45. Determine whether $(2, 7)$ is a solution for $3x - y = -4$.

[9.4] 46. Determine whether $(2.5, -0.5)$ is a solution for $y = -x + 2$.

[9.4] 47. Determine whether $\left(-4\dfrac{3}{4}, -1\dfrac{9}{10}\right)$ is a solution for $y = \dfrac{2}{5}x$.

[9.4] *For Exercises 48–55, find three solutions, then graph.*

    48. $2x + y = 6$ **49.** $3x - 6y = 12$

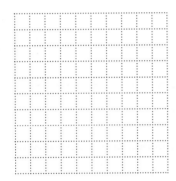 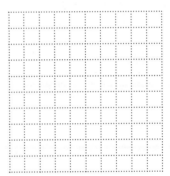

    50. $y = x + 3$ **51.** $y = -3x$

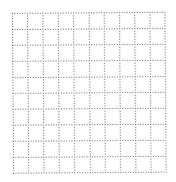 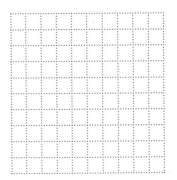

52. $y = \dfrac{2}{3}x$

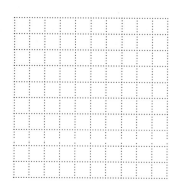

53. $y = -x - 4$

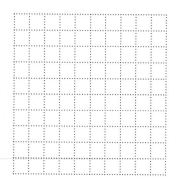

54. $y = 7$

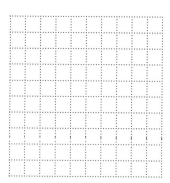

55. $x = -3$

[9.4] *For Exercises 56–59, find the coordinates for the x- and y-intercepts.*

56. $5x + y = 10$

57. $y = 4x - 1$

58. $y = \dfrac{1}{5}x$

59. $x = 6$

[9.5] 60. The linear equation $p = 0.4r - 12,000$ describes the profit for a company, where r is the revenue in dollars.

   a. If the company makes $430,680 in revenue, what is the profit?

   b. If the company has a profit of $80,250, what was the revenue?

   c. Graph the equation for profit.

[9.5] 61. The coordinates of the vertices of a figure are $(0, -1)$, $(0, 4)$, $(3, 4)$, and $(7, -1)$. Find the centroid of the figure.

[9.5] 62. The coordinates of the vertices of a figure are $(-2, 5)$, $(-1, -3)$, $(4, -3)$, and $(3, 5)$. Find the area of the figure.

1. _____

[9.1] **1.** Calculate the mean, median, and mode of the set of test scores.

88	85	78	62	94	52
80	70	90	76	84	98
84	74	96	80	96	82

2. _____

[9.2] **2.** Below is a list of prices of houses for sale in a particular area. Calculate the mean, median, and mode of the prices.

$76,500	$88,700	$96,400
$98,200	$110,000	$105,000
$94,500	$90,250	$102,000

3. _____

[9.1] **3.** Steve has the following test scores: 84, 88, 95, and 85. His instructor will allow him to be exempt from the final exam if he has a test average (mean) of 90 or better. What must he score on the fifth test to have a test average (mean) of 90?

[9.2] *For Exercises 4–6, use the pie chart. The chart shows the percentage of a family's net monthly income.*

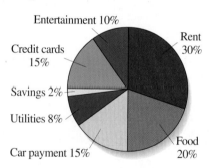

4. _____

4. What percent of the family income goes toward credit cards?

5. _____

5. If this family has a net monthly income of $2206.25, how much goes toward food?

6. _____

6. How much of the $2206.25 monthly income goes toward the credit cards and car payment combined?

[9.2] *For Exercises 7–9, use the bar graph in Figure 9.12.*

Figure 9.12 **Full-time-equivalent students per staff member in public and private degree-granting institutions: 1976 and 1999**

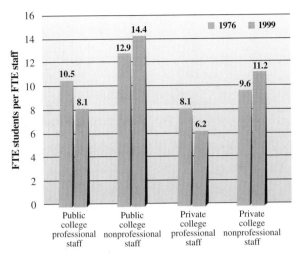

Source: U.S. Department of Education

7. Which type of college had the highest number of students per professional staff member?

8. How many FTE (full-time-equivalent) students were there per professional staff member at private colleges with professional staff in 1999?

9. How many FTE students were there per professional staff member at public colleges with professional staff in 1976?

7. _____

8. _____

9. _____

[9.2] *For Exercises 10–12, use the line graph in Figure 9.13.*

Figure 9.13 Total private average weekly work hours: 1985–2003 (seasonally adjusted)

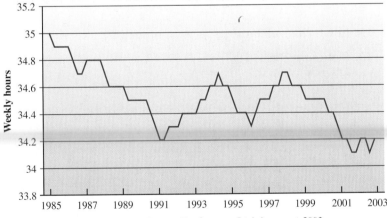

Source: Bureau of Labor Statistics, Current Employment Statistics survey 2003

10. What year(s) had the lowest number of private average weekly work hours?

11. What was the average number of weekly work hours at the beginning of 2002?

12. What was the average number of weekly work hours at the beginning of 1998?

10. _____

11. _____

12. _____

13. _____

[9.2] **13.** The following are the results of a poll. Construct a pie chart showing each response as a percentage of total responses.

Strongly agree: 226

Agree: 121

Neutral: 80

Disagree: 65

Strongly disagree: 48

14. _A:_ _____

_B:_ _____

_C:_ _____

_D:_ _____

[9.3] **14.** Determine the coordinates of each point in the rectangular system shown.

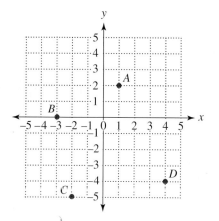

15. _____

[9.3] **15.** In which quadrant is $(-16, 48)$ located?

[9.3] **16.** Plot the points $A\,(-4, 2)$, $B\,(2, 1)$, $C\,(0, -3)$, and $D\,(-3, -5)$.

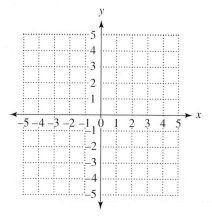

17. _____

[9.3] **17.** Calculate the midpoint of $(4, -2)$ and $(-3, -7)$.

18. _____

[9.4] **18.** Determine whether $(5, -3)$ is a solution for $-x + 2y = -11$.

[9.4] **19.** Find the coordinates of the *x*- and *y*-intercepts for $3x - 5y = 6$.   19. _____

For Exercises 20–23, graph.

[9.4] **20.** $y = -3x$

[9.4] **21.** $y = \dfrac{1}{5}x - 3$

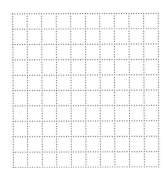

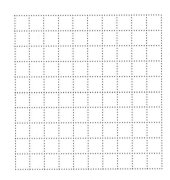

[9.4] **22.** $x + y = 7$

[9.4] **23.** $3x - y = 2$

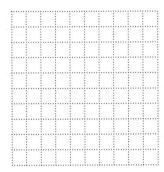

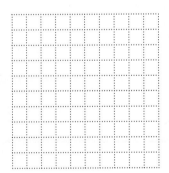

[9.5] **24.** The equation $v = -9.8t + 40$ describes the velocity in meters per second of a model rocket *t* seconds after being launched.

   a. What is the initial velocity of the rocket?

   b. What is the velocity of the rocket 3 seconds after being launched?

   c. After how many seconds will the rocket stop in midair before returning to Earth?

   d. What is the velocity of the rocket 10 seconds after being launched?

24. a. _____

b. _____

c. _____

d. _____

[9.5] **25.** The coordinates of the vertices of a figure are $(4, 3)$, $(-2, 3)$, $(-2, -5)$, and $(4, -5)$.

   a. Find the centroid of the figure.

   b. Find the area of the figure.

25. a. _____

b. _____

For Exercises 1–12, answer true or false.

[5.1] **1.** 0 is a rational number.

[3.6] **2.** 41 is a prime number.

[2.3] **3.** $5 - 8 = 8 - 5$

[3.7] **4.** $6^0 = 0$

[3.2] **5.** $5x^2 + 2$ is a monomial.

[3.2] **6.** The degree of 9 is 1.

[3.6] **7.** 12 is the GCF of 36 and 24.

[4.2] **8.** $4x = 9(x + 1.5)$ is a linear equation.

[5.1] **9.** $-9.2 \div 0 = 0$

[7.2] **10.** $\dfrac{2.5}{9} = \dfrac{6}{21.6}$

[8.1] **11.** $\dfrac{3}{8} > 37\%$

[9.3] **12.** $(-5, -3)$ is in quadrant III.

[1.5] **13.** State the order of operations agreement.

[2.2] **14.** Explain in your own words how to add two numbers that have the same sign.

[2.2] **15.** Explain in your own words how to add two numbers that have different signs.

[2.3] **16.** Explain in your own words how to write a subtraction statement as an equivalent addition statement.

[2.4] **17.** When multiplying or dividing two numbers that have the same sign, the result is _____.

[2.4] **18.** When multiplying or dividing two numbers that have different signs, the result is _____.

[4.3] **19.** Explain in your own words how to solve a linear equation in one variable.

[9.4] **20.** Explain in your own words how to graph a linear equation.

[1.1] **21.** Write 24,607 in expanded form.

[6.1] **22.** Write the word name for 4607.09.

[6.1] **23.** Round 79,804.652 to the nearest:

    a. ten thousand

    b. thousand

    c. whole number

    d. tenth

    e. hundredth

[1.2]
[1.3] **24.** Estimate each calculation by rounding so that there is only one nonzero digit.

    a. $21,459 + 6741$

    b. $729,105 - 4519$

    c. 86×17

    d. $2219 \div 45$

[6.3] **25.** Write 27,500,000,000 in scientific notation.

[6.3] **26.** Write 3.53×10^6 in standard form.

27. Graph on a number line.

[2.1] **a.** $-(-3)$

[3.2] **28.** What is the degree of each expression?

  a. $-4y$

  b. $15x^2$

[5.1] **b.** $5\dfrac{1}{4}$

  c. $4x^3 - 9x + 10x^4 - 7$

[6.1] **c.** -7.2

For Exercises 29–41, simplify.

[2.4] **29.** $(-3)^4$

[2.4] **30.** -3^4

[5.3] **31.** $\left(\dfrac{1}{6}\right)^2$

[5.3] **32.** $\left(-\dfrac{1}{2}\right)^3$

[2.5] **33.** $9 - 4|3 - 8| - \sqrt{49}$

[2.5] **34.** $19 - (6 - 8)^3 + \sqrt{100 - 36}$

[2.5] **35.** $\dfrac{14 + (18 - 2 \cdot 4)}{4^2 - 10}$

[5.6] **36.** $9\dfrac{3}{4} - \left(-2\dfrac{2}{3}\right)$

[5.7] **37.** $-6\dfrac{1}{8} \div \dfrac{3}{16} - \dfrac{3}{4}$

[6.5] **38.** $-11.7 \div 1.8 + 2.6(7.5)$

[6.5] **39.** $\left(\dfrac{3}{4}\right)^2 (-8.4)$

[6.5] **40.** $\dfrac{3}{5} - (0.2)^3$

[6.4] **41.** $\sqrt{(27)(0.03)}$

[6.4] **42.** Approximate $\sqrt{126}$ to the nearest hundredth.

43. Write as a decimal number.

[6.4]   **a.** $\dfrac{3}{8}$

[6.4]   **b.** $\dfrac{2}{3}$

[8.1]   **c.** 8%

[8.1]   **d.** $24\dfrac{2}{5}\%$

[8.1]   **e.** 4.5%

44. Write as a fraction.

[5.1]   **a.** $5\dfrac{1}{6}$

[6.1]   **b.** 0.145

[8.1]   **c.** 6%

[8.1]   **d.** $5\dfrac{1}{2}\%$

[8.1]   **e.** 16.4%

[8.1] 45. Write as a percent.

    a. $\dfrac{1}{4}$

    b. $\dfrac{4}{9}$

    c. 0.65

    d. 0.035

    e. 2.3

[3.1] 46. Evaluate each expression using the given values.

    a. $t^3 - 5t; t = -4$

    b. $-xy + 2\sqrt{x + y}; x = -4, y = 20$

[6.2] 47. Combine like terms and write the resulting polynomial in descending order of degree.
$$9x^3 - 15.4 + 8x - 12x^3 + 4x^2 + x^4 - 4.9 - x$$

[6.2] 48. Add. $(5n^2 + 8n - 9.3) + (4n^2 - 12.2n - 3.51)$

[6.2] 49. Subtract. $\left(x^4 - \dfrac{1}{3}x^2 + 9.6\right) - \left(4x^3 + \dfrac{3}{5}x^2 - 14.6\right)$

50. Multiply.

[5.3]   **a.** $\left(\dfrac{5}{8}a^2\right)\left(-\dfrac{4}{9}a\right)$

[3.5]   **b.** $(m - 9)(m + 9)$

[6.3]   **c.** $(4.5t - 3)(7.1t + 6)$

[3.6] 51. Find the prime factorization of 630.

[3.7] 52. Find the GCF of $30n^4$ and $45n^2$.

[5.5] 53. Find the LCM of $36x^3$ and $24x$.

[3.7] 54. Divide. $\dfrac{30y^5}{-6y^2}$

[3.7] 55. Factor. $28b^5 + 24b^3 - 32b^2$

56. Simplify.

[5.3]   **a.** $-\dfrac{9x^3}{10} \cdot \dfrac{25}{12x}$

[5.4]   **b.** $\dfrac{-10}{9n^2} \div \left(\dfrac{-5}{12n}\right)$

[5.6]   **c.** $\dfrac{3}{5} - \dfrac{y}{4}$

For Exercises 57–64, solve and check.

[4.2] **57.** $m - 25 = -31$

[4.3] **58.** $-6x = 54$

[4.3] **59.** $-4x + 19 = -1$

[5.8] **60.** $\dfrac{3}{5}k - 7 = \dfrac{1}{4}$

[6.6] **61.** $6.5h - 16.4 = 9h + 3.2$

[6.6] **62.** $2.4(n - 5) + 8 = 1.6n - 14$

[7.2] **63.** $\dfrac{u}{14} = \dfrac{-3}{8}$

[7.2] **64.** $\dfrac{4}{5} = \dfrac{1\frac{3}{5}}{a}$

$\begin{bmatrix} 7.3 \\ 7.4 \end{bmatrix}$ **65.** Convert.

    a. 12 feet to inches

    b. 78 feet to yards

    c. 2.4 miles to feet

    d. 4.8 meters to centimeters

    e. 12.5 kilometers to meters

    f. 4500 millimeters to meters

$\begin{bmatrix} 7.3 \\ 7.4 \end{bmatrix}$ **66.** Convert.

    a. 30 square feet to square yards

    b. 4 square yards to square feet

    c. 504 square inches to square feet

    d. 0.6 square meters to square centimeters

    e. 0.085 square kilometers to square meters

    f. 42,000 square centimeters to square meters

$\begin{bmatrix} 7.3 \\ 7.4 \end{bmatrix}$ **67.** Convert.

    a. 9.5 pounds to ounces

    b. 4500 pounds to American tons

    c. 0.23 kilograms to grams

    d. 9600 hectograms to metric tons

$\begin{bmatrix} 7.3 \\ 7.4 \end{bmatrix}$ **68.** Convert.

    a. 4 gallons to pints

    b. 68 cups to quarts

    c. 400 milliliters to liters

    d. 0.9 centiliters to cubic centimeters

[7.5] **69.** Convert.

    a. 77°F to degrees Celsius

    b. -20°C to degrees Fahrenheit

For Exercises 70–100, solve.

$\begin{bmatrix} 6.6 \\ 7.5 \end{bmatrix}$ **70.** A parallelogram has an area of 411.4 square meters. If the base is 24.2 meters, what is the height? What is the height in feet?

[1.3] **71.** An alarm system has a three-digit code. How many possible codes are there?

[2.6] **72.** A family has the following assets and debts. Calculate their net worth.

Assets	Debts
Savings = $1282	Credit card balance = $2942
Checking = $3548	Mortgage = $83,605
Furniture = $21,781	Automobile 1 = $4269
Jewelry = $5745	

[2.6]
[6.4] **73.** A person weighs 148 pounds. What is their mass in slugs? (*Hint:* The acceleration due to gravity is 32.2 feet per second per second.) Convert to kilograms.

[6.5] **74.** Write an expression in simplest form for the area of the triangle shown. Calculate the area if $h = 4.2$ inches.

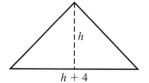

[2.3]
[6.5] **75.** A company's revenue is expressed by the polynomial $3.5x + 600$, and the cost is expressed by $0.2x + 4000$, where x represents the number of units sold.

   a. Write an expression in simplest form for the profit.

   b. If the company sells 1500 units in one week, what is the net?

[4.5] **76.** The sum of two positive integers is 123. One integer is three more than twice the other. What are the integers?

[4.5] **77.** The sum of the angles in any triangle is 180°. Suppose we have a triangle with the second angle measuring 10° more than the first and the third angle measuring 7° less than the first. What are the three angle measurements?

[4.5] **78.** Jeremy has some $5 bills and $10 bills in his wallet. If he has a total of 17 bills worth a total of $140, how many of each bill is in his wallet? (Use a four-column table.)

[5.3] **79.** On Sandra's phone bill, 24 out of the 32 long-distance calls were in state; $\frac{3}{4}$ of the in-state calls were to her parents. What fraction of all her long-distance calls were to her parents? How many calls were to her parents?

[5.6] **80.** A poll is taken to assess the president's approval rating. Respondents can answer four ways: excellent, good, fair, or poor. $\frac{1}{6}$ said "excellent," $\frac{3}{5}$ said "good," and $\frac{1}{8}$ said "fair."

   a. What fraction of the respondents said "excellent" or "good?"

   b. What fraction said "poor"?

[5.8] **81.** Alan and Jessica pass each other going in opposite directions. If Alan walks at 3 miles per hour and Jessica at 2 miles per hour, how long will it take them to be $\frac{1}{4}$ miles apart? (Use a four-column table.)

[6.5] 82. Find the area of the shaded region.

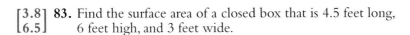
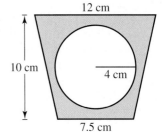

[3.8]
[6.5] 83. Find the surface area of a closed box that is 4.5 feet long, 6 feet high, and 3 feet wide.

[6.5] 84. Find the volume of a can that is 3.5 inches tall and is 3 inches in radius.

[6.5] 85. Find the volume of a ball that is 12 inches in diameter.

[6.5] 86. What is the volume of a cone that is formed from a blowtorch flame with a 0.8-centimeter radius and a 12-centimeter height?

[6.6] 87. A steel support beam on a tower needs to be replaced. It is known that the connecting joint is 20 feet above the ground, and the base of the support beam is 8 feet from a point directly below the connecting joint. How long is the beam?

[7.1] 88. What is the probability of winning a drawing if you entered your name 41 times and the total entries are 650?

[7.1] 89. Which is the better buy?
15.5 ounces of cereal at $2.89 or 20 ounces of the same cereal at $3.45

[7.2] 90. The triangles shown are similar. Find the missing side lengths.

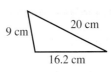
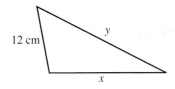

[8.5] 91. Because Coly is an employee of a store, he gets a 20% discount when he purchases merchandise. What would be his cost for an item that is normally $75.95?

[8.5] 92. A restaurant increases its price on a dish from $8.95 to $9.75. What was the percent increase?

[8.4] 93. Katrina answered 45 out of 70 questions correctly. What percent of the total questions did she answer correctly?

[8.5] 94. How much interest is earned if $680 is invested at 4.2% simple interest rate for half a year?

[8.6] 95. $3200 is invested at 6% APR compounded semiannually. What will be the balance after two years?

[9.1] 96. Find the mean, median, and mode for the following set of ages of participants in a survey.

28, 21, 45, 65, 54, 38, 39, 35, 36, 32, 40

[9.2] **97.** Use the following data to construct a pie chart showing the percentages of a total house payment that was spent on each item.

Principal = $158.25

Interest = $496.78

Taxes = $125.50

Insurance = $38.50

[9.5] **98.** Find the centroid and area of a figure with vertices $(2, 4)$, $(-5, 4)$, $(-2, -3)$, and $(2, -3)$.

[9.4] **99.** Graph.

    a. $x - y = 6$

    b. $y = -3x + 2$

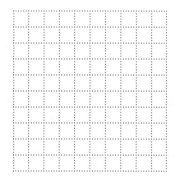
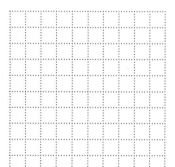

    c. $y = -\dfrac{1}{2}x$

    d. $y = 2$

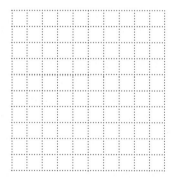
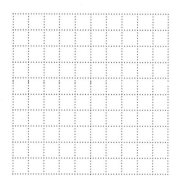

[9.5] **100.** The linear equation $v = -32.2t + 580$ describes the velocity in feet per second of a rocket t seconds after being launched.

    a. What is the initial velocity of the rocket?

    b. What is the velocity of the rocket after 5 seconds?

    c. How many seconds after launch will the rocket stop before returning to Earth?

GLOSSARY

Bracketed numbers following each definition indicate the section in which the term is covered.

Absolute value: A given number's distance from zero. **[2.1]**

Acceleration: A measure of how quickly an object increases or decreases its speed: measured in feet per second per second or meters per second per second. **[2.6]**

Addends: Numbers that are added. **[1.2]**

Addition: The arithmetic operation that combines amounts. **[1.2]**

Addition/subtraction principle of equality: We can add or subtract the same amount on both sides of an equation without affecting its solution(s). **[4.2]**

Additive inverses: Two numbers whose sum is zero. **[2.1]**

Algebraic equation: An equation that contains variables. **[3.1]**

Algebraic expression: An expression that contains variables. **[3.1]**

Amortize: To pay off a loan or debt in installments. **[8.6]**

Annual percentage rate (APR): An interest rate that is used to calculate the compound interest. **[8.6]**

Area: The total number of square units that completely fill a shape. **[1.3]**

Arithmetic average: The sum of all given numbers divided by the number of numbers (see mean). **[9.1]**

Associative property of addition: $(a + b) + c = a + (b + c)$, where a, b, and c are any numbers. **[1.2]**

Associative property of multiplication: $(a \cdot b) \cdot c = a \cdot (b \cdot c)$, where a, b, and c are any numbers. **[1.3]**

Average: A statistic that describes the middle, or central tendency, of a set of data, like mean, median, and mode. **[9.1]**

Average rate: A measure of the rate at which an object travels a total distance in a total amount of time. **[2.6]**

Axis: A line used for reference in a graph. **[9.2]**

Back-end ratio: The ratio of the total monthly debt payments to gross monthly income. **[7.6]**

Balance technique: If we add or remove an amount on one side of an equation, we must add or remove the same amount on the other side to keep the equation balanced. **[4.2]**

Base: The number that is repeatedly multiplied. **[1.3]**

Base-10 system: A numeral system that has 10 symbols where each place value is a power of 10. **[1.1, 6.1]**

Base unit: A basic unit; other units are named relative to it. **[7.4]**

Binomial: A polynomial that has exactly two terms. **[3.2]**

Capacity: A measure of the amount of liquid a container holds. **[7.3]**

Cartesian coordinate system: Two perpendicular axes that describe the position of a point in a plane (see rectangular coordinate system). **[9.3]**

Centroid: A point that is the center of gravity of a figure (the mean of x-coordinates of all vertices, the mean of y-coordinates of all vertices). **[9.5]**

Circle: A collection of points that are equally distant from a central point, called the center. **[5.3]**

Circumference: The distance around a circle. **[5.3]**

Coefficient: The numerical factor in a monomial. **[3.2]**

Combinatorics: The branch of mathematics that deals with counting total combinations or arrangements of items. **[1.3]**

Commission: A portion of sales earnings that a salesperson receives. **[8.4]**

Commutative property of addition: $a + b = b + a$, where a and b are any numbers. **[1.2]**

Commutative property of multiplication: $a \cdot b = b \cdot a$, where a and b are any numbers. **[1.3]**

Complementary angles: Two angles whose sum is 90°. **[4.5]**

Complex fraction: An expression that is a fraction with fractions in the numerator and/or denominator. **[5.4]**

Composite form: A geometric form that combines basic forms. **[6.5]**

Composite number: A natural number that has factors other than 1 and itself. **[3.6]**

Compound interest: Interest that is calculated based on principal and prior earned interest. **[8.6]**

Congruent angles: Angles that have the same measurement.   **[4.5]**

Conjugates: Binomials that differ only in the sign separating the terms.   **[3.5]**

Constant: Any symbol that does not vary in value. **[1.2]**

Coordinate: A number in an ordered pair that refers to the position on a number line in a coordinate system.   **[9.3]**

Cost: Money spent on production, operation, labor, and debts.   **[2.3]**

Cubic unit: A 1 × 1 × 1 cube.   **[1.6]**

Current: A measure of electricity moving through a wire; measured in amperes, or amps, A.   **[1.2, 2.6]**

Decimal notation: A base-10 notation for expressing fractions.   **[6.1]**

Deduction: An amount that is subtracted from income.   **[8.5]**

Degree: (monomial) The sum of the exponents on all variables in a monomial; (polynomial) the greatest degree of all the terms that make up the polynomial.   **[3.2]**

Denominator: The number written in the bottom position in a fraction.   **[1.5, 5.1]**

Diameter: The distance across a circle along a straight line through the center.   **[5.3]**

Difference: The answer in a subtraction problem. **[1.2]**

Digit: A symbol used to represent a number.   **[1.1]**

Distributive property: $a(b + c) = ab + ac$ and $a(b - c) = ab - ac$, where a, b, and c are any numbers.   **[1.3]**

Dividend: The number to be divided in a division problem. The dividend is written to the left of a division sign. In fraction form, the dividend is written in the numerator (top).   **[1.4]**

Division: Repeated subtraction of the same number. **[1.4]**

Divisor: The number that divides in a division problem. When the division sign is used, the divisor follows the division sign. In fraction form the divisor is written in the denominator (bottom).   **[1.4]**

Ellipsis: Three periods that mean that the pattern continues forever.   **[1.1]**

Equation: A mathematical statement that contains an equal sign.   **[1.1, 3.1]**

Equilateral triangle: A triangle with all three sides of equal length.   **[4.5]**

Equivalent fractions: Fractions that name the same number.   **[5.1]**

Estimate: A quick approximation.   **[1.2]**

Expanded form: A number written as a sum of all digits multiplied by their place value.   **[1.1]**

Expanded notation: see expanded form.   **[1.1]**

Exponent: A symbol written to the upper right of a base number that indicates how many times to use the base as a factor.   **[1.3]**

Exponential form: A number with an exponent. **[1.3]**

Expression: A collection of constants, variables, and operations.   **[3.1]**

Factored form: A number or expression written as a product of factors.   **[3.7]**

Factorization: A number written as a product of factors.   **[3.6]**

Factors: Numbers that are multiplied.   **[1.3]**

FOIL: A mnemonic used to describe the order in which two binomials are multiplied: **F**irst **O**uter **I**nner **L**ast.   **[3.5]**

Force: Something that accelerates an object (a push or pull); measured in newtons, N or pounds, lb.   **[2.6]**

Formula: An equation that describes a procedure. **[1.3]**

Fraction: A number that describes a part of a whole. **[5.1]**

Front-end ratio: The ratio of the total monthly house payment to gross monthly income.   **[7.6]**

GCF: (see greatest common factor)

Grade point: A numerical value assigned to a letter grade.   **[9.1]**

Grade point average (GPA): The sum of the total grade points earned divided by the total number of credit hours taken.   **[9.1]**

Greatest common divisor: see greatest common factor **[3.6]**

Greatest common factor (GCF): The largest number that divides all given numbers with no remainder.   **[3.6]**

Hypotenuse: The side directly across from the 90° angle in a right triangle.   **[6.6]**

Improper fraction: A fraction in which the absolute value of the numerator is greater than or equal to the absolute value of the denominator.   **[5.1]**

Indeterminate: An expression whose value cannot be determined, for example $\frac{0}{0}$ or 0^0.   **[1.4, 3.1]**

Inequality: A mathematical statement that contains an inequality symbol.   **[1.1]**

Installment: A partial payment made to pay off a loan. **[8.6]**

Integers: A set of numbers that contains all whole numbers and the negative counting numbers. ... $-3, -2, -1, 0, 1, 2, 3, ...$ **[2.1]**

Interest: An amount of money that is a percent of the principal. **[8.6]**

Interest rate: A percent used to calculate interest. **[8.6]**

Inverse operations: Operations that undo each other. **[1.2]**

Irrational number: A number that cannot be expressed in the form $\frac{a}{b}$, where a and b are integers and $b \neq 0$. **[5.3, 6.4]**

Isosceles triangle: A triangle with two sides of equal length. **[4.5]**

Key words: Specific words in a sentence that translate to mathematical operations. **[1.2]**

LCD: (see least common denominator)

LCM: (see least common multiple)

Least common denominator (LCD): The LCM of the denominators. **[5.5]**

Least common multiple (LCM): The smallest natural number that is divisible by all the given numbers. **[5.5]**

Legs: The sides that form the 90° angle in a right triangle. **[6.6]**

Like terms: Monomials that have the same variables raised to the same exponents. **[3.2]**

Linear equation: An equation that is made of polynomials or monomials that are at most degree 1. **[4.2]**

Loss: A negative net (or when revenue is less than cost). **[2.3]**

Lowest terms: A fraction is in lowest terms when the greatest common factor of its numerator and denominator is 1. **[5.2]**

Mass: A measure of how much material or matter (atoms and molecules) makes up an object: measured in slugs or grams. **[2.6]**

Mean: The sum of all given numbers divided by the number of numbers (see arithmetic average). **[9.1]**

Median: The middle score in an ordered set of scores. **[9.1]**

Midpoint of two points: The point that lies halfway between the two points on a straight line. **[9.3]**

Minuend: The first number in a subtraction problem. **[1.2]**

Mixed number: An integer combined with a fraction. **[5.1]**

Mode: The score that occurs most often in a set of scores. **[9.1]**

Monomial: An algebraic expression that is a constant or a product of a constant and variables that are raised to whole-number powers. **[3.2]**

Multiple: A number that is divisible by a given number. **[5.1]**

Multiplication: Repeated addition of the same number. **[1.3]**

Multiplication/division principle of equality: We can multiply or divide both sides of an equation by the same non-zero amount without affecting its solution(s). **[4.3]**

Multiplicative inverses: Numbers whose product is 1. (see reciprocals) **[5.4]**

Multivariable polynomial: A polynomial with more that one variable. **[3.2]**

Natural numbers: 1, 2, 3, ... **[1.1]**

Net: Money remaining after subtracting costs from revenue (or money made minus money spent). **[2.3]**

Net pay: Pay left after all deductions are subtracted for the gross pay. **[8.5]**

Node: A wire connection in a circuit. **[1.2]**

Nonlinear equation: An equation that is not linear. **[4.2]**

Numbers: Amounts or quantities. **[1.1]**

Numeral: see digit **[1.1]**

Numerator: The number written in the top position of a fraction. **[1.5, 5.1]**

Numeric equation: An equation that contains only constants. **[3.1]**

Numeric expression: An expression that contains only constants. **[3.1]**

Ordered pair: A pair of numbers where the order matters, such as those describing a point in a coordinate system. **[9.3]**

Origin: The point where axes of a coordinate plane intersect. **[9.3]**

Parallel lines: Lines that never intersect. **[1.6]**

Parallelogram: A four-sided figure with two pairs of parallel sides. **[1.6]**

Percent: Ratio representing some part out of 100. **[8.1]**

Perfect square: A number that has a whole number square root. **[1.4]**

Perimeter: The total distance around a shape. **[1.2]**

Period: A group of places in the place value system, for example "billions period." **[1.1]**

Pi (π): An irrational number that is the ratio of the circumference of a circle to its diameter, usually approximated as $\frac{22}{7}$ or 3.14. **[5.3]**

Place value system: A system that uses a combination of numerals to represent the numbers greater than 9. **[1.1]**

Polynomial: A monomial or an expression that can be written as a sum of monomials. **[3.2]**

Polynomial in one variable: A polynomial with only one variable. **[3.2]**

Power: see exponent **[1.3]**

Prime factorization: A number written as a product of only prime factors. **[3.6]**

Prime number: A natural number other than 1 that has exactly two different factors, 1 and the number itself. **[3.6]**

Principal: An initial amount of money. **[8.6]**

Principal square root: The positive square root. **[2.4]**

Product: The answer in a multiplication problem. **[1.3]**

Profit: A positive net (when revenue is greater than cost). **[2.3]**

Proportion: An equation in form $\frac{a}{b} = \frac{c}{d}$, where $b \neq 0$ and $d \neq 0$. **[7.2]**

Proportional ratios: Ratios that name the same number and have equal cross products. **[7.2]**

Pythagorean theorem: The sum of the areas of the squares of the legs is the same as the area of the square on the hypotenuse; $a^2 + b^2 = c^2$, where a and b are the lengths of the legs of a right triangle, and c is the length of the hypotenuse. **[6.6]**

Quadrant: One of four regions created by the intersection of the axes in the coordinate plane. **[9.3]**

Quotient: The answer in a division problem. **[1.4]**

Radical sign: $\sqrt{}$ **[1.4]**

Radicand: The number under the radical sign. **[1.4]**

Radius: The distance from the center to any point on the circle. **[5.3]**

Rate: A unit ratio comparing two different measurements. **[7.1]**

Ratio: A comparison between two quantities using a quotient. **[7.1]**

Rational expression: A fraction that is a ratio of monomials or polynomials. **[5.2]**

Rational number: A number that can be expressed in the form $\frac{a}{b}$, where a and b are integers and $b \neq 0$. **[5.1, 6.1]**

Real numbers: The set of all rational and irrational numbers. **[6.4]**

Reciprocals: Two numbers whose product is 1. (see multiplicative inverses) **[5.4]**

Rectangular array: A rectangle formed by a pattern of neatly arranged rows and columns. **[1.3]**

Rectangular coordinate system: Two perpendicular axes that describe the position of a point in a plane (see Cartesian coordinate system). **[9.3]**

Reduce: To create an equivalent fraction by dividing both the numerator and denominator by the same number. **[5.1]**

Related sentence: A mathematical equation that relates the same pieces of a given equation using the inverse operation. **[1.2]**

Remainder: The amount left over after dividing two whole numbers. **[1.4]**

Resistance: The resistance of a wire to the flow of electricity through it: measured in ohms, Ω **[2.6]**

Revenue: Income (money made). **[2.3]**

Right angle: An angle that measures 90°. **[1.6]**

Right triangle: A triangle that has one right angle. **[6.6]**

Scientific notation: A notation composed of a decimal number whose absolute value is greater than or equal to 1, but less than 10, multiplied by 10 raised to an integer exponent. **[6.3]**

Set: A group of elements. **[1.1]**

Similar figures: Figures that have congruent angles and proportional side lengths. **[7.2]**

Simple interest: Interest calculated using only the original principal and the amount of time the principal earns interest. **[8.6]**

Simple percent sentence: A **percent** of a **whole amount** is a **part** of the whole amount. **[8.2]**

Simplest form: An equivalent expression written with the fewest symbols and smallest numbers possible. **[3.2, 5.1]**

Simplify: To write an equivalent expression with fewer symbols or smaller numbers. **[5.1]**

Solution: A number that makes an equation true when it replaces the variable(s) in the equation. **[1.2, 4.1, 9.4]**

Solve: To find the solution or solutions to an equation. **[4.1]**

Square: In geometry, a rectangle with all sides equal in length; in algebra, to multiply a number by itself. **[1.4]**

Square root: The base number that can be squared to make a given number. **[1.4]**

Square unit: A 1×1 square. **[1.3]**

Standard form: A number written using the place value system. **[1.1]**

Statistic: A number used to describe some characteristic of a set of data. **[9.1]**

Subset: A set within a set. **[1.1]**

Subtraction: An operation of arithmetic that can be interpreted as 1. take away, 2. difference, or 3. missing addend. **[1.2]**

Subtrahend: The number following a minus sign in a subtraction problem. **[1.2]**

Sum: The answer in an addition problem.   **[1.2]**

Supplementary angles: Two angles whose sum is 180°.   **[4.5]**

Surface area: The total number of square units that completely cover the outer shell of an object.   **[3.8]**

Term: see monomial   **[3.2]**

Theoretical probability of equally likely outcomes: The ratio of the number of favorable outcomes to the total number of possible outcomes.   **[7.1]**

Trapezoid: A four-sided figure with one pair of parallel sides.   **[5.7]**

Trinomial: A polynomial that has exactly three terms. **[3.2]**

Undefined: When no numeric answer exists, as in $n \div 0$ when $n \neq 0$.   **[1.4, 3.1]**

Unit fraction: A fraction with a value equivalent to 1. **[7.3]**

Unit price: The price for each unit of an item.   **[6.3]**

Unit ratio: A ratio in which the denominator is 1. **[7.1]**

Up-scale: To create an equivalent fraction by multiplying both numerator and denominator by the same non-zero number.   **[5.1]**

Variable: A symbol that can vary or change in value. **[1.2]**

Vertex: A point where two lines join to form an angle. **[9.5]**

Voltage: The electrical pressure created by the current: measured in volts, V.   **[2.6]**

Volume: A total number of cubic units that completely fill an object.   **[1.6]**

Weight: Force due to gravity trying to accelerate the object toward the ground: measured in pounds, lb., or newtons, N.   **[2.6]**

Weighted average: A mean that takes into account that some scores in the data set have more weight than others.   **[9.1]**

Whole numbers: 0, 1, 2, 3, ...   **[1.1]**

x-intercept: A point where a graph intersects the x-axis.   **[9.4]**

y-intercept: A point where a graph intersects the y-axis.   **[9.4]**

INDEX OF APPLICATIONS

Furniture dimensions, 393, 536
Housing development, 395, 686
Ladders and stairs, 463, 470, 475, 482, 485, 496
Landscaping, 45, 47, 65, 67, 148, 199, 257, 275, 280, 282, 326, 338, 429, 503, 508, 549
Mining, 102, 109
Oil exploration, 123
Paving, 65, 70
Piping and tubing, 393, 489, 548
Pyramids, 71, 368, 427, 453, 455, 510
Roofing, 269, 275, 463, 488, 496, 548
Room dimensions, 21, 68, 186, 199, 407, 515, 588
Security and surveillance, 275, 282, 701
Siding, 71
Structural supports, 376, 398, 429, 462, 463, 469, 470, 703
Tools, 536, 703
Water tower, 454
Windows and doors, 253, 270, 282, 284, 485
Wiring, 48, 395, 462

Consumer

Comparison shopping, 494, 553, 555, 559, 703
Candies, 498
Cell phones, 461
Clothing, 393, 594, 596
Consumer goods prices, 493, 547, 553, 555, 633
Customers in a store, 34
Food and beverage consumption, 34, 508, 559, 635, 648
Food and beverage servings, 336, 338, 393, 453, 508, 536, 625
Internet use, 460
Long-distance calling, 47, 326, 425, 428, 459, 466, 467, 480, 499, 552, 555, 632, 702
Mail and stamps, 47, 496
Meal menus, 173
Meal spending, 34
Price discounts, 393, 593–601, 620, 622, 703
Purchases, 416, 417, 428, 440, 482, 495, 499
Restaurant tipping, 599, 600
Sales tax, 590, 591, 593, 596, 599–601, 617, 620, 622

Economics

Currency exchange, 277
Gross domestic product (GDP), 5
Gross national debt, 9, 87, 427
Household income, 645
Stock market, 148, 277, 326, 355, 425, 499, 549, 552
Taxes, 283, 424, 428, 601

Education

Answer keys, 35
Bachelor's degrees by field of study, 648
Class scheduling, 83
College enrollment, 689
College funding, 47, 586, 646
College ID cards, 200
College/university programs, 302
Course pass/failure rate, 302, 584
Grade calculations, 629, 635, 636
Grade distribution, 319, 357, 585, 621, 632–634, 681, 682
Grade point average (GPA), 630, 631, 636, 649, 686
Group size, 199
International achievement in science, 642
Issues in a public school, 649
Number of words in a textbook, 427

Salary by field of study, 647
School safety, 583
Student loans, 20, 440, 480
Student-to-faculty ratio, 492, 499, 552, 555, 695
Test scores, 395, 581, 583, 585, 586, 616, 628, 681, 682, 686, 694, 703
Writing instruments, 536

Environment

Craters, 438, 441, 469
Grand Canyon, 510
Lakes and rivers, 469, 510
Mt. Everest, 9, 92
Rainfall, 634
Sahara Desert, 537
South Pole, 537
Temperature, 86, 99, 102, 108, 111, 144, 282, 484, 537, 635, 680, 690
Tides, 140

Finance

Account balances, 80, 83, 87, 94, 95, 102, 104, 105, 109, 111, 112, 130, 144, 148, 412, 413, 415, 416, 480, 679
Bank account interest, 602–605, 607, 612, 613, 617, 620, 622, 625, 681, 703
Bank service charges, 120, 123, 146
Bills and coins, 277, 282, 368, 453, 455, 464, 465, 470, 480, 601, 702
Budgeting, 13, 247, 588, 622, 639, 694
Credit card balances, 86, 95, 96, 98, 102, 111, 123, 130, 144, 231, 245, 247, 282, 468
Credit card interest, 607, 608, 613
Debt or loan balance, 92, 105, 123
Home purchase, 247, 413, 416
Income tax deductions, 248
Installment payments, 120, 123, 144, 231, 437, 440, 459, 468, 480, 482, 609, 611, 614, 618, 620, 622, 704
Investment planning, 47
Loan amortization, 609, 611, 614, 618, 620, 622
Loan and mortgage interest, 603, 605
Loan and mortgage qualifications, 492, 498, 539–541, 545–547, 554, 556
Net (profit or loss) from resale of goods, 111, 130, 131, 137, 146, 148
Net pay, 248, 594–596, 600, 601, 646
Net worth, 98, 102, 286, 397, 701
Paycheck amount, 92
Paycheck deductions, 417, 594–596, 600, 646
Personal spending, 20
PITI payments, 540, 546, 554

Geometry

Angle measures, 277, 379
Area of a circle, 362, 367, 393, 395, 445, 453, 468, 469, 478, 482
Area of a composite form, 63–67, 70, 71, 78, 81, 112, 286, 367, 393, 449, 455, 478, 482, 484, 558, 625, 703
Area of a parallelogram, 56–58, 60–62, 81, 148, 211, 213, 218, 229, 231, 286, 310, 429, 559, 624, 672, 677, 681, 693, 697, 701, 704
Area of a rectangle, 32, 35, 47, 48, 54, 57, 58, 60, 62, 80, 81, 83, 123, 173, 176, 183, 186, 218, 227, 247, 257, 282, 338, 429, 440, 553, 555, 681
Area of a square, 43, 45, 81, 200, 429

Government

Health/Life Science

Hobbies/Entertainment

Household

Labor

Miscellaneous

Firefighting, 470
Food banks, 17
Fund-raising, 17, 379
Identification cards, 83
Jewelry, 286, 453
Liberty Bell weight, 327
Noah's ark, 537
Real estate, 112
Sculpture design, 218
Stationery, 275
Time (clocks, watches), 379, 511
Tower of Pisa, 218

Physics

Cable and rope strength, 99, 100, 133, 137, 144
Center of gravity, 675, 676, 681, 685, 697, 704
Electrical circuits, 18, 21, 80, 134, 135, 139, 144, 258, 270, 286, 440
Free-falling objects, 219, 227, 232, 397, 673, 678
Force on cable or rope, 99, 103
G force, 139
Galileo's gravity experiment, 218
Marbles, 215
Particle accelerator, 327
Speed of light, 9, 427
Speed of sound, 338, 379, 520

Social Sciences

Archaeology, 248
Daily schedule, 587
Education level, 641
Household population, 318
Illegal drugs, 587
Substance abuse, 585
Teenage smoking, 291
Unemployment, 641–644

Sports

Baseball, 470, 587
Basketball, 282, 586, 634
Billiards, 276
Bowling, 455
Cycling, 377, 395, 488, 493, 496, 547, 591
Fitness, 133, 326, 393
Football, 27, 32, 92, 291, 426, 521, 536, 586, 587
Golf, 633, 682
Kite flying, 559
Running, jogging, and marathons, 393, 395, 398, 536, 542, 547, 554, 625
Skydiving, 215, 219
Swimming pool dimensions, 173, 282
Swimming, 380, 469, 536
Tennis, 227
Track and field, 338, 493, 536
Walking and hiking, 380, 393, 428, 554, 625, 702

Statistics/Demographics

Age distribution, 306, 625, 628, 633, 703
Bachelor's degrees by field of study, 648
Birth lengths, 682
Birth weights, 633
Daily living allowance, 587
Education level, 641, 642, 687, 688
Gender ratios, 495, 552
Height distribution, 634
Home state demographics, 498
House prices, 686, 694
Income distribution, 5, 686
Life expectancy, 34
Medical experiments, 318
Population, 9, 427, 497
Product preference, 318
Rented and owned households, 645
Salary of college graduates distribution, 647
Surveys and opinion polls, 306, 309, 319, 326, 355, 357, 379, 393, 485, 580, 583, 585, 587, 620, 640, 646, 648, 696
Unemployment, 641–644
Work-to-home distance, 326

Technology

Binary codes, 35, 80, 148
Computer hardware, 257, 470, 555
Computer memory, 35, 379
Computer microchips, 258, 310
Computer software, 35, 590
Electronics, 245, 266, 429, 437
Speakers and amplifiers, 583
Lasers, 275, 276, 282
Microphones, 284

Transportation

Airplanes, 17, 71, 92, 139, 338, 379, 428, 440, 453
Average driving speed, 136, 140, 144, 287, 338, 493, 499, 552, 555
Busses, 71, 136, 522
Driving distance, 144, 146, 428
Driving time, 257
Maps, 45, 553, 555
Multiple-car situations, 377, 541, 545, 547, 554, 556,
Parking facilities, 34, 81
Reaction time, 253, 380
Road maintenance, 428
Satellite dimensions, 218
Search and rescue, 62
Ships, 140
Street intersection, 34
Submarines, 86, 87, 99, 102, 108, 120, 130, 135, 140, 455
Titanic, 92, 139, 143
Traffic accidents, 380, 520, 637, 638
Traffic lights, 34
Traffic signs, 536
Trains, 135
Travel planning, 257, 368, 503, 508, 620
Tunnels, 92

INDEX

A

Absolute value, 88–89
Absolute zero, 537
Addend(s)
  defined, 10
  missing, 14
Addition
  of decimal numbers, 408–409, 410–411
  defined, 10
  of fractions
    with different denominators, 349
    with same denominator, 346–347
  of integers, 95–100
  key words for, 12
  of mixed numbers, 350–351, 353
  of polynomials, 411
    in one variable, 167–168
  properties of, 10–11
  of rational expressions, 350
    with same denominator, 348–349
  of whole numbers, 10–13
Addition/subtraction principle of equality, 239–245, 412, 456
Additive inverses, 89–91, 328
Algebra, origin of word, 245
Al-Khowarizmi, Mohammed ibn Musa, 245
American measurement, 512–518
  converting between metric system and, 531–535
  units
    of area in, 514–515
    of capacity in, 515–516
    of length in, 512–513
    of speed in, 517–518
    of time in, 517
    of weight in, 516
Amortization
  defined, 609
  solving problems involving, 608–611
Angle(s)
  complementary, 269
  congruent, 268
  right, 56
  supplementary, 269
  symbol for, 268
Annual percentage rate (APR), 605
Applications, solving. *See* Solving application(s)
APR (annual percentage rate), 605
Archimedes, 450
Area
  of circle, 362–363
  of composite form, 449–450
  defined, 31
  of figure, calculating, using coordinates of its vertices, 677
  measurement of, 31, 32, 57
    American units of, 514–515
    metric units of, 526–528
  problem solving and, 63–67, 212–213

of rectangle, expression for, 183
surface, 213–214
total, 63–67
of trapezoid, 361, 445–446
of triangle, 319–320, 445
Arithmetic average, 628
Arithmetic mean, 628
Associative property
  of addition, 11
  of multiplication, 23
Average
  arithmetic, 628
  grade point (GPA), 629–631
Average rate
  defined, 135
  solving problems involving, 135–136
Axis, 641

B

Back-end ratio, 498, 538–539, 540–541
Balance technique, 239–242
Bar graphs, problem solving using, 640–642
Base, 29
Base unit, 521
Base-10 notation, 400
Base-10 system, 2, 521
Binary code, 28
Binary digits (bits), 28
Binomial, 159
Bits (binary digits), 28
Brahmagupta, 86

C

Calculation, 58
Calculator. *See* Scientific calculator
Capacity
  American units of, 515–516
  converting between American and metric systems, 532
  defined, 515
  metric units of, 524–525
Cartesian coordinate system, 651–655
Cavendish, Henry, 134
Celsius scale, 20
Central tendency, measures of, 628
Centroid
  calculating, 674–676
  defined, 675
Charts, problem solving using, 639–640
Circle(s)
  area of, 362–363
  circumference of, 322–323
  defined, 320
  radius and diameter of, 320–321
Circumference, 322–323
Coefficient
  decimal, 436
  defined, 156

Formula
  defined, 31
  solving for missing number in, 58–59
  use of, 55
Fraction(s)
  comparing, 296–297
  complex, 329–330
  defined, 290
  with different denominators, adding and subtracting, 349
  dividing, 328–330, 335–336
  eliminating from equation using least common denominator (LCD), 369–372
  equivalent. *See* Equivalent fractions
  graphing on a number line, 291–292
  improper. *See* Improper fraction(s)
  key words associated with, 372–373
  multiplying, 311–313, 317–319
  raised to a power, simplifying, 316–317
  reducing, 295
  with same denominator, adding and subtracting, 346–347
  simplifying polynomials containing, 363–365
  solving equations involving, 333
  square root of, finding, 332–333
  unit, 512
  upscaling, 295
  using order of operations to simplify expressions containing mixed numbers and, 359–360
  writing as decimals, 433
  writing as percent, 564–565
  writing percent as, 562–563
Free-body diagram, 100
Front-end ratio, 498, 538–540

G

Galilei, Galileo, 214
Gauss, Carl Friedrich, 450
Geometry, applications in, 442–450
Giga, 528
Grade point average (GPA), 629–631
Graphing
  applications with, 673–677
  of decimals on a number line, 402–403
  of fractions on a number line, 291–292
  of horizontal and vertical lines, 663–664
  of integers, 87–88
  of linear equations, 659–665
Greater than symbol, using to make true statement, 5, 295–297, 403–404
Greatest common factor
  defined, 193
  factoring out of polynomial, 206–208
  finding
    by listing, 192–194
    of set of monomials, 196–197
    using prime factorization, 194–196
Group, working with, 487
Grouping symbols, 49

H

Help, getting, 399
Horizontal lines, graphing, 663–664
Hypotenuse, 461

I

Improper fraction(s)
  defined, 297
  simplifying within mixed numbers, 306–307
  writing as mixed numbers, 297–299
  writing mixed numbers as, 299–300
Increase, percent of, 590–593, 596–598
Inequality, 5
Integer(s), 85–148
  adding, 95–100
  defined, 86
  dividing, 116–117
  graphing, 87–88
  multiplying, 113–115
  subtracting, 104–105, 107–109
Intercept(s)
  defined, 664
  x-, 664–665
  y-, 664–665
Interest
  compound, 605–608
  defined, 602
  simple, 602–605
  solving problems involving, 602–611
Interest rate, 602
Inverse(s), 14
  additive, 89–91, 328
  multiplicative, 328
Irrational number, 322, 434
Isosceles triangle, 267–268

K

Kelvin, William Thomson, 20
Kelvin scale, 20
Key words
  for addition, 12
  associated with fractions, 372–373
  defined, 12
  for division, 43
  for equal sign, 259–260
  for multiplication, 25
  for subtraction, 17
Kilogram (kg), 425

L

LCD. *See* Least common denominator
LCM. *See* Least common multiple
Learning styles, 85
Least common denominator (LCD)
  defined, 341
  using to eliminate fractions from equations, 369–372
  working fractions as equivalent fractions with, 341–342
  writing rational expressions as equivalent expressions with, 343
Least common multiple (LCM)
  defined, 339
  finding, 339–341
Legs, 461
Length
  American units of, 512–513
  converting between American and metric systems and, 531–532
  metric units of, 521–524
Less than symbol, using to make true statement, 5, 295–297, 403–404
Like terms
  combining, 411
  defined, 157
Line graphs, problem solving using, 642–644
Linear equation(s)
  defined, 239
  graphing, 659–665
  solving, 239–242, 673–674

ANSWERS

Exercise Set 1.1

1. 1, 2, 3, . . .   **3.** No, zero is not a natural number.   **5.** 1. Write the name of the digits in the left-most period. 2. Write the period name followed by a comma. 3. Repeat steps 1 and 2 until you get to the ones period. We do not follow the ones period with its name.   **7.** 0   **9.** 5   **11.** Thousands place   **13.** Hundred thousands place
15. $2 \times 10,000 + 4 \times 1000 + 3 \times 100 + 1 \times 10 + 9 \times 1$ **17.** $5 \times 1,000,000 + 2 \times 100,000 + 1 \times 10,000 + 3 \times 1000 + 3 \times 100 + 4 \times 1$ **19.** $9 \times 10,000,000 + 3 \times 1,000,000 + 1 \times 10,000 + 4 \times 1000 + 8 \times 1$ **21.** 8792
23. 6,039,020   **25.** 40,980,109   **27.** seven thousand, seven hundred sixty-eight   **29.** two hundred ninety million, eight hundred ten thousand   **31.** one hundred eighty-six thousand, one hundred seventy-one   **33.** <   **35.** >   **37.** =
39. 5,652,992,000   **41.** 5,653,000,000   **43.** 6,000,000,000   **45.** 5,652,992,500   **47.** 30,000   **49.** 900,000
51. 9000   **53.** 93,000,000 miles

Exercise Set 1.2

1. Changing the order of addends does not affect the sum.   **3.** the total distance around the shape   **5.** Write a related subtraction sentence, subtracting the known addend from the sum.   **7.** Estimate: 9000 Actual: 8849
9. Estimate: 101,000 Actual: 100,268   **11.** Estimate: 17,000 Actual: 16,014   **13.** Estimate: 15,200 Actual: 15,246
15. Estimate: 181,400 Actual: 181,383   **17.** Estimate: 5400 Actual: 5352   **19.** Estimate: 34,200 Actual: 34,161
21. Estimate: 45,000 Actual: 45,334   **23.** Estimate: 20,000 Actual: 26,519   **25.** Estimate: 170,000 Actual: 166,488
27. $x = 4$ **29.** $t = 21$ **31.** $n = 18$ **33.** $u = 155$ **35.** $b = 225$ **37.** 1810 tickets   **39.** 590 ft.   **41.** 2100 K
43. $16   **45.** $6611   **47.** 33 A   **49.** 1720 sq. ft.

Review Exercises:

1. 6   **2.** $3 \times 100,000,000 + 7 \times 1,000,000 + 4 \times 100,000 + 9 \times 10,000 + 1 \times 1000 + 2 \times 10 + 4 \times 1$
3. one million, four hundred seventy-two thousand, three hundred fifty-nine   **4.** <   **5.** 23,410,000

Exercise Set 1.3

1. Changing the factor order will not affect the product.   **3.** $a(b + c) = ab + ac$ **5.** the total number of square units that completely fill the shape   **7.** 352   **9.** 0   **11.** Estimate: 400 Actual: 432   **13.** Estimate: 4500 Actual: 4914
15. Estimate: 80,000 Actual: 93,726   **17.** 44,940   **19.** 995,330   **21.** 8,550,756   **23.** 241,356,618   **25.** 330,050
27. $2 \cdot 2 \cdot 2 \cdot 2 = 16$ **29.** $1 \cdot 1 \cdot 1 \cdot 1 \cdot 1 \cdot 1 = 1$ **31.** $3 \cdot 3 \cdot 3 \cdot 3 \cdot 3 = 243$ **33.** $10 \cdot 10 \cdot 10 \cdot 10 \cdot 10 \cdot 10 \cdot 10 = 10,000,000$ **35.** 42,719,296   **37.** 130,691,232   **39.** 9^4 **41.** 7^5 **43.** 14^6 **45.** $2 \times 10^4 + 4 \times 10^3 + 9 \times 10^2 + 2 \times 1$ **47.** $9 \times 10^6 + 1 \times 10^5 + 2 \times 10^4 + 8 \times 10^3 + 2 \times 10$ **49.** $4 \times 10^8 + 7 \times 10^6 + 2 \times 10^5 + 1 \times 10^4 + 9 \times 10^2 + 2 \times 10 + 5 \times 1$ **51.** 7954 intersections; 63,632 traffic lights   **53.** 36,792,000 heartbeats in a year, 2,722,608,000 heartbeats in a lifetime   **55.** 740 mg   **57.** $35 each week; $1750 each year   **59.** 18,560,000,000 oz.
61. 112 combinations   **63.** 32 binary numbers; No, it cannot recognize $26 + 10 = 36$ binary numbers.
65. 128 locations   **67.** 624 ft.^{2}; No, he has overestimated.   **69.** 300 ft.^{2}

Review Exercises:

1. sixteen million, five hundred seven thousand, three hundred nine   **2.** $2 \times 10,000 + 3 \times 1000 + 5 \times 100 + 6 \times 1$ **3.** 56,988   **4.** 831,724   **5.** $n = 13$

Exercise Set 1.4

1. the dividend   **3.** 1   **5.** a base number that can be squared to equal the given number   **7.** 1; because $26 \cdot 1 = 26$
9. 0; because $49 \cdot 0 = 0$ **11.** Indeterminate, because $x \cdot 0 = 0$ is true for any number   **13.** Undefined, because there is no number that can make $x \cdot 0 = 22$ true   **15.** no   **17.** yes   **19.** yes   **21.** yes   **23.** no   **25.** yes   **27.** yes
29. yes   **31.** no   **33.** 426   **35.** 207 r3   **37.** 217   **39.** 246 r3   **41.** 307 r5   **43.** 230   **45.** 1900
47. undefined   **49.** 1601 r12   **51.** $x = 6$ **53.** $m = 8$ **55.** $t = 0$ **57.** $n = 7$ **59.** $v = 42$ **61.** $h = 20$
63. $2854   **65.** 14 ml   **67.** 125 pieces of paper   **69.** 40 stamps; there are 1500 cents in $15 and $1500 \div 37 = 40$ r20. We round down since we cannot buy part of a stamp.   **71a.** 9600 boxes   **71b.** 1200 bundles
71c. 50 pallets   **71d.** 2 trucks; $50 \div 28 = 1$ r22. Since all the pallets cannot fit on one truck, we round up to 2.
73. 18 ft.   **75.** 22 sections   **77.** 10   **79.** 13   **81.** 0   **83.** 1   **85.** 7   **87.** 14   **89.** 37   **91.** 214
93. 6 in. by 6 in.   **95a.** 144 ft.^{2} **95b.** 12 ft. by 12 ft.

Review Exercises:

1. $t = 177$ **2.** 60 ft.   **3.** Estimate: 14,800 Actual: 15,414   **4.** 625   **5.** $4 \times 10^4 + 9 \times 10^3 + 6 \times 10^2 + 2 \times 1$

Exercise Set 1.5

1. 1. Grouping symbols 2. Exponents 3. Multiply or divide from left to right 4. Add or subtract from left to right
3. Subtract 4 from 10 within the parentheses. We work within grouping symbols before simplifying any other operations.
5. 22   **7.** 45   **9.** 141   **11.** 14   **13.** 1   **15.** 35   **17.** 22   **19.** 49   **21.** 24   **23.** 5   **25.** 25   **27.** 99
29. 81   **31.** 37   **33.** 29   **35.** 83   **37.** 47   **39.** 30   **41.** 51   **43.** 6   **45.** 10   **47.** 13   **49.** 1354
51. Mistake: subtracted before multiplying; Correct: 18.   **53.** Mistake: squared the addends; Correct: 50.

Review Exercises:

1. 70,000   **2.** 104,187,127   **3.** 1250 r3   **4.** 56 ft. of molding; $112   **5.** 192 tiles

Exercise Set 1.6

1. Replace the variables with the corresponding given values, then solve for the unknown variable. **3.** the total number of cubic units that completely fill an object **5.** 94 cm **7.** 128 in. **9.** 64 km **11.** 228 m² **13.** 144 in.² **15.** 112 km² **17.** 280 ft.³ **19.** 216 in.³ **21.** 210 km³ **23.** 9 m **25.** 62 in. **27.** 20 in. **29.** 16 ft. **31.** 18 in. **33.** 4 ft.

Review Exercises:

1. $12,833 **2.** 224 ft.² **3.** $512 **4.** 120 combinations **5.** 300 pieces of paper

Exercise Set 1.7

1. Develop an understanding of the problem. **3.** Add the areas of each shape. **5.** $288 **7.** $28 **9.** $5525 **11a.** 138 posts; $276 **11b.** 4360 ft.; 4 rolls of barbed wire; $180 **11c.** 552 nails; 6 boxes of nails; $18 **11d.** $549. This is considerably less than what the company will charge him. **11e.** He must consider the time it will require, the price of the tools, and whether building a fence is something he knows how to do. **13.** 104 cm² **15.** 70 sheets; $1050 **17.** 648 ft.²; 2 gallons **19.** 14,400 ft.²; $28,800. There are rooms such as bathrooms and closets that do not require carpeting. **21.** 8,400,000 ft.³ **23.** 35,280 ft.³

Review Exercises:

1. $2 \times 10^6 + 4 \times 10^5 + 8 \times 10^3 + 7 \times 10 + 3 \times 1$ **2.** $x = 1929$ **3.** Estimate: 35,000,000 Actual: 32,183,756 **4.** $y = 203$ **5.** 53

Chapter 1 Review Exercises

1. false **2.** true **3.** false **4.** true **5.** true **6.** false **7.** Replace x with 8 and verify that the sum is 17. **8.** Any number multiplied by 0 should be 0, not 14. **9.** Mistake: added before multiplying Correct: 38 **10.** The square of a number is the number multiplied by itself. The square root of a number is the base that can be squared to equal the given number. **11.** $5 \times 1,000,000 + 6 \times 100,000 + 8 \times 10,000 + 9 \times 100 + 1 \times 1$ **12.** $4 \times 10,000 + 2 \times 1000 + 5 \times 100 + 1 \times 10 + 9 \times 1$ **13.** 98,274 **14.** 8,020,096 **15.** 700,928,006 **16.** forty-seven million, six hundred nine thousand, two hundred four **17.** nine thousand, four hundred twenty-one **18.** one hundred twenty-three million, four hundred five thousand, six hundred **19.** < **20.** > **21.** 5,690,000 **22.** 3,000,000 **23.** Estimate: 53,000 Actual: 52,721 **24.** Estimate: 39,680 Actual: 39,673 **25.** Estimate: 480,000 Actual: 485,716 **26.** Estimate: 7700 Actual: 7707 **27.** $x = 25$ **28.** $y = 189$ **29.** $706 **30.** 22 A **31.** Estimate: 1,500,000 Actual: 1,409,437 **32.** Estimate: 600,000 Actual: 403,500 **33.** 128 **34.** 125 **35.** 10^5 **36.** 7^8 **37.** 64 numbers **38.** 288 ft.² **39.** Estimate: 4000 Actual: 4127 **40.** Estimate: 2666 r2 Actual: 3209 r17 **41.** $b = 24$ **42.** $k = 306$ **43.** 2 ml each minute **44.** $2438 **45.** 14 **46.** 15 **47.** 4 **48.** 60 **49.** 84 **50.** 10 **51.** 168 m² **52.** 378 ft.³ **53.** 16 ft. **54.** 15 ft. by 15 ft. **55.** 9000 ft. **56.** 89,500 ft.²

Chapter 1 Practice Test

1. 5 **2.** $4 \times 10^7 + 8 \times 10^6 + 2 \times 10^5 + 1 \times 10^4 + 9 \times 10^2 + 7 \times 1$ **3.** > **4.** sixty-seven million, one hundred ninety-four thousand, two hundred ten **5.** 2,800,000 **6.** 68,190 **7.** 425,709 **8.** $x - 8$ **9.** 246,687 **10.** 64 **11.** 1607 r23 **12.** $y = 18$ **13.** 14 **14.** 34 **15.** 22 **16.** 18 **17.** 1 **18.** 1 r16 **19.** 162 ft.² **20.** $18 **21.** 260 different identification cards **22.** 83 classes **23.** 290 ft.; $1740 **24.** 336 ft.³ **25.** 123 ft.²

Exercise Set 2.1

1. ...−3, −2, −1, 0, 1, 2, 3,... **3.** positive **5.** zero **7.** positive **9.** +450 **11.** +215,000 **13.** −220 **15.** +40 **17.** +100 **19.** +29,028 **21.** −75,243 **23.**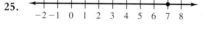

25.  **27.**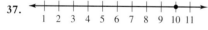

29.  **31.**

33. **35.**

37. **39.** < **41.** > **43.** < **45.** > **47.** > **49.** < **51.** >

53. 26 **55.** 21 **57.** 18 **59.** 10 **61.** 0 **63.** 14 **65.** 18 **67.** 2004 **69.** 8 **71.** 0 **73.** 47 **75.** 377 **77.** 18 **79.** −61 **81.** 0 **83.** −8 **85.** 43 **87.** −6 **89.** 14 **91.** 0 **93.** −63 **95.** 4 **97.** −12 **99.** 87 **101.** the absolute value of negative four; 4 **103.** the additive inverse of negative eight; 8 **105.** the additive inverse of the additive inverse of fourteen; 14 **107.** the additive inverse of the absolute value of negative five; −5 **109.** the additive inverse of the absolute value of the additive inverse of negative four; −4

Review Exercises:

1. 94,265 **2.** 864,488 **3.** 278,313 **4.** 2031 r12 **5.** $482

Exercise Set 2.2

1. Adding a debt of $6 to a debt of $24 increases the debt to $30, so the result is −30.   **3.** add; keep the same sign   **5.** 23
7. −23   **9.** −15   **11.** 46   **13.** −76   **15.** 10   **17.** −14   **19.** 14   **21.** −18   **23.** 56   **25.** 138   **27.** −72
29. 61   **31.** 27   **33.** −21   **35.** −21   **37.** 0   **39.** 34   **41.** 32   **43.** 0   **45.** −59   **47.** −$33
49. −$55,551   **51.** $29   **53.** −78°F   **55.** −78 ft.   **57.** 257 lb.

Review Exercises:

1. $4 \times 10,000,000 + 2 \times 1,000,000 + 5 \times 100,000 + 6 \times 10,000 + 1 \times 1000 + 9 \times 1$ **2.** two million, four hundred
seven thousand, six   **3.** $7265   **4.** 55,445   **5.** $n = 13$

Exercise Set 2.3

1. Change the operation sign from − to + and change the subtrahend (second number) to its additive inverse.   **3.** Answers
will vary, but will have the form of a positive number subtracting a negative number, as in $8 - (-2)$.   **5.** $8 + (-25) = -7$
7. $-15 + (-18) = -33$ **9.** $20 + 8 = 28$ **11.** $-14 + 18 = 4$ **13.** $-15 + 8 = -7$ **15.** $0 + 5 = 5$
17. $-21 + (-19) = -40$ **19.** $-4 + 19 = 15$ **21.** $31 + (-44) - 13$ **23.** $35 + 10 = 45$
25. $-28 + 16 = -12$ **27.** $0 + (-18) = -18$ **29.** $-27 + 16 = -11$ **31.** $t = -6$ **33.** $d = 15$
35. $u = 13$ **37.** $m = 14$ **39.** $h = -5$ **41.** $k = -12$ **43a.** −$119   **43b.** −$139   **45.** Net: $2,222,570; profit
47. net: −$21,958; loss   **49.** 46°F   **51.** 55°C   **53.** $67   **55.** 1396 ft.

Review Exercises:

1. 30,305   **2.** 243   **3.** 4 mi.^{2} **4.** $y = 8$ **5.** 15

Exercise Set 2.4

1. positive   **3.** positive   **5.** two   **7.** Find the additive inverse of the principal square root of n, which is the negative
square root of n.   **9.** −32   **11.** −63   **13.** −8   **15.** −65   **17** 0   **19.** 32   **21.** 0   **23.** −15   **25.** 168
27. −380   **29.** −35   **31.** 54   **33.** 120   **35.** 1   **37.** 9   **39.** 49   **41.** −64   **43.** 81   **45.** 64   **47.** −64
49. −1,000,000   **51.** −1   **53.** −9   **55.** −12   **57.** −3   **59.** 8   **61.** 0   **63.** −31   **65.** −5   **67.** undefined   **69.** 31   **71.** −7   **73.** $x = 3$ **75.** $x = -2$ **77.** $t = -4$ **79.** $m = 9$ **81.** $a = 12$ **83.** no solution
85. $c = -17$ **87.** $m = 0$ **89.** $d = -5$ **91.** $g = 9$ **93.** 9   **95.** 8   **97.** 7   **99.** not an integer   **101.** −11
103. 0   **105.** −$642   **107.** −$119   **109.** $46

Review Exercises:

1. 60 ft.   **2.** 138 ft.^{2} **3.** 504 ft.^{3} **4.** 9   **5.** 40

Exercise Set 2.5

1. 1. Grouping symbols   2. Exponents   3. Multiply or divide from left to right   4. Add or subtract from left to right
3. Subtract 8 from 3 within the parentheses. We work within grouping symbols before simplifying any other operations.
5. −2   **7.** −15   **9.** 6   **11.** −6   **13.** −1   **15.** 10   **17.** 84   **19.** −29   **21.** 7   **23.** 4   **25.** 2   **27.** 16
29. −32   **31.** −9   **33.** 9   **35.** −1   **37.** 21   **39.** 38   **41.** −19   **43.** −60   **45.** 53   **47.** −24   **49.** 0
51. −5   **53.** −32   **55.** −40   **57.** −1   **59.** −22   **61.** 23   **63.** undefined   **65.** −17   **67.** 22   **69.** −64
71. no integer solution   **73.** −4   **75.** −5   **77.** −1   **79.** 3   **81.** −2   **83.** undefined
85. $(-2)^4 = (-2) \cdot (-2) \cdot (-2) \cdot (-2)$ and $-2^4 = -[2 \cdot 2 \cdot 2 \cdot 2]$ **87.** The exponent is odd.   **89.** Mistake: subtraction was performed before the multiplication; Correct: 58.   **91.** Mistake: exponent was evaluated before doing the operations within parentheses; Correct: −21.   **93.** Mistake: the brackets were eliminated before performing the operations inside; Correct: 41.   **95.** Mistake: square roots were found before subtracting in the radical; Correct: 12.

Review Exercises:

1. −$23   **2.** −$213; loss   **3.** 73 ft.   **4.** −$705

Exercise Set 2.6

1. N represents the net, R represents revenue, and C represents cost   **3.** F represents force, m represents mass, and a represents acceleration   **5.** d represents distance, r represents rate, and t represents time   **7.** −$15,030; loss   **9.** $4576; profit
11. 2912 lb.   **13.** −1890 N   **15.** No, the force on the cable will be 2720 lb.   **17.** 78 lb.   **19.** 156 lb.   **21.** 55 lb.
23. 165,818 kg   **25.** 800 kg   **27.** 1110 lb.   **29.** −63 V   **31.** −44 A   **33.** 22 ft./sec.   **35.** 51,180 mi.
37. 65 mph

Review Exercises:

1.   **2.** −5   **3.** 15   **4.** $k = -110$ **5.** $m = -4$

Chapter 2 Review Exercises

1. true   **2.** false   **3.** false   **4.** true   **5.** false   **6.** true   **7.** subtract and keep the sign of the number with the greater absolute value   **8.** change the operation sign from minus to plus and change the subtrahend to its additive inverse
9. negative   **10.** negative   **11.** −13,000   **12.** +212   **13a.**

13b.
number line with point at 5, marks from −5 to 5

14a. <   **14b.** >   **14c.** >   **15.** 41   **16.** 16   **17.** −27
18. 17   **19.** −26   **20.** −12   **21.** 9   **22.** −26   **23.** −12   **24.** −7   **25.** −37   **26.** 6   **27.** −5
28. −54   **29.** 60   **30.** 36   **31.** −30   **32.** 81   **33.** −1000   **34.** −6   **35.** 7   **36.** undefined   **37.** not an
integer   **38.** −7   **39.** −12   **40.** −32   **41.** −15   **42.** 43   **43.** 16   **44.** −2   **45.** $x = -41$ **46.** $n = 5$
47. $k = -9$ **48.** $h = 8$ **49.** −$261   **50.** 0 lb.; The block is not moving.   **51.** $719,600; profit   **52.** 31°F
53. $70   **54.** −576 lb.   **55.** 36V   **56.** $178   **57.** Yes, because the weight of the boulder is 2400 lb.   **58.** 195 mi.
59. 60 mph

Chapter 2 Practice Test

1.
number line with points at −4 and 8, marks from −5 to 8

2. 26   **3.** −18   **4.** −12   **5.** −45   **6.** −14   **7.** −35

8. −16   **9.** $k = 44$ **10.** −$452   **11.** −108   **12.** −84   **13.** 4   **14.** $n = -9$ **15.** −64   **16.** −4

17. 9 and −9 (or $\pm$ 9)   **18.** −$70   **19.** −$1742; loss   **20.** 138 kg   **21.** 124 mi.   **22.** −2

23. −1   **24.** 5   **25.** 0   **26.** undefined   **27.** −30

Chapters 1–2 Cumulative Review

1. true   **2.** false   **3.** false   **4.** true   **5.** false   **6.** true   **7.** 1. Grouping symbols   2. Exponents   3. Multiplication or division from left to right.   4. Addition or subtraction from left to right.   **8.** Mistake: square roots were found before adding in the radical; Correct: −23   **9.** associative; multiplication   **10.** Addends can exchange places: 2 + 3 = 3 + 2.
11. $5 \times 1,000,000 + 6 \times 100,000 + 8 \times 10,000 + 9 \times 100 + 1 \times 1$ **12.** 50,836,009

13. four hundred nine million, two hundred fifty-four thousand, six   **14.**
number line with point at −3, marks from −5 to 5

15. >   **16.** >   **17.** −23,000,000   **18.** 57,000   **19.** 16   **20.** 5   **21.** −8   **22.** −4   **23.** 93   **24.** −43
25. −50   **26.** −84   **27.** 96   **28.** −144   **29.** 243   **30.** 64   **31.** −64   **32.** 201 r2   **33.** −15   **34.** 11
35. −19   **36.** −48   **37.** $x = 25$ **38.** $y = -37$ **39.** $a = 102$ **40.** $c = -7$ **41.** perimeter: 74 m; area: 300 m^2
42. 36 in.^{3} **43.** $27   **44.** −$15,210; loss   **45.** 21,325 ft.^{2} **46.** 128 numbers   **47.** 5 ml   **48.** 8 hr.
49. 1875 slugs   **50.** $5250

Exercise Set 3.1

1. An equation has an equal sign, whereas an expression does not.   **3.** when the divisor is 0 with a nonzero dividend
5. equation   **7.** expression   **9.** expression   **11.** equation   **13.** equation   **15.** 7   **17.** −7   **19.** −1   **21.** 14
23. −18   **25.** 7   **27.** −16   **29.** 11   **31.** −3   **33.** −39   **35.** 40   **37.** 6   **39.** −6   **41.** 10   **43.** 17
45. 13   **47.** 5   **49.** 7   **51.** −2   **53.** If $x = 10$, the expression is undefined.   **55.** If $y = 0$, the expression is undefined.   **57.** If $a = -4$ or $a = 7$, the expression is undefined.

Review Exercises:

1. $3 + (-7) + (-9)$ **2.** 2   **3.** 300,000   **4.** $2 \times 100 + 7 \times 10 + 9 \times 1$ **5.** 1, because 7 raised to the first power is 7.   **6.** 64

Exercise Set 3.2

1. an expression that is a constant or a product of a constant and variables, each raised to whole number powers; Examples will vary, but three possibilities are $12, 3x,$ and $-7y^3$.   **3.** the sum of the exponents of the variables   **5.** an expression that is a sum of monomials; Examples will vary, but one example is $3x^2 + x - 9$.   **7.** the greatest degree of all its terms
9. monomial; product of a constant with variables   **11.** not a monomial; not a product   **13.** not a monomial; not a product   **15.** monomial; a product of a constant with variables   **17.** monomial; a product of a constant −1 and variables
19. monomial; a constant   **21.** coefficient: 3; degree: 8   **23.** coefficient: −9; degree: 1   **25.** coefficient: 8; degree: 0
27. coefficient: 1; degree: 3   **29.** coefficient: −1; degree: 6   **31.** coefficient: −1; degree: 0   **33.** like   **35.** not like
37. like   **39.** like   **41.** not like   **43.** like   **45.** first term: $5x^2$, coefficient: 5; second term: $8x$, coefficient: 8; third term: -7, coefficient: -7 **47.** first term: $6t$, coefficient: 6; second term: -1, coefficient: -1 **49.** first term: $-6x^3$, coefficient: -6; second term: x^2, coefficient: 1; third term: $-9x$, coefficient: -9; fourth term: 4, coefficient: 4   **51.** binomial
53. monomial   **55.** none of these   **57.** trinomial   **59.** monomial   **61.** binomial   **63.** 3   **65.** 6   **67.** 12
69. $14t^6 - 8t^4 + 9t^3 + 5t^2 - 1$ **71.** $y^5 - 18y^3 - 10y^2 + 12y + 9$ **73.** $9a^5 + 7a^3 + 2a^2 - a - 6$

Review Exercises:

1. −2   **2.** −16   **3.** −7   **4.** −26   **5.** No, because subtraction is not commutative.   **6.** Yes, because addition is commutative.   **7.** Yes, because multiplication is commutative.   **8.** 28 m

Exercise Set 3.3

1. Add or subtract the coefficients and keep the variables and their exponents the same.   **3.** $12x$ **5.** $4y$ **7.** $5m$
9. $-5n$ **11.** $2x$ **13.** $-16a$ **15.** $-3t$ **17.** $-7q$ **19.** $13a^2$ **21.** $-7y^2$ **23.** $2j^3$ **25.** 0   **27.** $6y^3$
29. $-8m^5$ **31.** $11x^2 + 2x + 4$ **33.** $-13a^3 + 6a + 4$ **35.** $-5c^2 - 4c$ **37.** $-7n^2 - 2n + 7$ **39.** $7m^3 + m^2 - 4$
41. $8x^4 - 3x^2 + 8$ **43.** $-7t^5 + 6t^4 + 9t^3 - 4t - 2$ **45.** $-13y^4 - 8y^3 - 9y^2 + 18y$

Review Exercises:
1. $4 \times 100 + 7 \times 10 + 2 \times 1$ **2.** 18 ft.   **3.** 6   **4.** 541   **5.** -112 **6.** 341   **7.** distributive property

Exercise Set 3.4
1. Combine all the like terms.   **3.** $5x + 6$ **5.** $9y - 5$ **7.** $13x - 2$ **9.** $5t + 2$ **11.** $2x^2 + 6x + 10$
13. $15n^2 - 12n + 3$ **15.** $7p^2 - 3p - 4$ **17.** $-8b^3 - b^2 - 9b + 11$ **19.** $3a^3 + 5a^2 + a - 2$ **21.** $2x^3 + 3x^2 - 2$
23. $9t^3 - 5t^2 + t + 4$ **25.** $5m^4 + 3m^3 + 6m^2 + m - 3$ **27.** $-x^5 - 9x^4 - 6x^3 - 9x^2 + 16$ **29.** $4x + 7$
31. $8a + 4$ **33.** Mistake: combining $4y^2$ with $-2y^2$ does not equal $-2y^2$; Correct: $10y^3 + 2y^2 - 6y$ **35.** Mistake: unlike
terms were combined;   Correct: $8x^6 + x^5 - 9x^4 + 7x^3 - 4$ **37.** $2x + 7$ **39.** $3t$ **41.** $8n - 6$ **43.** $3x - 7$
45. $6x^2 - x - 11$ **47.** $12a^2 - 11a - 6$ **49.** $-4x - 10$ **51.** $4m^2 - 18m + 4$ **53.** Mistake: did not change
all signs in the subtrahend;   Correct: $2x^2 - 2x$ **55.** Mistake: adding $-9t^2$ to $-9t^2$ does not equal 0;   Correct:
$-5t^3 - 18t^2 + 4t + 5$

Review Exercises:
1. -682 **2.** 81   **3.** -32 **4.** 128   **5.** 375 m^2 **6.** 20

Exercise Set 3.5
1. add   **3.** multiply   **5.** Use the distributive property.   **7.** binomials that differ only in the sign separating the terms;
Examples will vary, but one pair is $x + 5$ and $x - 5$.   **9.** x^7 **11.** t^{10} **13.** $21a^3$ **15.** $10n^7$ **17.** $-24u^6$ **19.** $8y^7$
21. $90x^7$ **23.** $-36m^4$ **25.** $60y^9$ **27.** Mistake: multiplied the exponents; Correct: $45x^7$ **29.** $4x^6$ **31.** $49m^{10}$
33. $-32y^{30}$ **35.** $256x^{20}$ **37.** $25y^{12}$ **39.** $-125v^{18}$ **41.** $48x^{13}$ **43.** Mistake: did not raise 3 to the 4th power, added
exponents of x; Correct: $81x^{16}$ **45.** $8x + 12$ **47.** $15x - 35$ **49.** $-15t - 6$ **51.** $-28a + 36$ **53.** $18u^2 + 24u$
55. $-6a^2 - 2a$ **57.** $-10x^4 + 16x^3$ **59.** $16x^3 + 24x^2 - 32x$ **61.** $-5x^4 + 6x^3 - 9x^2$ **63.** $-6p^4 - 8p^3 + 10p^2$
65. $x^2 + 6x + 8$ **67.** $m^2 + 2m - 15$ **69.** $y^2 - 7y - 8$ **71.** $12x^2 - 7x - 10$ **73.** $12x^2 - 23x + 5$
75. $12t^2 - 17t - 5$ **77.** $2a^2 - 19a + 35$ **79.** $x + 7$ **81.** $2x - 5$ **83.** $-2 - 8b$ **85.** $9 + n$ **87.** $x^2 - 9$
89. $25t^2 - 36$ **91.** $4m^2 - n^2$ **93.** Mistake: did not square the initial x, and 5 should be negative;   Correct: $6x^2 - 13x - 5$
95. $2x^3 + 7x^2 - 14x + 5$ **97.** $8y^3 - 10y^2 - 15y + 18$ **99.** $2a^3 + 2a^2 - 4a$ **101.** $-2x^4 - 5x^3 - 3x^2$
103. $-27q^4 + 3q^2$ **105.** $24y^2 + 8y$ **107.** $3x^3 + 3x^2 - 6x$

Review Exercises:
1. 423   **2.** -4 **3.** 11 ft.   **4.** 5 ft.   **5.** $-4x$

Exercise Set 3.6
1. a natural number other than 1 that has exactly two factors, 1 and the number; Examples will vary, but three possibilities are 2, 3,
and 5.   **3.** the greatest number that evenly divides all the given numbers   **5.** composite   **7.** prime   **9.** neither
11. prime   **13.** neither   **15.** composite   **17.** composite   **19.** prime   **21.** No, 2 is an even prime number.
23. $2^4 \cdot 5$ **25.** $2^2 \cdot 3 \cdot 13$ **27.** $2^2 \cdot 67$ **29.** $2^3 \cdot 5^2$ **31.** 7^3 **33.** $3 \cdot 5^2 \cdot 13$ **35.** $2 \cdot 3^3 \cdot 7$ **37.** $2^3 \cdot 7 \cdot 17$
39. 1, 2, 3, 4, 5, 6, 10, 12, 15, 20, 30, 60   **41.** 1, 3, 9, 27, 81   **43.** 1, 2, 3, 4, 5, 6, 8, 10, 12, 15, 20, 24, 30, 40, 60, 120
45. 24: 1, 2, 3, 4, 6, 8, 12, 24;   60: 1, 2, 3, 4, 5, 6, 10, 12, 15, 20, 30, 60;   GCF = 12   **47.** 81: 1, 3, 9, 27, 81;   65: 1, 5, 13, 65
GCF = 1   **49.** 72: 1, 2, 3, 4, 6, 8, 9, 12, 18, 24, 36, 72;   120: 1, 2, 3, 4, 5, 6, 8, 10, 12, 15, 20, 24, 30, 40, 60, 120;   GCF = 24
51. $140 = 2^2 \cdot 5 \cdot 7$; $196 = 2^2 \cdot 7^2$;   GCF $= 2^2 \cdot 7 = 28$ **53.** $130 = 2 \cdot 5 \cdot 13$; $78 = 2 \cdot 3 \cdot 13$;
GCF $= 2 \cdot 13 = 26$ **55.** $336 = 2^4 \cdot 3 \cdot 7$; $504 = 2^3 \cdot 3^2 \cdot 7$;   GCF $= 2^3 \cdot 3 \cdot 7 = 168$ **57.** $99 = 3^2 \cdot 11$;
$140 = 2^2 \cdot 5 \cdot 7$;   GCF = 1   **59.** $60 = 2^2 \cdot 3 \cdot 5$; $120 = 2^3 \cdot 3 \cdot 5$; $140 = 2^2 \cdot 5 \cdot 7$;   GCF $= 2^2 \cdot 5 = 20$
61. $64 = 2^6$; $160 = 2^5 \cdot 5$; $224 = 2^5 \cdot 7$;   GCF $= 2^5 = 32$ **63.** 10 in. by 10 in.   **65.** 8 ft.; 5 in 40-ft. trench;
4 in 32-ft. trench; and 3 in 24-ft. trench   **67.** 10   **69.** $2x^2$ **71.** h^3 **73.** 1   **75.** $8x^7$ **77.** $6n^2$ **79.** $7x^3$

Review Exercises:
1. $5 \times 1{,}000{,}000 + 7 \times 100{,}000 + 8 \times 10{,}000 + 4 \times 1000 + 2 \times 100 + 9 \times 1$ **2.** 2019 r5   **3.** 128
4. No, there are only 260 possible combinations.   **5.** 6 ft. by 6 ft.

Exercise Set 3.7
1. subtract   **3.** Divide each term in the polynomial by the monomial.   **5.** x^7 **7.** m^6 **9.** 1   **11.** $4a^3$ **13.** $-5t^4$
15. $20n^8$ **17.** $-3x^4$ **19.** $-4b^2$ **21.** -14 **23.** $3a + 2$ **25.** $-7c + 4$ **27.** $3x + 2$ **29.** $-9d + 7$
31. $4a^3 + 6a$ **33.** $4x^3 - 2x^2 + x$ **35.** $-2a^3 + 3a^2 - 4a$ **37.** $5a - 4$ **39.** $-8 - 4c^3 + 10c^7$ **41.** $2a^2$
43. $-7x^4$ **45.** $4u$ **47.** a^5 **49.** $2x + 3$ **51.** $5a - 1$ **53.** $3 - 5z$ **55.** $2t^3 - 3t^2 + 4$ **57.** $5n^2 - 4n - 2$
59. $9m$ **61.** $4t^2$ **63.** $4(2x - 1)$ **65.** $10(a + 2)$ **67.** $2n(n + 3)$ **69.** $x^2(7x - 3)$ **71.** $4r^3(5r^2 - 6)$
73. $3y^2(2y + 1)$ **75.** $4x(3x^2 + 5x + 8)$ **77.** $3a^3(3a^4 - 4a^2 + 6)$ **79.** $7m^5(2m^3 + 4m + 1)$
81. $10x^3(x^6 - 2x^2 - 4)$

Review Exercises:
1. 20 ft.   **2.** 144 m^2 **3.** 108 in.^{3} **4.** $2x^3 - 9x^2 - 17$ **5.** $x^2 - 2x - 24$

Exercise Set 3.8
1. a measure of the total number of square units that cover the outer shell of an object   **3.** The formula is used to calculate
the height of a freefalling object, where h represents the height in feet after falling for t seconds from an initial height of h_0 feet.
5. $4x + 14$ **7.** $3n - 6$ **9.** $8b - 10$ **11a.** 90 in.   **11b.** 122 cm   **13a.** 30 ft.   **13b.** 18 yd.   **15a.** 70 in.
15b. 150 ft.   **17.** $2n^2 + 8n$ **19a.** 120 km^2 **19b.** 384 ft.^{2} **21.** $d^3 + 5d^2 - 6d$ **23a.** 54 in.^{3} **23b.** 220 ft.^{3}

25. 32 ft.^{2} **27.** 1536 panels   **29.** 116 ft. after 2 sec.; 36 ft. after 3 sec.   **31.** 4256 ft.   **33a.** $35r + 75s - 245$
33b. \$11,905 profit   **35a.** $R = 5s + 9m + 15l$ **35b.** $C = 2s + 4m + 7l$ **35c.** $N = 3s + 5m + 8l$ **35d.** \$87 profit

Review Exercises:
1. 15   **2.** -1 **3.** 0   **4.** $-x^3 - 21x^2 + 14$ **5.** 1   **6.** $-12a - 28$

Chapter 3 Review Exercises
1. false   **2.** true   **3.** true   **4.** false   **5.** false   **6.** false   **7.** coefficients; variables   **8.** add; base   **9.** subtract;
divisor's; dividend's   **10.** multiply   **11.** 40   **12.** -13 **13.** 8   **14.** -2 **15.** Yes, it is a product of a constant with
variables raised to whole-number exponents.   **16.** No, it contains subtraction.   **17.** coefficient: 18;   degree: 1
18. coefficient: 1;   degree: 3   **19.** coefficient: -9;   degree: 0   **20.** coefficient: -3;   degree: 6   **21.** They have the
same variables raised to the same exponents.   **22.** The same variables are raised to different exponents.   **23.** binomial
24. none of these   **25.** monomial   **26.** trinomial   **27.** 6   **28.** $5a^6 + 7a^4 - a^2 - 9a + 13$ **29.** $-a^2 + 2a - 1$
30. $7m^3 + 9m - 7$ **31.** $-5x^7 + 3x^2 + 13$ **32.** $3x + 5$ **33.** $4y^4 + 2y^3 - 10y - 4$ **34.** $3a^2 - 5a$
35. $13h^3 - 5h^2 - 5h - 3$ **36.** m^7 **37.** $-10x^5$ **38.** $-12y^5$ **39.** $6t^8$ **40.** $125x^{12}$ **41.** $4y^6$ **42.** $64t^9$
43. $-8a^{12}$ **44.** $6x - 8$ **45.** $-16y^2 + 20y$ **46.** $15n^3 - 3n^2 + 21n$ **47.** $a^2 - 2a - 35$ **48.** $2x^2 - x - 3$
49. $10y^2 - 21y + 8$ **50.** $9t^2 - 16$ **51.** $u^3 - 3u^2 - 7u + 6$ **52.** $-3p^4 + 5p^3 + 2p^2$ **53.** $3y + 4$ **54.** $-7x - 2$
55. composite   **56.** prime   **57.** $2^3 \cdot 3^2 \cdot 5$ **58.** $2^3 \cdot 3 \cdot 5^2 \cdot 7$ **59.** 28   **60.** 1   **61.** $=12x^5$ **62.** $6a^2$ **63.** r^6
64. $-4x^3$ **65.** $5x^4 - 4x^2 + 2$ **66.** 1   **67.** $-9m^5$ **68.** $3 + 4x^2$ **69.** $4xy^4 - 3y^2 - 2$ **70.** $14b$
71. $6(5x + 2)$ **72.** $3n^3(3n - 5)$ **73.** $6x(3x^2 + 4x - 6)$ **74a.** $16x - 12$ **74b.** 132 m   **75a.** $m^2 - 2m - 15$
75b. 33 in.^{2} **76a.** $48n^3 - 32n^2$ **76b.** 2560 ft.^{3} **77.** 86 ft.   **78a.** $43n + 43b - 105$ **78b.** \$31,414 profit

Chapter 3 Practice Test
1. -43 **2.** binomial   **3.** coefficient: -1; degree: 4   **4.** 4   **5.** $-a^4 + 5a^2$ **6.** $5y^3 - 10y - 4$
7. $6y^5 + 5y^3 - y^2 - 4$ **8.** $20u^7$ **9.** $-32a^{15}$ **10.** $-2t^4 + 6t^2 + 14t$ **11.** $v^2 - 36$ **12.** $1u^4 - 2u^3 + 2u^2$
13. $6u + 6$ **14.** composite   **15.** $2^3 \cdot 5 \cdot 17$ **16.** 36   **17.** $12h^5$ **18.** 1   **19.** m^2 **20.** $-8x^4 + 9x^2 - 11x$
21. $5x^3$ **22.** $4(3x - 5)$ **23.** $2y^2(5y^2 - 9y + 7)$ **24a.** $2n^2 - 7n - 4$ **24b.** 45 cm^2 **25a.** $84a + 91b$
25b. \$21,217 profit

Chapter 3 Cumulative Review Exercises
1. false   **2.** true   **3.** false   **4.** true   **5.** true   **6.** false   **7.** Grouping symbols, Exponents, Multiplication or division
from left to right, Addition or subtraction from left to right   **8.** negative   **9.** negative   **10.** Multiply every term in the
second polynomial by every term in the first polynomial (FOIL).   **11.** two million, four hundred eighty thousand, forty-five
12. 700,000,000
13.

14. $-110,000$ **15.** 140,000   **16.** -1 **17.** 1   **18.** 5   **19.** 6
20. -9 **21.** -144 **22.** -38 **23.** 108   **24.** -125 **25.** -81 **26.** -103 **27.** 20   **28.** -15 **29.** 108
30. 8   **31.** $-x^3 + 16x^2 - x - 12$ **32.** $4y^3 - 9y^2 + 21y - 6$ **33.** $63x^4$ **34.** $2b^2 - 13b - 24$ **35.** $2^3 \cdot 3^2 \cdot 5$
36. $10x^2$ **37.** $6n^3$ **38.** $-2x^2 + 4x - 6$ **39.** $6m^2(2m^2 - 3m + 4)$ **40.** $x = 6$ **41.** $x = 8$ **42.** 255 cm^2
43. $16w$ **44.** $5y^3 + 5y^2$ **45.** \$282,600   **46.** 560 combinations   **47.** 10 ml   **48.** $-\$279$ **49.** 548 ft.^{2}
50. 16 ft.   **51a.** $113b + 158a - 345$ **51b.** \$19,711 profit

Exercise Set 4.1
1. An expression has no equal sign, whereas an equation does.   **3.** to find its solution   **5.** equation   **7.** expression
9. expression   **11.** -5 is not a solution.   **13.** -9 is not a solution.   **15.** 3 is not a solution.   **17.** 8 is a solution.
19. 5 is not a solution.   **21.** -3 is a solution.   **23.** -3 is not a solution.   **25.** 0 is a solution.   **27.** 3 is not a solution.
29. -2 is not a solution.

Review Exercises:
1. $5x - 12$ **2.** $-2y - 8$ **3.** $5m + 15$ **4.** $-3m + 4$ **5.** $x = 7$

Exercise Set 4.2
1. an equation made of polynomials that are, at most, degree 1   **3.** The same amount can be added to or subtracted from
both sides of an equation without affecting its solution(s).   **5.** Add that term to both sides of the equation.   **7.** linear
9. not linear   **11.** not linear   **13.** linear   **15.** not linear   **17.** linear   **19.** $n = 6$ **21.** $x = 8$ **23.** $x = 12$
25. $y = -1$ **27.** $x = 7$ **29.** $m = 4$ **31.** $n = -7$ **33.** $y = 11$ **35.** $n = -4$ **37.** $t = -12$ **39.** \$1473
41. 70 cc   **43.** He must sell \$2465 more. Yes, midway through the month he has sold \$8035.   **45.** Length is 10 ft. Total
area of the dining room is 80 ft.^{2}, which is enough for the square table and chairs.   **47.** 26 cm

Review Exercises:
1. -40 **2.** -8 **3.** -9 **4.** -2 **5.** $x = 3$

Exercise Set 4.3
1. Both sides of an equation can be multiplied or divided by the same amount without affecting its solution(s).   **3.** $x = 7$
5. $a = -9$ **7.** $b = -11$ **9.** $a = 11$ **11.** $y = 6$ **13.** $x = 7$ **15.** $b = -7$ **17.** $u = -4$ **19.** $x = 11$
21. $k = -3$ **23.** $h = 4$ **25.** $h = -3$ **27.** $t = -8$ **29.** $x = 2$ **31.** $u = 1$ **33.** Mistake: did not write the
minus sign to the left of 11   Correct: $-1 = x$ **35.** Mistake: did not change minus sign in $x - 8$ Correct: $x = -3$
37. 18 in.   **39.** 7 hr.   **41.** 72 in.   **43.** 4 hr.   **45.** 5 sec.   **47.** 615 chips

Review Exercises:
1. six million, seven hundred eighty-four thousand, two hundred nine   **2.** 1   **3.** 43   **4.** $2^4 \cdot 3 \cdot 5$
5. $6x^3(4x^2 - 5x + 3)$

Exercise Set 4.4
1. Answers will vary; three possibilities are add, sum, plus.   **3.** Answers will vary; three possibilities are multiply, times, product.   **5.** $n + 5 = -7; n = -12$ **7.** $n - 6 = 15; n = 21$ **9.** $x + 17 = -8; x = -25$
11. $-3y = 21; y = -7$ **13.** $9b = -36; b = -4$ **15.** $5x + 4 = 14; x = 2$ **17.** $-6m - 16 = 14; m = -5$
19. $39 - 5x = 8x; x = 3$ **21.** $17 + 4t = 6t - 9; t = 13$ **23.** $2(b - 8) = 5 + 9b; b = -3$
25. $6x + 5(x - 7) = 19 - (x + 6); x = 4$ **27.** $-8(y - 3) - 14 = -2y - (y - 5); y = 1$
29. Mistake: incorrect subtraction order   Correct: $n - 7 = 15$ **31.** Mistake: multiplied 2 times x instead of the sum
Correct: $2(x + 13) = -9$ **33.** Mistake: incorrect subtraction order on the left side of the equation
Correct: $16 - 6n = 2(n - 4)$

Review Exercises:
1. $6xy^5$ **2.** $9b^2(2b^3 - 3b + 6)$ **3.** $8w$ **4.** 48 ft.   **5.** 9 small shrubs; $94

Exercise Set 4.5
1. a triangle with all three sides of equal length   **3.** two angles whose sum is $180°$ **5.** width; $w + 5$ **7.** The numbers are 15 and 19.   **9.** The numbers are 8 and 24.   **11.** 120 ft. long by 30 ft. wide   **13.** 80 ft. long by 30 ft. wide
15. The base is 16 m and the sides are 27 m.   **17.** The length of the rectangle is 4 ft. and its width is 2 ft. The sides of the triangle are 4 ft.   **19.** The angles are $18°$ and $72°$.   **21.** The angles are $33°$ and $147°$.   **23.** The angles are $34°$, $112°$, and $34°$.   **25.** The second angle is $92°$, the third is $45°$.   **27.** The angles are $59°$, $69°$, and $52°$.   **29.** 16 small bottles, 11 large bottles   **31.** 6 $10 bills, 13 $5 bills   **33.** The bills are worth $2 and $6.   **35.** The bills are worth $7 and $9.

Review Exercises:
1. -3 is not a solution.   **2.** No, because it has a degree higher than 1.   **3.** $m = -7$ **4.** $n = -5$ **5.** $x = 3$

Chapter 4 Review Exercises
1. false   **2.** true   **3.** false   **4.** true   **5.** true   **6.** false   **7.** add; subtract; same   **8.** multiply; divide; same
9. 1. Simplify a. Distribute b. Combine 2. addition/subtraction 3. multiplication/division   **10.** Replace the variable(s) in the original equation with the solution and verify that it makes the equation true.   **11.** 5 is not a solution.   **12.** -4 is a solution.   **13.** 1 is a solution.   **14.** -2 is not a solution.   **15.** -3 is a solution.   **16.** 5 is not a solution.
17. 3 is a solution.   **18.** -1 is not a solution.   **19.** $n = 8$ **20.** $y = -2$ **21.** $k = -9$ **22.** $m = 3$
23. $x = -7$ **24.** $h = -9$ **25.** $t = 3$ **26.** $y = -2$ **27.** $m = 1$ **28.** $v = -6$ **29.** $x = 3$ **30.** $y = -5$
31. $43°F$ **32.** $197   **33.** 4 ft.   **34.** 8 ft.   **35.** $15 - 7n = 22; n = -1$ **36.** $5x - 4 = 3x; x = 2$
37. $2(n + 12) = -6n - 8; n = -4$ **38.** $12 - 3(x - 7) = 6x - 3; x = 4$ **39.** The numbers are 15 and 27.
40. The numbers are 6 and 30.   **41.** 24 m wide and 70 m long   **42.** The base is 60 in.; the two equal sides are 98 in.
43. The angles are $65°$ and $115°$.   **44.** The angles are $98°$ and $82°$.   **45.** 8 twenty-dollar bills and 17 ten-dollar bills
46. 19 large bags and 8 small bags

Chapter 4 Practice Test
1. Equation; there is an equal sign.   **2.** Not linear, because the degree is higher than 1.   **3.** -4 is not a solution.
4. 2 is a solution.   **5.** $n = 8$ **6.** $m = -6$ **7.** $y = 4$ **8.** $k = 1$ **9.** $x = 3$ **10.** $t = -21$ **11.** $u = -2$
12. $k = 9$ **13.** $83   **14.** 400 yd.   **15.** The number is 8.   **16.** $x = -6$ **17.** The numbers are 11 and 33.
18. The two equal sides are 105 in. and the base is 48 in.   **19.** The two angles are $105°$ and $75°$.   **20.** 8 model A guitars and 4 model B guitars

Chapters 1–4 Cumulative Review Exercises
1. true   **2.** false   **3.** false   **4.** false   **5.** true   **6.** false   **7.** positive   **8.** add   **9.** subtract
10. Replace the variables with the given values, then calculate.   **11.** $3 \times 10,000 + 6 \times 1000 + 9 \times 10 + 7 \times 1$

12.   **13.** 30,000   **14.** 30   **15.** -8 **16.** 0   **17.** 5   **18.** 19

19. 97   **20.** -26 **21.** -61 **22.** -16 **23.** -22 **24.** -391 **25.** -7 **26.** $8t^3 + 14t^2 - 25t - 7$
27. $6x^3 - 7x^2 - x - 7$ **28.** $-54a^6$ **29.** $9x^2 - 25$ **30.** $2^2 \cdot 3 \cdot 5 \cdot 7$ **31.** $36m^2$ **32.** $-7k$
33. $5n^2(4n^3 + 3mn - 2)$ **34.** $x = 39$ **35.** $t = -6$ **36.** $b = 22$ **37.** $y = 4$ **38.** $n = -7$ **39.** 222 hours
40. -4 A   **41.** $-$67,181   **42.** $50,710   **43.** 216 in.^{2} **44.** $x^2 - 4x - 12$ **45.** 7 ft.   **46.** 5 in.
47. 62 mph   **48.** The numbers are 32 and 64.   **49.** 9 ft. wide and 13 ft. long   **50.** 9 small blankets and 7 large blankets

Exercise Set 5.1
1. a number that can be expressed in the form $\frac{a}{b}$, where a and b are integers and $b \neq 0$.   **3.** undefined   **5.** Divide the denominator into the numerator. Write the results in the form: quotient $\frac{\text{remainder}}{\text{original denominator}}$ **7.** $\frac{1}{3}$ **9.** $\frac{5}{8}$ **11.** $\frac{2}{3}$
13a. $\frac{5}{16}$ **13b.** $\frac{15}{16}$ **13c.** The sense of smell seems to contribute more information about foods than taste.   **13d.** If she is trying the same foods in the same order, she may be guessing better the second time.   **15.** $\frac{17}{800}$ survive, $\frac{783}{800}$ do not survive.

17. $\frac{249}{258}$ passed, $\frac{9}{258}$ did not pass. Very few of Mrs. Jones' students do not pass the course.   **19.**

21.

23.

25.

27. 23   **29.** 0

31. 1   **33.** undefined   **35.** 15   **37.** 7   **39.** 30   **41.** -30 **43.** $>$ **45.** $>$ **47.** $>$ **49.** $=$ **51.** $4\frac{2}{7}$
53. $21\frac{1}{4}$ **55.** $-12\frac{4}{5}$ **57.** $-12\frac{7}{8}$ **59.** $\frac{31}{6}$ **61.** $\frac{11}{1}$ **63.** $-\frac{79}{8}$ **65.** $-\frac{29}{20}$

Review Exercises:
1. 4   **2.** $2^3 \cdot 3 \cdot 5 \cdot 7$ **3.** 12   **4.** $8x^2$ **5.** $8x^2(5x^3 - 7)$

Exercise Set 5.2
1. greatest common factor; 1   **3.** $\frac{5}{6}$ **5.** $\frac{-2}{5}$ **7.** $\frac{1}{2}$ **9.** $-\frac{3}{5}$ **11.** $\frac{3}{4}$ **13.** $\frac{-3}{4}$ **15.** $\frac{6}{7}$ **17.** $\frac{7}{9}$ **19.** $\frac{-11}{17}$
21. $\frac{-3}{4}$ **23.** $\frac{5}{12}$ **25.** $\frac{12}{31}$; 12 out of every 31 women can be expected to use the product. 248 may not be a large enough
sample.   **27.** $\frac{212}{1001}$ **29.** $\frac{16}{77}$ **31.** $3\frac{3}{4}$ **33.** $-3\frac{1}{3}$ **35.** $4\frac{1}{7}$ **37.** $-5\frac{1}{6}$ **39.** $\frac{5x}{16}$ **41.** $\frac{-x^2}{y}$ **43.** $-\frac{2n}{5m^3}$ **45.** $\frac{1}{4t^3u}$
47. $\frac{3ac}{5b}$ **49.** $\frac{-b^4}{2a^6c^5}$

Review Exercises:
1. $-15{,}015$ **2.** 72   **3.** $35x^5$ **4.** $-32x^{15}$ **5.** 99 m^2

Exercise Set 5.3
1. 1. Divide out any numerator factor with any like denominator factor.   2. Multiply numerator by numerator and denominator
by denominator.   3. Simplify.   **3.** The formula is used to calculate the area of a triangle where A represents area, b represents
the base, and h represents the height.   **5.** the distance around a circle   **7.** an irrational number that is the ratio of the circumference of a circle to its diameter; It can be approximated using the fraction $\frac{22}{7}$.   **9.** $\frac{8}{35}$ **11.** $-\frac{1}{54}$ **13.** $\frac{15}{28}$
15. $\frac{49}{1000}$ **17.** $\frac{4}{5}$ **19.** $-\frac{1}{3}$ **21.** $\frac{13}{20}$ **23.** $-\frac{3}{10}$ **25.** $\frac{2}{5}$ **27.** $\frac{21}{50}$ **29.** $\frac{1}{2}$ **31.** Estimate: 6   Actual: $5\frac{3}{5}$
33. Estimate: 28   Actual: $23\frac{1}{3}$ **35.** Estimate: 80   Actual: 74   **37.** Estimate: -3 Actual: $-3\frac{1}{2}$ **39.** Estimate: 42
Actual: $40\frac{4}{5}$ **41.** Estimate: -15 Actual: $-13\frac{1}{2}$ **43.** $\frac{2x^2}{3}$ **45.** $\frac{x^2}{6}$ **47.** $\frac{2xy^4}{45}$ **49.** $-\frac{5k^3}{12}$ **51.** $\frac{10x^2y}{7}$ **53.** $-\frac{3m}{10n^3}$
55. $\frac{25}{36}$ **57.** $-\frac{27}{64}$ **59.** $\frac{1}{64}$ **61.** $\frac{x^3}{8}$ **63.** $\frac{8x^6}{27}$ **65.** $\frac{m^{12}n^4}{81p^8}$ **67.** 1560 pieces   **69.** $\frac{5}{8}$ of the employees   **71.** $\frac{2}{3}$ of her
long-distance calls; 20 calls   **73.** $12\frac{3}{4}$ in.   **75.** $-147\frac{49}{50}$ N   **77.** 51 ft.^{2} **79a.** $4\frac{1}{2}$ m   **79b.** $28\frac{2}{7}$ m   **81a.** $1\frac{53}{220}$ mi.
81b. $3\frac{9}{10}$ mi.

Review Exercises:
1. -204 **2.** 16   **3.** $3x(3x - 4)$ **4.** $x = -5$ **5.** 2 ft.

Exercise Set 5.4
1. two numbers whose product is 1; Examples will vary, but one pair is $\frac{2}{3}$ and $\frac{3}{2}$.   **3.** Write each mixed number as an
improper fraction, then follow the process for dividing fractions.   **5.** $\frac{3}{2}$ **7.** 6   **9.** $-\frac{1}{15}$ **11.** $-\frac{4}{x}$ **13.** 3   **15.** $\frac{9}{25}$
17. $-\frac{8}{5}$ **19.** $\frac{1}{24}$ **21.** $\frac{3}{2}$ **23.** -18 **25.** Estimate: 10;   Actual: $11\frac{5}{8}$ **27.** Estimate: 3;   Actual: $2\frac{8}{15}$ **29.** Estimate: $-\frac{3}{10}$;
Actual: $-\frac{1}{4}$ **31.** Estimate: 2;   Actual: $1\frac{23}{27}$ **33.** $\frac{14x^2}{5}$ **35.** $\frac{15m^3}{16n^2}$ **37.** $-\frac{9x^4}{8y^5}$ **39.** $\frac{8}{9}$ **41.** $\frac{11}{6}$ **43.** 6   **45.** 9
47. $x = 16$ **49.** $y = -3\frac{1}{3}$ **51.** $n = -\frac{9}{5}$ **53.** $m = \frac{2}{7}$ **55.** 30 doses, should be 28 doses; 2 extra doses   **57.** $\frac{7}{8}$ cup
59. $91\frac{60}{61}$ ft.   **61.** $4\frac{6}{13}$ ft.   **63.** $989\frac{109}{110}$ ft.   **65.** $45\frac{3}{4}$ mph

Review Exercises:
1. $2 \cdot 3^3 \cdot 7$ **2.** $60x^3y$ **3.** $=$ **4.** $9t^3 + 9t^2 - 13t - 16$ **5.** $16x(2x + 1)$

Exercise Set 5.5
1. the smallest natural number that is divisible by all the given numbers   **3.** 1. Find the prime factorization of each given
number.   2. Write a factorization that contains each prime factor the greatest number of times it occurs in the factorizations.
3. Multiply to get the LCM.   **5.** 4   **7.** 30   **9.** 36   **11.** 60   **13.** 90   **15.** 72   **17.** 252   **19.** 364   **21.** 1800
23. 3360   **25.** 1680   **27.** $24xy$ **29.** $30y^3$ **31.** $16mn$ **33.** $36x^2y^3$ **35.** $\frac{9}{30}$ and $\frac{25}{30}$ **37.** $\frac{21}{36}$ and $\frac{11}{36}$ **39.** $\frac{3}{60}$ and $\frac{34}{60}$
41. $\frac{27}{36}, \frac{6}{36}$, and $\frac{28}{36}$ **43.** $\frac{7}{12x}$ and $\frac{9x}{12x}$ **45.** $\frac{9}{16mn}$ and $\frac{6n^2}{16mn}$ **47.** $\frac{21}{30y^3z}$ and $\frac{25y^2z^2}{30y^3z}$ **49.** $\frac{2y^2z}{36x^2y^3}$ and $\frac{-15x}{36x^2y^3}$

Review Exercises:
1. -12 **2.** 11   **3.** $4x^3 + 39x^2 - 16x - 3$ **4.** $m = -8$ **5.** $x = 53$ **6.** $a = -\frac{4}{5}$

Exercise Set 5.6
1. 1. Add or subtract numerators and keep the same denominator. 2. Simplify.   **3.** Multiply the numerator and denomina-
tor of $\frac{3}{4}$ by 3 and multiply the numerator and denominator of $\frac{5}{6}$ by 2.   **5.** $\frac{5}{7}$ **7.** $-\frac{2}{3}$ **9.** $\frac{6}{17}$ **11.** $-\frac{1}{5}$ **13.** $\frac{12}{x}$ **15.** $\frac{2x^2}{3}$
17. $-\frac{2m}{n}$ **19.** $\frac{4x^2 - 3x + 1}{7y}$ **21.** $\frac{5n^2 - 3}{5m}$ **23.** $1\frac{2}{15}$ **25.** $\frac{5}{18}$ **27.** $\frac{37}{60}$ **29.** $1\frac{25}{36}$ **31.** $\frac{23x}{24}$ **33.** $\frac{3}{16m}$ **35.** $\frac{13h + 16}{20h}$
37. $\frac{6 - 7n}{9n^2}$ **39.** $10\frac{4}{9}$ **41.** $7\frac{1}{2}$ **43.** $7\frac{1}{2}$ **45.** $10\frac{5}{24}$ **47.** $9\frac{3}{4}$ **49.** $9\frac{1}{6}$ **51.** $3\frac{7}{24}$ **53.** $\frac{1}{2}$ **55.** $4\frac{5}{12}$ **57.** $-10\frac{1}{2}$

59. $-3\frac{5}{8}$ 61. $-6\frac{5}{8}$ 63. $x=\frac{1}{10}$ 65. $n=1\frac{1}{20}$ 67. $b=-\frac{11}{12}$ 69. $t=-\frac{7}{10}$ 71. $\frac{11}{16}$ in.   73a. $\frac{29}{40}$ 73b. $\frac{13}{120}$
75. $68\frac{1}{2}$ in.   77. 4 ft.

Review Exercises:
1. 10   2. 3   3. 51   4. $10x^2+3x-5$ 5. $-y^3-y^2-4y-5$ 6. $-54b^4$ 7. $2x^2+x-15$ 8. $-4x^2y^5$
9. $3y^3+3y^2$

Exercise Set 5.7
1. a four-sided figure with one pair of parallel sides   3. the side lengths of the parallel sides in a trapezoid   5. $1\frac{1}{2}$
7. 4   9. $1\frac{15}{16}$ 11. $-2\frac{3}{4}$ 13. $-2\frac{1}{4}$ 15. $-\frac{1}{8}$ 17. $-24\frac{5}{6}$ 19. 10   21. $\frac{45}{64}$ 23. $-1\frac{9}{40}$ 25. $\frac{2}{3}$ 27. $52\frac{1}{4}$ m²
29. $30\frac{1}{4}$ ft.²   31. 616 in.²   33. $4\frac{51}{56}$ m²   35. $240\frac{9}{10}$ m²   37. $\frac{1}{6}x^2-\frac{27}{4}x$ 39. $6y^3+\frac{11}{5}y^2+y-\frac{5}{6}$
41. $\frac{1}{10}t^3+\frac{5}{12}t^2-\frac{2}{3}$ 43. $-\frac{1}{10}m^4$ 45. $\frac{1}{2}t^2-\frac{5}{12}t$ 47. $x^2-\frac{1}{4}x-\frac{1}{8}$

Review Exercises:
1. $x=5$ 2. $n=11$ 3. $6(n+2)=n-3; n=-3$ 4. 4 hr.   5. 6 ten-dollar bills and 9 five-dollar bills

Exercise Set 5.8
1. Multiply both sides of the equation by the LCD of the fractions.
3.

Object 2   Object 1   10t represents a greater distance than 5t, so object 2 travels the greater distance.
10t   Start 5t

5. $x=-\frac{1}{4}$ 7. $p=\frac{1}{2}$ 9. $c=-\frac{7}{30}$ 11. $a=\frac{2}{3}$ 13. $f=3\frac{1}{9}$ 15. $b=\frac{1}{4}$ 17. $x=\frac{4}{9}$
19. $a=2\frac{1}{10}$ 21. $n=-4\frac{3}{4}$ 23. $x=10\frac{2}{5}$ 25. $\frac{3}{8}n=4\frac{5}{6}; 12\frac{8}{9}$ 27. $y-8\frac{7}{10}=-2\frac{1}{2}; y=6\frac{1}{5}$
29. $3\frac{1}{4}+2n=-\frac{1}{6}; n=-1\frac{17}{24}$ 31. $\frac{3}{4}(b+10)=1\frac{1}{6}+b; b=25\frac{1}{3}$ 33. $4\frac{3}{5}$ cm   35. \$22,530   37. 1200 gal. each
39. $\frac{3}{5}$ 41. 45° and 135°   43. $\frac{100}{11,127}$ hr.   45. $\frac{1}{9}$ hr.

Review Exercises:
1. <   2. $\frac{3}{8x}$ 3. $-\frac{14}{45}$ 4. $\frac{10a}{9b^3}$ 5. $6\frac{13}{24}$

Chapter 5 Review Exercises
1. true   2. true   3. false   4. true   5. false   6. false   7. multiplying; dividing   8. Write the numerator and denominator in prime factored form, then divide out all common primes.   9. Change the divisor to its reciprocal, then multiply.   10. Find the LCD. Rewrite fractions to have the LCD. Add numerators, keep the common denominators.
11a. $\frac{1}{2}$ 11b. $\frac{1}{4}$

12a. [number line: 4, $4\frac{1}{5}$, 5]   12b. [number line: -3, $-2\frac{5}{8}$, -2]   13a. $4\frac{4}{9}$ 13b. $-7\frac{1}{4}$ 14a. $\frac{20}{3}$

14b. $-\frac{11}{2}$ 15a. 18   15b. undefined   15c. 0   15d. 1   16a. $\frac{3}{7}$ 16b. $-\frac{4}{5}$ 17a. $-\frac{4m^3}{13n}$ 17b. $\frac{3x}{10y^3}$ 18a. >

18b. <   18c. =   19a. $\frac{4}{15}$ 19b. -12 20a. $\frac{35n}{12p^2}$ 20b. $-\frac{h^2}{5}$ 21a. $1\frac{5}{9}$ 21b. 7   22a. $\frac{4}{3b^2}$ 22b. $-\frac{4x^2z^2}{5}$
23a. $\frac{5}{3}$ 23b. 5   24a. 168   24b. $60x^2y$ 25a. $1\frac{1}{30}$ 25b. $11\frac{7}{24}$ 25c. $-\frac{1}{5}$ 25d. $-2\frac{5}{6}$ 26a. $\frac{n}{2}$ 26b. $-\frac{2}{3x}$
26c. $\frac{5}{12h}$ 26d. $\frac{15-14a}{24a}$ 27a. $3\frac{5}{8}$ 27b. $-6\frac{7}{12}$ 28a. $\frac{1}{64}$ 28b. $\frac{9}{25}x^2y^6$ 29a. $\frac{1}{10}x^2-2x-2$ 29b. $\frac{5}{8}n^2-\frac{2}{3}n-2$
29c. $\frac{23}{9}y^3-2y^2-\frac{5}{6}y+\frac{1}{15}$ 29d. $-\frac{1}{4}b^4$ 29e. $\frac{5}{4}x^2+\frac{77}{8}x-3$ 30a. $y=4\frac{2}{15}$ 30b. $n=-3\frac{5}{9}$ 30c. $m=11\frac{1}{2}$
30d. $n=-9\frac{1}{3}$ 31. 9 ft.²   32. $110\frac{1}{4}$ cm²   33. $1\frac{3}{4}$ in.   34. $28\frac{2}{3}$ cm   35. $7\frac{6}{7}$ ft.   36. $12\frac{4}{7}$ ft.²   37. 220 cm²
38. $\frac{1}{6}$ 39. 18 servings   40. $\frac{7}{20}$ 41. $\frac{1}{4}(n+2)=\frac{3}{4}n-\frac{3}{5}; n=2\frac{1}{5}$ 42. $2\frac{2}{9}$ m   43. $24\frac{1}{2}$ ft. and $10\frac{1}{2}$ ft.   44. $\frac{7}{75}$ hr.
45. $11\frac{13}{17}$ sec.

Chapter 5 Practice Test
1. $\frac{3}{8}$ 2. [number line: -4, $-3\frac{1}{4}$, -3]   3a. =   3b. >   4. $6\frac{1}{6}$ 5. $-\frac{37}{8}$ 6a. -16 6b. undefined

6c. 0   6d. 1   7a. $\frac{3}{5}$ 7b. $-\frac{3x^2y}{10}$ 8a. $-11\frac{5}{8}$ 8b. $\frac{4}{3b^2}$ 9a. $2\frac{1}{2}$ 9b. $-\frac{8m^3n}{15}$ 10. $36t^3u$ 11a. $\frac{11}{20}$ 11b. $-1\frac{5}{8}$
12a. $-\frac{x}{3}$ 12b. $\frac{11}{12a}$ 13a. $\frac{4}{25}$ 13b. $\frac{1}{64}$ 14. $2\frac{9}{16}$ 15a. $-\frac{3}{2}n-3$ 15b. $2m^2+m-\frac{3}{8}$ 16. $m=6$ 17. $22\frac{1}{2}$ m²
18. 62 in.²   19. $14\frac{13}{14}$ in.   20. $19\frac{9}{14}$ in.²   21. $\frac{5}{8}$ 22. 8 pieces   23. $\frac{5}{24}$ chose C.   24. $\frac{2}{3}(n+5)=\frac{1}{2}n-\frac{3}{4}; n=-24\frac{1}{2}$
25. $\frac{1}{18}$ hr.

Chapters 1–5 Cumulative Review Exercises
1. false   2. false   3. false   4. false   5. true   6. true   7. multiply   8. Find the prime factorization. Use the smallest exponent of those primes that are common to all factorizations.   9. added -9 and -3 to get -6
10. Find LCD. Rewrite. Add or subtract the numerators and keep the LCD. Reduce.   11. four million, five hundred eighty-two thousand, six hundred one   12. [number line: $-10\ -9\ -8\ -7\ -6\ -5\ -4\ -3\ -2\ -1\ 0$]   13. 850,000   14. 35,000
15. 1   16. $2^4 \cdot 3^2 \cdot 5^2$ 17. $9x$ 18. -15 19. -15 20. 36   21. -111 22. $20\frac{1}{2}$ 23. $\frac{9}{10}$ 24. $\frac{39}{46}$
25. $-\frac{3x^2}{8}$ 26. $3\frac{2}{3}$

27. $\frac{12 + 10x}{15x}$ **28.** $124\frac{2}{3}$ **29.** $-b^3 + 15b^2 - \frac{11}{12}b$ **30.** $-7x^3 - 7x + 26$ **31.** $32x^5$ **32.** $x^2 - 49$ **33.** t^4
34. $6m(3m^2 + 4m - 5)$ **35.** $x = -16$ **36.** $a = -27$ **37.** $n = \frac{4}{5}$ **38.** $y = -1$ **39.** 10,000 combinations
40. $-\$58,946$ **41.** 28 ft. **42a.** $10y - 2$ **42b.** 78 ft. **43.** $\frac{1}{2}h^2 + \frac{3}{2}h$ **44.** 14 in. **45.** 2 ft.
46. $71\frac{2}{3}°$ and $108\frac{1}{3}°$ **47.** width is 6 ft., length is 18 ft. **48.** 8 and 17 **49.** 359 small boxes and 298 large boxes
50. $\frac{2}{5}$ hr.

Exercise Set 6.1

1.

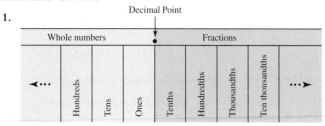

Decimal Point

| Whole numbers | | | | Fractions | | | |

3. and **5.** $\frac{1}{5}$ **7.** $\frac{1}{4}$ **9.** $\frac{3}{8}$ **11.** $\frac{6}{25}$ **13.** $1\frac{1}{2}$ **15.** $18\frac{3}{4}$ **17.** $9\frac{5}{8}$ **19.** $7\frac{9}{25}$ **21.** $-\frac{1}{125}$ **23.** $-13\frac{3}{250}$
25. ninety-seven thousandths **27.** two thousand fifteen millionths **29.** thirty-one and ninety-eight hundredths
31. five hundred twenty-one and six hundred eight-thousandths **33.** four thousand one hundred fifty-nine and six tenths
35. negative one hundred seven and ninety-nine hundredths **37.** negative fifty thousand ninety-two hundred-thousandths

39.  **41.** 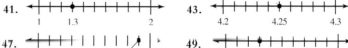 **43.**
 0   0.8   1   1   1.3   2   4.2   4.25   4.3

15. 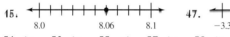 **47.** **49.**
 8.0   8.06   8.1   -3.3   -3.21 -3.2   -19.02 -19.017   -19.01

51. > **53.** > **55.** < **57.** > **59.** > **61.** < **63.** 610.3 **65.** 610.2832 **67.** 610 **69.** 1 **71.** 1.0
73. 0.951 **75.** -408.1 **77.** -408.0626 **79.** -410

Review Exercises:
1. $\frac{41}{100}$ **2.** -13 **3.** $-7a^3 - 3a^2 + 15$ **4.** $4x^2 + 21x + 1$ **5.** $x = -17$ **6.** 57 ft.

Exercise Set 6.2
1. 1. Stack the numbers so that the place values align. Line up the decimal points. 2. Add the digits in the corresponding place values. 3. Place the decimal point in the sum so that it aligns with the decimal point in the addends. **3.** 27.31 **5.** 159.71
7. 59.7511 **9.** 333.317 **11.** 841.45 **13.** 784.28 **15.** 0.01789 **17.** 10.11 **19.** -4.317 **21.** -0.9985
23. -190.02 **25.** -73.25 **27.** 0.9952 **29.** $-12.81y^2 + 5y - 0.93$ **31.** $4.5a^3 + 9a^2 - 1.5a - 10.6$
33. $10.2x^2 - 0.9x + 8.4$ **35.** $10.01a^3 - 4.91a^2 + 7.5a + 5.99$ **37.** $0.4n^2 + 10.8n - 3.6$ **39.** $2.8k^3 + 2.5k^2 + 0.06$
41. $-1.42t^4 - 2.4t^3 + 0.95t^2 - 0.2t + 1.45$ **43.** $2.91x^3 - 11.1x^2 + 2.6x - 4.45$ **45.** $m = 10.83$ **47.** $y = -5.81$
49. $k = -3.22$ **51.** \$46.83 overdrawn **53.** The client owes \$421.94. **55.** \$9.38 **57.** \$1919.89 **59.** 22.3 m

Review Exercises:
1. -300 **2.** $\frac{21}{100}$ **3.** 306 **4.** $2\frac{32}{39}$ **5.** 11 **6.** $-18x^7$ **7.** $y^2 - 2y - 15$ **8.** $x = -9$

Exercise Set 6.3
1. Place the decimal point in the product so that the product has the same number of decimal place values as the total number of decimal place values in the factors. **3.** No, because 24.7 is greater than 10. The absolute value of the decimal number needs to be greater than or equal to 1 but less than 10. **5.** 0.54 **7.** 10.585 **9.** 702.5922 **11.** 6.1945 **13.** 196
15. -15.08 **17.** -0.152 **19.** 0.00435 **21.** 0.81 **23.** 9.261 **25.** 0.00000081 **27.** -0.064
29. 300,000,000 m/sec.; three hundred million m/sec. **31.** 2,140,000 lt-yr.; two million, one hundred forty thousand lt-yr.
33. 293,000,000 people; two hundred ninety-three million people **35.** 2.3×10^6 blocks **37.** 3.54×10^9 mi.
39. 5.3×10^9 nucleotide pairs **41.** $5.76x^8$ **43.** $-0.016a^3$ **45.** $0.25t^6$ **47.** $0.9y^2 + 1.3y + 19$
49. $-0.136a^3 + 6.8a - 12.92$ **51.** $x^2 + 0.92x - 0.224$ **53.** $6.4x^2 - 18.32x - 3.66$ **55.** $33.64a^2 - 81$
57. $14.7k^2 - 4.27k + 0.28$ **59.** \$102.96 **61.** \$3.84 **63.** \$1827.20 **65.** \$3.75 **67.** 0.0294 N; The force is
downward. **69.** 701.96 lb.; The force is downward. **71.** 56.25 ft. **73.** 23.04 cm^2 **75.** 12,551.1 ft.^{2}
77. 0.0081 m^3

Review Exercises:
1. 205 **2.** 13 **3.** $6x^4$ **4.** $t = -9$ **5.** $x = 9$ **6.** $a = 3$ **7.** radius: 7 in.; circumference: ≈ 43.96 in.

Exercise Set 6.4
1. 6.2 **3.** Divide the denominator into the numerator. **5.** 23.85 **7.** 0.024 **9.** 600 **11.** 4.8 **13.** -40
15. 125 **17.** -0.0206 **19.** 814.5 **21.** 178.18 **23.** 1.06 **25.** 0.6 **27.** 0.45 **29.** -0.4375 **31.** $0.4\overline{3}$
33. 13.25 **35.** -17.625 **37.** $104.\overline{6}$ **39.** $-216.\overline{571428}$ **41.** 0.04 **43.** 0.5 **45.** ≈ 4.90 **47.** ≈ 14.14
49. 1.3 **51.** ≈ 0.09 **53.** $25.5x^2$ **55.** $-6.4m^2n$ **57.** $0.15ac$ **59.** $b = 6.02$ **61.** $h = -1.6$ **63.** $n = 0.6$
65. $x = -3.5$ **67.** \$613.32 **69.** 40.5 rads **71.** 400 Ω **73.** ≈ 1.71 hr. **75.** ≈ 6829.5 mi. **77.** ≈ 6.69 mi.
79. ≈ 1.04 ft.

Review Exercises:
1. -5 **2.** 18   **3.** -180 **4.** 56 in.^{2} **5.** 680 cm^2

Exercise Set 6.5
1. Writing the decimals as fractions is better in this case because the decimal equivalent of $\frac{2}{3}$ is $0.\overline{6}$, which would have to be rounded for the calculation and rounding would lead to an inaccurate result.   **3.** cylinder: $V = \pi r^2 h$, cone: $V = \frac{1}{3}\pi r^2 h$; The volume of a cone is one-third of the volume of a cylinder with the same radius and height.   **5.** 2.64   **7.** 11.73   **9.** 42.99
11. -22.24 **13.** -28 **15.** -117.69 **17.** $-9.7\overline{2}$ **19.** -0.51 **21.** 0.29   **23.** -4.775 **25.** $-0.3\overline{8}$ **27.** 40
29. ≈23.6 **31.** 2.25×10^{23} **33.** 9.75   **35.** 1.0609   **37.** 57.6 cm^2 **39.** 1.296 m^2 **41.** ≈86.5 in.^{2}
43. ≈0.9 cm^3 **45.** 168,750 ft.^{3} **47.** ≈92.106 in.^{3} **49.** $\approx2.45 \times 10^{11}$ mi.^{3} **51.** ≈2.71 cm^2 **53.** ≈134.375 cm^2
55. $\approx151,475.69$ ft.^{3} **57.** $\approx2,582,957.6$ m^3 **59a.** 0.000027 in.^{3} **59b.** 37,037 grains

Review Exercises:
1. $x = 7$ **2.** $y = 20\frac{1}{4}$ **3.** 300 ft.^{2} **4.** 720 cm^2 **5.** $\frac{1}{2}(x + 12) = \frac{3}{4}x - 9$; $x = 60$ **6.** 8 of the 5-dollar bills and 12 of the 10-dollar bills

Exercise Set 6.6
1. Multiply both sides by an appropriate power of ten as determined by the decimal number with the most decimal places.
3. a triangle with one angle that measures 90°, which is a right angle   **5.** the longest side in a right triangle   **7.** $n = 0.2$
9. $y = 0.6$ **11.** $t = -0.12$ **13.** $k = -6.2$ **15.** $n = -0.58$ **17.** $x = 3.3$ **19.** $t = 0.5$ **21.** $x = -11$
23. $n = -24$ **25.** \$6.57   **27.** \$188.16   **29.** \$33.42   **31.** \$327.00   **33.** $\approx$\$586.55   **35.** $p = 0.8$
37. $t = -3.8\overline{3}$ **39.** $y = 2.52$ **41.** ≈100 km   **43.** 5 mi.   **45.** ≈14.1 ft.   **47a.** 59.96 ft.   **47b.** No, the person is 63 ft. above ground.   **49.** ≈91.9 ft.   **51.** 5 half-dollars and 17 quarters   **53.** 42 of the 12-oz. drinks and 23 of the 16-oz. drinks

Review Exercises:
1. twenty-four billion, nine hundred fifteen million, two hundred four   **2.** 46,300,000   **3.** $13\frac{11}{24}$ **4.** $4\frac{9}{20}$ **5.** $4x^6$
6. Mistake: divided by 3 rather than by -3 Correct: $x = 2$

Chapter 6 Review Exercises
1. true   **2.** false   **3.** false   **4.** true   **5.** false   **6.** true   **7.** Mistake: decimal point placed incorrectly in the product; Correct: 0.0312   **8.** Line up decimal points, add the digits, then bring down the decimal point directly below its position in the addends.   **9.** denominator; numerator   **10.** Write all digits without the decimal point in the numerator, then write the last place value as the denominator.   **11a.** twenty-four and thirty-nine hundredths   **11b.** five hundred eighty-one and four hundred fifty-nine thousandths   **11c.** two thousand nine hundred seventeen ten-thousandths   **12a.** $\frac{4}{5}$ **12b.** $\frac{3}{125}$
13c. $-2\frac{13}{20}$ **13a.** $>$ **13b.** $<$ **14a.** 32   **14b.** 31.8   **14c.** 31.81   **14d.** 31.806   **15a.** 88.015   **15b.** 7.916
15c. -6.2 **15d.** -6.12 **15e.** 23.118   **15f.** 20.49   **16a.** $11.1y^4 - 12.2y^2 - 8y + 3.8$
16b. $5.4a^2 + 5.2a + 10.33$ **16c.** $-2.7x^3 - 98.1x^2 + 2.1x - 4.2$ **17a.** 17.544   **17b.** -14.505 **17c.** 1.44
17d. -0.008 **17e.** $0.025h^2k^3$ **17f.** $-3.78x^8y^2$ **17g.** $-0.027x^9$ **17h.** $1.44y^6$
17i. $0.12n^3 - 0.28n^2 + 0.2n - 3.6$ **17j.** $-3.08a + 5.39b$ **17k.** $9a^2 - 14.45a + 3.05$ **17l.** $1.69x^2 - 17.64$
18a. 5.25   **18b.** 0.4   **18c.** 8.4   **18d.** 600   **18e.** $7.95x^2$ **18f.** $-0.35a^3c$ **19a.** 0.8   **19b.** 1.1   **20a.** ≈12.17
20b. ≈1.26 **21a.** 0.6   **21b.** 0.16   **21c.** 6.25   **21d.** -4.18 **22a.** 3,580,000,000   **22b.** $-420,000$
23a. 9.2×10^9 **23b.** -1.03×10^5 **24a.** -12.97 **24b.** -3.05 **24c.** 0.27   **24d.** -2.16 **25.** radius: 7.5 cm; circumference: ≈47.1 cm;   area: ≈176.625 cm^2 **26a.** ≈46.12 in.^{2} **26b.** 0.15 m^2 **27a.** ≈105.975 in.^{2}
27b. ≈1445.43 cm^2 **28a.** ≈42.39 in.^{3} **28b.** ≈20.096 cm^3 **28c.** ≈904.32 m^3 **28d.** 22,696.8 m^3
28e. ≈39.978 in.^{3} **29.** ≈7.536 in.^{3} **30a.** $x = 1.54$ **30b.** $y = -0.64$ **30c.** $y = 3$ **30d.** $n = -0.02$
30e. $b = 0$ **30f.** $k = 0.8$ **30g.** $a = -1.05$ **30h.** $x = 8.1$ **30i.** $x = 0.75$ **30j.** $n = 5$ **31.** \$713.18
32. \$47.58 each payment; \$47.60   **33.** 28.6   **34.** \$9.82   **35.** 11.62 ft.   **36.** 90 min.   **37.** 10 nickels and 19 dimes
38. 9 larger-size and 15 smaller-size figurines

Chapter 6 Practice Test
1. fifty-six and seven hundred eighty-nine thousandths   **2.** $\frac{17}{25}$ **3.** $<$ **4. a.** 2   **b.** 2.1   **c.** 2.09   **d.** 2.092
5. 41.991   **6.** 7.916   **7.** -38.745 **8.** 0.3   **9.** 6.02   **10.** 0.9   **11.** 3.83   **12. a.** 2,970,000   **b.** -3.56×10^7
13. 5.1344   **14.** 0.256   **15.** $-2.7x^3 - 98.1x^2 + 2.1x - 4.2$ **16a.** $1.5x^4$ **16b.** $9a^2 - 14.45a + 3.05$ **17.** $t = 4.1$
18. $k = 0.6$ **19a.** ≈75.36 ft.   **19b.** ≈452.16 ft.^{2} **20a.** ≈33.493 cm^3 **20b.** 777.6 ft.^{3} **21.** 785.5 ft.^{2} **22.** \$7.03
23. $\approx$\$178.73   **24.** ≈19.36 ft.   **25.** 8 round-layered and 4 sheet cakes

Chapters 1–6 Cumulative Review Exercises
1. true   **2.** true   **3.** false   **4.** false   **5.** true   **6.** false   **7.** undefined   **8.** coefficients   variables   **9.** Mistake: missing the middle term;   Correct: $x^2 + 5x + 6$ **10.** Change the divisor to its reciprocal, then multiply.   **11.** twenty-nine and six thousand eighty-one ten-thousandths   **12.**

13. 2.02   **14.** 40   **15.** 6   **16.** $2^3 \cdot 3 \cdot 5 \cdot 7$ **17.** 168   **18.** 7   **19.** -52 **20.** $-\frac{4}{5}$ **21.** $\frac{3x}{4y^3}$ **22.** $-\frac{31}{54}$
23. 0.25   **24.** $\frac{8}{5}n$ **25.** $12\frac{1}{8}$ **26.** $\frac{2x + 9}{12}$ **27.** 3.98   **28.** 2.258   **29.** $1.6x^3 + \frac{3}{4}x^2 + 3.55$
30. $4.9m^3 - 6.7m^2 + 18.6$ **31.** $-12x^2y$ **32.** $4y^2 + 5y - 6$ **33.** 768,000,000; seven hundred sixty-eight million
34. $5n(6n^4 + 3n - 5)$ **35.** $y = 24.7$ **36.** $k = -6.8$ **37.** $x = \frac{7}{8}$ **38.** $n = 7$ **39.** 32°F   **40.** \$1,400,900; profit

41. $1939.33 **42a.** $8w^3 - 4w^2$ **42b.** 1584 cm^3 **43.** 2.2 ft. **44.** $\approx$81.25 in.^{2} **45.** $x = 2$ **46.** $\frac{1}{2}$ of the employees **47.** $\frac{7}{30}$ had no opinion. **48.** The base is 15 ft.; the two equal sides are 27.5 ft. **49.** 19.08 ft. **50.** 9 larger-size and 11 smaller-size bowls

Exercise Set 7.1

1. a comparison between two quantities using a quotient **3.** Divide the denominator into the numerator. **5.** a rate which describes the price for one quantity unit **7a.** $\frac{31}{66}$ **7b.** $\frac{35}{66}$ **7c.** $\frac{31}{35}$ **7d.** $\frac{35}{31}$ **9a.** $\frac{3}{2}$ **9b.** $\frac{2}{3}$ **9c.** $\frac{3}{5}$ **9d.** $\frac{2}{5}$ **11.** $\frac{7}{8}$ **13.** $\frac{14}{9}$ **15a.** $\frac{7}{30}$ **15b.** $\frac{17}{30}$ **15c.** 1 **15d.** $\frac{1}{2}$ **17a.** $\frac{1}{3}$ **17b.** $\frac{3}{5}$ **17c.** $\frac{13}{186}$ **17d.** $\frac{2}{165}$ **19.** $\frac{1}{13}$ **21.** $\frac{2}{13}$ **23.** $\frac{1}{5}$ **25.** 0 **27.** $\frac{7}{13}$ **29a.** $\frac{1}{5}$ **29b.** $\frac{28}{125}$ **29c.** $\frac{72}{125}$ **29d.** $\frac{97}{125}$ **31.** $\approx$0.16; debt of $0.16 for every $1 of income **33.** $\approx$0.26; $0.26 is paid for mortgage for every $1 of income **35.** $\approx$16.9; There are about 17 students for every faculty member. **37.** $\approx$12.08; The stock is selling at $12.08 for every $1 of annual earning. **39.** 68.2 mi./hr. **41.** 12¢/min. **43.** $7.50/lb. **45.** 16-oz. can **47.** bag containing 16 diapers **49.** two 15.5-oz. boxes

Review Exercises:
1. 120 **2.** $4\frac{1}{2}$ **3.** 12.375 **4.** $x = 8.6$ **5.** $y = 7\frac{1}{2}$

Exercise Set 7.2

1. an equation in the form $\frac{a}{b} = \frac{c}{d}$, where $b \neq 0$ and $d \neq 0$ **3.** 1. Equate the cross products. 2. Use the multiplication/division principle of equality to isolate the variable. **5.** yes **7.** no **9.** no **11.** yes **13.** no **15.** yes **17.** no **19.** $x = 5$ **21.** $n = -7.5$ **23.** $n = 6$ **25.** $m = -35$ **27.** $h = 25$ **29.** $b = -25$ **31.** $d = 14\frac{1}{4}$ **33.** $t = 14\frac{2}{3}$ **35.** $\approx$34.3 gal. **37.** 36 servings **39.** 1168.8 oz. **41.** $2\frac{1}{4}$ cups **43.** $3\frac{1}{4}$ in. **45.** $1633\frac{1}{3}$ days **47.** $c = 8.96$ cm **49.** $a = 3\frac{11}{18}$ in.; $b = 3\frac{1}{4}$ in.; $c = 4\frac{29}{48}$ in. **51.** 9.9 m **53.** $480\frac{17}{18}$ ft. **55.** $11,994.\overline{6}$ m

Review Exercises:
1. $\frac{5}{27}$ **2.** 3.7 **3.** 1344 **4.** 5.875 **5.** $20\frac{1}{4}$

Exercise Set 7.3

1. a fraction with a ratio equivalent to 1 **3.** 12; 3; 5280 **5.** 8; 2; 2; 4 **7.** 156 ft. **9.** 7.5 ft. **11.** 34,320 ft. **13.** 66 in. **15.** 36,608 yd. **17.** 2.75 yd. **19.** 270 ft.^{2} **21.** 13,608 in.^{2} **23.** 18.75 yd.^{2} **25.** 18 c. **27.** 20 oz. **29.** 15 pt. **31.** 4.25 qt. **33.** 51.2 oz. **35.** 5.1 gal. **37.** 196 oz. **39.** 11.25 lb. **41.** 0.6 T **43.** 8400 lb. **45.** 2.5 hr. **47.** 1.5 min. **49.** 630 sec. **51.** 135 min. **53.** 720 sec. **55.** $\approx$0.082 yr. **57.** 1440 min. **59.** 37,869,120 min. **61.** 12.36 mi./min. **63.** No, the driver and the officer are not in agreement because the officer's estimate converts to $54.\overline{54}$ mi./hr., not 40 mi./hr. **65.** $25,014.\overline{27}$ mi./hr.

Review Exercises:
1. 840 **2.** 950 **3.** 0.329 **4.** 4.56 **5.** 0.009 **6.** 4.8

Exercise Set 7.4

1. a basic unit; other units are named relative to it **3.** meter **5.** a measure of the amount of matter in an object **7.** 450 cm **9.** 70 m **11.** 3.8 km **13.** 9.5 m **15.** 15.4 dam **17.** 1.12 hm **19.** 120 ml **21.** 5 l **23.** 0.0095 dal **25.** 80 cc **27.** 12 ml **29.** 400 cc **31.** 12,000 mg **33.** 9 g **35.** 50 cg **37.** 3.6 t **39.** 106 kg **41.** 6.5 t **43.** 18,000 cm^2 **45.** 0.22 m^2 **47.** 5 mm^2 **49.** 0.0443 hm^2

Review Exercises:
1. 1.008 **2.** $21\frac{2}{3}$ **3.** $6.5x^2 - 18.2x - 5.6$ **4.** $481y^2 - 280y + 700$ **5.** $x = 2.8$

Exercise Set 7.5

1. 2.54 cm **3.** $F = \frac{9}{5}C + 32$ **5.** $\approx$91.4 m **7.** $\approx$42.2 km **9.** 9.525 mm **11.** $\approx$437.6 yd. **13.** $\approx$24.84 mi. **15.** $\approx$17.72 in. **17.** $\approx$0.473 l **19.** $\approx$591.25 ml **21.** $\approx$0.79 gal. **23.** $\approx$0.68 oz. **25.** $\approx$61.29 kg **27.** $\approx$340.5 g **29.** $\approx$0.0176 oz. **31.** $\approx$121 lb. **33.** 130°F **35.** 42.8°F **37.** -459.67°F **39.** 37°C **41.** $38.\overline{8}$°C **43.** $-20.\overline{5}$°C

Review Exercises:
1. 0.28 **2.** $0.118\overline{3}$ **3.** $-\frac{4}{3}x - \frac{7}{20}y$ **4.** $t = \frac{2}{3}$ **5.** $m = 10.88$

Exercise Set 7.6

1. the ratio of the total monthly house payment to gross monthly income **3.** a payment that includes principal, interest, taxes, and insurance **5.** front-end ratio $\approx$0.22; back-end ratio $\approx$0.33; The Wu family would most likely qualify. **7.** back-end ratio $\approx$0.42; The Rivers family would most likely not qualify. **9.** front-end ratio $\approx$0.31; back-end ratio $\approx$0.45; The Bishop family would most likely not qualify. **11.** $1344 **13.** $533.60 **15.** $1252.80 **17.** $686.75 **19.** $973.75 **21.** 9:30 P.M. **23.** 1:33 P.M. **25.** 522.5 mg **27.** $\approx$5 tablets **29.** 10 to 11 drops/min. **31.** 400 units/hr.

Review Exercises:
1. $\frac{7}{25}$ **2.** $\frac{2}{13}$ **3.** $\approx$$0.14/oz. **4.** 25 yd. **5.** 8500 mg

Chapter 7 Review Exercises

1. false **2.** true **3.** true **4.** false **5.** true **6.** false **7.** favorable; possible **8.** 1 **9.** denominator; numerator
10. denominator **11.** $\frac{11}{16}$ **12.** $\frac{25}{108}$ **13.** $\frac{2}{13}$ **14.** $\frac{1}{3}$ **15.** ≈ 16.4 **16.** ≈ 7.09 **17.** 70.2 mi./hr.
18. 9 ¢/min. **19.** 4.5 ¢/oz. **20.** 3.9 ¢/oz. **21.** 32-oz. cup **22.** two 12-oz. cans **23.** $k = -4.5$
24. $m = 9\frac{3}{13}$ **25.** 29.9 m² **26.** 280 mi. **27.** 432 in. **28.** 2.5 lb. **29.** 192 oz. **30.** 135 min. **31.** $24.\overline{1}$ yd.²
32. 79,200 mi./hr. **33.** 50 dm **34.** 260 g **35.** 0.95 t **36.** 75 cc **37.** ≈ 6.09 m **38.** ≈ 19.31 km
39. ≈ 65.83 kg **40.** ≈ 0.79 gal. **41.** $32.\overline{2}$°C **42.** 24.8°F **43.** 12 to 13 drops/min. **44.** 92.8 mg **45.** ≈ 0.28
46. ≈ 0.40 **47.** \$571.30 **48.** \$1722 **49.** 3 hr. **50.** 11:32 A.M.

Chapter 7 Practice Test

1. $\frac{19}{26}$ **2.** $\frac{11}{2289}$ **3.** $\frac{1}{5}$ **4.** ≈ 30.8 **5.** $70.\overline{48}$ mi./hr. **6.** 14 ¢/min. **7.** ≈ 4 ¢/oz. **8.** 20-oz. can
9. $n = -5.85$ **10.** \$574.50 **11.** 900 mi. **12.** 168 in. **13.** 320 oz. **14.** 4.5 gal. **15.** 2.5 hr. **16.** $27.\overline{7}$ yd.²
17a. 5.8 cm **17b.** 0.42 km **18a.** 0.024 kg **18b.** 91,000 g **19a.** 0.028 m² **19b.** 80 dm² **20a.** 0.65 l
20b. 800 cc **21.** ≈ 22.85 m **22.** $-1.\overline{1}$°C **23.** ≈ 0.45 **24.** 15 to 16 drops/min. **25.** 13 hr.

Chapters 1–7 Cumulative Review Exercises

1. false **2.** false **3.** true **4.** false **5.** true **6.** false **7.** reciprocal **8.** LCM **9.** cross **10.** Cross
multiply, then isolate the variable **11.** forty-nine thousand, eight hundred two and seventy-six hundredths

12. 9.14×10^8 **13.**

14. 41.33 **15.** -1 **16.** 5 **17.** -91 **18.** $-2\frac{8}{13}$

19. $5\frac{9}{10}$ **20.** 30.75 **21.** -31.2 **22.** $\frac{9}{16}$ **23.** 3 **24.** 0.0081 **25.** 3.04 **26.** $8.3y^3 - 8.2y^2 - y + 10.2$
27. $\frac{1}{2}a^2 - \frac{65}{12}a + 7$ **28.** $2^5 \cdot 3 \cdot 5^2$ **29.** $20x$ **30.** $6x^4$ **31.** $8m(4m^3 + 3m - 2)$ **32.** $\frac{3}{16x^4}$ **33.** $40k^2$ **34.** $\frac{18 + 5x}{9x}$
35. $n = 18.4$ **36.** $x = 14\frac{6}{11}$ **37.** $b = -16.32$ **38.** 66,000 ft. **39.** 480 g **40.** 140°F **41.** 126 m³
42. ≈ 349.45 ft.² **43.** \$50,710 **44.** 6.2 m **45.** $y^2 - 4y - 45$ **46.** length = 14 m; width = 11 m **47.** $\frac{3}{13}$
48. 16-oz. can **49.** ≈ 313 l **50.** ≈ 85.7 m

Exercise Set 8.1

1. a ratio representing some part out of 100 **3.** Write the percent amount in decimal form, then divide by 100, which
causes the decimal point to move two places to the left. **5.** $\frac{1}{5}$ **7.** $\frac{3}{20}$ **9.** $\frac{37}{250}$ **11.** $\frac{3}{80}$ **13.** $\frac{91}{200}$ **15.** $\frac{1}{3}$
17. 0.75 **19.** 1.25 **21.** 0.129 **23.** 0.0165 **25.** 0.534 **27.** $0.1\overline{6}$ **29.** 50% **31.** 60% **33.** 37.5%
35. $16.\overline{6}\%$ **37.** $66.\overline{6}\%$ **39.** $44.\overline{4}\%$ **41.** 96% **43.** 80% **45.** 9% **47.** 120% **49.** 2.8% **51.** 405.1%
53. $66.\overline{6}\%$ **55.** $163.\overline{63}\%$

Review Exercises:

1. 14.4 **2.** 1620 **3.** 3.84 **4.** $x = 1333.\overline{3}$ **5.** $y = 38$ **6.** The number is 32.

Exercise Set 8.2

1. whole amount; part out of the whole **3.** divide **5.** $40\% \cdot 350 = c$; 140 **7.** $c = 15\% \cdot 78$; 11.7
9. $150\% \cdot 60 = c$; 90 **11.** $c = 16\% \cdot 30\frac{1}{2}$; 4.88 **13.** $110\% \cdot 46.5 = c$; 51.15 **15.** $c = 5\frac{1}{2}\% \cdot 280$; 15.4
17. $66\frac{2}{3}\% \cdot 56.8 = c$; 37.8̄6 **19.** $c = 8.5\% \cdot 54.82$; 4.6597 **21.** $12.8\% \cdot 36,000 = c$; 4608 **23.** $35\% \cdot w = 77$; 220
25. $4800 = 40\% \cdot w$; 12,000 **27.** $605 = 2.5\% \cdot w$; 24,200 **29.** $105\% \cdot w = 48.3$; 46 **31.** $47.25 = 10\frac{1}{2}\% \cdot w$; 450
33. $p \cdot 68 = 17$; 25% **35.** $2.142 = p \cdot 35.7$; 6% **37.** $p \cdot 24\frac{1}{2} = 7\frac{7}{20}$; 30% **39.** $p = \frac{21}{60}$; 35%
41. $\frac{25}{35} = p$; $\approx 71.4\%$ or $71\frac{3}{7}\%$ **43.** $\frac{43}{65} = p$; $\approx 66.15\%$ or $66\frac{2}{13}\%$

Review Exercises:

1. 1280 **2.** 575 **3.** $86\frac{5}{8}$ **4.** $x = 10.5$ **5.** $n = 62.5$ **6.** 360 units

Exercise Set 8.3

1. Use the form percent $= \dfrac{\text{part}}{\text{whole}}$, where the percent is expressed as a fraction with a denominator of 100. **3.** $\dfrac{30}{100} = \dfrac{c}{560}$; 168

5. $\dfrac{12}{100} = \dfrac{c}{85}$; 10.2 **7.** $\dfrac{125}{100} = \dfrac{c}{90}$; 112.5 **9.** $\dfrac{15}{100} = \dfrac{c}{9\frac{1}{2}}$; 1.425 **11.** $\dfrac{102}{100} = \dfrac{c}{88.5}$; 90.27 **13.** $\dfrac{12\frac{1}{2}}{100} = \dfrac{c}{440}$; 55

15. $\dfrac{33\frac{1}{3}}{100} = \dfrac{c}{87.6}$; 29.2 **17.** $\dfrac{6.5}{100} = \dfrac{c}{22,800}$; 1482 **19.** $\dfrac{16.9}{100} = \dfrac{c}{2450}$; 414.05 **21.** $\dfrac{65}{100} = \dfrac{52}{w}$; 80

23. $\dfrac{80}{100} = \dfrac{4480}{w}$; 5600 **25.** $\dfrac{4.5}{100} = \dfrac{1143}{w}$; 25,400 **27.** $\dfrac{120}{100} = \dfrac{65.52}{w}$; 54.6 **29.** $\dfrac{5\frac{1}{2}}{100} = \dfrac{13.53}{w}$; 246

31. $\dfrac{P}{100} = \dfrac{24}{80}$; 30% **33.** $\dfrac{P}{100} = \dfrac{2.226}{74.2}$; 3% **35.** $\dfrac{P}{100} = \dfrac{18\frac{1}{5}}{30\frac{1}{3}}$; 60% **37.** $\dfrac{P}{100} = \dfrac{76}{80}$; 95% **39.** $\dfrac{P}{100} = \dfrac{32}{45}$; $71.\overline{1}\%$

41. $\dfrac{P}{100} = \dfrac{17}{19}$; $\approx 89.47\%$ **43.** $\dfrac{P}{100} = \dfrac{80}{75}$; $106.\overline{6}\%$

Review Exercises:
1. 268.8 **2.** 1.24 **3.** $\frac{5}{12}$ **4.** 250 people disagreed **5.** $\frac{3}{5}$ were females that agreed.

Exercise Set 8.4
1. 1. Determine whether the percent, the whole, or the part is unknown. 2. If needed, write the problem as a simple percent sentence. 3. Translate the sentence to an equation (word-for-word or proportion). 4. Solve for the unknown.
3. unknown; 4000; 2400 **5.** $60\% \cdot 800 = c$ or $\frac{60}{100} = \frac{c}{800}$; 480 ml **7a.** $90\% \cdot 30 = c$ or $\frac{90}{100} = \frac{c}{30}$; 27 questions **7b.** 10% **7c.** 3 questions **9.** $80\% \cdot 2.0 = c$ or $\frac{80}{100} = \frac{c}{2.0}$; 1.6 million people **11.** $10\% \cdot 3106 = c$ or $\frac{10}{100} = \frac{c}{3106}$; $310.60 **13.** $8\% \cdot 4249 = c$ or $\frac{8}{100} = \frac{c}{4249}$; commission: $339.92; gross pay: $699.92 **15.** $25\% \cdot 16,850 = c$ or $\frac{25}{100} = \frac{c}{16,850}$; $4212.50 **17.** $750 = 6\% \cdot w$ or $\frac{6}{100} = \frac{750}{w}$; $12,500 **19.** $39 = 42\% \cdot w$ or $\frac{42}{100} = \frac{39}{w}$; about 93 shot attempts **21.** $43 = 38\% \cdot w$ or $\frac{38}{100} = \frac{43}{w}$; about 113 employees **23.** $p = \frac{58}{65}$ or $\frac{P}{100} = \frac{58}{65}$; about 89% **25.** $p = \frac{48}{110}$ or $\frac{P}{100} = \frac{48}{110}$; $43.\overline{63}\%$ **27.** $p = \frac{3}{6.3}$ or $\frac{P}{100} = \frac{3}{6.3}$; about 47.6% **29.** $p = \frac{7}{24}$ or $\frac{P}{100} = \frac{7}{24}$; $29.1\overline{6}\%$ **31.** $p = \frac{810}{899}$ or $\frac{P}{100} = \frac{810}{899}$; about 90.1% **33.** $p = \frac{2.06 \times 10^{16}}{1.41 \times 10^{18}}$ or $\frac{P}{100} = \frac{2.06 \times 10^{16}}{1.41 \times 10^{18}}$; about 1.46% **35a.** Mortgage: ≈23.3%; Childcare: ≈19.6%; Car payment 1: ≈7.2%; Car payment 2: ≈6.8%; Credit card payments: ≈3.7%; Utilities: ≈7%; Groceries: ≈11.7%; Entertainment: ≈6.4% **35b.** about 85.6% **35c.** about 14.4%

Review Exercises:
1. $6.1x - 2.6$ **2.** $x = 14$ **3.** $y = 30.6$ **4.** $m = 1.6$ **5.** $2550

Exercise Set 8.5
1. percent; initial amount; amount of increase or decrease **3.** 25%; unknown; $80 **5.** Sales tax is $9.30; total amount is $195.25. **7.** Sales tax is $1587; total amount is $28,037. **9.** $30,636 **11.** $9.75 **13.** $25,200 **15.** $29,250 **17.** $42.95 **19.** Discount amount is $19.80; final price is $46.19. **21.** Discount amount is $245; final price is $454.99. **23.** $1977.08 **25.** $22.43 **27.** $2596 **29.** $146.97 **31.** $2570 **33.** $379.51 **35.** 7% **37.** 3.5% **39.** 8% **41.** 17.6% **43.** 22.3% **45.** 15% **47.** 33.6%

Review Exercises:
1. $x = 3750$ **2.** $y = \frac{1}{3}$ **3.** 1.1664 **4.** $\frac{1}{125}$ **5.** five $10 bills, eleven $5 bills

Exercise Set 8.6
1. an initial amount of money **3.** $I = Prt$, where I represents the simple interest, P represents principal, r represents the interest rate, and t represents the time the principal earns interest. **5.** $B = P\left(1 - \frac{r}{n}\right)^{nt}$, where B represents the final balance, P represents the principal, r represents the APR, t represents the time in years, and n represents the number of compound periods per year. **7.** $120; $4120 **9.** $38.50; $388.50 **11.** $7901.25; $20,151.25 **13.** $96; $2496 **15.** $22.68; $2022.68 **17.** $4798.84 **19.** $1946.25 **21.** $5832 **23.** $943.82 **25.** $15,142.40 **27.** $518.01 **29.** $1910.48 **31.** $412.41 **33.** $1646.80 **35.** $571.04 **37.** $872.23 **39.** $706.15 **41.** $311.38 **43.** $423.05 **45.** $744.27 **47.** $853 **49.** $926.92

Review Exercises:
1.

2. $0.8\overline{3}$ **3.** 77.75 **4.** 6 **5.** $x = -1\frac{13}{15}$

Chapter 8 Review Exercises
1. false **2.** false **3.** true **4.** false **5.** false **6.** true **7.** divide **8.** multiply **9.** Translate *is* to an equals sign. Translate *of* to multiplication if preceded by a percent, and division if preceded by a whole number. **10.** Write the percent as a fraction with a denominator of 100. Set this fraction equal to the part out of the whole. **11a.** $\frac{2}{5}$ **11b.** $\frac{13}{50}$ **11c.** $\frac{13}{200}$ **11d.** $\frac{49}{200}$ **12a.** 0.16 **12b.** 1.5 **12c.** 0.032 **12d.** $0.40\overline{3}$ **13.** 54% **14.** 130% **15.** 37.5% **16.** $44.\overline{4}\%$ **17.** $c = 15\% \cdot 90$; 13.5 **18.** $12.8\% \cdot w = 5.12$; 40 **19.** $12.5 = p \cdot 20$; 62.5% **20.** $p = \frac{40}{150}$; $26.\overline{6}\%$ **21.** $\frac{40.5}{100} = \frac{c}{800}$; 324 **22.** $\frac{15}{100} = \frac{10\frac{1}{2}}{w}$; 70 **23.** $\frac{P}{100} = \frac{8.1}{45}$; 18% **24.** $\frac{P}{100} = \frac{16}{30}$; $53.\overline{3}\%$ **25.** $35\% \cdot 600 = c$ or $\frac{35}{100} = \frac{c}{600}$; 210 ml **26.** $p = \frac{540}{2000}$ or $\frac{P}{100} = \frac{540}{2000}$; 27% **27.** $c = 6\% \cdot 285.75$ or $\frac{6}{100} = \frac{c}{285.75}$; $17.15; $302.90 **28.** $25\% \cdot 56.95 = c$ or $\frac{25}{100} = \frac{c}{56.95}$; $42.71 **29.** $180 = 35\% \cdot w$ or $\frac{35}{100} = \frac{180}{w}$; 514.29 mi. **30.** $1102.5 = 3.5\% \cdot w$ or $\frac{3.5}{100} = \frac{1102.5}{w}$; $31,500 **31.** $p = \frac{27.68}{86.50}$ or $\frac{P}{100} = \frac{27.68}{86.50}$; 32% **32.** $p = \frac{3.50}{12}$ or $\frac{P}{100} = \frac{3.50}{12}$; ≈29.2% **33.** $28.80 **34.** $8104 **35.** $600 **36.** $6298.56 **37.** $2033.79 **38.** $917.91

Chapter 8 Practice Test
1. $\frac{6}{25}$ **2.** 0.042 **3.** 0.125 **4.** $\frac{163}{400}$ **5.** 26% **6.** 120% **7.** 40% **8.** $55.\overline{5}\%$ **9.** $c = 15\% \cdot 76$; 11.4

10. $6.5\% \cdot w = 8.32$; 128　　**11.** $\frac{P}{100} = \frac{14}{60}$; $23.\overline{3}\%$　　**12.** $\frac{P}{100} = \frac{12}{32}$; 37.5%　　**13.** $80\% \cdot 30 = c$ or $\frac{80}{100} = \frac{c}{30}$; 24 students
14. $25\% \cdot 1218 = c$ or $\frac{25}{100} = \frac{c}{1218}$; \$304.50　　**15.** $p = \frac{345.75}{2786.92}$ or $\frac{P}{100} = \frac{345.75}{2786.92}$; about 12.4%　　**16.** $c = 5\% \cdot 295.75$
or $\frac{5}{100} = \frac{c}{295.75}$; sales tax: \$14.79; total amount: \$310.54　　**17.** $c = 70\% \cdot 84.95$ or $\frac{70}{100} = \frac{c}{84.95}$; \$59.47
18. $2.5\% \cdot f = 586.25$ or $\frac{2.5}{100} = \frac{586.25}{f}$; \$23,450　　**19.** $p = \frac{1.75}{8.75}$ or $\frac{P}{100} = \frac{1.75}{8.75}$; 20%　　**20.** \$300　　**21.** \$816
22. \$2400　　**23.** \$2590.06　　**24.** \$1248.72　　**25.** \$901.17

Chapter 1–8 Cumulative Review Exercises

1. true　　**2.** true　　**3.** true　　**4.** true　　**5.** false　　**6.** false　　**7.** Multiply every term of the first polynomial by every term in the second polynomial and combine like terms.　　**8.** Find a common denominator. Rewrite. Add numerators and keep the common denominator. Simplify.　　**9.** Multiply both sides by an appropriate power of 10 as determined by the decimal number with the most decimal places.　　**10.** Multiply the given measurement by unit fractions so that the undesired units divide out.　　**11.** 2.75×10^{10}　　**12.** 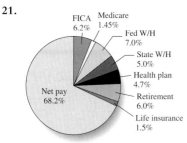　　**13.** 1,600,000　　**14.** 1　　**15.** 20
16. $14\frac{5}{12}$　　**17.** $-57\frac{1}{6}$　　**18.** 64.975　　**19.** -24.5　　**20.** 0.39　　**21.** 1.5　　**22.** ≈ 8.49　　**23.** 0.125　　**24.** 71%
25. $x^4 - 4x^3 - \frac{14}{15}x^2 + 24.2$　　**26.** $46.02x^{14}$　　**27.** $24.8x^2 + 27x - 5$　　**28.** $2^2 \cdot 3^3 \cdot 5$　　**29.** $120xy^2$　　**30.** $6x^4$
31. $8m(4m^3 + 3m - 2)$　　**32.** $\frac{8m}{3n}$　　**33.** $\frac{12 - 5y}{20}$　　**34.** $y = 6$　　**35.** $x = -4.6$　　**36.** $n = 3\frac{6}{13}$　　**37.** 152 oz.
38. 0.08 l　　**39.** $4.\overline{4}°C$　　**40.** 12.2 ft. or about 3.7 m　　**41.** $A = \frac{1}{2}h^2 + \frac{1}{2}h$; 45 in.^{2}　　**42.** 40 and 83　　**43.** ≈ 81.76 cm^2
44. ≈ 18.1 in.^{3}　　**45.** ≈ 2143.6 in.^{3}　　**46.** $\frac{1}{40}$　　**47.** $x \approx 16.1$ cm; $y \approx 19.3$ cm　　**48.** 0.025 hr. = 1.5 min.
49. $85.\overline{3}\%$　　**50.** \$3601.63

Exercise Set 9.1

1. the sum of all given numbers divided by the number of numbers　　**3.** 1. Arrange the scores in order from least to greatest. 2. Find the mean of the middle two scores.　　**5.** mean = \$26,841.6; median = \$26,975; mode = \$28,700
7. mean = \$78.25; median = 80; mode = 82　　**9.** mean = 1.82 m; median = 1.85 m; modes = 1.8 m and 2.1 m
11. mean = 1.25 in.; median = 1.1 in; mode = 0.6 in.　　**13.** year 1: mean = \$126.76; median = \$131.26; no mode
year 2: mean = \$131.95; median = \$134.08; no mode　　**15.** 90　　**17.** 106　　**19.** 3.192　　**21.** 2.667　　**23.** 86.6 or B

Review Exercises:

1. 75%　　**2.** 0.125　　**3.** $11\frac{7}{20}$　　**4.** $-14x - 6y + 12$　　**5.** $b = -15$

Exercise Set 9.2

1. portions　　**3.** a line used for reference in a graph　　**5.** 106,406,950　　**7.** \$52,161　　**9.** Anaheim-Santa Ana, CA
11. Charlotte, NC　　**13.** $\approx 314,145$　　**15.** endowment income　　**17.** 3.8%　　**19.** \$29.1 billion

21.

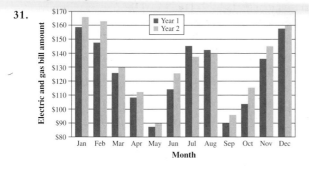

23. engineering　　**25.** education　　**27.** $\approx \$40,000$　　**29.** $\approx \$28,000$

31.

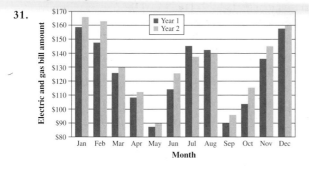

33. business and management　　**35.** $\approx 60,000$ degrees　　**37.** $\approx 25,000$ degrees　　**39.** $\approx 37\%$　　**41.** 1985

43. use of drugs ≈ 13%; lack of discipline ≈ 17%; lack of financial support ≈ 23%

45.

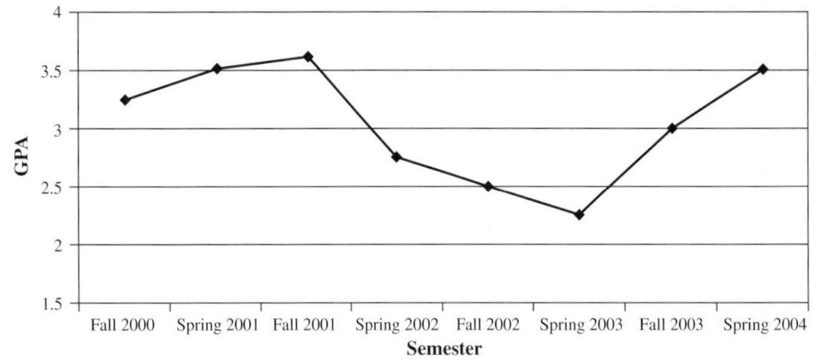

Review Exercises:

1.

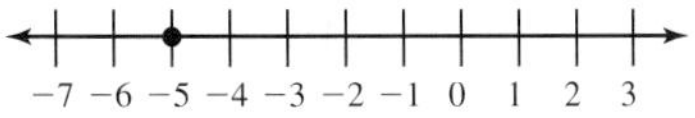

2.

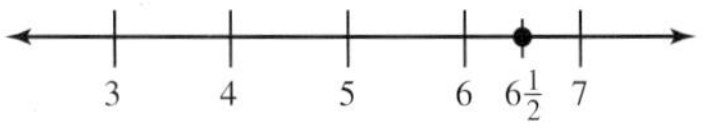

3. 7 **4.** 16

5. c ≈ 9.849 in.

Exercise Set 9.3

1.

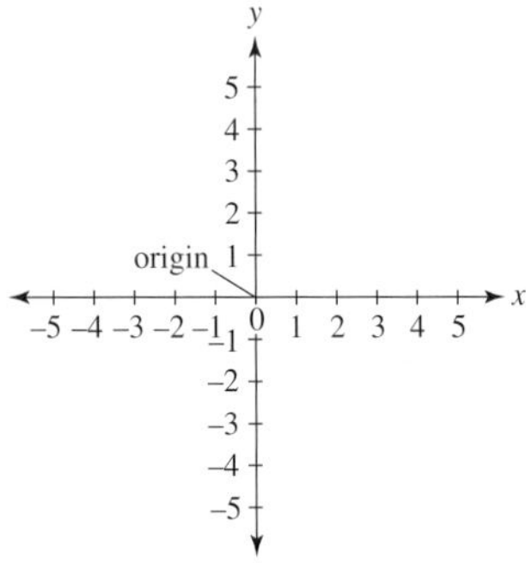

3. Beginning at the origin, move to the right or left along the x-axis the amount indicated by the first coordinate. From that position on the x-axis, move up or down by the amount indicated by the second coordinate. **5.** Both coordinates are negative. **7.** A (2, 3); B (−3, 0); C (−1, −3); D (3, −4) **9.** A (−4, 1), B (0, 3), C (−3, −3); D (3, −5)

11.

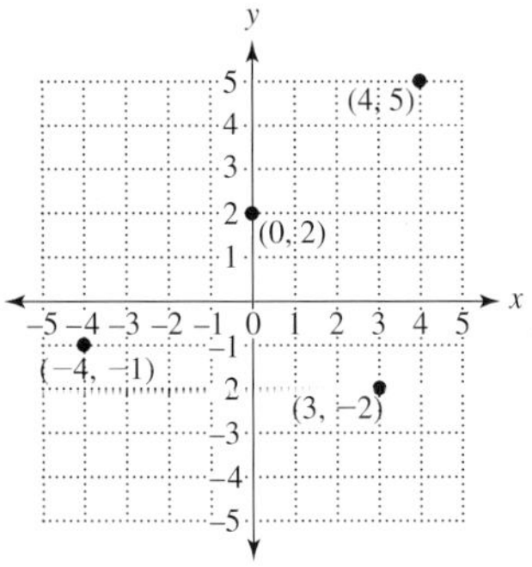

13.

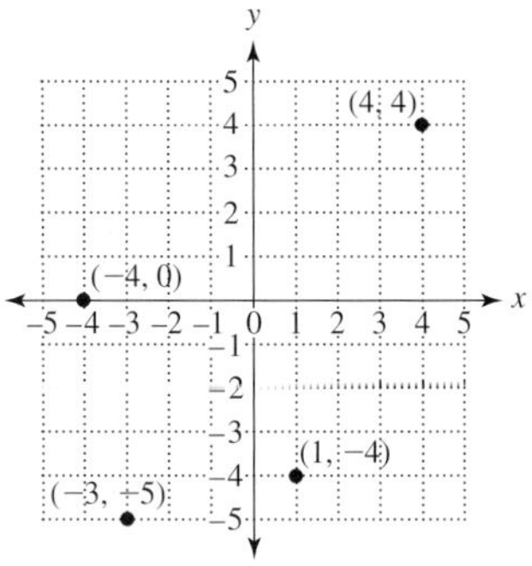

15. II **17.** I **19.** IV **21.** III **23.** (4, 3) **25.** $\left(5, \frac{1}{2}\right)$ **27.** $\left(-\frac{1}{2}, 1\right)$ **29.** $\left(-\frac{7}{2}, \frac{3}{2}\right)$ **31.** $\left(-7, -\frac{5}{2}\right)$
33. (−1.5, −3.4) **35.** $\left(4\frac{1}{8}, 5\frac{1}{4}\right)$

Review Exercises:

1.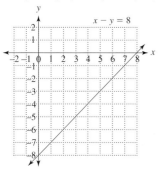
-4 $-3\frac{1}{2}$ -3

2. 2   **3.** -36 **4.** $x = 7.5$ **5.** $y = -\dfrac{9}{10}$

Exercise Set 9.4

1. Answers will vary, but $x + y = 5$ is one possibility.   **3.** 1. Choose a value for one of the variables. 2. Replace the corresponding variable with your chosen value. 3. Solve the equation for the other variable.   **5.** a straight line
7. a point where a graph intersects the y-axis   **9.** 0   **11.** yes   **13.** no   **15.** yes   **17.** no   **19.** no   **21.** no

23. $(0, -8)$, $(4, -4)$, and $(8, 0)$

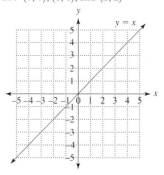

25. $(1, 4)$, $(2, 2)$, and $(3, 0)$

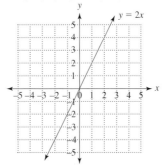

27. $(0, 0)$, $(1, 1)$, and $(2, 2)$

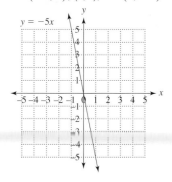

29. $(0, 0)$, $(1, 2)$, and $(2, 4)$

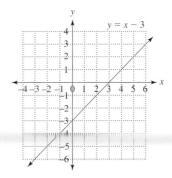

31. $(-1, 5)$, $(0, 0)$, and $(1, -5)$

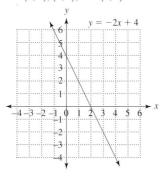

33. $(0, -3)$, $(1, -2)$, and $(2, -1)$

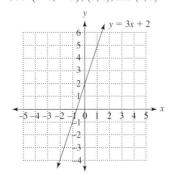

35. $(0, 4)$, $(1, 2)$, and $(2, 0)$

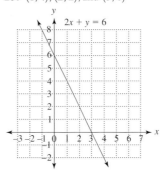

37. $(-1, -1)$, $(0, 2)$, and $(1, 5)$

39. $(0,0), (2,1),$ and $(4,2)$

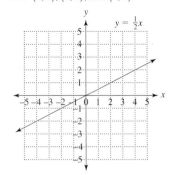

41. $(0,4), (3,2),$ and $(6,0)$

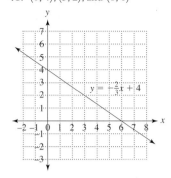

43. $(0,-8), (1,-4),$ and $(2,0)$

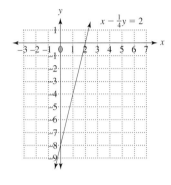

45. $(0,-2.5), (3,-1.3),$ and $(5,-0.5)$

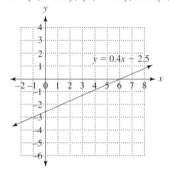

47. $(0,-5), (1,-5),$ and $(4,-5)$

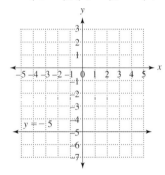

49. $(7,0), (7,2),$ and $(7,5)$

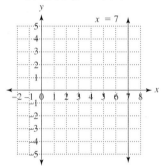

51. $(12,0), (0,6)$ **53.** $(4,0), (0,-5)$ **55.** $(-8,0), (0,6)$ **57.** $(0.2,0), (0,0.65)$ **59.** $(0,0)$ **61.** $\left(\dfrac{1}{4}, 0\right), (0,1)$

63. $(25,0), (0,-5)$ **65.** $(9,0),$ no y-intercept   **67.** no x-intercept, $(0,4)$ **69.** no x-intercept, $(0,-3)$

Review Exercises:
1. -64.4 **2.** 40 m   **3.** $45\dfrac{1}{2}$ ft.^{2} **4.** 5.67 in.^{2} **5.** 55.5 ft.^{2}

Exercise Set 9.5
1. When the object stops, its velocity is 0. Therefore, we replace v with 0 and solve for t.   **3.** a square

5a. 600 ft./sec. **5b.** 503.4 ft./sec. **5c.** ≈ 18.6 sec. **5d.**

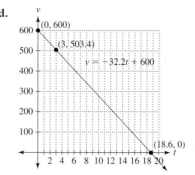

7a. $15,000 **7b.** $187,500 **7c.**

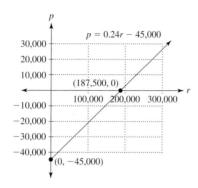

9a. $300 **9b.** $367.50 **9c.** $570 **9d.**

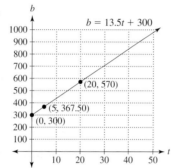

11a. 32°F **11b.** $-17\frac{7}{9}$°C **11c.** 104°F **11d.**

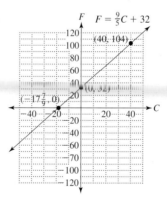

13a. $T = 9h + 58$ **13b.** $80.50   **13c.** $4\frac{2}{3}$ hr.   **13d.**

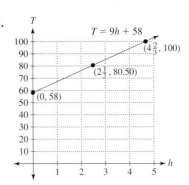

15. $(2, 2.5)$; 20 square units   **17.** $(0.5, -2)$; 42 square units   **19.** $(-3.5, -1)$; 30 square units   **21.** $(3.5, 3)$; 30 square units   **23.** $(5, 2)$; 30 square units   **25.** $(1.5, -2.5)$; 63 square units   **27.** $(3.25, 3.5)$; 45.5 square units   **29.** $(0.5, 2.5)$; 27 square units   **31.** $(2, 1)$; 54 square units   **33.** $(1.5, 1)$; 24 square units

Review Exercises:
1. mean ≈ 79.2; median $= 83$; modes $= 70$ and 88 **2.** 9 people   **3.** $440.05   **4.** $\frac{1}{2}$ **5.** The numbers are 25 and 81.

Chapter 9 Review Exercises
1. false   **2.** true   **3.** false   **4.** false   **5.** true   **6.** true   **7.** Divide the sum of all scores by the number of scores.
8. List the given numbers in order from least to greatest, then locate the middle number.   **9.** Find at least two solutions. Plot those solutions on the rectangular coordinate system. Draw a straight line through the points.   **10.** To find the x-intercept, replace y with 0, and solve for x. To find the y-intercept, replace x with 0, and solve for y.   **11.** mean ≈ 81.3; median $= 84$; mode $= 94$ **12.** mean $\approx 127,847.27$; median $= 127,100$; no mode   **13.** 93   **14.** ≈ 3.233 **15.** 23%
16. 65%   **17.** Canada   **18.** Italy   **19.** Japan   **20.** Canada   **21.** 17.5%   **22.** 15.7%   **23.** 177.0 million people
24. 30.8 million people   **25.** 10.6 million people   **26.** 4.8 million people   **27.**

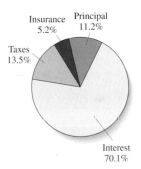

28. $\approx 19,000   **29.** $\approx 41,000   **30.** women with no high school   **31.** men with a master's degree
32. women with a bachelor's degree   **33.**

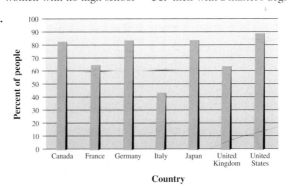

34. $\approx 5,700,000$ people   **35.** $\approx 2,000,000$ people   **36.** $\approx 4,000,000$ people   **37.** under 22 years old

38. $\approx 2,000,000$ people

39.

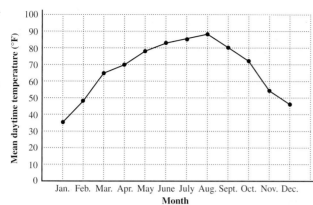

40. A $(3, 4)$; B $(-3, 2)$; C $(-2, 0)$; D $(-2, -4)$; E $(0, -5)$; F $(4, -2)$

41.

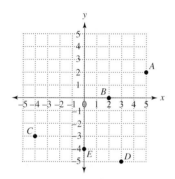

42a. II **42b.** III **42c.** I **42d.** on x-axis **43.** $(4, 5)$ **44.** $\left(1, -\dfrac{1}{2}\right)$ **45.** no **46.** yes **47.** yes

48. $(0, 6)$, $(2, 2)$, and $(3, 0)$ **49.** $(0, -2)$, $(2, -1)$, and $(4, 0)$ **50.** $(-1, 2)$, $(0, 3)$, and $(1, 4)$

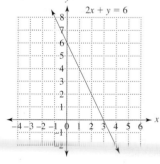

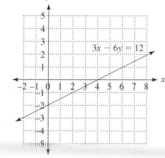

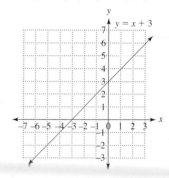

51. $(-1, 3)$, $(0, 0)$, and $(1, -3)$ **52.** $(-3, -2)$, $(0, 0)$, and $(3, 2)$ **53.** $(-2, -2)$, $(-1, -3)$, and $(0, -4)$

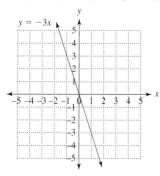

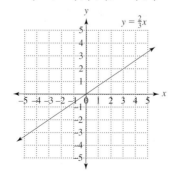

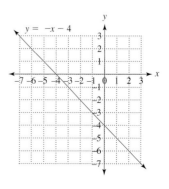

54. $(0, 7)$, $(2, 7)$, and $(5, 7)$

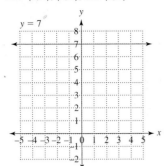

55. $(-3, 0)$, $(-3, 4)$, and $(-3, 5)$

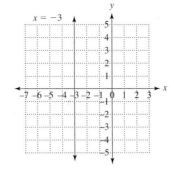

56. $(2, 0)$; $(0, 10)$ **57.** $\left(\frac{1}{4}, 0\right)$, $(0, -1)$ **58.** $(0, 0)$ **59.** $(6, 0)$; no y-intercept

60a. $160,272 **60b.** $230,625 **60c.**

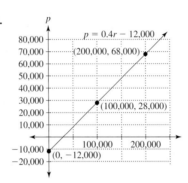

61. $(2.5, 1.5)$ **62.** 40 square units

Chapter 9 Practice Test

1. mean ≈ 81.6; median $= 83$; modes $= 80, 84,$ and 96 **2.** mean $\approx \$95,727.78$; median $= \$96,400$; no mode **3.** 98

4. 15% **5.** $441.25 **6.** $661.88 **7.** public college **8.** 6.2 **9.** 10.5 **10.** 2001 and 2002 **11.** 34.2

12. 34.7 **13.**

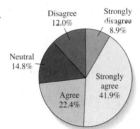

14. A: $(1, 2)$; B: $(-3, 0)$; C: $(-2, -5)$; D: $(4, 4)$ **15.** II

16.

17. $\left(\frac{1}{2}, -\frac{9}{2}\right)$ **18.** yes **19.** $(2, 0)$; $\left(0, -\frac{6}{5}\right)$

20.

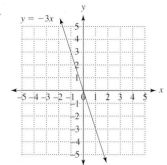

21.

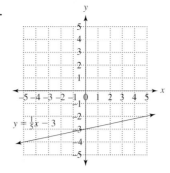

22.

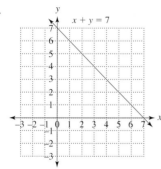

23.

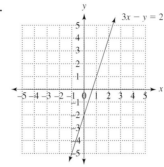

24a. 40 m/sec.   **24b.** 10.6 m/sec.   **24c.** ≈ 4.1 sec.   **24d.** −58 m/sec.   **25a.** $(1, -1)$ **25b.** 48 square units

Chapters 1–9 Cumulative Review Exercises

1. true   **2.** true   **3.** false   **4.** false   **5.** false   **6.** false   **7.** true   **8.** true   **9.** false   **10.** true   **11.** true
12. true   **13.** 1. grouping symbols; 2. exponents; 3. multiply/divide from left to right; 4. add/subtract from left to right
14. Add the absolute values and keep the same sign.   **15.** Subtract the absolute values and keep the sign of the number with
the greater absolute value.   **16.** Change the operation from a minus sign to a plus sign. Change the subtrahend to its additive
inverse.   **17.** positive   **18.** negative   **19.** Simplify by clearing parentheses, fractions, and decimals and combining like
terms. Use the addition/subtraction principle of equality so that all variable terms are on one side of the equation and all
constant terms are on the other side. Use the multiplication/division principle of equality to clear any remaining coefficients.
20. Find at least two solutions. Plot the solutions on the rectangular coordinate system. Draw a straight line through the points.
21. $2 \times 10{,}000 + 4 \times 1000 + 6 \times 100 + 7 \times 1$ **22.** four thousand six hundred seven and nine hundredths
23a. 80,000   **23b.** 80,000   **23c.** 79,805   **23d.** 79,804.7   **23e.** 79,804.65   **24a.** 27,000   **24b.** 695,000
24c. 1800   **24d.** 40   **25.** 2.75×10^{10} **26.** 3,530,000

27a.

27b.

27c.    **28a.** 1   **28b.** 2   **28c.** 4   **29.** 81   **30.** −81   **31.** $\dfrac{1}{36}$

32. $-\dfrac{1}{8}$ **33.** −18   **34.** 35   **35.** 4   **36.** $12\dfrac{5}{12}$ **37.** $-33\dfrac{5}{12}$ **38.** 13   **39.** −4.725   **40.** 0.592   **41.** 0.9

42. ≈ 11.22   **43a.** 0.375   **43b.** $0.\overline{6}$ **43c.** 0.08   **43d.** 0.244   **43e.** 0.045   **44a.** $\dfrac{31}{6}$ **44b.** $\dfrac{29}{200}$ **44c.** $\dfrac{3}{50}$

44d. $\dfrac{11}{200}$ **44e.** $\dfrac{41}{250}$ **45a.** 25%   **45b.** $44.\overline{4}\%$ **45c.** 65%   **45d.** 3.5%   **45e.** 230%   **46a.** −44   **46b.** 88

47. $x^4 - 3x^3 + 4x^2 + 7x - 20.3$ **48.** $9n^2 - 4.2n - 12.81$ **49.** $x^4 - 4x^3 - \dfrac{14}{15}x^2 + 24.2$ **50a.** $-\dfrac{5}{18}a^3$

50b. $m^2 - 81$ **50c.** $31.95t^2 + 5.7t - 18$ **51.** $2 \cdot 3^2 \cdot 5 \cdot 7$ **52.** $15n^2$ **53.** $72x^3$ **54.** $-5y^3$

55. $4b^2(7b^3 + 6b - 8)$ **56a.** $-\dfrac{15x^2}{8}$ **56b.** $\dfrac{8}{3n}$ **56c.** $\dfrac{12 - 5y}{20}$ **57.** $m = -6$ **58.** $x = -9$ **59.** $x = 5$

60. $k = 12\dfrac{1}{12}$ **61.** $h = -7.84$ **62.** $n = -12.5$ **63.** $u = -5.25$ **64.** $a = 2$ **65a.** 144 in.   **65b.** 26 yd.

65c. 12,672 ft.   **65d.** 480 cm   **65e.** 12,500 m   **65f.** 4.5 m   **66a.** $3\dfrac{1}{3}$ yd.^{2} **66b.** 36 ft.^{2} **66c.** 3.5 ft.^{2}

66d. 6000 cm^2 **66e.** 85,000 m^2 **66f.** 4.2 m^2 **67a.** 152 oz.   **67b.** 2.25 T   **67c.** 230 g   **67d.** 0.96 t

68a. 32 pt.   **68b.** 17 qt.   **68c.** 0.4 l   **68d.** 9 cc   **69a.** 25°C   **69b.** -4°F   **70.** 17 m; $\approx$ 55.8 ft.

71. 1000 combinations   **72.** $-\$58,460$ **73.** $\approx$ 4.6 slugs; $\approx$ 67.3 kg   **74.** $A = \dfrac{1}{2}h^2 + 2h$; 17.22 in.^{2}

75a. $P = 3.3x - 3400$ **75b.** \$1550   **76.** The integers are 40 and 83.   **77.** 59°, 69°, and 52°   **78.** 11 \$10 bills

and 6 \$5 bills   **79.** $\dfrac{9}{16}$; 18 calls   **80a.** $\dfrac{23}{30}$ **80b.** $\dfrac{13}{120}$ **81.** $\dfrac{1}{20}$ hr. = 3 min.   **82.** $\approx$ 47.26 cm^2 **83.** 117 ft.^{2}

84. $\approx$ 98.91 in.^{3} **85.** $\approx$ 904.32 in.^{3} **86.** $\approx$ 8.04 cm^3 **87.** $\approx$ 21.54 ft.   **88.** $\dfrac{41}{650}$ **89.** 20 oz.

90. $x = 21.6$ cm; $y = 26.\overline{6}$ cm   **91.** \$60.76   **92.** $\approx$ 8.9%   **93.** $\approx$ 64.3%   **94.** \$14.28   **95.** \$3601.63

96. mean = $39.\overline{36}$; median = 38; no mode   **97.**

98. $(-0.75, 0.5)$; 38.5 square units

99a.

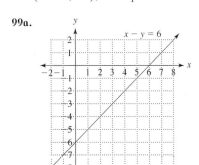

99b.

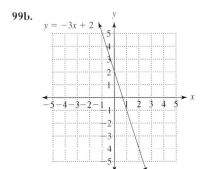

99c.

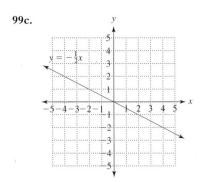

99d.

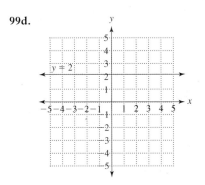

100a. 580 ft./sec.   **100b.** 419 ft./sec.   **100c.** $\approx$ 18 sec.